Boiler Operations
Questions and Answers

Boiler Operations

Questions and Answers

P Chattopadhyay

Process Engineer
HFC (Haldia Division), Haldia
West Bengal

Boston, Massachusetts   Burr Ridge, Illinois
Dubuque, Iowa   Madison, Wisconsin   New York, New York
San Francisco, California   St. Louis, Missouri

McGraw-Hill

A Division of The McGraw·Hill Companies

Library of Congress Cataloging-in-Publication Data

Chattopadhyay, P.
    Boiler operations : questions and answers / P. Chattopadhyay.

       p.   cm.
    Includes bibliographical references and index.
    ISBN 0-07-460296-9
    1. Steam-boilers—Examinations, questions, etc.   I. Title.
TJ288.C44   1995
621.1′83′076—dc20   95-5964
                                         CIP

First published as *Boiler Operation Engineering* © 1994,
Tata McGraw-Hill Publishing Company Limited

3 4 5 6 7 8 9 0 BKM BKM 9 0 9

ISBN 0-07-460296-9

To
my revered parents,
my beloved frau, HONEY
& our son, RAHUL

Preface

The production of steam and its utilization have undergone radical changes over the years through the pioneering efforts of scientists and engineers in the field of fuel and combustion technology, boiler operation, and power generation.

Though there are many excellent texts dealing separately with these subjects, engineers and operators in charge of boiler operation may find it difficult to obtain a single book covering the various aspects of boiler operation technology. I hope this book will fulfil their requirements and also be advantageous to students undergoing courses in power plant engineering and fuel combustion technology.

The book is based on four interrelated disciplines in the production and utilization of steam:

- Water treatment and water conditioning
- Combustion of fuels and the physico-chemical principles involved
- Boilers and steam generation
- Utilization of steam for power generation

The subject matter is presented in a question-answer form. Real-life problems have been supplemented with study matter pertaining to the basic concepts, and numerical calculations, wherever necessary, have been incorporated for better understanding of the subject.

However, with the growing concept of efficient combustion of fuel, utilization of low calorie fuels, and better pollution control measures, the methods of production and utilization of steam are becoming more and more complex: the old ones phasing out to give way to the new. Therefore, in a sense the book is incomplete. And I will be grateful to those who will lend it completeness by giving constructive criticism and suggestions for further improvement of the book.

I sincerely acknowledge my indebtedness to Mr Debu Roy, my friend and colleague for checking the numerical calculations in the manuscript. I am grateful to my wife, Honey, for ferreting out a good many typographical mistakes and omissions from the typescript.

P CHATTOPADHYAY

Contents

x Contents

1

Boilers

Q. *What is a boiler?*

Ans. Broadly speaking, a boiler is a device used for generating.

(a) **steam** for power generation, process use or heating purposes, and (b) **hot water** for heating purposes.

However, according to the Indian Boiler Act, 1923, a boiler is a closed pressure vessel with capacity exceeding 22.75 litres used for generating steam under pressure. It includes all the mountings fitted to such vessels which remain wholly or partly under pressure when steam is shut-off.

Q. *What is the difference between a steam boiler and a steam generator?*

Ans. Technically speaking, a steam boiler consists only of the containing vessel and convection heating surfaces, whereas a steam generator covers the whole unit, encompassing waterwall tubes superheaters, air heaters and economizers.

Q. *How are boilers classified?*

Ans. Boilers are classified on the basis of:
1. Mode of circulation of working fluid
2. Type of fuel
3. Mode of firing
4. Nature of heat source
5. Nature of working fluid
6. Position of the furnace
7. Type of furnace
8. Boiler size
9. Materials of construction
10. Shape of tubes and their spatial position
11. Content of the tubes
12. Steam pressure
13. Specific purpose of utilization
14. General shape
15. Manufacturer's trade name

16. Special features.

Q. *What is circulation?*

Ans. It is the motion of the working fluid in the evaporating tubes. This motion is effected by head or pressure difference in the working fluid between the downcomer and uptake (riser) tubes.

The circulation may be natural or forced and the circulation circuit formed by the heated and unheated tubes may be a closed or open hydraulic system. A once-through boiler represents an open-hydraulic system (Fig. 1.3). While in natural (Fig. 1.1) and forced multiple circulation boilers (Fig. 1.2), the circulation circuit is a closed hydraulic system. In combined-circulation (Fig. 1.4) boilers, the plant operates on closed hydraulic system at the start-up and is switched over to an open hydraulic system after attaining the specified load.

Q. *How are boilers classified on the basis of mode of circulation of working fluid?*

Ans.
1. Natural circulation boiler
2. Forced (i.e., positive) circulation boiler

Q. *What is natural circulation?*

Ans. The natural convection of water set up in the closed hydraulic system of heated and unheated tubes of the waterwall.

Q. *How is it accomplished?*

Ans. The natural convection current is induced to water due to a difference in density resulting from difference in temperature.

The baffle separates out the heated riser from the unheated downcomer and therefore creates a temperature difference between the two tube systems.

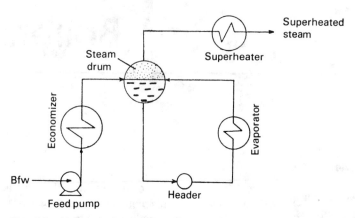

Fig. 1.1 *Natural circulation. It is a closed circuit in which the working fluid circulates by virtue of its density difference*

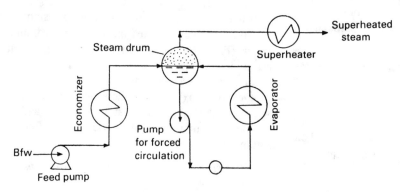

Fig. 1.2 *Multiple forced circulation. It is a closed hydraulic system in which the working fluid is circulated by forced circulation pump*

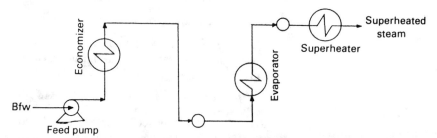

Fig. 1.3 *Open-hydraulic circuit. This system is adopted for once-through boilers*

Saturated water flows down the unheated downcomer and receives heat in the riser whereupon a part of it gets converted into steam. The difference in densities of saturated water in the downcomer and the steam – water mixer in the riser brings about natural circulation. (Fig. 1.5)

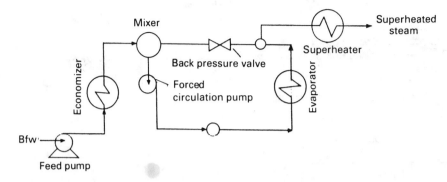

Fig. 1.4 *Combined circulation. It operates on closed hydraulic system at low loads and open hydraulic system beyond specified load*

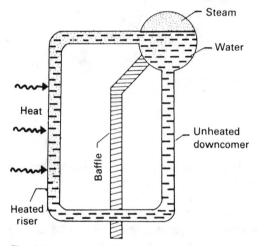

Fig. 1.5 *Natural circulation mechanism*

Q. *What is its limitation?*

Ans. It is applicable to all subcritical boilers, i.e., all those which are operating at a pressure less than critical pressure.

Q. *What is forced circulation?*

Ans. If the working fluid is forced through the boiler circuits by an external pump, the ensuing circulation is called positive or forced circulation.

Q. *What are the advantages of forced circulation over natural circulation?*

Ans.
1. Steam generation rate is higher
2. Greater capacity to meet load variation
3. Quicker start-up quality from cold
4. Lower scaling problem due to high circulation velocity
5. More uniform heating of all parts reduces the danger of overheating and thermal stresses
6. Smaller tube diameter and hence lighter tubes
7. Greater freedom in arrangement of furnace, boiler component and tube layout
8. Operating temperature and pressure can be made to deviate from the designed values.

Q. *What is circulation ratio?*

Ans. It is the ratio of the mass flow rate of circulating water (G_{fw} t/h) to the rate of steam generation (G_s t/h)

$$k = G_{fw}/G_s$$

Q. *What is its value for natural circulation?*

Ans. It usually ranges from 4 to 30.

Q. *What is the value of circulation ratio for forced circulation?*

Ans. It ranges from 3 to 10.

Q. *What is the value of circulation ratio for once-through steam boilers?*

Ans. Unity.

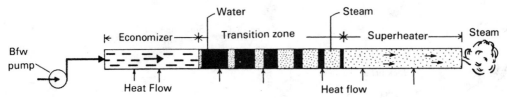

Fig. 1.6 *Transformation of BFW to steam in once-through boiler*

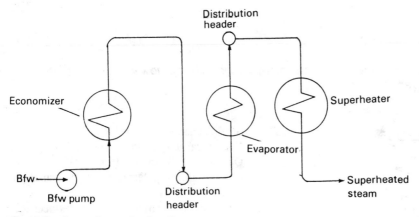

Fig. 1.7 *Once-through circuit*

Q. *Why?*

Ans. In such units, (Fig. 1.6 and 1.7) the entire feed-water is continuously converted to steam as it passes through the evaporating surfaces, i.e.,

$$G_{fw} = G_s$$

Q. *Can once-through boilers operate at subcritical as well as supercritical pressures?*

Ans. Yes.

Q. *What is the difference between a closed hydraulic system and an open hydraulic system?*

Ans. The former features a drum that acts both as a reservoir to provide working fluid circulation and separator to separate water from steam, while the latter has no drum and the working fluid passes through the evaporating tubes only once.

Q. *How can boilers be classified on the basis of tube shape and position?*

Ans. Depending on the form of tubular heating surface, boilers may be classified as
   1. Straight tube boiler
   2. Bent tube boiler.
   Depending on the inclination of tubular heating surface, boilers may be classified as
   1. Horizontal Boiler (Fig. 1.9)
   2. Vertical Boiler (Fig. 1.8)
   3. Inclined Boiler.

Q. *What is the difference between a horizontal boiler and a vertical boiler?*

Ans. The difference basically lies in the geometric position of the boilers.
   A horizontal boiler has its principal axis horizontal or slightly inclined while that of a vertical boiler is perpendicular to the horizontal plane.

Q. *What is a single tube boiler?*

Ans. A boiler having only one firetube. (Cornish or simple vertical boiler)

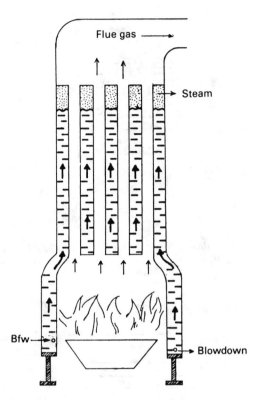

Fig. 1.8 Vertical boiler

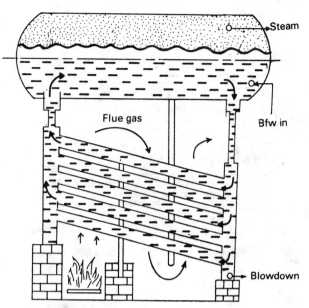

Fig. 1.9 Horizontal boiler

Q. *What is a multitube boiler?*

Ans. Boiler having two or more fire or watertubes.

Q. *How can boilers be classified on the basis of use?*

Ans. This is done on the basis of the nature of service they perform. Customarily boilers are called
  (a) **stationary**: these are land based boilers
  (b) **mobile**: these are mounted on marine vessels and steam locomotives.

Q. *What is a stationary boiler?*

Ans. This boiler, as its name implies, is not required to be transported from one place to another.

Q. *What are mobile boilers?*

Ans. Locomotive and marine boilers, which are moved from place to place, are mobile boilers.

Q. *How may stationary boilers be classified further?*

Ans. They can be classified further depending on the specific service they meet:
  1. Stationary boilers for central station (district) heating
  2. Stationary boilers for process steam generation
  3. Stationary boilers for power generation.
Stationary boilers used for heating are often classified as:
  1. Residential boilers
  2. Commercial boilers.

Q. *How can boilers be classified on the basis of furnace position?*

Ans. Depending on the relative location of the furnace to the boiler, the boiler classification can be made by:
  1. Externally fired furnace
  2. Internally fired furnace.

Q. *What is the difference between externally fired and internally fired boilers?*

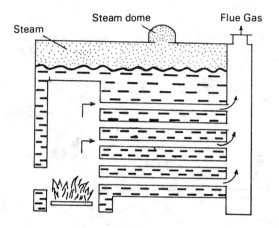

Fig. 1.10 (a) *Locomotive boiler : Gas flow circuit*

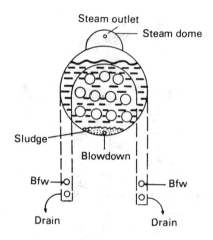

(b) *Locomotive boiler : water circulating system*

Ans. In the case of externally fired boilers, the combustion of fuel takes place in a chamber outside the boiler shell while in the case of internally fired boilers, the combustion chamber is provided inside the boiler shell.

Q. *How can boilers be classified on the basis of tube contents?*

Ans. Depending on whether the flue gas or the water is in the tube side, the boiler can be classified as:
1. Firetube boiler
2. Watertube boiler.

Q. *Give some examples of firetube boilers.*

Ans.
1. Locomotive boiler
2. Cochran boiler
3. Cornish boiler.

These have hot flue gas in the tube side and BFW in the shell side.

Q. *Cite some examples of watertube boilers.*

Ans.
1. Babcock and Wilcox boiler
2. Stirling boiler
3. La Mont boiler
4. Yarrow boiler.

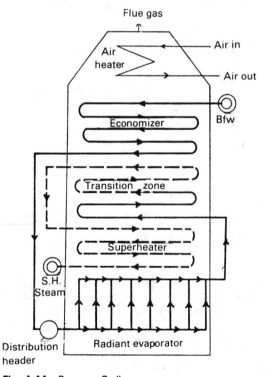

Fig. 1.11 *Benson Boiler*

All these have BFW in the tube side and hot flue gas in the shell side.

Q. *What are the comparative advantages and disadvantages of firetube and watertube boilers?*

Ans.

Parameters	Firetube boilers	Watertube boilers
1. Rate of steam generation	Less rapid	More rapid
2. Suitability for power plants	Unsuitable	Suitable. All major power plants are based on these.
3. Operating steam pressure	Limited to 25 kgf/cm^2	Can well exceed 125 kgf/cm^2
4. Chances of explosion	Less	More
5. Risk of damage due to explosion	Much more	Much less
6. Water treatment	Not very necessary as minor scaling would not go far enough to cause overheating and tube-bursting	Required as scaling will lead to tube-bursting
7. Floor space requirement	Much	Less
8. Cost and construction problem	Higher	Much less
9. Transportation	Inconvenient due to large size of the shell	Comparatively easier
10. Skill required for efficient operation	Less	More

Q. *How many types of watertube boilers are there?*

Ans. Two. Straight tube and bent tube boilers.

Q. *On what grounds are bent tube boilers more favourable than straight tube boilers?*

Ans.

1. Bent tube boilers lend greater economy in fabrication and operation than straight tube boilers. These are due to use of welding, improved quality steel, waterwall construction and new manufacturing techniques.
2. Bent tube boilers afford greater accessibility for inspection, cleaning and maintenance due to more spacious layout of tubes.
3. They have a higher steam generation rate than straight tube boilers
4. They produce drier steam than straight tube boilers.

Q. *What are the essential qualities of a good boiler?*

Ans.

1. It should be capable of quick start-up
2. Should meet large load fluctuations
3. Occupy less floor space
4. Should afford easy maintenance and inspection

5. Should essentially possess the capacity of producing maximum steam with minimum fuel consumption
6. Light and simple in construction
7. Various joints should be accessible for inspection and should be away from direct flame impact
8. Tubes should be sufficiently strong to resist wear and corrosion
9. Mud and other deposits should not collect on heated plates
10. The velocity of water and that of flue gas should be a minimum.

Q. *What basic factors will you consider in selecting a boiler?*

Ans.

1. Power required to be generated
2. Operating pressure
3. Fuel – quality and type
4. Water availability and its quality
5. Probable load factor
6. Location of the power house or process plants
7. Cost of operation and maintenance
8. Cost of installation and erection
9. Availability of floor space.

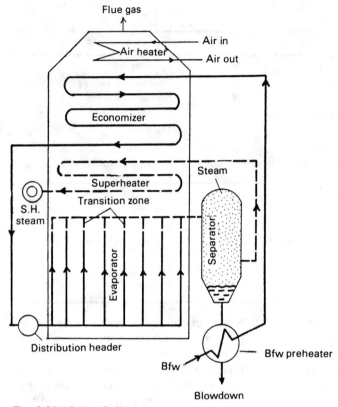

Fig. 1.12 *Sulzer Boiler*

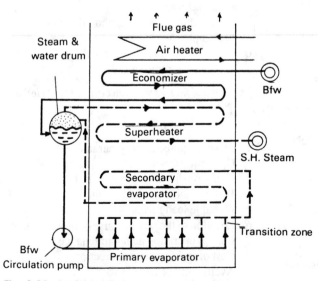

Fig. 1.13 *La Mont Boiler*

Q. How can boilers be classified on the basis of pressure?

Ans. According to the American Society of Mechanical Engineers' "Boilers and Pressure Vessel Code" known as ASME Boiler Code, boilers may be differentiated as

1. **Low pressure boilers** — operating steam pressure not exceeding 1.021 atm[*] (103.427 kN/m^2) and operating pressures of hot water boiler not exceeding 10.893 atm* (1.10 MN/m^2) and temperature 394°K[*].
2. **Power boilers** — all boilers with operating pressure and temperature exceeding those of LP boilers stated above
3. **Miniature boilers** — include all fired pressure vessel having the following parameters
   (a) ID of shell $\not>$ 406.4 mm[*]
   (b) Gross volume (casing and insulation included) $\not>$ 0.14158 m^{3*}
   (c) Water heating surface $\not>$ 1.858 m^{2*}
   (d) Allowable working pressure $\not>$ 6.8 atm (689.7 kN/m^{2*}).

(*Note* * refers to converted unit)

Q. How can classification of boilers be effected on the basis of materials of construction?

Ans. According to ASME Boiler Code Material Specifications

(a) Low Pressure Heating Boilers can be constructed of cast iron or steel
(b) Miniature Boilers may be constructed of copper, stainless steel, etc.
(c) Power Boilers are constructed of special steels.

Q. How are boilers classified on the basis of size?

Ans. Based on their size and rating, the Steel Boiler Institute (USA) has classified boilers into three categories:

Category (I): These are commercial boilers. Twenty two sizes have been standardized. Their:

Heating surface = 11.98 – 331.756 m^{2*}

Gross heat output = (45333 – 1260) $kcal/s^{2*}$
$$= (10827 – 300) \ kJ/s^*$$

Category (II): These are residential boilers. Seventeen sizes.

Heating Surface = 1.486–27.313 m^{2*}
Gross heat output = upto 126 kcal/s (30 kJ/s)[*]
Category (III): Oil fired boilers. Fourteen sizes. Gross heat output range is of up to 30 kJ/s[*]

For low-pressure cast-iron heating boilers, the Institute of Boiler and Radiator Manufacturers (USA), standardized cast-iron boilers into 33 sizes. Their steam generation rate is upto 3143.44 kg/h[*].

(*Note* * refers to converted unit)

Q. How may boilers be classified on the basis of firing?

Ans.

1. **Fired boilers** in which supplied heat comes from the combustion of fuel.
2. **Non-fired boilers** receive heat other than that produced by the burning of fuel.

Q. How can classification of boilers be done depending on the heat source?

Ans. Depending on the nature of heat source, boilers can be classified into

1. **Fuel fired boiler**—these derive their heat energy by combustion of fuel which may be solid, liquid or gaseous.
2. **Waste heat boilers**—recover heat from the hot waste gases of other chemical reactions.
3. **Electrical powered boilers**—generate steam by the application of electrical energy.
4. **Nuclear powered boilers**—utilize the energy of controlled thermonuclear fission reactions to generate steam.

Q. How can boilers be designated on the basis of the nature of fuel used?

Ans.

1. **Coal fired**
   (a) Pulverized coal fired
   (b) Stoker fired
   (c) Hand fired
2. **Gas fired**
3. **Oil fired**
4. **Wood fired**
5. **Bagasse fired** (Fluidized bed bagasse fired furnace for boilers have recently been developed).

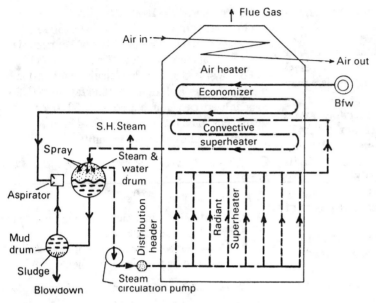

Fig. 1.14 Loeffler Boiler

Q. *How can boilers be described in terms of the type of furnace?*

Ans.
1. Dutch oven boiler
2. Open boiler
3. Scotch boiler
4. Screened boiler
5. Twin boiler.

Q. *How is the general shape factor harnessed to classify a boiler?*

Ans. Depending on their shapes and design features, boilers can be classified as follows:

(A) **Watertube boilers**
1. Horizontal straight tube:
    (a) boxed header type
    (b) sectional header type
2. Bent tube boilers:
    (a) mono-drum type
    (b) bidrum type
    (c) tridrum type
    (d) quadridrum type.

If the drum is parallel to the tubes, it is called longitudinal drum type boiler; if across the tubes, it is called a cross-drum type.

(B) **Firetube boilers**
   (a) short fire box type
   (b) compact
   (c) vertical tube type
   (d) horizontal type
   (e) locomotive
   (f) Scotch type.

Q. *How can the boiler type be indicated by the manufacturer's trade name?*

Ans. Many a manufacturer applies his trade name to the particular type of boiler he designs, e.g.
1. Benson boiler
2. La Mont boiler
3. Sulzer boiler
4. Velox boiler
5. Loeffler boiler
6. Babcock boiler
7. Wilcox boiler
8. Yarrow boiler
9. Manning boiler
10. Thornycraft boiler

Q. *Sometimes special features of boilers are used to classify them. How is this done?*

Ans. Depending upon the placement and firing operation of burners, boilers may be classified as:

differential fired type

tangential fired type.

Similarly there is the anthratube boiler, which is a completely self-contained anthracite burning unit.

Different variations exist, such as tube-in-tube boiler, top-fired boiler, and so on.

There are also:

dual circulation boiler,

boiler with gas recirculation

pressurized and supercharged boilers.

Q. *How are boilers generally categorized?*

Ans. They may be categorized into 4 (four) general types:

1. **Steel boilers:**
   (a) Firetube type
   (b) Watertube type
   (c) Shell type
2. **Cast-iron boilers**
3. **Special design boilers**
4. **Nuclear powered boilers.**

Watertube boilers may be further subdivided into:

1. Horizontal straight tube type
2. Bent tube type:
   (a) Natural circulation type
   (b) Forced circulation type.

Q. *What are the inherent advantages of horizontal straight tube boilers if the tube sizes are small?*

Ans.

1. Tube replacement is easier
2. Draught (draft) loss is low
3. Better end-to-end visibility through tubes before and after cleaning
4. Greater accessibility to all components for inspection and cleaning
5. Low head room.

Q. *However, smaller straight tube horizontal boilers are associated with certain disadvantages. What are they?*

Ans.

1. Access to internal components uses more time and labour. Inadequacy in design adds to the grievance of the operators

2. Because of relatively low circulation rates and poor circulatory distribution, the steam generation rate is sharply impaired
3. If steaming rate is raised, the separation of steam from water in the steam drum becomes inadequate because a limited surface area is available for disengagement.

Q. *Briefly describe the horizontal straight tube boiler.*

Ans. These boilers are made up of banks of straight tubes laid out in a staggered arrangement at an angle 5° to 15° to the horizontal, which are expanded into headers at the ends. The tubes are 75–100 mm in diameter with length not exceeding 600 cm. Smaller diameter tubes are selected when greater tube spacing is required to meet steam demand at higher pressure. Providing flat surfaces for these tube connections, the header may be a box or sectional type connected to the drum by means of circulation tubes — downtakes and uptakes. The drum may be of the longitudinal or cross type.

These boilers may be oil, gas, coal, bagasse or wood fired. Pulverized coal firing can also be incorporated.

When the boiler is fired, steam and water rise along the inclined tubes to the front headers and then pass to the drum through circulation tubes (uptakes). The water then circulates down through downtakes to the rear header and finally to the tubes to complete the cycle.

Q. *Why are the tubes, in horizontal straight tube boilers, inclined at an angle of 5 – 15° to the horizontal?*

Ans. To ensure better circulation.

Q. *Briefly describe a bent tube boiler.*

Ans These are multidrum watertube boilers fitted with two, three or four drums, and usually having one lower drum called mud drum that serves as a blowdown for removal of sludge and concentrations of salts. The remaining drums placed at the top of the boiler are called steam and water drums.

The tubes may be inclined or vertical, arranged in banks, or forming waterwalls backed with furnace wall refractories.

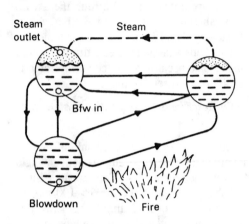

Fig.1.15A *Inclined bent-tube boiler*

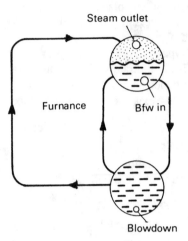

Fig. 1.15B *Vertical bent-tube boiler*

Baffles are placed in the gas path to ensure even heating of watertubes. The drums are protected from the radiant heat of fire.

The mode of steam-water circulation is as shown in Fig. 1.15.

The boilers lend themselves to a wide range of fuel burning. They can be fired with oil, gas, coal, bagasse or wood. For higher capacity of steam generation ($45t/h$) pulverized coal or crushed coal (cyclone furnace) firing is adopted.

Bent tube boilers have high steam generation rate and are quick to respond to fluctuating loads.

Q. *What are the advantages of bent tube boilers over straight tube boilers?*

Ans.

1. Steaming rate is higher
2. Deliver more drier steam
3. More responsive to load fluctuations
4. Lend greater economy in fabrication and operation
5. Afford greater accessibility for inspection, cleaning and maintenance
6. Greater design flexibility as the tubes enter the drum radially
7. The bent tubes allow free expansion and contraction
8. Boiler capacity can be increased without increasing the drum diameters.

Q. *Why do bent tube boilers have a quicker response to load fluctuation?*

Ans. They have very good evaporating flexibility because of the relatively small water volume for the steam generating capacity.

Q. *When are vertical tubes preferred to inclined tubes in designing bent tube boilers?*

Ans. This is given primary design consideration when bent tube boilers are fired with coal having low-melting ash or particularly abrasive ash.

Q. *What is a waste heat boiler?*

Ans. It is a special purpose boiler designed to generate steam by

(a) removing the generated heat, as called for, by the chemical processes involving exothermic reactions viz, the partial gasification of fuel oil; synthesis of ammonia $N_2 + 3H_2 \rightarrow 2NH_3 + 22$ kcal ; oxidation of sulphur dioxide to trioxide ($2SO_2 + O_2 \rightarrow 2SO_3 + 45$ kcal) ; conversion of ammonia to nitric oxide ($4NH_3 + 5O_2 \rightarrow 4NO + 6H_2O + 3O_5$ kcal)

(b) recovering heat that is
   (i) evolved as an integral part of the process and would otherwise go waste, such as from an open-hearth furnace
   (ii) a by-product of chemical process, e.g., black liquor recovery

(iii) made available by burning wastes, e.g., wood scraps.

Q. *Does it serve any purpose in addition to producing useful steam?*

Ans.
1. It reduces air and water pollution
2. It lowers the flue gas temperature, reducing the maintenance of flues, fans, and stacks.

Q. *How may waste heat boilers be classified?*

Ans. They can be classified as:
1. Watertube boiler
2. Gastube boiler
3. Bent tube boiler
4. Positive circulation boiler
5. Supercharged boiler
6. Three-drum-low-head boiler
7. Waterwall bidrum type boiler.

Q. *Where are these generally employed?*

Ans.

Waste Heat Boiler Type	Application
1. Watertube	For clean or dust laden flue gas
2. Gastube	For relatively clean flue gases
3. Bent tube	Handles heavily dust laden gases
4. Positive circulation	For clean, low-temperature gases
5. Supercharged	Gas turbine exhaust
6. Three-drum-low-head boiler	Suitable for light dust loadings
7. Waterwall bidrum type	For gases with suspended sticky particles.

Q. *What should be the design considerations in the selection of a waste heat boiler?*

Ans.
1. Heat load and temperature of the gases available for waste heat recovery for steam generation
2. Chemical nature and corrosiveness of the gases
3. Available draught (draft)

4. Dust load and its nature in the gases
5. Whether the gases are under pressure or suction
6. Available space
7. Requirement for a start-up furnace, gas preheating emergency use or added capacity, etc.
8. Location for the outlet in the case of flue gases.

Q. *In case of lower gas temperature what modifications should be made in designing a waste heat boiler?*

Ans. This drawback is offset by
(a) providing high gas velocity
(b) reducing the diameter of the heat exchanger tubes
(c) increasing the number of such tubes to increase the magnitude of convective heat transfer.

Q. *What is the usual mass flowrate in the WHB (Waste Heat Boiler) of flue gases being discharged to the atmosphere after waste heat recovery?*

Ans. 29 000–39 000 kg/h· m^2.

Q. *In some cases a large quantum of dust particles is found laden in the waste gases available for steam generation. What problems are associated with dust load?*

Ans. Dust laden waste gases, if they contain abrasive dust particles, may cause severe damage to boiler tubes by erosion. Therefore, waste gases loaded with such particles require low gas velocities and that decreases the heat transfer rate in the WHB.

If the gases contain sticky or tacky particles they will deposit on heat transfer tubes, impair the heat exchange process and reduce the efficiency of the WHB.

Q. *How can the dust load in waste gases be controlled?*

Ans. By
(a) installing dust collectors
(b) preventing by-passing of tubes with baffles
(c) providing abrupt change in the direction of gas flow path to settle dust

(d) minimizing bridging between tubes by adequate spacing.

Q. *What is a gastube WHB?*

Ans. Such boilers are shell- and -tube type boilers with hot gases flowing in the tube-side and BFW in the shell-side.

They are usually single pass in arrangement and absorb only convection heat from the hot gases.

They have a high weight-to-heat output ratio.

They are usually found in applications for gas pressure 27–35 atm. and temperature up to 1255°K (982°C).

The external surface of the boiler is hot insulated.

The boiler tubes are of smaller diameter and more closely spaced than direct-fired waste heat boilers.

Soot blowers with nozzles directed towards the tube ends are used to clean the deposits.

Gastube boilers are used in the case of gases with light dust loadings, for obvious reasons.

Q. *How many types of gastube WHBs are there?*

Ans. Two types : Horizontal and Vertical.

Q. *Horizontal gastube waste heat boilers are set at an angle of 15° to the horizontal. Why?*

Ans. This is to
(a) ensure steam collection at the high point
(b) allow suspended solids in the water to settle down by gravity to the lowest point
(c) enhance circulation.

Q. *In which case is a vertical gastube WHB preferred to its horizontal counterpart?*

Ans. It is preferred when it is a prime necessity to save floor space.

Q. *A vertical gastube WHB with shell diameter 2.5 m is rarely found. What is the cause of such a limitation?*

Ans. This is due to limitations inherent in transportation and drum attachments.

Q. *Why are watertube boilers most frequently used for waste heat recovery?*

Ans.
1. They can work successfully at higher pressures
2. Since water is circulated in the tube-side which can be readily cleaned, watertube WHBs are not so susceptible to damage from poor feedwater quality
3. Better capacity to withstand the shock due to fluctuation of gas temperature
4. The furnace wall can be adequately cooled by applying water- wall tubes. This imparts long life to the refractory lined WHB interior wall
5. The slagging and erosion problems can be minimized by varying the tube size and spacing
6. Dust particles may be recovered
7. Lends itself to a more economic arrangement.

Q. *What are the primary considerations in designing a watertube WHB?*

Ans. The design considerations should primarily focus on two major problems. These are:
(a) problems due to sticky dust particles
(b) problems due to heavy dust loadings

The cooling and subsequent elimination of sticky particles is necessary before their entry into the convection shaft of the WHB, otherwise they will deposit on the external surfaces of watertubes and impair the heat transfer characteristics.

Hot gases heavily laden with dust particles may entail erosion problems for tubes as well as reduced heat transfer due to surface deposition of dust particles on watertubes.

Q. *What are the reasons that led to the application of bent tube boilers for waste heat recovery?*

Ans.
1. These boilers can tolerate heavy dust loadings. The vertical tubes collect less dust
2. Greater flexibility in tube size, spacing and arrangement
3. Tube damage problem is minimized by more positive circulation
4. Allows pendant superheater installation
5. Less space is required for tube removal.

Q. *What is the chief advantage of a positive circulation waste heat boiler?*

Ans. The unit is more compact and light as extremely small diameter tubes (30–40 mm) are used, which can be arranged without regard to natural circulation requirements.

Q. *Name one such boiler.*

Ans. Positive circulation La Mont Waste Heat Boiler.

Q. *What are the applications of waste heat boilers?*

Ans. Waste heat boiler finds its application in:
1. **Steel mills** — use all types of small boilers:
    (a) horizontal and vertical gastube WHB
    (b) horizontal straight tubular WHB
    (c) Bent watertube type boilers.
   Waste heat boilers are fitted to open-hearth, forge and continuous heating furnaces.
   Coke oven and blast furnace gases being heavily dust laden need special handling in the burner.
2. **Cement kilns** — need a WHB that must be specially designed to handle extremely dust laden gas. Hoppers are provided under the boiler and economizer to remove dust continuously.
   As much as 20–40 tons of cement dust is recovered per day from a single kiln.
   Two-drum and three-drum WHBs are particularly used. These are fitted with economizers, superheaters and soot blowers.
3. **Ore roaster**—A typical WHB for recovering waste heat from ore roasters is a three-drum, low-head boiler fitted with hoppers under the sections of gas path to collect gas-borne ore particles as they settle when the gases makes low-velocity turns around the baffles.
4. **Lead and Zinc smelters** — These need a kind of WHB that is capable of handling a gas whose temperature is as high as 1450–1480°K and which is laden with solids in a semimolten or sticky form. Usually a vertical watertube WHB is used, which must cool the gas down to 1000–1030° K in the radiant chamber to condense out the metal (Zn, Pb) vapour from the hot gas before its entry to superheater and convection shaft.
5. **Paper making**–uses a WHB to
    (a) generate process steam by burning the waste liquor
    (b) recover the salt cake
    (c) eliminate stream pollution.
   The waste liquor is dehydrated to produce char which is burnt in a large heap in a reducing atmosphere in the recovery furnace. The furnace temperature is as high as 1500–1530° K. Gas velocities are kept low to avoid fouling of the heat absorbing surfaces which comprise waterwall tubes laid on the refractory lined boiler furnace. The flue gases after passing through the economizer go through the evaporator to concentrate the black liquor.

Q. *How are marine boilers classified?*

Ans.
1. **Sectional header boiler**—either horizontal straight tube type or the sectional express type
2. **Drum type**—these are bent tube boilers having a double furnace or simple furnace
3. **Positive circulation boiler**
4. **Thermonuclear steam generator**

Q. *How are marine boilers typified?*

Ans. Depending upon the specific usuage of steam they produce, the marine boilers are divided into:
1. **Main boilers** — used for ship propulsion
2. **Auxiliary boilers** — used for running the auxiliary systems abroad the ship.

Q. *What are these auxiliary systems?*

Ans.
1. Turbogenerators for electric power generation
2. Water distillation
3. Fuel oil heating
4. Space heating
5. Cooking
6. BFW turbines

Q. *What factors must be considered in designing a marine boiler?*

Ans.
1. The boiler must be compact in size and minimum in weight

2. Maximum operating efficiency to be imparted by ensuring
   (a) proper combustion and its control
   (b) efficient heat recovery
   (c) adequate insulation
   (d) higher operating steam pressure and temperature.
3. Special design features must be effected to
   (a) outweigh the effect of ships's rolling, pitching and vibration on steam generation
   (b) eliminate gas leakage in the fireroom
   (c) minimize heat loss to the fireroom
4. Operation reliability is to be attained by
   (a) correct designing and construction of boiler
   (b) proper installation of boiler unit, superheater and auxiliaries
5. Simplicity in design and operation
6. Complete accessibility to boiler internals for inspection, cleaning and repairing must be ensured to minimize outage time
7. Flexible enough to maintain constant steam temperature and pressure over a wide range of steam loads
   This necessitates:
   (a) large steam drum
   (b) large ratio of heat generating to contained water volume.
8. Fuel and its efficient utilization.
   Pulverized coal is seldom used because of space limitation.
   Almost all oceanliners use oil fired boilers as
   (a) oil burns more cleanly than coal
   (b) oil leaves no ash
   (c) oil needs 55% of the storage capacity needed for coal
   (d) oil burning devices require less space.

Q. *What is called an 'M' boiler?*

Ans. Because of its shape, the double-furnace-single-uptake unit is referred to as an 'M' boiler. It is fitted with a separately fired superheater and extended surface economizer.

Q. *What is the range of steam generation capacity of the 'M' boiler?*

Ans. 45 to 115 *t/h.*

Q. *What are the advantages of an 'M' boiler?*

Ans.
1. Close control over superheater temperature ensures greater efficiency
2. Generation of steam at almost any desired temperature and pressure condition can be effected at any load condition by varying the rate of combustion in the twin furnaces
3. Desuperheater is not required
4. Requires less maintenance
5. Ensures good fuel economy at high efficiency under normal conditions
6. Compact. Less floor space is required.

Q. *What are its main drawbacks?*

Ans.
1. Unequal distribution of heating surfaces
2. Unbalanced combustion characteristics
3. Complicated in operation in earlier design
4. It is difficult to light off the superheater fire due to high furnace pressure and this becomes particularly dangerous when the boiler load is considerable.
5. Since the saturated auxiliary steam is taken directly from the steam drum, under conditions of low evaporation rate the superheater may receive insufficient steam, causing superheater starvation.

Q. *What is a 'D' boiler?*

Ans. It is a single-furnace, single-uptake marine boiler having two drums. The water-cooled furnace is shaped in the form of a 'D' and hence its name.

Q. *What are the advantages of a 'D' boiler over an 'M' boiler?*

Ans.
1. More compact and light for a given steam output
2. More reliable in operation
3. Higher steam generation efficiency
4. More economical at normal load
5. Operation is more simplified. Maintenance is much less and outage time is low

6. Superheated steam temperature can be better controlled by varying
   (a) the number of burners in line
   (b) feedwater temperature
   (c) the amount of excess air.

Q. *What are the disadvantages of a 'D' boiler?*

Ans.
1. Requires greater care during cold start-up
2. Efficiency goes down at higher steam generation rate.

Q. *Naval boilers have been found to install all tubes straight. What are its advantages?*

Ans. Straight tubes:
   (a) lend themselves to greater ease of inspection, cleaning and replacement if necessary
   (b) require a small number of spares.

Q. *What are the disadvantages associated with all-tubes- straight installation?*

Ans.
1. It weakens the drum structure
2. It delimits the number of tubes that can be installed for a specified drum diameter and length.

Q. *Modern high pressure naval steam boilers are designed with all tubes bent to an arc of a circle. What are the advantages of this design?*

Ans.
1. Maximum number of tubes can be provided for a given drum length and diameter
2. Drilling holes for tube fitting on the drum is easy
3. Ease of fitting tubes to the drum.
4. Permits operation at higher boiler pressure.

Q. *The bent tube design has certain disadvantages. What are they?*

Ans.
1. Tube cleaning becomes more difficult and time consuming
2. Tube replacement is troublesome
3. Because of variation in length, size and curvature, a large number of spare tubes must always be kept ready.

Q. *What are the usual methods of installing tubes in marine boilers?*

Ans.
1. Straight tube design, known as Yarrow design. All tubes are straight
2. Foster design, in which all tubes are bent to an arc of a circle at their entry to the drum
3. Bent tube design, in which all tubes are bent at right angles to the drum surface at their entry to the drum.

Q. *What types of superheaters are generally used in marine boilers?*

Ans. Convective type superheaters.

Q. *Why not radiant or radiant-convective superheaters?*

Ans. Marine boilers have a low steam generation rate. At a low steaming rate, the radiant and radiant-convective types of superheaters will not operate satisfactorily.

Q. *Marine Boiler's superheaters are plagued with excessive slagging requiring modification in superheater design to carry-out the deslagging operation. How does this problem arise, if almost all marine boilers are oil fired?*

Ans. As the fuel oil tanks, during the operation of the marine boiler, get gradually depleted, seawater is pumped into the tanks for ballast. As a result of this, the seawater salinity is imparted to the fuel oil whose sulphur and vanadium contents combine with sodium chloride during combustion to form slag which deposits on superheater tubes.

Q. *What constitutes the furnace wall of a marine boiler?*

Ans. It is constructed of a refractory wall cooled by waterwall tubes facing the furnace core. The refractory wall is backed up with insulating material and then a metal casing.

Q. *Is any draught (draft) installed in a marine boiler?*

Ans. Yes. Both forced and induced draught (draft) find their way into all marine boilers. Either F.D. or I.D. is used.

Q. *Which type of draught (draft) is more prevalent in marine boilers?*

Ans. Forced draught (draft).

Q. *Why?*

Ans. F.D. fans are smaller in size and hence fit comfortably in the limited space available for marine installation.

Q. *Why is I.D. less favoured?*

Ans. I.D. Fans occupy more vertical space, which is at a premium in marine installations.

Q. *In which cases are both forced and induced draughts (drafts) used in marine boilers?*

Ans. In cases of coal-fired marine boilers.

Q. *Why is the use of a desuperheater in marine boilers not favoured?*

Ans. To eliminate the complications in installation arising out of the fittings, valves and pipings are necessary.

Q. *Is any reheater installed in marine boiler units?*

Ans. Reheaters are rarely installed in marine boilers.

Q. *Why?*

Ans.
1. It adds to the weight and size of the boiler unit
2. It imparts operational complexities.

Q. *What are the primary advantage of steam reheating?*

Ans. When the reheat cycle is incorporated, the overall plant efficiency increases. The superheated steam may become wet after being expanded through turbine stages which tells upon the body of the turbine blades by eroding its surface and thereby reducing its power output. To avoid this problem, steam, after its partial expansion in the turbine, is reheated to super-heated steam and is again delivered to the turbine.

Q. *When is it more economic to adopt a reheat cycle?*

Ans. Reheat cycle becomes more economic to generate high pressure superheated steam for its expansion in the turboalternator for the generation of electric power.

Q. *Why is H.P. steam favoured?*

Ans. HP steam possesses higher enthalpy content than MP or saturated steam and hence a higher quantum of heat drop, during expansion over turbine blades, is available from HP steam before it becomes wet.

Q. *What do you mean by utility boilers?*

Ans. These are boilers that burn coal, oil and natural gas to provide steam to generate electricity.

Q. *What are the primary design criteria for central-station steam generators and auxiliaries?*

Ans. There are three main factors—efficiency, reliability and cost. At present, preference is given to equipment reliability to improve station availability.

Q. *The present day trend is to reduce the size of the utility boilers, i.e., to limit the average size of the units between 300 and 400 MW or even 150 to 250 MW. Why?*

Ans.
1. These smaller sized units require less floor space. They can be fitted comfortably to the available space of existing utility central stations.
2. They require less erection time. A 300 MW unit will take about 6 – 12 months less erection time than a unit twice the size.
3. Economies of scale do not favour large units.
4. Financial risk is limited.

Q. *Why are supercritical once-through units preferred to subcritical drum-type units for large units — 800 MW and above?*

Ans. For very large sizes, the supercritical once-through units are more efficient (i.e., more fuel efficient) than subcritical drum-type boiler units, for the same amount of heat input. Hence supercritical boilers are favoured in those countries (viz. Japan) where the import fuel bill is high and efficient heat

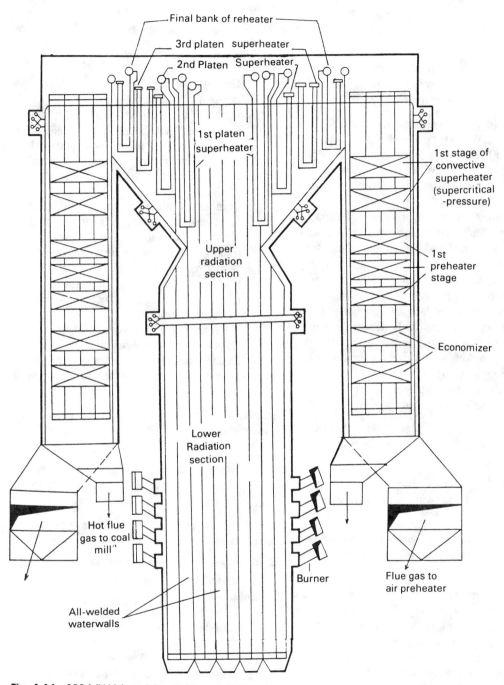

Fig. 1.16 *800 MW Monobloc Unit*
Steam generating cap. 2650 t/h
Steam temperature: 545°C/545°C
Steam pressure: 25.5 MN/m²
T – Shaped in layout, it has its boiler structure suspended from boiler room

utilization overwhelms the higher construction cost of these units.

Q. *Why, is the present trend to prefer once-through boilers to drum-type boilers for 500 MW and higher capacity monobloc power units?*

Ans. Monobloc units with drum-type boiler for large capacity (500 MW) power plants need a high-capacity, high-pressure boiler drum. These giant boiler drums (boiler drum of a 500 MW boiler unit may weigh as much as 200 ton) are considerably larger in dimension and have much thicker wall cross-sections.

The operating experience of these high-capacity, high-pressure drum-type boilers reveals that at varying operating modes (load conditions) in which the unit is shutdown and started up too frequently, non-uniform temperature fields appear in the wall cross-section, producing high thermal stresses that bring about cracks of corrosion fatigue in the boiler drum.

However, in once-through boilers there are no such heavy metal elements as the drum. They are much lighter in weight and can be shutdown and started up more rapidly. For these reasons, once-through boilers are recommended for units with the duty of supplying semi-peak and peak loads.

Q. *Why is the furnace of single-housing boilers for 300MW monobloc powerplants made of prismatic (open) shape without constriction?*

Ans. The purpose is to bring down the average heating intensity of waterwalls in the flame core zone to a safe level.

Q. *Is this the only way to reduce the average heating intensity of waterwalls in the flame core zone to a safe level?*

Ans. No. Recirculation of combustion products drawn at relatively low temperature, i.e., from the convective gas duct (usually downstream of economizer) and reintroducing the same, with the aid of fan, into the furnace will decrease the heat absorption in the lower radiation zone and stabilize temperature conditions of waterwalls. Therefore, the average heating intensity of waterwalls in the flame core zone is reduced, whereupon the possibility of high-temperature corrosion is minimized.

Q. *Of late, steam boilers with gas-tight enclosures are gaining ground. Why?*

Ans.

1. Gas tight waterwalls ensure a substantial increase in economic efficiency
2. It extends the reliability of the boiler units
3. Heat losses to waste gases are reduced as there is no air in-leakage in the furnace and the ducts
4. Low auxiliary power consumption in the supply of air and discharge of combustion products
5. Combustion under optimal conditions with least excess air ratio is possible. This eliminates low-temperature corrosion and fouling of heat-transfer surfaces
6. Heavy refractory lining is replaced by light heat insulation as a result of which
   (a) the mass of boiler structure and foundation is curtailed
   (b) loss of heat decreases
   (c) quicker start-up/shutdown.
7. Easier removal of slag and soot from the furnace by water without the risk of damaging the lining.

Q. *Why is the load-carrying boiler structure predominantly combined with the building structure in modern high-capacity boiler design?*

Ans. To achieve a considerable saving in metal. For example, such a design of an 800MW monobloc unit results in the saving of 1500 ton of metal.

Q. *Briefly describe the boiler of a superthermal power station.*

Ans. One typical example is the TGMP-1202 boiler of suspended design (Russian) for a 1200MW monobloc unit.

It is a supercharged boiler where the mass flowrate of working fluid flowing by single-pass mode in the furnace waterwalls is as high as 2000 kg/m^{2}s.

It is a gas and fuel oil fired supercritical pressure where the burners are laid out in two fronts and two tiers (see Fig. 1.17) and FO burners are used for startup.

The furnace carries membrane sections of unified width and has large depths so that flames do not lick

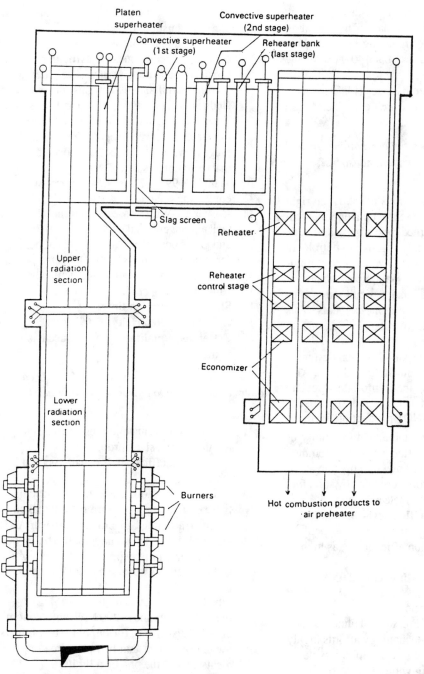

Fig. 1.17 *1200MW Monobloc Unit TGMP-1202*
Gas and fuel oil fired boiler of suspended design generates 3950 ton/h of
supercritical-pressure superheated steam

the waterwalls which are of all-welded type. The reliability of the waterwalls is enhanced by providing the boiler with a water recirculating system which operates when the load is less than 40–50% of the designed load.

In the horizontal convective duct is laid a supercritical pressure superheater which combines platen sections and two convective stages.

Q. *Why are fluidized-bed boilers gaining ground for utility power and cogeneration industries?*

Ans.

1. **Fuel flexibility**—A variety of solid fuels right from low cost solid fuels to conventional fossil fuels can be burned. Fuels burned include many types of coals, biomass, petroleum cokes and other wastes, viz., anthracite mining waste (culm), oil shale, spent shale, tar sands, coal slurries and agricultural wastes using a variety of limestone sorbents. This means flexible, reactive capabilities of fuel market fluctuation.

2. **Lower installation cost**—Modular design allows for lower cost on-site installation. Also modular designs accommodate specific modifications without major redesign cost.

3. **Low emissions**—NO_x production is minimum and they have the added capability of removing SO_2 from flue gas during combustion. These units are known to operate at the lowest emission profile obtained by solid fuel combustor low particulate emission.

4. **High combustion efficiency**—Simple design and fuel flexibility allow users to achieve excellent combustion efficiencies, as high as 98%.

5. **High availabilities**—in excess of 95% is possible.

6. **High turndown ratio**

7. **Combined cycle**—Pressurized Fluidized Bed (PFB) combustion technology affords highly efficient gas-turbine-based combine cycle

8. **Minimum building volume**

9. **Advanced control system**—Allows accurate control of temperature of combustion products in the furnace, minimizing NO_x production, particulate formation and complete combustion of fuels.

10. **Burning hazardous wastes**—Fluidized bed combustion techniques are particularly useful for firing hazardous wastes. Their flexibility permits simultaneous combustion of solids, sludges, slurries and gases in the same unit. These units can attain higher destruction efficiencies at lower temperatures for a wider range of waste materials than most other thermal oxidation systems. Destruction efficiency as high as 99.99% has been reported for fuels containing organic hazards.

11. **Cofiring**—With refuse derived fuel (RDF) as well as conventional pulverized coal, fuel oils and gas is possible. This is an added bonanza for fluidized combustion systems using low-cost fuels to meet higher load demands.

12. **Low operating cost**

13. **Short construction time**

14. **Low cost of electricity**

15. **Better ash quality**—Ash produced is non-toxic

Q. *What are the disadvantages of fluidized-bed boilers as compared to pulverized coal fired boilers?*

Ans.

1. The feed stream of solid fuels must be closely monitored and sized properly.

2. Combustion efficiency is usually poor at low loads.

3. When burning low grade fuels, particularly those containing salts, care must be taken to maintain accurate bed temperature over AFT. Otherwise the salts may melt at the bed temperatures or may produce eutectics in the bed causing bed seizure and loss of fluidization.

4. Inbed tube erosion is higher.

5. For utility units above 200 MW, any economy of scale disappears for bubbling-bed boilers owing to the fact that bed plan area requirements exceed the practical limits.

   CFB is handicapped by the same problem. It is height-limited as unit output increases due to the large amount of solids that must be circulated through the furnace. After this height-limit is reached, the designer must increase the bed plan area to accommodate greater heat-transfer surface requirements. And from a practical point of view this is not affordable.

6. Pulverized coal-fired conventional boilers register higher efficiency than fluidized-bed boilers of units above 200 MW. The efficiency gain is about 1 ½%.

7. P.C. fired boilers require at least half the limestone requirement of fluidized bed boilers for equivalent sulphur arrest.

8. The most nagging drawback with fluidized bed boilers is that they produce a substantial quantity of solid waste (ash) that must be safely disposed of. Some experts believe that the quantity is even greater than that from a conventional pulverized- coal-fired-boiler befitted with a wet limestone scrubber at equivalent conditions.

Q. *Why is the quantity of ash produced by a FBC unit greater than a conventional pulverized-coal-fired boiler fitted with wet-limestone-scrubber at equivalent conditions?*

Ans. This is mainly due to the fact that whereas a wet scrubber can operate at a calcium-to-sulphur ratio close to 1.0, an FBC unit on the other hand may require anywhere from 1.5 to 3.0.

Q. *How can fluidized-bed boilers be classified?*

Ans. All fluidized-bed boilers can be categorized into two main groups, depending on the mode of operation of the fluidized bed.

These are:
1. Atmospheric Fluidized Bed (AFB) boilers
2. Pressurized Fluidized Bed (PFB) boilers

Atmospheric Fluidized Bed boilers can be of two types:
1. Atmospheric Bubbling Bed boilers (Fig. 1.18)
2. Atmospheric Circulating-Bed boilers (CFB) (Fig. 1.19)

AFB boilers can be subdivided into
(a) Bubbling-Bed with inbed tubes
(b) Bubbling-Bed without inbed tubes
(c) Circulating-Fluidized-Bed with external heat exchangers
(d) Circulating-Fluidized-Bed without external heat exchangers.

A distinction can also be made on the basis of fluidizing velocity of air, which is the fundamental distinguishing feature of fluidized-bed-combustion units.(Fig.1.20)

Bubbling-beds have lower fluidization velocities (1.2–3.6 m/s) so that particles do not escape from the

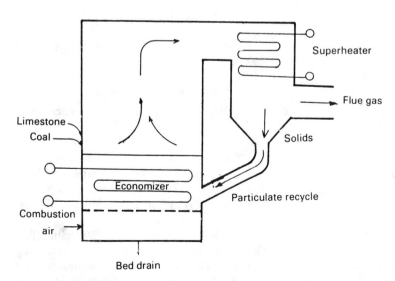

Fig. 1.18 Bubbling-bed boiler. It has inbed evaporator tubes and superheater tubes in the convective shaft

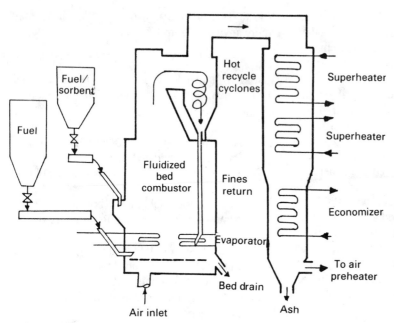

Fig. 1.19 *Circulating Fluidized Bed. Modular design for cost effective field installation and advanced emission control is possible*

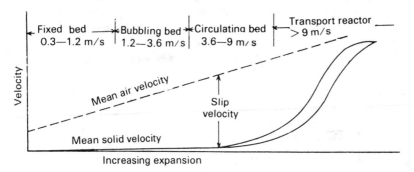

Fig. 1.20 *Fluidizing velocity of air is the fundamental distinguishing feature of fluidized bed combustion units. Bubbling beds have low air velocity while the CFBS have higher*

bed (elutriation) into the convective passes. CFBs apply higher velocities (3.6–9 m/s) and in fact promote solids elutriation.

Q. *What are the principal attributes of atmospheric-fluidized-bed (AFB) units?*

Ans.

    1. In a typical AFB combustion unit, solid, liquid or gaseous fuels (particularly coal plus solid waste fuels, viz., culm, municipal wastes, etc.), together with inert material—viz., silica, alumina or ash—and/or a sorbant such as limestone are kept suspended in the combustion chamber by primary air distributed below the combustor floor.

    2. Turbulence is promoted by fluidization that turns the entire mass of solids into something like a liquid. Owing to improved mixing, heat is generated at a substantially lower and more uniformly distributed temperature—typically

810°–870°C than a stoker fired unit or a pulverized coal fired boiler. Thus given the properly sized combustion chamber, the heat release rate will be at a level comparable to a conventional boiler, but at a lower temperature and theoretically without any loss of efficiency.

3. The operating temperature (810°–870°C) of an FBC unit is well below the formation of thermal NO_x

4. Staged combustion can be applied to minimize the formation of fuel-bound NO_x as well

5. The operating temperature range is such that the reactions of SO_2 with a suitable sorbent, commonly limestone, are thermally balanced

6. The fluidization mechanism leads to less volatilization of alkali compounds

7. Added turbulence minimizes the chances of a hot spot on the boiler and shell surfaces

8. Less sensitivity to quantity and nature of ash

9. Smaller furnace volume.

Q. *What is the layout of evaporator tubes in AFB units?*

Ans. The evaporator tubes serve as a suitable heat sink in the combustion area to control temperature. They are laid out in the traditional pattern of waterwall enclosures. There may be different designs as, for instance, where the boiler tubes are located within the bed itself.

Q. *Does this mean that only watertube design is possible?*

Ans. No; both watertube and firetube designs are possible. Firetube boilers are more popular for larger sizes.

Q. *How is steaming rate controlled in AFB units?*

Ans. This is effected by manipulating the primary bed parameters
  (a) bed height
  (b) bed temperature
  (c) inventory
plus controlling the superficial gas velocity. Indeed the manipulation of gas velocity accomplishes most of the control.

Q. *Why are the bubbling fluidized bed boilers so called?*

Ans. The combustors of such boilers are provided with a grid upon which is supported a dense bed of fuel and limestone. The bed is maintained in a fluid-like state by air blown up through the grid and the bed from the bottom of gridholes. As the air flows upwards it mixes the limestone and fuel and at the same time keeps the bed in suspension. The action of air creates extreme turbulence in the bed material and produces the effect of a bubbling fluid.

Q. *Can't it be made amenable to limited load fluctuation?*

Ans. Yes; limited load fluctuation can be made possible.

Q. *How?*

Ans. By lessening the amount of air passing through the bed.

Q. *Why?*

Ans. This lowers the bed depth and retards combustion.

Q. *What is bed-slumping?*

Ans. Reducing bed depth is sometimes called bed-slumping.

Q. *Why must the velocity of air through the bed be kept always above the minimum fluidizing velocity?*

Ans. To avoid incomplete combustion which results in smoking. If the velocity of air through the bed is below the minimum fluidizing velocity, it will affect the removal of sulphur dioxide.

Q. *How many types of bubbling-bed boilers are there?*

Ans. Two:
  (a) single-stage bubbling-bed units
  (b) two-stage bubbling-bed units

Q. *What is the concept behind two-stage bubbling-bed units?*

Ans. The basic idea is to separate the combustion process from the absorption process so that each can be optimized without the design compromises inherent in the single-bed unit. For instance, a higher combustion temperature can be attained in a lower bed since absorption is no longer a limiting factor.

Q. *What is the world's largest bubbling-bed boiler?*

Ans. It is the 160 MW bubbling-bed boiler of Tennessee Valley Authority. Till date (1989) it is the largest AFB bubbling-bed unit.

Q. *What are its important features?*

Ans.
1. It supports a single-level bed divided into 12 compartments
2. It has under-bed feed system
3. Its primary superheater and evaporator are fitted in the bed
4. Its finishing superheater and reheater are fitted in a split convection pass
5. Main-steam temperature control by spray water
6. Reheat-steam temperature control by gas-flow dampers

Q. *What is the world's largest circulating fluidized bed (CFB) unit?*

Ans. It is the 150MW unit of the Texas-New Mexico Power Company.

Q. *What is its fuel?*

Ans. Texas lignite

Q. *What are its important features?*

Ans.
1. Dense-phase fuel injection in the combustor
2. Single combustor system fitted with water-cooled walls, grate and plenum
3. Four refractory-lined cyclones for recirculation of unburnt fuel
4. Two numbers of external fluidized-bed heat exchangers (FBHE) each with watercooled walls
5. One FBHE sports a superheater and evaporator while the other has one reheater and superheater

Q. *What are the advantages of CFB units over bubbling-bed systems?*

Ans.
1. Larger residence time of fuel that enhances combustion efficiency
2. Greater amount of absorption of SO_2 and other acid gases because of longer residence time
3. Use can be made of less-prepared fuel and sorbent
4. Free from many problems associated with underbed and/or overbed feeding of the fuel as encountered in bubbling-bed combustion system.

Q. *What is the common disadvantage of a CFB unit?*

Ans. Higher electricity bill because of greater fan power is required to maintain the higher velocity through bed.

Q. *What is PFBC?*

Ans. It means Pressurized Fluidized Bed Combustion

Q. *What are the two key features that distinguish PFBC units from AFB units?*

Ans.
1. Combustion takes place at pressures higher than the atmospheric
2. It couples two cycles—conventional steam-turbine cycle with the gas-turbine cycle

Q. *How does PFBC work?*

Ans. Fluidized bed combustion takes place at elevated pressures.

It incorporates a gas turbine that drives an air compressor which supplies combustion air at high pressure to the fluidized bed combustor. The resulting flue gas, also at high pressure, is cleaned of suspended solids and directed to drive the gas turbine. (Fig. 1.21)

Evaporator tubes laid out in the fluidized-bed combustor abstract the heat generated, producing steam which drives a steam turbine to produce 80% of the units' power.

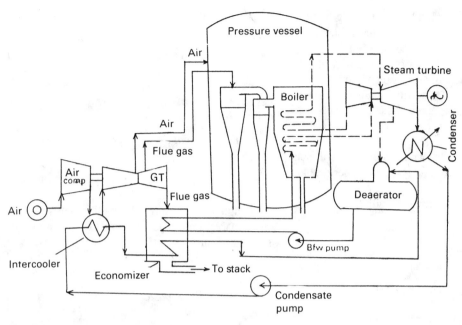

Fig. 1.21 *Pressurized fluidized bed boiler*
It operates on steam turbine and gas turbine cycle combined to achieve high overall efficiency

Q. *What is the paramount advantage of a PFBC unit over conventional and atmospheric fluidized-bed boilers?*

Ans. PFBC units offer higher efficiency than conventional and AFB units.

Q. *What is the operational efficiency of a PFBC unit?*

Ans. PFBC can operate at 40–42% efficiency.
  The designed efficiency of the 80MW PFBC unit of the Tidd Power Station, USA is 40%. Another 320MW PFBC unit with 42% efficiency is planned. Other commercial units that will come to life in future will have still greater efficiencies.

Q. *What primary factors are being considered in the selection of industrial boilers today?*

Ans.
1. Fuel flexibility over the life of the boiler unit
2. Capability to burn low-quality fuels or derived fuels
3. Possible hook-up for cogeneration facility
4.

4. Capability to fulfill ever-increasing emission standards
5. Optimizing the existing equipment regarding the efficiency, performance and service life
6. Turndown and part-load capability

Q. *How are industrial boilers classified?*

Ans.
1. Watertube boiler
2. Firetube boiler
3. Heat recovery boiler
4. Electrode boiler
5. Packaged boiler
6. Field erected boiler
7. Fluidized-bed boiler
8. Refuse-fired boiler
9. Unfired boiler

Q. *Briefly describe watertube boilers used for industrial purposes?*

Ans. Only a few years ago all industrial watertube boilers would generate most of the steam in the convective bank of waterwall tubes. With the growth of demand of steam at higher pressure and temperature, and to meet the industry's need for higher capacity and higher efficiency, most of the steam is now-a-days generated in the radiant section of the waterwall tubes where heat transfer occurs principally through radiation. The design features of these radiant boilers are basically the same as those used in utility central stations.

Fuels burned range from conventional fossil fuels to a wide range of waste and low-grade energy sources. These include:

(a) Industrial and municipal sludges
(b) Woodwaste
(c) Petroleum coke
(d) Coal-mining waste
(e) Liquid industrial wastes
(f) Blast furnace gas
(g) Gas from sewage-sludge digesters

Q. *In which cases are packaged boilers used?*

Ans. Packaged boilers suit those industries where the steam requirement is modest.

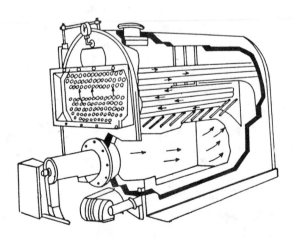

Fig. 1.22 *A Typical two-pass packaged boiler. It is a fire-tube type boiler that burns liquid fossil fuels*

Q. *What fuels are used for packaged boilers?*

Ans. Packaged boilers are almost exclusively designed for liquid and gaseous fuels.

Also, pulverized solid fuel can be used to generate steam to the highest practical limit of 45 t/h if the solid fuel has high calorific value and the fuel is pulverized to ultra-fine particles.

Q. *Is the practical limit of a packaged boiler 45 t/h steam?*

Ans. Yes; because of the size that is commensurate with higher steam demand.

Though packaged units of capabilities upto 272 t/h are available, boilers larger than 112 t/h cannot be supplied by rail.

Note: Packaged boilers are shop-assembled units.

Q. *What are the basic design characteristics of packaged boilers?*

Ans.
1. Maximum use of vertical or near-vertical waterwall tubes in the radiant as well as convective zone of the furnace
2. Use of economizer to hike up overall efficiency
3. Maximum use of natural circulation of working fluid to increase heat absorption

Q. *What are the usual structural configurations of packaged boilers?*

Ans. Most of the packaged boiler units come in the form of A, O and D type configurations.

Q. *What are the distinguishing features of each such type of packaged unit?*

Ans. The D-type boiler is the most flexible among the three. It has
(a) only two drums
(b) large volume of combustion space to fit a superheater or economizer into it

The O-type boiler is a symmetrical unit. It exposes the least amount of heat to the radiant surfaces.

The A-type packaged boiler is a tridrum type unit. It has two small lower drums plus one large upper drum where separation of steam from water takes place.

Q. *What are the ranges of temperature and pressure in which the packaged boilers, in general, operate?*

Ans. Operating pressure : 8.5–68 atm.
Operating temperature : 510°C (maxm.)

Q. *Can these boiler units be hooked up for cogeneration purposes?*

Ans. Yes; only the largest units can lend themselves for cogeneration facilities.

Q. *Why does this limitation exist?*

Ans. Low superheat temperature is available due to space limitation.

Q. *Is water carryover into the steam a problem with packaged units?*

Ans. Yes.

Q. *What measures are taken to counter it?*

Ans. Steam separators are installed in the steam drum.

Q. *What type of separators are used?*

Ans. Right from the simplest ones (e.g., baffle and chevron types) to highly sopisticated centrifugal types are used.

The simplest types are used for separating low pressure saturated steam while for high pressure steams (to be used in steam turbines where water droplet carryover in steam is detrimental) centrifugal separators are used.

Q. *What types of superheaters are used in packaged boilers?*

Ans.
1. Radiant type
2. Radiant-convective type

Q. *What is the important difference between the working characteristics of a radiant type and a radiant-convective type superheater?*

Ans. Radiant type superheater superheats the steam to higher-than-design temperatures at low loads.

Radiant-convective superheater maintains a relatively steady superheat temperature over the entire load range.

Q. *What are field erected boilers?*

Ans. These are predominantly watertube boilers erected in the field.

Fuels used range from coal, solid waste (wood-waste, municipal waste and industrial waste) to oil, gas and gaseous or liquid wastes. (Fig. 1.23)

When coal and solid biomass fuels are used, stoker firing is the dominant choice.

The mode of firing may be of the suspension type if pulverized coal or biomass fuel in the form of dry fines, e.g., sawdust, sanderdust and rice hulls are available. Suspension firing and fluidized bed combustion are usually the choices for the larger units.

Q. *In which case are fluidized-bed field erected boilers preferred to suspension fired boilers?*

Ans. Though for larger units, fluidized-bed boilers are too expensive, they are preferred when stringent restrictions on the emission of NO_x, SO_x and particulates are imposed.

Q. *What are the usual operating temperatures pressures and capacities field erected boilers?*

Ans. Steam generating capacity 20 t/h to 180 t/h
Operating (Steam) temperature 450–500°C
Operating Steam pressure 61–102 atm.

Q. *Why is close control of furnace-exit gas temperature necessary in the case of refuse-fired boilers?*

Ans. This temperature monitoring and control is imperative to minimize slag and flyash deposition.

Q. *What types of fluidized-bed boilers are used as industrial units?*

Ans.
1. Bubbling-bed boilers
2. Circulating fluidized-bed boilers are the fluidized-bed units mostly used to meet the demands of industries. They are all watertube boilers.

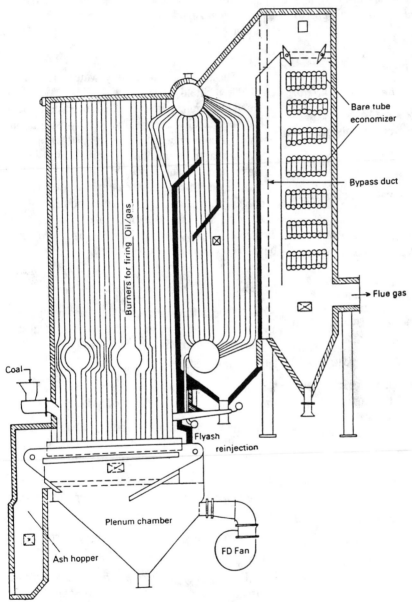

Fig. 1.23 *Field-erected spreader stoker boiler unit*
It burns a wide variety of coal and solid wastes in addition to gas and liquid fuels.
Unit capacity is 115 ton/hr

Q. *How are the waterwalls arranged in the bubbling-bed boiler?*

Ans. The waterwall tubes are oriented either horizontally or vertically within the bed.

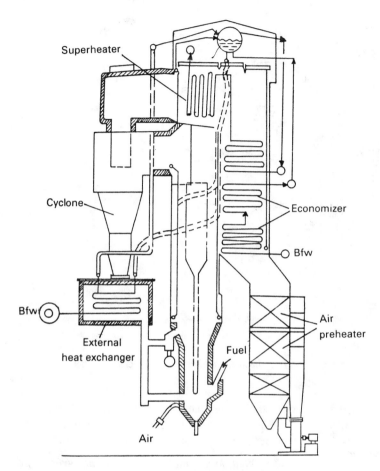

Fig. 1.24 *CFB fitted with external heat exchanger*

Q. *What is the chief advantage of vertically laid waterwall tubes over horizontally laid tubes?*

Ans. Horizontal steam-generating tubes are associated with more erosion problems than the tubes disposed vertically.

Q. *What measures are adopted to minimize erosion of in-bed tubes?*

Ans. Tubes are
  (a) fitted with studs or fins
  (b) coated with protective metal coating
  (c) chromized to combat erosion.

Q. *What parameters are manipulated to control the steaming rate?*

Ans.
  1. Bed height
  2. Bed temperature
  3. Solids inventory
  4. Superficial gas velocity

Q. *What are the basic characteristics of circulating fluidized-bed boilers?*

Ans.
  1. They generally do not have in-bed tubes
  2. They use hot-solids separation device (cyclones) to minimize erosion of heat absorption surfaces
  3. Many CFBs have an external heat exchanger (EHE) (Fig. 1.24)

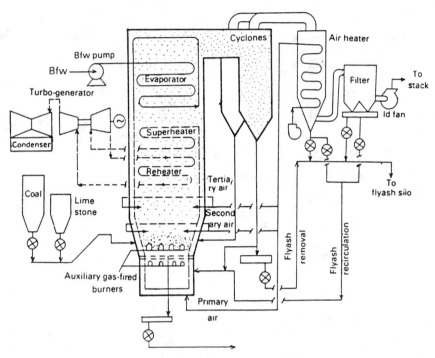

Fig. 1.25 *Circulating fluidized bed-boiler. It has no EHE. Its superheater is placed in the dense phase flow*

Some CFBs do not have EHE. The superheater is placed in the dense-phase flow at the top of the furnace (Fig.1.25) or in the convective shaft downstream of the particle separator.

Q. *What is EHE?*

Ans. It is a refractory lined box into which are arranged tube bundles.

Q. *What is the purpose of fitting an EHE?*

Ans. It cools the solids returning from cyclone separator. Changes in load conditions and fuel properties alters the heat absorption rate in the furnace. The EHE helps to compensate for variations in the rate of heat-absorption.

Heat transfer in the EHE increases when absorption in the combustor decreases and vice-versa.

Q. *Are the CFBs provided with sootblowers?*

Ans. Though the requirements of sootblowing for CFBs are much less than for other solid-fuels fired boilers, such units are provided with sootblowers.

Q. *What do you mean by the heating surface of a boiler?*

Ans. It is the surface area of boiler tubes exposed to the hot gases of combustion in the furnace space in order to transfer heat to the working fluid to generate steam.

Q. *How many types of heating surface may a boiler have?*

Ans.

1. **Radiant Heating Surface**—waterwalls receiving direct radiant flame heat
2. **Convective Heating Surface**—located in the convective shaft of the furnace and receives heat entirely by convection
3. **Radiant-Convective Heating Surface**—heated partly by radiant heat and partly by convective heat

Q. *In which type of boiler—watertube or firetube— is the heating surface more for tubes with the same specifications?*

Ans. Watertube boilers.

Q. *Why?*

Ans. The effective heating surface of watertube boilers is the external surface area of the tubes exposed to furnace gases while the heating surface of firetube boilers is the internal surface area of the tubes. For tubes having the same dimension, the internal surface area of the tube is less than its external surface area. Therefore, watertube boilers will have greater heating surface than firetube boilers.

Q. *Why is a safety relief valve installed close to a pressure reduction valve in a steam line?*

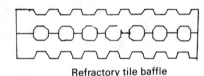

Refractory tile baffle

(a)

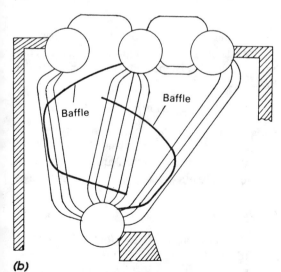

(b)

Fig. 1.26 *(a) Refractory tile*
(b) Baffle arrangement in a bent-tube boiler

Ans. It is an added safety to the steam line to avoid sudden steam pressure development in the steam main in case the pressure reducing valve fails to operate properly, i.e., the valve may get stuck-up in the open position, impairing the downstream pressure of steam (L.P.). The safety valve fitted closely to the PRV is set at a pressure slightly greater than L.P. steam pressure and it goes off if there is an accidental pressure development in the low pressure side.

Q. *What are boiler baffles?*

Ans. The flow of hot combustion gas over the banks of boiler tubes is controlled by what are called baffles.[Fig. 1.26]

Q. *Why are baffles placed in the path of hot flue gases in the convective shaft of watertube boilers?*

Ans. To put into effect a more even distribution and transfer of convective heat from hot combustion gases to the convective heating surfaces of the boiler unit.

Moreover, they maintain proper gas velocity, facilitate flyash and soot deposition and reduce draft loss.

Q. *Upon what factors does baffle arrangement depend?*

Ans.
1. Allowable draft (draught) loss
2. The layout of boiler heating surface and its dimension
3. Gas outlet position
4. Fuel
5. Fuel burning equipment and operating parameters
6. Baffle supporting device

Q. *What factors should be considered in designing a baffle?*

Ans. The baffle design should be such that it
(a) ensures best distribution of heating surface with respect to the gaseous products of combustion
(b) ensures proper flue gas velocity
(c) maintains cross-flow of flue gas past the boiler tubes
(d) deposits flyash and dust where it is accessible for easy removal

(e) eliminates dead pockets

(f) avoids high draught (draft) loss

Q. *When may the furnace baffles open out?*

Ans. Due to crack or formation of patch holes.

Q. *What will happen in the case of baffle failure?*

Ans. Combustion gases will be short circuited when baffles open out. A modified heat redistribution will take place. Some parts will get more overheated than the others. Tubes may fail. Excessive flue gas temperature at the stack outlet will occur and this will bring down the overall thermal efficiency of the boiler.

Q. *Why is a slag screen installed?*

Ans. The slag screen, as its name implies, is installed to overcome clogging of boiler tubes as a result of cooling and adhering of molten slag on the tube surface.

Q. *Where is it installed?*

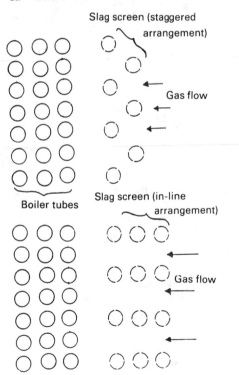

Fig. 1.27 *Slag screen is installed upstream of boiler tubes*

Ans. At the entry to the convective shaft of the boiler furnace.

Q. *Why?*

Ans. It is required that hot flue gases and molten slag, before their entry to the convective shaft of the furnace, should be cooled to a point at which they will not stick to the boiler tube surfaces. To avoid clogging, the watertube boilers are given the facility of slag screens installed at the entry.

Q. *What are the drawbacks of a slag screen?*

Ans. It
1. limits the maximum distance for effective tube cleaning by soot blowing. This imposes restrictions on tube bank design.
2. reduces the accessibility for the repairing or replacement of faulty boiler tubes.

Q. *In what pattern can a slag screen be arranged?*

Ans. Two:
1. in-line arrangement
2. staggered arrangement.[Fig. 1.27]

Q. *Which one is better?*

Ans. In-line arrangement.

Q. *Why?*

Ans. In-line arrangement has wider tube spacings. It reduces slagging. The wider tube spacing facilitates manual cleaning, washing and replacement of boiler tubes.

Q. *Why is the steam drum provided?*

Ans. It is provided to carry out the process of separation and steam purification at higher pressures with the aid of baffles and other devices called steam internals as well as to provide an adequate amount of steam storage.

Q. *What are drum internals?*

Ans. It includes all apparatus within a steam drum, viz., baffles and various devices to effect the separation of steam from steam-water mixture, the various chemical and feedwater admission pipelines and blowdown lines.

Q. *What should be the basic qualities of drum internals?*

Ans. They should

1. provide highly pure steam to prevent superheater tubes from burn-outs and the steam turbine from fall-out of the nominal capacity

2. maintain purity of steam at a steady rate despite the fluctuation of boiler water concentrations over a wide range

3. maintain steam purity despite the fluctuations of water level occurring during normal operation

4. ensure steam-free water to the downcomer tubes for maximum circulation

5. ensure minimum pressure drop through the separators

6. provide maximum accessibility for inspection of the drum

7. be basically simple in design to minimize the installation and removal time.

Q. *What fundamental steps are involved during the process of steam separation and purification in the steam drum?*

Ans. The process involves three steps:
1. separation
2. steam washing
3. steam scrubbing

Separation is the primary step while steam washing and steam scrubbing are secondary.

Q. *What is steam separation?*

Ans. It is the process of eliminating bulk masses of water droplets from steam.

Q. *How can it be achieved?*

Ans. This can be effected by any of the following three methods:
1. gravity separation
2. abrupt change in direction of flow pattern
3. centrifugal separation
4. using baffles
5. impact against a plate

Q. *What is a dry pipe?*

Ans. It is a perforated or slotted pipe placed in the highest part of the boiler to provide commercially dry (97% dry) steam. It is a capped pipe fitted with a drain to prevent accumulation of water and perforated openings at the top that allow the entry of steam but discourage the entry of water. The openings are drilled holes (20 mm) dia or slots (7.5 mm) wide. (Fig. 1.28)

Q. *What are baffles?*

Ans. Baffles are steam drum internals that break up the steam-water jets entering the drum.

Q. *How many types of baffles are in use?*

Ans.
1. Deflector-plate baffle (Fig. 1.29)

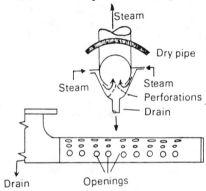

Fig. 1.28 *Dry pipe*

Fig. 1.29

Fig. 1.30

Fig. 1.31

Fig. 1.32

Fig. 1.33

Fig. 1.34

Fig. 1.35

Fig. 1.36

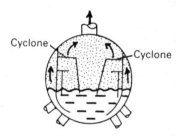

Fig. 1.37 *Cyclone separator*

2. Offset deflector-plate baffle (Fig. 1.30)
3. Slotted deflector-plate baffle (Fig. 1.31)
4. V-baffle (Fig. 1.32)
5. Perforated plate and V-baffle (Fig. 1.33)
6. Angle-iron deflector baffle (Fig. 1.34)
7. Hydraulic barrage baffle (Fig. 1.35)
8. Compartment baffle (Fig. 1.36)

Q. *Why are deflectors or compartment baffles necessary?*

Ans. For low steam release rates a V-baffle or single vertical baffle is adequate, but for higher release rates deflectors or compartment baffles are necessary for separation.

Q. *When is a hydraulic barrage installed?*

Ans. These types of baffles are installed to control excessive foam or spray formation.

Q. *How does it perform this job?*

Ans. It creates a successful high velocity screen or curtain through which steam must pass in order to exit. As a result, the water particles losing their kinetic energy by successive impacts with the 'curtains' fall off, and steam escapes dry.

Q. *What is a cyclone separator?*

Ans. This is an effective device to harness steam-water separation by utilizing centrifugal forces.[Fig. 1.37]

Q. *How does it do so?*

Ans. It imparts to the steam-water mixture, as it enters the drum, a rotary motion which develops a centrifugal force that destroys the foam bubble, eliminates spray and separates out solid particles.

Q. *What is steam washing?*

Ans. It is the process of rinsing the separated steam with steam condensate or fresh, relatively clean feedwater.

Q. *What is its purpose?*

Ans. Its purpose is to condense out the dissolved silica from steam as well as to wash out impurities carried over by steam.

Q. *How much of the silica carryover by steam can be eliminated by the process of steam washing?*

Ans. As much as 90% with a proper washer.

Q. *What are the ways to reduce the silica content of steam leaving the washer?*

Ans.

1. Increasing the stage efficiency of the washer
2. Adding more washer stages in series
3. Increasing the ratio (by weight) of wash water to steam
4. Washing the steam with water with lower silica content
5. Reducing the silica content of steam entering the washer

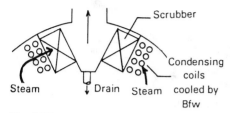

Fig. 1.38

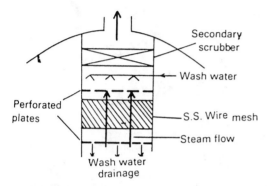

Fig. 1.39

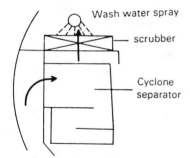

Fig. 1.40

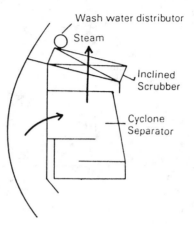

Fig. 1.41

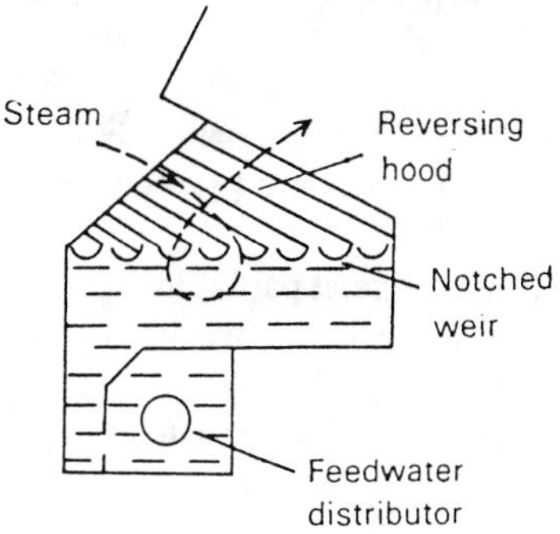

Fig. 1.42

Q. *How many types of steam washers are in vogue?*

Ans.

1. Condensing type (Fig. 1.38)
2. Wire-mesh type (Fig. 1.39)
3. Spray-type (Fig. 1.40)
4. Wetted scrubber type (Fig. 1.41)
5. Hood type (Fig. 1.42)

Q. *How does the condensing type steam washer function?*

Ans. In this device steam is passed over condensing coils cooled by feedwater flowing through them. Upon contact with the relatively cold metal surface of the coil, a portion of the steam is condensed and this condensate provides a washing liquid for the rest of the saturated steam. This steam condensate is of higher quality than feedwater and hence possesses a higher affinity for silica. Thus this kind of steam washer is quite effective for removing vapourized silica as well as the solids carried in the moisture and other impurities.

Q. *What are the basic features to be kept in mind while designing steam washers?*

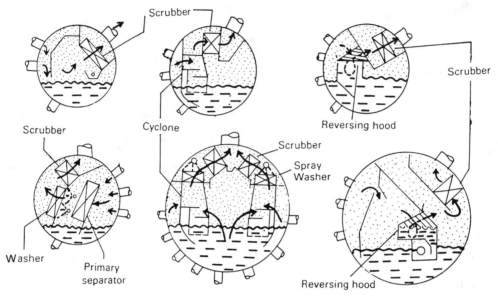

Fig. 1.43 *Various arrangements of steam scrubbers*

Ans.
1. They must provide for intimate mixing of washing liquid (feedwater/condensate) with steam
2. They must be effective at partial load
3. They should have a minimal tendency to adhere to sludge
4. Multistage washing for higher efficiency

Q. *What is steam scrubbing?*

Ans. It is analogous to filtration. The wet steam is passed between closely spaced corrugated plates or screens that collect the water droplets carried by steam. The mass of water collected on the screens is dripped by gravity to the water below in the boiler drum. (Fig. 1.43)

The scrubber eliminates most of the water droplets added to steam from spray and splatter. Also, it breaks up and eliminates dry foam.

Q. *How much water carryover by steam can be arrested by the scrubber?*

Ans. It can slash the water carryover by steam down to 1/30th of the 1% moisture at full load.

Q. *On which factors does the efficiency of a steam scrubber depend?*

Ans. It depends upon such variables as:
1. the surface area of the scrubber
2. time of contact
3. velocity and pressure of steam
4. the shape of steam passage (sinuous passage)

Q. *What should be the velocity of steam at low pressure?*

Ans. 2.45–3 m/s.

Q. *What should be the steam velocity at high pressure?*

Ans. 0.6 m/s

Q. *The scrubber is not highly effective in removing small quantities of fine mists suspended in steam as more efficient scrubbers are impractical. Why?*

Ans. More efficient scrubbers cannot be designed because of the limitations imposed by such factors as plugging, corrosion and pressure drop. A scrubber covered with a heavy accumulation of wet foam may break and give rise to a high degree of carryover.

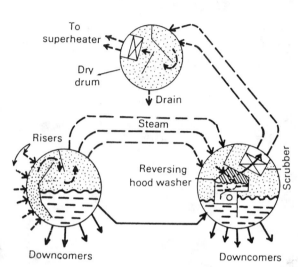

To superheater

Dry drum

Drain

Steam

Risers

Reversing hood washer

Scrubber

Downcomers

Downcomers

Fig. 1.44 *Dry drum arrangement*

Q. *A scrubber fitted together with a washer is more efficient. Why?*

Ans. As the steam passes through the washer, the moisture droplets suspended in the steam get enlarged lending themselves more amenable to fall-out upon contact with the metal surfaces as the steam moves through the scrubber. This increases the effectiveness of the scrubber and helps reduce the carryover.

Q. *What is a dry drum?*

Ans. It is an additional drum fitted occasionally above the boiler drum proper as an extra safeguard from carryover. As no boiler water is held up within its space, it is called a dry drum. (Fig. 1.44)

Q. *For what purpose is it set up?*

Ans. A dry drum, sufficiently large and provided with adequate drainage facility is a good protection against priming and foamover.

Q. *What are the basic elements of consideration in designing a boiler drum?*

Ans.
1. Limitations of space
2. Even distribution of steam
3. Pressure drop across the unit
4. Drainage facility
5. Problem of corrosion
6. Accessibility to drum internals
7. Problem of dirt deposition
8. Cleaning
9. Weight
10. Strength
11. Method of installation.

Q. *Why is feedwater treatment added to the boiler rather than to the feedwater itself?*

Ans. This is to eliminate the problem of plugging of washer by sludge resulting from chemical reactions.

Q. *In some marine boilers, water baffles—called swash plates—are installed in the lower half of athwartship (across-ship) steam drums. Why?*

Ans. These plates reduce the surging of boiler water from end to end in the drum as the ship rolls in choppy seas.

Q. *Why is a dual circulation boiler installed?*

Ans. For boilers operating at pressures above 4137 kN/m^2 (40.96 atm.), the solubility of silica in steam becomes considerable and may cause blade deposition in the steam turbines. The dual circulation boiler is designed to eliminate this silica carryover in particular.

Q. *How does it work?*

Ans. It works on the principle of stage evaporation. It has two separate heat absorbing sections—the heavy duty primary section is composed of furnace waterwall tubes and furnace circulatory system while the low duty secondary section is constituted of a boiler convection tubes bank which is moderately heated.

These two separate sections, each having its own independent circulating system, are functionally identical to two separate boilers mounted in one setting.

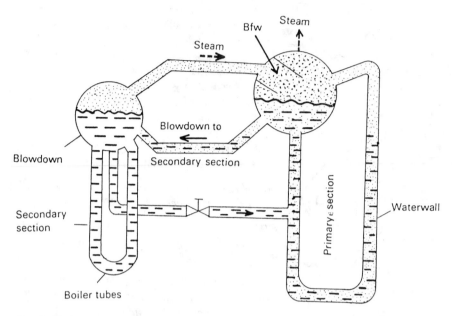

Fig. 1.45 *Dual circulation boiler*

BFW is delivered to the primary section while continuous blowdown from this section serves as feedwater to the secondary section. (Fig. 1.45)

In comparison to the steam output of the unit, the blowdown from the primary section is considerable, rendering a low-silica- feedwater maintained in the primary section while a much higher concentration of silica is maintained in the secondary section where the steam generation rate being comparatively low, the possibility of foaming and excessive carryover is precluded. Since the primary section contributes to the lion's share of overall steam output, a low concentration of silica in the boiler water of this section reduces the carryover of silica by steam. Moreover, the generated steam in the secondary section being scrubbed by feedwater, a part of it gets condensed and the silica contamination in steam leaving this section is greatly reduced.

The drum internals of these two sections are totally segregated by a baffle arrangement that does not allow the feedwater and condensate to dilute the feed to the secondary section or the high-silica water of the secondary section to contaminate the feedwater in the primary section.

Q. *What are the distinct advantages of a dual circulation steam generation unit?*

Ans.

1. Marked reduction in carryover of silica as well as total dissolved solids in steam
2. Permits higher impurity level in the intake feed
3. Low blowdown as high concentrations are maintained in the secondary section. This also minimizes heat loss and feedwater make-up
4. Lower concentration of silica and TDS in BFW in the high duty furnace section greatly reduces the danger of scaling
5. Permits higher heat absorption rate over a prolonged period
6. Outage time and maintenance cost are minimal
7. Needs smaller economizer
8. Reduces treatment chemical cost

Q. *What are the upper pressure limitations of boilers?*

Ans.

Steel Firetube Boiler	10–17 atm
Steel Watertube Boiler	No effective limitations. Boilers for operation at 340 atm have been built
Cast-Iron Boiler	1 atm steam
	2 atm hot water (although hot water heaters may operate at 11 atm if the temperature does not exceed 120°C)

Q. *What are the effects of expansion and contraction?*

Ans. When heated the boilers expand. When cooled they contract. The magnitude of expansion depends on the source of the heating effect, i.e., whether the heat source is a radiant flame or relatively cooler flue gases. The heating and cooling is not uniform and may give rise to hot spots within the boiler. Hence, a boiler designer should give necessary consideration to expansion.

Q. *What about the expansion problem in steel boiler?*

Ans. Properly designed steel boilers present little expansion difficulty and is relatively free from stresses due to expansion and contraction. The connecting pipelines should be adequately provided for expansion and movement.

Q. *What is the effect of expansion or contraction stresses on tubes?*

Ans. Leakage develops where the tubes are bent into the drum or tubesheet.

Q. *What about the expansion and contraction problem in a cast-iron boiler?*

Ans. Thermal stresses induced due to expansion or contraction will warp sections of the cast-iron boiler. And deformed sections take a permanent shape. Replacement may be difficult.

Q. *How does flame impingement affect the boiler?*

Ans. As the flame impinges on waterwalls, it becomes cooled and the flame temperature drops which in turn may result in incomplete combustion and soot formation.

Q. *Does a draft affect boiler operation?*

Ans. The draft problems that are encountered when a chimney alone is used are eliminated if draft fans are installed. Practical experiences show that a draft fan is far more economical than a chimney of adequate height in all cases except those of small installations.

Q. *So a chimney can be eliminated if draft fans are installed. Is it?*

Ans. No. Some kind of chimney or stack is necessary despite the installation and operation of draft fans. This is because a chimney/stack of appropriate height will permit discharge of flue gases to such a level that the dilution of obnoxious gases (air pollutants) by the atmospheric air will be within acceptable limits.

Q. *What is the need for a large water capacity?*

Ans. If a boiler has large water capacity, it will function smoothly and better against a relatively steady load or slowly varying load.

Q. *For sudden load changing conditions, what effect will the boiler water capacity have?*

Ans. If sudden load-changes are prevalent, boilers having less or a small water capacity will be more responsive to quicker load fluctuations.

Q. *Compare watertube and firetube boilers in the light of water capacity?*

Ans. Watertube boilers have a smaller water capacity than firetube boilers. Therefore watertube boilers are more responsive to a rapid load change.

The larger water capacity of firetube boilers imparts stability but does not permit rapid load fluctuation.

Q. *How does water circulation affect a boiler?*

Ans. Good circulation is a desirable feature for all boilers. Inadequate circulation of working fluid will reduce its heat removal rate efficiency, causing higher metal temperature.

Firetube boilers usually have poor circulation compared to watertube boilers. If properly designed, watertube boilers have good circulating characteristics that ensure higher heat removal efficiency.

Q. *What are the steaming characteristics of a firetube boiler?*

Ans. Firetube boilers in general, are not fast steamers. The steaming rate of a firebox type boiler is much faster than that of externally fired boilers.

However, the steaming characteristics of a firetube boiler vary with differing design and with the ratio of heating surface to water volume.

Q. *What are the steaming characteristics of a packaged boiler?*

Ans. Packaged boilers have quicker steaming rate than firetube boilers because of a large heating surface with smaller water volume.

Q. *What are the steaming characteristics of watertube boilers?*

Ans. Fast.

Q. *What are the steaming characteristics of cast-iron boilers?*

Ans. Slow steaming rate, except those having smaller water volume. This is because a large quantity of metal is involved in a cast-iron boiler.

At the start-up, the steaming rate is low but thereafter such boilers have relatively stable steaming characteristics.

Q. *What are the effects of low-water in the case of a steel boiler?*

Ans. Replacement of exposed tubes or crown sheets is required. This means a considerable downtime loss as the boiler is shutdown and kept out of service till the repair jobs are completed.

Q. *What are the effects of low-water operation in the case of cast-iron boilers?*

Ans. It occasionally results in cracked sections.

Though little difficulty is encountered in disconnecting or replacing the cracked sections of an external header type boiler, total shutdown of a push-nipple type boiler must be taken until a replacement section is obtained. Warpage may sometime invite replacement of the boiler.

Q. *How can sludge be removed from the boiler?*

Ans. Periodic blowdown is the method most practised to remove sludge deposited on the mud drum or mud ring.

If it has become solid cake, a high-pressure water hose is harnessed to dislodge the cake during shutdown periods. If it still remains, it is chipped out.

Due to lack of adequate clearance, it is usually difficult to remove sludge from the bottom of Scotch type boilers.

Q. *How can the boiler tube scales be removed?*

Ans. Incrustation of scales on the outside surface of a boiler tube is removed as follows:
1. The boiler is emptied
2. The boiler is then carefully heated
3. Cold water is sprayed on the tubes to induce thermal stress in the solidified scale and remove the ruptured scales.

Scales inside a boiler tube are turbined (reamed) out carefully with special equipment.

Q. *Why does soot deposit in the boiler flue passes?*

Ans. Improper combustion is the cause of soot formation in the boiler furnace. Combustion-gas-borne soot gets deposited in the boiler flue passes.

Q. *What is the effect of soot deposition in the flue gas passage of a boiler?*

Ans. It impairs heat transfer characteristics and thereby reduces the overall efficiency of the boiler. Tests conducted at the Bureau of Standards have confirmed that reduction of boiler efficiency by 8% over an 8-day operating period was principally due to the rapid buildup of soot deposits.

Q. *Compare the soot blowing characteristics of a steel firetube boiler and a steel watertube boiler?*

Ans. Annual or biannual soot blowing may be adequate for steel firetube boilers because they have large tubes and ample furnace volume. However such infrequent soot cleaning is not recommended.

Steel watertube boilers should be cleaned regularly. Large installations have soot blowers.

Q. *What is the effect of soot on cast-iron boilers?*

Ans. Initially there is some soot buildup and with that there occurs initial efficiency reduction. After that there is little additional soot development because of the large size of passageways. And so no further efficiency drop takes place.

Annual soot cleaning may be adequate.

Q. *What about the relative ease of servicing of cast-iron boiler, steel boiler and package boiler?*

Ans. For cast-iron boilers, the waterside is inaccessible whereas the fireside is accessible for inspection and maintenance by simply opening the doors. The flue ducts are large and lend themselves to easy cleaning.

The waterside of watertube steel boilers and packaged units is accessible for inspection and maintenance jobs through manholes, handholes or washout plugs. The mere opening of doors makes the fireside accessible to inspection and maintenance.

Accessibility to tubes for firetube boilers is gained through hinged, front flue doors. Soot cleaning is accomplished through the doors set in the rear smoke box.

If the firetube boilers are fitted with smaller steel tubes they present more difficulty in cleaning than cast-iron boilers because the latter provide ample cleaning space.

Watertube boilers of power plants have handholes for tube rolling.

The waterside of a packaged firetube unit is accessible for inspection through manholes. The rear wall is to be unbolted and removed to get access to the fireside. Hence firetube packaged boilers are not generally considered easily accessible for maintenance and cleaning.

Q. *Why do the present day boiler designers not recommend maximum flue gas velocities in boilers?*

Ans. It is due to deteriorating fuel quality. With the degraded variety of fuel burned in the combustion chamber, the metal loss in the convection pass caused by flyash mounts. This erosion problem being proportional to an exponential function of flue-gas velocity, the latter is a major consideration in boiler design.

Q. *Why is chromizing of waterwall, superheater and reheater tubes done?*

Ans. This strengthens the tubes in combating flyash erosion problems.

Q. *How does it work?*

Ans. Chromium, a powerful corrosion resistant metal, diffuses into the base metal of the tubes of the heat exchangers, viz., evaporators, superheaters, etc., thereby protecting against fireside grooving and corrosion, with the effect that tube life is extended dramatically.

Q. *Why are internally ribbed tubes laid out in the high-heat-absorption zones of natural circulation drum-type boilers operating at high supercritical pressures?*

Ans. It ensures high availability of the boilers.

Q. *How?*

Ans. When internally ribbed tubes are arranged in the high-heat-absorption zones of the combustion chamber, they provide a far greater margin of safety against tube failure from departure from nucleate boiling (DNB) conditions. They impart a centrifugal action to the working fluid that forces water droplets towards the waterwall tube surface and inhibits the formation of a steam film.

Q. *Is it true then that the DNB is more possible with smooth tubes than with internally ribbed tubes?*

Ans. Yes.

Q. *What do you mean by cycling service?*

Ans. It means boilers must be designed to cycle on and off. For example, a typical requirement may be nightly or weekend shutdowns.

Q. *In what way does a boiler designed for cycling service differ from one designed for base-load duty?*

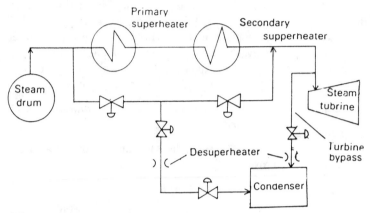

Fig. 1.46

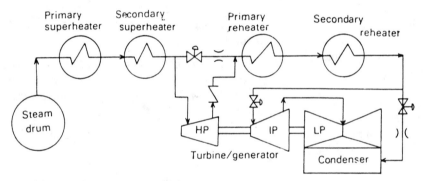

Fig. 1.47 *European bypass system*

Ans. They differ in the arrangement of superheater and reheater surfaces. This change is required to permit operation at rated steam temperatures at all loads.

Q. How is the start-up system for cycling units designed to allow quicker matching of final steam-to-turbine temperature?

Ans. The arrangement is as shown.[Fig. 1.46]

A part of the steam from the steam drum flows straight to the condenser via the desuperheater while the remaining steam passes through the superheaters and becomes superheated at a temperature higher than that which would result if the entire steam were passed through the superheater. This allows more rapid final steam-to-turbine temperature matching than is experienced by a boiler without this system.

Q. Why do the spray systems of some superheaters and reheaters on some cycling boilers have twice as much capacity as those of non-cycling boiler designs?

Ans. This extra spray capability endows the cycling system with a greater control flexibility for maintaining the boilertube metal temperature within recommended limits.

Q. What is the European by-pass system?

Ans. It is the full by-pass system.[Fig. 1.47]

Q. What are the advantages of this full-bypass system?

Ans. This system bestows:
1. Virtually unlimited turbine temperature-matching capabilities

2. Positive reheater cooling protection at all loads

3. Allows the boiler to operate continuously from 60% to 100% (depending on boiler size) of Maximum Continuous Rating (MCR) after the turbine trips.

Q. *What factors affect the combustion efficiency of boilers?*

Ans. There are three factors that affect combustion efficiency:

1. **Sensible Heat Losses:** it is the total heat carried away by hot dry flue gases.
2. **Hydrogen Losses:** it is the heat lost in vaporizing the moisture present in the fuel and flue gases.
3. **Combustible Losses:** it is the heat lost due to incompletely burnt products of combustion.

Q. *What parameters are to be monitored to calculate the foregoing three factors determining the combustion efficiency?*

Ans.

1. Net stack temperature
2. All the flue gas components

Q. *Why is the monitoring of net stack temperature required?*

Ans. The importance of monitoring net stack temperature can be realized from the fact that if it is high, it indicates that a large amount of heat is being wasted to the atmosphere. If, on the other hand, stack temperature is kept too low, it may result in coldend corrosion.

Q. *What components will you get in flue gas analysis?*

Ans. Oxygen, carbon dioxide, water vapour, sulphur dioxide, sulphur trioxide, oxides of nitrogen, and unburnt combustibles, viz., CO, H_2 and hydrocarbons.

Q. *What is the fundamental indicator of good combustion?*

Ans. The presence of minimum O_2, maximum CO_2 and nil combustibles in the flue gas.

Q. *What is the disadvantage of using excess air to reduce the combustibles content in flue gas to zero?*

Ans. It will

1. lower the furnace temperature
2. tax excess fan horsepower
3. result in heat loss due to heating of the induced air that does not enter into the combustion process.

Q. *Why is the measurement of draft essential?*

Ans. It assists in setting the damper correctly for various loads.

Excessive draft can result in increase of net stack temperature and lower the percentage of CO_2 in the flue gas.

If, on the other hand, the draft is inadequate, it will result in insufficient combustion air and smoky operation.

Q. *What is the conventional practice to determine the correct draft for introducing minimum excess air required for smokeless operation?*

Ans. This is obtained by a trial and error method, as the precise relationship between damper position and the amount of draft is not known. So the correct draft for minimum excess air compatible with smokeless combustion is determined for various loads by a trial method using a flue gas analyzer to monitor the levels of excess air, as dictated by the damper position and the amount of draft.

Q. *Though close monitoring of net stack temperature, oxygen and total combustibles in the flue gas is essential to control the combustion efficiency, yet more emphasis is given to monitor the combustibles at very close levels. Why?*

Ans. The combustion efficiency curve shows that the rate of energy loss in case of excess fuel, i.e., higher combustibles is much greater than in the case of excess air, i.e., higher percentage of oxygen. [Fig. 1.48]

The rate of energy loss in the former case is about six times the rate of energy loss due to higher oxygen level.

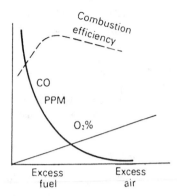

Fig. 1.48 *Effect of excess fuel and excess air on combustion efficiency*

Q. *Why it is required that fuel firing equipment operate under optimal combustion conditions?*

Ans. It will ensure:
1. better heat transfer rate
2. increase in combustion efficiency
3. decrease in fan power loss
4. decrease in air pollution
5. reduction in downtime, e.g., reduction in soot blowing.

Q. *What are the disadvantages in monitoring the flue gas with Orsat Apparatus?*

Ans. The apparatus is cumbersome. The monitoring of flue gas with it is a time consuming process. It also suffers from inaccuracies. It calls for special skills of the personnel to minimize human error.

Q. *Do you recommend the use of stand-alone analyzers for monitoring flue gases?*

Ans. No.

Q. *Why?*

Ans. These stand-alone systems may sometimes prove their worth under specific circumstances, yet their use for monitoring total combustion efficiency suffers from certain specific limitations.

Q. *What are these specific limitations?*

Ans. Take the case of **CO_2-Analyzer**.
It needs
1. higher maintenance

2. greater precision of measurements.

Moreover, the readings it produces are a function of the carbon/hydrogen ratio of the fuel burnt. These factors in combination make CO_2 monitoring difficult.

Take the case of **On-line CO-Analyzer**.

It uses infrared techniques that suffer from interference problems at elevated temperatures. Also, the presence of particulate solids in the flue gases impart inaccuracy. They interfere with the reading by blocking the light source. Besides analyzers measuring CO only do not indicate the magnitude of fuel wastage due to the presence of other combustibles.

Oxygen-Analyzers by themselves cannot measure the combustion efficiency.

Q. *Then what instrument do you recommend for flue gas monitoring?*

Ans. Portable, multivariable combustion analyzer.

Q. *Any specific example?*

Ans. Model ENERAC-2000. It is manufactured and marketed by Energy Efficiency Systems, Inc. USA.

Q. *What are the important and striking features of a good portable combustion analyzer?*

Ans. A good portable analyzer should have the following features:
It should monitor
1. net stack temperature
2. excess oxygen
3. total combustibles.

Besides, some comprehensive analyzers have some striking features, such as:
1. draft measurement and excess air calculation
2. data logging and storing capability
3. autocalibration facilities
4. direct calculation of energy savings from computer programs by virtue of its interfacing capabilities with computer input data
5. add-on modules with NO_x, SO_x measurement

(This enables pollution control)
6. quicker flue gas analysis. What Orsat apparatus completes in hours, Enerac-2000 finishes within minutes

7. multifuel switches are provided to determine the combustion efficiency of as many as 15 different fuels
8. can be lined up with any fuel firing equipment.

Q. *What are the viable techniques, that if implemented, may result in a 5–10% savings of coal in a coal fired boiler?*

Ans.

1. Control of excess air

It has been observed that for every 20–30% excess air, the boiler efficiency drops by 0.7%

2. Soot Blowing

Fouling of heat exchanger surfaces by soot will bring about a rise in flue gas temperature. And a 4.5°C rise in flue gas temperature will cut down boiler efficiency by 0.2%.

3. Uses of high pressure BFW heaters
4. Ensuring proper fuel/air distribution in the burners
5. Avoiding low combustion air temperature
6. Avoiding low primary air velocity
7. Ensuring adequate fuel/air mixing
8. Minimizing unburnt loss by increasing fineness of coal
9. Controlling Blowdown Loss. 16% of blow-down means a total heat loss of 0.42%, making it imperative to maintain blowdown as per the optimum requirement.
10. Reduction of unburnt gases in the flue gas
11. Minimizing condenser losses
12. Efficient Steam Management. This means maintaining rated steam parameters at the turbine inlet and the back pressure of the turbine exhaust.
13. Prevention of spontaneous coal combustion
14. Prevention of coal loss by natural forces like air/wind, rain, etc. during transportation and storage
15. Proper use and application of bowl mills for accurate sizing of coal.

Q. *What may be the cause of a fall in condenser vacuum in the surface condenser?*

Ans.

1. Deposition of foulants on the surface condenser's tubes

2. Low circulation flow
3. High air ingress in the condenser
4. High cooling water inlet temperature.

Q. *If a thermal powerplant burns inferior coal of a quality different from the designed coal specification of the boilers, what harm will it cause?*

Ans. Use of inferior quality of coal in boilers may lead to

1. operational problems
2. excessive use of coal
3. malfunctioning of coal handling plants
4. chocking of chutes
5. poor functioning of dust extraction plants
6. drastic cut in the availability of plant
7. reduction in plant load factor (PLF)

Q. *What factors may lead to excessive fuel oil consumption in coal fired boilers of thermal power stations?*

Ans.

1. Burning of inferior quality coal
2. Operating the plant at part load
3. Maloperation of the equipment
4. Frequent shutdown/start-up of the plant
5. Problems in stablization as well as full-load operation of new units.

Q. *What is the drawback of operating a boiler at part load condition?*

Ans. Despite being under part load of steam generation, the running equipment consume almost the same amount of energy as in a full load condition.

Q. *How can start-up and shutdown losses be cut down?*

Ans. A fast start-up is highly desirable, as it cuts down the energy consumption during this period.

A few thermal powerplants follow an open valve boiler start-up. This saves half of coal start-up costs.

Q. *What is the turn-down ratio of a steam control valve?*

Ans. It is the ratio of normal design flow to minimum controllable flow.

T = Normal design flow/Minimum controllable flow

Q. *What is the rangeability ratio of a steam control valve?*

Ans.   It is the ratio of maximum controllable flow to minimum controllable flow.

R = Maximum controllable flow/Minimum controllable flow

Note: The above definitions hold good for gun and nozzle also.

Generally, $T \approx 70\% \ R$

2

High Pressure Boilers

Q. *Why is high pressure steam in growing demand in power plants?*

Ans. To increase the efficiency of the plant and reduce the cost of production of electricity.

Q. *How?*

Ans. Higher the temperature and pressure of the steam, higher is its heat content and therefore, greater the enthalpy drop available in the expansion turbine. This means the efficiency of the power plant will be higher.

In high pressure steam generation plants greater economy in the production of electricity is obtained by making use of high temperature flue gases.

Q. *What are the unique features of high pressure boilers?*

Ans.

1. **Method of Water Circulation:** Can be both natural as well forced circulation types. The use of natural circulation is limited to subcritical boilers with operating pressure upto 140 kgf/cm^2, while for critical (steam pressure 218 atm) and supercritical boilers forced circulation is used.
2. **Type of Tubing:** Most of the high pressure boilers are of the water-tube type in which water is circulated through tubes arranged parallel to each other.
3. **Improved Method of Heating:** Greater heat transfer is achieved by improved methods of heating:
   (a) by evaporating water at pressure higher than the critical pressure of steam and thus saving the latent heat

(b) by heating of water by mixing the superheated steam to obtain the highest heat transfer coefficient
   (c) by increasing the velocity of water through the tubes and raising the flue gas velocity above sonic level, with the effect that the overall heat transfer coefficient is enhanced.

Q. *Why is natural circulation limited to subcritical boilers?*

Ans. Natural circulation works on the basis of density difference between steam and water.

At critical pressure (218 atm) and saturation temperature (374°C) of steam, the density difference of steam and water is zero and so natural circulation ceases. Hence natural circulation is operative in the case of subcritical boilers.

Q. *Why are the water tubes arranged in parallel?*

Ans. To reduce the pressure loss and to get better control over the quality of the steam.

Note: There would have been a large pressure drop due to friction if the water flow were to take place through one continuous tube.

Q. *What are the advantages of high pressure boilers?*

Ans.

1. Chances of scale formation on tube walls are very low due to high velocity of feedwater.
2. Quick start-up from cold is possible if power from an external source is available.
3. Increased efficiency (40 – 42%) of the plant due to higher enthalpy of steam output. As a result, more work output is available from the expansion turbine. (Fig. 2.1)

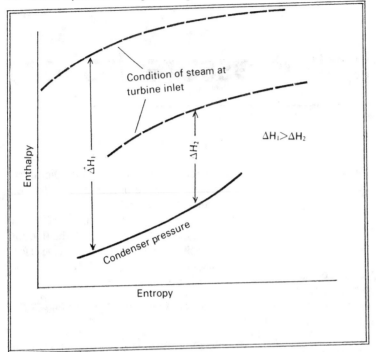

Fig. 2.1 *Higher enthalpy drop is available from steam at higher pressure*

4. Fit for meeting variable load quickly, as steam can be raised quickly.
5. Light-weight tubes with better heating characteristics can be used to reduce the cost, erection time and space requirement for tube layout.
6. When forced circulation is adopted, there is a greater degree of freedom in the layout of the furnace, tubes and boiler compartments.
7. Danger of overheating and thermal shock is minimized due to all parts being uniformly heated.

Q. *Who first introduced the forced circulation boiler?*

Ans. Credit goes to La Mont for introducing the forced circulation boiler for the first time (1925).

Q. *Briefly describe a La Mont boiler.*

Ans. It consists primarily of three circuits:

(a) **Air circuit** — Cold air is blown by a blower through the air preheater where it gets heated up in the process of heat exchange with flue

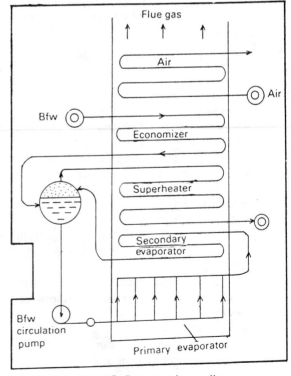

Fig. 2.2 *La Mont Boiler—a schematic representation*

gases. Hot air is then directed to the combustion chamber to be used as combustion air.

(b) **Water circuit**—Deaerated BFW is pumped through the economizer—a coil type heater—by a BFW pump and fed to the steam separating drum from which it is forced through the tubes of radiant evaporators and convective evaporator by a circulating pump. Together with steam it returns to the steam drum to complete the cycle.

(c) **Steam circuit**—The steam separated in the steam drum is saturated steam. It is superheated in the superheater coil and fed to the expansion turbine to generate power or drive compressors.

Note: This boiler is adaptable for all pressures from subcritical to supercritical; capacity exceeding 50 t/h.

Q. *What is the ratio of BFW circulation to tonnage of steam produced?*

Ans. Water circulation is 8 to 10 times the weight of the steam produced.

Q. *Why is such a high circulation ratio maintained?*

Ans. To prevent the evaporator tubes from overheating.

Q. *What is the main drawback of a La Mont boiler?*

Ans. Reduced heat transfer coefficient due to formation and attachment of steam bubbles on the inner surface of watertubes.

Q. *How can this be eliminated?*

Ans. By operating the boiler at critical pressure (218 atm).

Q. *Why?*

Ans. At this pressure the density of water is equal to that of steam. Hence no bubble will form.

Q. *What is the minimum steam generating pressure of a Benson boiler?*

Ans. 225 atm.

Q. *When did such a boiler go into operation first?*

Ans. In the year 1927 in Germany by Siemens-Schuckert-Merke.

Note: It was first built in England (1923) by Mark Benson, a Czechoslovakian.

Q. *Briefly describe a Benson boiler.*

Ans. It consists mainly of two circuits (Fig. 2.3)

(a) **Air circuit:** Cold air blown by the blower is heated up in the air preheater by hot flue gas by indirect contact heat transfer.

The resulting hot air is directed to the combustion chamber to be used as combustion air.

(b) **Water-steam circuit:** Forced circulation is introduced to generate superheated steam in this open hydraulic system.

A high pressure feed pump feeds BFW (Boiler Feed Water) to the economizer, radiant evaporator, convective evaporator and superheater, thus converting feedwater into superheated steam in a once-through cycle.

Q. *What is the major problem associated with a Benson boiler?*

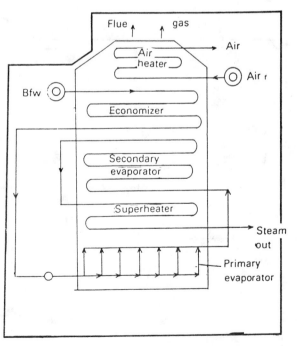

Fig. 2.3 *Benson boiler—a schematic representation*

Ans. Salt deposition in the transformation zone where all remaining water is converted into steam.

Q. *How can this be avoided?*

Ans. By flushing out the boiler after every 4000 working hours.

Q. *What are the maximum operating temperature and pressure achieved so far in a commercial Benson boiler?*

Ans. Temperature : 650°C; Pressure : 500 atm

Q. *What are the advantages of a Benson boiler?*

Ans.

1. There is no boiler drum, which means a reduction in the overall weight of the boiler and a cut in capital investment.
2. There is no expansion joint as the boiler is of the forced circulation type. Since all parts are welded at their sites, it is easier and quicker to erect a Benson boiler
3. Start-up is quicker because of welded joints

4. Needs comparatively less floor-space than other boilers of the same capacity
5. Better and more efficient protection of furnace walls as the high pressure boiler tubes have smaller diameters and are closely spaced
6. Economic operation at partial load or overload is possible
7. Inherently free from the problem of bubble formation which occurs in the case of natural circulation as the load drops abruptly
   This feature makes Benson boilers very suitable for grid power stations
8. Blowdown loss is less, hardly 4% of natural circulation boilers of the same capacity
9. Minimum explosion hazard as tubes are of small diameter with very little storage capacity.

Q. *Briefly describe the water-steam circuit of a Schmidt–Hartmann boiler.*

Ans. The water-steam system of this type of boiler comprises two circuits—primary circuit and secondary circuit. (Fig. 2.4)

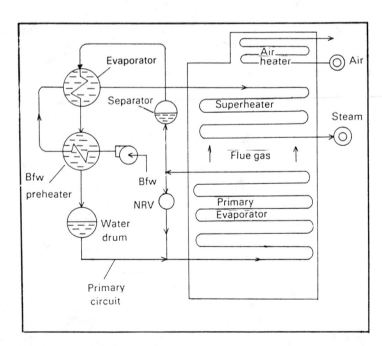

Fig. 2.4 *Air and steam-water circuits of a Schmidt-Hartmann boiler*

High pressure (100 atm) steam produced from distilled water in the primary evaporator of the primary circuit is allowed to exchange its heat with impure water in the evaporator and BFW preheater in the secondary circuit to generate steam (60 atm) which is passed through the superheater to produce superheated steam.

Q. *What is the nature of circulation in the primary circuit of a Schmidt–Hartmann boiler?*

Ans. Natural circulation.

Q. *Will it be sufficient to effect the necessary rate of heat transfer as to overcome the thermo-siphon head of about 2 to 10 m of water column?*

Ans. Yes.

Q. *What is the rate of heat transfer from the primary circuit to the secondary circuit?*

Ans. About 10000–12000 kJ/m^2 h°C

Q. *What are the advantages of Schmidt–Hartmann boilers?*

Ans.
1. Chances of overheating are minimal as the highly heated components carry pure water with no risk of salt deposition at all
2. Accepts wide load fluctuations without undue priming or abnormal increase of pressure in the primary circuit
3. Evaporation proceeds without priming
4. Salt deposition due to evaporation of impure water in the evaporation drum can be quickly and easily brushed off.

Q. *What is the speciality of a Velox boiler?*

Ans. Here the steam generation is based on the principle that the rate of heat transfer from hot flue gas to water is much higher at supersonic velocity than at subsonic velocity of the flue gas. Hence compressed air (2.5 atm) is used to produce flue gas at supersonic velocity (200 to 250 m/s).

Q. *What is the degree of heat release in the combustion chamber of a Velox boiler?*

Ans. $(30–42) \times 10^6$ kJ/m^3 of the combustion chamber.

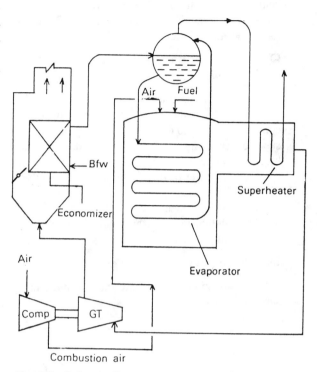

Fig. 2.5 *Schematic representation of Velox boiler*

Q. *Briefly describe the Velox boiler.*

Ans. Air is compressed to 2.5 atm by an air compressor driven by a gas turbine to produce flue gas at supersonic velocity in the combustion chamber. The gas turbine is driven by hot (500°C) flue gas from the superheater exit. (Fig. 2.5)

The combustion chamber is vertical and cylindrical, lined with tubes of 10 cm dia.

Inside each tube is another concentric tube of 2.5 cm dia through which hot flue gas passes at supersonic velocity while BFW flows through the annulus between the two tubes.

The mixture of water and steam produced passes through a separator that separates steam which is fed to the superheater heated by hot flue gas to produce superheated steam.

The flue gas (500°C) from the superheater is allowed to expand in the gas turbine to drive the air compressor. The turbine exhaust gas is passed countercurrently through the economizer to preheat the BFW and then finally discharged at 90°C and 120 m/s to the atmosphere.

Q. *What is the basic factor that limits the capacity of a Velox boiler?*

Ans. With the increase of the capacity of Velox boilers, greater is the power requirement of the air compressor, so as to deliver more combustion air at supersonic velocity. Since a gas turbine alone may not be able to cope with this power demand of the air compressor, power from an external source needs to be supplied.

That is why the size of the Velox boiler is limited to 100 t/h and the power requirement of the air compressor at this velocity is 450 kW.

Q. *What is a supercritical boiler?*

Ans. Boiler producing steam above critical pressure, i.e., above 218 atm.

Q. *What is the basic difference between subcritical and supercritical boilers?*

Ans. Subcritical boilers operate at pressures below 218 atm while supercritical boilers operate at pressures above 218 atm.

A subcritical boiler has three distinct sections—economizer, evaporator and superheater while a supercritical boiler has only an economizer and a superheater.

Q. *What are the advantages of supercritical boilers over subcritical ones?*

Ans.
1. Higher heat transfer rate
2. More flexible in accepting load variation
3. Greater ease of operation
4. Higher thermal efficiency (40 – 42%) of power generating stations can be achieved by using supercritical steam

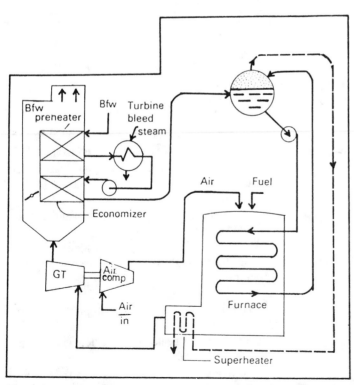

Fig. 2.6 *Schematic diagram of a Supercharged boiler*

5. The absence of two-phase mixture in super-critical boilers minimizes the problem of erosion and corrosion

6. Steadier pressure level.

Q. *What is a supercharged boiler?*

Ans. A boiler where combustion is carried out under pressure in the combustion chamber by supplying compressed air. (Fig. 2.6)

Q. *What are the direct utilities of combustion under pressure?*

Ans. The flue gas is also under pressure (about 5 atm) which is utilized to drive the gas turbine to run the air compressor coupled with it to supply the compressed air for combustion.

Due to pressurized combustion, the overall heat transfer coefficient is very high.

surface of a conventional boiler. And that means a drastic saving in material cost

3. Quicker start-up possible. A 150 t/h super-charged boiler can be brought to full load in less than half an hour

4. Better response to control

5. The power available from the gas turbine can be utilized to run the auxiliaries

Q. *What is the main drawback of such a pressurized combustion system?*

Ans. The entire gas passage should be carefully made leak-proof.

Q. *Draw the pressure distribution in the gas-air path of a boiler with balanced draft and one with a supercharged furnace.*

Ans. See Fig. 2.7

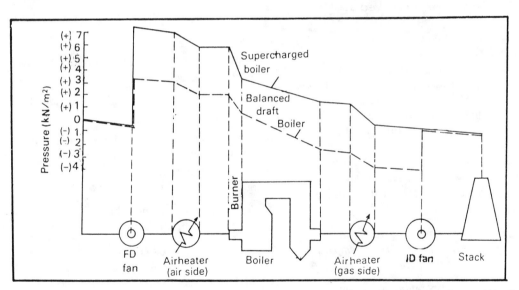

Fig. 2.7 *Pressure distribution in the gas-air path of balanced draft and Supercharged boilers*

Q. *What are the specific advantages of super-charged boilers over other conventional boilers?*

Ans.

1. Much higher overall heat transfer coefficient
2. The heat transfer surface of a supercharged boiler is barely 25–30% of the heat transfer

Q. *How many heat zones are there in most common-ly used pulverized coal fired boilers?*

Ans. Three.

Q. *What are they?*

Ans. One radiation zone (Zone I) and two convection zones (Zones II and III). (Fig. 2.8)

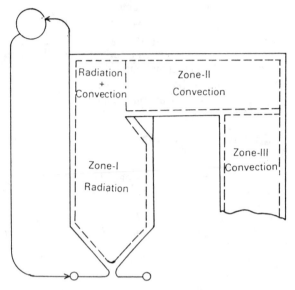

Fig. 2.8 *Different zones of pulverized coal fired boilers*

In the bottom and middle portion of Zone I, most of the heat transfer takes place by radiation from the yellow flames. Hence this zone is called the radiation zone. This effect of radiation is reduced in the top portion of Zone I where secondary air is introduced, with the effect that a considerable amount of heat transfer is contributed by convection. Here heat transfer takes place due to both radiation and convection.

Heat transfer in zones II and III takes place predominantly by convection. The former is called the high temperature zone and the latter is called the low temperature zone.

Q. *Why should the flue gas temperature in zones II and III be lower than the ash fusion temperature (AFT)?*

Ans. If the flue gas temperatures in the zones II and III are higher than the AFT (Ash Fusion Temperature), then the convective mode of heat transfer will bring about slag deposition on metal tube surfaces carrying water or steam, as the surface temperature of the tubes is always less than AFT.

Q. *How much of the total heat generated is absorbed in Zone I?*

Ans. About half of the total heat.

Q. *How can this be increased?*

Ans. By firing pulverized coal with low AFT and by reducing the excess air supply.

Q. *Why can more heat transfer by radiation be effected using low AFT coal?*

Ans. Firing pulverized coal with low AFT will not deposit slag on the water-tube surfaces in this zone and as a result more heat transfer will be effected through the clean heat transfer surface.

Q. *Why is more heat transfer by radiation essential when using low AFT coal?*

Ans. Firing pulverized coal with low AFT in hot turbulent air coupled with low excess air produces a very high flame temperature at which the ash always remains in a molten state. The evaporator tube surface temperatures are always less than this AFT and that entails the risk of solidification of molten ash on the surface of the evaporator tubes. In order to avoid slagging of evaporator tubes, the use of convection heat transfer should be avoided as long as the gas temperature is higher than AFT. And therefore, heat transfer by radiation is essential. Even though the radiant surface is more expensive for the same degree of heat transfer, the heat transfer must be by radiation.

Q. *If there is a fall in excess air supply, why does heat transfer in zone I increase?*

Ans. In this zone, heat transfer occurs predominantly by radiation. Introduction of excess air will entail a convection current, flame temperature will drop and that will reduce the amount of heat transfer by radiation. Conversely, with the supply of less excess air, the heat transfer by radiation in zone I will increase.

Q. *Why is the evaporator, not the superheater located in zone I?*

Ans. Reduction in furnace size demands the lowest possible surface temperature of tube metals. The evaporator tubes always offer lower surface temperature than superheater tubes carrying superheated steam. Hence the evaporator, not the superheater, is the most suitable component to be located in zone I.

Q. *Why is the superheater located in zone II?*

Ans. For a medium pressure boiler (40–60 atm and 440 – 460°C), the heat absorbed by the superheater does not exceed one-fifth of the total heat absorption of the boiler and the evaporator tubes occupy the entire surface of the furnace. The superheater is of the convective type. Hence zone II is the most preferred location for the superheater, as the heat transfer in this zone is mainly of the convective type.

For higher pressure boilers (100 atm/540°C), the superheater is partly convective and partly radiative. The convective part is placed in zone II while the radiant platen section is suspended in the top portion (radiant + convection) of zone I. These two superheaters are joined in series.

3

Boiler Auxiliaries

Q. *What do you mean by boiler auxiliaries?*

Ans. These are the devices incorporated in the boiler circuit to boost up the efficiency and performance of the steam generation plant and assist in the systematic and adequate operation of the boiler unit for prolonged periods.

Q. *What are they?*

Ans. Usually a boiler is fitted with the following auxiliaries (accessories):
1. Air preheater
2. Economizer
3. Superheater
4. Desuperheater
5. Boiler feed pumps
6. Forced draft and induced draft fans
7. Mechanical separator
8. Equipment tanks
    (a) Feedwater tank
    (b) Deaerator
    (c) Continuous blowdown expander
    (d) Drainage expander
9. Chemical dosing system
10. Soot blowers and wall blowers
11. Pressure reduction valve
12. Pulverizers and fuel firing system
13. Ash handling system.

Q. *What is a boiler feed pump?*

Ans. A pump that feeds boiler water (polished water) to the steam drum via the economizer (for preheating).

Q. *What are the two most important criteria of a boiler feed pump?*

Ans. It must be absolutely positive and reliable **under** all variable operating conditions.

Q. *What kind of pump should it be?*

Ans. It may be:
(a) direct acting pump driven by its own cylinders
(b) a reciprocating pump driven by a motor or belted to the machinery
(c) a centrifugal pump—turbo-driven or electrically driven by motor.

Q. *How are direct acting pumps classified?*

Ans. They may be classified as simplex, duplex and triplex pumps.

Q. *What is the basic difference between these three?*

Ans. The simplex pump has one pump cylinder and one engine cylinder.

In duplex and triplex pumps there are two and three pump cylinders respectively.

Q. *What kinds of reciprocating pumps can be used?*

Ans. Both plunger driven and piston driven.

Q. *Among the three kinds of pumps—direct acting, reciprocating and centrifugal—which one will you prefer as a boiler feedwater pump?*

Ans. Centrifugal pump.

Q. *Why?*

Ans. A centrifugal pump is capable of:
1. delivering steady flow of BFW
2. supplying the largest quantity of BFW under a given head
3. accepting load variations most easily
4. trouble free and smooth operation; less floor space in required and maintenance cost is low.

Q. *Why are both turbo-driven and motor driven centrifugal BFW pumps installed?*

Ans. During start-up, the motor driven pumps are lined up to establish circulation.

When the unit is in full operation and steam is available, one or two turbo-driven pumps may be lined up to spare corresponding number(s) of motor-driven pumps for reserve, rest or repairing or saving of electricity, which is costlier than steam. Motor driven pumps will continue to circulate BFW through the boiler unit when the latter trips.

Q. *Suppose both steam and electricity are available. Which centrifugal pump would you prefer to switch on?*

Ans. Turbo-driven.

Q. *Why?*

Ans. Steam is cheaper than electricity.

Q. *Why is the FD fan installed before the air preheater and not after?*

Ans. This reduces the volume of air handled by the FD fan. The volume of air increases considerably after the air preheater.

Q. *The size of the ID fan is always larger than the FD fan. Why?*

Ans. ID fan is to handle a fluid (flue gas) with higher specific volume than the fluid (air) handled by the FD fan. Greater the specific volume of the gas handled, larger will be the size of the fan.

Q. *Why are FD fans installed with impellers having blades inclined backwards?*

Ans.
1. Backward curved blades endow the fan with greatest efficiency
2. Air handled by FD fans is clean, i.e., free from grit, soot, etc., which might otherwise impair the performance of an FD fan fitted with backward curved blades.

Q. *The efficiency of the backward curved blades is higher than forward curved blades. Why?*

Ans. The former offer minimum resistance to the rotation of the impeller.

Q. *Why are backward curved blades with radial tips used in ID fans?*

Ans. ID fans handle flue gas containing soot and grit. The radial tips make the wheel self-cleaning and prevent dust from clogging the blades.

Q. *Economizers are so designed that the water temperature is not raised within 30°C of the boiler temperature. Why?*

Ans. This is to avoid water hammer due to sudden condensation of pockets of steam.

Q. *BFW should not enter the economizer below a certain temperature.* Why?

Ans. Before being delivered to the boiler drum, BFW is preheated in the economizer, by indirect heat exchange with the flue gas,—either in crossflow or counterflow with the BFW. However, BFW is not allowed to enter the economizer below a minimum temperature, for, this may cause a sharp drop in the temperature of the flue gas. If the flue gas temperature drops to its dew point (140–150°C), it will give rise to condensation of sulphuric acid resulting from the reaction of oxides of sulphur with water vapour produced due to fuel burning. Acid mist condensing on the economizer and air preheater tubes will cause severe corrosion of the metal and even lead to tube failure.

Q. *Why were earlier economizer tubes made of cast iron?*

Ans. Earlier economizer tubes of low pressure boilers were made from cast iron which exhibits good resistance to acid corrosion as inflicted by SO_2/SO_3 + water vapour. Hence the chances of attack on cast iron by acid mist near dew point were narrow.

Q. *Why may a sub-economizer be a necessity?*

Ans. The one economizer may be by-passed for desuldging (blowdown) without putting the boiler off operation.

Q. *Why are fins used all over the economizer tubes?*

Ans. A bare tube economizer is beset with the disadvantage that there are always some dead spaces behind the tube in the gas flow direction. To avoid this problem, fins are planted. Due to increased surface area, overall heat-transfer increases.

[Finned economizer tubes are called gilled tubes.]

Q. *What should be the ideal design of an economizer?*

Ans. Flue gas will pass vertically downwards in countercurrent flow with the boiler feedwater moving upwards for highest thermal efficiency.

Q. *What kind of heat exchanger is the superheater?*

Ans. It is a surface heat exchanger.

Q. *What kind of function does it accomplish?*

Ans. It produces superheated steam by first bringing wet steam to saturation point and then raising its temperature to the degree of superheat according to design.

Q. *Why is a higher degree of superheat a necessity?*

Ans. Since the work done per kg of superheated steam in expanding in the turboalternator is proportional to the adiabatic heat drop, so the greater the degree of superheat, the greater is the enthalpy drop available from the superheated steam without being wet, thereby increasing the life and efficiency of the turboalternator.

Q. *Why is forced circulation a necessity?*

Ans. At higher pressures beyond 140 atm $(14MN/m^2)$, the density of steam, at saturation temperature, approaches that of water and as such natural circulation, which is based on the density difference between BFW and steam increasingly becomes inoperative. And that makes it imperative to install, in the steam-water circuit, a circulation pump to force water through the evaporating tubes of the boiler.

Q. *Why are extra precautions necessary to bring a vertical-coil superheater in line?*

Ans. Vertical coils are non-drainable and hence extra precautions are required during boiler start-up, to drive away any entrapped condensate.

Q. *What should be the temperature difference between the gas-side and steam-side of superheaters?*

Ans. 140°C for effective heat transmission which primarily depends on mass velocity and temperature difference.

Q. *Which will influence more the effective heat transfer—a slight rise in mass velocity or a slight rise in gas temperature?*

Ans. Slight rise in mass velocity.

Run	$\Delta \dot{M}$	$\Delta \Theta$	Product ($\Delta \dot{M}$ $\Delta\Theta$)
1st	2	3	6
2nd	(2 + 1)	3	9
3rd	2	(3 + 1)	8

Q. *Why must expansion loops be provided in the supply piping?*

Ans. To avoid thermal strain due to differential expansion between equipment piping and the boiler.

Q. *Why is it imperative to carry out the soot blowing operation when the boiler is operating at a stable rate?*

Ans. This is done so as to prevent any fire from being blown out as the soot blowing is going on.

(Soot blowing at 30–40% of the rated load of a boiler is sufficient for purging; best result: 75% load of boiler.)

Q. *Why is it not recommended to carry out the soot blowing when any boiler is out of service?*

Ans. To avoid the risk of explosion due to leftover unburnt combustibles in the passages.

Q. *How many safety valves, at least, are required for a boiler?*

Ans. Two.

Q. *Why?*

Ans. If one gets out of order (i.e., stuck up), the other will save the boiler by blowing the excess steam.

Q. *What should be the total capacity of the safety valves?*

Ans. Equal to the design steam flow of the boiler.

Q. *What is the critical pressure of steam?*

Ans. It is 218 atm or 225 kgf/cm^2, at which the volume of steam is equal to the volume of water.

Q. *What is wire drawing?*

Ans. It refers to the process of expansion of steam when no energy is expended. Steam, as it expands through any orifice or valve, does not perform any external work and as such does not undergo any loss of its heat energy.

Q. *Why does dry, saturated steam get superheated after wire drawing?*

Ans. The total heat content of steam during wire drawing is constant:
    Total heat before wire drawing = Total heat after wire drawing
    Due to expansion through the orifice, the steam pressure falls but not its ΔH. As such the dry, saturated steam gets superheated.

Q. *In a boiler where coal is fired on grates, slightly negative pressure is maintained over the grate and slightly positive pressure (5–7.5 cm of water) below it. Why?*

Ans. To avoid leakage of furnace gases to the surroundings. Positive pressure below the grate will force the entire combustion air up through the coal bed on the grate to ensure better and more efficient combustion of the fuel while the negative pressure above the grate will augment the flow of combustion air through the coal bed and eliminate the risk of positive pressure build-up above the coal bed, thus reducing the chances of flue gas leakage to the surroundings.

Q. *Why does the fuel bed have to be maintained in such a way that the bed-resistance remains uniform throughout?*

Ans. Otherwise there will be a channeling of combustion air leading to incomplete combustion. Combustion air will follow the path of less resistance across the coal bed and therefore, the fuel-air mixture will not be uniform throughout, leading to uneven combustion.

Q. *What is a clinker?*

Ans. It is a solidified mass of fused ash.

Q. *Why is it unwanted in a stoker furnace?*

Ans. It obstructs the flow of combustion through the coal bed.

Q. *What is the temperature inside the bowl mill of the pulverizer unit?*

Ans. 75°–90°C.

Q. *If it is raised above this range what will happen?*

Ans. Volatile matter of pulverized coal will quickly evaporate and there is a good chance of an explosion.

Q. *What determines this temperature range?*

Ans. Coal's volatile matter.

Q. *If the ash content of the coal is high, will you increase or decrease the temperature?*

Ans. Temperature should be raised.

Q. *Which bearing—motor bearing or exhauster bearing of the bowl mill—is likely to be damaged first?*

Ans. Exhauster bearing.

Q. *Why?*

Ans. Coal particles erode into the fan blades causing a defective balancing which will entail vibration of the journal, impairing the bearing.

Q. *One bowl mill feeds four burners placed at different locations in the boiler unit of SGP of HFC Haldia Division. How is the same flowrate maintained all through?*

Ans. By placing orifices in each tube feeding pulverized coal to the burner, the flow through all tubes is made nearly constant. Flow is restricted most in the case of the shortest tube.

Boiler Mountings and Accessories

4

Q. *What do you mean by boiler mountings?*

Ans. These are fittings primarily intended for the safety of the boiler and control of the steam generation process completely.

Q. *What are they?*

Ans. According to the Indian Boilers Act, 1923, they are:
1. Pressure gauge
2. Safety valves (2 Nos.)
3. Water level indicators (2 Nos.)
4. Feed check valve
5. Steam stop valve
6. Fusible plug
7. Blow-off cock
8. Manholes and mudholes.

Q. *What is a pressure gauge?*

Ans. It is a device fitted to indicate the pressure of a fluid.

Q. *How does it work in a steam boiler?*

Ans. The steam enters a hollow tube of elliptical cross-section that gets circular under steam pressure. This results in a slight movement of the other end of the tube which is attached to a rack and pinion system via a rod and a lever. The pinion greatly magnifies the movement and a pointer fitted to the pinion moves on a scale to indicate the pressure applied.

Q. *What is a safety valve?*

Ans. It is a device that lets out the excess steam when the steam pressure in the boiler, steam header or pipeline exceeds the working pressure.

Q. *How does a safety valve work?*

Ans. It works automatically. When the steam pressure exceeds the working pressure, it automatically vents some steam from the system with the effect that the system pressure returns to the normal working limit.

Q. *In general, how many types of safety valves are used in practice?*

Ans. Four types.

Q. *What are they?*

Ans.
1. Spring loaded safety valve
2. Lever safety valve
3. Dead weight safety valve
4. High steam and low water safety valve.

Q. *What is a Ramsbottom safety valve?*

Ans. It is a spring loaded safety valve.

Q. *How does it work?*

Ans. The valve is loaded with a helical spring (Fig. 4.1) that pulls two valves to press upon the respective valve seats. It is the compressive strength of the spring that holds the valves tightly upon the valve-seats. When the steam pressure exceeds the normal working limit, the force of the steam acting on the valve seats exceeds the compressive force of the spring, the valves get lifted from their seats and release the excess steam till the pressure falls below the preset limit, after which the valves again rest on their seats.

Q. *What is its limitation?*

Ans. The spring may get stuck-up.

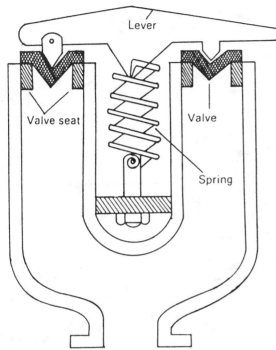

Fig. 4.1 *Ramsbottom Safety Valve*

Q. *How does the lever safety valve work?*

Ans. The valve is loaded by a lever (Fig. 4.2). One end of it is fulcrumed while the other end carries a weight. The loading force pressing the valve against its seat is transmitted by virtue of a short strut joined to the lever arm.

When the steam pressure of the system exceeds the normal working limit, the valve is lifted from its seat, allowing some steam to escape and the pressure to drop to its normal value.

Q. *What is the casing material of a lever safety valve?*

Ans. Generally, cast iron.

Q. *What is the material of construction of these valves?*

Ans. Usually made of gun metal.

Q. *What is the material of construction of the lever?*

Ans. Mild steel or wrought iron.

Q. *What is a dead-weight safety valve?*

Ans. It is a safety valve that uses a weight sufficiently heavy to keep the valve on its seat below the allowable pressure limit.[Fig. 4.3]

Q. *How does it operate?*

Ans. The weight exerted upon the valve is enough to press it tightly upon the valve seat against the normal steam pressure.

When the steam pressure exceeds the normal limit, it lifts the valve up together with its weight, allowing

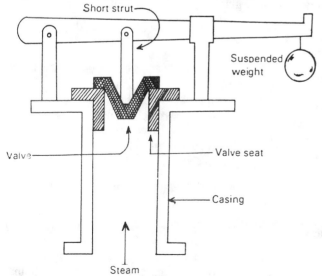

Fig. 4.2 *Lever safety valve*

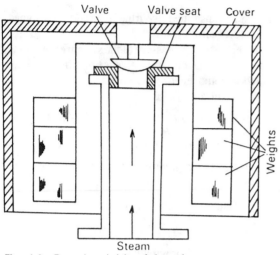

Fig. 4.3 *Dead-weight safety valve*

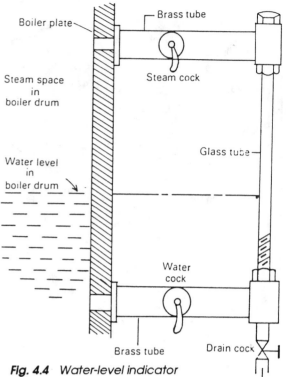

Fig. 4.4 *Water-level indicator*

excess steam to escape and bringing the pressure down to its normal value.

Q. *Can you prescribe this type of safety valve for high pressure boilers?*

Ans. No.

Q. *Why?*

Ans. Large amounts of weights will be required to balance the steam pressure.

Q. *What is the function of a water-level indicator?*

Ans. To indicate the level of water in the boiler.

Q. *How does it work?*

Ans. It is a vertical hard glass tube (Fig. 4.4) whose upper and lower ends are connected to steam space and water space respectively, with brass tubes. Therefore, the same pressure of steam acts upon the water in the boiler and the water in the gauge glass, which therefore indicates the same level of water as that in the boiler.

Q. *If the water cock, during operation, is kept open and the steam cock closed, what will happen?*

Ans. Water from the boiler will channel into the gauge glass and fill it indicating a false, high water level in the boiler.

Q. *Why?*

Ans. Since there is no steam inlet to the gauge glass, a differential pressure is set up between the water in the boiler and that in the gauge glass. Boiler water pressure being higher, it will force its way up the gauge glass.

Q. *During the line-up of a gauge glass, what will you do?*

Ans. The following sequence of operations is performed:
1. Drain cock is opened
2. Water cock is crack opened
3. Steam cock is crack opened
4. Drain cock is fully closed
5. Steam cock and water cock are fully opened.

Q. *What will happen in case of breakage of the gauge glass?*

Ans. The gauge glass is provided with valves which are arranged with steel balls inside such that in the event of any accidental breakage of gauge glass, steam and water will gush forth abruptly and in that process they will position the balls in such a way that flow of steam and water ceases automatically.

Q. *Why is a boiler normally provided with two dependable gauge glasses to work on?*

Ans. If one fails due to accidental breakage, the other one will continue to be in-line and will show the water level.

Q. *Why should the steam connection of the gauge glass from the steam drum not be lagged whereas the water connection of the gauge glass should be lagged?*

Ans. This will keep an all time small flow of condensate through the gauge glass with the effect that the gauge glass will remain active all the time.

Q. *What is the function of a feed check valve?*

Ans. To control the flow of BFW from the feed pump to the boiler and to prevent the backflow of water from the boiler to the pump when it is not working or its pressure is less than boiler pressure.

Q. *Where is it fitted?*

Ans. It is placed downstream of the BFW pump and near the boiler end of the discharge pipe.

Q. *How does it work?*

Ans. During normal operation, the valve (Fig. 4.5) is lifted off the valve seat due to the discharge pressure of the water from the pump and that allows water to pass into the boiler. When the pump is stopped, the boiler water at high pressure will rush back and press the valve tightly upon its seat.

Also, the flow can be controlled with the aid of a handwheel that can regulate the lift of the valve by means of a spindle connected to it.

Q. *What is the function of a steam stop valve?*

Ans. To stop or allow the flow of steam from the boiler to the steam pipe.

Q. *What is a junction valve?*

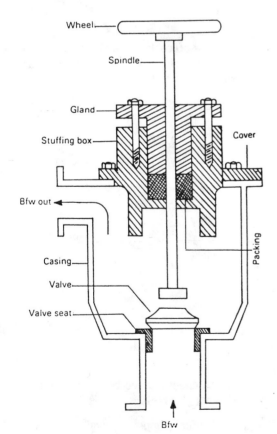

Fig. 4.5 *Feed check valve*

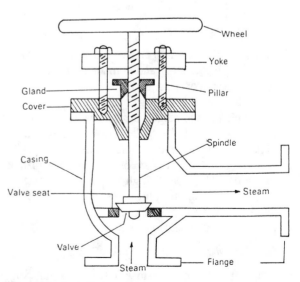

Fig. 4.6 *Junction valve*

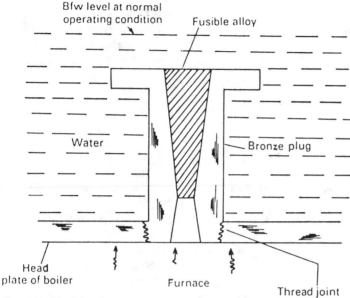

Fig. 4.7 Fusible plug

Ans. It is a steam stop valve (Fig. 4.6) mounted on the boiler, at the junction of the boiler and steam pipe, to control the flow of steam from the boiler.

Q. *What is the function of the fusible plug?*

Ans. To prevent the boiler from being damaged due to overheating when the water level falls below the lowest permissible limit in the boiler.

Q. *How does it work?*

Ans. A bronze plug tapered within, it is stuffed with an easily fusible alloy and is mounted on the head plate of the boiler. [Fig. 4.7]

Under normal operating conditions, the upper end of the plug is submerged in water while its bottom remains exposed to the boiler furnace located immediately below the plug. If the water level falls to expose dry the top surface of the plug, the fusible material will melt due to high temperature and steam will gush out of the plug hole into the boiler to extinguish the fire, or will let out a sound of warning to the boiler attendant to expedite action.

Q. *For what purpose is a blow-off cock fitted?*

Ans. It is imperative to empty the boiler periodically for cleaning and inspection. It is for this purpose that a blow-off cock (Fig. 4.8) is fitted. It will let out any sediment along with a portion of BFW when it is opened.

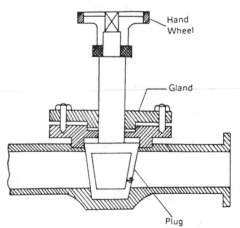

Fig. 4.8 Blow-off cook

Q. *Where is it fitted?*

Ans. It is fitted directly to the boiler shell or to a pipe connected with the boiler.

Q. *What is the utility of a fusible plug?*

Ans. It is a safety device installed in almost all firetube boilers at the lowest safe water level in the boiler drum. If the boiler water level goes below this level, the fusible plug of tin will melt (at 560°K) and allow steam to pass through making a warning noise for low water level.

Q. *Why are all-welded boilers now preferred to rivetted boilers?*

Ans. Because of the limitation posed by the ligament—a section of solid plate between rivet/tube holes—rivetted boilers are less favoured. Though ligament efficiency as high as 85% can be achieved in the case of rivetted boilers, welded construction parts have better joint efficiency imparted by improved welding technology and high quality weld rods.

5

Boiler Operation Inspection and Maintenance

Q. *What principal characteristics are taken into consideration in describing the operating conditions of boilers?*

Ans.

1. Average efficiency of a boiler for a particular operating period
2. Net efficiency of the boiler at rated load
3. Availability factor, i.e., the ratio of operation time and reserve time to the calendar time
4. Operation factor, i.e., the actual operating time of the boiler to the length of the calendar time (month, year) considered
5. Capacity factor which is the ratio of the total steam generated during operation time to the probable steam generation during calendar time at the rated steam generation capacity
6. Average and maximum time of a campaign (i.e., the operating time to failure)

Q. *What do you mean by boiler failure?*

Ans. It refers to any incident that disturbs the operating ability of a boiler.

Q. *When is the boiler expected to have the highest values of the indicated characteristics?*

Ans. Boiler operating in base regime.

Q. *What do you mean by a monobloc unit?*

Ans. The boiler and the steam turbine taken as a whole are called a monobloc unit.

Q. *What do you mean by manoeuvrability of monobloc units?*

Ans. This concept includes:

1. start-up and shutdown characteristics
2. operating load range

3. the dynamic properties
4. the characteristics at sudden surging of load and load-sheddings.

Q. *What do you mean by load control range?*

Ans. It is the load range in which the automatic control system of the boiler responds quickly to the load variations without the interference of the operating personnel.

In this range, the number of burners in line is not changed nor is any adjustment made of automatic regulators, and the composition of the automatic auxiliary equipment remains the same.

Q. *What is the allowable load range?*

Ans. It includes loads right from the lowest load at which the boiler can function steadily to the load at the lower limit of control range.

Q. *How does the load vary in this range?*

Ans. It varies slowly.

Q. *What changes are effected during load variation in the allowable load range?*

Ans.

1. Change of composition of the auxiliary equipment
2. Change in the number of burners in line
3. Some automatic regulators are switched off.

Q. *What is the steady regime of boiler operation?*

Ans. In this region the steam parameters vary insignificantly at any load.

Q. *What is the unsteady regime of boiler operation?*

Ans. In this region load variation and fluctuation of steam parameters occur due to internal or external disturbances.

Q. *Which factors are responsible for internal disturbances?*

Ans. Variation in:
1. flowrate of BFW
2. temperature of BFW
3. fuel consumption rate
4. combustion air flowrate, etc.

Q. *Which factors cause external disturbances?*

Ans. Variation in:
1. steam pressure in the steam main
2. load of the turbo-alternator
3. the degree of opening of start-up and shut-down device.

Q. *What do you mean by the acceleration characteristic of a boiler?*

Ans. It is a characteristic by virtue of which the boiler is able to change its load quickly.

Q. *What are the principal types of boiler shutdown?*

Ans.
1. Emergency shutdown
2. Shutdown for repairing jobs with cooling of the whole or part of the boiler unit
3. Shutdown for repairing jobs or to reserve without cooling of the boiler and steam pipelines.

Q. *In which cases must a boiler be shutdown immediately?*

Ans.
1. Explosion in the furnace damaging the brick-work or pressure parts
2. Flame extinction in the furnace
3. Deformation of pressure parts that might invite explosion and endanger the operating personnel
4. Failure to ensure reliable boiler operation because of bad visibility, fire and danger of explosion
5. Non-permissible rise of superheated steam temperature

6. Failure of feed pumps
7. Failure of both water-level gauges for drum type boilers and feedwater flowmeters for once-through boilers
8. When the water level in the drum drops below the safety mark or in the case of once-through boilers, the supply of feedwater is interrupted for more than 30s
9. Rupture of tubes in the water-steam path
10. Fuel burning on the heat recovery zone. This is accompanied by abnormal rise of temperature of the flue gases
11. Inadmissible pressure drop of gas or fuel oil behind the control valve
12. When there is no steam flow through the steam reheater.

Q. *What do you mean by normal shutdown to hot standby?*

Ans. It is the normal shutdown for a relatively short spell of time while maintaining the existing pressure and temperature conditions substantially.

Q. *What procedure is adopted, over and above the normal shutdown, for this purpose, in the case of pulverized fuel fired drum type boilers?*

Ans. The normal shutdown procedure is followed, except for the following:
1. Drum pressure should not be reduced in line with unit load reduction
2. Water level should be kept in sight in the gauge glass. Whenever water-level drops out of sight fresh make-up is to be taken to restore the water level in the gauge glass
3. Hot air shut-off gate valve should be closed as each pulverizer is taken out of service. When the coal-air temperature drops to 45°C, the pulverized coal feeder is stopped. The pulverizer is to be run until empty and then stopped.
4. ID and FD fans are to be kept in line at least five minutes at 30% air flow, after all fires are out, to purge the unit
5. All gas ducts, secondary air duct and windbox dampers are to be closed

Q. *What procedures are adopted in the case of a pulverized coal fired drum type boiler for normal shutdown to cold?*

Ans.

1. Load on the unit is gradually reduced and so is the firing rate
2. Steam temperature control, combustion control and feedwater control are put on auto till a point is reached when these are switched over to 'manual' for better control
3. When the feeder rating on pulverizers is dropped below 50%, stabilizing fuel oil guns are put in line
4. The pulverizers are taken out of service when the minimum rating is reached. The pulverizers are run for two minutes to get totally empty
5. The second pulverizer is taken out of service when the feeder rating on this pulverizer reaches 40%
6. Soot blowers are put in line
7. Along with fuel reduction air flow is reduced to and maintained at 30% of the maximum
8. As the load is being reduced, expansion movement of the unit is checked
9. When the last pulverizer is taken shutdown, fuel oil guns are removed from service
10. As the unit goes off line, the superheater outlet vents and drains are wide opened
11. Desuperheating isolating valves are closed
12. ID and FD fans are kept in line for five minutes at 30% air flow to purge the unit after the fires are extinguished

Q. *What steps should be taken in case an emergency fuel trip occurs in the pulverized coal fired boiler?*

Ans.

1. The system should be purged with air for five minutes with air flowrate at pre-trip value
2. If during the fuel-trip, the oil guns were in line, they should be checked to ensure complete closure of individual nozzle shut-off valve
   The oil guns are retracted, removed, cleaned and reinstalled.
3. If pulverizers were in line during the fuel-trip, all pulverizers are to be emptied of coal.

Q. *In which cases does the water level fall out of sight of the water gauge?*

Ans.

1. Failure of the BFW supply.
2. Momentary fluctuations due to extraordinary changes in load.
3. Neglect of the operator.

Q. *What may happen due to low water in the boiler?*

Ans. It may be anything from leakage to explosion. Of course, it depends greatly upon the type of boiler, the rate of combustion and just how low the water level is.

Q. *If you find the water level cannot be restored immediately, what will you do?*

Ans.

1. All fuel supply should be cut off immediately
2. All steam being discharged from the unit should be shut off
3. High air flowrate is to be maintained to cool down the furnace quickly
4. Steam pressure is to be reduced gradually, if pressure parts damage is suspected, by opening the superheater start-up drains
5. Inspection and repairing of leaks after draining of the boiler, are to be carried out.

Q. *In the case of high water level what will happen?*

Ans. It may lead to carryover of BFW and cause priming.

Q. *What necessary steps will you take in the case of high water level?*

Ans.

1. Water level is to be brought down immediately by opening the drain valves of the waterwall system
2. Steam generation rate is to be reduced.

Q. *How will you detect priming?*

Ans. It is indicated by rapid fluctuations in the outlet steam temperature.

Q. *What will you do in such a case?*

Ans.

1. Rate of steam generation is to be reduced
2. Water level, if abnormally high, is to be reduced
3. Alkalinity and TDS of BFW are to be checked
4. Investigation of the condition of the drum internals, when opportunity comes, is to be carried out.

Q. *How will you detect tube failure in a boiler?*

Ans. Tube leakages, if minor, can be detected by the loss of working fluid from the system.

It can be detected by the noise produced by the leak.

It can also be suspected in case of loss of boiler water chemicals.

Q. *What procedures will you adopt in case a leakage has been detected?*

Ans.

1. Boiler is to be shutdown and cooled
2. Boiler drum is to be drained
3. Inspection and detection of leakages are to be carried out
4. Leaks are to be repaired and leaky tubes may be replaced.

Q. *What steps are to be taken in case an economizer fails?*

Ans. Boilers having no economizer by-pass are to be shutdown immediately.

For boilers befitted with an economizer by-pass, the by-pass should be taken in line and economizer taken out of service for repairing.

Q. *During steam raising what precautions should be taken for non-drainable superheaters?*

Ans.

1. Temperature of the superheater tubes should not be allowed to exceed the higher allowable limit
2. No abnormal temperature difference between any two parts should be allowed
3. During start-up, the tube metal temperature is kept below the temperature which the tubes attain at maximum designed capacity

4. The firing rate should be controlled to avoid accumulation of condensate in the superheater coils.

Q. *Why is the safety valve of superheater set at a lower pressure than the safety valve of the boiler drum?*

Ans. It is a safety measure to protect the superheater from starvation.

If the safety valve of the boiler drum is set at a pressure lower than that of the superheater, then in case the boiler safety valve blows there will not be adequate steam flow to the superheater. Therefore, the superheater coils will starve.

Q. *What will happen to the superheater in the case of starvation?*

Ans. The superheater coils will get overheated and this may result in warping of tubes or in extreme cases, tube failure.

Q. *What do you mean by boiler load?*

Ans. It refers to the total demand (including radiation) imposed upon the boiler by the equipment and its connecting piping system when there is the greatest requirement (usually at the time of start-up).

Q. *What is net load?*

Ans. This means total load plus other such loads as hot water. This is also inclusive of radiation at design temperature.

Q. *What is design load?*

Ans. It is the net load together with the piping tax.

Q. *What is gross load?*

Ans. It refers to the design load and pickup allowance combined.

Q. *What do you mean by boiler rating?*

Ans. It refers to the manufacturer's stated capacity of a boiler capable of handling the boiler load.

The boilers are rated on the basis of catalog heating surface area or by performance test or by both.

Power boilers are rated on the basis of boiler horse power together with a suggested overfiring rate. It is important to note that the catalog heating surface forms the very basis of boiler horse power.

Q. *In which units may ratings be expressed?*

Ans.

1. Equivalent evaporation (i.e. kg of steam generated per hr from and at 100°C)
2. Kg of steam per hour for specified pressure and temperature
3. For utility plants by their turbine capability in kilowatts or megawatts
4. Rate of heat generation (kcal/hr) in the furnace.

Q. *What do you mean by gross boiler output?*

Ans. It is the heat (steam or hot water) available from a boiler working under the limiting conditions specified by the rating code.

Q. *What do you mean by the 'heat available from a boiler'?*

Ans. It is the heat output of the boiler, i.e., the total load the boiler will carry.

Q. *Is it the same as the maximum boiler nozzle output?*

Ans. Not necessarily

Q. *Why?*

Ans. Usually a boiler is capable of greater capacity, which may exceed some of the limiting conditions as imposed by the rating code.

That is why gross boiler output should never be equated to a rating.

Q. *What do you mean by maximum boiler output?*

Ans. This refers to the greatest amount of heat that can be developed at the boiler nozzle for a brief spell. It has no connection to normal operating limitations.

Q. *What is nominal rating?*

Ans. It is the theoretical heating load a boiler is designed to handle. Of course, it does not necessarily indicate the correct boiler rating.

Q. *What is net boiler rating?*

Ans. It is the actual heating load the boiler is capable of handling.

Q. *What parameters are included in this 'actual heating load' that a boiler is capable of handling?*

Ans.

1. All radiation at design temperature.
2. Estimated amount of maximum heat requirement of a connected water heater or other connected apparatus.

Q. *What is piping tax?*

Ans. It is an arbitrary allowance to compensate for the heat losses by a normal amount of insulated piping.

Q. *What is pickup allowance?*

Ans. It is an arbitrary allowance to compensate for the extra load imposed during warming up periods. That is why it is also called warming-up allowance. For instance, if during the warming-up period, the air inlet temperature is 10°C, then radiation at this temperature has approximately 17.5% greater output than at 18°C.

Q. *What is piping and pickup?*

Ans. This corresponds to the total amount of heat allowed, apart from net load, to provide for piping, pickup and contingencies.

(This is also an arbitrary allowance to compensate for piping, pickup and contingencies, in addition to the net load.)

Q. *How much of the net boiler rating is maintained as piping and pickup allowance?*

Ans. Steel Boiler Institute as well as Mechanical Contractors Association of America have established:

Pickup and Piping Allowance

Mechanically fired boilers ►	33.33% of net boiler rating
Hand fired boilers ►	50% of net boiler rating

Q. *What are ECR and MCR of a boiler?*

Ans. ECR stands for Economic Continuous Rating, i.e., it is the boiler load at which the boiler runs with optimum efficiency.

MCR stands for Maximum Continuous Rating, i.e. it is the maximum boiler load at which the boiler can run without any problem.

Q. *What is the value of ECR in terms of MCR?*

Ans. Generally ECR of a boiler is 75–80% of its MCR.

Q. *In which case—MCR or ECR—the will boiler run with higher efficiency?*

Ans. ECR.

Q. *What is the peak load of a boiler?*

Ans. It is the maximum load at which the boiler can run for a short period, normally a couple of hours, in a day's service without problems.

Q. *Why has a provision been made for it?*

Ans. To accommodate a sudden increase in load demand over a brief period of time.

Q. *What is overloading of a boiler?*

Ans. It is the running of a boiler at a load higher than its MCR value for a considerable length of time.

Q. *What do you mean by system load?*

Ans. It is the total power produced, at any particular instant, by all the plants of a power generating system.

Q. *What type of units are suitable for base load supply?*

Ans. Large monobloc units operating on supercritical steam parameters usually supply a base load.

Q. *What type of units are suitable for semi-peak and peak-load supply?*

Ans. Subcritical-pressure plants are particularly suitable for this duty.

Q. *Why?*

Ans.
1. They are well suited for load variations
2. They have faster startup
3. They require less time to attain the installed load.

Q. *Why are the monobloc units of supercritical pressure not assigned to supply semi-peak and peak loads?*

Ans.
1. These steam generating giants have large boilers and heavy metal parts. Therefore, they require a longer time for boiler firing and heating up of pipelines.
2. They consume a greater amount of heat during startup. So it is uneconomic to resort to frequent startup and shutdown of these units.
3. They are less convenient for control.
4. Since they are high-pressure boilers, their valves and fittings are prone to wear due to frequent load fluctuations and frequent startup.
5. These high-pressure boilers have thicker tubes that are more liable to develop high thermal stresses (causing corrosion fatigue) owing to frequent load variations.

Q. *What is the lower limit of the load control range?*

Ans. Operating experience has established the following lower limits of the load control range:

Fuel Burned/Boiler Type	Lower Limit of Load Control Range
Fuel oil/Gas/High-volatile coal	40–50% of rated load
Lean coal	50–60% of rated load
Slag-bottom boiler	60–75% of rated load

Q. *What is the lowest load for stable operation of a boiler?*

Ans. It is usually 30 – 40% of the rated load.

Q. *Why is it advisable to select initial steam pressure equal to 13 MN/m^2 instead of 16 MN/m^2, for the operation of a steam turbine?*

Ans. It has been established from past experience that approximately the same economic efficiency characteristics are obtained irrespective of whether the initial steam pressure is 13 MN/m^2 or 16 MN/m^2.

Moreover, the manoeuvrability of the power plants at these steam pressures is roughly the same.

Past experience shows positive results in the operation of steam- turbine plant at a pressure of 13 MN/m^2.

Hence for a manoeuvring power plant it is advisable to select initial steam pressure equal to 13 MN/m² instead of 16 MN/m².

Q. *For boiler plants operating at semi-peak and peak load conditions the temperature of live and reheated steam is kept in the region of 530–535°C. Why?*

Ans. The tubes for boilers for semi-peak and peak load duty are made of pearlitic steel. For reliable operation under variable load, the temperature of live and reheated steam is kept at 530–535°C.

Q. *Why are austenitic steels not used to enable the tubes to operate at higher temperatures?*

Ans. Austenitic steels are much more expensive (five to eight times) than pearlitic steels. If austenitic steels are used, it will greatly enhance the capital expenditure.

Q. *Why is the lower limit of superheated steam temperature kept at 480–500°C in a thermal powerplant?*

Ans. This is to protect the final stages of the turbine from erosion damage. These stages of the turbine tend to be operated in the wet region of steam expanding through the blades of the previous stages. Water particles striking on the blade profile can cause impingement, corrosion/erosion.

Q. *Why is it that a manoeuvrable boiler unit should be compact?*

Ans. A manoeuvrable boiler unit should be compact so that it contains a small mass of metal. With a lower mass of metal, less time is required for startup and changeover from one thermal mode to another.

Q. *Why are manoeuvrable boiler units usually fuel-oil fired?*

Ans. A manoeuvrable boiler unit should essentially possess the capacity of rapid changeover from one thermal state to another. In the case of fuel-oil fired boilers this basic condition is easily met, as the fuel oil combustion is efficient as well as stable over a wide range of operating loads.

Q. *Why are the convective elements of manoeuvrable boilers made with a horizontal arrangement of tubes?*

Ans. A manoeuvring boiler unit should be drainable for maintenance convenience. This being the case, the tubes are arranged horizontally in the convective heat transfer elements.

Q. *Why is the working fluid in a manoeuvring once-through boiler circulated at high mass velocity?*

Ans. To ensure:
1. proper cooling over a wider range of loads
2. reliable operation of a boiler at low loads.

Q. *It is true that the efficiency of a monobloc unit and a thermal power station is a function of the initial steam parameters. What technical difficulties inhibit the possibility of increasing the temperature and pressure of the steam infinitely?*

Ans.
1. The first deterring factor is the availability and cost of the special steels that will operate reliably at higher parameters of superheated steam for prolonged periods.
2. The cost of metal in the superheater and pipelines increases sharply with the increase of temperature.
3. For generating subcritical-pressure steam, natural circulation drum type boilers are usually employed. And with the increase of pressure, the driving circulation head decreases so that the highest pressure in the drum is established at 17 MN/m² so as to ensure reliable circulation. Therefore, a higher pressure may tell upon reliability.
4. The most deterrent factor in selecting higher steam pressure is the cost of the boiler drum. For a high-capacity drum-type boiler, higher steam pressure means a drum of larger dimension and wall thickness. With the addition of more metal, the cost of the drum mounts. It becomes very heavy and bulky, which presents considerable difficulties in mounting and operating the boiler.

Q. *What are the consequences of overloading a boiler?*

Ans. It will:

1. affect the design circulation velocity
2. result in higher flue gas temperature
3. bring about higher superheated steam temperature, that may lead to superheater coil failure by bulging or overheating.

A boiler is never guaranteed for running at overload.

Q. *Why is it not advisable to run a boiler at part load?*

Ans. Constant running of a boiler at part load will cut short its life due to rapid wear and tear.

Q. *What are the effects of running a monobloc unit at derated pressure?*

Ans. If the boiler of a monobloc unit having turbo-alternator in line runs at a derated pressure:

1. the pressure of the steam inlet to turbine will be low and that will consequently reduce the output of the turbine. Of course, a steam pressure drop of 5 to 10% will not affect the TA output much
2. the specific volume of steam will increase, but as the boiler and piping design are constant, the rate of steam generation will drop and steam supply will be lower
3. the overall thermal efficiency of the monobloc unit will be less.

Q. *Why is CO_2% in the flue gases measured?*

Ans. To detect whether the fuel combustion within the furnace is complete or not.

Q. *What are the reasons for black smoke emission through the chimney of a spreader stoker?*

Ans.

1. Incomplete combustion of fuel in the furnace and disintegrated carbon particles being released without being completely burned in the furnace.
2. Carryover of coal fines due to excess of its percentage in the coal-feed.
3. Low supply of secondary air that would otherwise complete the combustion.

Q. *Where is the problem of coal segregation encountered?*

Ans. In a stoker fired furnace.

Q. *What is it?*

Ans. Clogging of coal feeder or chain grate due to large size of coal (in the former case) or higher percentage of coal fines in the feed (later case).

Q. *Why does coal segregation occurs?*

Ans. Segregation will occur if

1. coal contains a higher percentage of fines
2. the design of the coal bunker is not proper and the bunker sides are not slopped to 40°–60° for proper rolling of coal-feed
3. the coal size is more than 23 cm.

Q. *Where is the fire slippage encountered?*

Ans. In a stoker fired furnace

Q. *Why?*

Ans. If the quality of coal is poor or if the stoker runs at a higher speed due to overloading of the boiler, the proper combustion of coal may be impaired. Therefore, the arch may get cooled and have less radiant heat, with the effect that the fire gradually advances.

Q. *How will you operate a blow-off valve hooked up in series with a quick opening valve?*

Ans. The quick opening valve should be opened first followed by the blow-off valve.

During closing, the blow-off valve is to be closed first and then the quick opening valve.

Q. *What do you mean by the firing regime of a boiler?*

Ans. Heating the water-steam and gas-air paths, whereupon the process of steam generation sets in and the temperature and pressure of the steam is gradually raised to the specified value. The firing regime is completed when steam is generated at the rated capacity and the steam parameters attain the specified values.

This is also called the start-up regime.

Q. *Does the boiler start-up depend on the normal state of the equipment?*

Ans. Yes.

Q. *How many kinds of start-up are there?*

Ans.
1. Hot start-up
2. Warm start-up
3. Cold start-up

Q. *How are these distinguished?*

Ans. These are distinguished by the length of idle time preceding start-up.

For a monobloc unit:

Mode Of Start-Up	Preceding Idle Time
Hot	6–10 h
Warm	6–10 h to 70–90 h
Cold	70–90 h

For conventional boilers, the difference between hot and warm start-ups is determined by the preceding idle time of 6–10 h or more, while the difference between warm and cold start-ups is by the loss of gauge pressure in the boiler and the temperature drop of the hottest elements down to 150°C.

Q. *How can the load of a pulverized coal fired drum type boiler unit be decreased?*

Ans.
1. Load is gradually reduced on the unit and the firing rate is reduced.
2. If the unit load decreases to a point when the feeder rating on the pulverizers is reduced to below 50% of the maximum, stabilizing fuel oil firing is introduced and one pulverizer is taken out of service as follows:
   - (a) The pulverizer coal feeder rating of the selected pulverizer is gradually reduced to a minimum.
   - (b) When the minimum rating (approximately 25%) is reached, the hot air shut-off gate valve is closed.
   - (c) When the pulverizer coal-air temperature drops to 45°C the coal feeder is taken out of service.
   - (d) The pulverizer is run for about two minutes to make it completely empty and then stopped.

Q. *How can the load on the above boiler unit be raised?*

Ans.
1. The load is increased gradually and the firing rate is raised.
2. If the unit load is increased to a point where the feeder rating of the pulverizers in line exceeds 80% of the maximum, then the second pulverizer is taken in line, as follows:
   - (a) Pulverizer is started
   - (b) Pulverizer is brought to operating temperature by opening the hot air shut-off gate valve
   - (c) Pulverizer coal-feeder is started at minimum feeder rating
   - (d) After the coal ignition has been established, feeder rating is increased until the pulverizer loading equals the other pulverizers in line

Q. *A boiler is designated by*

$$TT \frac{5}{6.5} \, VU \, 40 \, \frac{137}{91} \cdot \frac{1}{5} \cdot \frac{40}{48} \cdot \frac{7}{18}$$

What does each symbol or numeral mean?

Ans. TT stands for Tilting Tangential type burners (for corner firing)

5 represents	furnace width in metre
6.5 represents	furnace depth in metre
VU40 represents	type of boiler series
137 represents	internal dia (in cm) or steam drum
91 represents	internal dia (in cm) of water drum
5 represents	size of boiler bank tubes (in cm)
40 represents	number of rows wide boiler bank tubes
18 represents	number of rows deep boiler bank tubes
7 represents	distance between the centres of steam drum and water drum (in m)
18 represents	distance between waterwall system bottom ring header and water drum centres (in m).

Q. *How can a bidrum type pulverized fuel fired balance draft boiler be grouped into parts?*

Ans. All the important parts can be grouped under three heads:

1. **Pressure parts**	steam drum ; water drum ; evaporator, superheater, desuperheater, boiler banks, economizer, boiler valves and fittings and integral parts	
2. **Non-pressure parts**	air heater, gas duct, dust collectors, boiler support, fuel firing system, insulation and refractory materials	
3. **Rotating parts**	pulverizers, exhausters, feeder, F.D. and I.D. fans, soot blowers	

Q. *What is the function of an exhauster?*

Ans. It sucks in a pulverizerd coal and air mixture from the coal mill and delivers it to the burners.

Q. *Why is inspection of boilers and their auxiliaries to be carried out on a regular basis?*

Ans. The inspection of boilers is carried out according to preplanned schedules and on a regular basis to detect defects, locate deterioration of material, abnormal wear, etc. so that these can be rectified to avoid serious damage.

Q. *When is such inspection carried out?*

Ans. Both during operation and shutdown.

Q. *At what intervals should the boiler maintenance engineer carry out a check of the boiler during operation, in case the boiler is being put into operation after a long period of repair, adjustments, reserve or overhaul, etc.?*

Ans. At least once in a quarter.

Q. *How should he carry out this check?*

Ans. He should inspect, during the loading of the boiler, to ascertain the following:
- (a) feed system, signals, interlocks, regulating devices, instruments, auxiliaries, furnaces and boilers functioning properly and satisfactorily
- (b) the boiler operation is in conformity with the instructions specified by the boiler suppliers, Govt. regulations, safety rules, etc., with respect to working parameters, viz. steam temperature, pressure, water level, draft losses, etc.

- (c) equipment is kept clean, tidy and in workable condition
  Floors and passages are clean;
  Adequate provision of fire-fighting equipment exists.
- (d) skilled and well-informed operators are manning the boiler.

Q. *Why is it essential to examine thoroughly the pressure parts of a boiler?*

Ans. This thorough examination is done to detect the presence of scales, deposits, corrosion and pitting, etc., on the surfaces of the pressure parts. It is carried out both internally and externally.

Q. *At what intervals, is internal inspection of a boiler normally carried out?*

Ans. Once an year.

Q. *What steps precede internal inspection of a boilers?*

Ans.
1. The boiler is to be shutdown and cooled carefully.
2. ID and FD fans should be run to complete thorough ventilation of the boiler.
3. The boiler is then to be drained through waterwall drain valve.
4. Feedlines, steam lines and drainage expanders are to be disconnected from the boiler to avoid ingestion of hot water or steam in the boiler.
5. All furnace inspection doors and manholes are to be kept open.
6. All manholes on drum and heater should be kept open.

Q. *How is internal inspection carried out?*

Ans.
1. Checking the presence of deposits on the tube outer surfaces inside the drum and the exposed portions of the waterwall tubes, steam lines, headers, feedlines, etc.
2. Collecting the samples of these deposits and getting them analyzed in the lab.
3. Arranging for cleaning of the boiler.

(a) The external surfaces of tubes, drum and waterwall tube are cleaned by using powerful water-jets.

(b) Deposits which are hard to dislodge are brushed off:

Components	Look for
1. Drum inner surface	Presence of development of cracks, corrosion, pitting, thinning of sections. Cleanliness of waterwalls
2. Tubes and headers	Blockage, thinning out, deposits, etc.
3. Other surfaces of waterwall, superheater, economizer tubes	Deformation, increase of dia, bulging out due to corrosion, thinning out due to erosion, obstruction to free thermal expansion due to deposits in expansion gaps.

Q. *Why is boiler preservation a necessity?*

Ans. The internal surfaces of boilers are prone to corrosion by leftover water after operation or by atmospheric oxygen when they are out of service. Hence boiler preservation is required to protect the internal surfaces from corrosion.

Q. *How many methods are there for the preservation of boilers?*

Ans. There are two methods:
1. preservation by wet method
2. preservation by dry method

Q. *For which categories of boilers, is preservation by wet method recommended?*

Ans. Boilers which are kept ready for stand-by service and may be required for sudden demands of operation are preserved by the wet method.

The wet method is also recommended for large boiler units with multiple circuits that are very difficult to drain and dry out thoroughly.

Q. *Why are boilers for stand-by service preserved by the wet method?*

Ans. This renders them available and ready for service quickly. Also this is a more practical method.

Q. *For which category of boilers, is the dry method of preservation recommended?*

Ans. This method is recommended for boilers which are scheduled to be kept out-of-service for a long period and are not expected to be put into operation at short notice.

Q. *Why?*

Ans. As sufficient time allowance can be made for the preparation of the boiler prior to placing it in service.

Q. *Is the wet method also followed for this category of boilers?*

Ans. Yes.

Q. *What is the procedure for the preservation of external surfaces?*

Ans. The boiler should be drained completely.

Obeying the relevant cleaning standards, the external as well as internal surfaces of all pressure parts and metallic surface should be cleaned thoroughly.

Using alkaline water, the whole boiler pressure parts surfaces as well as the gas side surfaces of the air heater should be thoroughly washed. BFW is recommended for this purpose.

Next, the boiler is to be filled with deaerated polish water and the unit fired to raise about 7 kg/cm^2 pressure. This pressure should be maintained till the setting is completely dry, which may require intermittent firing.

The setting, inclusive of the stack should be isolated, i.e. closed to prevent, as far as possible, the infiltration of air from outside.

Periodic inspection is to be followed to guard against condensate on and corrosion of external surfaces of pressure parts. This is particularly important when the wet method is used.

Finally, the safety valve escape piping should be sealed off (closed) so that no rainwater can ingress into the system.

Q. *Why is alkaline water used to wash the pressure parts as well as the gas side surfaces of the air heater?*

Ans. To remove sulphur bearing flue dust, since sulphur is soluble in caustic alkali.

Q. *How is the preservation of boilers by wet method carried out?*

Ans.

1. The boiler should be filled with feedwater to the normal water level through the condenser. The water pH should be 10.5–11.

2. Hydrazine to the extent of 200 ppm is to be dosed with the water. The unit is to be steamed in service to ensure uniform concentration of boiler water throughout the unit and to eliminate dissolved oxygen from water.

3. The boiler pressure should be stepped down gradually and the water level raised as high as is permissible for safe operation while delivering some steam to the line.

   But before a vaccum, due to cooling off is produced, the boiler should be filled with deaerated water until the water spills over and fills the superheater.

4. If the superheater is non-drainable, it must be backfilled with deaerated water from the outlet drain prior to filling the drum to overflowing into the superheater.

5. Boiler connections should be checked for leakage.

6. Analysis of boiler water should be carried out frequently. If the hydrazine concentration in water happens to drop below 50 ppm, the water in the drum should be lowered to the normal operating level and an appropriate quantity of chemicals should be dosed to bring back 200 ppm of N_2H_4 or 300–400 ppm Na_2SO_3. The boiler should be steamed to circulate chemicals to uniform concentration.

Q. *If hydrazine is not available to dose in the filling water, what should be done?*

Ans. Sodium sulphite (Na_2SO_3) is another good alternative. This chemical should be added to the filling water to raise the sulphite concentration to 300–400 ppm. It acts as a deoxygenerator.

Q. *What precautions should be taken for wet preservation?*

Ans.

1. Fully demineralized and deoxygenated water should be taken.

2. Wet preservation should obviously be avoided if there is any possibility of the ambient temperature dropping to freezing point.

3. The drum aircock should be connected to the surge tank (capacity ≈ 0.45–0.5 m^3) located 300 to 450 cm above the drum. This will ensure positive pressure within the boiler, thus barring the entry of oxygen from outside. This surge tank should be covered and closed with hydrazine.

4. Water should be regularly circulated, i.e. once a week by
   (a) using circulating pumps
   (b) steaming the boiler by providing connections on the feedline

5. Boilers meant for stand-by service should be emptied after every two or three weeks down to the bottom of the steam drum. The drum should be inspected for any signs of deterioration and then the boiler refilled to the normal operating level, steamed and dosed with hydrazine or sodium sulphite in appropriate amounts.

   Boilers to be preserved for a prolonged period should be completely emptied after every eight weeks for similar inspection.

6. Nitrogen blanket in the steam space should be maintained by opening the nitrogen supply so that positive nitrogen pressure is always kept on the steam space.

Q. *Why is the periodic circulation of preserving liquid required at regular intervals during the wet preservation of boilers?*

Ans. It has been found from experience that if the preservation liquid is kept static, it does not prevent corrosion after two months or so.

Q. *What type of active corrosion is usually formed in case of static storage over two months?*

Ans. Active pitting corrosion is usually found.

Q. *How is the preservation of the internal surfaces of a boiler carried out by the dry method?*

Ans.

1. The boiler should be filled with deaerated polish water and fired to raise the pressure to 7 kg/cm^2.

    After that firing is discontinued and the boiler pressure is allowed to drop to 2–3 kg/cm^2.

    The water is to be drained out of the unit.

2. The drum blowdown, air cocks, gauge glass and waterwall box drains, excluding superheater box drains should be opened.

3. All drum manhole doors and all superheater drains should be opened and hot air from portable blowers should be directed into the boxes to complete the drying of internal surfaces.

4. After thorough drying, trays containing silica gel on activated alumina (1.3 kg/m^3 or 0.25 kg/m^2 of heating surface) should be placed inside the drum. Alternatively, anhydrous calcium chloride or unslaked lime (5.1 kg/m^3 or 0.098 kg/m^2 of heating surface) may be used as a dessicant.

5. Pressure parts are then to be closed airtight so that no moisture/air leaks in. Any line through which moisture can ingress into the unit should be disconnected.

6. Once a month inspection should be carried out and the dessicant trays replenished as necessary.

7. Alternatively, nitrogen can be used for dry storage. This nitrogen must be as pure as is practically possible.

Before injecting the nitrogen, the boiler should be freed from as much water as possible. The boiler should be purged with nitrogen by connecting the lower header drains to the nitrogen supply line. Thereafter, a positive pressure of nitrogen should be maintained in the boiler.

Q. *When anhydrous calcium chloride is used as a dessicant, what portion of the tray should be filled, before placing it in the boiler drum?*

Ans. Less than 3/4th of the tray.

Q. *Why?*

Ans. Anhydrous calcium chloride is a dessicant and upon absorbing water it gradually dissovles in it, swells in volume and thereby entails the risk of spillover. The liquid is corrosive.

Q. *How much pure nitrogen is preferred when dry preservation using nitrogen is recommended?*

Ans. The oxygen impurity in it should not exceed 200 ppm.

Q. *Why?*

Ans. Traces of oxygen together with isolated pockets of water on the internal surfaces of boilers can lead to severe pitting corrosion.

Q. *What precautions should be taken during dry preservation?*

Ans.

1. Inspection personnel going inside the drum must wear an oxygen mask to prevent suffocation.

2. The steam drum should always be kept under positive nitrogen pressure when nitrogen is used as preserving agent.

3. The drum should be inspected regularly, i.e. at least once every month.

4. Dessicants should be replenished as required.

5. The superheater box drain should be kept closed.

6

Boiler Calculations

Q. *What is equivalent evaporation?*

Ans. It is the quantity of water evaporated from and at 100°C to produce dry saturated steam at 100°C by absorbing the same amount of heat as used in the boiler under actual operating conditions.

$$M_{eq} = M_{act} (H - H_{w_1})/539$$

where M_{eq} = equivalent evaporation

M_{act} = actual mass of steam generated per unit mass of fuel burnt

H = total specific enthalpy of steam under operating conditions, kcal/kg

H_{w_1} = specific enthalpy of feedwater, kcal/kg

Latent heat of dry, saturated steam at 100°C is 539 kcal/kg.

Q. *What is factor of evaporation?*

Ans. It is the factor to be multiplied with the quantity of steam generated under working conditions to get the equivalent evaporation.

Equivalent evaporation = Actual evaporation × (f)

or $M_{eq} = M_{act} (f)$

or $M_{act} (H - H_{w_1})/539 = M_{act} f$

∴ $f = (H - H_{w_1})/539$

Q. *What is boiler efficiency?*

Ans. It is the ratio of the heat load of the generated steam to the heat supplied by the fuel over the same period.

Heat load of generated steam

= $G_s (H - H_{w_1})$ kcal/s

where G_s = rate of steam generation, kg/s

Rate of heat supplied by fuel = $G_f \times (CV)_f$ kcal/s

where G_f = rate of fuel burning, kg/s

$$\eta_{boiler} = \frac{G_s (H - H_{w_1})}{G_f (CV)_f}$$

$$= M_{act} (H - H_{w_1})/(CV)_f$$

where G_s/G_f = actual evaporation = M_{act}

Q. *What is economizer efficiency?*

Ans. It is defined as the ratio of the heat absorbed by the BFW in the economizer to the heat supplied by the flue gases in the economizer, the temperature of flue gases being reckoned above the temperature of the air supplied to the boiler

$$\eta_{econ} = \frac{M_{act} \Delta \Theta}{M_{f_1} C_p (\Theta_f - \Theta_{air})}$$

where $\Delta \Theta$ = rise in BFW temperature in the economizer

M_{f_1} = mass of flue gases per unit mass of fuel

C_p = specific heat of flue gases

Θ_f = flue gas temperature at inlet to economizer

Θ_{air} = temperature of air delivered to the boiler

Problem 6.1 A boiler generates 4.5 t of superheated steam (500°C, 90 kgf/cm² abs.) per ton of coal feed.

The BFW temperature = 45°C

What is the equivalent evaporation from and at 100°C per ton of coal?

Solution

STEAM
90 kgf/cm² abs. ——— Specific Enthalpy
500°C = 809 kcal/kg

Sensible heat of feedwater at 45°C = 45 kcal/kg

Heat required to produce 4.5 t steam (90 kgf/cm² abs., 500°C)

$$= 4.5 \times 10^3 \times (809 - 45)$$

$$= 3\ 438 \times 10^3 \text{ kcal}$$

Latent heat of dry, saturated steam at 100°C

$$= 539 \text{ kcal/kg} = 539 \times 10^3 \text{ kcal/t}$$

Therefore, equivalent evaporation from and at 100°C

$$= \frac{3\ 438 \times 10^3}{539 \times 10^3} \frac{\text{kcal}}{\text{kcal/t}} = 6.378 \text{ t per ton of coal}$$

Ans.

Problem 6.2 A steam boiler generates 7.5 tons of steam per ton of coal burned. Calculate the equivalent evaporation from and at 100°C per ton of coal from the following data:

Steam pressure = 10 kgf/cm². abs.
Dryness fraction = 0.95
Feedwater temperature = 50°C

Solution Working formula

$$M_{eq} = \frac{M_{act}(H - H_{w_1})}{539}$$

M_{act} = 7.5 t/t of coal

H_{w_1} = 50 kcal/kg = 50 × 10³ kcal/t

H_w = 181.3 × 10³ kcal /t

x = 0.95

L = 483 × 10³ kcal/t

$$H = H_w + xL = [181.3 + 0.95 (483)] \times 10^3$$

$$= 640 \times 10^3 \text{ kcal/t}$$

$$M_{eq} = 7.5 (640.15 - 50) \times 10^3/539$$

$$= 8.211 \text{ t of steam/t of coal}$$

Ans.

Problem 6.3 A boiler is working at 14 bar and evaporates 8.5 kg of water per kg of coal fired from BFW entering at 39°C. Determine the equivalent evaporation from and at 100°C if the steam is 0.96 dry at the stop valve.

Solution The equivalent evaporation from and at 100°C is

$$M_{eq} = M_{act}(H - H_{w_1})/L_{100}$$

Now

M_{act} = 8.5 kg steam per kg of coal

H_{w_1} = 163.4 kJ/kg

H_w = 830 kJ/kg (at 14 bar)

x = 0.96

L = 1957.7 kJ/kg (at 14 bar)

$$H = H_w + xL$$

$$= 830 + 0.96 (1957.7)$$

$$= 2709.39 \text{ kJ/kg}$$

L_{100} = 2257 kJ/kg

$$M_{eq} = 8.5 (2709.39 - 163.4)/2257$$

$$= 9.588 \text{ kg steam/kg of coal}$$

$$= 9.59 \text{ kg steam/kg of coal}$$

Ans.

Problem 6.4 A boiler produces 220 t of dry saturated steam per hour at a pressure 60 kgf/cm², abs. from feedwater at a temperature of 120°C.

Coal consumption = 1200 t/day
Calorific value of coal = 4200 kcal/kg
1% of coal escapes unburnt.

Determine

(a) the equivalent evaporation per ton of coal fired
(b) the efficiency of the boiler
(c) the overall efficiency of the boiler

Solution

Step (I) Heat Load of Steam/Ton

Enthalpy of dry, saturated steam at 60 kgf/cm² abs.
= 665.4 kcal/kg

Enthalpy of water at 120°C = 120 kcal/kg

Therefore, heat required to raise 1 ton of steam

$$= 10^3 (665.4 - 120) = 545.4 \times 10^3 \text{ kcal}$$

Step (II) Equivalent Evaporation

Coal consumption = 1200 t/day = 50 t/h

Steam generated per ton of coal fired

$$= 220/50 = 4.4 \text{ t}$$

Therefore, equivalent evaporation

$$= \frac{4.4\,(545.4 \times 10^3)}{539 \times 10^3}$$

= 4.452 ton of steam/t of coal

Ans.

Step (III) Boiler Efficiency

Energy output = 220 (545.4 × 10³) kcal/h
Coal charged to the boiler = 50 t/h
Actual coal burnt = 50 (1 – 1/100) = 49.5 t/h
Therefore, energy input = 49.5 × 10³ (4200) kcal/h

$$\eta_{boiler} = \frac{220\,(545.4 \times 10^3)}{49.5 \times 10^3 \times 4200}$$

= 0.577 i.e. 57.7%

Ans.

Step (IV) Overall Efficiency of the Boiler

$$\left[\eta_{boiler}\right]_o = \frac{220\,(545.4 \times 10^3)}{50 \times 10^3 \times 4200}$$

= 0.5713

= 57.13%

Ans.

Problem 6.5 A boiler consumes 224 tons of coal to produce 1864 tons of steam per day. The steam is dry, saturated at 90 atm. abs. Calculate the boiler thermal efficiency, and the equivalent evaporation per ton of coal if the calorific value of coal is 5400 kcal/kg of coal, the specific enthalpy of feedwater being 425.036 kJ/kg of water.

Solution
Step (I) Rate of Evaporation

Mass of steam produced = 1864 ton
Mass of coal consumed = 224 ton
Actual evaporation capacity = 1864/224

= 8.321 t/t of coal

Step (II) Equivalent Evaporation
Evaporation capacity, M_a = 8.321 t/t of coal
Sp. enthalpy of dry, satd. steam (90 atm. abs.),

H = 2705 kJ/kg

= 2705 × 10³ kJ/t
Sp. enthalpy of BFW, H_w = 425.036 kJ/kg

= 425.036 × 10³ kJ/t

Equivalent evaporation, $M_e = M_a\,(H - H_w)/L$

= 8.321 (2705 – 425.036) × 10³/(2257 × 10³)

= 8.405 ton steam/ton of coal

Ans.

Step (III) Boiler Thermal Efficiency
Working Formula: Boiler thermal efficiency

= energy to steam/energy from fuel

Energy to steam

= 8.321 (2705 – 425.036) × 10³ kJ

Energy from fuel

= 5400 × 10³ kcal/t of coal

= 5400 × 4.1868 × 10³ kJ/t of coal

Boiler thermal efficiency

$$= \frac{8.321\,(2705 - 425.036) \times 10^3}{5400 \times 4.1868 \times 10^3}$$

= 0.8391

= 83.91%

≈ 84%

Ans.

Problem 6.6 A boiler generates 7.5 tons of steam per hour at 18 bar (1 bar = 10⁵ N/m². The steam temperature is 598K and the feedwater temperature is 328K.

When fired with oil of calorific value 47 250 kJ/kg, the boiler plant achieves an efficiency of 85%.

The generated steam is fed to drive a turbine which develops 0.75 MW and exhausts at 1.8 bar, the dryness fraction of the steam being 0.97.

Determine the rate of fuel consumption and the fraction of enthalpy drop, through turbine, converted to useful work.

If the turbine exhaust is directed for process heating, estimate the heat transfer available per ton of exhaust steam above 322.4 K.

Solution
Step (I) Energy to Raise Steam
Specific enthalpy of generated steam

= 3106 – 0.84(3106 – 3083)

= 3 086.6 kJ/kg (by interpolation)
Specific enthalpy of BFW at 328 K

$= 230.274$ kJ/kg

Specific energy to raise steam
$= 3086.68 - 230.274$

$= 2\,856.4$ kJ/kg.

Step (II) Rate of Fuel Oil Consumption

Rate of steam generation $= 7.5$ t/h $= 7500$ kg/h

Sp. energy to raise steam $= 2856.4$ kJ/kg

Energy input to steam/h $= 7500\ (2\,856.4)$ kJ

$$\text{Boiler efficiency} = \frac{\text{Energy to steam/h}}{\text{Energy from fuel/h}}$$

$$= 85\%$$

$$= 0.85$$

Rate of fuel consumption (Energy from oil/h)

$= 7500\ (2856.4)/\ (0.85)(47250)$

$= 533.408$ kg

Ans.

Step (III) Rate of Sp. Enthalpy Drop in Turbine

Sp. enthalpy of exhaust steam,

$$H_2 = H_w + x \cdot L$$

$$= 490.7 + 0.97\ (2211)$$

$$= 2635.37 \text{ kJ/kg}$$

Sp. enthalpy of inlet steam, $H_1 = 3086.68$ kJ/kg

Sp. enthalpy drop in turbine,.

$$\Delta H = H_1 - H_2$$

$$= 3086.68 - 2635.37$$

$$= 451.31 \text{ kJ/kg of steam}$$

Steam feed $= 7.5$ t/h $= 7500/3600$ kg/s

Rate of sp. enthalpy drop in turbine

$= 451.31\ (7\,500/3\,600)$ kJ/s

$= 940.229$ kJ/s

Step (IV) Fraction of Enthalpy Converted to Useful Work

Energy output from turbine

$= 0.75$ MW

$= 0.75 \times 10^3$ kW

$= 0.75 \times 10^3$ kJ/s

Rate of enthalpy drop in turbine $= 940.229$ kJ/s

Fraction of enthalpy drop converted to useful work

$= 0.75 \times 10^3 / 940.229$

$= 0.7976 \approx 0.8$

Step (V) Heat Transfer Available in Exhaust Steam Above 322.4 K

Sp. enthalpy of exhaust steam $= 2635.37$ kJ/kg

Sp. enthalpy of water at 322.4 K $= 207$ kJ/kg

Heat transfer available in exhaust steam above 322.4 K

$= 2635.37 - 207$

$= 2428.37$ kJ/kg

Ans.

Problem 6.7 The following observations were made in the case of a boiler fitted with an economizer:

Rate of steam generation $= 5$ t/t of coal

Equivalent evaporation from and at 100°C
$= 5.5$ t/t of coal

Boiler feedwater temp. inlet to economizer
$= 100$°C

Temperature of BFW inlet to boiler $= 180$°C

Temperature of air supplied to the boiler $= 30$°C

Temperature of flue gases entering the economizer
$= 400$°C

Weight of flue gases produced per ton of dry coal
$= 15$ t

Mean specific heat of flue gases $= 0.20$ kcal/kg °C

Calorific value of coal $= 5400$ kcal/kg

Determine

(a) the boiler efficiency

(b) the economizer efficiency

(c) the combined efficiency of the whole plant

Solution

Step (I) Heat Output

Steam generated from and at 100°C $= 5.5$ t/t of coal burnt

Therefore, heat output $= 5.5 \times 10^3\ (539)$ kcal/t of coal burnt

Step (II) Heat Input

Calorific value of coal $= 5400$ kcal/kg

Therefore, heat input $= 5400 \times 10^3$ kcal/t of coal burnt

Step (III) Boiler Efficiency

$$\eta_{boiler} = \frac{\text{Heat Output}}{\text{Heat Input}} = \frac{5.5 \times 10^3\ (539)}{5400 \times 10^3}$$

$$= 0.5489 \text{ i.e. } 55\% \text{ (approx)}$$

Ans.

Step (IV) Heat of Flue Gases

Heat of the flue gases entering the economizer

$$= 15 \times 10^3 \, (0.20) \, (400 - 30) \, \text{kcal/t}$$
$$= 111 \times 10^4 \, \text{kcal/t of coal}$$

Step (V) Heat Absorbed by BFW in the Economizer

Heat absorbed by BFW in the economizer

$$= 5 \times 10^3 \, (180 - 100) \, \text{kcal/t of coal}$$
$$= 40 \times 10^4 \, \text{kcal/t of coal}$$

Step (VI) Economizer Efficiency

$$\eta_{econ} = (40 \times 10^4) / (111 \times 10^4) = 0.3603 \text{ i.e. } 36\%$$

Ans.

Step (VII) Combined Efficiency

Heat absorbed in the boiler $= 5.5 \times 10^3 \, (539) \, \text{kcal/t}$ of coal

Heat absorbed in the economizer $= 40 \times 10^4 \, \text{kcal/t}$ of coal

Total heat absorbed in boiler and economizer combined

$$= 5.5 \times 10^3 (539) + 40 \times 10^4$$
$$= 336.45 \times 10^4 \, \text{kcal/t of coal}$$

Energy released by burning coal $= 5400 \times 10^3$ kcal/t of coal

$$\eta_{comb} = \frac{336.45 \times 10^4}{5400 \times 10^3} = 0.623 \text{ i.e. } 62.30\%$$

Ans.

Problem 6.8
A boiler produces steam at 90 kgf/cm² abs. at the rate 150 t/h from the feedwater at 120°C. The steam is dry, saturated. What is the boiler horse power?

Solution

| Steam, 90 kgf/cm² abs. Dry, saturated | — Total heat = 655.7 kcal/kg |

Sensible heat of BFW at 120°C = 120 kcal/kg
Equivalent evaporation from and at 100°C

$$= 150 \times 10^3 \, (655.7 - 120) / 539$$
$$= 149.08 \times 10^3 \, \text{kg/h}$$

Therefore, boiler horse power

$$= 149.08 \times 10^3 / 15.653$$
$$= 9524.15$$

Ans.

Boiler Horse Power is a very commonly used unit for measuring the capacity of a boiler. ASME (American Society for Mechanical Engineers) defines a unit boiler horse power as the boiler capacity to evaporate 15.653 kg of BFW per hour from and at 373 K into dry, saturated steam or equivalent in heating effect.

Boiler h.p. = Equivalent evaporation from and at 373°K per hour/15.653

Problem 6.9
A boiler generates 6.5 t of steam per ton of coal fired.

The steam is at 18 kgf/cm² gauge
The boiler feedwater temperature = 110°C downstream of deaerator
Boiler efficiency = 75%
Factor of evaporation = 1.15
C_p of steam = 0.55 kcal/kg °C
Determine
(a) the temperature of the steam
(b) the degree of superheat, if any
(c) the equivalent evaporation per ton of coal burnt
(d) the calorific value of coal

Solution

Step P (I) Steam Parameters

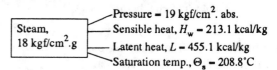

Steam, 18 kgf/cm².g	Pressure = 19 kgf/cm². abs.
	Sensible heat, H_w = 213.1 kcal/kg
	Latent heat, L = 455.1 kcal/kg
	Saturation temp., Θ_s = 208.8°C

Step (II) Degree of Superheat

Total heat of the steam $= H_w + L + C_p \, \Delta\Theta$

where, $\Delta\Theta$ = degree of superheat
Sp. enthalpy of feedwater $= H_{fw}$

Therefore, the factor of evaporation

$$= \frac{H_w + L + C_p \, \Delta\Theta - H_{fw}}{539}$$

or $\quad 1.15 = \dfrac{213.1 + 455.1 + 0.55 \, (\Delta\Theta) - 110}{539}$

$\therefore \quad \Delta\Theta = 112.09°C$

Ans.

Step (III) Superheated Steam Temperature

$$\Delta\Theta = 112.09°C$$

or, $\Theta - \Theta_s = 112.09°C$

or $\Theta = 208.8 + 112.09 = 320.89°C \approx 321°C$

Ans.

Step (IV) Heat Output

Heat required to generate steam

$= 6.5 \times 10^3 (213.1 + 455.1 + 0.55\, \Delta\Theta - 110)$ kcal/t of coal

$= 6.5 \times 10^3 (558.2 + 0.55 \times 112.09)$ kcal/t of coal

$= 4029.021 \times 10^3$ kcal/t of coal

Step (V) Heat Input

Calorific value of coal $=$ CV kcal/kg
Energy released per ton of coal burnt

$= 10^3 \times CV$ kcal

Step (VI) Boiler Efficiency

$\eta_{boiler} = \dfrac{\text{Heat output}}{\text{Heat input}} = \dfrac{4029.021 \times 10^3}{10^3 \times CV}$

or $0.75 = 4029.021 / CV$;

$\therefore$ CV $= 5372$ kcal/kg of coal

Ans.

Step (VII) Equivalent Evaporation

$M_{eq} = \dfrac{6.5 (213.1 + 455.1 + 0.55 \times 112.09 - 110)}{539}$

$= 7.474$ t/t of coal burnt

Ans.

Problem 6.10 The following observations were made during the trial run of a boiler.

Steam generation rate $= 16$ t/h
Feedwater temperature $= 30°C$
Steam quality $= 0.9$ dry
Steam pressure $= 15$ kgf/cm^2 abs.
Coal consumption $= 2.5$ t/h
Calorific value of coal $= 6540$ kcal/kg
Ash + unburnt coal collected from beneath the grates $= 0.2$ t/h (Calorific value $= 700$ kcal/kg)
Weight of flue gases $= 15$ t/t of coal fired
Flue gas temperature $= 350°C$
Average specific heat of flue gases $= 0.25$ kcal/kg °C
Ambient air temperature $= 25°C$

Calculate

(a) the boiler efficiency
(b) the percentage of heat loss to the flue gases
(c) the percentage of heat loss to the ash
(d) the percentage of heat loss unaccounted for

Solution

Step (I) Steam Parameters

Steam, 15 kgf/cm^2 abs.	Sensible Heat, $H_w = 200.7$ kcal/kg
	Latent Heat, $L = 466$ kcal/kg

Step (II) Heat Output Rate

Rate of steam generation $= 16$ t/h
Heat output rate $= 16 \times 10^3 (H_w + x L - H_{fw})$

$= 16 \times 10^3 (200.7 + 0.9 \times 466 - 30)$

$= 9441.6 \times 10^3$ kcal/h

Step (III) Heat Input Rate

Coal consumption $= 2.5$ t/h
Calorific value of coal $= 6540$ kcal/kg
Heat input rate $= 2.5 \times 10^3 \times 6540$ kcal/h

Step (IV) Boiler Efficiency

$\eta_{boiler} = $ Heat output rate/Heat input rate

$= 9441.6 \times 10^3 / (2.5 \times 10^3 \times 6540)$

$= 0.5774$ i.e. 57.74%

Ans.

Step (V) Heat Load of Flue Gases

Flue gases generated $= 15$ t/t of coal
Heat load of flue gases

$= 15 \times 10^3 \times 0.25 (350 - 25)$

$= 1218.75 \times 10^3$ kcal/t of coal

Step (VI) Heat Generated By 1 Ton of Coal

Heat produced by 1t of coal $= 10^3 \times 6540$ kcal

Step (VII) Percentage of Heat Loss to Flue Gases

$= [1218.75 \times 10^3 / 6540 \times 10^3] (100)$

$= 18.63\%$

Ans.

Step (VIII) Percentage of Heat Loss to Ash

Ash + unburnt coal collected $= 0.2$ t/h
Heat loss due to ash + unburnt coal

$= 0.2 \times 10^3 \times 700$ kcal/h

Heat generated in the furnace

$= 2.5 \times 10^3 \times 6540$ kcal/h

Therefore, percentage of heat loss to ash

$= \dfrac{0.2 \times 10^3 \times 700}{2.5 \times 10^3 \times 6540} (100) = 0.85\%$

Ans.

Step (IX) Percentage of Unaccounted Heat

Useful heat = 57.74%

Heat lost to flue gases = 18.63%

Heat lost to ash = 0.85%

Total accounted heat = 57.74+18.63+0.85

= 77.22%.

∴ Unaccounted heat = 100 − 77.22 = 22.77%

Ans.

Problem 6.11 A boiler generates 7.5 t of steam per hour at pressure 1.8 MN/m² and temperature 325°C from feedwater at 49.4°C. When fired with oil of calorific value 45 MJ/kg, the boiler attains an efficiency of 78%. The steam (325°C) is fed to a turbine that develops 650 kW and exhausts at 0.18 MN/m², the dryness fraction of steam being 0.95.

Determine

(a) the mass of oil fired per hour

(b) the fraction of the enthalpy drop through the turbine which is converted to useful work

Also determine the heat transfer available per kg of exhaust steam above 49.4°C, if the turbine exhaust is used for process heating.

Solution

Step (I) Specific Enthalpy of Generated Steam

$H = 3106 - 0.84 (3106 - 3083)$

= 3086.45 kJ/kg (by interpolation)

Step (II) Specific Enthalpy of BFW (49.4°C)

$H_{fw} = 206.9$ kJ/kg

Step (III) Heat Output

Energy required to generate steam

= 3086.45 − 206.9

= 2 879.55 kJ/kg

The rate of steam generation = 7.5 t/h

∴ Heat output = 7.5 × 10³ × 2 879.55 kJ/h

Step (IV) Rate of Oil Burning

Let the mass of oil fired be $\overset{\bullet}{M}$ kg/h

Heat input = $\overset{\bullet}{M}$ × 45 × 10³ kJ/h

Boiler efficiency = 78%

0.78 = Heat output/Heat input

$= 7.5 \times 10^3 \times 2\,879.55/\overset{\bullet}{M} \times 45 \times 10^3$

∴ $\overset{\bullet}{M} = 615$ kg/h

Ans.

Step (V) Specific Enthalpy Drop in Turbine

Specific enthalpy of exhaust steam

$= H_w + x L = 490.7 + 0.95 (2210.8)$

= 2590.96 kJ/kg

∴ Specific enthalpy drop in turbine

= 3086.45 − 2590.96

= 495.45 kJ/kg

Step (VI) Rate of Enthalpy Drop in Turbine

Rate of steam fed to turbine

= 7.5 t/h

= 7.5 × 10³/3600 kg/s

Specific enthalpy drop in turbine = 495.45 kJ/kg

∴ Rate of enthalpy drop in turbine

= 495.45 (7.5 × 10³/3600) kJ/s

= 1032.187 kJ/s

Step (VII) Fraction of Enthalpy Drop Converted to Useful Work

Energy output from turbine

= 650 kW

= 650 kJ/s

Energy input to turbine = 1032.187 kJ/s

∴ Fraction of enthalpy drop converted to useful work

= 650/1032.187

= 0.629

Ans.

Step (VIII) Heat Transfer from Exhaust Steam

The net heat available, for process heating, from exhaust steam above 49.4°C = 2590.96 − 206.9 = 2384.06 kJ/kg

Ans.

Problem 6.12 A steam generation plant supplies 8500 kg of steam per hour at pressure 0.75 MN/m². The steam is 0.95 dry.

Feedwater temperature = 41.5°C

Coal consumption = 900 kg/h

Calorific value of coal = 32450 kJ/kg

Determine

(a) the boiler efficiency
(b) the equivalent evaporation from and at 100°C
(c) the saving in fuel consumption, if by installing an economizer it is estimated that the feedwater temperature could be raised to 100°C, assuming that other conditions remained unchanged and the efficiency of the boiler increases by 6%.

Solution

Step (I) Steam Generation Per Ton of Coal

Rate of steam generation = 8500 kg/h
Coal consumption = 900 kg/h
Therefore, steam generation per kg of coal

= 8500/900

= 9.44 kg

∴ Steam generation/ton of coal

= 9440 kg

= 9.44 t

Step (II) Specific Enthalpy of Steam Raised

$H = H_w + x L = 709.3 + 0.95 (2055.5)$

= 2662.025 kJ/kg

Step (III) Energy Required to Generate Steam

Specific enthalpy of steam raised

= 2662.025 kJ/kg

Specific enthalpy of BFW = 173.9 kJ/kg

Heat output = 2662.025 − 173.9

= 2488.125 kJ/kg of steam

Step (IV) Boiler Efficiency

$$\eta_{boiler} = \frac{\text{Heat output/kg of coal}}{\text{Heat input/kg of coal}} = \frac{2488.125}{32450} (9.44)$$

= 0.7238 i.e. 72.38%

Ans.

Step (V) Equivalent Evaporation

Steam raised per kg of coal = 9.44 kg
Energy required to generate this steam

= 9.44 (2488.125) kJ/kg coal

Specific enthalpy of evaporation from and at

100°C = 2256.9 kJ/kg.

Equivalent evaporation = 9.44(2488.125)/2256.9

= 10.40 kg/kg of coal

Ans.

Step (VI) Energy Required to Generate Steam Under New Conditions

Specific enthalpy of BFW at 100°C = 419.1 kJ/kg
Energy required to generate steam when economizer is incorporated

= 2662.025 − 419.1

= 2242.925 kJ/kg

Energy to steam/h

= 2242.925 × 8 500 kJ

Step (VII) Rate of Coal Consumption when Economizer is Fitted

Energy output = 2242.925 × 8500 kJ/h

Energy input = $\overset{\bullet}{M}$ × 32450 kJ/h

Boiler efficiency = 72.38 + 6 = 78.38%

∴ $0.7838 = \dfrac{2242.925 \times 8500}{\overset{\bullet}{M} \times 32450}$

∴ $\overset{\bullet}{M}$ = 749.57 kg/h

Step (VIII) Saving in Fuel Consumption

Initial fuel consumption rate = 900 kg/h
Modified fuel consumption rate when economizer is fitted = 749.57 kg/h

Saving in fuel consumption = 900 − 749.57

= 150.43 kg coal/h

Ans.

Problem 6.13 The following observations were made during the trial run of a boiler:

Rate of steam generation = 5 t/h

Steam quality: dry, saturated

Steam pressure = 10 kgf/cm² gauge

Average specific heat of steam = 0.55 kcal/kg. °C

Feedwater temperature = 85°C

Room temperature = 25°C

Atmospheric pressure = 1 kgf/cm²

Fuel consumption = 650 kg coal/h

Calorific value of coal = 7500 kcal/kg of coal

Moisture content of coal = 2.5%

Fuel contains: C – 86%; H – 5%; Ash – 9%

Flue gas temperature = 300°C

Mean sp. heat of flue gases = 0.25 kcal/kg °C

Analysis of dry flue gases:

CO_2 – 10% ; O_2 – 8% ; N_2 – 82%

Produce a complete heat balance sheet taking 1 kg dry coal as the basis.

Solution

Step (I) Energy to Steam Per kg of Dry Coal

Steam pressure = 10 kgf/cm^2 gauge

$$= 11 \text{ kgf/cm}^2 \text{ abs.}$$

Sensible heat of steam at 11 kgf/cm^2 abs.

= 185.7 kcal/kg

Latent heat of evaporation at 11 kgf/cm^2 abs.

= 478.4 kcal/kg

Specific enthalpy of dry, saturated steam generated

= 185.7 + 478.4

= 664.1 kcal/kg

Specific enthalpy of feedwater = 85 kcal/kg

Coal consumption

= 650 kg/h

Dry, coal consumption = 650 (100 – 2.5)/100

$$= 650 \times 0.975 \text{ kg/h}$$

Energy to steam/kg coal

= (664.1 – 85) (5000)/(650 × 0.975)

= 4568.836 kcal

Step (II) Flue Gas Analysis

Basis: 100 m^3 of dry flue gas

Con-stituent	Volume m^3	Mol.wt.	Proportional Mass	Remarks
CO_2	10	44	44(10) = 440	Carbon content = 440(12/44) = 120
O_2	8	32	32(8) = 256	
N_2	82	28	28(82) = 2296	
	100	104	2 992	

Step (III) Coal Analysis

Basis: 1kg dry coal

Constituent	Chemical Reaction during combustion	Remarks
C	$C + O_2 \rightarrow CO_2$ (12)　　(44)	Wt. of dry flue gas produced = (2992/104)(86/100) = 24.741 kg/kg of coal
H	$H_2 + \frac{1}{2} O_2 \rightarrow H_2O$ (2)　　(18)	Wt. of water vapour produced = (18/2)(5/100) = 0.45 kg/kg of dry coal

Wt. of moisture fired

= 0.025/0.975

= 0.02564 kg/kg of dry coal

Total wt. of water vapour in flue gases

= 0.45 + 0.02564

= 0.4756 kg/kg of dry coal.

Step (IV) Heat Load of Water Vapour

= 0.4756 [638.8 + 0.55(300 – 90) – 25]

= 344.5 kcal/kg of dry coal

where 638.8 kcal/kg = total heat of water vapour at 1 kgf/cm^2 abs. to which flue gases are discharged

Step (V) Heat Load of Dry Flue Gases

Heat loss to flue gases

= 24.741 (0.25) (300 – 25)

= 1700.944 kcal/kg of dry coal

Step (VI) Heat Balance

Basis: 1 kg of dry coal

Heat Input	%	Heat Expenditure	%
Total heat supplied = 7500 kcal	100	Heat consumed in steam formation = 4568.836 kcal	60.92
		Heat lost to flue gas = 1700.944 kcal	22.70
		Heat lost to vapour = 344 kcal	4.60
		Heat unaccounted for = 886.220 kcal	11.80
7 500 kcal	100	7 500 kcal	100.00

Problem 6.14 During the trial run of a boiler the following data were recorded:

Coal consumption	83.1 t
Steam produced	606 t
Boiler:	
Steam pressure	1.461 MN/m^2
	= 14.42 atm
Steam temperature	470 K
Superheater:	
Superheated steam temperature	610 K
Economizer:	
Water inlet temperature	353 K
Water outlet temperature	400 K
Air heater:	
Air inlet temperature	320 K
Air outlet temperature	380 K
Flue gas inlet temperature	503 K
Flue gas outlet temperature	405 K

Coal analysis (by weight)		Flue gas analysis (dry basis)	
C	62.5%	CO_2	13.2% (by volume)
H	4.25%	O_2	4.85% (-do-)
O	5.11%	N_2	81.95% (-do-)
N	1.2%		
S	0.85%		
Ash	9.85%		
Moisture	16.24%		
Total	100%		

Gross calorific value = 30550 kJ/kg (dry coal)
Boiler house temperature = 298 K
Enthalpy of dry, saturated steam at 1.461 MN/m^2 = 2791 kJ/kg.

Substances	Specific Heats (kJ/kg °K)
Dry flue gas	1.005
Water vapor in flue gas	2.095
Water	4.187

Determine
(a) theoretical air requirements per kg of coal
(b) actual air supplied per kg of coal fired
(c) weight of flue gas per kg of coal burned
(d) thermal efficiencies of
   (i) boiler
(ii) superheater
(iii) air heater
(iv) economizer
(e) heat lost in the flue gas
Summarize the overall result on the basis of 1 kg coal burnt.

Solution
Step (I) Theoretical Air Requirements
   Basis: 100kg coal

Constituent Element	% by Weight	Molecular Weight	kmol	kmol of O_2 required for complete combustion
C	62.5	12	5.208	5.208
H	4.25	2	2.125	2.125/2
				= 1.0625
O	5.11	32	0.159	(−) 0.159
N	1.2	28	0.042	
S	0.85	32	0.027	0.027
Ash	9.85			
Moisture	16.24	18		
				$\Sigma = 6.1385$

There, theoretical air requirement

= 6.1385 (100/21)

=29.23 kmol/100 kg coal

= 29.23 (28.9) kg/100 kg coal

= 844.744 kg/100 kg coal

= 8.447 kg/kg of coal

Ans.

(*cf* The average molecular weight of air is 28.9)

Step (II) Actual Air Supplied
   100 kg coal contains 5.208 kmol of C
   100 kmol of dry flue gas contains 13.2 kmol of C
   Therefore, the amount of flue gas produced

= 5.208 (100/13.2)

= 39.45 kmol/100 kg coal

Let x mol of air be supplied per 100 kg of coal burnt,
   therefore by nitrogen balance we get,

$$79x/100 + 0.042 = \frac{81.95}{100} (39.45)$$

$\therefore$ $x = 40.87$ kmol/100 kg coal

Therefore, the weight of air supplied

= 40.87 (28.9)

= 1 181.14 kg/100 kg coal

= 11.81 kg/kg coal

Ans.

Note % excess air = (11.81 − 8.447) (100/8.447)

= 39.81%

Step (III) Weight of Flue Gas
Basis: 100 kmol

Flue Gas Constituent	kmol	Molecular Weight	Weight	Weight in 39.45 kmol of Flue Gas
CO_2	13.2	44	13.2(44)	13.2(44)(39.45/100) = 229.125
O_2	4.85	32	4.85(32)	4.85(32)(39.45/100) = 61.226
N_2	81.95	28	81.95(28)	81.95(28)(39.45/100) = 905.219
				$\Sigma = 1195.57$ kg

Water produced due to combustion of hydrogen content of coal

= 2.125 kmol = 2.125 (18) = 38.25 kg

Free moisture = 16.24 kg

Therefore, the total weight of wet flue gas

= 1195.57 + 38.25 + 16.24

= 1 250.06 kg/100 kg coal

= 12.50 kg/kg coal

Ans.

Step (IV) Thermal Efficiencies
1. Boiler
Total heat content of steam at 1.461 MN/m^2

= 2 791 kJ/kg

Total heat content of water charged to boiler

= 4.187 (400 − 273)

= 531.749 kJ/kg

Therefore, the net heat transferred to steam

= 2791 − 531.749

= 2259.251 kJ/kg

Rate of steam generation/t of coal

= 606/83.1 t/t

= 7.292 t/t of coal (or kJ/kg coal)

Therefore, the heat transferred to steam/kg of coal burned

= 7.292 (2259.251)

= 16474.458 kJ/kg coal

Gross calorific value of coal as fired

= 30550 (100 − 16.24)/100

= 25588.68 kJ/kg coal

Therefore, thermal efficiency of the boiler

= 16474.458/25588.68

= 0.6438 i.e. 64.38%

Ans.

2. Superheater
Net heat transferred to steam in the superheater

= 7.292 (2.095) (610 − 470)

= 2138.74 kJ/kg coal

Therefore, thermal efficiency of the superheater

= 2138.74/25588.68

= 0.0835 i.e. 8.35%

Ans.

3. Air Heater
Weight of air charged to the boiler

= 11.81 kJ/kg coal

Heat absorbed by air in the air heater

= 11.81 (1.005)(380 − 320)

= 712.143 kJ/kg coal

Therefore, the efficiency of the air heater

= 712.143/25588.68

= 0.0278 i.e. 2.78%

Ans.

4. Economizer
Heat transferred to BFW

= 7.292 (4.187)(400 − 353)

= 1434.985 kJ/kg coal

Therefore, thermal efficiency of the economizer

= 1434.985/25 588.68

= 0.0560   i.e.   5.6%

Ans.

Step (V) Heat Lost to the Flue Gas

Weight of the dry flue gas

= 11.95 kJ/kg coal

Enthalpy of the dry flue gas

= 11.95 (1.005) (405 − 298)

= 1285 kJ/kg coal

Water content in flue gases when 100 kg coal burnt

= 38.25 + 16.24

= 54.49 kg

Therefore, the weight of water vapour/kg of coal burnt

= 0.5449 kg

Therefore, the enthalpy of water in the flue gases

= 0.5449 [2.095 (405 − 311*) + 2411.2*

$\qquad$ + 4.187 (405 − 298)]

= 1665.29 kJ/kg coal

* [*Dew Point of Wet Flue Gas*]

Flue Gas Constituent	kmol		% Composition
CO_2	(13.2/100)(39.45) =	5.2	12.24
O_2	(4.82/100)(39.45) =	1.9	4.47
N_2	(81.95/100)(39.45) = 32.33		76.13
H_2O	54.49/18 = 3.02		7.11
	= Σ 42.46		Σ = 99.95

Therefore, the vapour pressure of water vapour

= (7.11/99.95) (101.3)

= 7.20 kN/m²

which corresponds to the dew point 311°K and latent heat of evaporation 2411.2 kJ/kg

Therefore, the total heat lost to flue gas

= 1285 + 1665.29

= 2950.29 kJ/kg coal

Hence, the percentage of heat lost to the flue gas

= (2950.29/25588.68)(100)

= 11.52%

Tabulation of Result (Basis: 1 kg coal burned)

Unit	Heat recovered (kJ)	% GCV wet coal	% Efficiency
Boiler	15762	61.59 ⎤	
Air Heater	712	2.78 ⎪	78.34
Superheater	2139	8.36 ⎬	
Economizer	1435	5.61 ⎦	
Heat to flue gas	2950	11.53 ⎤	21.66
Heat unaccounted	2591	10.13 ⎦	
		Σ = 100.00	

Problem 6.15 A boiler generates 6000 kg steam per hour at 10 kgf/cm² from BFW at 40°C. The steam is 0.97 dry. The boiler is fired with coal at the rate of 700 kg/h using 16 kg of air (at 15° C) per kg of coal fired.

Assuming the boiler efficiency to be 70%, determine

(a) excess air coefficient

(b) flue gas temperature leaving the boiler

Given

The coal is composed of carbon and hydrogen besides its ash content 12%.

Combustion	Heat Of Combustion
$C + O_2 \rightarrow CO_2$	8075 kcal/kg of carbon
$H_2 + O_2 \rightarrow H_2O$	34500 kcal/kg of hydrogen

Specific heat of flue gas = 0.25 kcal/kg °C

18% of total heat generated by coal is lost to substances other than coal.

Solution

Step (I) Heat Content of Generated Steam

$H = H_w + x L - H_{fw}$

$\qquad$ = 181 + 0.97 (482) − 40 = 608.54 kcal/kg

Step (II) Calorific Value of Coal

It can be determined from the boiler efficiency relationship.

η_{boiler} = Heat output/Heat input

Now Heat Output = 6000 (608.54) kcal/h

Heat input $= 700 \times CV$ where $CV =$ calorific value of coal

or $\quad 0.70 = 6000 \,(608.54)/(700 \times CV)$

$\therefore \quad CV = 7\,452$ kcal/kg

Step (III) Carbon and Hydrogen Contents of Coal
Basis: 1 kg coal
Therefore, the coal contains $(1 - 0.12)$, i.e.

0.88 kg of C + H per kg of coal fired.

If x be the part of C/kg of coal, then

$x\,(8075) + (0.88 - x)(34500) = 7452$

$\therefore \quad x = 0.8669$

Carbon	+	Hydrogen
1 kg Coal 0.8669 kg		0.0131 kg

Step (IV) Theoretical Air Requirement for Complete Combustion
Basis: 1 kg coal

Element	Combustion Reaction	Weight	O_2 Requirement
C	$C + O_2 \rightarrow CO_2$ (12) (32)	0.8669 kg	0.8669(32/12) = 2.3117 kg
H	$2H_2 + O_2 \rightarrow 2H_2O$ (4) (32)	0.0131 kg	0.0131(32/4) = 0.1048 kg
			Σ = 2.4165 kg

Since air contains 23% O_2 by mass, the stoichiometric (theoretical) air requirement for complete combustion

$= 2.4165\,(100/23)$

$= 10.506$ kg

Step (V) Excess Air Coefficient
Excess air coefficient = Actual air/minimum air

$= 16/10.506$

$= 1.523$

Ans.

Step (VI) Enthalpy of Flue Gas
Mass of coal + mass of air = Mass of flue gas

$(1 - 0.12)$ kg + 16 kg = Mass of flue gas

or Mass of flue gas = 16.88 kg.

Heat Generated (kcal/h)	Heat Received by (kcal/h)
by coal: 700(7452) = 5216400	Flue gas: 16.88(700)(0.25)($\Delta\Theta$) = 2954($\Delta\Theta$)
	Steam: 6000(608.54) = 3651240
	Substances other than flue gas: 700(7452)(0.18) = 938952

By heat balancing

$2954\,(\Delta\Theta) + 3651240 + 938952 = 5216400$

$\therefore \quad \Delta\Theta = 211.98°C \approx 212°C$

Hence the temperature of the flue gas at the boiler outlet

$= 212 + 15 = 227°C$

Ans.

Problem 6.16 A boiler is fired with coal having following percentage composition by mass:

C—85%; H—5%; S—1%; O—2.5%; Incombustible—6.5%.

Determine the boiler efficiency from the given data:

Excess air supplied = 40%
Flue gas temperature at boiler exit = 170°C
Ambient air temperature = 25°C
Specific heat of flue gas = 0.25 kcal/kg °C
Specific heat of steam = 0.48 kcal/kg°C

Combustion	Heat of Combustion
$C + O_2 \longrightarrow CO_2$	8075 kcal/kg
$S + O_2 \longrightarrow SO_2$	2220 kcal/kg
$H_2 + O_2 \longrightarrow H_2O$	34500 kcal/kg

Unaccounted heat loss = 18%

Solution
Step (I) Calorific Value of Coal

$CV = 8075(C) + 2220(S) + 34500(H - O/8)$

where, C, S, H & O stand for carbon, sulphur, hydrogen and oxygen percentage.

$= 8075(0.85) + 2220(0.01)$
$\qquad + 34500(0.05 - 0.025/8)$

$= 8503$ kcal/kg

Step (II) Stoichiometric Oxygen

Basis: 1 kg fuel

Element	Combustion Reaction	Oxygen Required Per Kg of Fuel
C	$C + O_2 \longrightarrow CO_2$ (12) (32)	(32/12)(0.85) = 2.2666 kg
S	$S + O_2 \longrightarrow SO_2$ (32) (32)	(32/32)(0.01) = 0.01 kg
H	$2H_2 + O_2 \longrightarrow 2H_2O$	(32/4) (0.05) = 0.4 kg
		$\Sigma = 2.6766$ kg

Since the fuel contains 0.025 kg oxygen/kg of fuel, the actual mass of O_2 requirement per kg of coal burnt

$$= 2.6766 - 0.025$$
$$= 2.6516 \text{ kg}$$

Step (III) Air Supplied

Theoretical mass of air requirement

$$= 2.6516 \ (100/23)$$
$$= 11.5289 \text{ kg}$$

40% excess air supplied.

Hence the actual air supplied

$$= 1.4 \ (11.5289)$$
$$= 16.14 \text{ kg/kg of coal}$$

Step (IV) Mass of Flue Gas

Mass of combustibles per kg of coal

$$= 1 - 0.065$$
$$= 0.935 \text{ kg}$$

Fuel + Air = Flue Gas

0.935 kg   16.14 kg = Flue Gas

Hence the total mass of flue gas (inclusive of water vapour) produced per kg of coal burnt

$$= 0.935 + 16.14$$
$$= 17.075 \text{ kg}$$

Step (V) Mass of Dry Flue Gas

$$2H_2 + O_2 \longrightarrow 2H_2O$$
$$(4) \qquad\qquad (2 \times 18)$$

Mass of water produced/kg of H_2 burned

$$= 2 \ (18)/4$$
$$= 9 \text{ kg}$$

Mass of water produced/kg of coal (H-content : 0.5%) burned

$$= 9 \ (0.05) \text{ kg}$$
$$= 0.45 \text{ kg}$$

Mass of dry flue gas produced/kg of coal burnt

$$= 17.075 - 0.45$$
$$= 16.625 \text{ kg}$$

Step (VI) Heat Balance

Basis: 1 kg coal

Heat Evolved	Heat Lost To
8 503 kcal	Flue gas (dry) = 16.625 (0.25) (170–25) = 602.65 kcal
	Steam (1 atm press.) generated from fuel burning $= 0.45 \ [H + C_p \ (\Delta\Theta) - \Theta_{air}]$ $= 0.45 \ [639 + 0.48 \ (170 - 100) - 25]$ = 291.42 kcal
	Unaccounted sources = (18/100) (8 503) = 1 530.54 kcal
	Total = 2 424.61 kcal

Heat utilized = 8503 – 2424.61

$$= 6078.39 \text{ kcal/kg coal}$$

Step (VII) Boiler Efficiency

$$\eta_{boiler} = \frac{\text{Heat utilized}}{\text{Heat generated}}$$
$$= \frac{6078.39}{8503}$$
$$= 0.7148 \text{ i.e. } 71.48\%$$

Ans.

Problem 6.17 A water tube boiler operates 8400 h/year at 80% efficiency. The unit rated at 27215 kg/h operates at 7.82 atm.

It burns natural gas for six months of the year and No.2 fuel oil for the rest.

Average annual boiler loading is 60% with an input of 11347303 kcal/h.

Without Economizer

Natural gas consumption = 1274.25 Nm³/h ⎱ at 60%
Fuel oil (No.2) consumption = 1.459 m³/h ⎰ load

After Adding An Economizer
BFW flowrate (including blowdown) at 60% load

= 17145 kg/h

Feedwater temperature at the economizer inlet

= 105°C

Feedwater temperature at the economizer outlet

= 136°C

Fluegas temperature at economizer inlet = 260°C
Fluegas temperature at economizer outlet = 149°C
Determine
(a) the fuel saving using the economizer
(b) total annual fuel cost without installing the
economizer
(c) total annual saving of fuel after installing the
economizer
(d) the payback months, if the economizer cost is
Rs. 600,000 installed

Given: Natural gas cost = Rs. 1.06 per Nm³ of
gas

Fuel oil (No.2) cost = Rs 1 255 per m³ of F.O.

Solution The addition of an economizer to a watertube boiler system reduces fuel cost.

The fuel saving using the economizer is

$$S = \frac{H \times 100}{I}$$

where S = fuel saving in percent

H = heat recovered, kcal/h = $\dfrac{F \times \Delta\Theta}{B}$

F = BFW flowrate, kg/h

$\Delta\Theta = \Theta_2 - \Theta_1$ = temperature difference of BFW
before and after the economizer

Θ_1 = BFW temperature at economizer inlet, °C

Θ_2 = BFW temperature at the economizer outlet,
°C

B = boiler efficiency

(a) $H = (17145)(136 - 105)/0.8$

$= 664368.78$ kcal/h

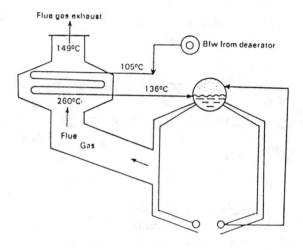

Flue gas exhaust

149°C

Bfw from deaerator

105°C

260°C

136°C

Flue
Gas

Fig. 6.1 *Figure to the Problem 6.17*

$$\therefore \quad S = \frac{664368.78 \times 100}{11347303} = 5.85\%$$

Ans.

(b) Total operating period = 8400 h/year
Natural gas burned for 4200 h and F.O. burned for
rest 4200 h over the year.
Annual cost of natural gas

$$= 1274.25 \left(\frac{Nm^3}{h}\right) \times 4200 \left(\frac{h}{year}\right) \times 1.06 \left(\frac{Rs}{Nm^3}\right)$$

= Rs. 5672961.

Annual cost of fuel oil

$$= 1.459 \left(\frac{m^3}{h}\right) \times 4200 \left(\frac{h}{year}\right) \times 1255 \left(\frac{Rs}{m^3}\right)$$

= Rs. 7690389

$\therefore$ Total annual cost of fuel prior to installation
of economizer

= Rs. 5672961 + Rs. 7690389

= Rs. 13,363,350/-

Ans.

(c) After the installation of the economizer, a 5.85% saving in fuel results.

Annual saving in natural gas

= Rs. 5672961 × 5.85/100

= Rs. 331868/-

Annual saving in fuel oil cost

= Rs. 7690389 × 5.85/100

= Rs. 449888/-

∴ Total annual saving of fuel cost after installation of the economizer

= Rs. 331868 + Rs. 449888 = Rs. 781756/-

Ans.

(d) The payback is

$$P = \frac{E \times 12}{A}$$

where P = payback months
E = installed economizer cost, Rs
A = annual fuel savings with economizer, Rs

$$\therefore \quad P = \frac{Rs.\ 600\ 000}{Rs.781\ 756} \times 12$$

$$= 9.21 \text{ months}$$

Ans.

Problem 6.18 A waste heat boiler is hooked up with a diesel generator set to produce steam from waste heat.

At 60% DG, set load, saturated steam of 10 kg/cm² g is produced in the waste heat boiler at the rate of 40 ton/day.

Average electric energy generated per day varies from 65 to 70 MWH.

Estimate

(a) the economics of incorporating the waste heat boiler

(b) the payback period of the waste heat boiler

Given

The cost of purchased electrical energy from the grid = Rs 1.12/KWH

Cost of generated electricity after the installation of WHB = Rs 0.95/KWH

(This includes overheads and depreciation charges)

Number of operating days per year = 270

Cost of waste heat boiler installed

= Rs 3,000 000/-

Maintenance and overhead expenses

= 12% of the cost of WHB

Rate of interest payable = 20% on the principal amount

DG Set Load	Steam Generation Rate
60%	40 t/day
100%	80 t/day

1 ton of coal generates 4.5 t of saturated steam at 10 kg/cm² g

Cost of coal = Rs 750 per ton

Solution

(a) Economics of Incorporating WHB

The cost of electricity purchased from the grid

= Rs 1.12/KWH

The cost of generated electricity after the installation of WHB = 0.95/KWH

Monetary savings per unit electricity generated

= Rs (1.12 – 0.95)/KWH = Rs 0.17/KWH

Average electricity generation/day

$$= \frac{65 + 70}{2} \text{ MWH} = 67.5 \text{ MWH}$$

Number of working days per year = 270

Therefore, monetary savings with respect to power purchased from the grid

$$= 0/17 \left(\frac{Rs}{KWH} \right) \times 67.5 \times 1000 \left(\frac{KWH}{day} \right)$$

$$\times 270 \left(\frac{days}{year} \right) = Rs\ 3,098\ 250/\text{- per year}$$

(b) Payback Period Of WHB

Cost of waste heat boiler

= Rs 3,000,000/-

Maintenance and overheads

= 12% of capital cost of WHB

= Rs 360,000/-

Interest on principal amount

$= 20\%$ of Rs.3,000,000/–

$=$ Rs 600,000/–

Total steam generated on 100% load $=$ 80 t/day

4.5t of coal generate 1 ton of steam

Amount of coal saved $=$ 80/4.5 $=$ 17.777 t/day

Monetary savings, on the basis of coal, per year

$$= 750 \left(\frac{Rs}{ton}\right) \times 17.777 \left(\frac{ton}{day}\right) \times 270 \left(\frac{day}{year}\right)$$

$=$ Rs 3599842/–

WHB is an energy saving equipment. So it qualifies for 100% depreciation in the 1st year.

Approximate savings in corporate taxes (@ 55%) per year

$=$ Rs 3000000 $\times 0.55$

$=$ Rs 1650000/-

Net Savings per year

$=$ Rs (3599842 – 360000 – 600000 + 1650000)

$=$ Rs 4289842/-

Payback period $= \dfrac{Rs\ 3000000}{Rs\ 4289842} \times 12$ months

$= 8.39$ months

Ans.

Problem 6.19 Determine the

(a) rate of fuel consumption in kg/h

(b) efficiency of WHB of problem 6.18

Given

1 lt. of fuel generates 4.025 KWH of electrical energy

Specific gravity of liquid fuel $=$ 0.90

Exhaust gas flowrate and temperature at 68% load are 7.55 kg/s and 325°C respectively.

Flue gas temperature at WHB inlet $=$ 320°C

Flue gas temperature at WHB outlet $=$ 170°C

Average feedwater temperature to the boiler

$=$ 75°C

Specific heat of flue gas $=$ 0.26 kcal/kg °C

Assume 5% radiation loss suffered by the flue gas in the WHB.

Solution The determination of efficiency of the waste heat boiler is to be made on the basis of heat balance.

(a) Rate of Fuel Consumption

Average electric energy generated per day

$=$ 67500 KWH

$\therefore$ Average fuel consumption per day

$=$ 67500/4.025

$=$ 16770 lt.

$\therefore$ Mass rate of fuel consumption

$$= 16770 \left(\frac{lt}{day}\right) \times 0.9 \left(kg\frac{kg}{lt}\right) \times \frac{1}{24} \frac{1}{(h/day)}$$

$=$ 628.875 kg/h

Ans.

(b) Flowrate of Flue Gas at 60% Load

$= 7.55 \times 3600 = 27180$ kg/h

(c) Useful Heat of Flue Gas

Total heat rejected by hot flue gas in the WHB

$$= 27180 \left(\frac{kg}{h}\right) \times 0.26 \left(\frac{kcal}{kg°C}\right) \times (320 - 170)\ (°C)$$

$= 1060020$ kcal/h

Heat lost to radiation

$= 1060020 \times 5/100$ kcal/h

$= 53001$ kcal/h

Useful heat available for steam generation

$= 1060020 - 53001$

$= 1007019$ kcal/h

$=$ Heat input rate

(d) Heat Output

Average steam (10 kg/cm^2 g and saturated) generation rate

$= 40$ t/day

$= 40 \times 1000/24$ kg/h

$= 1666.66$ kg/h

Average feedwater temperature $=$ 75°C

Heat required to generate 1666.66 kg steam (10 kg/cm^2 g and saturated)

$= 1666.66 \times (183 - 75) + 1666.66 \times 478.4$ kcal

$= 977329$ kcal

$\therefore$ Heat output rate $=$ 977329 kcal/h

Heat input rate $=$ 1007019 kcal/h

∴ Efficiency of waste heat boiler

$$= \frac{977329}{1007019} \times 100\%$$

$$= 97.05\%$$

Ans.

BOILER HEAT BALANCE CALCULATIONS

Basis: 1kg fuel

Heat Input

(A) H_1 = Gross calorific value of fuel, kcal

(B) H_2 = Heat input of fuel

$$= c \cdot (\Theta_f - \Theta_r) \cdot \text{kcal}$$

c = specific heat of fuel, kcal/kg °C

Θ_f = temperature of fuel, °C

Θ_r = reference temperature, °C

(C) H_3 = Heat input of air

$$= M_a c_a (\Theta_a - \Theta_r), \text{kcal}$$

M_a = mass of input dry air/kg fuel, kg air/kg fuel

c_a = specific heat of humid air

$$= 0.24 + 0.46H, \text{kcal/kg dry air °C}$$

H = humidity of air, kg moisture/kg dry air

Θ_a = air temperature, °C

Total heat input, $H_i = H_1 + H_2 + H_3$, kcal

Heat Output

(A) **Heat consumed in generating steam**

1. **Economizer** $H_4 = M_w (h_{ew} - h_{fw})$, kcal

M_w = mass of feedwater per unit mass of fuel, kg/kg fuel

h_{ew} = enthalpy of water at economizer outlet, kcal/kg

h_{fw} = enthalpy of feedwater, kcal/kg

2. **Evaporator (Boiler)** $H_5 = M_s (h_s - h_{ew})$, kcal

M_s = mass of steam generation per unit mass of fuel, kg steam/kg fuel

h_s = enthalpy of steam generated, kcal/kg

3. **Superheater** $H_6 = M_s (H_{ss} - h_s)$, kcal

h_{ss} = enthalpy of superheated steam, kcal/kg

(B) **Heat lost in flue gases**

$$H_7 = \Sigma \, n_i \, \bar{c}_{p_i} \, (\Theta_{fg} - 25), \text{kcal}$$

n_i = number of moles of i-th component present in the flue gas produced due to combustion of 1 kg fuel.

$\bar{c}_{p_i}$ = the mean specific heat of i-th component at Θ_{fg}

Θ_{fg} = flue gas temperature, °C

(C) **Heat loss due to evaporation**

1. Moisture is formed due to combustion of hydrogen in the fuel. Loss of heat to evaporate this moisture

$$H_8 = M_m L, \text{kcal}$$

M_m = mass of moisture formed by burning of hydrogen per kg of fuel, kg H_2O/kg fuel

L = latent heat of evaporation of the moisture at the dew point of the flue gases, kcal/kg

2. Heat loss due to evaporation of moisture present in the fuel

$$H_9 = M_{mf} L, \text{kcal}$$

M_{mf} = mass of moisture present in the fuel, kg/kg fuel

(D) **Heat Loss due to incomplete combustion of carbon as carbon monoxide.**

$$H_{10} = \left[\frac{CO}{CO + CO_2} \right] \times C \times 5\,636.7 \text{ kcal}$$

CO = % (by volume) of carbon monoxide in the flue gas

CO_2 = % (by volume) of carbon dioxide in the flue gas

C = carbon burnt per kg fuel burnt, kg/kg fuel

(E) **Heat loss due to unburnt carbon**

$$H_{11} = M_c (7\,837.5), \text{kcal}$$

M_c = mass of unconsumed carbon in refuse, kg/kg fuel

(F) **Heat loss due to blowdown**

$$H_{12} = M_{bl} (h_{bw} - h_{fw}), \text{kcal}$$

M_{bl} = mass of blowdown water, kg/kg fuel

h_{bw} = enthalpy of boilerwater, kcal/kg

(G) **Unaccounted heat loss**

$$H_{13} = H_i - (H_4 + H_5 + H_6 + H_7 + H_8 \\ + H_9 + H_{10} + H_{11} + H_{12})$$

Problem 6.20 A stoker-fired watertube boiler burns coal at the rate of 4 t/h to generate steam of 30 kg/cm^2 abs and 430°C at the rate of 30 t/hour. Evaluate the boiler performance from the following data:

(a) Component Proximate analysis of coal
   Ash   12.7% (by weight)
   Moisture   7.9% (by weight)

(b) Gross calorific value of coal = 6 250 kcal/kg

(c) Component   Flue Gas Analysis
   CO_2 12.85%
   O_2 6.5%
   N_2 rest

(d) Carbon present in the cinder as unburnt combustible = 2.75%

(e) The feedwater temperature = 90°C

(f) Flue gas temperature at economizer outlet = 150°C
   Flue gas pressure at economizer outlet = 755 mmHg

(g) Air temperatures at burner inlet: 30°C DB and 22°C WB. Ignore the presence of sulphur and oxygen in coal.

Solution The boiler performance, i.e. the overall thermal efficiency of the boiler is to be evaluated on the basis of heat balance.

Basis: 100 kmol of dry flue gas.

1. Oxygen Supplied with Combustion Air

N_2 in the flue gases

= 100 – (12.85 + 6.5)

= 80.65 kmol

100 kmol air contains 79 kmol N_2 and 21 kmol O_2

∴ O_2 supplied for combustion

= (21/79) × 80.65 = 21.438 kmol

2. Water Vapour Produced During Combustion

$$C + O_2 \longrightarrow CO_2$$
1 kmol 1 kmol   1 kmol

1 kmol of CO_2 requires 1 kmol of O_2 for combustion.

∴ 12.85 kmol of CO_2 require 12.85 kmol of O_2 for combustion.

Therefore, the oxygen utilized for hydrogen burning of fuel

= 21.438 – (12.85 + 6.5)

= 2.088 kmol

$$2H_2 + O_2 \longrightarrow 2H_2O$$
2 kmols 1 kmol   2 kmols

∴ Hydrogen burnt = 2 (2.088) = 4.176 kmol

∴ Water produced = 4.176 kmol

3. Unburnt Carbon for 100 kmol of Dry Flue Gas

Carbon retained in the cinder

$$= \frac{2.75}{100} \times 4 \times 1000 \text{ kg/h}$$

= 110 kg/h

(FC + VM) in coal

= 100 – (ash% + moisture%) in coal

= 100 – (12.7 + 7.9) = 79.4%

∴ $\dfrac{\text{unburnt carbon}}{(\text{FC + VM} - \text{unburnt carbon})}$

$$= \frac{2.75}{79.4 - 2.75} = 0.0358$$

Carbon in the flue gas

= carbon in the CO_2 in the flue gas

= 12.85 kmol

= 12.85 × 12

= 154.2 kg

Hydrogen in the flue gas

= 4.176 kmol

= 4.176 × 2

= 8.352 kg

Total burnt combustible

= 154.2 + 8.352

= 162.552 kg

Carbon unburnt for 100 kmol of dry flue gas

= 0.0358 × 162.552

= 5.8193 kg

= 0.4849 kmol

4. Excess Air

$$C + O_2 \longrightarrow CO_2$$

1 kmol   1 kmol   1 kmol

1 kmol of C requires 1 kmol of O_2 for combustion

∴ 0.4899 kmol of C requires 0.4899 kmol of O_2 for combustion

∴ Oxygen required to burn that unburnt carbon = 0.4899 kmol

∴ Excess O_2 supplied

= 6.5 − 0.4899 = 6.0151 kmol

Total stoichiometric O_2 required

= 12.85 + 2.088 + 0.4899

= 15.4229 kmol

Therefore, excess air supplied

$$= \frac{6.0151}{15.4229} \times 100$$

= 39%.

5. Moisture Content of the Flue Gas

100 kg coal contains 7.9 kg free moisture and [100 − (12.7 + 7.9)] i.e. 79.4 kg combustibles

Free moisture appearing with the combustion of 162.552 kg combustibles

$$= \frac{7.9}{79.4} \times 162.552$$

= 16.173 kg = 16.173/18 kmol = 0.8985 kmol

From psychrometric chart, humidity of air at 30°C DB and 22°C WB

= 13.4 gm/kg dry air

$$= \frac{13.4}{18 \times 1000} \text{ kmol } H_2O/\frac{1}{29} \text{ kmol dry air}$$

$$= \frac{0.02158 \text{ kmol water}}{\text{kmol dry air}}$$

Therefore, the total moisture entering the combustion zone

= 0.02158 × kmol of air containing 21.438 kmol O_2

$$= (0.02158) \times \left[\frac{100}{21} \times 21.438 \right]$$

= 2.203 kmol

Therefore, total moisture in the flue gases

= (4.176 + 0.8985 + 2.203) kmol

= 7.2775 kmol

6. Composition of Flue Gas

Component	kmol	mol%
CO_2	12.85	11.97
O_2	6.5	6.05
N_2	$\frac{79}{21} \times 21.438 = 80.647$	75.179
H_2O	7.275	6.781
	107.272	99.980

7. Heat Input Rate

Rate of fuel burning = 4 ton coal/h

Gross clorific value of fuel = 6250 kcal/kg

Heat input rate = 4 × 1000 × 6250

= 25000000 kcal/h

Ignoring the heat input of air (at 30°C), the net heat input rate = 25000000 kcal/h

8. Heat Output Rate

[A] Heat absorbed in generating superheated steam

$$H_4 + H_5 + H_6$$

= 30,000 × (787.8 − 90.04) kcal/h

= 20932800 kcal/h

where 787.8 kcal/kg = enthalpy of steam at 30 kgf/cm^2 abs and at 430°C = h_{ss}

90.04 kcal/kg = enthalpy of water at 90°C

[B] Combustibles left in the cinder (as C)

= 110 kg/h

Calorific value of carbon (GCV = NCV)

= 94.05 kcal/mol

$$= \frac{94.05}{12} \times 1000 \text{ kcal/kg}$$

Heat lost in the combustibles

$$H_{11} = \frac{94.05}{12} \times 1000 \times 110 \text{ kcal/h}$$

= 862125 kcal/h

[C] Total free moisture evaporated from coal

$$= \frac{7.9}{79.4} \times 4000 \times \frac{79.4 - 2.75}{100}$$

= 305.055 kg/h

Now, partial pressure of water vapour in the flue gas

= 755 × mol fraction of water vap. in the flue gas

= 755 (7.275/107.272) mm Hg

= 51.20 mm Hg

Dew point of flue gases ≈ 38°C

Latent heat of water at 38°C = 575.83 kcal/kg

Heat lost due to evaporation of free moisture

= 305.055 × 575.83 = 175660 kcal/h

[D] Heat loss due to evaporation of moisture formed due to combustion of hydrogen in the coal burnt

$$H_8 = \left[\frac{4.176}{162.552}\right](4000)\left[\frac{79.4 - 2.75}{100}\right](18)(575.83)\text{kcal/h}$$

$$\left[\frac{\text{kmol } H_2O}{\text{kg combustibles}}\right]\left[\frac{\text{kg coal}}{h}\right]\left[\frac{\text{kg combustibles}}{\text{kg coal}}\right]$$

$$\left[\frac{\text{kg } H_2O}{\text{kmol } H_2O}\right]\left[\frac{\text{kcal}}{\text{kg}}\right] = 816408 \text{ kcal/h}$$

[E] Heat lost in flue gases (H_7) is evaluated on the basis of mean specific heat data of flue gas components:

Component	Mean Sp. Heat in the Range 25°–150°C
CO_2	9.5 kcal/kmol °C
H_2O	8.12 kcal/kmol °C
O_2	7.12 kcal/kmol °C
N_2	7.00 kcal/kmol °C

Therefore, for 107.272 kmol of flue gas

$$n_i c_{p_i} = 12.85 (9.5) + 7.275 (8.12) + 6.5 (7.12)$$

$$+ 80.647 (7) \text{ kcal/ °C}$$

$$= 791.957 \text{ kcal/°C}$$

Heat lost in flue gases = 791.957 (150 – 25)

= 99,994.62 kcal

Therefore, the rate of heat loss of flue gases

$$= 99,994.62 \ (4,000/162.552)\left(\frac{79.4 - 2.75}{100}\right)\text{kcal/h}$$

$$= 1,886,064 \ \frac{\text{kcal}}{h}$$

9. Heat Balance

Heat Input Rate (kcal/h)	Heat Output Rate (kcal/h)		
25,000,000	Steam generation	– 20932800	83.73%
	Heat loss due to unburnt combustibles	– 862125	3.44%
	Heat loss due to evaporation of free moisture	– 175660	0.70%
	Heat loss due to evaporation of moisture formed due to combustion of hydrogen in the fuel	– 816408	3.26%
	Heat lost to flue gases	– 1886064	7.54%
	Unaccounted heat loss (by difference)	– 326943	1.30%
	Total – 25 000 000		

(10) Overall Thermal Efficiency of the Boiler

$$= \frac{\text{Heat output rate (steam generation)}}{\text{Heat input rate (fuel combustion)}}$$

$$= \frac{20\ 932\ 800}{25\ 000\ 000} \times 100 = 83.73\%$$

7

Draught

Q. *What is draught (draft)?*

Ans. The difference in pressure of the combustion products within a boiler furnace and the cold air outside is known as (draught or draft).

Draught is also the differential pressure between the air column (chimney height) outside and the hot flue gas column (chimney height) inside the chimney.

$$= H (d_a - d_f) g \text{ kgf/cm}^2$$

Q. *Is there any other definition of draught?*

Ans. It is the difference in the weight of a cold air column of chimney height (H) and that of flue gas of the same column of chimney height.

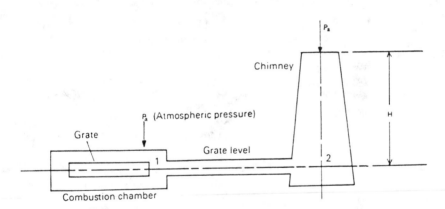

Fig. 7.1

Equipment	Level	Pressure
Combustion chamber	Grate	$p_1 = p_a +$ pressure of cold air of column H where $d_a =$ density of air $= p_a + H d_a g$
Chimney	Grate	$p_2 = p_a +$ pressure of hot flue gas of column H $= p_a + H d_f g$

Since $p_1 > p_2$, the draught produced

$$= p_1 - p_2$$
$$= (p_a + H d_a g) - (p_a + H d_f g)$$

Q. *How?*

Ans. Draught $= H (d_a - d_f) g \text{ kgf/cm}^2$

Now, if unit area of cross-section of the chimney is considered, then

$$H d_a g = H \underbrace{\text{(Area of Cross-section)}}_{\text{Volume}} d_a g$$

$$= \text{wt. of air column of height } H$$
$$= W_a$$

$$H d_f g = H \text{ (Area of Cross-section)} d_f g$$
$$= \text{wt. of flue gas column of height } H$$
$$= W_f$$

Draught $= (W_a - W_f)$ kgf

Q. *Why is draught a necessity?*

Ans.

1. To force adequate air through the combustion chamber to assist in the proper combustion of fuel
2. To draw out the resulting hot flue gas from the combustion chamber
3. To vent the products of combustion to atmosphere after necessary heat recovery in superheater, economizer and air heater.

Q. *How is draught classified?*

Ans. Draught can be broadly classified into two classes—natural draught and artificial draught.

Artificial draught is of two types—steam jet draught and mechanical draught.

Steam jet draught can be of induced as well as forced draught type.

Mechanical draught can be of induced, forced and balanced draught types.

Q. *What is natural draught?*

Ans. When the draught is generated with the help of the chimney only.

Q. *What are the advantages of natural draught?*

Ans.

1. Simple in design and construction
2. Low capital investment (lower than that for artificial draught)
3. No maintenance cost
4. Needs no power input for operation
5. High dispersion of flue gas as chimney height is very high
6. Long life.

Q. *What is artificial draught?*

Ans. The draught produced by steam jet or mechanical means like fans and blowers.

Q. *Why is artificial draught required?*

Ans. It is because natural draught will not be sufficient to generate enough static draught (25 to 350 mm of water column) as is required by large steam generation plants.

Moreover, natural draught has one serious limitation:

Variation of draught with the variation of the climatic conditions.

Q. *How does natural draught vary with the climatic conditions?*

Ans. It decreases with the increase of ambient air temperature and increases with the fall of the latter.

For maximum discharge, the draught (draft) produced (in mm of water column),

$$h_w = 176.5H/T_{air}$$

where T_{air} is the absolute temperature (K) of inlet air and H is the height of the chimney.

Q. *What are the specific advantages of artificial draught?*

Ans.

1. No need for large chimney, which is both uneconomic and inconvenient to build
2. Higher operating efficiency (about 70%) than natural draught (hardly 1%)
3. Functionally independent of climatic conditions
4. Any grade of fuel—superior or inferior variety—can be economically burnt
5. Eliminates the need for high temperature of the exhaust flue gas as required in the case of natural draught. About 20% of heat released by fuel is lost to the flue gases in the natural draught
6. Draught can be regulated as per the requirement of the furnace.

Q. *What are the major disadvantages of artificial draught?*

Ans.

1. High investment
2. High operating and maintenance cost

Q. *How many types of steam jet draught are there?*

Ans. Two: induced draught and forced draught (draft).

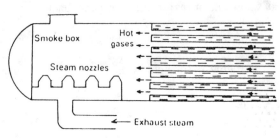

Fig. 7.2 Induced draught produced by steam

Ans.

1. Simplicity in design and operation
2. Low installation cost
3. Has the capability of accepting low grade fuel
4. Occupies very little space.

Q. *What are the disadvantages of such a system?*

Ans.

1. Operate only when steam is available
2. Draught produced is very low (20–30 mm of water column).

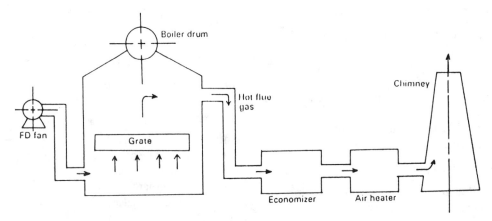

Fig. 7.3 Forced draught (mechanical system)

Q. *How is induced draught by steam produced?*

Ans. Exhaust steam is directed into the smoke box through nozzles. (Fig. 7.2) Steam emerging from these nozzles at high velocity creates a partial vacuum that drags out the air through the furnace space to chimney.

Q. *Where is such system employed?*

Ans. Steam locomotive boilers.

Q. *What is the pressure of exhaust steam?*

Ans. 1.5 to 2 kgf/cm^2

Q. *What is turbine draught?*

Ans. Forced draught produced by steam is also called turbine draught.

Q. *What are the advantages of steam jet draught?*

Q. *How does the forced draught mechanical system operate?*

Ans. A blower or forced draught (F.D.) fan is installed upstream of the boiler and air is forced to pass through the furnace, flues, economizer, air heater and the chimney. (Fig. 7.3)

Q. *Why is it called forced draught?*

Ans. Air is forced to flow through the entire system under pressure.

Q. *It is also called positive draught system. Why?*

Ans. The air pressure throughout the system is positive, i.e. above the atmospheric pressure.

Q. *If the chimney is eliminated, will it affect the forced draught system?*

Ans. Not very much.

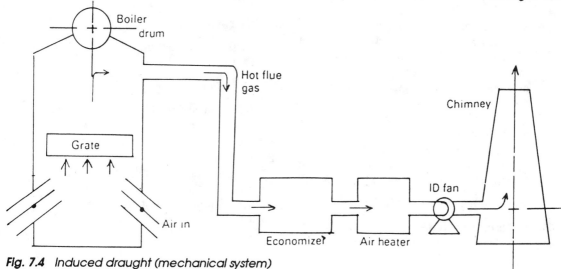

Fig. 7.4 *Induced draught (mechanical system)*

Q.   *Then why is it installed?*

Ans.   To prevent the contamination of stack gas in the ground level. Its chief function is to discharge the flue gas at higher levels.

Q.   *How does an induced draught mechanical system operate?*

Ans.   In this case a blower or induced draught fan is installed downstream of the air heater to suck out the air from the furnace and let it out, via economizer and air heater, to the atmosphere through the chimney. (Fig. 7.4).

Q.   *What is the total draught produced by this system?*

Ans.   It is the sum of the draughts produced by the I.D. (Induced Draught) fan and the chimney.

Q.   *Induced draught is used, in general, when economizer and air heater are incorporated in the system. Why?*

Ans.   Overall draught generated is independent of the temperature of the hot flue gas and as such the flue gas can be discharged after extracting as much heat as possible in the economizer and air heater.

Q.   *What is the most suitable location of the I.D. fan?*

Ans.   It should be located at a place where the flue gas temperature is lowest, i.e. downstream of the air heater.

Q.   *Why?*

Ans.   Since the flue gas can be discharged as cold as possible, it is convenient to locate the I.D. fan where most of the heat of the flue gas has been recovered. This will also mean reduced size of the I.D. fan, as lower the gas temperature lower will be the specific volume of the gas handled.

Q.   *What are the advantages of forced draught (draft) over the induced draught?*

Ans.

1. The size of the F.D. fan is smaller and its power requirement is lower than the I.D. fan since the temperature (hence the specific volume) of the fluid (air) handled by the F.D fan is lower than that of the fluid (flue gasses) handled by the I.D. fan. The size of the I.D. fan is 1.3 times the size of the F.D. fan.

2. For I.D. fans, a separate cooling water system should be included to cool the fan bearings as it will have to withstand the high temperature of the flue gases.

3. Since during forced draught, the furnace is always at pressure above the atmospheric, there is no chance of leakage of cold air into it. Whereas in the case of induced draught, there is always ingestion of cold air from outside into the furnace causing dilution of combustion.

4. There is more uniform and greater penetration of combustion air through the furnace grate in the case of forced draught than in the case of induced draught. Better the penetration and distribution of the air, greater is the rate of firing.

5. Whenever firing is to be done in the case of an I.D. system, there is always a rush of cold air into the furnace space as the doors are opened. This means draught loss and loss in heat transmission efficiency of the surfaces.

Q. *Then why is not an F.D. used alone?*

Ans. In this case, the furnace cannot be opened for firing or for inspection, as the high pressure air inside

Q. *Then what is the solution?*

Ans. Use of balanced draught, i.e. I.D. and F.D. combined.

Q. *How will it function?*

Ans. The forced draught fan will supply the combustion air for proper and complete combustion of fuel and will overcome the fuel bed resistance in the case of stoker grate.

The induced draught will remove the flue gas plus excess air from the furnace, maintaining the pressure inside the furnace just below the atmospheric. (Fig. 7.5)

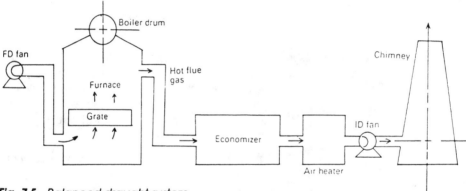

Fig. 7.5 *Balanced draught system*

the furnace will try to gush out suddenly. And this may blow out the fire completely, stopping the furnace.

Q. *What is the pressure distribution throughout this system?*

Ans. Fig. 7.6.

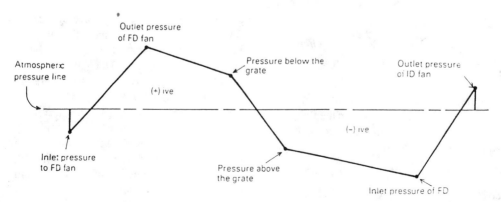

Fig. 7.6 *Pressure distribution profile in a balanced draught system*

Q. *What is the pressure inside the furnace?*

Ans. It is near atmospheric and somewhat below it.

Q. *What is its effect?*

Ans. This eliminates the danger of blowing out of flames as well as the danger of inrushing of air into the furnace when the furnace doors are opened for inspection.

Q. *What is the pressure of air below the grate?*

Ans. Above atmospheric.

Q. *Why is it necessary?*

Ans. To flush the fuel bed with adequate and uniform supply of air for proper combustion.

Q. *What is the pressure of air above the grate?*

Ans. Below atmospheric.

Q. *Why is it necessary?*

Ans. To drag out the products of combustion as quickly as possible from the zone of combustion.

Q. *How can the efficiency of the fans (ID/FD) be computed?*

Ans. If the fan generates a draught of h_w mm of water $(= h_w$ kgf/m$^2)$ and discharges V m^3/min of gas, then the work done by the fan

$$= h_w \, V \, \text{kgf . m/min}$$

$$\text{H.P. of the fan} = \frac{h_w \, V}{4500}$$

If η_1 and η_2 be the transmission efficiency (between the prime mover and the fan) and mechanical efficiency (of the fan), then

$$\text{B.H.P. of primemover} = \frac{h_w \, V}{4500 \, \eta_1 \, \eta_2}$$

This is the required relationship from which the mechanical efficiency of the fan can be computed.

Q. *How the volumetric flowrate (V m^3/min) of ID or FD is determined?*

Ans. For an F.D. fan, the gas handled is air. Hence the volume of air supplied per minute

$$V_a = \frac{W \, M_a}{\rho_a} \text{ m}^3$$

where W = kg of coal burnt per minute

M_a = kg of air supplied per kg of coal

ρ_a = density of air at temperature T_a

Since the pressure difference between atmospheric air and discharged air of F.D. fan is small, by applying Charles' Law

$$\rho_1 \, T_1 = \rho_2 \, T_2$$

or

$$1.293(273) = \rho_a \, T_a$$

$$\therefore \quad \rho_a = \frac{1.293 \, (273)}{T_a}$$

$$\therefore \quad V = \frac{W \, M_a \, T_a}{1.293 \, (273)} = W \, M_a \, T_a / 353$$

For I.D. fan by following mass balance

$$\text{Fuel} \quad + \quad \text{Air} \quad \longrightarrow \quad \text{Flue Gas}$$

$$W \qquad W \cdot M_a \qquad W + W \cdot M_a$$

$$= W \, (1 + M_a)$$

Therefore, the volume of flue gas handled

$$V_f = \frac{W \, (1 + M_a)}{\rho_f}$$

where ρ_f = density of flue gas at flue gas temperature T_f, at the inlet of I.D. fan

Now during combustion,

$$C \; + \; O_2 \; \longrightarrow \; CO_2$$

$$1 \text{ vol.} \qquad 1 \text{ vol.}$$

the volume of air required for combustion is equal to the volume of flue gas produced, ignoring the reaction $2H_2 + O_2 \rightarrow 2H_2O$ as the H-content of coal is too small.

Therefore, volume of $W \, (1 + M_a)$ kg of flue gas

$$= \frac{W \, M_a \, T_f}{1.293 \, (273)}$$

$$\therefore \rho_f = \frac{W(1+M_a)}{\dfrac{WM_a T_f}{(273 \times 1.293)}} = \frac{1+M_a}{M_a} \frac{273 (1.293)}{T_f}$$

$$\therefore V_f = \frac{W(1+M_a)}{1} \frac{T_f}{273(1.293)} \cdot \frac{M_a}{1+M_a}$$

or

$$V_f = \frac{W M_a T_f}{353} \ m^3$$

Q. *Show that the ratio of B.H.P. of the prime mover of an F.D. fan to the B.H.P. of the prime mover of an I.D. fan is the ratio of absolute temperature of air to flue gas, provided both fans produce the same draught (draft) and have the same efficiency.*

Ans. B.H.P. of the prime mover of F.D. Fan

$$= \frac{h_w V_a}{4500 \ \eta_1 \ \eta_2}$$

$$= \frac{h_w W M_a T_a}{4500 \ \eta_1 \ \eta_2 \ 353}$$

B.H.P. of the prime mover of I.D. fan

$$= h_w V_f / 4500 \ \eta_1 \ \eta_2$$

$$= h_w W M_a T_f / 4500 \ \eta_1 \ \eta_2 (353)$$

$$\frac{\text{B.H.P. of the prime mover of F.D.Fan}}{\text{B.H.P. of the prime mover of I.D. fan}} = \frac{T_a}{T_f}$$

Q. *An I.D. fan requires more power, generally 50% more power, than an F.D. fan. Why?*

Ans. Generally, $T_f = 1.5 T_a$ and as such

$$\frac{\text{B.H.P of the prime mover of I.D. fan}}{\text{B.H.P of the prime mover of F.D. fan}} = \frac{T_f}{T_a}$$

$$= \frac{1.5 \ T_a}{T_a}$$

$$= 1.5$$

I.D. fan requires 1.5 times the power required by F.D. fan, i.e. an I.D. fan draws 50% more power than an F.D. fan.

Q. *Why is the control of air supply to the combustion chamber of thermal power plants necessary?*

Ans. To meet the variable load demand.

Q. *What is system resistance characteristic?*

Ans. Interpolation of various static pressures developed within the furnace at varying air flows yields a certain curve called system resistance characteristic. [Fig. 7.7]

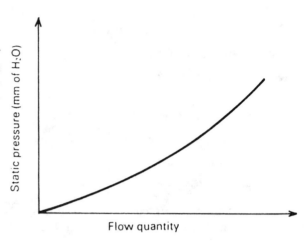

Fig. 7.7 *System resistance characteristic*

Q. *How many resistance curves are there for a particular boiler system?*

Ans. Only one.

Q. *What is the total resistance along the boiler system?*

Ans. It is the sum of the pressure losses in fittings, filters, ducts, fuel bed, economizer, air heater and many others.

Q. *How does this total resistance vary with the velocity and quantity of airflow into the boiler system?*

Ans. The total resistance increases with the increase of velocity and quantity of air flow (Q)

$$h_1 = K Q^2$$

where h_1 = head loss due to total resistance

Q. *How can the supply of air be controlled to meet the variable load demand?*

Ans. Two ways:
1. Damper control
2. Speed control

Q. *Show how air flow can be adjusted by damper control?*

Ans. Let for the fan operating at speed N_1 (rpm), point A marks the point of operation at full load whence the quantity of air supply is Q_a. [Fig. 7.8]

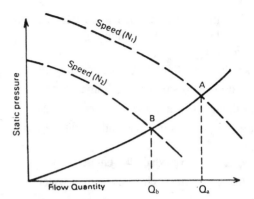

Fig. 7.9 Speed Control

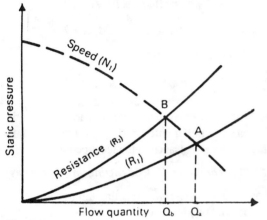

Fig. 7.8 Damper Control

Now, if the load is to be reduced, the flow of air is to be reduced, say to Q_b. This is determined by increasing the system resistance, i.e. by partly closing the dampers. This brings about a new resistance curve (R_2) which intersects the fan characteristic at point B. Now quantity of airflow (Q_b) corresponding to this point is the desired value.

Q. *How can the flow quantity of air be adjusted by speed control?*

Ans. This is achieved by changing the fan rpm to bring about the desired change in the fan characteristic.

If the airflow is to be reduced from Q_a to Q_b, the speed of the fan is reduced from N_1 to N_2 rpm so that the new fan characteristic intersects the resistance curve (R_1) at point B corresponding to airflow Q_b [Fig. 7.9]

Q. *Between damper control and speed control which is more economical?*

Ans. Speed control.

Q. *Why?*

Ans. The power (B.H.P.) requirement in the case of speed control is less than that required in the case of damper control to bring about the desired change in airflow quantity.[Fig. 7.10]

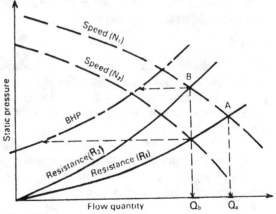

Fig. 7.10

Q. *But most of the steam generation plants use inlet louvers or discharge dampers to control airflow while keeping motor speed constant. Why?*

Ans. This is because speed control with slip ring motors or multiple winding induction motors is too expensive to favour speed control system.

Q. *What is the total draught required to produce the desired airflow in the furnace and discharge the flue gas out of the chimney?*

Ans. It is the arithmetic sum of draught losses in the air and gas loops in the boiler unit

$$h_t = h_{fb} + h_e + h_d + h_v$$

where h_t = total draught loss (in mm of water)

h_{fb} = draught loss in fuel bed (mm of water)

h_e = draught loss in equipment (mm of water)

h_d = draught loss in ducts and chimney (mm of water)

h_v = head loss to produce the required exit velocity of the flue gas from the chimney (mm of water)

Q. *What factors influence the fuel bed resistance responsible for draught loss?*

Ans. Fuel bed thickness, size of coal and rate of combustion.

Q. *If a forced draught stoker furnace is used to fire anthracite coal and bituminous coal separately, in which case will the draught loss be higher for the same rate of combustion (kg of coal/m^2 of grate area)?*

Ans. Draught loss will be higher in the case of forced draught stoker firing anthracite coal.

Q. *Why?*

Ans. Anthracite coal contains much less volatile matter than bituminous coal. Hence it is more compact than the bituminous variety. So for the same rate of combustion, it will require greater quantity of airflow than bituminous coal.

Since the total resistance of the equipment increases with the increase in velocity and quantity of airflow, draught loss will be higher in the case of anthracite bed than its bituminous counterpart.

Q. *What is the velocity head loss?*

Ans. The velocity head loss (h_v) is given by

$$h_v = \frac{v^2}{2g}$$

where v = velocity of the flue gas at the exit of the chimney.

Q. *What factors influence it?*

Ans. Air temperature and natural air velocity at the chimney height.

Q. *What is the head loss in the ducts and chimney?*

Ans. It is the draught loss to overcome friction between air/flue gas and gas ducts/chimney.

It is given by the Fanning equation:

$$h_d = \frac{4fL\,v^2}{2g\,d}$$

where f = friction factor of the duct and chimney

L = length of the duct and chimney

v = velocity of air/flue gas flow

d = hydraulic diameter of the duct $= 2\left[\dfrac{\pi r^2}{P}\right]$

P = wetted perimeter

Q. *What is a chimney?*

Ans. It is a long, vertical cylindrical construction made of steel or masonry or concrete to discharge the flue gas of a boiler sufficiently high into the atmosphere to avoid nuisance to the local people.

Q. *What are the main loads acting on the chimney?*

Ans. Its own dead weight acting through the centre of mass and wind pressure.

Q. *What is the wind pressure normally taken to design a chimney?*

Ans. It is 0.015 kgf/cm^2.

Q. *In which case is a steel chimney preferred?*

Ans. Generally preferred for short exhaust stacks.

Q. *Why?*

Ans. For economy.

Q. *How are they erected? Are they a monobloc system or otherwise?*

Ans. In general, they are made of several sections erected in the field and welded or riveted along the horizontal joints.

Q. *What is the major drawback of a steel chimney?*

Ans. Combustion of high-sulphur fuel liberates acidic oxides like SO_2 and SO_3 producing acid mist (H_2SO_3, H_2SO_4) at dew point. These acids act upon the steel surface leading to a severe corrosion problem.

Q. *How can this be prevented?*

Ans.
1. Lining the steel chimney with acid-resistant brick (silica brick)
2. Cladding the inner surface of the chimney with aluminium.

Q. *What is known as cold air inversion?*

Ans. It marks the phenomenon when the flue gas pressure inside the chimney is less than the air pressure outside the chimney.

Q. *When is it said to occur?*

Ans. When several boilers working on partial load are connected to a common chimney.

Q. *Why is a steel chimney particularly favoured in the case of a gas turbine power plant?*

Ans. A gas turbine attains its full load in less than a minute and as such the chimney has to withstand a thermal shock resulting from the increase in temperature of 450–500°C (723–773 K) during this period.

The thin walled steel chimney having high thermal conductivity is best fitted for accepting such a load.

Q. *Instead of common bricks, perforated bricks are used nowadays to construct brick chimneys. Why?*

Ans. The perforations aid
(a) structural stability
(b) heat insulating properties of the construction with the effect that maximum draft performance is gained.

Q. *In which cases are reinforced concrete chimneys favoured?*

Ans. In cases where the life of the chimney is singled out as the most important and prime element of consideration and where high thermal stresses are absent.

Q. *What is the main drawback of such concrete chimneys?*

Ans. They are susceptible to spalling due to high thermal shock. Acid/water will ingress through the cracks developed due to thermal stress and cause splitting.

Q. *In which cases are glass-reinforced plastic chimneys favoured?*

Ans. Under the following operating conditions, a glass-reinforced plastic chimney is most favourable:
1. low flue gas temperature
2. low thermal stress
3. highly corrosive emissions.

Q. *How can chimney height be calculated from draught (draft)?*

Ans. Draught is the difference between the pressures exerted by the air column and flue gas column of the same chimney height *(H)* measured from the boiler grate level, i.e.

$$\text{draught} = p_{air} - p_{fg}$$

[See steps I to IV on p. 112]

Q. *How can this equation be expressed in terms of mm of water?*

Ans. Since $1 \text{ kg/m}^2 \equiv 1$ mm of water column, so draught measured in terms of water column comes out as

$$h = 353H \left[\frac{1}{T_1} - \frac{W+1}{W} \frac{1}{T} \right] \text{mm of water column}$$

Q. *How can draught be expressed in terms of the height of the column of the flue gas?*

Ans. If h_{fg} be the height of a column of hot flue gases equivalent to the draught pressure of p kgf/m^2, then

$$h_{fg} = p \,/\, \text{Density of hot flue gases}$$

$$= \frac{1.293 HT_0 \left[\dfrac{1}{T_1} - \dfrac{W+1}{W} \dfrac{1}{T} \right]}{1.293 \left(\dfrac{W+1}{W} \right) \dfrac{T_0}{T}}$$

$$= H \left[\frac{W}{W+1} \cdot \frac{T}{T_1} - 1 \right] \cdot m$$

Q. *How can the chimney diameter be determined?*

Ans. Velocity of flue gases in the chimney

$$v = \sqrt{2g \, (h_{fg} - h_L)} \quad \text{where } h_L = \text{draught loss in chimney}$$

Step (I) *Now*

$$\text{Fuel} \quad + \quad \text{Air} \quad \longrightarrow \quad \text{Flue Gas } (T°K)$$

$$\text{Mass} \quad 1 \text{ kg} \quad W \text{ kg} \quad (W+1) \text{ kg}$$

Step (II)

Gas	Temperature	Volume	Remarks
Air (1 kg)	$T_0\text{K} - 273°\text{K}$	$\dfrac{RT_0}{P} = \dfrac{29.27(273)}{1.023\times10^4} - 0.7734 \text{ m}^3$	
-do-	$T_1°\text{K}$ (air temp. outside chimney)	$\dfrac{0.7734\, T_1}{T_0}\text{ m}^3$	Applying Charle's Law as pressure difference between the furnace and chimney is too low
Air (W kg)	-do-	$0.7734 \dfrac{WT_1}{T_0}\text{ m}^3$	
Flue Gas $(W+1)$ kg	$T°\text{K}$	$0.7734 \dfrac{WT}{T_0}\text{ m}^3$	The volume of air required for complete combustion $\;C + O_2 \longrightarrow CO_2$ $\qquad$ 1 vol. $\qquad$ 1 vol. is practically equal to the volume of the products of combustion

Step (III)

Gas	Mass	Volume	Density	Pressure
Air	W kg	$0.7734\,W\,(T_1/T_0)$	$W/0.7734W\,(T_1/T_0)$ $= 1.293\,(T_0/T_1)\text{ kg/m}^3$	$P_{\text{air}} = 1.293\,H\,(T_0/T_1)\,g,\ \text{kgf/m}^2$
Flue Gas	$(W+1)$ kg	$0.7734\,W\,(T/T_0)$	$(W+1)/0.7734W\,(T/T_0)$ $= 1.293\dfrac{W+1}{W}\dfrac{T_0}{T}$	$P_{\text{fg}} = 1.293\,H\dfrac{W+1}{W}\dfrac{T_0}{T}\,g\text{ kgf/m}^2$

Step (IV)

$$\text{Draught (Draft)} = P_{\text{air}} - P_{\text{fg}} \ = 1.293 T H\left[1/T_1 - \frac{W+1}{W}(1/T)\right]g\ \text{kgf/m}^2 \ = 353\,H\left[\frac{1}{T_1} - \frac{W+1}{W}\frac{1}{T}\right]g$$
kgf/m^2

where $T = 273\text{K}$

With the help of this equation chimney height can be calculated.

$$= 4.43 \sqrt{h_{fg} (1 - h_L/h_{fg})}$$

$$= 4.43 \sqrt{(1 - h_L/h_{fg})} \sqrt{h_{fg}} = C \sqrt{h_{fg}} \qquad \text{(I)}$$

where $C = \text{constant} = 4.43 \sqrt{1 - h_L/h_{fg}}$

Again, mass flowrate of flue gases through chimney

$$\overset{\bullet}{M} = \text{volume flowrate} \times \text{density}$$

$$= (vA) \times \rho_{fg} = C \sqrt{h_{fg}} \, A \, \rho_{fg} \qquad \text{(II)}$$

From the known values of v, ρ_{fg} and $\overset{\bullet}{M}$ the value of A (area of cross- section) of the chimney can be found out from equation (II). When A is known, the diameter of the chimney can be easily determined.

Q. *What is the condition for maximum discharge through a chimney?*

Ans. For maximum discharge to take place, the chimney draught expressed in terms of column of hot flue gases should be equal to the height of the chimney.

Q. *Why is it said that chimney draught is created at the cost of thermal efficiency of the boiler?*

Ans. The chimney draught is a function of the temperature

$$\text{Draught} = 353H \left[\frac{1}{T_1} - \frac{W+1}{W} \left(\frac{1}{T} \right) \right] g \text{ kgf/m}^2$$

of the hot flue gases leaving the chimney. As is evident from the above equation, higher the temperature of the flue gases, greater will be the draught.

So a sizeable portion of the heat generated in the boiler furnace goes to heat up the flue gases to the desired temperature to maintain the desired draught level. This heat could have otherwise been utilized for heating the BFW or preheating the combustion air to increase the efficiency of the boiler.

Q. *How can chimney efficiency be determined?*

Ans. This can be done with the aid of the following equation:

$$\eta_{chim} = \frac{H \left[(W/W+1) (T/T_1) - 1 \right]}{4.187 \, C_p \, (T - T_2)}$$

where $T_2 = $ temperature of flue gases in artificial draught.

Q. *What is the magnitude of chimney efficiency, in general?*

Ans. It comes out to be a fraction of a per cent.

Problem 7.1 The height of a chimney of a steam generation plant is 35 m. Determine

(a) the draught produced by the chimney
(b) the available draught from the following parameters:
   Flue gas temperature = 300°C,
   Ambient air temperature = 25°C
   Mass of air supplied = 20 kg/kg of fuel,
   Available draught (draft)
       = 0.82 times natural draught

Solution
Step (I) Chimney Draught Theoretical

H	T_1	T	W	$h = 353 \, H \left\{ \dfrac{1}{T_1} - \left[\dfrac{W+1}{W} \right] \dfrac{1}{T} \right\}$, mm of H_2O
35 m	$25 + 273 = 298°K$	$300 + 273 = 573K$	20 kg/kg of fuel	$353 \, (35) \, \{1/298 - (21/20) \, (1/573)\}$, mm of H_2O $= 18.82$ mm of H_2O \hfill *Ans.*

Step (II) Available Draught

Available draught = 0.82 (Natural draught)
= 0.82 (18.82) mm of H_2O
= 15.43 mm of H_2O \hfill *Ans.*

Problem 7.2 Determine the height of the chimney required to produce a static draught of 18 mm of H_2O when the mean temperature of the flue gases is 550°K and ambient air temperature is 300°K.
   Barometric pressure = 76 cm of Hg

Characteristic gas constant for air = 29.27 kgf m/kg°K

Characteristic gas constant for flue gas = 27 kgf m/kg°K

Solution

Step (I) Density of Air at 300°K

$$\rho_{air} = p_{air}/R_{air} \ (300)$$

$$= \frac{1.032 \times 10^4}{29.27 \times 300}$$

$$= 1.1752 \ \text{kg/m}^3$$

Step (II) Density of Flue Gases at 550°K

$$\rho_{fg} = p_{fg}/R_{fg} \ (550)$$

$$= 1.032 \times 10^4/(27) \ (550)$$

$$= 0.6949 \ \text{kg/m}^3$$

Step (III) Chimney Height

Let the height and area of cross-section of the chimney be H and A respectively

Therefore, weight of the air column of chimney height $= h \, \rho_{air} \, A$

∴ Weight of flue gas column of chimney height

$$= H \rho_{fg} A$$

∴ Draught (Draft)

$$= H \rho_{air} A - H \rho_{fg} A = HA \ (\rho_{air} - \rho_{fg})$$

or $18A = HA \ (1.1752 - 0.6949)$

∴ $H = 37.47$ m

Ans.

Problem 7.3 A boiler consumes 25 t coal per hour. The flue gases temperature is 575°K while the ambient air temperature is 300°K. If the draught produced by a 40 m high concrete chimney is 20 mm of H_2 O, calculate

(a) the air supplied per ton of coal burnt
(b) the draught (draft) in terms of hot flue gas
(c) the flowrate of hot flue gases through the chimney
(d) the velocity of the flue gases in the chimney
(e) the diameter of the chimney at the base

Assume 0.0666 times the draught is available for producing the velocity.

Solution

Step (I) Combustion Air Requirement

T	T_1	H	h	Combustion Air Requirement
575°K	300°K	40 m	20 mm of H_2 O	W

Working Formula:

$$h = 353H \left[\frac{1}{T_1} - \frac{W+1}{W} \right] \frac{1}{T}$$

or $20 = 353 \ (40)\{1/300 - [(W + 1)/W \ ] \ (1/575)\}$

or $\dfrac{W+1}{W} = 1.102$

∴ $W = 9.775$ t/t of coal fired

Ans.

Step (II) Draught (Draft) in terms of Hot Flue Gas

$$h_{fb} = H \ [(W/W+1) \ (T/T_1) - 1] \ \text{m}$$

$$= 40[(9.775/10.775) \ (575/300) - 1]$$

$$= 29.55 \ \text{m}$$

Ans.

Step (III) Mass Flowrate of Flue Gases

Coal consumption = 25 t/h

Air supplied $= 25 \ (9.775) = 244.375$ t/h

Flue gas produced $= 25 + 244.375 = 269.375$ t/h

Step (IV) Density of Flue Gases

$$\rho_{fg} = 1.293 \left[\frac{W+1}{W} \right] \left[\frac{T_0}{T} \right]$$

$$= 1.293 \ (10.775/9.775) \ (273/575)$$

$$= 0.676696 \ \text{kg/m}^3$$

$$= 0.676696 \times 10^{-3} \ \text{t/m}^3$$

Step (V) Volumetric Flowrate of Flue Gases

Mass Flowrate	Density	Volumetric Flowrate
269.375 t/h	0.676696×10^{-3} t/m^3	$269.375/(0.676696 \times 10^{-3})$ $= 398.0736 \times 10^3$ m^3/h $\approx 398 \ 074$ m^3/h

Step (VI) Draught (Draft) available for Velocity producing

$= 29.55 (0.0666)$

$= 1.968$ m of hot flue gas column

Step (VII) Velocity of Flue Gases in the Chimney

$v = \sqrt{2g\ (1.968)}$ m/s $= 6.21$ m/s

Ans.

Step (VIII) Chimney Diameter

Area of cross-section of chimney at the base × Velocity of flue gases = Volumetric flowrate of flue gases

$$\frac{\pi}{4} D^2 v = 398\ 074 \times \frac{1}{3600}$$

or $D^2 = \frac{398\ 074}{3\ 600} \times \frac{4}{\pi} \times \frac{1}{6.21}$

$\therefore\quad D = 4.762$ m

Ans.

Problem 7.4 Calculate the static draught that would be produced by a chimney 30 m high for the maximum discharge through the chimney. The ambient air temperature is 27°C.

Solution

Step (I) Condition for Maximum Draught

For maximum draught in a given chimney

$$\frac{T}{T_1} = 2\frac{W+1}{W}$$

where T = abs. temp. of flue gases in the chimney
T_1 = abs. temp. of ambient air
W = mass of air required per kg of coal burnt

Step (II) Static Draught for Maximum Discharge

The draught equation is

$$h = 353H\left[1/T_1 - \frac{W+1}{W}(1/T)\right] \text{ mm of water}$$
column

For maximum discharge,

$$h_{max} = 353H\left[\frac{1}{T_1} - \frac{W+1}{W}\frac{W}{W+1}\frac{1}{2T_1}\right]$$

$= 353H/2T_1$

$= 353(30)/[(2)\ (273+27)]$

$= 17.65$ mm of water column

Ans.

Problem 7.5 The chimney of a steam generation plant is 35 m high producing a natural draught of 30m column of hot flue gases.

Estimate:
(a) the temperature of the flue gases leaving the chimney
(b) the heat lost to flue gases if the average temperature of the flue gases leaving the boiler in artificial draught system 160°C. The mean specific heat of flue gases is 0.25 kcal/kg°C.
(c) the chimney efficiency
(d) the amount of total heat spent in producing natural draught
(e) the temperature of flue gases in the chimney for maximum discharge.

Given
The calorific value of coal = 6500 kcal/kg
Combustion air supplied = 20 kg/kg of coal burnt
Ambient air temperature = 27°C

Solution

Step (I) Temperature of Flue Gases leaving the Chimney

This can be estimated with the help of the following equation:

$$h_{fg} = H\left[\frac{W}{W+1}\frac{T}{T_1} - 1\right]$$

or $30 = 35\ [\ (20/21)\ (T/273 + 27) - 1\ ]$

or $T = 585°K$ i.e. 312°C

Ans.

Step (II) Extra Heat Loss to Flue Gases above 160°C

Extra heat carried away by 1 kg of flue gases

$= 1(0.25)\ (312 - 160)$ kcal

$= 38$ kcal

Ans.

Step (III) Chimney Efficiency

Static head = 30 m

Maximum energy which this static head will impart to 1 kg of flue gases = $30 \times 1 = 30$ kgf. m

Chimney Efficiency

$$= \frac{30}{38 \times (427)}$$

$= 1.84 \times 10^{-3}$ i.e.   0.18%

Ans.

[where, 1 kcal = 427 kgf. m of mechanical work]

Step (IV) Amount of Heat Spent for Creating Draught

Extra heat carried away for creating draught

= wt. of flue gas × sp. heat × temperature difference

$= (20 + 1)(0.25)(312 - 160)$

= 798 kcal/kg of fuel

Heat produced by burning coal = 6500 kcal/kg of fuel

∴   % heat spent in creating draught

$= [798/6\ 500] \times 100 = 12.27\%$

Ans.

Step (V) Flue Gas Temp. for Maximum Discharge in Chimney

For maximum discharge,

$$\frac{T}{T_1} = 2\frac{W+1}{W}$$

or $T = 2\left(\frac{21}{20}\right)(273 + 27)$

$= 630°K$ i.e.   357°C

Ans.

8

Primary Fuels

Q. *What is fuel?*

Ans. Fuel is a substance which, upon burning in air or oxygen, liberates heat.

Q. *How are fuels classified?*

Ans. Fuels can be classified according to
   (a) the physical state in which they exist in nature—solid, liquid and gaseous
   (b) the mode of their procurement—natural and synthetic
   (c) their calorific value

Primary Fuels	Secondary Fuels
Solid	**Solid**
1. Wood	1. Coke
2. Peat	2. Charcoal
3. Coal	3. Briquettes
4. Lignite	
5. Oil shale	
Liquid	**Liquid**
1. Petroleum	1. Gasoline
	2. Naphtha
	3. Diesel oil
	4. Kerosene
	5. Fuel oils
	6. Shale oil
	7. Coal-tar
	8. Colloidal fuel
	9. Alcohols
Gaseous	**Gaseous**
1. Natural gas	1. Coal gas
	2. Producer gas
	3. Water gas
	4. Oil-gas
	5. Hydrogen
	6. Acetylene
	7. Butane

Q. *What are primary fuels?*

Ans. These are those fuels which occur naturally, such as wood, peat, coal, lignite, petroleum and natural gas.

Q. *What are secondary fuels?*

Ans. These are fuels produced from primary fuels by some treatment processes.

Q. *What is coal?*

Ans. Coal is the stratified rock of decaying trees and vegetation that underwent morphological changes by anaerobic biodegradation through millions of years under heat and pressure of the earth crust above them.

Q. *Which conditions led to the formation of peat?*

Ans. Conditions which were not truly anaerobic.

Q. *How did the degree of alkalinity in the surrounding clay affect the quality of coal?*

Ans. The nature of alkalinity of surrounding clay was the deciding factor in determining the type of coal resulting from the decaying trees and vegetation heaped under tons of the earth's crust many millions years ago.

During anaerobic decomposition of cellulose of the wood, CO_2 and H_2O were liberated and the absorption in the surrounding clay of carbon dioxide produced led the decay to proceed further.

As the decomposition proceeded further, hydrogen and oxygen were gradually consumed from the solid mass and eliminated with the effect that the residue became gradually enriched in carbon and depleted of H and O content in its successive stages of transformation.

Material	%C	Parts of		
		C	H	O
Cellulose	44.6	100	14	111
Wood	50	100	12	88
Peat	60	100	10	57
Lignite	62	100	7.9	54
Brown coal	69.5	100	7.8	36
Bituminous coal	78.8	100	6	21
Anthracite	91	100	4.7	5.2
Graphite	100	100	nil	nil

Therefore, with the propagation of decomposition, more and more CO_2 as well as H_2O were liberated and the quality of the coal improved. Since the absorption of carbon dioxide by the surrounding clay allowed the decay to proceed further, higher alkalinity (absorbing CO_2) led to the formation of bituminous and anthracite variety of coal while lower alkalinity led to the formation of lignite variety.

Q. *How does the volatile matter content determines the rank of coal?*

Ans. Higher the volatile matter content of the coal, the lower is the rank of the coal and vice versa.

Q. *With the growth of the rank, more and more oxygen is expelled from the coal and with that its calorific value increases. Why?*

Ans. The presence of oxygen in coal tends to reduce its calorific value because the oxygen is combined with hydrogen in coal as water. The heat of this chemical reaction has already been liberated and is, therefore, not available during the combustion of coal.

So, as oxygen is excluded from coal, its rank gradually increases and it becomes enriched with carbon and consequently the calorific value of the resulting coal increases.

Coal	Parts of			Calorific Value
	C	H	O	
Peat	100	10	57	18670 kJ/kg
Lignite	100	7.8	54	21000 – 25600 kJ/kg
Brown coal	100	7.8	36	27250 kJ/kg
Bituminous	100	6	21	32125 kJ/kg
Anthracite	100	4.7	5.2	32550 kJ/kg
Graphite	100	–	–	32 910 kJ/kg

Q. *How was petroleum formed?*

Ans. Petroleum was formed by anaerobic decay of marine plant and animal life.

Dead plants and animals of aquatic origin deposited at the bottom of shallow seas were covered with a thick layer of silt and set to decay under near-anaerobic conditions. The steady accumulation of mud above the decaying remains of organic matter led to an increase of temperature and pressure that favoured the formation of liquid and gaseous hydrocarbons.

Q. *What is the ultimate analysis of coal?*

Ans. It is the chemical analysis of coal to determine the percentage of more important chemical elements in coal, which are basically carbon, hydrogen, nitrogen and sulphur and occasionally phosphorus and chlorine.

Q. *How are the percentages of hydrogen and carbon determined?*

Ans. By burning a fixed weight of the coal sample to a fixed weight of residue (ash) in a strongly oxidized atmosphere and determining the amounts of CO_2 and H_2O resulting from combustion.

The carbon dioxide is absorbed in a preweighed solution of KOH while water vapour is absorbed in a preweighed mass of anhydrous $CaCl_2$.

Q. *How is the sulphur content in coal determined?*

Ans. By Eschka method. In this process a fixed weight of coal sample is strongly heated with a mixture of sodium carbonate and magnesium oxide. The sulphur content of the coal gets converted into sulphites and sulphides of sodium and magnesium. The residue is washed into a beaker, oxidized to sulphate by bromine water and the sulphur is then precipitated as barium sulphate. The weight of the sulphur is then equal to 13.73% of the precipitated barium sulphate.

Q. *How is the nitrogen content of coal determined?*

Ans. By Kjeldahl method. In this method, a fixed weight of coal sample is oxidized with concentrated sulphuric acid whereupon the nitrogen content of coal is converted into ammonium sulphate. This is distilled off and titrated with a quantity of standard alkali and the amount of ammonia and hence the percentage of nitrogen is obtained by back titration.

Q. *How is the moisture content of coal determined?*

Ans. Accurately weighed quantity of coal sample is heated in a clean, dry crucible placed in an oven maintained at 373–383°K (100–110°C) for one hour. The crucible is then allowed to cool in a desicator and reweighed. This process, i.e. heating, cooling and weighing is repeated until the final weight becomes constant.

$$\text{Therefore, \% of moisture} = \frac{\text{Loss in weight}}{\text{Weight of coal}} \times 100$$

Q. *How is the ash content of coal determined?*

Ans. Accurately weighed quantity of coal sample taken in a dry, clean, preweighed crucible is completely combusted in air at 730°C (1003°K) for about an hour so that all its combustible material is burnt off leaving the ash as residue. The ash with crucible is weighed after cooling in a desicator.

$$\therefore \quad \text{\% of ash} = \frac{\text{Weight of residue}}{\text{Weight of coal}} \times 100$$

Q. *How is the oxygen content of coal determined?*

Ans. It is not determined directly. It is calculated as the sum of the percentage of the elements plus moisture and ash subtracted from 100.

$$O(\%) = 100 - [C + H + N + S + H_2O + Ash]\%$$

Q. *Why is this quantity called oxygen plus errors?*

Ans. This is because all the experimental errors in the determination of the percentages of all the elements (C, H, N, S) and moisture and ash get included the percentage weight of oxygen of coal.

Q. *What is the utility of ultimate analysis?*

Ans. The results of ultimate analysis are of prime importance.

The concentrations of carbon, hydrogen and oxygen are particularly required in the classification of coal as well as in combustion calculations.

The concentrations of chlorine and vanadium are vital in considering the problems of corrosion and boiler deposits.

The sulphur and phosphorus contents have significant bearing on the metallurgy of iron and steel. The presence of sulphur poses severe problems of corrosion due to formation of clinker and corrosive gases like SO_2 and SO_3.

Q. *How does the sulphur content of coal exert influence upon the metallurgy of iron and steel?*

Ans. In the metallurgy of iron and steel, coke is required as the reducing agent to reduce FeO and Fe_2O_3 to iron. This coke results from the carbonization of coal, during which 65–70% of the sulphur content of coal remains in the coke, 20% escapes as H_2S and the rest gets its way to tar and aqueous liquor. Coke thus obtained is contaminated coke unfit for metallurgical purposes.

[Coal must be washed in scrubbers and purifiers to reduce its sulphur content to 0.5 g/m^3 and then upon carbonization it will produce metallurgical grade coke containing less than 0.6% sulphur.]

Q. *The phosphorous content in coal is usually very low (0.1%). So how can this coal influence the metallurgy of iron and steel?*

Ans. Through the phosphorous content coal is very low, the coke produced from it by carbonization process should not exceed 0.01% phosphorous for the coke being suitable metallurgical purposes. As the phosphorous will react with iron to form hard, brittle iron phosphide.

Q. *What is the sulphur content of coal?*

Ans. It varies depending upon the nature of coal. Usually sulphur content of coal ranges from 0.5 to 3%. Sometimes it may be as high as 5–6%.

Q. *How does it occur in coal?*

Ans. It occurs in three forms:
(a) as pyrites FeS_2
(b) as organic sulphur combined with carbon atoms of giant coal molecule
(c) as gypsum, $CaSO_4$

Q. *Does the presence of pyrites in coal pose any harmful effect?*

Ans. Yes. The effect may be disastrous as the resulting clinker having relatively low fusion point, forms a eutectic with the firebars (in Stoker furnace) leading to severe corrosion.

Q. *What is proximate analysis?*

Ans. It is the chemical analysis to determine the percentages of such important components of coal as moisture, volatile matter, ash and fixed carbon.

Q. *What is fixed carbon?*

Ans. It is a hypothetical term and is not a precise constituent of coal. It does not imply the existence of uncombined carbon in coal material nor does it bear any relationship with the total carbon as determined by ultimate analysis. It is obtained by subtracting from 100 the sum of the percentages of moisture, volatile matter and ash content of the coal sample.

Q. *What is the significance of proximate analysis of coal?*

Ans. Its significance lies in the classification of coal on the basis of its percentage content of volatile matter and calorific value. (Fig. 8.1)

Q. *How is moisture present in coal?*

Ans. It exists in three forms:
  (a) surface moisture
  (b) inherent moisture
  (c) combined moisture

Q. *How is surface moisture acquired by coal?*

Ans. During washing and scrubbing of coal for storage and cleaning.

Q. *How is inherent moisture ingressed in coal?*

Ans. It is the absorbed moisture as water is absorbed through capillaries in the coal substance and makes its way throughout the body of coal.

Q. *How is combined moisture present in coal?*

Ans. In the same way as it is present in hydrates. It is held in loose combination.

Q. *How does the percentage moisture affect the quality of coal?*

Ans. The lower the rank of a coal, the higher is the moisture percentage of coal and vice versa.

Q. *The presence of moisture diminishes the gross calorific value of coal. Why?*

Ans. The oxygen content of coal is present as moisture that has consumed from coal its hydrogen content equal to the 1/8th of the weight of the oxygen content.

Hence the amount of free hydrogen available for combustion will be less than the total hydrogen content of coal. That explains why the gross calorific value or higher calorific value diminishes with the increase of moisture content in coal.

Q. *If it is so then why as much as 15% of water is added to the coal feed in large boiler plants?*

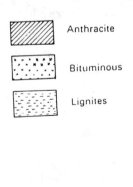

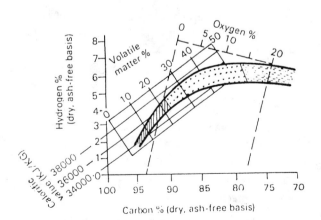

Fig. 8.1 *Coal classification on the basis of Proximate Analysis of coal*

Ans. The purpose is to disperse the fine material in the combustion zone of the furnace.

Q. *What is volatile matter in coal and how is it determined?*

Ans. It is defined as the percentage loss in weight when a finely ground coal specimen is heated in a closed vessel at 1200°K for 40 minutes. After cooling and drying in a desicator the residue (coke) is weighed. From this experiment the coke percentage is determined. Moisture percentage is determined in a separate experiment.

Therefore, % volatile matter

= 100% − (coke + moisture) %

Q. *What are the basic components of volatile matter?*

Ans. There are three basic components of volatile matter:

1. **Gases:** containing CO, H_2, CH_4, C_2H_6 and H_2S
2. **Tar:** a complex mixture of such hydrocarbons as benzene, toluene, napthalene, xylene, phenols, cresols, anthracine and free carbon
3. **Ammoniacal liquor:** aqueous condensate of nitrogen and sulphur compounds plus cyanides.

Q. *Does the ammoniacal liquor result from free or inherent moisture content of coal?*

Ans. No. It results from the decomposition of combined moisture, i.e. decomposition of hydrates.

Q. *What is the effect of volatile matter in the combustion process?*

Ans. To produce smoke.

Q. *What is smoke?*

Ans. It is essentially the aerosols of tar and finely divided carbon particles.

Q. *What is the overall effect of volatile matter in coal?*

Ans. They affect the design of furnace volume and arrangement of heating surfaces.

Q. *How does it affect the combustion volume?*

Ans. Coal feed having high volatile percentage will not be suitable in a furnace with small combustion volume if the burning rate is high.

Burning of low volatile coal requires forced draught and therefore reduced combustion volume will increase the combustion efficiency.

Q. *What is the percentage of volatile matter in coal?*

Ans. It varies with the origin and rank of coal. It may range from as low as 3% to as high as 50%.

Q. *Is coal a homogeneous substance?*

Ans. No; coal is not homogeneous. Basically coal is a heterogeneous hydrocarbon containing mineral matter and water as impurities.

Q. *How is coal classified?*

Ans. (A) Coal can be classified according to the purpose for which the varieties of coal are employed, e.g.

1. **Gas Coal:** coal having high percentage of volatile matter, hydrogen, but a lesser amount of oxygen
2. **Coking Coal:** coal having the property of swelling and yielding high percentage of coke upon carbonization
3. **Smithy coal:** coal having high degree of swelling characteristics
4. **Steam coal:** coal with low swelling, low ash content, producing short flame, with high calorific value.

(B) Coal can be classified according to some visual superficial differences:

1. Dry coal 2. Lean coal 3. Fatty coal 4. Short-flame coal 5. Long-flame coal.

(C) Solid fuels including coal can be classified according to their fixed carbon, volatile matter and oxygen content.

Fuel	Fixed Carbon (%)	Volatile (%)	Oxygen (%)
Wood			43.16
Peat	29.95	70	53.55
Lignite	47	53.12	19.61
Sub-bituminous	59.5	40.52	17.01
Bituminous	69.81	30.21	5.19
Semi-bituminous	83.22	16.81	2.65
Semi-anthracite	91	9.93	2.16
Anthracite	93.84	6.19	2.13

(D) Frazer made an attempt to classify different varieties of coal according to the ratio of fixed carbon to volatile matter. But arbitrary results were obtained.

(E) **Paar's Method of Classification:** It is a simplified method classifying American coals on the basis of calorific value of sulphur-free coal and volatile matter.[Fig. 8.2]

The calorific value of sulphur-free coal was calculated as

$$C.V. = \frac{C.V.\ of\ dry\ coal - (5000)\ S}{1 - [1.08\ A - 0.55\ S]}$$

and the volatile matter as

$$V.M. = 100\ \frac{Fixed\ carbon\ of\ dry\ coal}{1 - [1.08\ A - 0.55\ S]}$$

where A = ash content of the coal
S = sulphur content of the coal

(F) **Ralston's Classification:** is based on carbon-hydrogen and oxygen content of coal and on ash-moisture-sulphur-nitrogen free basis.

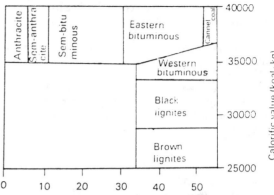

Fig. 8.2 *Parr's method of coal classification*

$$BTU = 16\,780 - \frac{17.23}{x + 0.98}\ per\ pound\ of\ coal$$

where $x = \dfrac{C}{O + A}$

C = carbon %
O = oxygen %

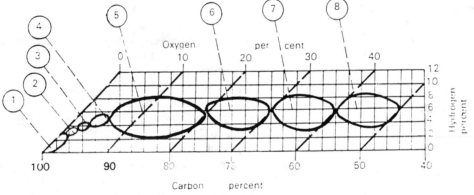

Fig. 8.3 *Ralston's method of coal classification*

Plotting three parameters on a graph (Fig. 8.3) he obtained eight ellipses that belonged to:

Graphite: ellipse No. (1); **Anthracite:** ellipse No. (2); **Semi- anthracite:** ellipse No. (3); **Semi-bituminous:** ellipse No. (4); **Bituminous + Sub-bituminous:** ellipse No. (5); **Lignites:** ellipse No. (6); **Peats:** ellipse No. (7); **Wood + Plants:** ellipse No. (8).

(G) **White's Classification:** coal is classified according to the ratio of carbon to the sum of oxygen and ash content, i.e. according to calorific values.

He drew his curve Fig. 8.4 by plotting calorific value against **carbon/(oxygen + ash)** ratio, on the basis of the following mathematical expression:

A = ash %

(H) **GRÜNER's Classification:** This classification is based on the ultimate analysis of coals on a dry-ash-free basis, according to their carbon, hydrogen and oxygen contents and is correlated against the coke-yielding capacity of coals (See table at the bottom of next page).

However, it must be borne in mind that the nature of coals is so diverse in composition, varying both geologically and geographically, that none of the above mentioned classifications is exact or universally adequate.

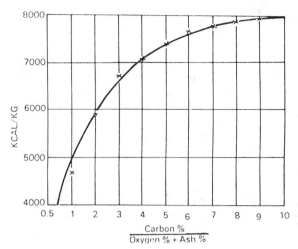

Fig. 8.4 *White's classification of coal*

Coals with low volatile content have slower burning characteristics but generate high fuel bed temperature. Hence forced draught is required. Therefore, grates should be so designed and fuel feed so controlled that the grate is protected by adequate ash.

Coals with high volatile content, on the other hand, have a high rate of burning and hence require a large combustion chamber for combustion of the volatiles. Such coals are very useful for thermal power plants to meet the sudden increase of load. They quickly liberate their combustible volatile gases that rapidly burn off to increase the furnace temperature to boost up steam productivity.

Also, important factors of consideration are sizing, caking characteristics of coal, grindability, resistance to degradation, ash fusion temperature and sulphur content of coal.

Q. *How are Indian coals classified?*

Ans. Coal is graded on the basis of ash and moisture content and is classified into Slack Grade - I and Slack Grade - II

Grade	Moisture + Ash	Calorific Value
Slack Grade-I	19–24%	5940–6340 kcal/kg
Slack Grade-II	24–28%	5340–5940 kcal/kg

Q. *What factors are considered in selecting coals for thermal power plants?*

Ans. Firing qualities of coal are of prime importance and influence the design of the combustion chamber, type of combustion equipment and layout of heat-transfer parts of thermal power plants.

Q. *What is coal preparation?*

Ans. Processing run-of-mine coal to remove the inert material and moisture.

Q. *Why is it necessary?*

Ans. Run-of-mine coal is not suitable for combustion.

Q. *What are the specific steps involved in the preparation of coals?*

Ans.
1. Removal of clay and dirt
2. Drying of coal
3. Sizing of coal
4. Removal of sulphur

GRÜNER's Classification

Type Of Coal	Coking	Coke	Flame	C	H	O+N+S	% VM at 1200°K	% Moisture
Lignite	non	—	long and smoky	60–77	5	20–36	45	20
Bituminous	non		long	75–80	5	15–20	40–45	10–20
	coking			80–85	6	10–15	30–40	5–15
	hard-coking			85–90	5–6	5–11	18–32	5
Semi-bituminous	very weakly	80-85	short smokeless	90–92	4–4.5	4–5	10–20	2–3
Semi-anthracite	non	82–90	-do-	90–93	4–4.5	3–5.5	8–15	2
Anthracite	non	95	very little flame	92–94	3–4	3–4.5	7–8	—

Q. *How is coal freed from clay and dirt?*

Ans. In two ways : dry cleaning and wet cleaning.

Q. *What is dry cleaning?*

Ans. Run-of-mine coal is screened and coal below 80 mm size is subjected to dry cleaning.

The coal is charged to vibrating screens and the mechanical energy imparted to the screens dislodges the fine clay substances adhered to the coal surface. The final condition of the coal is dictated by the degree of vibration imparted to the screens.

Q. *What is the limitation of this process?*

Ans. Its only handicap is the moisture content of the coal. If it is less than 3%, then coal can be cleaned by the dry process. But unfortunately, most of the coals mined and brought to the surface are too wet to fit dry cleaning.

Q. *What is wet cleaning of coal?*

Ans. Coal is washed by sprinkling water over the bed of coal. Shale clay and coal dust are washed away with the result that the final product contains much less incombustibles.

Q. *What is the drawback of this system of coal washing?*

Ans. Moisture content of washed coal increases.

Q. *How can coal be dried?*

Ans. By
 (a) steam heating
 (b) fuel gas drying
 (c) oil dehydration

Q. *How much moisture can be expelled by steam drying?*

Ans. Moisture content of coal can be reduced from 15% to 4–5%

Q. *What is the pressure of the steam generally employed?*

Ans. 30 ata.

Q. *Why is drying of coal necessary?*

Ans.
 1. **For better combustion result:** as a sizeable fraction of heat of combustion is expended in evaporating the moisture content of coal (about 1.5% of the liberated heat of combustion goes to evaporate 10% moisture content)
 2. **To boost up the capacity of pulverizing mill:** the capacity of pulverizer decreases by 2.5% per 1% increase of moisture content of coal
 3. **To reduce the power consumption of auxiliaries:** such as I.D., F.D. fans and pulverizing mills.

Q. *What is coal sizing?*

Ans. Coal lump is broken and screened to maintain the uniformity of coal size.

Q. *Why is it necessary?*

Ans. For higher efficiency utilization.

Q. *How?*

Ans. Though a great deal of work is still required to determine precisely the relationship between the size of coal and its efficiency of utilization, since combustion is a surface reaction and accelerates with the increase of surface area exposed, as the size of the lump is reduced the rate of combustion increases.

Q. *How is sulphur removed from coal?*

Ans. Only the pyritic sulphur can be removed from coal. Crushed coal with high sulphur content is washed in scrubbers to remove the pyrites.

Q. *Why can organic sulphur not be eliminated from coal?*

Ans. Organic sulphur, unlike iron pyrite (which forms a heterogeneous constituent in the coal matrix), is in the compound state with the coal hydrocarbon. Hence it is inseparable from coal and no physical process is known to eliminate it.

Q. *How much of total sulphur content in coal is distributed in pyrite and organic forms?*

Ans. Pyrite accounts for 50–80% of total sulphur load. Organic sulphur accounts for upto 40% of the total sulphur load.

Q. *How can coal be transported from coal mines to thermal power stations?*

Ans. By four ways:
1. Transportation by rail
2. Transportation by road
3. Transportation by sea or river
4. Transportation by pipelines

Q. *Which is the speediest method?*

Ans. Transportation by pipelines.

Q. *How is this carried out?*

Ans. Coal is pulverized and made into a slurry with water. This slurry is pumped to transport coal through pipelines.

Q. *What are the drawbacks of this system?*

Ans. It requires
(a) prohibitively large capital investment
(b) huge quantity of water
(c) drying of coal before subjecting it to combustion chamber

Q. *How much of water per ton of coal is needed for transportation of coal by pipelines?*

Ans. About 1 t/t of coal.

Q. *Can you mention a thermal power plant where such a system is operative?*

Ans. Mohave generating station in UK with an installed capacity of 1580 MW power generation. A 437 km pipeline transports coal to this power station.

Q. *In which cases is the transportation of coal by road economical?*

Ans. For small and medium capacity thermal power stations situated within 30 to 50 km of the coal mine belt.

Q. *Outline in brief the in-plant coal transportation system.*

Ans. Coal dumped in the coal yard by trucks/railway wagons are spread and compacted upon the vibrating screen feeder by bulldozers.

From the vibrating screen feeder, coal is fed to a belt conveyor via a hopper that feeds coal to the crusher. Crushed coal is screened in a vibrating separator oversized lumps are recycled to the crusher while the correct-sized coal is transported by a conveyor belt via a magnetic separator which removes the magnetic materials from coal. The iron particles cling to the belt as it travels around the magnetized pulley and the coal falls off on another conveyor belt that carries them to storage bins.[Fig. 8.5]

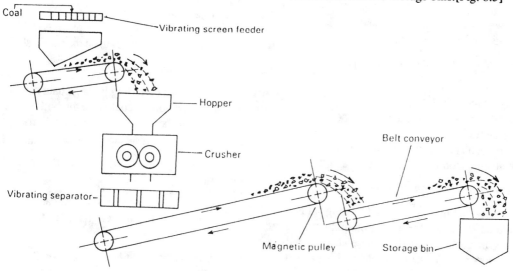

Fig. 8.5 *In-plant coal transportation system (A schematic layout)*

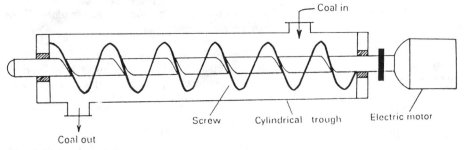

Fig. 8.6 *Screw conveyor*

Q. *Apart from a belt conveyor what are other similar equipment for the transportation of coal?*

Ans.
1. Flight conveyor
2. Screw conveyor
3. Grab bucket conveyor
4. Bucket elevator
5. Skip hoist, etc.

Q. *What are the advantages of a belt conveyor?*

Ans.
1. Most economical system in transportation of large amount of coals. Hence most suitable for large thermal power stations.
2. Power consumption per tonnage of coal transported is less than any other type of conveyor
3. Low operating and maintenance cost
4. Load can be easily varied

Q. *What are its disadvantages?*

Ans.
1. High capital investment
2. Unfit for shorter distance transport and at greater elevation
3. Long belt is required if coal is to be elevated at an elevation higher than 20°.

Q. *How does a screw conveyor works?*

Ans. A screw mounted on two ball bearings at two ends is rotated within a cylindrical trough, by means of a driving mechanism fitted at one end. Coal fed through a hopper into a trough at one end is carried by a screw drive at the other end. (Fig. 8.6)

Q. *What is the diameter of the screw?*

Ans. It usually varies from 15 to 50 cm.

Q. *What is the maximum capacity of a screw conveyor?*

Ans. 125 t/h

Q. *What are its advantages?*

Ans.
1. The operation is dust free
2. Simple and compact equipment
3. Requires minimum floor space
4. Investment cost is low

Q. *What are its disadvantages?*

Ans.
1. High power consumption per unit tonnage of coal transported.
2. High wear and tear
3. Length of the conveyor cannot exceed 30 m due to high torsional strain on the screw-shaft.

Q. *How does a bucket elevator function?*

Ans. Buckets fitted on infinitely long chain passing over two wheels scoop up coal from the bottom and discharge them at the top outlet as the chain moves over the wheels rotated by an electric motor. [Fig. 8.7]

Q. *What is most distinct advantage of a bucket elevator system?*

Ans. It is most advantageous for vertical lifts of solid materials and is extensively used for this purpose.

Q. *What is the capacity of bucket elevators?*

Ans. Their capacity is about 60–80 t/h

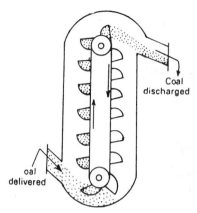

Fig. 8.7 *Bucket elevator*

Q. *What is the speed of the chain?*

Ans. 30–75 m/min

Q. *What are the maximum height and maximum inclination of a bucket elevator?*

Ans. 30.5 m and 60° (to the horizontal)

Q. *Why is storage of coal a necessity?*

Ans. To safeguard the steam generation based plants from total shutdown in the case of failure of normal supplies of coal, i.e. to ensure a steady supply of electricity.

To meet seasonal fluctuations of market prices of coal.

Q. *What is known as the auto-ignition point of coal?*

Ans. It is the temperature at which coal spontaneously ignites in air.

Q. *Therefore it is desirable to store as much coal as the factory site will permit. Is it?*

Ans. Obviously not.

Q. *Why?*

Ans. There should be an optimum level of coal storage.

Q. *What is that?*

Ans. It is roughly 10% of the total amount of the annual coal requirement of the steam generation plant.

Q. *Why should we go for an optimum level?*

Ans. Storage of coal above the optimum level is unnecessary as it would mean:

1. greater risk of oxidation and spontaneous combustion
2. greater loss of volatile matter
3. deterioration of coal quality due to weathering
4. blockage of large amount of capital
5. mounting interest due to large amount of capital locked in heaps of coal

Q. *What is the mechanism of spontaneous combustion of coal?*

Ans. It is thought to proceed via unstable intermediates known as peroxides, the formation of which can be exemplified through a simple example like ethane:

$$C_2H_6 \boxed{O_2} CH_3 \cdot O \longrightarrow O \cdot CH_3 \Longrightarrow H_2O + CH_3CHO \boxed{3/2\ O_2} CO_2 + H_2O + HCHO \boxed{O_2}$$

Ethane   Ethane peroxide   Acetaldehyde   Formaldehyde

$$H_2O + O$$

$$\underset{R_1}{\overset{H}{\diagdown}} C = C \underset{R_2}{\overset{H}{\diagup}} \xrightarrow{O_2} \underset{R_1}{\overset{H\ O-O\ H}{C-C}} \underset{R_2}{} \longrightarrow \underset{R_1}{CHO} + \underset{R_2}{CHO} \xrightarrow{[O]} R_2 CO_2 H \longrightarrow R_2H + CO_2 + H_2O$$

$$\downarrow [O]$$

$$R_1CO_2H \longrightarrow R_1H + CO_2 + H_2O$$

6. greater insurance charge
7. higher handling and reclamation cost

Q. *What factors influence the spontaneous combustion of coal?*

Ans.

1. Large amount of coal fines
2. Higher percentage of combined moisture
3. Rise in ambient temperature

Q. *Why is it advantageous to store screened coal?*

Ans. It will eliminate the fine coal particles that would otherwise lead to spontaneous combustion of coal.

Q. *Does sulphur play any part in spontaneous combustion?*

Ans. No. There is no evidence to support it.

Q. *How can it be determined?*

Ans. By **Wheeler method:**

A sample of coal encased in a sand-bath is steadily heated up while air is blown at a fixed rate through the coal under test.[Fig. 8.8]

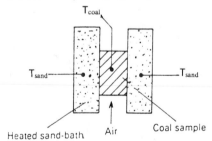

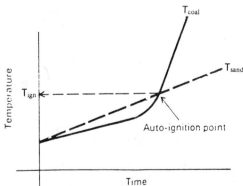

Fig. 8.8 *Temperature-time profile in coal ignition*

The temperature of the coal sample and the sand-bath are noted at periodic intervals and plotted in **Temp. vs. Time** coordinates.

The two curves intersect at a point which corresponds to the autoignition point of the coal specimen.

Q. *What is its value for coal?*

Ans. Low ranking coals $\Rightarrow$ 430–475°K
Higher ranking coals $\Rightarrow$ 500–575°K

Q. *What do you mean by the term 'rational use of coal'?*

Ans. It has a dual meaning.
Firstly, it means minimizing the avoidable losses.
Secondly, it means selection of right grades of coal for right industries.

Q. *What are these avoidable losses and how can they be minimized?*

Ans. Firstly, a huge amount of coal is lost to the manufacture of soft-coke and hard coke in stack burning and beehive ovens respectively. For instance, more than 130 million litres of tar are burnt in the open every year in the Jharia coalfield belt alone. This amounts to a loss of 200,000 tons of coal-equivalent every year. This considerable loss can be avoided by adopting a low-temperature carbonization technique of coal.

Secondly, the mines burn their coking grade coal to run their installations instead of transporting coal from neighbouring mines. This is a misuse and can easily be avoided by the process of mutual exchange of coal.

Thirdly, the steam locomotives of railways use first grade coal, even coking coal, for steam generation. Whereas low grade coal, coal dust converted into briquettes, could serve the purpose well.

About 20% of the coal retrieved from mine goes to dust and probably 10% of it is retrieved for use. And this 10% is equivalent to 2.8 million tons of coal per year. This loss can be minimized by setting up coal-briquetting plants close to the coal-mines.

Finally, an appreciable quantity of coke is lost in dust during the manufacture of soft-coke. For example, if 2 million tons of soft-coke is produced annually and the formation of dust is 20%, then only 10% of the dust can feasibly be utilized and the balance dust accounts for a loss of about 360 000 tons of coke per year. This huge loss can be safely avoided by briquetting the dust.

Q. *What do you mean by scientific utilization of coal?*

Ans. Efficient utilization of coal by adopting suitable methods and apparatuses ensuring complete combustion of coal with minimum excess air, minimizing heat loss and modernizing fuel consuming installations.

Q. *How can heat loss be minimized?*

Ans. By adopting waste heat recovery system and application of better and adequate amount of insulation.

Q. *How can savings of coal be ensured?*

Ans.
1. By complete combustion of coal with minimal excess air
2. Allowing minimum heat loss, i.e. utilizing total heat liberated for useful purposes alone
3. Complete recovery of byproducts of coal carbonization processes
4. Allocation of right grade of coal for specific consumers
5. Upgradation of low ranking coal with the help of such matrix as waste molasses.

Q. *What is the primary liquid fuel occurring naturally?*

Ans. Crude petroleum.

Q. *Is it directly used as fuel?*

Ans. No.

Q. *Why is it called fuel then?*

Ans. Though itself a combustible material, crude petroleum is not used, for the sake of efficient utilization, as fuel straightaway. Nor is it wise to burn it that way as crude petroleum provides the source of valuable materials for petrochemical industry and manufactured products, such as synthetic fibres, plastics, drugs, paints and detergents.

It is the source of a range of gaseous and liquid fuels that distill off the fractionating column at different ranges of temperature when crude petroleum is fractionated.

Q. *How are liquid fuels classified?*

Ans. They are classified according to the mode of procurement, viz.—natural or crude oils and artificial or manufactured oils.

Q. *What are natural oils?*

Ans. Distilled natural oils are:
petrol, benzene, petroleum spirit, kerosene, benzol, fuel oils, diesel fuels, gas oils.

Q. *What are artificial oils?*

Ans. Distilled artificial oils are:
natural-gas oil, shale-oil, tar-oil, coal-tar etc.

Q. *What is the composition of crude petroleum?*

Ans. It contains such major components as
1. **Hydrocarbons**—paraffins, olefins, aromatics and cyclic compounds.
2. **Oxygen compounds**—carboxylic acids and phenols
3. **Sulphur compounds**—mercaptans, thioethers and thiophenes
4. **Inorganic compounds**—mineral matters and organometallic salts.

Q. *Give specific examples of each of these groups of compounds.*

Ans.
Paraffins: n-octane; iso-octane; n-butane
Olefines: propylene; isobutene
Aromatics: napthalene; pyridine
Cylic compounds: cyclopentane
Carboxylic acids: napthenic acids
Phenols: phenols
Mercaptan: ethyl mercaptan
Thiophenes: thiophene

Q. *Does crude petroleum bear any important metal?*

Ans. Yes. It is vanadium in compound state (available in Mexican and Venezuelan crude).

Q. *What factors are considered for the gradation of*

petroleum?

Ans. The following physico-chemical properties are considered:
1. Specific gravity
2. Viscosity
3. Congealing point
4. Flash point
5. Calorific value
6. Specific heat
7. Sulphur content
8. Moisture and sediment contents

Q. *What is congealing point?*

Ans. It is the temperature at which the crude becomes so pasty that it remains in place and does not flow out for one minute from a test glass inclined at 45°.

Q. *What is the specific gravity of petroleum?*

Ans. It is the ratio of the weight of a given volume of petroleum to the weight of the same volume of water at a fixed temperature.

Q. *What is this fixed temperature?*

Ans. Normally it is 288°K.

Q. *What is the most convenient method of measuring the specific gravity of liquid fuels?*

Ans. Use of hydrometer.

Q. *What is °API?*

Ans. It is the scale of measuring specific gravity introduced by the American Petroleum Industry (API).

$$°API = \frac{141.5}{Sp.Gr. \ at \ 288°K} - 131.5$$

Q. *Why is the knowledge of specific gravity of crude important?*

Ans.
1. To assess the volume of a given weight of material (or vice versa) for transportation and storage
2. To calculate the volume associated with the corresponding weight of oil from the calorific value which is normally mentioned on a weight basis
3. To enable a rough check on the consistency of quality in a series of batches.

Q. *Is there any relation between the calorific value of petroleum and specific gravity?*

Ans. United States Bureau of Mines gives an empirical relation between the gross calorific value of petroleum and its specific gravity as

$$HCV = 51 \ 916 - 8792 \ \rho^2 \ kJ/kg$$

where ρ = sp. gr. of oil at 288°K

Q. *Why is viscosity factor an important consideration?*

Ans. Viscosity of a liquid fuel is a measure of its resistance to flow. Its importance lies in the fact that it affects such things as the rate of flow of liquid fuels through pipelines, atomization of fuel oils at the burners and the performance and wear of diesel pumps.

Q. *What is viscosity?*

Ans. It is a specific property of a fluid and is the force required to displace one square metre of imaginary plane surface of the fluid at a rate of 1 m/s with respect to the second plane separated by a 1 m distance from the first plane and parallel to it.

Unit: $N \ s/m^2$

Q. *With which equipment, can the viscosity of liquid fuels be conveniently measured?*

Ans. By
1. Ostwald viscometer
2. Ubbelohde suspended level viscometer
3. Redwood viscometer (U.K.)
4. Saybolt-Furol viscometer (USA)
5. Engler viscometer (Continental)

Q. *What factors influence the viscosity of liquid fuels?*

Ans. Viscosity decreases with the increase of temperature and increases with the fall of temperature.

Q. *What is viscosity index scale?*

Ans. It is an arbitrary scale based on the viscosity-temperature relationship of liquid fuels.

Q. *Why is a high viscosity index desirable for lube*

oils?

Ans. Their service condition demands that they to remain viscous at the operating temperature to prevent metal to metal contact between the rotating shaft and the bearings.

Q. *Why is viscosity index important for fuel oils?*

Ans. To enable us to know the temperature to which the fuel oil must be heated before it can be conveniently pumped.

Q. *What is the effect of pressure on the viscosity of oils?*

Ans. It is felt only at pressures beyond $96500 \, kN/m^2$ (983 kgf/cm^2), such as may be encountered in a bearing.

Viscosity increases very rapidly with the increase of pressure in excess of 98 000 kN/m^2, according to the relationship:

$$\log (\eta_1/\eta_2) = C (P_1 - P_2)$$

where η_1 and η_2 are dynamic viscosities of oil at pressures P_1 and P_2 respectively.

C is the constant depending on the chemical structure of the oil.

Q. *What is the flash point of a liquid fuel?*

Ans. It is the temperature at which a liquid fuel, when tested in a standard apparatus, gives off just sufficient vapour to create, in the air space above it, an explosive mixture that will flash if brought into contact with a flame.

Q. *Why is the knowledge of flash point important?*

Ans. It gives an indirect measure of volatility of the fuel and serves as an indication of the fire hazards associated with the fuel in bulk, i.e. during storage and application.

Q. *Give two examples of equipment which are used to determine the flash point of liquid fuel?*

Ans.
1. Pensky–Martin's apparatus used for liquid fuels having flash point higher than 322°K
2. Abel apparatus used for those liquid fuels

whose flash point is below 322°K.

Q. *How is it determined?*

Ans. The liquid fuel under test is heated in a metal cup closed by a lid. At certain intervals a small test flame is introduced through an opening in the lid. The temperature of the liquid fuel at which the insertion of flame causes the vapor-air mixture above the oil in the cup to ignite is the flash-point.

Q. *What is the fire point?*

Ans. If the test flame in the above case is introduced when the liquid fuel is heated above its flash point, a temperature is reached at which the bulk of the liquid fuel burns continuously. This temperature is the fire point of the liquid fuel under test.

Q. *What is the significance of the fire point?*

Ans. Uncertain.

Q. *How many types of flash points are there?*

Ans. Two. Open flash point and closed flash point.

Q. *What is open flash point?*

Ans. It is the flash point of a liquid fuel determined in an open crucible when a test flame is periodically applied above the surface.

Q. *What is closed flash point?*

Ans. It is the flash point of a liquid fuel determined in a closed crucible as mentioned earlier.

Q. *What is the difference between these two?*

Ans. The open flash point of a liquid fuel is a few degrees higher than its closed flash point.

Q. *Why?*

Ans. It is because when open flash point is determined ignition takes place provided there is sufficient vapour in excess air rather than in a restricted space as with the closed flash point.

Q. *What is the octane number of a liquid fuel?*

Ans. It is an empirical rating of the antiknock quality of a liquid fuel and is a measure of its suitability for spark ignition engines.

It is the percentage of iso-octane in a mixture of

iso-octane (octane value 100) and n-heptane (octane value 0)-that will give the same knocking characteristics as the fuel under test in a standard spark-ignition engine.

Q. *What is knock and why does it occur in spark-ignition engines?*

Ans. Knock is the explosion occurring in the cylinder of spark-ignition engines.

It is caused by the secondary ignition of unburnt fuel after normal spark ignition, with the effect that a fast moving flame propagates along the cylinder. It gives rise to pressure waves that vibrate against the cylinder walls and we hear knocking.

Q. *How is the octane number of a liquid fuel determined if its octane number is greater than 100?*

Ans. In this case the performance of that fuel is matched to the performance of a mixture of iso-octane and tetraethyl lead.

The octane number of the fuel is then

= 100 + quantity of tetraethyl lead in the mixture

Q. *What is the cetane number of a liquid fuel?*

Ans. It is the proportion of cetane (n-hexadecane with cetane value 100) in a mixture of n-hexadecane and alfa-methyl napthalene (cetane value 0), that will give the same ignition delay after the injection of the fuel as the liquid fuel specimen in the same standard compression ignition engine.

Q. *For which fuels is cetane number used?*

Ans. Diesel fuels.

Q. *What is the cetane number of high-speed-diesel fuel?*

Ans. It ranges between 52 and 55.

Q. *What is diesel index?*

Ans. It is the product of the aniline point and specific gravity of the liquid fuel concerned:
Diesel Index = Aniline Point (°K) × Sp. Gr. (°API)

Q. *What is the aniline point?*

Ans. It is the temperature at which equal volumes of liquid fuel and aniline are just miscible.

Q. *What is its significance?*

Ans. It indicates the paraffinicity of the fuel, i.e. ignition characteristics of the fuel.

Q. *What are the advantages of liquid fuels over solid fuels?*

Ans. These are:
1. Require less excess air for complete combustion
2. Higher heat generation per unit weight of fuel consumed because of higher calorific value
3. Better control of fuel consumption
4. Require no ash disposal system
5. More cleanliness of operation
6. Require less floor space for storage and handling.

Q. *Why is low sulphur content of fuel oil desirable?*

Ans. Higher sulphur content of fuel oil leads to corrosion problems. For example, fuel oil gun tip made of tungsten carbide gets quickly damaged during partial oxidation reaction in the gasifier

$$S + O_2 \rightarrow SO_2$$

(from F.O.)

$$WC + SO_2 \rightarrow WS + CO_2$$

The tungsten sulphide so produced is a brittle material and breaks off the gun-tip easily causing serious gun damage.

Q. *How are gaseous fuels classified?*

Ans. They are classified into two classes:
1. Natural gaseous fuel
2. Manufactured gaseous fuel.

Q. *What are natural gaseous fuels?*

Ans. The fuels belonging to this category are:
1. **Natural gas:** a colourless, odourless gas, composed of mostly methane and accumulating in the upper parts of oil of gas wells. Suitable for domestic and industrial consumption.
2. **Liquified petroleum gas:** is the primary flash distillation product of crude petroleum. Also it is produced by fractionating natural

gas. It is mainly propane and butane, which are easily liquified. Used as both a domestic and industrial fuel.

3. **Refinery oil gas:** composed of some light gases obtained during processing of crude petroleum. These are light hydrocarbons with a small percentage of hydrogen and carbon monoxide. They cannot be economically converted into liquid product. Used as inplant source of fuel.

Q. *What are manufactured gaseous fuels?*

Ans. These are:
1. Coal gas
2. Producer gas
3. Water gas
4. Carburetted water gas
5. Oil gas
6. Town gas
7. Blast furnace gas

Q. *What is coal gas?*

Ans. It is the gaseous product of carbonization of coal.

Q. *What are its chief constituents?*

Ans. Methane, hydrogen and carbon monoxide.

Carbonization	Gas Composition
Low Temperature	CH_4–46%; H_2–28%; CO–12%; N_2–7%; CO_2–4%; H_2O–2%; O_2–1%
High Temperature	H_2–52%; CH_4–25%; CO–5.6%; CO_2–2.5%; C_xH_y–2.1%; N_2–12.5%; O_2–0.2%

Q. *Upon what variables does the quality of coal gas depends?*

Ans.
1. Temperature of carbonization
2. The type of plant, i.e., whether vertical retort or horizontal retort

Q. *What are the chief constituents of producer gas?*

Ans. Nitrogen, carbon monoxide and hydrogen

Q. *At what temperature is the carbonization of coal carried out?*

Ans. Low Temperature Carbonization: 775–900°K
High Temperature Carbonization: 1250–1600°K

Q. *What are the applications of coal gas?*

Ans. Used as fuel in gas fired boilers and commercial purposes.

Q. *What is producer gas?*

Ans. It is the gaseous fuel resulting from the complete gasification of the combustible material in solid fuels by air-steam mixture.

$$C + O_2 \rightarrow CO_2 \, ; \, CO_2 + C \rightleftharpoons 2CO$$

A mixture of air and steam is passed continuously through a heated bed of coke/coal for continuous generation of producer gas of uniform composition. (See Table below)

Q. *Where is producer gas extensively used?*

Ans.
1. Firing open hearth furnaces in steel industry
2. Heating of retorts in coal carbonization.

Q. *What is water gas?*

Ans. It is the gaseous fuel resulting from total gasification of the combustible material in solid fuels by steam (superheated)

$$C + H_2O \longrightarrow CO + H_2$$

Steam is blown through a bed of incandescent coke/coal. As the above endothermic reaction proceeds, the coal bed temperature drops. When it falls to 900°C, air is blown through the bed for some time to bring the bed temperature back to the desired level (1000–1100°C)

$$C + O_2 \, (Air) \rightarrow CO_2 + Q \text{ cals of heat}$$

Solid Fuel	Gas Composition (%)						
	N_2	CO	H_2	CO_2	CH_4	O_2	Calorific Value
Coke	56–57	26–27	10–12	5–6	0.3–0.5	—	4650–4700 kJ/m³
Coal	46–47	28–30	15–16	3–5	2–3	0.2	6450–6500 kJ/m³

Q. *What are its chief constituents?*

Ans. Carbon monoxide and hydrogen.

Solid Fuel	Gas Composittion							
	CO	H_2	CO_2	N_2	CH_4	C_xH_y	O_2	Calorific Value (kJ/Nm^3)
Coke	40–45	50–52	4–5	3–4	0.5–0.6	—	—	11200–12000
Coal	28–30	52–54	6–8	4–5	6–7	0.6–0.7	0.2	12450–13250

Q. *What are the chief usages of water gas?*

Ans. Used as
1. Industrial fuel
2. Synthesis gas in chemical processing

Q. *What are the advantages of gaseous fuels over solid fuels?*

Ans.
1. Require less excess air for complete combustion
2. Smokeless combustion is possible
3. No ash formation
4. Higher calorific value
5. More cleanliness of operation
6. Better control of flame length and nature of flame (oxidizing or reducing)
7. Affords greater economy than coal or fuel oil at higher operating temperatures
8. Provides quicker furnace light-up
9. Better response to boiler load variations.

9

Principles of Combustion

Q. *Does the combustion of a fuel in a furnace involve purely chemical processes?*

Ans. No. It involves a number of complex physical and chemical processes.

Q. *What physical process is most important in the study of kinetics of fuel combustion?*

Ans. It is the aerodynamic factor, i.e. the process of mixing the fuel and oxidant.

Q. *Which chemical factors are essentially required in the study of the kinetics of fuel combustion?*

Ans. Temperature and concentration of the reacting substances.

Q. *What kind of reactions are involved in the fuel combustion?*

Ans. Mainly the exothermic reactions: burning of carbon, hydrogen and sulphur in the fuel

$$C + O_2 \longrightarrow CO_2 + 408.86 \text{ kJ/mol}$$

$$2H_2 + O_2 \longrightarrow 2H_2O + 286.22 \text{ kJ/mol}$$

$$S + O_2 \longrightarrow SO_2 + 292.25 \text{ kJ/mol}$$

At higher temperatures, i.e. in the flame core some endothermic reactions may occur

$$N_2 + O_2 \longrightarrow 2NO - 180 \text{ kJ/mol}$$

$$C + CO_2 \longrightarrow 2CO - 7.25 \text{ MJ/kg}$$

The last reaction takes place on the incandescent surface of the carbon particles under conditions of oxygen deficiency.

Q. *How can the rate of homogeneous and heterogeneous combustion be expressed?*

Ans. For homogeneous combustion (i.e. combustion of gaseous fuel and atomized liquid fuel), the rate of combustion at a constant temperature at any particular moment is the product of the concentrations of the reacting species

$$r = k\, C_{fuel}^{x}\, C_{o_2}^{y}$$

where k is the rate constant which depends on the temperature and chemical nature of the reagents.

C_{fuel} = concentration of fuel

C_{o_2} = concentration of oxygen

x, y are the moles of fuel and oxygen involved in the stoichimetric chemical equation.

For heterogeneous combustion, i.e. the combustion of solid fuel, the concentration of combustible substance is constant and hence the rate of this reaction is dependent only on the concentration of oxygen on the surface of the solid fuel

$$r = k\, C_{o_2}^{s}$$

where $C_{o_2}^{s}$ = concentration of oxygen on the fuel surface.

Q. *Why is the rate of formation of carbon monoxide higher than the rate of formation of carbon dioxide on the surface of burning coal?*

Ans. It is because the activation energy for the formation of CO

$$C + \frac{1}{2} O_2 \longrightarrow CO \quad ; \quad E_{co} = 60 \text{ kJ/mol}$$

is much less than that for the formation of CO_2

$$C + O_2 \longrightarrow CO_2 \quad ; \quad E_{co_2} = 140 \text{ kJ/mol}$$

With higher activation energy the molecular bonds of the reactants break off less easily and consequently the reaction rate is lower.

Q. *What is activation energy?*

Ans. It is the minimum energy required to excite the reactant molecules to such an activated state that these molecules undergoing collisions break their old molecular bonds and rearrange themselves into molecules of new substances.

All collisions do not result in chemical reaction. Only collisions between reactant molecules excited by activation energy E result in chemical reaction. This activation energy is given by

$$k = k_o \exp\left(-\frac{E}{RT}\right)$$

Q. *Why does an excess of fuel (rich mixture) or a fuel deficiency (lean mixture) cause the rate of reaction to decrease?*

Ans. It is because of lower heat evolution per unit volume.

Since during the process of combustion, there is a steady supply of fuel and oxidant (air) in the combustion zone, the concentrations of the reagents are practically invariable in time. When these conditions prevail, the highest reaction rate is attained at a nearly stoichiometric ratio of the concentrations of the reagents. Therefore, any deviation from the stoichiometric ratio will bring about a decrease in heat evolution per unit volume.

Q. *Can combustion take place at any arbitrary concentration of fuel in the air-fuel mixture?*

Ans. No.

Q. *Why?*

Ans. There exists a lower concentration limit below which combustion is impossible and at the same time there exists an upper concentration limit when any further increase of the concentration of the fuel prevents combustion. That is, there exists a range of concentrations between these two limits at which combustion is possible.[Fig. 9.1]

Q. *What is explosive combustion?*

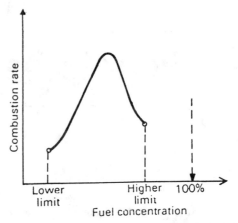

Fig. 9.1 *Combustion is possible within two defined limits of fuel concentration*

Ans. If a gaseous fuel-air mixture prepared for combustion completely fills in a particular volume, an ignition source will trigger the oxidation reaction propagating at a high rate all over the volume. This will result in a sharp increase of temperature and pressure. This type of combustion is called explosive combustion.

Q. *What do you mean by upper and lower explosive limits of a gaseous fuel-air mixture?*

Ans. The fuel gas-air mixture is capable of exploding in the whole range of concentrations between these two limits.

Q. *What is ignition temperature?*

Ans. The temperature above which a self-sustained oxidation reaction between fuel and oxidant is possible.

Q. *How will you determine graphically the ignition temperature of a gaseous fuel-oxidant mixture?*

Ans. Heat generated in the initial stage of combustion can be determined with the help of the equation

$$Q_g = k_o \exp(-E/RT)\, C_f^x\, V\, q \tag{I}$$

where C_f = concentration of gaseous fuel

V = volume of prepared gaseous fuel-air mixture

q = thermal effect of the reaction per unit mass

And the quantity of heat removed from the reaction zone is determined by the equation

$$Q_r = \alpha A (T - T_s) \qquad \text{(II)}$$

where α = heat-transfer coefficient
A = surface area of surrounding walls
T = temperature of gaseous fuel-air mixture
T_s = temperature of surrounding walls

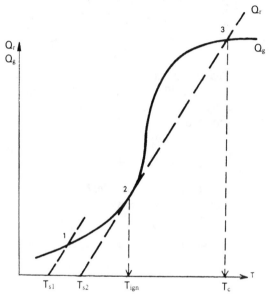

Fig. 9.2

As evident from equation (I), the heat generated at the initial stage of oxidation of fuel is described by an exponential curve while the heat removal at any stage can be represented by a straight line inclined at an angle to the X-axis.[Fig. 9.2]

In the beginning let the fuel-air mixture and surrounding wall temperature be T_{s_1}. Due to heat evolution in the fuel combustion, the mixture will be heated to a temperature, say, T_1. Now $T_1 > T_{s_1}$.

Initially, $Q_g > Q_r$.

But at the point (1), $Q_g = Q_r$ and therefore further preheating of the fuel gas-air mixture becomes impossible. This is the zone of slow oxidation.

Now, let the temperature of the surrounding walls be raised to T_{s_2}. At this point, again $Q_g > Q_r$ initially, and temperature of the mixture rises. Finally, the point (2) is reached where $Q_g = Q_r$, but in contrast to point (1) this point is unstable. Only a slight increase of

temperature shoots Q_g up to greater than Q_r, with the effect that heat evolution will increase more rapidly than heat removal. The temperature corresponding to this point (2) is called ignition temperature.

Q. *How can the extinction temperature be determined from such a curve?*

Ans. The point A corresponds to a state of stable combustion. If heat is extracted more forcibly, i.e. along the line LM, the combustion temperature will drop to a point B. At this point, the high temperature oxidation process interrupts for as long as Q_r becomes greater than Q_g.[Fig. 9.3]

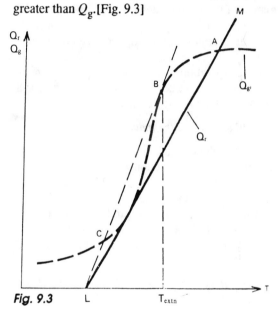

Fig. 9.3

If the temperature of the gas mixture at B is slightly decreased it will bring about a sharp drop in Q_g and since $Q_r > Q_g$ all along from B to C, combustion will cease to exist.

The temperature of the fuel gas-air mixture corresponding to the point B is called extinction temperature.

Q. *Why is extinction temperature always higher than ignition temperature?*

Ans. It is because of the lower concentrations of gaseous fuel and air in the zone of active combustion than their initial concentration at ignition.

Q. *It has been experimentally established that the rates of combustion reaction are much higher than the rates calculated on the basis of law of mass action and Arrhenius' law. Why?*

Ans. The rates of combustion reactions, using the law of mass action and Arrhenius theory, are derived by considering the number of active molecules of the initial substances entering into oxidation reactions.

However, in fact, reactions do not occur immediately between the original molecules, i.e. combustion reaction is not a one step process but proceeds through a number of intermediate stages that involve the formation of such active molecular fragments as atoms and radicals like H, O, OH, etc. Since radicals and individual atoms can form free-valency particles, these intermediate reactions, as a result, have low levels of activation energy. Such reactions, therefore, proceed at a high rate and therefore account for the substantial high rates of combustion reactions.

Q. *What is called the induction period?*

Ans. It is the time taken by the reacting molecules to generate active reaction centres in the form of charged particles and the time taken by these particles to accumulate in the medium.

These active centres are produced due to partial destruction of the original molecules due to collision with other molecules having higher energy than the atomic bond energy of the original molecules.

Q. *What is the mechanism of combustion of hydrogen?*

Ans. The combustion of hydrogen at temperatures above 500°C is an explosive chain reaction that involves the following stages:

1. **Nucleation:** formation of active centres

$$H_2 + M^* \longrightarrow H + M$$

$$H_2 + O_2^* \longrightarrow 2OH$$

where M^* and O_2^* are high-energy molecules

2. **Propagation:** entering of the active centres—radicals and atoms—into reactions with the surrounding molecules producing final reaction products and an ever greater number of active centres

$$OH \xrightarrow{\ H_2\ } H_2O + H$$

$$H + O_2 \longrightarrow \begin{array}{c} + \\ O \end{array}$$

$$O \xrightarrow{\ H_2\ } OH \xrightarrow{\ H_2\ } H_2O + H \\ + \\ H$$

3. **Ceasation:** quenching of the active centres as the reaction products accumulate and the concentration of the starting material becomes lower.

$$H + H \longrightarrow H_2$$

$$OH + H \longrightarrow H_2O$$

Q. *What is the rate expression of hydrogen burning by considering the law of mass action and Arrhenius law?*

Ans. The stoichiometric equation for the combustion of hydrogen is

$$2H_2 + O_2 \longrightarrow 2H_2O$$

And accordingly, the theoretical rate expression for the combustion of hydrogen is

$$r_{H_2O} = k_o \exp(-E/RT) \, C_{H_2}^2 \, C_{O_2}$$

Q. *What is the actual reaction rate of hydrogen combustion?*

Ans. It is described by the equation

$$r_H = 10^{11} \exp(-E'/RT) \, C_H \, C_{O_2} \, T^{\frac{1}{2}}$$

where E' = activating energy of H-atoms (reaction centres) and O_2 molecules reacting

[*Note: E' < E*]

Q. *What are the decisive factors affecting the reaction rate of hydrogen combustion?*

Ans. Concentrations of hydrogen atoms and oxygen molecules excited by an activation energy E'.

Q. *What is the mechanism of hydrocarbon combustion?*

Ans. W. Bone suggested the following mechanism of hydroxylation:

1. NUCLEATION: formation of active centres, H, OH and O

$$CH_4 \xrightarrow{heat} CH_3 + H$$

$$H + O_2 \longrightarrow OH + O$$

2. PROPAGATION: formation of intermediate hydroxylated compounds that in turn burn or disintegrate thermally

$$CH_4 + OH \longrightarrow \underset{+H}{CH_3OH} \xrightarrow{OH} \underset{+H}{CH_2(OH)_2} \longrightarrow$$

$$\underset{\underset{H}{+}}{\overset{H}{H-C}} = O \xrightarrow{OH} \begin{array}{l} \overset{H+H}{\underset{|}{OH-C}} = O \longrightarrow H_2O + CO \\ \quad\quad \downarrow OH \\ HO-C = O \longrightarrow H_2O + CO_2 \\ \quad | \\ \quad OH \end{array}$$

According to Bakh, combustion of hydrocarbons proceeds via unstable peroxide intermediates

1. NUCLEATION:

$$R \cdot CH = CH \cdot R' \xrightarrow{heat} H$$

$$H + O_2 \longrightarrow OH + O$$

2. PROPAGATION:

$$R \cdot CH = CH \cdot R' \xrightarrow{O_2} \underset{\underset{O\text{---}O'}{|\quad\quad|}}{R \cdot CH\text{---}CH \cdot R'} \xrightarrow{OH} \underset{\underset{O\text{------}O}{|\quad\quad\quad|}}{\overset{H}{\underset{|}{R\text{---}C\text{---}OH}} \quad \overset{H}{\underset{|}{\cdot C\text{---}R'}}}$$

$$\underset{+\ H}{\underset{\underset{H}{|}}{\overset{OH}{\underset{|}{RH + CO_2 \longleftarrow R\text{---}C}} = O}} \xleftarrow{OH} \overset{H}{\underset{|}{R\text{---}C}} = O \longleftarrow \overset{H}{\underset{|}{\underset{OH}{R\text{---}C\text{---}OH}}} \xleftarrow{H} \overset{H}{\underset{|}{\underset{O\cdot}{R\text{---}C\text{---}OH}}} + \overset{H}{\underset{\|}{\underset{O}{C\text{---}R'}}}$$

Q. *How does the reaction rate of combustion of gaseous fuel change with time?*

Ans. During the induction period—a short time during which a sufficiently large number of active centres (atoms and radicals) accumulate—the reaction rate is almost unnoticeable.

Following this period, the reaction rate increases rapidly due to the propagation of a large number of parallel chain reactions over the entire volume.

Finally an equilibrium is reached between the appearance and disappearance of the active centres whereupon no more rate increase is observed. This is the maximum rate and the combustion will proceed at this rate provided there is a steady supply of combustion material and oxidant to the combustion zone.

Q. *Why is liquid fuel evaporated prior to combustion?*

Ans. So far as liquid fuel is concerned, both ignition and combustion temperatures are higher than the boiling point of the individual fuel fractions.

Hence heat is supplied to evaporate the liquid fuel, then its vapours are mixed with air, preheated to ignition temperature and ignited.

Q. *What is the mechanism and combustion characteristics of a liquid fuel droplet?*

Ans. The combustion characteristics of atomized liquid fuel (i.e. fuel droplets) in stagnant air involve a diffusion mechanism.

Step (I) Formation of a Vapour Cloud—As soon as the liquid fuel droplet comes in contact with the air, it evaporates till the partial pressure of fuel vapour in the air is equal to the vapour pressure of the liquid at that temperature. As a result, a part of liquid from the surface of the droplet evaporates to form a vapour cloud surrounding the remaining liquid droplet.[Fig. 9.4]

Step (II) Diffusion of Vapour Cloud—The vapour cloud that forms around the droplet diffuses into the environment, spreading radially outward while the oxygen of the air is diffusing into the body of the vapour cloud. These two diffusions take place in the opposite direction—the fuel-vapour diffusing away from the fuel droplet and the oxygen towards the droplet.

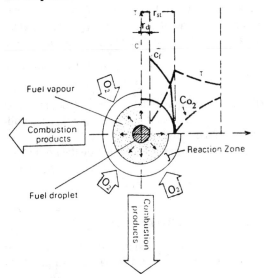

Fig. 9.4 *Mechanism of liquid fuel droplet combustion*

Step (III) Combustion—At a certain radial distance (r_{st}) from the centre, the stoichiometric relationship between fuel vapours and oxygen is established and combustion takes place producing a spherical combustion front around the liquid droplet.

The zone with $r > r_{st}$, contains primarily combustion products mixed with oxygen while the zone with $r < r_{st}$, fuel vapours prevail, their concentration progressively increasing towards the liquid droplet.

Q. *What is the magnitude of r_{st}?*

Ans. It is 4 to 10 times of r_d—the radius of the liquid droplet.

Q. *On which factors does r_{st} depends?*

Ans. It depends primarily on the droplet size and temperature of the combustion zone.

Q. *On which factors does the rate of combustion of a liquid fuel droplet depend?*

Ans. It depends on three factors:
1. rate of evaporation of droplet from its surface
2. rate of chemical reaction (oxidation) in the combustion zone
3. rate of oxygen diffusion to the combustion zone.

Q. *Why is the rate of combustion of a liquid droplet predominantly determined by evaporation from its surface?*

Ans. The rate of combustion of a liquid droplet basically depends on the rate of oxygen diffusion to the combustion zone. Now, the quantity of oxygen diffusing through the spherical cloud of fuel vapour surrounding the liquid fuel droplet is directly proportional to the square of the sphere-diameter. Hence a slight shift in the combustion zone from the surface of the droplet, that means, the increase of the rate of evaporation from the surface of the liquid droplet will noticeably increase the mass flow rate of oxygen to the combustion zone.[Fig. 9.5]

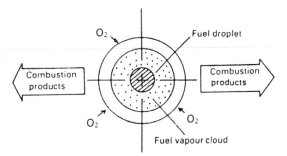

Fig. 9.5

Q. *Why is the combustion rate of liquid fuels increased by atomizing the fuel just before burning?*

Ans. Atomization of liquid fuels brings about substantial increase of the total surface available for evaporation, increasing the intensity of evaporation per unit area of their surface.

With the increase of the rate of evaporation the diameter of the vapour cloud surrounding the liquid droplets increases. Since the quantity of oxygen diffused through the vapour cloud is proportional to the square of the diameter of the cloud sphere, atomization increases the rate of oxygen diffusion through the sphere and thereby increases the combustion rate of liquid fuels.

Q. *For streamline airflow (Re << 1) how can it be shown that the rate of heat exchange between a fuel droplet and the surrounding air increases as the size of the droplet decreases?*

Ans. For air flow at low Reynolds number Re << 1, the heat flow through a spherical surface suspended in air is determined by λ, the conductivity of vapour film in the boundary layer.

Under these conditions, the heat transfer coefficient α is given by Sokolsh's formula

$$Nu = \frac{\alpha d}{\lambda} = 2$$

$$\therefore \quad \alpha = \frac{2\lambda}{d} = \frac{\lambda}{r}$$

where,

d = droplet diameter

From this equation it follows that with the reduction of droplet size, the rate of heat exchange between a droplet and the surrounding air increases.

Q. *What stages are involved in the combustion of pulverized solid fuel (coal)?*

Ans. **Stage (I) THERMAL PREPARATION:** Evaporation of residual moisture and separation of volatiles, as the fuel particles are heated up (400–600°C) in a few tenths of a second.

Stage (II) BURNING OF VOLATILES AND HEATING OF COKE PARTICLES: The volatiles are then ignited and they burn up within a very short period (0.2–0.5 s) liberating enough quantity of heat to bring coke particles to their ignition temperature.

Stage (III) COMBUSTION OF COKE PARTICLES: Coke particles, intensely heated by the combustion of volatiles, start to burn as soon as the ignition temperature is reached. This reaction takes place at elevated temperature (800–1000°C) and takes up 1/2 to 2/3 of the total time of combustion (1–2.5 s).

Q. *What is the temperature profile of pulverized coal combustion?*

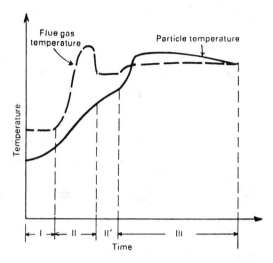

Fig. 9.6 *Temperature–time profile in pulverized coal combustion*

Ans. The temperature conditions of combustion of finely divided coal particles can be best represented graphically.[Fig. 9.6]

Stage I representing the thermal preparation zone is accomplished at 400–600°C within a fraction of a second. All the residual moisture and volatiles are intensively evolved at this stage.

Stage II is the zone of burning volatiles as well as heating of coke particles by the heat liberated due to combustion of volatiles. If the volatile yield of coal (brown coal, oil shales, pit) is high, the liberated heat of volatile combustion will ignite the coke particles. If the yield of volatiles is low, the coke particles must be heated up externally (stage II') to ignition temperature.

The final stage is the combustion of coke particles that takes place in the temperature range 800–1000°C.

Q. *What kind of process is the final stage of pulverized coal combustion, i.e. combustion of coke?*

Ans. It is a heterogeneous process.

Q. *On which factor is its rate principally dependent?*

Ans. Supply of oxygen to the heated surface of the coke particles.

Q. *What is the mechanism of burning of coke particles?*

Ans. The probable mechanism of reaction between carbon and oxygen is as follows:

1. ADSORPTION OF OXYGEN MOLECULES from the gas volume on the surface of the carbon particles to form a complex carbon-oxygen intermediate C_xO_y.

2. THERMAL DISSOCIATION OF C_xO_y into CO and CO_2 whose proportions vary with reaction temperature.

At about 1200°C, the overall reaction can be represented as

$$4C + 3O_2 \longrightarrow 2CO + 2CO_2$$

where the CO/CO_2 ratio is 1:1.

But at 1700°C, the resulting reaction can be written in the form:

$$3C + 2O_2 \longrightarrow 2CO + CO_2$$

whence CO/CO_2 is equal to 2:1.

The carbon particle is immediately surrounded by a sheath of hot gas film consisting of primary reaction products which are being continuously removed from the particle surface to the environment.[Fig. 9.7]

In this process, carbon monoxide diffusing away from the carbon particle encounters oxygen molecules diffusing in the opposite direction and reacts with it within the boundary gas-film to produce CO_2. As a result, the concentration of oxygen across the boundary film falls off on approaching the surface of the particle.

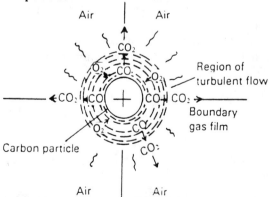

Fig. 9.7 *Mechanism of coke particle combustion*

Q. *How does the concentration of gaseous substance at the surface of burning carbon vary?*

Ans. This can be best interpreted through graphical representation.[Fig. 9.8 and 9.9]

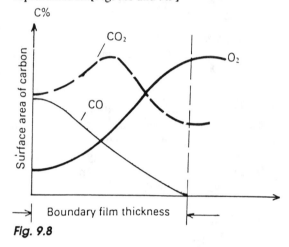

Fig. 9.8

For carbon burning at moderate temperatures (Fig. 9.8), the oxygen concentration across the boundary gas layer decreases sharply on approaching the surface of the carbon particles, while the concentration of CO_2 increases upto a certain radial distance, as more CO is getting converted to CO_2 by diffusing O_2 and is maximum at a point when stoichiometric concentrations of O_2 and CO are attained for the reaction

$$2CO + O_2 \longrightarrow 2CO_2$$

Beyond this, the CO_2 concentration towards the carbon particle falls off because of conversion of some CO_2 to CO

$$C + CO_2 \longrightarrow 2CO$$

For high temperature combustion, (Fig. 9.9) oxygen concentration across the boundary layer decreases more sharply on approaching the carbon particles as the carbon monoxide produced consumes all the oxygen supplied. No oxygen can reach the coke particles' surface. Under these conditions of oxygen deficiency, the endothermic reduction

$$CO_2 + C \longrightarrow 2CO - q$$

will occur on the surface of the coke particles increasing the carbon monoxide concentration on and adjacent to the surface of the particles.

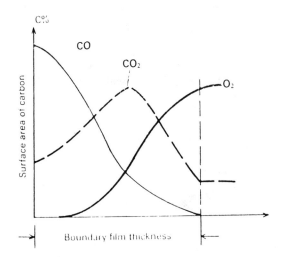

Fig. 9.9

Q. *How can the combustion of carbon particle be represented through chemical reactions?*

Ans. It involves four reactions out of which two are primary and the rest secondary.

Primary Reactions
$$1.\ C + O_2 \longrightarrow CO_2 + Q_1 \text{ (MJ/mol)}$$
$$2.\ C + \frac{1}{2}O_2 \longrightarrow CO + Q_2 \text{ (MJ/mol)}$$

Secondary Reactions
$$3.\ CO + \frac{1}{2}O_2 \longrightarrow CO_2 + Q_3 \text{ (MJ/mol)}$$
$$4.\ C + CO_2 \longrightarrow 2CO - Q_4 \text{ (MJ/mol)}$$

Q. *Is the overall reaction exothermic or endothermic?*

Ans. Exothermic. It is because $Q_4 = 0.57Q_3$ which implies that despite the endothermic reaction (reaction No. 4), the temperature of combustion is maintained at a high level due to higher heat evolution in the volume.

Q. *What process accelerates the burning-off of coke particles?*

Ans. The combustion of carbon (coke) particles proceeds via partial gasification of carbon to CO and its after-burning to CO_2

$$C \xrightarrow{\ O_2\ } CO \xrightarrow{\ O_2\ } CO_2$$

So, as more and more CO burns of into CO_2, more carbon undergoes partial gasification to produce CO shifting the foregoing chain reaction to the right and thus, in the process, accelerating the burning-off of coke particles.

Q. *Is the total rate of combustion of pulverized solid and liquid fuels determined only by the rate of chemical reactions proper?*

Ans. It is determined not only by the rate of chemical reactions proper but also by the rate of oxygen supply to the combustion zone.

Q. *How does the oxygen supply take place?*

Ans. It takes place by turbulent and gas diffusion mechanism.

Q. *What is this mechanism?*

Ans. When the carbon particle burns, it is immediately surrounded by a boundary film consisting of predominantly CO near the particle surface and predominantly O_2 and CO_2 further from the surface. [Fig. 9.10]

The CO produced due to partial gasification of carbon diffuses out from the surface towards the film boundary as more and more CO is oxidized to CO_2 when the stoichiometric ratio of $CO : O_2$ is reached within the boundary layer

$$C \xrightarrow{\ O_2\ } CO \xrightarrow{\ O_2\ } CO_2$$

Therefore, as this process continues, oxygen concentration gets constantly depleted on approaching

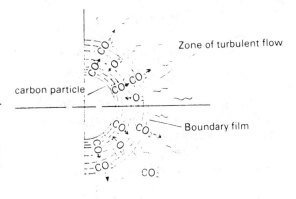

Fig. 9.10 *Carbon particle combustion (Film theory)*

the particle surface. Thus a concentration gradient of oxygen is established across the film—concentration of O_2 at the film surface ($C_{O_2}^f$) being greater than concentration of O_2 at the particle surface ($C_{O_2}^s$). This concentration difference is the drive for molecular diffusion of O_2 across the gas film towards the particle as long as combustion proceeds. Beyond this gas film, intensive turbulent mass transfer occurs.

Q. *How can the rate of oxygen diffusing per unit surface area of the carbon particles be determined?*

Ans. This can be determined by Fick's law:

$$K_d = \alpha_d [\ C_{O_2}^b - C_{O_2}^s\ ]\ \frac{kmol}{m^2}\ s^{-1}$$

α_d = rate of mass transfer = D/δ m/s
D = coefficient of molecular diffusion, m^2/s
δ = boundary layer thickness, m
$C_{O_2}^b$ = oxygen concentration in the bulk of main flow
$C_{O_2}^s$ = oxygen concentration on the surface of carbon

particles

Q. *When will this rate of diffusion be maximum?*

Ans. For $C_{O_2}^s = 0$ whence $\left[K_d\right]_{max} = \alpha_d\ C_{O_2}^b$

Q. *If all the oxygen diffusing reacts on the particle surface, what would be the rate of reaction per unit surface area in terms of oxygen consumption?*

Ans. It will be

$$K_r = k \cdot C_{O_2}^s$$

Q. *When will this reaction rate be maximum?*

Ans. If $C_{O_2}^s = C_{O_2}^b$ whence $\left[K_r\right]_{max} = k\ C_{O_2}^b$

[Since the rate of oxygen supplied through the boundary film is equal to the rate of oxygen consumed in the surface reaction, so

$$K_d = K_r = K_s \text{ (say)}$$

Solving the two equations above,

$$K_s = \frac{\alpha_d\ k}{\alpha_d + k}\ C_{O_2}^b = k_r\ C_{O_2}^b$$

where

$$k_r = \frac{\alpha_d\ k}{\alpha_d + k}$$

Q. *How do K_d and K_r vary with temperature?*

Ans. This can be best represented graphically as shown in Fig. 9.11.

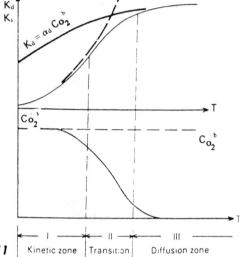

Fig. 9.11 | Kinetic zone | Transition zone | Diffusion zone

Q. *How can these be interpreted at low and high temperatures?*

Ans. At temperatures less than 1000°C, the rate of surface reaction

$$2C + O_2 \longrightarrow 2CO$$

is rather slow. And the oxygen consumption, as a consequence, becomes a very small fraction of the total oxygen supplied to the surface, i.e.

$$k \ll \alpha_d$$

Under these circumstances

$$k_r \approx k$$

and

$$C_{O_2}^b \approx C_{O_2}^s$$

$$\therefore \qquad K_s = k\ C_{O_2}^s$$

which implies that the total reaction rate is limited by the kinetics of the chemical reacting on the coke particle's surface.

At high temperatures (i.e. above 1400°C), the rate constant of the surface reaction shoots up rapidly and eventually exceeds the maximum rate of oxygen supply to the surface.

$$\therefore \qquad k \gg \alpha_d$$

And as such the total reaction rate is determined by the rate of oxygen supply:

$$K_s = \alpha_d \, C_{o_2}^b$$

Under these conditions, the rate of the surface reaction is so high that all the oxygen reaching the coke particle surface through the diffusion mechanism reacts instantaneously and its concentration at the surface becomes virtually equal to zero. In this zone, the surface reaction rate approaches a maxima and varies very little despite any increase of temperature.

Q. *Into how many zones, can the regions of constant-size coke particle burning be divided?*

Ans. Kinetic combustion zone
Transition combustion zone
Diffusion combustion zone

Q. *What is kinetic combustion zone?*

Ans. The temperature region (800–1000°C) for combustion of coke particles is called kinetic combustion zone.

Q. *Why?*

Ans. During this temperature range, the total reaction rate is determined by the kinetics of the chemical reacting on the surface.

Q. *What is the diffusion combustion zone?*

Ans. The temperature range from 1400°C onwards is called the diffusion combustion zone.

Q. *Why is it so called?*

Ans. During this temperature range, the overall reaction rate is determined by the rate of oxygen diffusing to the particle surface. The reaction rate varies very little with the increase of temperature but is retarded considerably if the oxygen supply is insufficient.

Q. *What chemical reactions take place in this diffusion combustion zone in the case of coke particle burning?*

Ans. In this zone the rate of surface reaction is too high and the oxygen supplied to the surface by diffusion reacts instantaneously, with the effect that its concentration at the coke particle surface becomes nearly zero.

Due to oxygen deficiency at the surface a part of CO_2 is reduced to CO on the white hot coke surface

$$CO_2 + C \longrightarrow 2CO$$

This CO diffusing outwards meets the oncoming O_2 diffusing inwards across the boundary film and burns to produce CO_2

$$2CO + O_2 \longrightarrow 2CO_2$$

Q. *How is the reaction rate in diffusion combustion zone affected by gas flow around the particles and particle size?*

Ans. The rate of combustion increases with the increase of gas flow around the coke particles as well as with the decreasing size of the particles.

Q. *What is the transition zone?*

Ans. The temperature range from 1000°C to 1400°C is called transition zone.

Q. *Why?*

Ans. During this temperature region, the rate of surface reaction on the coke particles is determined simultaneously by the rate of oxygen supply as well as the kinetics of the chemical reaction on the surface.

Q. *How is the transition combustion zone affected by the size of the coke particles?*

Ans. If the coke particles are large, the transition zone shifts to the lower limit of temperature while with smaller particles it appears at a higher temperature.

Q. *What is flame core zone?*

Ans. It is the zone of the furnace within which intensive combustion of fuel occurs to a burn-off intensity $\Psi_f = 0.85$–0.90.

It is a zone of high temperature and the best portion of heat transfer from this zone to surrounding water-walls takes place by radiation.[Fig. 9.12]

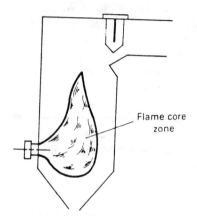

Flame core zone

Fig. 9.12 *Flame core zone*

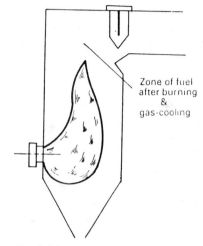

Zone of fuel
after burning
&
gas-cooling

Fig. 9.13

Q. *What is the size of the flame core zone with respect to the furnace space?*

Ans. It occupies 20 to 33% of the furnace space volume.

Q. *What about the rest of the furnace space volume?*

Ans. The remaining portion of the furnace space volume is known as zone of fuel afterburning and gas cooling.

Q. *Why does afterburning (burn-off) take place deep in the diffusion region of the fuel afterburning and proceed slowly?*

Ans. It is because in this zone(Fig.9.13)there is low concentration of leftover fuel and oxidant and gas flow turbulence is weak.

Q. *What is resolved flame length?*

Ans. It is equal to the:

horizontal distance from the burner-end to the furnace axis (*AB*) + vertical distance from the burner level to the level of the horizontal gas duct (*BC*) + horizontal distance to the furnace outlet (*CD*)

i.e. $L_{flm} = AB + BC + CD$ [Fig. 9.14]

Q. *What is the importance of resolved flame length?*

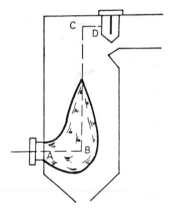

Fig. 9.14

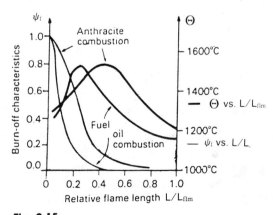

Fig. 9.15

Ans. Fuel afterburning (burn-off) and temperatures in a furnace space are functionally related to the resolved flame length (L_{flm}).

Q. *How do these two parameters—burn-off degree (Ψ_f) and furnace temperature (Θ) vary with the resolved flame length?*

Ans. These can be best interpreted through curves obtained by extrapolating test results of Ψ_f and Θ at different values of relative flame length.[Fig. 9.15]

From the curves, it is evident that burn-off of anthracite coal is completed at relative flame length $L/L_{flm} = 0.35$ while that of fuel-oil is completed at $L/L_{flm} = 0.25$

Q. *Why is burn-off of fuel oil completed at such a small value of relative flame length?*

Ans. It is because, in this case, the flame core disappears in the initial horizontal portion (*AB* in the above figure) of the resolved flame length.

Q. *Why does the combustion of larger fractions of pulverized coal take place mainly in the diffusion region?*

Ans. When pulverized coal is used as a fuel, it consists of particles having a range of size distribution—the finest fractions get heated up quickly (within a few hundredths of a second), reach ignition point earliest and burn first. As they burn, the generated heat accelerates the heating of larger particles. But these begin burning when a large portion of oxygen has already been consumed. Therefore, the combustion of large particles, under oxygen deficiency, takes place predominantly in the diffusion region.

Q. *Why do larger fractions of pulverized coal account for the loss of fuel as the unburnt carbon in the flue gas exhaust?*

Ans. As stated above, the combustion of larger fractions of coal particles takes place when there is already a deficiency of oxygen, as the major part of the supplied oxygen has already been consumed up by the finest fractions. Therefore, only that fraction of a large carbon particle is burnt up from the surface that has received adequate oxygen penetrated through the gas film by molecular diffusion mechanism during the retention time of the particle in the furnace. This leaves a considerable portion of large size carbon particles unburnt and hence accounts for loss of fuel as unburnt carbon in the flue gas exhaust.

10

The Chemistry of Combustion

Q. *What are the important reactions that take place during the combustion of fuels?*

Ans. These are:

3. Draw and material balance.
4. What will be the specific weight of the flue gas?
5. Estimate the expected calorimetric temperature.

	kcal/mol	kcal/kg	kJ/kg
1. $H_2 + \frac{1}{2}O_2 \longrightarrow H_2O$ (vap.)	+ 57.81	28905	121 025 (of H_2)
2. $H_2 + \frac{1}{2}O_2 \longrightarrow H_2O$ (liq.)	+ 68.36	34180	143111 (of H_2)
3. $C + \frac{1}{2}O_2 \longrightarrow CO$ (gas)	− 29.43	2452.5	10 268.61 (of C)
4. $CO + \frac{1}{2}O_2 \longrightarrow CO_2$ (gas)	+ 68.22	2436.42	10201.32 (of CO)
5. $C + O_2 \longrightarrow CO_2$ (gas)	+ 97.65	8137.5	34071.71 (of C)
6. $CO_2 + C \longrightarrow 2CO$ (gas)	− 38.79	3232.5	13534.77 (of C)
7. $C + H_2O \longrightarrow CO + H_2$	− 28.38	2365	9902.25 (of C)
(gas) (gas)			
8. $CH_4 + 2O_2 \longrightarrow CO_2 + 2H_2O$	+ 192.40	12025	50348.67 (of CH_4)
(gas) (liq.)			
9. $C_2H_2 + 2.5\,O_2 \longrightarrow 2CO_2 + H_2O$	+ 312.40	12015.38	50308.41 (of C_2H_2)
(gas) (liq.)			
10. $C_2H_4 + 3O_2 \longrightarrow 2CO_2 + 2H_2O$	+ 345.80	12350	51709 (of C_2H_4)
(gas) (liq.)			
11. $H_2S + 1.5\,O_2 \longrightarrow H_2O + SO_2$	+ 124.85	3672	15375 (of H_2S)
(gas) (gas)			
12. $S + O_2 \longrightarrow SO_2$ (gas)	+ 69.8	2181	9133 (of S)

Problem 10.1 The composition of wood is as follows:

C—49.5%; H—6.5%; O—43%; N—0.4%; Ash—0.6%

1. What volume of air is required to burn 1 kg of wood?
2. What will be the volume of the products of combustion of 1 kg of wood?

Solution

Step (I) Composition of Wood

In Terms of % Wt.	In Terms of Mol. Volume
C – 49.5	C – 49.5/12 – 4.125
H – 6.5	H – 6.5/2 – 3.25
O – 43	O – 43/32 – 1.3437
N – 0.4	N – 0.4/28 – 0.01428
Ash – 0.6	

Step (II) Oxygen Requirement for Combustion

Now, if as per Dulong's theory, the oxygen present in a fuel would react with the hydrogen content of it in a 1:2 ratio to produce water, so 2×1.3437, i.e. 2.6874 mol. vol. of H will combine with the oxygen content of the fuel leaving $3.35 - 2.6874 = 0.5626$ mol. vol. hydrogen for combustion with air.

Therefore, above data modifies to:

Mol. Vol.	Oxygen Requirement for Combustion
C – 4.125	4.125 mol. $\boxed{C + O_2 \longrightarrow CO_2}$
	vol. $\boxed{\text{1 mol} \quad \text{1 mol}}$
H – 3.25 – 2.6874	0.5626/2 – 0.2813 mol. vol
– 0.5626	$\boxed{2H_2 + O_2 \longrightarrow 2H_2O}$
O – 0.000	$\boxed{\text{2 mol 1 mol}}$
N – 0.01428	
	O_2 requirement – 4.4063 mol. vol per 100 kg of wood

Step (III) Volume of Air Requirement

Air contains N_2 and O_2 in volume ratio 79:21
= 3.762
Oxygen requirement for combustion
= 4.4063 mol. vol.
Air requirement for combustion
= 4.4063 + 4.4063 (3.762)
    (Oxygen)   (Nitrogen)
= 20.9828 mol/100 kg of wood
Air required for combustion of 1 kg of wood
= (20.9828/100) (22.4) m^3
= 4.7 m^3 *Ans.*

Step (V) Material Balance

Basis: 100 kg fuel

Input			Output		
	Mol. Vol.	Weight (kg)	Flue gas	Mol. Vol.	Weight (kg)
(a) FUEL					
C	4.125	4.125(12) – 49.5	CO_2	4.125	4.125(44) – 181.5
H	0.5626	0.5626(2) – 1.1252	H_2O	3.25	3.25(18) – 58.5
H_2O	2.6874	2.6874(18) – 48.3732	N_2	16.5907	16.5907(28) – 464.53
N	0.01428	0.01428(28) – 0.399			– 704.53
(b) AIR					
O	4.4063	4.4063(32) – 141.00			
N	4.4063 × 3.762 –16.5765	16.5765(28) – 464.14			
		= 704.5418			

Calculation Error = 704.5396 – 704.5418
                       = – 0.0022

Step (VI) Specific Weight of the Combustion Products

Flue Gas Composition			Weight
CO_2 – 4.125 mol. vol.	4.125 (44/100)		– 1.815 kg
H_2O – 3.25 mol. vol.	3.25 (18/100)		– 0.5850 kg
N_2 – 16.5907 mol. vol.	16.5907 (28/100)		– 4.6453 kg
		Total	7.0453 kg

Therefore, the sp. wt. of flue gas = 7.0453/5.368
                              = 1.312 kg/m^3 *Ans.*

Step (IV) Volume of Combustion Products

Basis: 100 kg wood

Reactants	Chemical Reaction	Products of Combustion
C (4.125 mol)	$C + O_2 \longrightarrow CO_2$ (4.125 mol) 4.125 mol. vol.	CO_2 – 4.125 mol. vol.
H_2 (3.25 mol)	$H_2 + 0.5O_2 \longrightarrow H_2O$ (3.25 mol) 3.25 mol. vol.	H_2O – 3.25 mol. vol.
		N_2 – 4.4063(3.762) + 0.01428 – 16.5907 mol. vol.
		Vol. – 23.9657 mol. vol. – 23.9657 (22.4) Nm^3 – 536.8316 Nm^3

Hence, volume of combustion products for 1 kg wood burnt = 5.368 Nm^3 *Ans.*

Step (VII) Calorific Value of Wood

Reactant	Combustion Reaction	Heat of Combustion (kcal/kg)	Heat Absorbed/ Produced (kcal/kg)
C	$C + O_2 \rightarrow CO_2$	+ 8137.5	+ 8137.5(49.5/100) = + 4 028.06
H	$H_2 + 0.5O_2 \rightarrow H_2O$	+ 28905	+ 28 905 $\dfrac{1.1252}{100}$ = + 325.24
			Total = + 4 353.30

From this total heat developed, the quantity of heat of evaporation of 2.6874 mol. vol. of water should be subtracted.

Hence, the calorific value of wood

= 4353.30 – 2.6874(18/100)(586) = 4069.83 kcal/kg

Ans.

Step (VIII) Expected Calorimetric Temperature

This is done through a series of successive approximations.

Basis: 1 kg fuel. 1st Assumption : 2000°C

Gas	Heat Content at 2000°C	Quantity of Heat Reqd.	to develop 2000°C Temp
CO_2	562.954 kcal/kg	562.954(1.815)	– 1021.76 kcal
H_2O	1183.333 kcal/kg	1183.333(2.6874)(18/100)	– 572.41 kcal
N_2	541.785 kcal/kg	541.785(4.6453)	– 2516.75 kcal
H_2	7585 kcal/kg	7585(0.5626 × 2/100)	– 85.34 kcal
			Total = 4196.26 kcal

This is higher than the calorific value of wood calculated (4 069.83 kcal/kg). So let's have a second trial.

2nd Assumption : 1900°C

Gas	Heat Content at 1900°C	Quantity of Heat Reqd.	to develop 1900°C Temp
CO_2	531.613 kcal/kg	531.613 (1.815)	– 964.88 kcal
H_2O	1 094.833 kcal/kg	1 094.833 (2.6874) (18/100)	– 529.60 kcal
N_2	511.642 kcal/kg	511.642 (4.6453)	– 2376.73 kcal
H_2	7 163 kcal/kg	7 163 (0.5626) (2/100)	– 80.598 kcal
			Total = 3 951.81 kcal

1900°C	3951.81 kcal/kg of fuel
t°C	4069.83 kcal/kg of fuel
2000°C	4196.26 kcal/kg of fuel

$$\frac{2000 - 1\,900}{4\,196.26 - 3\,951.81} = \frac{t - 1\,900}{4069.83 - 3951.81}$$

$$t = 1900 + \frac{100}{244.45}(118.02)$$

$$= 1948.28°C \qquad\qquad Ans$$

Problem 10.2 A wood specimen has the following composition in terms of per cent weight:

C– 49.7; H–5.9; O – 43.5; N – 0.5; Ash – 0.4

1. Calculate the volume of air required to burn 1 kg of wood
2. Calculate the volume of combustion products per kg of fuel consumed
3. Draw the material balance
4. Determine the specific weight of the flue gas
5. Determine the calorific value of wood
6. What will be the expected calorimetric temperature?

Solution

Step (I) Composition of Wood

In Terms of % Wt.	In Terms of Mol. Vol.
C – 49.7	C – 49.7/12 – 4.141
H – 5.9	H – 5.9/2 – 2.95
O – 43.5	O – 43.5/32 – 1.359
N – 0.5	N – 0.5/28 – 0.017
Ash – 0.4	

Step (II) Oxygen Requirement for Combustion

Mol. Vol.	Combustion Reaction	O_2 Reqd. for Combustion
C – 4.141	C + O_2 $\longrightarrow$ CO_2 1 mol 1 mol 1 mol	4.141 mol. vol.
H – 2.95 – 2(1.359) – 0.232	H_2 + $0.5O_2$ $\longrightarrow$ H_2O 1 mol 0.5 mol 1 mol	$\dfrac{0.232}{2}$ – 0.116 mol. vol.
O – 0.000		
N – 0.017		
		O_2 requirement – 4.257 mol. vol. per 100 kg of wood.

Step (III) Volume of Air Requirement

Air contains 3.762 vol. N_2 per unit volume of O_2
Oxygen requirement for combustion
= 4.257 mol. vol. /100 kg of wood

Air requirement for combustion
= 4.257(Oxygen) + 4.257(3.762)(Nitrogen)
= 20.2718 mol. vol./100 kg of wood
Air requirement for combustion of 1 kg wood
= 20.2718 (22.4)/100 = 4.54 Nm^3 *Ans.*

Step (IV) Volume of Combustion Products

Basis: 100 kg wood

Reactants	Combustion Reactions	
C (4.141 mol)	C + O_2 $\longrightarrow$ CO_2 4.141 4.141 (mol) (mol)	CO_2 – 4.141. mol. vol.
H_2 (2.95 mol)	H_2 + $0.5O_2$ $\longrightarrow$ H_2O 2.95 2.95 (mol) (mol)	H_2O – 2.95 mol. vol.
		N_2 = 4.257(3.762) + 0.017 – 16.0318 mol. vol.
		Volume – 23.1228 mol. vol. – 23.1228(22.4) Nm^3 – 517.951 Nm^3

Hence the volume of combustion products for 1 kg of wood burnt = 5.179 Nm^3 *Ans.*

Step (V) Material Balance

Basis: 100 kg fuel

	Input		Output		
	Mol. Vol.	Weight (kg)	Flue Gas	Mol. Vol.	Weight (kg)
(A) Fuel					
C	4.141	49.692	CO_2	4.141	182.204
H	0.232	0.464	H_2O	2.95	53.1
H_2O	2(1.359)				
	– 2.718	48.924			
N	0.017	0.476	N_2	16.0318	448.89
(B) Air					
O_2	4.257	136.224			
N_2	4.257				
	(3.762)				
	–16.0148	448.415			
	Total – 684.195			Total – 684.194	

Calculation error = 684.194 – 684.195 = – 0.001

Step (VI) Specific Weight of Combustion Products

Flue Gas Composition	Weight (kg)	
CO_2 – 4.141 mol. vol	4.141(44/100)	– 1.822
H_2O – 2.95 mol. vol	2.95(18/100)	– 0.531
N_2 – 16.0318 mol. vol.	16.0318(28/100) – 4.489	
	Total – 6.842	

Therefore, the specific weight of flue gas

$$= 6.842/5.179 = 1.321 \text{ kg/Nm}^3$$

Ans.

Step (VII) Calorific Value of Wood

Reactant	Combustion Reaction	Heat of Reaction	Heat Evolved/ Absorbed (kcal/kg)
C	$C + O_2 \longrightarrow CO_2$	(+) 8137.5 kcal/kg	(+) $\dfrac{8137.5\,(49.7)}{100}$ = (+) 4044.34
H	$H_2 + 0.5O_2 \longrightarrow H_2O$	(+) 28905 kcal/kg	(+) $\dfrac{28905\,(0.464)}{100}$ – (+) 134.12
			(+) 4178.46

From this total heat developed, the quantity of heat of evaporation of 2(1.359), i.e. 2.718 mol. vol. of water should be subtracted.

Hence the calorific value of wood

$$= 4178.46 - 2.718(18/100)(586)$$

$$= 3891.766 \text{ kcal/kg of fuel}$$

Ans.

Step (VIII) Calorimetric Temperature of Combustion

1st Assumption: Let this temperature be 2000°C

Gas	Heat Content at 2000°C	Quantity of Heat Reqd. to develop 2000°C Temp.
CO_2	562.954 kcal/kg	562.954(1.822) – 1025.70 kcal
H_2O	1183.333 kcal/kg	1183.333(2.718)(18/100) – 578.93 kcal
N_2	541.785 kcal/kg	541.785(4.49) – 2432.61 kcal
H_2	7585 kcal/kg	7585(0.232)(2/100) – 35.19 kcal
		Total – 4072.43 kcal

2nd Assumption: Temperature 1900°C

Gas	Heat Content at 1900°C	Quantity of Heat Reqd. to develop 1900°C Temp
CO_2	531.613 kcal/kg	968.598 kcal
H_2O	1094.833 kcal/kg	535.636 kcal
N_2	511.642 kcal/kg	2297.272 kcal
H_2	7163 kcal/kg	33.236 kcal
		Total – 3834.742 kcal

1900°C	3834.74 kcal/kg of fuel
t°C	3891.766 kcal/kg of fuel
2000°C	4072.43 kcal/kg of fuel

$$\frac{2000 - 1900}{4072.43 - 3834.74} = \frac{t - 1900}{3891.766 - 3834.74}$$

or

$$t = 1900 + (5702.6/237.69) = 1923.99°C$$

Ans.

Problem 10.3 The percentage composition of straw in terms of weight is as given below:

C—35.5; H—5.5; O—39; N—0.5; Ash—3.75; Water—15.75

Determine:

(a) the volume of air required for complete combustion of 1 kg straw

(b) the volume of combustion products per kg of straw burnt
(c) the specific weight of flue gas
(d) the calorific value of straw
(e) the calorimetric temperature of combustion (expected value)

Also draw the material balance.

Solution

Step (I) Composition of Straw

In Terms of % Wt.	In Terms of Mol. Vol.
C – 35.5	C – 35.5/12 – 2.958
H – 5.5	H – 5.5/2 – 2.75
O – 39	O – 39/32 – 1.218
N – 0.5	N – 0.5/28 – 0.0178
Ash – 3.75	Ash – 3.75
Water – 15.75	Water – 15.75/18 – 0.875

Step (II) Oxygen Requirement for Combustion

Mol. Vol. (According to Dulong)	Combustion	O_2 Requirement for Combustion
C – 2.958	$C + O_2 \longrightarrow CO_2$	2.958 mol. vol.
H – 2.75 – 2(1.218) – 0.314	$H + 1/2\, O_2 \longrightarrow H_2O$	0.314/2 – 0.157 mol. vol.
O – 0.000		
N – 0.0178		
H_2O – 0.875		

Total O_2 requirement – 3. 115 mol. vol. per 100 kg of straw

Step (III) Air Requirement for Combustion

Air required

= 3.115 (Oxygen) + 3.115 (3.762) (Nitrogen)

= 14.833 mol. vol. for 100 kg of straw

Air required for 1 kg of straw

= 14.833 (22.4/100) = 3.323 Nm^3

Ans.

Step (IV) Volume of Combustion Products

Basis: 100 kg of straw

Reactant	Combustion Reaction	Product of Combustion
C (2.958 mol)	$C + O_2 \longrightarrow CO_2$ 2.958 mol 2.958 mol	CO_2 – 2.958 mol. vol.
H_2 (2.75 mol)	$H_2 + 0.5 O_2 \longrightarrow H_2O$ (2.75 mol) (2.75 mol)	H_2O – 2.75 mol. vol
		H_2O – 0.875 mol. vol.
		N_2 – 3.115(3.762) + 0.0178 mol. vol.
		– 11.736 mol. vol.

Total Volume

= 18.319 mol. vol.

= 18.319(22.4/100) Nm^3

= 4.103 Nm^3 per kg of straw burnt

Ans.

Step (V) Material Balance for Combustion

Basis: 100 kg of straw

Input		Output		
Mol. Vol.	Weight (kg)	Flue Gas	Mol. Vol.	Weight (kg)
(A) Fuel				
C 2.958	35.496	CO_2	2.958	130.152
H 0.314	0.628	H_2O	2.75 + 0.875 – 3.625	65.25
H_2O 2(1.218) + 0.875 – 3.311	59.598	N_2	– 11.736	328.608
N 0.0178	0.499			
(B) Comb. *Air*				
O_2 3.115	99.68			
N_2 3.115 (3.762) –11.7186	328.121			
Total – 524.022		Total – 524.01		

Calculation error = 524.022 – 524.01 = 0.012 kg

Step (VI) Specific Weight of Combustion Products

Flue Gas Composition	Mol. Vol.	Weight	
CO_2	2.958	2.958 (44/100)	– 1.3015 kg
H_2O	3.625	3.625 (18/100)	– 0.6525 kg
N_2	11.736	11.736 (28/100)	– 3.2860 kg
		Total	– 5.2400 kg

Therefore, the specific weight of flue gas

$$= 5.24/4.103$$

$$= 1.277 \text{ kg/m}^3$$

Ans.

Step (VII) Calorific Value of Straw

Reactant	Combustion Reaction	Heat of Reaction	Heat Produced/ Absorbed (kcal/kg)
C	$C + O_2 \longrightarrow CO_2$	(+)8137.5 kcal/kg	(+) $\dfrac{8137.5\,(35.5)}{100}$ = + 2888.812
H	$H_2 + 0.5O_2 \longrightarrow H_2O$	(+)28905 kcal/kg	(+) $\dfrac{28905\,(0.628)}{100}$ = (+) 181.523
		Total = (+) 3070.335	

From this heat developed, according to Dulong, the quantity of heat of evaporation of 2 (1.218) + 0.875 i.e. 3.331 mol. vol. of water should be subtracted.

Hence the calorific value of water

$$= 3070.335 - 3.331(18)(586)/100$$

$$= 2721.09 \text{ kcal/kg}$$

Ans.

Step (VIII) Calorimetric Temperature

1st Assumption: Let this temperature be 1800°C

Gas	Heat Content at 1800°C	Quantity of Heat Reqd. to develop 1800°C Temp.
CO_2	500.318 kcal/kg	500.318(1.3015) = 651.163 kcal
H_2O	1011.611 kcal/kg	1011.611(3.311)(18/100) = 602.899 kcal
N_2	481.82 kcal/kg	481.82(3.286) = 1583.260 kcal
H_2	6745.5 kcal/kg	6745.5(0.314)(2/100) = 42.361 kcal
		Total = 2879.683 kcal

2nd Assumption: Calorimetric temperature 1700°C

Gas	Heat Content at 1700°C	Quantity of Heat Regd. to develop 1700°C Temp
CO_2	469.09 kcal/kg	610.528 kcal
H_2O	933.388 kcal/kg	556.281 kcal
N_2	452.321 kcal/kg	1486.328 kcal
H_2	6332.5 kcal/kg	39.768 kcal
		2692.905 kcal

Hence the calorimetric temperature lies between 1800°C and 1700°C.

1800°C	2879.683	kcal/kg of fuel
t°C	2721.09	kcal/kg of fuel
1700°C	2692.90	kcal/kg of fuel

$$\frac{1800 - 1700}{2879.683 - 2692.90} = \frac{t - 1700}{2721.09 - 2692.90}$$

$$\therefore \quad t = 1715.09 \text{ °C}$$

Ans.

Problem 10.4 Bagasse (the residue of sugarcane) has the following average composition, by weight per cent, on moisture-free basis:

C — 45; H — 6; O — 46; Ash — 3

Calculate

(a) the volume of air required for complete combustion of 1 kg of bagasse
(b) the volume of combustion products per kg of bagasse burnt
(c) the specific weight of flue gas
(d) the calorific value of bagasse
(e) the calorimetric temperature

Also draw the material balance.

Solution

Step (I) Composition of Bagasse

In Terms of Weight %	In Terms of Mol. Vol.
C – 45	C – 45/12 – 3.75
H – 6	H – 6/2 – 3
O – 46	O – 46/32 – 1.4375
Ash – 3	

Step (II) Oxygen Requirement for Combustion

Mol. Vol. (according to Dulong)	Combustion Reaction	O_2 Requirement for Combustion
C – 3.75 H – 3 – 2 (1.4375) – 0.125 O – 0.000	$C + O_2 \longrightarrow CO_2$ $H_2 + 0.5O_2$ $\longrightarrow H_2O$	3.75 mol. vol. 0.125/2 – 0.0625 mol. vol.
	Total O_2 requirement – 3.8125 mol. vol. per 100 kg of bagasse	

Step (III) Air Requirement for Combustion

Air required = 3.8125 + 3.8125 (3.762)
$\qquad$ (Oxygen) $\qquad$ (Nitrogen)
= 18.155 mol. vol. per 100 kg of Bagasse

Air required for combustion of 1 kg bagasse

= 18.155 (22.4)/100 = 4.066 Nm^3 $\qquad$ *Ans.*

Step (IV) Volume of Combustion Products

Basis: 100 kg of bagasse

Reactant	Combustion Reaction	Product of Combustion
C (3.75) mol H_2 3 mol	$C + O_2 \longrightarrow CO_2$ (3.75) mol $\quad$ (3.75) mol $H_2 + 0.5O_2 \longrightarrow H_2O$ 3mol $\qquad$ 3mol	CO_2 – 3.75 mol. vol. H_2O – 3 mol. vol. N_2 – 3.8125 (3.762) – 14.342 mol. vol
	Total Volume – 21.092 mol. vol.	

Therefore, the volume of combustion products per kg of bagasse burnt

= 21.092 (22.4/100) = 4.724 Nm^3 $\qquad$ *Ans.*

Step (V) Material Balance for Combustion

Basis: 100 kg of bagasse

	Input		Output		
	Mol. Vol.	Weight (kg)	Flue Gas	Mol. Vol.	Weight (kg)
(A) Fuel					
C	3.75	45	CO_2	3.75	165
H	0.125	0.25	H_2O	3	54
H_2O	2.875	51.75	N_2	14.342	401.576
(B) Air					
O_2	3.8125	122			
N_2	14.342	401.576			
	Total	620.576		Total	620.576

Step (VI) Specific Weight of Combustion Products

Flue Gas Composition	Weight (kg)
CO_2 – 3.75 mol. vol	165
H_2O – 3 mol. vol	54
N_2 – 14.342 mol. vol.	401.576
Total – 620.576 per 100 kg of bagasse.	

Therefore, the specific weight of flue gas

= 6.205/ 4.724 = 1.313 kg/Nm^3 $\qquad$ *Ans.*

Step (VII) Calorific Value of Bagasse

Reactant	Combustion Reaction	Heat of Reaction (kcal/kg)	Heat Evolved/Absorbed (kcal/kg)
C	$C + O_2 \rightarrow CO_2$	(+) 8137.5	(+) 8137.5(45/100) – (+) 3661.875
H	$H_2 + 0.5O_2$ $\rightarrow H_2O$	(+) 28905	(+) 28905 (0.25/100) – (+) 72.262
		Total – (+) 3734.137	

According to Dulong, the quantity of heat of evaporation of 2.875 mol. vol. of H_2O should be subtracted from this calculated value.

Hence the calorific value of bagasse

= 3734.137 – 2.875 (18) (586)/100

= 3430.882 kcal/kg.
$\qquad$ *Ans.*

Step (VIII) Calorimetric Temperature

1st Assumption: Let this temperature be 1900°C

Gas	Heat Content At 1900°C	Quantity of Heat Reqd. to develop 1900°C Temp.
CO_2	531.613 kcal/kg	531.613 (1.65) – 877.161 kcal/kg of fuel
H_2O	1094.833 kcal/kg	1094.833 (51.75/100) – 566.576 kcal/kg of fuel
N_2	511.642 kcal/kg	511.642 (4.01576) – 2054.631 kcal/kg of fuel
H_2	7163 kcal/kg	7163 (0.25/100) – 17.907 kcal/kg of fuel
		Total – 3516.275 kcal/kg of fuel

2nd Assumption: Let this temperature be $1800°C$

Gas	Heat Content at 1800°C	Quantity of Heat Reqd. to develop 1800°C Temp.
CO_2	500.318 kcal/kg	(+) 825.524 kcal/kg of bagasse
H_2O	1011.611 kcal/kg	(+) 523.508 kcal/kg of bagasse
N_2	481.82 kcal/kg	(+) 1934.873 kcal/kg of bagasse
H_2	6745.5 kcal/kg	(+) 16.863 kcal/kg of bagasse
	Total	− 3300.768 kcal/kg of bagasse

Hence the calorimetric temperature ($t°C$) lies between $1900°C$ and $1800°C$

$$\frac{1900 - 1800}{3516.275 - 3300.768} = \frac{t - 1800}{3430.882 - 3300.768}$$

$$\therefore t = 1860.37°C$$

Ans.

Problem 10.5 The percentage composition of peat is given below in terms of weight:
    C—58; H—5.45; O—33.45; N—0.75;
Ash—2.35
    Calculate:
    (a) the volume of air required for the complete combustion of 1 kg of peat
    (b) the volume of flue gas per kg of peat burnt
    (c) the specific weight of flue gas
    (d) the calorific value of peat
    (e) the calorimetric temperature of combustion
Also draw the material balance.

Solution
Step (I) Composition of Peat

In Terms of Weight %	In Terms of Mol. Vol.
C − 58	C − 58/12 − 4.833
H − 5.45	H − 5.45/2 − 2.725
O − 33.45	O − 33.45/32 − 1.045
N − 0.75	N − 0.75/28 − 0.0267

Step (II) Oxygen Requirement

Mol. Vol.	Combustion Reaction	O_2 Requirement for Combustion
C − 4.833	$C + O_2 \longrightarrow CO_2$	4.833 mol. vol.
H − 2.725 − 2(1.045) − 0.635	$H_2 + 0.5O_2 \longrightarrow H_2O$	0.635/2 − 0.3175 mol. vol.

Therefore, total oxygen requirement = 5.1505 mol. vol. per 100 kg of peat combustion.

Step (III) Air Requirement for Combustion

Air required for combustion
    = 5.1505 + 5.1505 (3.762) mol. vol. per 100 kg of peat
    = 24.5266 mol. vol. per 100 kg of peat
Therefore, air required for combustion of 1 kg of peat
    = 24.5266 (22.4/100) Nm^3 − 5.494 Nm^3

Ans.

Step (IV) Volume of Flue Gas

Basis: 100 kg of peat

Reactant	Combustion Reaction	Product of Combustion
C 4.833 mol	$C + O_2 \longrightarrow CO_2$	− 4.833 mol. vol.
H 2.725 mol	$H_2 + 0.5O_2 \longrightarrow H_2O$	H_2O − 2.725 mol. vol.
		N_2 − 5.1505 (3.762) + 0.0267 − 19.4027 mol. vol.
		Σ Volume − 26.9607 mol. vol. per 100 kg of peat burnt

Hence the volume of flue gas/kg of peat burnt
    = 26.9607 (22.4/100) − 6.0392 Nm^3

Ans.

Step (V) Specific Weight of Flue Gas

Flue Gas	Composition	Weight
CO_2	4.833 mol. vol.	212.652 kg
H_2O	2.725 mol. vol.	49.05 kg
N_2	19.4027 mol. vol.	543.275 kg
		Σ − 804.977 kg per 100 kg of peat

Therefore, the specific weight of flue gas
    = 8.04977/6.0392 = 1.333 kg/Nm^3

Ans.

Step (VI) Material Balance

Basis: 100 kg of peat

Input			Output		
	Mol. Vol.	Weight (kg)	Flue Gas	Mol. Vol.	Weight (kg)
(A) Fuel					
C	4.833	57.996	CO_2	4.833	212.652
H	0.635	1.27	H_2O	2.725	49.05
H_2O	2.09	37.62	N_2	19.4027	543.275
N	0.0267	0.7476			
(B) Air					
O_2	5.1505	164.816			
N_2	19.376	542.528			
		$\Sigma = 804.977$			$\Sigma = 804.977$

Step (VII) Calorific Value of Peat

Reac-tant	Combustion Reaction	Heat of Reaction	Heat Produced/ Absorbed
C	$C + O_2 \longrightarrow CO_2$	(+) 8137.5 kcal/kg	$(+) \dfrac{8137.5 \ (58)}{100}$ $= + 4719.75 \ \dfrac{kcal}{kg}$
H	$H_2 + 0.5O_2$ $\longrightarrow H_2O$	(+) 28905 kcal/kg	$(+) \dfrac{28905(1.27)}{100}$ $= + 367.09 \ \dfrac{kcal}{kg}$
			$\Sigma = + 5086.843$ kcal per kg of peat

But according to Dulong, 2.09 (18/100) (586)

= 220.453 kcal would be used up on evaporation.

Hence, the calorific value of peat

= 5086.843 – 220.453 kcal/kg

= 4866.39 kcal/kg

Step (VIII) Calorimetric Temperature

1st Assumption: Let this temperature be 2000°C

Gas	Heat Content at 2000°C	Quantity of Heat Reqd. to develop 2000°C
CO_2	562.954 kcal/kg	562.954 (2.1265) = 1197.121 kcal/kg of peat
H_2O	1183.333 kcal/kg	1183.333 (37.62/100) = 445.17 kcal/kg of peat
N_2	541.786 kcal/kg	541.786 (543.275)/100 = 2943.387 kcal/kg of peat
H_2	7585 kcal/kg	7585 (1.27/100) = 96.329 kcal/kg of peat
		$\Sigma = 4682.007$ kcal/kg of peat

This value being too small, let us assume the calorimetric temperature to be 2100°C.

Gas	Heat Content at 2100°C	Quantity of Heat Reqd. to develop 2100°C Temp.
CO_2	594.25 kcal/kg	1263.672 kcal/kg of peat
H_2O	1277.722 kcal/kg	480.679 kcal/kg of peat
N_2	572.250 kcal/kg	3108.891 kcal/kg of peat
H_2	8011.50 kcal/kg	101.746 kcal/kg of peat
		$\Sigma = 4954.988$ kcal/kg of peat

2100°C	4954.988 kcal/kg of peat
t°C	4866.39 kcal/kg of peat
2000°C	4682.007 kcal/kg of peat

$$\frac{2100 - 2000}{4954.988 - 4682.007} = \frac{t - 2000}{4866.39 - 4682.007}$$

$$\therefore \quad t = 2067.54°C$$

Ans.

Problem 10.6 The ultimate analysis of a coal sample gives the following composition on the basis of weight per cent:

C—71; H—5.5; O—3.5; N—1.25; S—1.2; Ash + Water—17.55

Determine the

1. the volume of air required for complete combustion of 1 kg coal
2. volume of flue gas per kg of coal burnt
3. specific weight of flue gas
4. calorific value of coal
5. calorimetric temperature of combustion

Also draw the material balance.

Solution

Step (I) Composition of Coal

In Terms of Weight %	In Terms of Mol. Vol.
C – 71	C – 71/12 – 5.916
H – 5.5	H – 5.5/2 – 2.75
O – 3.5	O – 3.5/32 – 0.1094
N – 1.25	N – 1.25/28 – 0.0446
S – 1.2	S – 1.2/32 – 0.0375

Step (II) Oxygen Requirement

Mol. Vol. (according to Dulong)	Combustion Reaction	Oxygen Requirement for Complete Combustion (mol. vol.)
C – 5.916 H – 2.75 – 2(0.1094) – 2.5312 N – 0.0446 S – 0.0375	$C + O_2 \longrightarrow CO_2$ $H_2 + 0.5O_2 \longrightarrow$ H_2O $S + O_2 \longrightarrow SO_2$	5.916 (2.5312)0.5 – 1.2656 0.0375
		Σ – 7.2191 mol. vol. per 100 kg of coal

Step (III) Air Requirement for Combustion

Air required = 7.2191 + 7.2191(3.762)
       (Oxygen)   (Nitrogen)
= 34.3773 mol. vol. for combustion of 100 kg coal
Air required for 1 kg coal combustion
= 34.3773 (22.4/100) Nm^3
= 7.7005 Nm^3

                               Ans.

Step (IV) Volume of Flue Gas
Basis: 100 kg of coal

Reactant	Combustion Reaction	Product of Combustion	(Mol. Vol.)
C (5.916 mol)	$C + O_2$ $\longrightarrow CO_2$	CO_2	– 5.916
H (2.75 mol.)	$H_2 + 0.5O_2$ $\longrightarrow H_2O$	H_2O	– 2.75
S (0.0375 mol.)	$S + O_2$ $\longrightarrow SO_2$	SO_2	– 0.0375
		N_2 – 7.2191 (3.762) + 0.0446	– 27.2028
		Σ Volume – 35.9063 mol. vol. per 100 kg of coal burnt	

Hence the volume of flue gas per kg of coal burnt
= 35.9063 (22.4/100)
= 8.043 Nm^3

                               Ans.

Step (V) Specific Weight of Flue Gas

Flue Gas	Composition	Weight
CO_2	5.916 mol. vol.	5.916 (44) – 260.304 kg
H_2O	2.75 mol. vol.	2.75 (18) – 49.5 kg
SO_2	0.0375 mol. vol.	0.0375 (64) – 2.4 kg
N_2	27.2028 mol.vol	27.2028(28) – 761.678 kg
		Σ – 1073.882 kg/100 kg of coal burnt

Specific weight of flue gas
= [1073.882/100] 8.043
= 1.335 kg/Nm^3

                               Ans.

Step (VI) Material Balance
Basis: 100 kg of coal

Input			Output		
	Mol. Vol.	Weight (kg)	Flue Gas	Mol. Vol.	Weight (kg)
(A) *Fuel*					
C	5.916	70.99	CO_2	5.916	260.304
H	2.5312	5.062	H_2O	2.75	49.5
H_2O	0.2188	3.938	SO_2	0.0375	2.4
S	0.0375	1.2	N_2	27.2028	761.678
N	0.0446	1.25			
(B) *Air*					
O_2	7.2191	231.0112			
N_2	7.2191 (3.762)	760.4311			
	– 27.1582				
	Σ –	1073.883		Σ –	1073.882

Step (VII) Calorific Value of Coal

Reactant	Combustion Reaction	Heat of Reaction (kcal/kg)	Heat Evolved/Absorbed (kcal)
C	$C + O_2$ $\longrightarrow CO_2$	(+) 8137.5	(+) 8137.5 (71/100) – + 5777.625
H	$H_2 + 0.5O_2$ $\longrightarrow H_2O$	(+) 28905	(+) 28905 (2.5312 × 2)/ 100 – + 1463.286
S	$S + O_2$ $\longrightarrow SO_2$	(+) 2181.25	(+) 2181.25 (0.0375)(32)/100 – 26.175
			Σ – 7267.087 kcal/kg of coal

According to Dulong, the heat of evaporation of 0.2188 mol. vol. of water should be subtracted from this calculated value.

Hence the calorific value of coal

$$= 7267.087 - \frac{0.2188\,(18)\,(586)}{100} = 7244 \text{ kcal/kg}$$

Ans.

Step (VIII) Calorimetric Temperature of Combustion

1st Assumption: Let us take it to be 2400°C

Gas	Heat Content at 2400°C	Quantity of Heat Reqd. to develop 2400°C Temp.
CO_2	688.091 kcal/kg	688.091 (2.603) = 1791.10 kcal/kg of coal
H_2O	1602 kcal/kg	1602 (3.938/100) = 63.086 kcal/kg of coal
SO_2	473.062 kcal/kg	473.062 (2.4/100) = 11.353 kcal/kg of coal
H_2	9318 kcal/kg	9318 (5.062/100) = 471.677 kcal/kg of coal
N_2	665.571	$\dfrac{665.571\,(761.678)}{100}$ = 5069.508 kcal/kg of coal
		Σ = 7406.726 kcal/kg of coal

This figure is too much. So we need further approximation.

2nd Assumption: 2300°C

Gas	Heat Content at 2300°C	Quantity of Heat Reqd. to develop 2300°C Temp
CO_2	656.863 kcal/kg	1709.814 kcal/kg of coal
H_2O	1486.444 kcal/kg	58.536 kcal/kg of coal
SO_2	451.593 kcal/kg	10.838 kcal/kg of coal
H_2	8878 kcal/kg	449.404 kcal/kg of coal
N_2	634.143 kcal/kg	4830.127 kcal/kg of coal
		Σ = 7058.719 kcal/kg of coal

2400°C   7406.726 kcal/kg of coal
t°C   7244 kcal/kg of coal
2300°C   7058.719 kcal/kg of coal

$$\frac{2400 - 2300}{7406.726 - 7058.719} = \frac{t - 2300}{7244 - 7058.719}$$

$$\therefore \quad t = 2353.24°C$$

Ans.

Problem 10.7 The percentage composition of petroleum by weight is given below:

C— 84.5; H—12.8; O —1.5; S—1.2

Determine

(a) the volume of air required for complete combustion of 1 kg petroleum
(b) the volume of flue gas per kg of petroleum burned
(c) the specific weight of the flue gas
(d) the calorific value of petroleum
(e) the calorimetric temperature of combustion

Also draw the material balance.

Solution

Step (I) Composition of Petroleum

In Terms of Weight %	In Terms of Mol. Vol.
C – 84.5	C – 84.5/12 – 7.0416
H – 12.8	H – 12.8/2 – 6.4
O – 1.5	O – 1.5/32 – 0.0468
S – 1.2	S – 1.2/32 – 0.0375

Step (II) Oxygen Requirement

Mol. Vol. (according to Dulong)	Combustion Reaction	Oxygen Requirement For Combustion (Mol. Vol.)
C – 7.0416	$C + O_2 \longrightarrow CO_2$	7.0416
H – 6.4 – 2 (0.0468) – 6.3064	$H_2 + 0.5O_2 \longrightarrow H_2O$	3.1532
O – 0.000		
S – 0.0375	$S + O_2 \longrightarrow SO_2$	0.0375
		ΣO_2 – 10.2323 Mol. Vol./100 kg of petroleum burned

Step (III) Air Requirement

Air required = 10.2323 + 10.2323(3.762)
      (Oxygen)   (Nitrogen)

= 48.7262 mol. vol. for combustion of 100 kg of petroleum

Hence, the combustion air required for 1 kg of petroleum

= 48.7262 (22.4/100) Nm^3 = 10.9146 Nm^3

Ans.

Step (IV) Volume of Flue Gas

Basis: 100 kg of petroleum

Reactant	Combustion Reaction	Product of Combustion
C (7.0416 mol)	$C + O_2 \longrightarrow CO_2$	CO_2 – 7.0416 mol. vol.
H (6.4 mol)	$H_2 + 1/2O_2 \longrightarrow H_2O$	H_2O – 6.4 mol. vol.
S (0.0375 mol)	$S + O_2 \longrightarrow SO_2$	SO_2 – 0.0375 mol.vol
		N_2 – 10.2323 (3.762) – 38.4939 mol. vol.
		Σ Volume – 51.973 mol.vol.

Petroleum Burned	Flue Gas Produced
100 kg	51.973 mol. vol. – 51.973(22.4) Nm^3
1 kg	51.973(22.4)/100 – 11.6419 Nm^2
	Ans.

Step (V) Specific Weight of Flue Gas

Flue Gas	Composition	Weight (kg)
CO_2	7.0416 mol. vol.	309.8304
H_2O	6.4 mol.vol.	115.2
SO_2	0.0375 mol. vol.	2.4
N_2	38.4939 mol. vol.	1077.829
	Σ = 1505.2596 kg/100 kg of petroleum burned	

Therefore, the specific weight of flue gas

= 15.0525/11.6419 = 1.292 kg/Nm^3

Ans.

Step (VI) Material Balance

Basis: 100 kg of petroleum

	Input		Output		
	Mol. Vol.	Weight (kg)	Flue Gas	Mol. Vol.	Weight (kg)
(A) Fuel					
C	7.0416	84.4992	CO_2	7.0416	309.8304
H	6.3064	12.6128	H_2O	6.4	115.2
H_2O	2(0.0468)	1.6848	SO_2	0.0375	2.4
	– 0.0936				
S	0.0375	1.2	N_2	38.4939	1077.829

(B) Air		
O_2	10.2323	327.4336
N_2	38.4939	1077.8292
	Σ – 1505.2596	Σ – 1505.2596

Step (VII) Calorific Value of Petroleum

Reactant	Combustion Reaction	Heat of Combustion (kcal/kg)	Heat Evolved/ Absorbed (kcal)
C	$C + O_2 \longrightarrow CO_2$	(+) 8137.5	(+) 8137.5 (84.4992)/100 – 6876.122
H	$H_2 + 1/2 O_2 \longrightarrow H_2O$	(+) 28905	(+)28905 (12.6128)/100 – 3645.729
S	$S + O_2 \longrightarrow SO_2$	(+) 2181.25	(+) 2181.25 (1.2)/ 100 – 26.175
			Σ – 10548.026 kcal/kg of petroleum

Hence the calorific value of petroleum

= 10548.026 – [0.0936 (18) (586)/100]

= 10538.153 kcal/kg   Ans.

Step (VIII) Calorimetric Temperature of Combustion

1st Assumption: Let this temperature be 2300°C

Gas	Heat Content at 2300°C (kcal/kg)	Quantity of Heat Reqd. to develop 2300°C Temp. (kcal/kg of petroleum)	
CO_2	656.863	$656.863\left(\dfrac{309.83}{100}\right)$	– 2035.158
H_2O	1486.444	$1486.444\left(\dfrac{1.6848}{100}\right)$	– 25.0436
SO_2	451.593	$451.593\left(\dfrac{2.4}{100}\right)$	– 10.8382
H_2	8878	$8878\left(\dfrac{12.6128}{100}\right)$	– 1119.7643
N_2	634.143	$634.143\left(\dfrac{1077.829}{100}\right)$	– 6834.9771
		Σ – 10025.781	

This value is too small.

2nd Assumption: 2400°C

Gas	Heat Content at 2400°C		Quantity of Heat Reqd. to develop 2400°C
CO_2	688.091 kcal/kg		2131.912 kcal/kg of Petroleum burned
H_2O	1602	"	26.990 "
SO_2	473.062	"	11.353 "
H_2	9318	"	1175.260 "
N_2	665.571	"	7173.717 "
			$\Sigma - 10519.232$

3rd Assumption: 2500°C

Gas	Heat Content at 2500°C (kcal/kg)	Quantity of Heat Reqd. to develop 2500°C Temp. (kcal/kg of petroleum burned)
CO_2	719.318	2228.663
H_2O	1725.666	29.074
SO_2	494.531	11.868
H_2	9762	1231.261
N_2	697.321	7515.928
		$\Sigma - 11016.794$ kcal/kg of petroleum burned

2500°C   11016.794 kcal/kg of petroleum burned
t°C   10578.255 kcal/kg of petroleum burned
2400°C   10519.232 kcal/kg of petroleum burned

$$\frac{2500 - 2400}{11016.794 - 10519.232} = \frac{t - 2400}{10578.255 - 10519.232}$$

$$\therefore \quad t = 2411.86°C$$

Ans.

Problem 10.8 The percentage composition of producer gas is given below in terms of volume:

CO—20; H_2—15.5; N_2—54; CH_4—1.25; C_2H_4—1.3; O_2—0.4; CO_2—7.55

Determine the

(a) Theoretical volume of air required for complete combustion of 1 m³ of producer gas
(b) volume of flue gas per m³ of producer gas burned
(c) specific weight of the flue gas
(d) calorific value of producer gas
(e) calorimetric temperature of combustion

Also draw the material balance.

Solution

Step (I) Air Requirement for Combustion
Basis: 1 Nm³ of producer gas

Composition	Combustion Reaction	Oxygen Requirement
CO – 0.2 Nm³	$CO + 1/2\, O_2$ $\longrightarrow CO_2$	1/2 (0.20) – 0.1 Nm³
H_2 – 0.155 Nm³	$H_2 + 1/2\, O_2$ $\longrightarrow H_2O$	1/2 (0.155) – 0.0775 Nm³
CH_4 – 0.0125 Nm³	$CH_4 + 2O_2 \longrightarrow$ $CO_2 + H_2O$	2(0.0125) – 0.025 Nm³
C_2H_4 – 0.013 Nm³	$C_2H_4 + 3O_2$ $\longrightarrow 2CO_2 +$ $2H_2O$	3(0.013) – 0.039 Nm³
N_2 – 0.54 Nm³ O_2 – 0.004 Nm³ CO_2 – 0.0755 Nm³		
	$\Sigma\, O_2$ Requirement – 0.2415 Nm³ for burning 1 Nm³ of producer gas	

Since the gas itself contains 0.004 Nm³ of O_2, the theoretical volume of oxygen required for complete combustion of 1 Nm³ of producer gas

$$= 0.2415 - 0.004 = 0.2375\ Nm^3$$

Hence, the volume of air required

$$= 0.2375 + 0.2375\ (3.762)\ Nm^3$$

$$= 1.1309\ Nm^3\ \text{per Nm}^3\ \text{of producer gas burned}$$

Ans.

Step (II) Volume of Flue Gas
Basis: 1 Nm³ of producer gas

Reactant	Chemical Combustion	Volume of the Combustion Product	
		CO_2	H_2O
CO (0.2 Nm³)	$CO + 1/2\, O_2$ $\longrightarrow CO_2$	0.20 Nm³	—
H_2(0.155 Nm³)	$H_2 + 1/2\, O_2$ $\longrightarrow H_2O$	—	0.155 Nm³
CH_4 (0.0125 Nm³)	$CH_4 + 2O_2$ $\longrightarrow CO_2 + 2H_2O$	0.0125 Nm³	0.025 Nm³
C_2H_4 (0.013 Nm³)	$C_2H_4 + 3O_2$ $\longrightarrow 2CO_2 + 2H_2O$	0.026 Nm³	0.026 Nm³
		$\Sigma - 0.2385$ Nm³	$\Sigma - 0.206$ Nm³

Hence the total volume of combustion products per 1 Nm^3 of producer gas burned

$$= \underbrace{0.2385 + 0.0755}_{(CO_2)} + \underbrace{0.206}_{(Water)} + \underbrace{0.54 + 0.2375}_{(3.762)}$$

$$= 1.953 \ Nm^3$$

Ans.

Step (III) Specific Weight of Flue Gas

Basis: 1.953 Nm^3 of flue gas

Compos-ition	Volume (Nm^3)	Weight (kg)
CO_2	0.2385 + 0.0755 – 0.314	0.314 (44/22.4) – 0.6167
H_2O	0.206	0.206 (18/22.4) – 0.1655
N_2	0.54 + 0.2375 (3.762) = 1.433	1.433 (28/22.4) – 1.7912
		Σ – 2.5734

Therefore, the specific weight of flue gas

= 2.5734/1.953 kg/Nm^3 = 1.317 kg/Nm^3

Ans.

Step (IV) Calorific Value of Producer Gas

Reactant	Combustion Reaction	Heat of Combustion	Heat Evolved (kcal/Nm^3 of producer gas burned)
CO (0.2 Nm^3)	$CO + 1/2O_2 \longrightarrow CO_2$	2436.42 kcal/kg of CO	$\dfrac{2436.42 \ (0.2) \ (28)}{22.4}$ – 609.105
H_2 (0.155 Nm^3)	$H_2 + 1/2O_2 \longrightarrow H_2O$	28905 kg of H_2 kcal/kg	$\dfrac{28905 \ (0.155) \ (2)}{22.4}$ – 400.024
CH_4 (0.0125 Nm^3)	$CH_4 + 2O_2 \longrightarrow CO_2 + 2H_2O$	12025 kcal/kg of CH_4	$\dfrac{12025 \ (0.0125) \ (16)}{22.4}$ – 107.366
C_2H_4 (0.013 Nm^3)	$C_2H_4 + 3O_2 \longrightarrow 2CO_2 + 2H_2O$	12350 kcal/kg of C_2H_4	$\dfrac{12350 \ (0.013) \ (28)}{22.4}$ – 200.687
			Σ – 1317.182 kcal/Nm^3 of Producer Gas burned

Step (V) Calorimetric Temperature of Combustion

1st Approximation: Let this temperature be 1600°C

Flue Gas Constituent	Volume (Nm^3)	Specific Heat at 1600°C (kcal/Nm^3 C)	ΔH (kcal)
CO_2	0.314	0.539	0.314 (0.539) (1600) – 270.793
H_2O	0.206	0.435	0.206 (0.435) (1600) – 143.376
N_2	1.433	0.331	1.433 (0.331) (1600) – 758.916
			Σ – 1173 kcal

This value is too small. So let us have a second approximation.

2nd Approximation: 1800°C

Flue gas Constituent	Volume (Nm^3)	Specific Heat at 1800°C (kcal/Nm^3·C)	ΔH (kcal)
CO_2	0.314	0.545	0.314 (0.545) (1800) – 308.034
H_2O	0.206	0.452	0.206 (0.452) (1800) – 167.601
N_2	1.433	0.335	1.433 (0.335) (1800) – 864.099
			Σ – 1339.73 kcal

1600°C   1173.0   kcal/Nm^3 of Producer Gas burned
t°C   1317.18   kcal/Nm^3 of Producer Gas burned
1800°C   1339.73   kcal/Nm^3 of Producer Gas burned

$$\therefore \ \frac{1800 - 1600}{1339.73 - 1173} = \frac{t - 1600}{1317.18 - 1173}$$

$$\therefore \ t = 1772.95°C$$

Ans.

Material Balance

Basis: 1 Mol. Vol. of producer gas

Input			Output		
Mol. Vol.	Weight (kg)	Flue Gas	Mol. Vol.	Weight (kg)	
(A) Fuel					
CO 0.20	0.2(28) – 5.6	CO_2	0.314	0.314(44) – 13.816	
H_2 0.155	0.155(2) – 0.31	H_2O	0.206	0.206(18) – 3.708	
CH_4 0.0125	0.0125(16) – 0.20	N_2	1.433	1.433(28) – 40.124	
C_2H_4 0.013	0.013(28) – 0.364				
N_2 0.54	0.54(28) – 15.12				
O_2 0.004	0.004(32) – 0.128				
CO_2 0.0755	0.0755(44) – 3.322				
(B) Air					
O_2 0.2375	0.2375(32) – 7.6				
N_2 0.2375 × 3.762 – 0.8934	0.8934 (28) – 25.017				
	Total – 57.6613			Total – 57.648	

Problem 10.9 From the following percentage composition (by volume) of town gas, determine the calorific value of and calorimetric temperature of combustion.

Also draw the material balance

	H_2	CH_4	CO	CO_2	N_2
Town gas	45	24.5	18	5.5	7

Solution

Step (I) Calorific Value

Basis: 1 Nm^3 of town gas

Town Gas Constituent (Nm^3)	Combustion Reaction	Heat of Combustion (kcal/kg)	Heat Evolved (kcal/Nm^3 of Town Gas)
H_2 – 0.45	$H_2 + 1/2O_2 \longrightarrow H_2O$	28905	$\dfrac{28905\ (0.45)\ (2)}{22.4}$ – 1161.361
CH_4 – 0.245	$CH_4 + 2O_2 \longrightarrow CO_2 + 2H_2O$	12025	$\dfrac{12025(0.245)(16)}{22.4}$ – 2104.375
CO – 0.18	$CO + 1/2O_2 \longrightarrow CO_2$	2436.42	$\dfrac{2436.42\ (0.18)\ (28)}{22.4}$ – 548.194
CO_2 – 0.055			
N_2 – 0.07			
			Σ – 3813.931 kcal/Nm^3 of Town Gas burned

In order to determine the calorimetric temperature of combustion, it is necessary to know the composition of the products of combustion.

Step (II) Air Requirement for Combustion

Basis: 1 Nm^3 of Town Gas.

Town Gas Constituent	Combustion Reaction	Oxygen Requirement (Nm^3)	Air Requirement (Nm^3)
H_2 – 0.45 Nm^3	$H_2 + 1/2O_2 \longrightarrow H_2O$	0.5 (0.45) – 0.225	
CH_4 – 0.245 Nm^3	$CH_4 + 2O_2 \longrightarrow CO_2 + 2H_2O$	2 (0.245) – 0.490	0.805 + 0.805 (3.762) – 3.833 Nm^3
CO – 0.18 Nm^3	$CO + 1/2O_2 \longrightarrow CO_2$	0.5 (0.18) – 0.09	
		Σ – 0.805	

Step (III) Volume of Flue Gas

Basis: 1Nm^3 of Town Gas

Reactant	Combustion Reaction	Volume of the Products of Combustion	
		CO_2	H_2O
H_2 (0.45 Nm^3)	$H_2 + 0.5\ O_2 \longrightarrow H_2O$		0.45 Nm^3
CH_4 (0.245 Nm^3)	$CH_4 + 2O_2 \longrightarrow CO_2 + 2H_2O$	0.245 Nm^3	0.49 Nm^3
CO (0.18 Nm^3)	$CO + 0.5\ O_2 \longrightarrow CO_2$	0.18 Nm^3	—
		Σ – 0.425 Nm^3	Σ – 0.94 Nm^3

Hence the total volume of the flue gas produced due to combustion of 1 Nm³ of town gas

$$= 0.425 + 0.055(CO_2) + 0.94(H_2O) + 0.805\,(3.762) + 0.07(N_2)$$

$$= 4.5184\;Nm^3$$

Step (IV) Material Balance

Basis: 1 Mol. Vol. of Town Gas

Input		Output		
Mol. Vol.	*Weight (kg)*	*Flue Gas*	*Mol. Vol.*	*Weight (kg)*
(A) *Fuel*				
H_2 0.45	0.45(2) − 0.9	CO_2	0.425 + 0.055 − 0.48	0.48(44) − 21.12
CH_4 0.245	0.245(16) − 3.92	H_2O	0.94	0.94(18) − 16.92
CO 0.18	0.18(28) − 5.04	N_2	3.028 + 0.07	3.098(28) − 86.74
CO_2 0.055	0.055(44) − 2.42			
N_2 0.07	0.07(28) − 1.96			
(B) *Air*				
O_2 0.805	0.805(32) − 25.76			
N_2 0.805 × 3.762 − 3.028	3.028(28) − 84.784			
	Σ − 124.784			Σ − 124.784

Step (V) Calorimetric Temperature

1st Approximation: Let the calorimetric temperature of combustion be 2100°C

Flue Gas Constituent	Volume (m³)	Sp. Heat at 2100°C (kcal/m³.°C)	ΔH (kcal)
CO_2	0.48	0.555	0.48 (0.555) (2100) − 559.44
H_2O	0.94	0.49	0.94 (0.49) (2100) − 967.26
N_2	3.098	0.34	3.098 (0.34) (2100) − 2211.972
			Σ − 3738.672

This value is to low.

2nd Approximation: Calorimetric temperature of combustion be 2200°C

Flue Gas Constituent	Volume (m³)	Sp. Heat at 2200°C (kcal/m³.°C)	ΔH (kcal)
CO_2	0.48	0.56	0.48 (0.56) (2200) − 591.36
H_2O	0.94	0.505	0.94 (0.505) (2200) − 1044.34
N_2	3.098	0.343	3.098(0.343) (2200) − 2337.750
			Σ − 3973.45

This value of ΔH suggests that the calorimetric temperature of combustion should lie between 2100°C and 2200°C.

$$\frac{2200 - 2100}{3973.45 - 3738.672} = \frac{t - 2100}{3813.931 - 3738.672}$$

$$\therefore\quad t = 2132°C \qquad Ans.$$

Problem 10.10 A coal sample having the following composition by weight:

C—75%; H—5.25%; O—9.7%; N—1.38%; S—0.95%; Moisture—4.75%; Ash—2.97% is to be burned in air.

Determine the volume of air requirement to burn 1 kg of coal by stoichiometric method (b) mol method.

Solution

(A) Stoichiometric method

Basis: 1 kg coal

Fuel Consti- tuent	Combustion Reaction	O_2 Requirement (kg)	Remarks
C (0.75 kg)	C + O_2 → CO_2 (12) (32) (44)	0.75(32/12) − 2	Already 0.09 kg oxygen is available in c
H (0.0525 kg)	H_2 + 1/2O_2 → H_2O (2) (16) (18)	0.0525(16/2) − 0.42	Therefore, the oxygen to be supplied − 2.4295 − 0.097 − 2.3325 kg
S (0.0095 kg)	S + O_2 → SO_2 (32) (32) (64)	0.0095 (32/32) − 0.0095	
		Σ − 2.4295	

Therefore, the air to be supplied

$= 2.3325\,(100/23)$

$= 10.141$ kg/kg of coal burnt.

Mass of Air Requirement	Molecular Wt. of Air	Density of Air at NTP	Volume of Air Requirement
10.141 kg/kg of coal burnt	$\dfrac{79 \times 28 + 21 \times 32}{79 + 21}$ $- 28.84$	28.84/22.4 $- 1.2875$ kg/Nm^3	10.141/1.2875 $- 7.876$ Nm^3/kg of coal burnt

(B) Mol Method
Basis: 100 kg coal

Fuel Constituent	kmol	Combustion Reaction	Oxygen Required
C $- 75$ kg	75/12 $- 6.25$	$C + O_2 \longrightarrow CO_2$	(+)6.25 kmol
H $- 5.25$ kg	5.25/2 $- 2.625$	$H_2 + 1/2\,O_2$ $\longrightarrow H_2O$	(+)2.625(1/2) $- 1.3125$ kmol
S $- 0.95$ kg	0.95/32 $-$ 0.0297	$S + O_2 \longrightarrow SO_2$	(+)0.0297 kmol
O $- 9.7$ kg	9.7/32 $- 0.303$		$(-)$0.303 kmol
			$\Sigma - 7.2892$ kmol

Therefore, oxygen to be supplied

$= 7.2892\,(22.4)$ Nm^3

Hence, the volume of air required

$= 7.2892(22.4)\,(100/21)$

$= 777.5146$ Nm^3/100 kg of coal burnt

$= 7.775$ Nm^3/kg of coal burnt   *Ans.*

Problem 10.11 If the coal in Problem 10.10 is burned in 50% excess air, what will be the volumetric composition of the dry flue gas?

Solution

Coal (kg)	Stoichiometric Requirement		Nitrogen Supplied (kmol)	With 50% Excess Air	
	Oxygen (kmol)	Air (kmol)		Nitrogen Supplied	Oxygen Supplied
100	7.2892	7.2892 (100/21) $= 34.71$	34.71 $- 7.2892$ $= 27.421$	27.421 (0.5) $- 13.71$ kmol	7.2892 (0.5) $- 3.6446$ kmol

Flue Gas Composition (Dry Basis)	Volumetric Composition
$CO_2 - 6.25$ kmol	6.25/51.055 $- 12.24\%$
$SO_2 - 0.0297$ kmol	0.0297/51.055 $- 0.058\%$
$O_2 - 3.6446$ kmol	3.6446/51.055 $- 7.138\%$
$N_2 - 27.421 + 13.71$ $- 41.131$ kmol	41.131/51.055 $- 80.56\%$
Total $- 51.055$ kmol	

Problem 10.12 A liquid fuel containing only carbon and hydrogen is burnt to produce a flue gas of the following analysis (by volume)

CO_2—11%; O_2—7.5%; N_2—81.5%

Determine

(a) the composition of fuel by weight

(b) the excess air used

Solution

Step (I) Oxygen Content in Flue Gas (Dry)
Basis: 100 kmol of dry flue gas

Flue Gas Constituent	kmol	Oxygen Content (Kmol)	Remarks
CO_2	11	11	$C +$ $O_2 \longrightarrow CO_2$ (1 kmol) (1 kmol) (1 kmol)
O_2	7.5	7.5	Free oxygen
		$\Sigma = 18.5$	

Step (II) Actual Air Used
Basis: 100 kmol dry flue gas

Nitrogen Content	Actual Air Used	Oxygen in Air Supplied
81.5 kmol	81.5 (100/79) $- 103.164$ kmol	103.164 (21/100) $- 21.664$ kmol

Step (III) Hydrogen Content in the Fuel
Basis: 100 kmol dry flue gas

Oxygen Supplied	Oxygen Content in Flue Gas	Oxygen Consumed in Water Formation	Hydrogen Content in Fuel
21.664 kmol	18.5 kmol	21.664 $- 18.5$ $- 3.164$ kmol	2(3.164) $- 6.328$ kmol

Step (IV) Fuel Composition

Basis: 100 kmol fuel

Carbon Content	Hydrogen Content	C : H Ratio
11(12) – 132 kg	6.328(2) – 12.656 kg	132 : 12.656 – 91.25% : 8.75%

Step (V) Percentage Excess Air

Basis: 100 kmol flue gas

Flue Gas Consti- tuent	Oxygen Require- ment	Oxygen Supp- lied	Excess Oxygen Supplied	% Excess Air
CO_2 – 11 kmol	11 kmol	21.664 kmol	21.664 – 14.164 – 7.5 kmol	7.5(100)/ 14.164 – 52.95 Ans.
H_2O – 6.328	$\frac{1}{2}$ (6.328)	– 3.164 kmol		
	Σ – 14.164 kmol			

Problem 10.13 A liquid fuel contains:

C—84%; H—16%

It is burnt with a quantity of excess air so that flue gas contains:

CO_2—12.5%; CO—1.2%; O_2—0.75%; N_2—84.55%; H_2O—rest by volume

Determine

(a) the ratio of air to liquid fuel

(b) the percentage of excess air

Solution

Step (I) Flue Gas (Dry Basis)

Basis: 1 kmol of flue gas

Flue Gas Consti- tuent	kmol*	Weight (kg)	Weight/kg of Flue Gas	Wt. of Carbon/kg of Flue Gas
CO_2	0.125	0.125 (44) – 5.5	5.5/29.75 – 0.1848	0.1848 (12/44) – 0.0504 kg
CO	0.012	0.012 (28) – 0.336	0.336/29.75 – 0.01129	0.01129 (12/28) – 0.0048 kg
O_2	0.0075	0.0075 (32) – 0.24		
N_2	0.8455	0.8455 (28) – 23.674		
		Σ – 29.75		Σ – 0.0552

* (mol per cent – volume per cent)

Step (II) Air : Fuel Ratio

Now, wt. of C/kg of fuel = 0.84

Wt. of C/kg of dry flue gas = 0.0552

Therefore, the wt. of dry flue gas/kg of fuel

= 0.84/0.0552

= 15.217 kg

$$H_2 + \frac{1}{2} O_2 \longrightarrow H_2O$$
$$(2) \qquad\qquad (18)$$

Hence water vapour produced per kg of fuel burnt

= 18(0.16)/2

= 1.44 kg.

Therefore, total weight of flue gas

= 15.217 + 1.44

= 16.657 kg/kg of fuel

Now, 1 kg fuel + air supplied

= 16.657 kg of flue gas

Air supplied = 16.657 – 1 = 15.657 kg/kg of fuel

Therefore, air : fuel ratio = 15.657 : 1

Ans.

Step (III) Stoichiometric Air

Basis: 1 kg fuel

Fuel Consti- tuent	Combustion Reaction	Oxygen Requirement (kg)	Air Requirement (kg)
C – 0.84 kg	$C + O_2 \rightarrow CO_2$ 12 32	(0.84 × 32/ 12) – 2.24	2.24 × 100/23 – 9.739
H – 0.16 kg	$H_2 + 1/2\,O_2$ $\longrightarrow H_2O$ (2) (16)	(0.6 × 16/2) – 1.28	1.28 × 100/23 – 5.565
			Σ = 15.304

Step (IV) Percentage Excess Air

Basis: 1 kg Fuel

Air Supplied	Air Requirement	Excess air	% Excess Air
15.657 kg	15.304 kg	15.657 – 15.304 – 0.353	(0.353/15.304)(100) – 2.30 Ans.

Problem 10.14 The analysis of coal fired in a boiler and the flue gas produced are as follows:

Coal C—25%; H—5%; O—7.5%; Non-combustibles—62.5% (by weight)

Flue Gas CO_2—11%; CO—1.5%; N_2—83%; O_2— 4.5% (by volume)

Determine

(a) the weight of air supplied per kg of coal burned

(b) the per cent excess air

Solution
Step (I) Oxygen Requirement
Basis: 1 kg coal

Fuel Constituent	Combustion Reaction	Oxygen Requirement (kg)	Remarks
C (0.25 kg)	C + O$_2$ → CO$_2$ 12 32 44	0.25(32/12) – 0.666	Already 0.075 kg O$_2$ is available in coal
H (0.05 kg)	H$_2$ + 1/2O$_2$ → H$_2$O 2 16 18	0.05(16/2) – 0.4	Hence actual O$_2$ requirement – 1.066 – 0.075 – 0.9916 kg/kg of coal
O (0.075 kg)	—	—	
Ash (0.625 kg)	—	—	
		Σ – 1.066	

Step (II) Air Requirement
The theoretical amount of air requirement for combustion
= 0.9916(100/23)
= 4.311 kg/kg of coal
(cf. Air contains N_2 and O_2 in approximately 77 : 23 ratio by weight

Step (III) Air Supplied
Basis: 100 kg of fuel
Working in kmol we get, the overall combustion reaction as:

$$\frac{25}{12}C + \frac{5}{2}H + \frac{7.5}{32}O_2 + aO_2 + 3.76a\,N_2$$
$$\longrightarrow b\,CO_2 + d\,CO + e\,O_2 + f\,N_2 + g\,H_2O$$

Equating the coefficients of carbon:

$$b + d = \frac{25}{12} = 2.08 \qquad (I)$$

Since volume % = mol % so,

$$CO_2 = \frac{b}{b+d+e+f} = \frac{11}{100} = 0.11;$$

$$CO = \frac{d}{b+d+e+f} = \frac{1.5}{100} = 0.015$$

$$O_2 = \frac{e}{b+d+e+f} = \frac{4.5}{100} = 0.045;$$

$$N_2 = \frac{f}{b+d+e+f} = \frac{83}{100} = 0.83$$

From these relationships we get:

$$b/d = 0.11/0.015 = 7.33 \qquad (II);$$
$$f/d = 0.83/0.015 = 55.33 \qquad (III)$$

From equations (I) and (II) we get, $d = 0.249$
Therefore, from equation (III), $f = 13.77$
Again, $3.76\,a = f = 13.77$; ∴ $a = 3.66$
Therefore, air supplied per 100 kg fuel

= $a(32) + f(28)$

= 502.819 kg

Hence, air supplied per kg of fuel = 5.028 kg

Step (IV) Excess Air

Basis: 1 kg fuel

Air Supplied	Actual Air Requirement	Excess Air	% Excess Air
5.028 kg	4.311 kg	5.028 – 4.311 – 0.717 kg	(0.717/4.311)(100) – 16.63

Problem 10.15 A flue gas has the following analysis on dry basis CO_2—11%; CO—1.2%; O_2—7.2%; N_2—80.6% (by volume)

If the fuel contains 87.6% carbon by mass, determine the percentage of carbon burnt to carbon monoxide by mass.

Also determine the mass of dry flue gas per kg of fuel burnt.

Solution

Step (I) Total Carbon in Flue Gas

Basis: 100 kmol flue gas

Flue Gas Constituent	kmol	Weight (kg)	Carbon Content (kg)	C Per kg Dry Flue Gas (kg)
CO_2	11	11(44) = 484	484(12/44) = 132	146/3004.8
CO	1.2	1.2(28) = 33.6	33.6(12/28) = 14.4	= 0.04872 kg
O_2	7.2	7.2(32) = 230.4	—	
N_2	80.6	80.6(28) = 2256.8	—	
		= 3004.8	= 146.4	

Step (II) Per cent of Carbon Burnt to Carbon Monoxide

[14.4/146.4]/100 = 9.83%   *Ans.*

Step (III) Mass of Dry Flue Gas

As only the carbon content in the fuel burns to produce dry flue gas, so the dry flue gas produced per kg of fuel

= 0.876/0.04872 kg

= 17.98 kg   *Ans.*

Problem 10.16 A fuel (C—83%; H—16%; Non-combustibles—1%) was completely burnt with excess air in a furnace. On a testing analysis, the dry flue gas registered CO_2–7.98% by volume.

Calculate the amount of air supplied per kg of fuel and the products of combustion per kg of fuel.

Solution

Step (I) Oxygen Requirement

Basis: 100 kg of fuel

Let the supplied air contained *a* kmol of oxygen. Therefore, it was associated with 3.76(a) kmol of nitrogen (cf. N_2/O_2 = 79/21 = 3.76)

Computing the combustion reaction in kmol we get,

$$\frac{83}{12} C + \frac{16}{2} H_2 + a O_2 + 3.76a N_2$$

$$\longrightarrow b CO_2 + d O_2 + 3.76a N_2 + \frac{16}{2} H_2O$$

Equating the coefficients of carbon: $b = 83/12$
= 6.916

Equating the coefficients of oxygen:

$$a = b + d + \frac{8}{2}$$

$$= 6.916 + d + 4$$

$$= 10.916 + d \qquad (I)$$

Again, $\dfrac{b}{b + d + 3.76a} = \dfrac{7.98}{100} = 0.0798$

or

6.916/[6.916 + d + 3.76 (10.916 + d)] = 0.0798
[Putting the value of *a* from Eq. (I)]

$d = 8.13$ kmol

$a = 10.916 + 8.13 = 19.047$ kmol

Step (II) Air Supplied

Basis: 100 kg fuel

Oxygen Supplied	Nitrogen Supplied	Total Air Supplied
19.047 kmol	3.76(19.047) kmol	609.524 + 2005.268
= 19.047(32) kg = 609.524 kg	= 3.76 (19.047) (28) kg = 2005.2681 kg	= 2614.792 kg

Therefore, mass of air supplied per kg of fuel

= 26.147 kg   *Ans.*

Step (III) Combustion Products

Basis: 100 kg fuel

Combustion Products	kmol	Weight (kg)	Per kg Fuel
CO_2	6.916	6.916(44) = 304.304	3.04 kg
H_2O	8	8(18) = 144	1.44 kg
O_2	8.13	8.13(32) = 260.16	2.60 kg
N_2	3.76 × 19.047	3.76 × 19.047 × 28 = 2005.268	20.05 kg
Ash	—	1.00	0.01 kg

Problem 10.17 A fuel gas has the following analysis by volume:

H_2—51%; CH_4—22%; CO—15%; N_2—5%; CO_2—3%; C_2H_4—2%; O_2—2%

Calculate the minimum volume of air required for complete combustion of 1 m³ of the fuel gas.

If 5 m³ of air be supplied per m³ of the fuel gas, determine the percentage composition of the combustion products by volume.

Solution

Step (I) Oxygen Requirement and Combustion Products

Basis: 100 kmol of fuel gas

Fuel Gas Constituent	kmol	Combustion Reaction	O_2 Requirement (kmol)	Combustion Products	
				CO_2 (kmol)	H_2O (kmol)
H_2	51	$H_2 + 1/2O_2 \longrightarrow H_2O$	$51/2$ = 25.5	—	51
CH_4	22	$CH_4 + 2O_2 \longrightarrow CO_2 + 2H_2O$	$2(22)$ = 44	22	2×22 = 44
CO	15	$CO + 1/2O_2 \longrightarrow CO_2$	$15/2$ = 7.5	15	—
N_2	5			5 (Nitrogen)	
CO_2	3			3	—
C_2H_4	2	$C_2H_4 + 3O_2 \longrightarrow 2CO_2 + 2H_2O$	$3(2)$ = 6	4	4
O_2	2		(–)2		
			Σ = 81 kmol	Σ = 44(CO_2) + 5(N_2) + 99(H_2O)	

Step (II) Minimum Amount of Air Requirement

Fuel	O_2 Requirement	Air Requirement
100 kmol	81 kmol	
100 m³	81 m³	
1 m³	0.81 m³	0.81 (100/21) = 3.857 m³
		Ans.

Step (III) Excess Air

Theoretical air requirement = 3.857 m³/m³ of fuel
Supplied air = 5 m³/m³ of fuel
Therefore, excess air

= 5 – 3.857 = 1.14 m³/m³ of fuel

= 114 kmol/100 kmol of fuel

Step (IV) Percentage Composition of Flue Gas

Basis: 100 kmol of fuel gas

Combustion Products	Composition (kmol)	Percentage Composition
CO_2	44	44(100)/566.703 = 7.76%
N_2	$5 + \dfrac{385.7 \ (79)}{100}$ = 309.703	$\dfrac{309.703 + 90.06}{566.703}(100)$ = 70.54%
H_2O	99	73.5(100)/566.703 = 17.47%
N_2 (excess air)	114 (79/100) = 90.06	
O_2 (excess air)	114 (21/100) = 23.94	23.94 (100)/566.703 = 4.22%
	Σ = 566.703	

Problem 10.18 A liquid fuel upon combustion in excess air produces flue gas of following percentage composition (by volume) on dry basis:

CO_2—8.86%; CO —1.25%; O_2—7%; N_2—82.89%

The liquid fuel has the following percentage composition (by mass):

C—84%; H—15%; O—1%

Determine the mass of air supplied per kg of fuel burnt.

Solution

Step (I) Oxygen Requirement

Fuel Constituent	Mass kg of Fuel	Combustion Reaction	Oxygen Requirement Per kg Constituent (kg)	Fuel (kg)
C	0.84	$C + O_2 \longrightarrow CO_2$	32/12 = 2.666	2.666 (0.84) = 2.24
H	0.15	$H_2 + 1/2O_2 \longrightarrow H_2O$	$\dfrac{32 \ (1/2)}{2}$ = 8	8(0.15) = 1.2
O	0.01	—		(–) 0.01
				Σ = 3.43

Step (II) Theoretical Air Requirement

Theoretical oxygen requirement

= 3.43 kg/kg of fuel

Theoretical air requirement = 3.43(100/23)

= 14.91 kg/kg of fuel

Step (III) Flue Gas Analysis

Flue Gas Constituent	% by Volume	(% by Volume) (Mol Wt.)	% by Mass
CO_2	8.86	8.86(44) – 389.84	$\dfrac{389.84}{2969.76}$ (100) – 13.127
CO	1.25	1.25(28) – 35	$\dfrac{35}{2969.76}$ (100) – 1.178
O_2	7	7(32) – 224	$\dfrac{224}{2969.76}$ (100) – 7.542
N_2	82.89	82.89(28) – 2320.92	$\dfrac{2320.92}{2969.76}$ (100) – 78.151
		Σ – 2969.76	

Step (IV) Excess Air

Basis: 1 kg dry flue gas

Flue Gas Constituent	Mass (kg)	Mass of Carbon (kg)
CO_2	0.13127	0.13127 (12/44) – 0.0358
CO	0.01178	0.01178 (12/28) – 0.00504
		Σ – 0.04084

Mass of C/kg of fuel = 0.84 kg

Mass of dry flue gas/kg of fuel = 0.84/0.04084

= 20.563 kg

Mass of excess O_2/kg of fuel = 20.563(0.07542)

= 1.5509 kg

Mass of excess air/kg of fuel = 1.5509(100/23)

= 6.743 kg

Ans.

Step (V) Air Supplied

Stoichiometric air/kg of fuel = 14.91 kg

Mass of excess air/kg of fuel = 6.743 kg

Total air supplied = 14.91 + 6.743 = 21.653 kg

Ans.

EXCESS AIR (Alternative Method)

Nitrogen Balance Method

Nitrogen in dry flue gas = Total N_2 from air

Mass of N_2/kg fuel = 20.563 (0.78151) = 16.07 kg

Mass of air/kg fuel = 16.07(100/77) = 20.87 kg

Ans.

Problem 10.19 The fuel supplied to a boiler contains 79% carbon, 7% hydrogen, 8% oxygen and 6% ash (by mass) as fired. The supplied air is 50% in excess of the stoichiometric air. Determine the mass of dry flue gas per kg of fuel.

If the boiler house temperature is 25° C and the flue gas temperature is 300° C, calculate the energy carried away by the dry flue gas per kg fuel burned.

Gas	Specific Heat (kcal/kg °C)
CO_2	0.222
N_2	0.242
O_2	0.313
Air	0.237

Solution

Step (I) Stoichiometric Oxygen Required

Basis: 1 kg fuel

Fuel Constituent	Mass (kg)	Combustion Reaction	Oxygen Requirement (kg)
C	0.79	$C + O_2 \longrightarrow CO_2$ 12 32	(32/12) (0.79) – 2.106
H	0.07	$H_2 + 1/2O_2 \longrightarrow H_2O$ 2 16	(16/2) (0.07) – 0.56 (–)0.08
O	0.08		Σ – 2.586

Step (II) Stoichiometric Air Required

2.586(100/23) = 11.246 kg/kg of fuel

Step (III) Mass of Dry Flue Gas Per kg of Fuel

Basis: 1 kg fuel

Stoichio-metric Air (kg)	Actual Air Supplied (kg)	Total Mass of Flue Gas	Mass of Water Produced (kg)	Mass of Dry Flue Gas (kg)
11.246	11.246 (1.5) – 16.869	16.869(Air) + 1(Fuel) – 17.869 kg	0.07(18/2) – 0.63	17.869 – 0.63 – 17.239

Ans.

Step (IV) Energy Content of Dry Flue Gas
Basis: 1 kg fuel

Dry Flue Gas Constituent	Mass (kg)	Specific Heat (kcal/kg.°C)	$\Delta\Theta$ (°C)	Heat Content (kcal)
CO_2	$\dfrac{44(0.79)}{12}$ $= 2.896$	0.222	275	2.896(0.222)(275) $= 176.841$
N_2	$\dfrac{11.246(77)}{100}$ $= 8.659$	0.242	275	8.659(0.242)(275) $= 576.284$
Air	$\dfrac{11.246(50)}{100}$ $= 5.623$	0.237	275	5.623(0.237)(275) $= 366.479$
				$\Sigma = 1119.604$

Problem 10.20 The analysis of coal fired in a boiler shows the following composition by mass:

Carbon—81%; Hydrogen—9%; Oxygen—2%; Ash—8%

The rate of coal consumption in the boiler is 0.9 t/h and air supplied by a blower is 30% in excess of stoichiometric air.

Calculate:
(a) the intake air volume when the intake conditions at the blowers are $100 \, kN/m^2$ and 291°K. ($R_{air} = 0.287 \, kJ/kg°K$)
(b) the percentage composition (by mass) of dry flue gas

Solution
Step (I) Stoichiometric Oxygen
Basis: 1 kg coal

Coal Composition	Mass (kg)	Combustion Reaction	Oxygen Requirement (kg)
C	0.81	$C + O_2 \longrightarrow CO_2$ $\quad 12 \quad 32$	(32/12) (0.81) $= 2.16$
H	0.09	$H_2 + 1/2O_2 \longrightarrow H_2O$ $\quad 2 \quad\quad 16$	(16/2) (0.09) $= 0.72$
O	0.02		(−) 0.02
			$\Sigma = 2.86$

Step (II) Theoretical Air Requirement
Basis: 1 kg coal

Theoretical Oxygen Requirement	Theoretical Air Requirement
2.86 kg	2.86(100/23) $= 12.434$ kg

Step (III) Supplied Air
Basis: 1kg coal

Theoretical Air	Excess Air	Supplied Air
12.434 kg	12.434 (30/100) $= 2.73$ kg	12.434 + 3.73 $= 16.164$ kg

Hence the air supplied per second

$= 16.164 \,(0.9)\,(1000)/3600$ kg

$= 4.041$ kg

Step (IV) Volume of Supplied Air
Working Formula: $PV = mRT$

P	m	R	T	V (volume flowrate)
100 kN/m^2	4.041 kg/s	0.287 kJ/kg K	291° K	$\dfrac{4.041\,(0.287)\,(291)}{100}$ $= 3.375 \ m^3/s$

Step (V) Dry Flue Gas Composition
Basis: 1kg coal

Dry Flue Gas Composition	Mass (kg)	% Composition by Mass
CO_2	(44/12) (0.81) $= 2.97$	(2.97/16.274)(100) $= 18.25$ *Ans.*
O_2 ← from excess air	2.86 (0.3) $= 0.858$	(0.858/16.274)(100) $= 5.27$ *Ans.*
N_2 ←	$\dfrac{16.164\,(77)}{100}$ $= 12.446$	(12.446/16.274)(100) $= 76.476$ *Ans.*
	$\Sigma = 16.274$	

Problem 10.21 A boiler generates 2500 kg of steam per hour—the steam being 97% dry at pressure $1.4 \, MN/m^2$.

The boiler feedwater temperature = 328°K
Boiler efficiency = 72%

Calorific value of coal = 27500 kJ/kg
Coal composition by mass:
C—84%; H—6%; O—3%; Ash—7%
Air supply: 25% in excess of stoichiometric air.
Specific heat capacity of water = 4.187 kJ/kgK
Calculate:
(a) the feed rate of coal per hour
(b) the mass of air supplied per hour
(c) the percentage analysis of flue gases by mass

Solution

Step (I) Energy to Generate Steam

From steam table

Abs. Press. (MN/m^2)	Satd. Temp. $(°C)$	Sp. Enthalpy (kJ/kg) H_w	L
1.4	195	830	1957.7

Specific enthalpy of generated steam

$= 830 + 1957.7 (0.97)$ kJ/kg

$= 2728.970$ kJ/kg

Sp. Enthalpy of Generated Steam	Sp. Enthalpy of Feedwater	Energy to Generated Steam	Energy to Steam per Hour
2728.97 kJ/kg	4.187 (328 – 273) kJ/kg – 230.285 kJ/kg	2728.97 – 230.285 – 2498.685 kJ/kg	2498.685 (2500) kJ/h

Step (II) Mass of Coal Consumption per Hour

Energy Supplied to Steam per Hour (kJ/kg)	Boiler Efficiency	Energy Required from Coal Feed (kJ/kg)	Calorific Value of Coal	Feedrate of Coal per Hour (kg/h)
2498.685 (2500)	72%	2498.685 (2500)/0.72	$27500 \frac{kJ}{kg}$	$\frac{2498.685(2500)}{0.72(27500)}$ – 315.49 *Ans.*

Step (III) Theoretical Air Requirement

Basis: 100 kg coal

Coal Constituent	Mass (kg)	Combustion Reaction	Oxygen Requirement (kg)	Theoretical Air Requirement
C	84	$C + O_2 \rightarrow CO_2$ 12 32	(32/12)(84) – 224	269(100/23) – 1169.565 kg
H	6	$H_2 + \frac{1}{2}O_2$ 2 16 $\rightarrow H_2O$	(16/2)(6) – 48	
O	3		(–)3	
			Σ – 269	

Step (IV) Supplied Air

Theoretical Air per kg of Coal	Excess Air (kg)	Mass of Supplied Air per kg Coal	Coal Consumption	Air Supplied
11.6956 kg	25% (11.6956) – 2.9239	11.6956 + 2.9239 kg – 14.6195 kg	315.49 kg/h	14.6195 × 315.49 – 4612.31 kg

Step (V) Flue Gas Composition

Basis: 1 kg coal

Coal Constituent	Mass (kg)	Combustion Reaction	Flue Gas Produced			
			H_2O	CO_2	N_2	O_2
C	0.84	$C + O_2$ (12) $\longrightarrow CO_2$ (44)		$\frac{44(0.84)}{12}$ – 3.08 kg	(from total air supply) 14.62 × 0.77 –11.25 kg	(from excess air) 2.923 × 0.23 –0.67 kg
H	0.06	$H_2 + \frac{1}{2}O_2$ $\longrightarrow H_2O$	$\frac{18(0.06)}{2}$ – 0.54 kg			

Basis: 1 kg coal

Flue gas Composition	Mass (kg)	Per cent Composition
CO_2	3.08	3.08 (100/15.549) – 19.80
H_2O	0.54	0.54 (100/15.549) – 3.47
N_2	11.257	11.257 (100/15.549) – 72.39
O	0.67	0.67 (100/15.549) – 4.30
	Σ – 15.549	

Problem 10.22 From the two following combustion reactions:

$$C + 1/2O_2 \longrightarrow CO + 2452.5 \text{ kcal/kg of C}$$
(s) (g) (g)

$$C + O_2 \longrightarrow CO_2 + 8137.5 \text{ kcal/kg of C}$$
(s) (g) (g)

determine the calorific value of carbon monoxide. Check the result with the following:

$$CO + 1/2O_2 \longrightarrow CO_2 + 2436.42 \text{ kcal/kg of CO}$$
(g) (g) (g)

Solution

Step (I) The calorific value of carbon monoxide is determined with the help of the following combustion reaction

$$CO + \frac{1}{2}O_2 \longrightarrow CO_2 + Q$$

Replacing CO and CO_2 from 1st and 2nd reactions we get,

$$(C + \frac{1}{2}O_2 - 2452.5 \text{ kcal/kg of C}) + \frac{1}{2}O_2$$

$$= (C + O_2 - 8137.5 \text{ kcal/kg of C}) + Q$$

or $Q = (8137.5 - 2452.5)$ kcal/kg of C

$$= 5685 \text{ kcal/kg of C}$$

Step (II) Check

$$C + \frac{1}{2}O_2 \longrightarrow CO$$
(12) (28)

$$\therefore \quad Q = 5685 \frac{12}{28} \text{ kcal/kg of CO}$$

$$= 2436.4285 \text{ kcal/kg of CO which is in excellent agreement with the check-data.}$$

Q. *What is the Higher Calorific Value (HCV) or Gross Calorific Value of fuel?*

Ans. It is the total energy liberated by the complete combustion of 1 kg of solid/liquid fuel or 1 m³ of gaseous fuel and, in all cases, when the combustion products are cooled to the original fuel temperature. (The standard temperature is 15°C, usually)

Q. *What is the Lower or Net Calorific Value (LCV) of fuel?*

Ans. It is the net energy liberated by the complete combustion of 1 kg of solid/liquid fuel or 1 m³ of gaseous fuel and, in all cases when the combustion products are cooled to 100°C without the condensation of steam.

Q. *What is the difference between heat evolved at constant volume and heat evolved at constant pressure?*

Ans. Heat evolved at constant pressure
= Calorific value at constant pressure
Heat evolved at constant volume
= Calorific value at constant pressure + work done
= Calorific value at constant pressure + PV – work
= Calorific value at constant pressure + $\frac{n}{M}(1.985)$ T

(cf.: $PV = nRT = \frac{n}{M}(R_m)T$; where R_m = molar gas constant = 1.985 kcal/kmol°C
M = molecular weight of the gas)

Problem 10.23 Determine the higher calorific value of carbon when it is burnt to
(a) CO_2
(b) CO
at constant volume.
Given HCV at constant pressure for
$C + O_2 \longrightarrow CO_2$ is 8137.5 kcal/kg of °C
$C + 1/2 O_2 \longrightarrow CO$ is 2452.5 kcal/kg of °C

Solution

Step (I) $\boxed{C \longrightarrow CO_2}$
$C + O_2 \longrightarrow CO_2 + 8137.5 \text{ kcal/kg of C}$
(s) (g) (g)

Here, Δ volume $= 1 - 1 = 0$

$$[HCV]_{vol} = [HCV]_{press} = 8137.5 \text{ kcal/kg of C}$$
Ans.

Step (II) $\boxed{C \longrightarrow CO}$

$$C + 1/2\,O_2 \longrightarrow CO + 2452.5 \text{ kcal/kg of C}$$
$$(s) \quad (g) \qquad\qquad (g)$$

Here, Δ volume $= 1 - \dfrac{1}{2} = \dfrac{1}{2}$ kmolar volume

$$[HCV]_{vol} = 2452.5 + \left(\frac{1}{2}\right)\left(\frac{1.985}{12}\right)(273) \text{ kcal/kg } ^\circ C$$

$$= 2475.07 \text{ kcal/kg of } ^\circ C$$
Ans.

Problem 10.24 A fuel contains 86% carbon and the rest hydrogen. Determine its probable formula.
Estimate the
(a) $[HCV]_{press}$.
(b) $[HCV]_{vol}$.

Given:

$$C \longrightarrow CO_2 \ [HCV]_{press} = 8137.5 \text{ kcal/kg of C}$$
$$(s) \qquad (g)$$

$$H_2 \longrightarrow H_2O \ [HCV]_{press} = 34180 \text{ kcal/kg of H}$$
$$(g) \qquad (l)$$

Solution
Step (I) Formula of the Fuel
Let the formula of the fuel be C_xH_y.

$$x:y :: \frac{86}{12} : \frac{14}{1} = 7.16 : 14 = 1 : 2$$

C_xH_y is CH_2. Since no fuel can exists as such, C_xH_y becomes C_2H_4, which is ethylene.

Step (II) Calorific Value at Constant Pressure
Combustion reaction of ethylene is

$$C_2H_4 + 3O_2 \longrightarrow 2CO_2 + 2H_2O$$
$$(g) \qquad (g) \qquad (g) \qquad (l)$$

$$[HCV]_{press}^{C_2H_4} = [HVC]_{press}^{C \longrightarrow CO_2} + [HCV]_{press}^{H_2 \longrightarrow H_2O}$$

$$= 8137.5 \text{ kcal/kg of C} + 34180 \text{ kcal/kg of H}$$

Molecular weight of ethylene is

$$2 \times 12 + 4 \times 1 = 28$$

$$28 \text{ kg } C_2H_4 = 24 \text{ kg C} + 4 \text{ kg H}$$

$$[HCV]_{press}^{C_2H_4} = (8137.5)(24/28) + (34180)(4/28)$$
kcal/kg of C_2H_4

$$= 11857.85 \text{ kcal/kg of } C_2H_4$$
Ans.

Step (III) Calorific Value at Constant Volume

$$[HCV]_{vol}^{C_2H_4} = [HCV]_{press}^{C_2H_4} + [PV - workdone]$$

Now, $PV - \text{workdone} = \dfrac{n}{M}(1.985)\,T$

Because of combustion reaction,

$$n = \underset{(CO_2)}{[2]} - \underset{(O_2)}{[3} + \underset{(C_2H_4)}{1]} = -2$$

$$PV - \text{workdone} = -\frac{2}{28}(1.985)(273) \text{ kcal/kg of } C_2H_4$$

$$= -38.7075 \text{ kcal/kg of } C_2H_4$$

$$[HCV]_{vol}^{C_2H_4} = 11857 - 38.7$$

$$= 11818.3 \text{ kcal/kg of } C_2H_4$$
Ans.

Problem 10.25 A coal containing 81% C, with a calorific value 32400 kJ/kg is burnt to produce a flue gas of the following composition by volume:
CO_2—13%; CO—1.25%; O_2—5.5%; N_2—80.25%

Calculate the
(a) weight of carbon converted to CO per kg coal
(b) percentage heat lost due to incomplete combustion

$$C + O_2 \longrightarrow CO_2 + 34071.71 \text{ kJ/kg of C}$$
$$(s) \quad (g) \qquad (g)$$

$$C + \frac{1}{2}O_2 \longrightarrow CO + 10268.61 \text{ kJ/kg of C}$$
$$(s) \quad (g) \qquad\quad (g)$$

Solution

Step (I) Carbon $\longrightarrow$ CO

Fraction of coal C burnt to CO

$$= \frac{\%CO}{\% \, CO + \% \, CO_2} = \frac{1.25}{1.25 + 13}$$

$= 0.0877$ kg/kg of C burnt

Therefore, in burning 1 kg of coal, i.e. 0.81 kg C, the weight of carbon transforming to CO is

$= 0.0877(0.81)$

$= 0.071$ kg/kg of coal

Ans.

Step (II) Per cent Heat Loss

Due to conversion of C to CO, heat lost

$= 34071.71 - 10268.61$ kJ/kg of C

$= 23803.1$ kJ/kg of C

Heat lost per kg coal burnt $= 23803.1 \, (0.071)$

$= 1690$ kJ

$\therefore$ Heat Loss $= \dfrac{1690}{32400} \, (100) = 5.216$

Ans.

Problem 10.26 A solid fuel contains 74% carbon and 16% ash. The ash discharged from the furnace contains 20% carbon.

Estimate:

(a) the weight of carbon lost in the ash per kg of fuel
(b) the percentage carbon burned
(c) the heat lost by the incomplete combustion

Solution

Step (I) Carbon Associated with Ash

If it is assumed that the ratio of carbon associated with ash is the same both in the solid fuel and the ash discharged then,

$$\frac{x}{16} = \frac{20}{(100 - 20)}$$

where x is the carbon associated with ash in 100 kg solid fuel.

$\therefore$ $x = 4$ kg C/100 kg of fuel

Step (II) Carbon Lost in Ash

Fuel	C Lost in Ash	
100 kg	4 kg	
1 kg	0.04 kg	*Ans.*

Step (III) Per cent Carbon Burned

Fuel	Carbon Content	Effective Carbon
100 kg	74 kg	$74 - 4 = 70$ kg

$\therefore$ %C burnt $= (70/74) \, (100) = 94.59$

Ans.

Step (IV) Heat Lost Due to Incomplete Combustion

The loss of heat due to incomplete combustion is due to slippage of carbon into ash.

Thus heat lost $= (4/100) \, (8137.5)$ kcal/kg of fuel

$= 325.5$ kcal/kg of fuel.

Therefore, per cent heat loss due to incomplete combustion

$= (325.5/8137.5) \, (100) = 4\%$

Ans.

Problem 10.27 Hot air is blown over red-hot coke (C—84%; Ash—10%) to manufacture producer gas whose composition is as given below:

CO —26%; CO_2—5%; H_2—16%; N_2—51%; CH_4—2%

The calorific value of the gas is 5900 kJ/Nm^3 and the ash discharged from the plant contains 15% carbon.

Calculate:

(a) the mass of carbon gasified per kg of coke
(b) the volume of gas produced per kg of coke feed
(c) the heat available in the gas per kg of coke fired

Solution

Step (I) Mass of Carbon Gasified

Assuming the ratio of carbon associated with ash is the same in both the coke and the final ash discharged,

$$\frac{x}{10} = \frac{15}{100 - 15}$$

Where x is the carbon per 100 kg of coke

$x = 1.764$ kg C/100 kg coke

Coke	Total Carbon	Carbon Lost to Ash	Carbon Gasified
100 kg	84 kg	1.764 kg	84 – 1.764 – 82.236 kg

Mass of carbon gasified = 0.822 kg/kg of coke

Step (II) Volume of Producer Gas Produced
   Basis: 100 kmol of producer gas

Gas Composition	Kmol	Carbon Content (Kmol)
CO	26	26
CO_2	5	5
H_2	16	
N_2	51	
CH_4	2	2
		$\Sigma - 33$

Basis: 100 kg fuel

Fuel Composition	Carbon Content	Carbon Gasified	Kmol C Gasified
C	84 kg	82.236 kg	82.236/12 – 6.853

CARBON   PRODUCER GAS

33 kmol | generating | 100 kmol

∴   6.853" $100(6.853/33) = 20.766$ kmol

$$= 20.766 (22.4) \text{ Nm}^3$$

$$= 465.173 \text{ Nm}^3$$

Therefore, the volume of producer gas manufactured per kg of coke feed

$$= 4.6517 \text{ Nm}^3$$

Ans.

Step (III) Heat Available in the Gas
   Basis: 1 kg coke

Gas Produced	Calorific Value	Heat in the Gas
4.6517 Nm^3	5900 kJ/Nm^3	4.6517 (5900) kJ
		– 27445.22 kJ
		– 27.44 MJ *Ans.*

Problem 10.28 In a combustion chamber a mixture of air-benzene vapour is burned. Calculate the available heat in the air-benzene mixture at NTP taking theoretical minimum air for combustion of benzene (lower calorific value = 40363 kJ/kg).

Also determine the percentage change of volume due to combustion.

Solution
Step (I) Combustion of Benzene
   The chemical reaction accompanying the combustion of benzene is

$$C_6H_6 + \frac{15}{2} O_2 \longrightarrow 6 CO_2 + 3H_2O + 40363$$

$$\text{kJ/kg of benezene}$$

Now that this oxygen comes from air, the overall reaction is

$$C_6H_6 + (15/2) O_2 + 3.76 (15/2) N_2 \longrightarrow$$

$$6 CO_2 + 3H_2O + 3.76 (15/2) N_2 + 40363$$
$$\text{kJ/kg of benezene}$$

(cf. kmol of N_2 in air/kmol of O_2 in air = 3.76)

Step (II) Calorific Value of Air-Benezene Mixture
   Air - Benzene Vapour Mixture

Total Kmol of Reactants	Calorific Value
1 + 15/2 + 3.76 (15/2) – 36.7	40363 kJ/kg of C_6H_6 – 40363 × 78 kJ/kmol of C_6H_6 i.e. 40363 × 78 kJ/kmol of above air-benzene mix

Hence the available heat in 1 Nm^3 of air-benzene vapour mixture

$$= 40363 \times 78/(36.7 \times 22.4)$$

$$= 3829.69 \text{ kJ/Nm}^3$$

Ans.

Step (III) Volume Change Due to Combustion

Reactants (Total Kmol)	Products (Total Kmol)	Kmol	% Volume
36.7	6 + 3 + 3.76 (15/2) – 37.2	37.2 – 36.7 – 0.5	(0.5/36.7)100 – 1.362 *Ans.*

Problem 10.29 In a standard engine a mixture of air and n-heptane vapour is tested taking air: n-heptane ratio 20 : 1 by mass.

Determine the calorific value/m³ of the mixture at NTP.

Calorific value of n-heptane is 45220 kJ/kg

Also determine the percentage composition (by volume) of the dry exhaust gas assuming complete combustion of n-heptane.

Solution
Step (I) Combustion Reaction of n-Heptane

Taking 100 kg n-heptane and working in kmols we get,

$$\frac{100}{100} C_7H_{16} + a\, O_2 + 3.76a\, N_2$$

$$\longrightarrow b\, CO_2 + d\, O_2 + 3.76a\, N_2 + e\, H_2O$$

where a, b, c, d and e denote the number of kmols

Step (II) Oxygen Supplied

n-Heptane	Air	Oxygen
100 kg	20 (100) kg	20(100) (23/100) kg
		= 20(23) kg = 20 (23)/32 kmol
		= 14.375 kmol = a

Step (III) Calorific Value
Basis: 100 kg Fuel

Reactants	K mol	Volume at NTP
C_7H_{16}	1	69.425×22.4 m³
		= 1555.12 m³
O_2	14.375	
N_2	14.375 (3.76)	
	= 54.05	
	$\Sigma = 69.425$	

Calorific value of 1 kmol of fuel (100 kg n-heptane) + air mixture occupying 1555.12 m³ at NTP is
= 45220 (100) kJ

Calorific value of 1 Nm³ of air-heptane mixture
= 45220 (100)/1555.12 kJ
= 2907.814 kJ

Ans.

Step (IV) Percentage Composition (by Volume) of the Dry Exhaust Gas

By equating the coefficients of carbon and nitrogen, we get

$b = 7$ kmol

$3.76\, a = 3.76\, (14.375) = 54.05$ kmol

$d = a -$ kmol of stoichiometric O_2

Now, $C_7H_{16} + 11 O_2 \longrightarrow 7 CO_2 + 8H_2O$

$d = a - 11 = 14.375 - 11 = 3.375$ kmol

Basis: 100 kg fuel

Combustion Products	Kmol	Percent by Volume
CO_2	$b = 7$	7(100/64.425) = 10.86
O_2	$d = 3.375$	3.375 (100/64.425) = 5.24
N_2	3.76a = 54.05	54.05 (100/64.425) = 83.89
	$\Sigma = 64.425$	

Problem 10.30 In a certain coal fired furnace, the coal has the following composition by mass:

C—83.5%; H—6.5%; O—4%; Moisture—3.5%; Ash—2.5%

Determine the theoretical temperature of combustion and the mean specific heat of the flue gas if the combustion is complete with minimum air and 30% excess air.

Given: Calorific value of coal = 7550 kcal/kg

$[C_p]_{CO_2} = [C_p]_{N_2} = [C_p]_{O_2} = 0.241$ kcal/kg°C

$[C_p]_{H_2O} = 0.469$ kcal/kg °C

Air contains O_2 21% by volume and 23% by mass.

Solution:
Step (I) $N_2 : O_2$ Mol Ratio in Air

Since volume per cent = mol per cent, so
kmol of N_2/kmol of O_2 = 79/21 = 3.76

Step (II) Combustion Equation
Basis: 100 kg coal

$$\frac{83.5}{12} C + \frac{6.5}{2} H_2 + \frac{4}{32} O_2 + x\, O_2 + 3.76\, x\, N_2 + \frac{3.5}{18} H_2O$$

$$\longrightarrow y\, CO_2 + 3.76\, x\, N_2 + \frac{3.5}{18} H_2O + p\, H_2O$$

Now $p = 3.25$ kmol; $y = 83.5/12 = 6.958$ kmol

$x = $ minimum O_2 supplied $= y + p \cdot (1/2) - 1/8$

$= 6.958 + 3.25/2 - 1/8 = 8.458$ kmol

Step (III) Combustion Products

$N_2 = 3.76x = 3.76\,(8.458) = 31.802$ kmol

$H_2O = (3.5/18) + 3.25 = 3.444$ kmol

$CO_2 = y = 6.958$ kmol

Step (IV) Mass of Flue Gas

Basis: 100 kg coal

Flue gas produced:

N_2 – 31.802 kmol   – 31.802 (28)   – 890.456 kg
H_2O – 3.444 kmol   – 3.444 (18)   – 61.992 kg
CO_2 – 6.958 kmol   – 6.958 (44)   – 306.152 kg

Total – 1258.600 kg

Step (V) Percentage of Mass of Flue Gas Composition

N_2 = (890.456/1258.6)(100) = 70.75%
H_2O = (61.992/1258.6)(100) = 4.925%
CO_2 = (306.152/1258.6)(100) = 24.325%

Step (VI) Mean Sp. Heat of Flue Gas

$$M\,C_p = M_{CO_2}\,[C_p]_{CO_2} + M_{N_2}\,[C_p]_{N_2}$$

$$+ M_{H_2O}\,[C_p]_{H_2O}$$

or $1258.6\,(C_p) = 306.152\,(0.241) + 890.456\,(0.241)$
$$+ 61.992\,(0.469)$$

$\therefore$ C_p = 0.2522 kcal/kg°C

Ans.

Step (VII) Combustion Temperature

Calorific Value of fuel =

[mass of combustion products] × [average sp. heat]
× [combustion temperature (Θ_c)]

$\therefore$ Θ_c = 7550/[(1258.6/100)(0.2522)]= 2378°C

Ans.

cf. Combustion Products

= 1258.6/100 kg/kg of coal
    Calorific value

= 7550 kcal/kg of coal

Step (VIII) Minimum Oxygen and Air Requirement

Minimum O_2 required per kg of coal

= 8.458 (32)/100

= 2.70 kg

Minimum air requirement

= 2.70 (100/23)

= 11.76 kg

Step (IX) Percentage Mass of Flue Gas Composition with 30% Excess Air

Now, stoichiometric air = 11.76 kg

Therefore, 30% excess air

= 11.76 (30)/100

= 3.528 kg

Therefore, total mass of the flue gas per kg of fuel

= M_{CO_2} (= 3.06 kg) + M_{N_2} (= 8.90 kg) + $M_{air\ excess}$

(= 3.528 kg) + M_{H_2O} (= 0.62 kg) = 16.108 kg

Therefore, percentage mass of

CO_2 = (3.06/16.108)/100 = 18.99%

N_2 = (8.90/16.108)/100 = 55.25%

H_2O = (0.62/16.108)/100 = 3.85%

Air_{excess} = (3.528/16.108)/100 = 21.90%

Step (X) Mean Sp. Heat of Flue Gas with Excess Air

16.108 C_p = 3.06 (0.241) + 8.90 (0.241) + 3.528
(0.241) + 0.62 (0.469) = 4.0233

$\therefore$ C_p = 4.0233/16.108 = 0.2497 kcal/kg.°C.

Ans.

Step (XI) Theoretical Combustion Temperature with Excess Air

Θ_c = 7550/[(16.108) (0.2497)] = 1877°C

Ans.

Problem 10.31 A fuel oil consists of 84.5% carbon, 12.5 % hydrogen, 2.5% oxygen and 0.5% residual matter by mass.

Working from first principles, determine the minimum theoretical air required for complete combustion of 1 kg of this fuel oil.

Also estimate:

(a) the higher calorific value
(b) the lower calorific value of this oil

Given: 1. Sp. enthalpy of water vapour formed by combustion = 2445 kJ/kg at 25°C

2. C + O_2 ⟶ CO_2 + 33.8 MJ/kg of C referred to 25°C

3. H_2 + O_2 ⟶ H_2 + 144 MJ/kg of H referred to 25°C.

Solution

Step (I) Combustion of Carbon

$$C + O_2 \longrightarrow CO_2$$

$$12 + 32 = 44$$

$$1 + 32/12 = 44/12$$

Therefore, 1 kg C requires 32/12 kg O_2 to produce 44/12 kg CO_2

Step (II) Combustion of Hydrogen

$$2H_2 + O_2 \longrightarrow 2H_2O$$

$$4 + 32 = 36$$

$$1 + 8 = 9$$

Therefore, 1kg H_2 requires 8 kg O_2 to produce 9 kg H_2O.

Step (III) Theoretical Quantity of Air Requirement

F.O. Constituent	Constituent Mass kg/kg of fuel	Oxygen Required (kg)
C	0.845	0.845 (32/12) – 2.253
H₂	0.125	0.125(8) – 1
O₂	0.025	(–) 0.025
Residual mass	0.005	
		Total O₂ required – 3.228 kg/kg of F.O.

Therefore, theoretical air requirement

$$= 3.228/(23/100)$$

$$= 14.036 \text{ kg/kg of fuel oil}$$

Ans.

(cf. Air contains 23% O_2 by mass)

Step (IV) HCV of Fuel Oil

$$\text{HCV} = (0.845)(33.8) + (0.125)(144) \text{ MJ/kg}$$

$$= 46.561 \text{ MJ/kg of F.O.}$$

Ans.

Step (V) LCV of Fuel Oil

LCV = HCV – Sp Enthalpy of water vapour formed by combustion

$$= 46.561 - (9)(0.125)(2.445)$$

$$= 43.810 \text{ MJ/kg of F.O.}$$

Ans.

Problem 10.32 A certain coal sample was tested in a bomb calorimeter to determine its calorific value. The following readings were recorded:

Mass of coal sample = 1.253 g
Mass of water = 2.5 kg
Water equivalent of the apparatus = 765 g
Temperature rise of water = 2.85°C
Temperature correction for cooling = (+) 0.018°C
Estimate the calorific value of the coal sample.
Take the specific heat capacity of water
= 4.187 kJ/kg°K

Solution

The calorific value of coal can be determined by energy balance.

Energy liberated by coal sample = Energy gained by water and the apparatus

or $1.253 \times 10^{-3} \times [CV]_{coal}$

$$= (2.5 + 0.765)(2.85 + 0.018)(4.187)$$

$$\therefore \ [CV]_{coal} = \frac{39.207}{1.253 \times 10^{-3}}$$

$$= 31290.62 \text{ kJ/kg}$$

Ans.

Q. *Is this heating value available when coal is fired in a boiler?*

Ans. No.

Q. *Why?*

Ans. Water gives up its enthalpy in condensing in the bomb but not in the boiler. Moreover, the explosion in the bomb is a constant volume process whereas the combustion of coal in a boiler furnace is a constant pressure process.

Problem 10.33 During the determination of calorific value of a fuel gas by using a gas calorimeter, the following results were recorded:

Water collected = 0.8 kg
Inlet temperature = 20°C
Outlet temperature = 38°C
Gas consumed = 0.0035 m³
Gas pressure = 95 mm of H_2O column absolute
Gas temperature = 25°C
Barometric pressure = 780 mm Hg
Estimate the calorific value of gas in kJ/Nm³.
Normal m³ is measured at 0°C and 760 mm Hg

$$[C_p]_{H_2O} = 4.187 \text{ kJ/kg°K}$$

Solution

Step (I) Energy Liberated

The amount of energy released by 1 m^3 of the fuel gas under test condition

$$= 0.8\ (4.187)\ (38 - 20)/0.0035 = 17226.5\ kJ$$

Step (II) Gas Pressure

Gas pressure $= 780 + 95/13.6 = 786.985$ mm Hg

Step (III) Gas volume in Normal State

Applying the formula $\dfrac{P_1 V_1}{T_1} = \dfrac{P_2 V_2}{T_2}$, we get

$$V_N = \frac{786.985}{760} \left(\frac{273 + 20}{273 + 25} \right) = 1.0181\ m^3$$

Step (IV) Calorific Value

The calorific of the fuel gas $= 17226.5/1.0181$

$$= 16919.7\ kJ/Nm^3$$

$$= 16920\ kJ/Nm^3$$

Ans.

Problem 10.34 A fuel gas having the following composition:

H_2—54%; CH_4—25%; CO—9.5%; C_3H_6—3.5%; CO_2—3%; N_2— 4.5% and O_2— 0.5%

is burned in a boiler whereupon a flue gas of following composition is produced:

CO_2—8.5%; O_2—7%; N_2—84.5%

This gas heats up air from 290 K to 650 K in the air preheater and leaves it at 480 K.

Determine the temperature at which the flue gas enters the air preheater.

Take specific heats of air, dry flue gas and water vapour equal to 1.005, 1.048, 2.029 kJ/kg K respectively.

Solution

Step (I) Carbon Content in Fuel Gas

Basis: 100 kmol of fuel gas

Constituent	Mass of Constituent	Carbon Content
CH_4	25 kmol	25 kmol
CO	9.5 kmol	9.5 kmol
C_3H_6	3.5 kmol	3(3.5) – 10.5 kmol
CO_2	3 kmol	3 kmol
		$\Sigma - 48$ kmol

Step (II) Composition of Dry Flue Gas

Basis: 100 kmol of fuel gas

The amount of dry flue gas produced upon combustion of 100 kmol of fuel gas

$$= 48\ (100/8.5)$$

$$= 564.70\ kmol$$

(cf.: 100 kmol of flue gas contains 8.5 kmol C as CO_2)

Dry Flue Gas Composition	Mass (kg)	
CO_2	(8.5/100)(564.70)(44)	– 2111.978
O_2	(7/100)(564.70)(32)·	– 1264.928
N_2	(84.5/100)(564.70)(28)	– 13360.802
	Total – 16737.708	

Step (III) Hydrogen Content in the Flue Gas

Basis: 100 kmol of fuel gas

Constituent	Mass of Constituent (kmol)	Hydrogen Content (kmol)
CH_4	25	25(4/2) – 50
C_3H_6	3.5	3.5(6/2) – 10.5
H_2	54	54
		Total – 114.5

Step (IV) Mass of Water Vapour Produced

114.5 kmol of H_2 upon combustion will produce

$$= 114.5(18)\ kg\ water$$

$$= 2061\ kg\ water$$

Step (V) Mass of Air Supplied For Combustion

Now, x kmol Air + 100 kmol FUEL GAS combustion 564.70 kmol FLUE GAS

Therefore, by nitrogen balance,

$$(79/100)\ x + 4.5 = (84.5/100)(564.70)$$

$$\therefore\quad x = 598.318\ kmol/100\ kmol\ of\ flue\ gas$$

Therefore, the mass of air used for combustion

$$= 598.318\ (28.9) = 17291.39\ kg \approx 17291\ kg$$

[cf. Mol. Wt. of Air – 28.9]

Step (VI) Flue Gas Temp. Inlet to Air Preheater

By energy balance,

Heat lost by dry flue gas plus water vapour = Heat gained by the air in the air preheater

or $[16737.708\,(1.048) + 2061\,(2.029)]\,(T-480)$

$= 17291\,(1.005)\,(650-290)$

$\therefore\quad T - 480 = 287.98$

$\therefore\quad T = 767.98$ K

≈ 768 K

Ans.

Problem 10.35 Steam is generated in a waste-heat boiler at pressure 1.17 MPa by the flue gas obtained from the combustion of producer gas N_2—60%; CO—26%; CO_2—5%; H_2—9% in a rotary furnace.

The flue gases enter the waste-heat boiler at a temperature 925 K and leave it at 500 K.

Calculate:
(a) the mass of flue gas produced per m^3 of producer gas burned
(b) the mass of water evaporated in the waste heat boiler per m^3 of producer gas burned
(c) the percentage of heat recovered in the WHB
Given: Boiler efficiency = 74%
Feedwater temperature inlet to WHB = 350 K
Excess air used for combustion = 25%
Take: C_p of flue gas = 1.05 kJ/kg K
Enthalpy of steam at 1.17 MPa = 2781 kJ/kg
Calorific value of CO = 12.7 MJ/m^3
Calorific value of H_2 = 10.78 MJ/m^3

Solution
Step (I) Stoichiometric Air
Basis: 100 kmol of producer gas

$2CO + O_2 \longrightarrow 2CO_2$

2 kmol CO require 1 kmol of O_2 for complete combustion
$\therefore$ 26 kmol CO require 26/2 i.e. 13 kmol O_2 for complete combustion.

$$2H_2 + O_2 \longrightarrow 2H_2O$$

2 kmol of H_2 require 1 kmol of O_2 for complete combustion
9 kmol of H_2 require 9/2 i.e. 4.5 kmol of O_2 for complete combustion

Therefore, O_2 requirement for complete combustion = 13 + 4.5 = 17.5 kmol/100 kmol of producer gas
Therefore, stoichiometric air requirement

$= (13 + 4.5)\,(100/21)$

$= 83.333$ kmol/100 kmol of producer gas

Step (II) Air Supplied
Actual air supplied = 25% excess of stoichiometric air
$= (125/100)/(83.333)$

$= 104.166$ kmol/100 kmol of producer gas

This supplied air contains:

O_2: $104.166\,(21/100) = 21.875$ kmol

N_2: $104.166 - 21.875 = 82.291$ kmol

Step (III) Flue Gas
The flue gas produced contains:
CO_2: 26 kmol (from combustion of CO) +
5 kmol (free CO_2) = 31 kmol = 31 × 44 = 1364 kg
H_2O: 9 kmol (from the combustion of H_2)

$= 9(18) = 162$ kg

O_2: 21.875 kmol (as supplied) – 17.5 kmol (used in combustion)

$= 4.375$ kmol = 4.375 (32) = 140 kg

N_2: 82.291 kmol (as supplied) + 60 kmol (from the Producer Gas

$= 142.291$ kmol = 142.291(28) = 3984.148 kg

Therefore, the total mass of the flue gas produced

$= 1364 + 162 + 140 + 3984.148$

$= 5650.148$ kg per 100 kmol of producer gas burned

Step (IV) Flue Gas per m^3 of Producer Gas Burned
1 kmol of PROD. GAS occupies 22.4 m^3 at NTP

$\therefore$ 100 kmol of PROD. GAS occupy 2240 m^3 at NTP

$\therefore$ 5650.148 kg flue gas is produced due to combustion of 100 kmol. i.e. 2240 Nm^3 of producer gas

Therefore, 5650.148/2240 = 2.522 kg flue gas is produced per Nm^3 of producer gas burned

Ans.

Step (V) Per cent Heat Recovered in the WHB
Heat lost by flue gas in WHB

$= 2.522\,(1.05)\,(925 - 500)$ kJ/m^3 of PROD.GAS.

$= 1125.442$ kJ/m^3 of PROD.GAS.

Boiler Efficiency

$$= \frac{\text{Actual heat used to raise steam in WHB}}{\text{Total heat lost by flue gas in WHB}}$$

Therefore, actual heat utilized to generate steam

$= 0.74(1125.442) = 832.827$ kJ/m^3 of producer gas

Heat released by 1 m^3 of producer gas upon combustion

$= 0.26(12.7) + 0.09(10.78)$ MJ $= 4.272$ MJ $= 4272$ kJ

Therefore, % heat recovered in WHB

$= (832.827/4272)(100) = 19.49\%$ *Ans.*

Step (VI) Mass of Water Evaporated

Heat transferred to BFW $= 832.827$ kJ/m^3 of producer gas

Enthalpy of water $= 2781 - 4.187\,(350 - 273)$

$= 2458.60$ kJ/kg

(cf. $[C_p]_{H_2O} = 4.187$ kJ/kg K)

Mass of water evaporated $= 832.827/2458.60$

$= 0.338$ kg/m^3 of producer gas burned   *Ans.*

Problem 10.36 N-hexane vapour and dry air in stoichiometric ratio at pressure 2.5 kgf/cm^2 abs. and temperature 127°C are ignited in a closed vessel and after complete combustion, the products are cooled until condensation ensues.

Determine:
  (a) the volume per kg of the mixture before combustion
  (b) the pressure and temperature of water vapour produced
  (c) the pressure of the products when condensation just begins

Air contains 23% O_2 by mass and 21% by volume.

Solution:

Step (I) Molar Ratio of N_2 and O_2 in Air

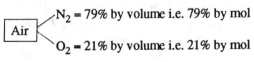

$N_2 = 79\%$ by volume i.e. 79% by mol

$O_2 = 21\%$ by volume i.e. 21% by mol

Therefore, the molar ratio of $N_2 : O_2$ in air

$= 79/21 = 3.76$

Step (II) Combustion of n-Hexane

The stoichiometric equation for chemical combustion of n-hexane can be represented as:

$$C_6H_{14} + 9\tfrac{1}{2}\,O_2 + 9\tfrac{1}{2}\,(3.76)\,N_2 \longrightarrow 6\,CO_2 + 7H_2O + 9\tfrac{1}{2}\,(3.76)\,N_2$$

Total kmol in the reaction mixture

$= 1 + 9\tfrac{1}{2} + 9\tfrac{1}{2}\,(3.76) = 46.22$

Step (III) Partial Pressure of the Reactants

Now, partial pressure $=$ [mol fraction] $\times$ [total pressure]

i.e. $p = x\,P$

Here, $P = 2.5$ kgf/cm^2 abs.

Component	Mol Fraction	Partial Pressure
C_6H_{14}	1/46.22	$(1/46.22)\,2.5$ $= 0.0541$ kgf/cm^2 abs.
O_2	9.5/46.22	$(9.5/46.22)\,(2.5)$ $= 0.5138$ kgf/cm^2 abs.
N_2	9.5(3.76)/46.22	$9.5\,(3.76/46.22)\,2.5$ $= 1.932$ kgf/cm^2 abs.

Step (III) Specific Volume of the Reaction Mixture

Applying the universal gas law equation we get,

$$PV = mRT = m \cdot (R_m/M)\,T$$

where $R_m =$ Molar gas constant

$= 848$ kgf/(kmol K)

$m = (PV)\,M/(R_m\,T)$; $M =$ molecular weight

Constituent	Partial Pressure (kgf/cm^2 abs.)	Volume (m^3)	Mol. wt.	Mass (kg)
C_6H_{14}	0.0432	1	86	$\dfrac{(0.0541\times10^4)(1)(86)}{848(127+273)}$ $= 0.1371$
O_2	0.5138	1	32	$\dfrac{(0.5138\times10^4)(1)(32)}{848(127+273)}$ $= 0.4847$
N_2	1.932	1	28	$\dfrac{(1.932\times10^4)(1)(28)}{848(127+273)}$ $= 1.5948$
				$\Sigma = 2.2166$

Therefore, the total mass of the reaction mixture

$= 2.2166 \text{ kg/m}^3$.

Sp. volume of the mixture

$= 1/2.2166$

$= 0.4511 \text{ m}^3/\text{kg}$

Ans.

Step (V) Mass of Combustion Products and Steam

Total mass of the combustion products, as obtained from the stoichiometric equation, is

$= 6(44) + 7(18) + 9\frac{1}{2}(3.76)(28) = 1390.16 \text{ kg/kmol}$

$\quad CO_2 \quad\quad H_2O \quad\quad N_2$

Mass of steam (water vapour) produced

$= 7(18)$

$= 126 \text{ kg}$.

Therefore, the mass of the water vapour per m^3 in the closed vessel

$= (126/1390.16)(2.2166) = 0.2009 \text{ kg/m}^3$

Step (VI) Specific Volume of Steam

Sp. volume of steam $= 1/0.2009 = 4.977 \text{ m}^3/\text{kg}$

From steam table:

Pressure (satd.) (kgf/cm².abs.)	Saturation Temp. (°C)	Sp. Volume (m³/kg)
0.34	71.6	4.74
0.32	70.2	5.01
Difference	Difference	Difference
– (+) 0.02	– (+)1.4	– (–) 0.27

Sp. Volume Diff.	Saturation Temp. Diff.	Press. (satd) Diff
– 0.27 m³/kg	(+) 1.4°C	(+) 0.02 kgf/cm² abs.
– 0.033	(+) (1.4)	(+) 0.02 (0.33/0.27)
	(0.033/0.27)	– (+) 0.0024 kgf/cm²
	– (+) 0.171°C abs.	

Steam Pressure Saturation Temp.

Therefore, for	0.32 + 0.0024	70.2 + 0.171 – 70.37°C
sp. volume of	– 0.3224	*Ans.*
steam 4.977	kgf/cm².abs.	
m³/kg.		

Step (VII) Pressure of the Products

Let P be the total pressure of the products mixture when condensation just starts.

Applying partial-pressure equation

$p_{steam} = x_{steam} P$

or $\quad 0.3224 = \dfrac{7}{6 + 7 + 9\frac{1}{2}(3.76)}(P)$;

$\therefore \quad P = 2.244 \text{ kgf/cm}^2 \text{ abs.}$

Ans.

Problem 10.37 Coal burnt in a stoker furnace contains 85% C, 5% H and rest non-combustibles. The ash collected during the trial runs of the boilers contains ash + unburnt fuel 13% of the supplied fuel and they contain 20% unburnt fuel.

The analysis of dry flue gas:·
CO_2—15%; CO—2.5% by volume
Determine
(a) the masses of excess O_2 and air supplied to the furnace per ton of coal burnt
(b) the furnace efficiency
Given:
1. Higher Calorific Value of coal – 7700 kcal/kg
2. $C + O_2 \rightarrow CO_2 + 8051$ kcal/kg of C burnt
3. $CO + O_2 \longrightarrow CO_2 + 2371$ kcal/kg of CO burnt
4. Air contains O_2 23% by mass & 21% by volume.

Solution

Step (I) Useful Carbon in the Fuel

Ash + Unburnt fuel – 13% of total fuel burnt

Carbon content left unburnt
$= 0.13(0.2)$

$= 0.026 \text{ kg/kg of fuel burnt}$

Therefore, useful carbon per kg of coal fired

$= 0.85 – 0.026$

$= 0.824 \text{ kg}$

Step (II) Air Supplied

Let x kmol of air be supplied per kg of coal burnt. Therefore, the combustion equation becomes,

$\dfrac{0.824}{12} C + \dfrac{0.05}{2} H_2 + 0.21 \, x \, O_2 + 0.79 \, x \, N_2 \longrightarrow$

$y \, CO_2 + z \, CO + 0.025 \, H_2O + p \, O_2 + 0.79 \, x \, N_2$

Since combustion is not complete, the flue gas contains some amount of CO as well as free O_2 due to excess air supply.

By carbon balance: $y + z = 0.824/12 = 0.0686$

$$(I)$$

The percentage of CO_2 and CO in the flue gas being 15% and 2.5%, $y/z = 15/2.5 = 6$, $y = 6z$

$$(II)$$

By O_2 balance: $0.21x = y + z/2 + 0.025/2 + p$

$$(III)$$

CO_2 in the dry flue gas:

$0.15 = y / [y + z + p + 0.79x]$ (IV)

From equations (I) and (II) we get, $6z + z = 0.0686$

$\therefore$ $z = 0.0098$ kmol

From equation (I) $y = 0.0686 - 0.0098 = 0.0588$ kmol

Substituting these values in equation (III), we get,

$0.21x - 0.0588 - 0.0098/2 - 0.025/2 = p$

or

$0.21x - 0.0762 = p$ (V)

Substituting the known values in equation (IV) we get,

$0.15 = 0.0588/[0.0588 + 0.0098 + 0.21x - 0.0762 + 0.79x]$

$= 0.0588/[x - 0.0076]$

or

$x = [0.0588/0.15] + 0.0076 = 0.3996$

$p = 0.21 (0.3996) - 0.0762 = 0.007716$ kmol
Therefore, excess O_2 in the flue gas per kg of coal

$= 0.007716(32) = 0.2469$ kg

Therefore, O_2 supplied

$= 0.21 (0.3996)$ kmol/kg of coal

$= 0.021 (0.3996) (32)$ kg/kg of coal

$= 2.6853$ kg/kg of coal

Ans.

Therefore, air supplied

$= 2.6853(100/23)$

$= 11.675$ kg/kg of coal

Ans.

Step (III) Heat Lost

Now, heat lost per kg of coal = heat lost due to incomplete combustion of C + heat lost due to unburnt fuel

Basis: 1 kg coal

Constituent of Flue Gas	Mass of Carbon
CO	$z - 0.0098$ kmol $- 0.0098(12) - 0.1176$ kg
CO_2	$0.824 - 0.1176 - 0.7064$ kg

Therefore from heat balance equation, heat lost/kg of coal

$= 0.1176 (8051 - 2371) + 0.026(8051)$

$= 877.294$ kcal,

Step (IV) Furnace Efficiency

$\eta_{fur} = (7700 - 877.294)/7700 = 0.8860$

i.e.   88.6%

Ans

Coal Pulverization

Q. *What are the advantages of pulverized coal firing systems?*

Ans.

1. Pulverization brings about a large increase of surface area per unit mass of solid fuel. Since combustion is a surface reaction, greater the extent of coal surface available, higher will be the rate of combustion. Herein lies the success of pulverized coal fired systems.
2. Less excess air is required for complete combustion because of greater surface area of fuel exposed
3. Higher combustion air temperatures ensure higher cycle efficiency
4. A good range of coal right from anthracite to peat can be successfully burnt
5. Better combustion control enables the system to respond quickly to extensive load variation
6. Better response to instrument control on auto
7. Large amount of heat release makes it very suitable for super thermal power stations where the rate of steam generation is as high as 2000 t/h.
8. Slagging and clinkering problems are low
9. Carry over of unburnt fuel to ash is practically nil
10. Ash handling problem low
11. Can operate successfully in combination with gas and oil fired systems
12. Cold start-up of boilers is very rapid and efficient
13. Less furnace volume is required
14. Banking loss low

Q. *How is the process of coal pulverization carried out?*

Ans. It is a two stage process:

I STAGE: raw and lump coal is crushed to a particle size not more than 15–25 mm in the crusher

II STAGE: crushed coal is delivered into raw coal bunkers and from here it is transferred to grinding mills that grind the feed into final particles of 200–300 mesh size. During grinding hot air is blown through the fuel to dry it to impart good fluidity to the coal dust.

Q. *What is a pulverization system?*

Ans. It is a family of equipment in which coal is ground, dried and fed to the burners of a boiler furnace.

Q. *How is a pulverization system classified?*

Ans. This classification is based on the method of delivery of pulverized fuel (coal) to the boiler furnaces and accordingly pulverization systems are of two types—central and individual.

Q. *What is a central system?*

Ans. In this system coal is pulverized on a centralized basis, the pulverized fuel is stored in a central bin wherefrom it is distributed through pipelines between the boilers. (Fig. 11.1).

Q. *What is an individual system?*

Ans. In this case each boiler is provided with its own pulverizing unit while certain provisions are made to transfer the pulverized coal to neighbouring boilers to increase the reliability of the fuel supply. (Fig. 11.2).

Q. *What are the advantages of a central system over an individual system?*

Ans.

1. Greater flexibility and better response to abrupt load variation

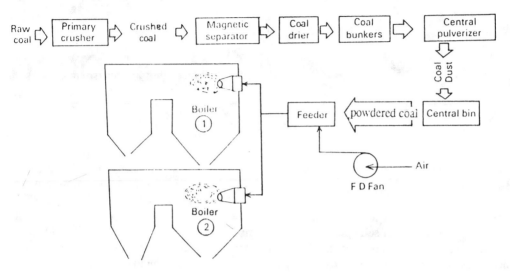

Fig. 11.1 *Schematic flow diagram of central pulverization system*

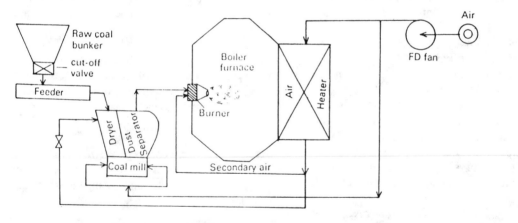

Fig. 11.2 *Schematic flow diagram of individual pulverization system*

2. Less power consumption per ton of coal pulverized

3. Operation of burners is independent of coal preparation

4. Pulverizer can be taken shutdown when there is enough reserve of pulverized coal

5. F.D. fan handles only air and so there is no erosion problem of fan blades

6. Affords better control over fineness of coal

7. Less manpower input

8. More efficient economically, especially when moist brown coal is pulverized

Q. *What are the disadvantages of a central system as compared to an individual system?*

Ans.

1. More expensive. Higher in first cost

2. Occupies more floor space

3. Higher power consumption of the auxiliaries. Hence overall power consumption per ton of

coal handled is higher than the individual system

4. Greater possibility of fire hazard due to storage of large amount of powdered coal
5. More intricate in operation and coal transportation becomes more complex
6. Operation and maintenance cost is higher
7. Drier is essential
8. Operation is less reliable

Q. *What are the advantages of an individual system over a central system?*

Ans.

1. Simpler in lay-out, design and operation
2. More economic
3. More reliable
4. Allows direct combustion control from the pulverizer
5. No drying unit is required
6. Affords better control of fuel feed
7. Lower maintenance charge

Q. *What are the disadvantages of an individual system with respect to a central system?*

Ans.

1. Lesser degree of flexibility
2. Poor performance of pulverizing unit at part load
3. If one pulverizing unit goes out of order, its corresponding boiler unit is to be taken shutdown
4. Greater erosion of F.D. fan blades as these handle both air and abrasive coal particles

Q. *How many types of individual pulverization system are there?*

Ans. Two types: closed system and open system.

Q. *How is this subdivision made?*

Ans. This is made on the basis of how the drying agent is utilized upon fuel drying.

In the closed system, the drying agent is directed into the boiler furnace together with the dried pulverized coal.

In the open system, it is used as dust carrier. The hot air is carefully cleaned from coal fines and ejected into the stack bypassing the boiler furnace.

Q. *Describe the individual pulverization system with closed fuel drying and direct dust blowing into the furnace.*

Ans. From the coal bunker, crushed coal is delivered to the grinding mill via coal feeder. At the same time hot air (523–673K), called primary air, is also directed into the mill to dry up the fuel and transfer it to the burners.

Coarse fractions of the pulverized coal are separated out in the dust separator, after which the fuel and air (which has picked up moisture from the fuel) mixture (353–400K) is supplied through pulverized fuel pipelines into the furnace burners to which secondary air is separately charged.

Q. *What factor(s) determines the quantity of primary air to be used for drying and transportation of pulverized coal?*

Ans. Fuel quality, particularly the moisture content.

Q. *What fraction of total combustion air is the primary air?*

Ans. It varies from 30 to 50% of the total consumption of air for combustion.

Q. *How does it vary with the moisture content of solid fuel?*

Ans. Greater the moisture content of the solid fuel, greater is the amount of primary air required for drying.

Q. *What is done in the case of extremely moist fuel?*

Ans. In this case, the primary air is blended with a part of the furnace gases to effect efficient drying.

Q. *Why?*

Ans. The hot primary air alone becomes economically inefficient to bring down the moisture content of extremely moist pulverized coal to the level required for efficient combustion. As the greater portion of pulverized coal enters the combustion zone at reduced temperature, the fuel combustion becomes unstable.

Q. *If the pulverization system is to become part and parcel of a boiler, what conditions should be fulfilled to ensure reliable operation?*

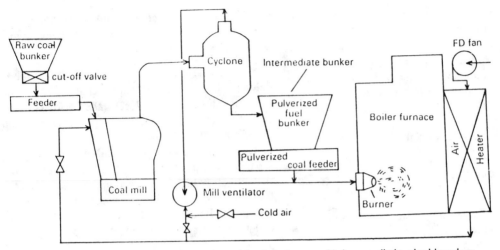

Fig. 11.3 *Pulverization system with closed fuel drying and intermediate dust bunker.*

Ans.
1. Number of grinding mills installed should be at least three
2. The number of mills in operation minus one mill must ensure at least 90% of the rated load of the boiler.

Q. *How is the productivity of a grinding mill related to the fuel consumption of the boiler?*

Ans. It is given by the relationship:

$$G_M \geq 0.9 B_F/(N-1)$$

where G_M = productivity of the grinding mill

B_F = fuel consumption of the boiler

N = number of installed mills attached to the boiler

Q. *The pulverized coal particles experience a resistance, i.e., suffer pressure drop during their journey from the mill to the burners. How is this problem overcome?*

Ans. By the head developed by the F.D. fan that ensures the operation under a slightly excessive pressure.

Q. *What is the pressure upstream of the mill?*

Ans. 1 to 2.5 kPa

Q. *But this would mean a hazard to safety and dust pollution. How can this problem be tackled?*

Ans. The only solution is to keep the equipment perfectly air-tight.

Q. *What are the discrete advantages of this direct blowing system?*

Ans.
1. Simple in operation
2. Pulverizing unit is compact
3. Consumption of electricity per unit ton of coal dust transportation is low
4. Fuel supply can easily be automatically controlled

Q. *How does the pulverization system with closed fuel drying and intermediate dust bunkers differ from the closed fuel drying and direct dust blowing system?*

Ans. The characteristic distinguishing feature is that in the former case the prepared pulverized coal is separated from the carrier air in a cyclone.

The coal dust is then led to an intermediate bunker from where it is fed to pulverized coal pipelines by a special feeder. (Fig. 11.3).

Now the moistened air (80–100°C, i.e., 353–373K) at the cyclone exit carries 10–15% of the finest coal particles. Since this can not be discharged through the stack, so it is blown by a mill ventilator into the primary air dust and gets distributed among the pulverized coal pipelines.

Q. *What is the discrete advantage of an intermediate dust bunker system over the direct dust blowing system?*

Ans. The primary advantage is that there is no need to match the mill productivity with the boiler productivity.

Q. *Why?*

Ans. It is due to the provision of an intermediate pulverized fuel bunker. Each mill, as such, can operate at optimal load.

Q. *What modifications are done in the case of less active coals with low yield of volatiles when the intermediate dust bunker system is in line?*

Ans. In this case the temperature of the pulverized coal-air mixture is raised to facilitate the ignition of coal dust by feeding 1/4 th to 1/5 th of the primary air into the air duct and then into the pulverized coal pipelines by an auxiliary hot blast fan, over and above 15–25% of the primary air directed into the pulverization system as usual.

Q. *Is this air sufficient for complete fuel combustion?*

Ans. No.

Q. *How is this corrected?*

Ans. Moistened primary air carrying out 10–15% coal fines from the cyclone exit is fed into the combustion zone through an annular channel around the main burners or through special discharge burners. (Fig. 11.4).

Q. *What are the disadvantages of an intermediate bunker system?*

Ans.
1. The equipment is bulky as well as intricate
2. Higher consumption of electric energy in dust transportation because of elevated hydraulic resistance
3. Storage of a large mass of pulverized fuel adds to fire and explosion hazard

Q. *How can the unit energy consumption for dust transportation be reduced?*

Ans. By using compressed air and a high fuel-air ratio.

In the conventional system, the fuel-air ratio is 0.4–0.6 but by using compressed air, in a recent technique, this ratio is increased to 30–60 (i.e., 30–60 kg dust per kg of primary air) imparting high fluidity to coal dust which is then conveniently transported through small dia (60–90 mm) pipelines.

Q. *How does the individual pulverization with open drying system operate?*

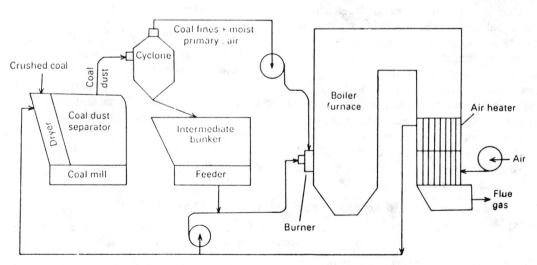

Fig. 11.4

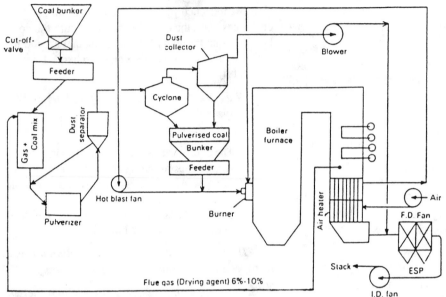

Fig. 11.5 *Individual pulverization with open drying system*

Ans. This system is adopted for solid fuels with high moisture content. The fuel is dried at elevated temperature by flue gas (400–450C, i.e., 673–723K) taken off in an amount of 6–10% of the gas volume from the flue gas duct downstream of the economizer.

The coal dust is transported from the pulverizing unit to the cyclone separator with this high temperature drying agent that exits the cyclone with 10% of the finest fuel fines. This moistened gas together with suspended coal fines is then directed to a system of multicyclones (a set of 150–250 small cyclone elements) and an electrostatic precipitator to separate out the flue-gas borne dust.[Fig. 11.5]

The separated dust flows by gravity through chutes into the intermediate pulverized fuel bunker from where it is fed to burners via pulverized fuelpipelines.

The drying agent leaving the dust collector is combined with the flue gas exit of the air heater and discharged into the air after being cleaned in the main electrostatic precipitator of the plant.

Q. *What are the advantages of the open system of fuel drying?*

Ans.

1. Quality of fuel improves because of high temperature drying

2. Efficiency of fuel combustion increases
3. Lower aerodynamic resistance and lower waste gas temperature as the volume of the combustion products is reduced in the flue gas ducts

Q. *What are its drawbacks?*

Ans.

1. Along with the discharged drying agent some fuel fractions are lost
2. Higher energy consumption to effect separation and purification of moist drying agent
3. With improper operation of the dust collectors or a high moisture content in the drying agent, a substantial amount of coal dust escapes into the atmosphere leading to air pollution

In spite of good operation 1–2% of the fuel is lost. Hence this system is limited to extremely moist fuel which cannot be otherwise burnt efficiently by conventional methods.

Q. *What determines the quality of pulverized coal?*

Ans. Milling fineness and the relative concentration of individual fractions.

Q. *How are these characteristics determined?*

Ans. By sieve analysis. A sample of pulverized coal is allowed to pass through 4–5 standard sieves with progressively decreasing mesh size which is the clear size of the mesh expressed in micrometers. And the total sieve residue R_x is determined by the total dust residue on a particular sieve of mesh size x micrometer and that on all other sieves above it with larger mesh size, expressed as a percentage of the initial sample mass.

Q. *Sometimes finer particles coalesce readily introducing an error in sieve analysis. How can this be avoided?*

Ans. By blowing the finest dust particles in an air classifier to grade it to size, after the screening operation is performed.

Q. *What is integral particle-size distribution curve (or total residue curve)?*

Ans. It is the curve obtained by graphically plotting the result of sieve analysis–the sieve mesh size x (μm) being laid off as the abscissae and the total sieve residue, R_x, as the ordinates.[Fig. 11.6]

Q. *What is the nature of the curve?*

Ans. It is an exponential curve described by the equation

$$\ln\left[\frac{R_x}{100}\right] = bx^n$$

or

$$R_x = 100 \exp(-bx^n)$$

where b and n are constants depending on the quality of fuel and grinding technique. (Fig. 11.6).

Q. *What is the importance of the polydispersity coefficient of dust?*

Ans. Its importance lies in its characterization of the structure of dust and its particle size distribution.

Q. *How?*

Ans. Differentiating the above equation with respect to x, the mesh size (μm), we get

$$y = -\frac{dR_x}{dx} = 100\, bnx^{n-1} \exp(-bx^n) = R_x\, bnx^{n-1}.$$

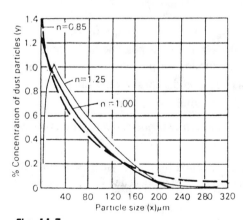

Fig. 11.7

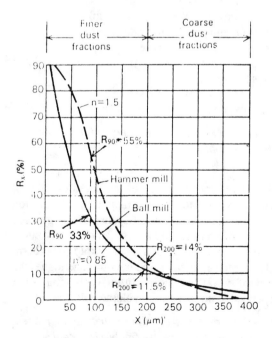

Fig. 11.6 *Integral particle-size distribution curve*

Now plotting y (% concentration of dust particles of the size x μm) vs. mesh size x (μm), several curves can be obtained depending on the value of the polydispersity coefficient (n) of the dust. (Fig. 11.7).

For $n > 1$, the curve shows a maxima in the zone $x = 15$–25 μm, thus showing that such a dust sample has a relatively small content of the finest fractions.

For $n \leq 1$, the curves show that the dust samples contain a higher quantity of the finest fractions.

Q. *What is the immediate impact of the concentration of coarse particles in boiler furnaces?*

Ans. Greater the concentration of the coarse particles (> 250 μm) in pulverized coal, greater is the heat loss with unburned carbon.

Q. *How can the energy consumption in pulverization be determined if the surface area of the dust produced is known?*

Ans. This can be computed with the help of Rittinger's equation:

$$E_{gr} = E^\circ_{gr} (A_d - A_c)$$

where E_{gr} = energy of grinding, kWh/kg

E°_{gr} = unit consumption of electric energy in grinding to produce 1 m^2 of ground surface

A_c = initial surface area of 1 kg of crushed coal m^2/kg

A_d = final surface area of 1 kg of produced dust m^2/kg

Since $A_d \gg A_c$ so $E_{gr} \approx E^\circ_{gr} A_d$

From this equation it is possible to determine the energy consumption for pulverization from the surface area of dust produced, since the value of E_{gr} for a wide range of solid fuel is known.

Q. *How is the actual surface area of the dust determined?*

Ans. It is obtained by multiplying the theoretical surface area of the dust produced with the shape factor

$$[A_d]_{actual} = k_s [A_d]_{th}$$

where k_s = shape factor

The theoretical surface area of dust (m^2/kg) is calculated with the help of the following equation

$$[A_d]_{th} = \frac{4.5 \times 10^5}{\rho} \frac{1}{n} \left[\ln \frac{100}{R_{90}} \right]^{1/n}$$

where ρ = density of ground fuel, kg/m^3
n = polydispersity coefficient

Q. *What is the value of the shape factor (k_s) for coal?*

Ans. It is usually taken as 1.75.

Q. *What is the density of pulverized coal?*

Ans. It varies from 1700 to 1800 kg/m^3.

Q. *What should be the value of polydispersity coefficient for pulverized fuel?*

Ans. $n > 1$

Q. *Why?*

Ans. In this case, the coal dust will contain a small concentration of very fine fractions and a small concentration of coarse fractions.

Q. *Why is the moisture content (% by weight) of pulverized fuel important?*

Ans. The moisture content of pulverized fuel should be at the recommended level.

Moisture content above this level will entail certain difficulties, such as:
  (a) lowering of boiler productivity
  (b) loss of fluidity
  (c) formation of slums in the bunkers
  (d) clogging of feeders and chutes
  (e) reduction in the ease of transport
Moisture content below the recommended level brings the possibility of:
  (a) self-ignition where stored
  (b) explosion when mixed with air

Q. *How does the explosion hazard originate with pulverized coal?*

Ans. Coal has the fundamental property of auto-ignition. Greater the degree of surface area of coal particles exposed to air, more rapid is its self-ignition even at lower temperature. Because of the gaseous products of combustion, a large amount of volume expansion occurs and that's how an explosion is born.

Larger the amount of fine fractions and higher the yield of volatiles, more intensively coal dust suspended in air in a closed volume will explode.

Q. *What factors primarily contribute to the explosion for a given quality of pulverized coal?*

Ans.
1. Temperature of coal dust-air mixture
2. Coal dust-air ratio

Q. *What is the most dangerous concentration of coal dust in air?*

Ans. It is within 0.3 to 0.6 kg coal dust/m^3 of air.

Q. *If the coal dust suspended in air in a closed volume undergoes auto-ignition, what parameters will undergo abrupt change?*

Ans. Temperature and pressure.

Q. *What preventive measure is taken?*

Ans. Safety (relief) valves are installed to let out a part of the coal dust-air mixture should the pressure rise abruptly.

Q. *What are the concentration limits of O_2 in the drying agent, below which fuel dust will not explode?*

Dust Sample	O_2 Content
Peat	16%
Oil shales	16%
Brown coals	18%
Coals (other)	19%

Q. *How can the oxygen concentration in the drying air be reduced?*

Ans. By blending hot air with the gaseous products of combustion.

Q. *What is the probability of explosion with the yield of volatiles?*

Ans. Greater the yield of volatiles, greater is the probability of explosion, lower the yield of volatiles, lower is the probability of explosion.

Q. *What is the safe limit of the yield of volatiles?*

Ans. A fuel having yield of volatiles less than 8% is explosion-safe.

Q. *Why must the temperature of coal-dust and air mixture downstream of the mill be strictly controlled?*

Ans. A high temperature of coal dust-air mixture makes it prone to explosion.

Q. *What should be this temperature limit?*

Ans. 70–80°C (343–353K) for coal with high volatiles.
120–130°C (393–403K) for coal with low volatiles.

Q. *What do you mean by the term coefficient of grindability?*

Ans. It is the ratio of unit energy (kWh) consumptions of a standard laboratory mill in grinding a reference solid fuel and the fuel under consideration, provided that both have the same initial particle size and the same ground dust characteristics i.e.

$$k_{gr} = E_r / E$$

Q. *How is the optimal value of grinding in terms of sieve residue R_{90} related to the yield of volatiles?*

Ans. $[R_{90}]^{opt} = 4 + 0.8n\, W_{vol}$
where n = polydispersity coefficient
W_{vol} = yield of volatiles

Q. *How many types of grinding mills are there? What is their principle and speed characteristics?*

Ans.

Grinding Equipment	Grinding Principle	Speed Characteristics
1. Ball-tube mill	Impact, abrasion	Low (15–30 rpm)
2. Roller mill	Crushing	Medium (50–80 rpm)
3. Hammer mill	Impact	High (750–1000 rpm)
4. Paddle-type mill	Impact	High (1400–1500 rpm)
5. Pulverizing fan	Impact	High (750–1450 rpm)

Q. *For grinding which type of fuel, are ball-tube mills preferred?*

Ans. Fuels with relatively low yield of volatiles.

Q. *For grinding which type of fuel, are hammer mills preferred?*

Ans. Fuels like brown coals, peat, oil shales.

Q. *In which cases are pulverizing fans preferred?*

Ans. Soft ($k_{gr} > 1.5$) and very moist brown coals.

Q. *Briefly describe a ball-tube mill.*

Ans. A ball-tube mill (or simply ball mill) consists of a large rotating drum (2–4 m in diameter and 3–10 m in length) partially filled with steel balls (30–60 mm dia). The inside surface of the drum is clad with armour plates and outside surface with heat and sound insulation. The drum is rotated by an electric motor via a speed reduction gear and a large driven wheel attached to the drum.

Crushed fuel and hot air are fed to the drum through the inlet pipe while the pulverized fuel-air mixture is taken out through the exit pipe via the classifier that retains the oversized particles to be ground again.

Dampers located in the exhaust fan inlet, dust control the mill output by varying the flow of air through the mill and hence the rate of fuel removed from the mill.

Q. *What is the energy consumption of a ball mill per ton of coal ground?*

Ans. 20–25 kWh.

Q. *How is the optimal rotational speed of a ball mill determined?*

Ans. It is 0.76 times the critical speed, i.e.

$$N_{opt} = 0.76 \, N_{cr}$$

where N_{cr} = critical rotational speed of the drum in revolutions per second

Q. *What is the critical rotational speed?*

Ans. It is the minimum rotational speed at which the balls stick to the drum wall as their weight is counterpoised by the centrifugal force due to rotation.

$$N_{cr} = 0.75 \, [D_d]^{-1/2}$$

where D_d = drum diameter

Q. *What is the mechanism of grinding in a ball mill?*

Ans. As the drum rotates, it lifts the balls to a certain height over the wall, imparting to them a potential energy which is expended as the balls detach from the wall and fall. The impact of the falling balls and the abrasion between them grind the solid fuel.

Q. *What are the principal factors upon which grinding capacity depends?*

Ans. Drum length and drum diameter.

Q. *What is the drying capacity of a grinding mill?*

Ans. It is the quantity of fuel that can be dried in the mill from the initial moisture content to the desired value.

Q. *Why it is advisable to run ball mills at full load?*

Ans. Energy consumption of a ball mill due to rotation is virtually independent of the mass of the fuel charged to the drum because of the large mass of the balls and drum. And as a result, as the quantity of charge to the drum decreases, the unit energy consumption for grinding E_{gr} increases

$$E_{gr} = E_m/B_m \text{ kW h/kg}$$

where E_m = power consumption for mill rotation, kW

B_m = mill productivity, kg/h

That is why it is always economical to run the ball mill at full load.

Q. *What are the advantages of a ball mill?*

Ans.
1. This type of mill can be successfully employed for pulverizing a wide range of solid fuels right from oil shales, pits to anthracite—the hardest variety of coal.
2. Infiltration of metallic objects occasionally present in the coal does not seriously affect the mill operation
3. The operation is simple
4. Low initial cost

Q. *What are its drawbacks?*

Ans.
1. Wear and tear of the armour plates and balls due to impact and abrasion
2. High operating cost, particularly for harder fuels

Unit energy consumption is up to 35 kWh/t for anthracite

Q. *How is the wear of the balls compensated?*

Ans. By introducing new balls periodically into the ball mill.

Q. *Briefly describe the operation of a hammer mill.*

Ans. A rotor with discs to which hammers are hinged rotates inside a steel casing clad with an armour plate (25–30 mm thick) inside.

Rotating hammers with a circumferential speed of 50 to 60 m/s strike the fuel lumps and crush them into smaller pieces which get pulverized by the abrasion in the gap between the hammers and casing.

The primary air fan induces through the pulverizer a flow of air that lifts the coal dust. The airborne fuel dust is subjected to a centrifugal dust separator to throw the oversize particles back into the grinding section while the finely divided fuel particles suspended in primary air are discharged through a centrally located dust-discharging duct.

Q. *For which type of solid fuels, is a hammer mill preferred to a ball mill?*

Ans. For fuels having $k_{gr} \geq 1.1$

Brown coals, oil shales, pits and coals with volatiles 28% belong to this category.

Q. *Why?*

Ans. With such kinds of fuel, the unit consumption of electric energy for grinding, in the case of hammer mill is 8–12 kWh per ton of ground fuel—which is 30–50% less than that required in the case of a ball mill.

Q. *Briefly describe the operation of a roller mill.*

Ans. A roller mill (also called bowl mill) consists of stationary rollers mounted on an electrically driven rotating bowl. Coal fed through the hopper gets pulverized by attrition as it passes between the sides of the rollers and bowl.

Hot primary air introduced into the pulverizer through the bottom of the bowl carries off coal-dust into the centrally located classifier fitted at the top. Coarse particles drop back into the bowl through the centre cone of the classifier while the fine coal dust-air mixture is led away to the burner.

Q. *What are the advantages of a roller mill (i.e., bowl mill) ?*

Ans.
1. Low unit energy consumption (12–15 kWh/t)
2. Small dimensions
3. Reduced noise level
4. The classifier can be adjusted to alter the degree of coal fineness while the mill is on
5. Since the mill operates at negative pressure, leakage of coal from the mill casing is practically nil

Q. *What are the disadvantages of a roller mill?*

Ans.
1. Sensitive to metallic objects should they get into the mill along with coal
2. Uneven wear of the grinding parts presents repairing complexities

Q. *In which cases are roller mills preferred?*

Ans. Systems:
(a) using direct dust blowing
(b) grinding moderately hard coals with relatively low moisture content and hard fractions

Q. *What auxiliary equipment would you have with a pulverizer?*

Ans.
1. Dust separator
2. Cyclone
3. Raw coal feeder
4. Dust duct
5. Pulverized coal bunkers

Q. *Why is a dust separator required?*

Ans. Its sole function is to separate dust into coarse and fine fractions. While the fine fractions are led away by primary air to the burners, the coarse particles are dropped out to the mill for regrinding.

Q. *How is the separation effected?*

Ans. The separation can be effected by three ways by using centrifugal, inertial and gravitational forces.

Q. *Briefly describe the operation of a centrifugal separator.*

Ans. It is fitted with two concentric cones. Coal dust-primary air mixture is directed to the bottom inlet of the separator with a flow velocity 15–20 m/s. The

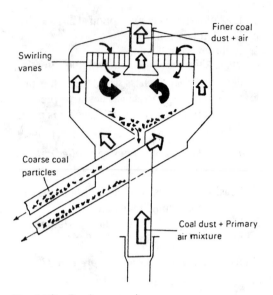

Swirling vanes

Finer coal dust + air

Coarse coal particles

Coal dust + Primary air mixture

Fig. 11.8 *Dust separator*

dust laden air experiences a volume expansion in the annular space between the cones, its flow velocity reduces to 1/3 rd, producing the effect of gravitational separation. Down along the walls of the cone, slide the coarsest coal particles and return to the mill. From the outer cone the coal dust-air mixture enters the inner cone through the inlet vanes which are tilted to impart a swirling motion to the dust-air flow. The coarser particles drop off by centrifugal effect and dust with necessary fineness is led through the central dust-duct. (Fig. 11.8).

12

Pulverized Coal Fired Furnaces

Q. *By how many methods can coal be burned in boiler furnaces?*

Ans. Three methods:
1. Flame combustion (Fig. 12.1a)
2. Cyclone combustion (Fig. 12.1b)
3. Fluidized-bed combustion (Fig.12.1c)

Q. *How rapid is the combustion?*

Ans. Combustion takes place within a very short period of 1–2 s.

Q. *What is cyclone combustion?*

Ans. A process in which combustion of particles is

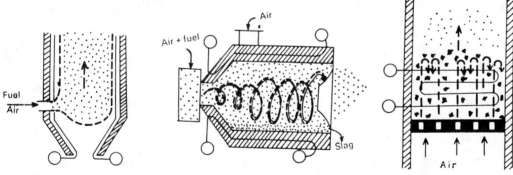

Fig. 12.1 *(a) Flame Combustion* *(b) Cyclone Combustion* *(c) Fluidized-bed combustion*

Q. *What is the basis of this classification?*

Ans. Aerodynamic characteristic of combustion system that determines the conditions of contact of the burning fuel with an oxidant.

Q. *Which one is the most popular in modern power engineering?*

Ans. Flame combustion.

Q. *What is flame combustion?*

Ans. Burning of pulverized coal in a suspended state in the combustion air in the furnace space is called flame combustion.

effected in the presence of intensive turbulent motion of air.

Q. *What is the difference between cyclone combustion and flame combustion?*

Ans. In cyclone combustion, fuel particles are subjected to a great turbulence by the combustion air supplied and they burn off more quickly. Whereas in the case of flame combustion, the fuel particles are so fine that they get easily airborne and as a result combustion takes place in a suspended state in the furnace space.

Only finely divided coal particles are suitable for flame combustion whereas the cyclone method is good for the combustion of coarse coal dust and even crushed coal.

Q. *Why are cyclone furnaces called slagging type furnaces?*

Ans. Due to cyclone combustion a high temperature is developed in such furnaces, with the effect that slags produced are in a molten state and liquid slag is discharged off. Hence such furnaces are called slagging type.

Q. *What is fluidized-bed combustion?*

Ans. Solid fuels (coal, bagasse, etc.) reduced to 1–6 mm size can be successfully burnt in a fluidized state over a grate at the bottom of which combustion air is blown through. The velocity of air is so controlled that the fuel particles are lifted off the grate and are reciprocated in the vertical plane. The finer and partially burnt fuel particles are carried off to the upper layer of the fluidized bed, whereupon their flow velocity decreases and they undergo complete combustion.

Q. *What is the thickness of such a bed?*

Ans. It varies. Usually it ranges from 0.5 to 1 m.

Q. *What is the specific characteristic of a fluidized bed?*

Ans. It expands in volume by 1.5–2 times during operation.

Q. *What is the bed temperature?*

Ans. 800–1000°C (1073–1 273K).

Q. *How are the boiler tubes placed in such a bed?*

Ans. They are placed in the form of in-line or staggered tube bundles arranged in and above the bulk of the fluidized bed.

Q. *What are the advantages of low furnace temperature in the fluidized bed?*

Ans.
1. Overheating is prevented
2. NO_x emission is reduced.
3. Introduction of such solid additions as limestone in the bed to neutralize SO_2 and SO_3 produced during combustion is possible.

Q. *How is a pulverized fuel fed furnace characterized geometrically?*

Ans. Geometrically a furnace can be characterized by its linear dimensions:

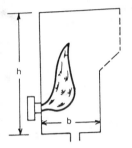

Fig. 12.2 (a)

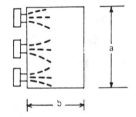

Fig. 12.2 (b)

front width a
depth b
height h
These parameters are determined on the basis of
(a) rated fuel consumption
(b) thermal and physico-chemical properties of fuel.

Q. *What is the cross-sectional area of the furnace normal to the path of the flue gas?*

Ans. $A_f = ab$, m^2

Q. *What is the principal thermal characteristic of a steam boiler furnace?*

Ans. It is the heat power of the furnace measured in kilowatt.

$$Q = B_f H$$

where Q = total heat released rate in the furnace, kW
B_f = rate of fuel consumption, kg/s

H = calorific value of fuel, kJ/kg

Q. *What is the heat release rate per unit cross-sectional area of the furnace?*

Ans. It is the total heat release rate in the furnace divided by the cross-sectional area of the combustion zone, i.e.

$$q = Q/A_f \quad kW/m^2$$

Q. *Upon which factors does this quantity q depend?*

Ans.

1. Kind of fuel
2. Type of burners
3. Arrangement of burners.

Q. *What is the value of q?*

Ans. The heat release rate per unit cross-sectional area of a furnace (q) ranges from 3500 to 6500 kW/m².

Q. *How can a furnace be characterized on the basis of burner arrangement?*

Ans. The furnace can be characterized on the basis of heat release rate per burner-tier if the burners are laid out in a number of tiers whence

$$q_t = Q_t/A_f$$

where Q_t = heat release rate of all burners in a tier, kW

q_t = heat release rate per unit area per tier

Q. *What is the value of q_t?*

Ans. It ranges from 1 200 to 2 400 kW/m².

Q. *What will happen if the values of q and q_t are allowed to go beyond their limiting value?*

Ans. It'll bring about intensive clinkering of waterwalls, particularly in the burner zone, and rising of the surface temperature of tube metal dangerously.

Q. *What is the value of the depth of a pulverized coal fired furnace?*

Ans. Its value ranges from 6 m to 10.5 m.

Q. *What factors exert an influence on the section of depth of furnace?*

Ans. The depth of the furnace is chosen such that
(a) burners can be arranged properly
(b) free flame-propagation does not bring about contact between flame tongues and waterwalls.

Q. *What is the value of the front-width of the pulverized coal fired furnace?*

Ans. It ranges from 9.5 m to 30 m.

Q. *On which factors does its value depend?*

Ans.

1. The quality and nature of fuel
2. The thermal power, i.e., steaming capacity of the boiler.

Q. *How can it be determined?*

Ans. It can be calculated with the aid of the formula:

1. $A_f = a \cdot b$ when A_f and b are known
2. $a = 0.67 \sqrt{G_s}$

where, G_s = steaming capacity of the boiler, t/h

Q. *How does the height of the furnace vary?*

Ans. It ranges from 15 m to 65 m.

Q. *What factors are to be considered in the determination of furnace height?*

Ans. Its value should be such that it will
(a) ensure complete combustion of the fuel along the flame length within the furnace
(b) allow sufficient space to layout the waterwalls on the furnace wall to cool the products of combustion to the specified temperature.

Q. *How is the furnace height determined to ensure complete combustion of fuel?*

Ans. It is determined on the basis of the following formula:

$$h_f = \bar{v}\,\tau$$

where $\bar{v}$ = average gas velocity in the furnace cross-section, m/s

τ = residence time of unit volume of flue gas in the furnace, s.

Q. *What is the allowable heat release rate of the furnace?*

Ans. It characterizes the energy release rate per unit volume, kW/m^3 of the furnace

$$q_v = Q/V_f = G_s H/V_f$$

where V_f = volume of the furnace, m^3

Q. *What is the value of the allowable heat release rate?*

Ans. Its value varies from $120\,kW/m^3$ for coal-fired dry-bottom furnaces to $210\,kW/m^3$ for slagging-bottom furnaces.

Q. *How does the allowable heat release rate vary with the residence time of the unit volume of the gas in the furnace?*

Ans. It increases with the decrease of and decreases with the increase of the latter as can be seen from the adjacent figure (Fig. 12.3).

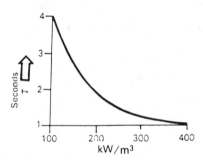

Fig. 12.3

Q. *Why does the increase in boiler capacity bring about an increase of Θ_2, which is the specified temperature to which combustion products are to be cooled by the waterwalls?*

Ans. With the increase of boiler capacity, the volume of the furnace increases and it increases more substantially than the surface area of the waterwalls. Therefore, the available unit surface area of waterwalls per m^3 of the furnace volume becomes lower, with the effect that combustion products leave the furnace space at a higher temperature than the specified temperature Θ_2.

Q. *How can the cooling surface area of a boiler be increased without altering the furnace dimensions?*

Ans. By inserting curtain walls (or platens).

Q. *What are these?*

Ans. These are additional tube-walls mounted in the furnace space dividing it into two or more sections.

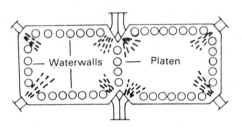

Fig. 12.4 (a)

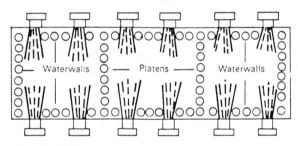

Fig. 12.4 (b)

Q. *In what respect does the curtain wall, i.e., platen differs from conventional waterwalls laid on the furnace wall?*

Ans. Conventional waterwalls arranged on the sides of the furnace receive heat from one side whereas curtain wall tubes receive heat from both sides and hence these are characterized by a higher heat release rate.

Q. *How can the desired intensity and complete combustion of pulverized coal be achieved?*

Ans. By proper supply of and intensive mixing of the pulverized fuel (primary airborne coal powder) with the secondary air in the burner assembly.

Q. *What is the function of a burner?*

Ans. A burner does not ignite a fuel. Its sole function is to produce two individual flows—one of fuel dust-air mixture and the other of secondary air—for ignition and active burning in the furnace space.

Q. *Why are coal dust-primary air mixture and hot secondary air introduced into the furnace space at different speeds and with different degrees of turbulence?*

Ans. To ensure good intermixing of the ignited fuel with the secondary air to complete the combustion. The result: Active burning takes place in the furnace space.

Q. *How many types of pulverized coal fired burners are there?*

Ans. Straight flow burner.
Turbulent (vortex) burner.

Q. *What are straight flow burners?*

Ans. Burners of this type produce long-range jet with low expansion angle and less turbulence of secondary and primary air flows.

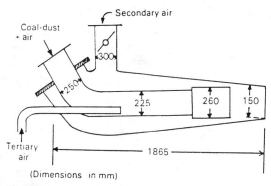

Fig. 12.5 *Straight flow burner*

Q. *What is the main drawback of this type of burner?*

Ans. Poor intermixing of primary air-coal dust flow with the secondary air flow. So a single burner is not enough to bring about efficient combustion.

Q. *What is done for efficient combustion?*

Ans. A series of burners are compounded to produce an assembly of straight-flow burners. [Fig. 12.6]

As a result of this, the jets from various burners will interact with one another to effect complete combustion in the furnace space.

Q. *For what kind of solid fuels, are straight-flow burners employed?*

Ans. These are employed with high-reactive solid fuels:
brown coals
high-volatile coals
peat
oil shales, etc.

Q. *What is the velocity of coal-dust-primary air flow at the burner exit?*

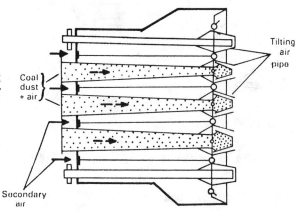

Fig. 12.6 *An assembly of straight flow burners*

Ans. 20–28 m/s.

Q. *What is the optimal value of secondary air?*

Ans. 35–45 m/s.

Q. *What are vortex burners?*

Ans. As the name implies, the vortex burners feed the coal dust-primary air mixture as well as secondary air in the form of jets with high turbulence. [Fig. 12.7]

The jet of ignited fuel-air mixture gets intensively mixed with the whirl of secondary air as the fuel jet propagates in the form of a cone-shaped expanding flame.

Q. *How many types of vortex burners are there?*

Ans. Four types
(a) Two-scroll burners
(b) Straight-scroll burners

(c) Scroll-vane burners

(d) Vane-type burners.

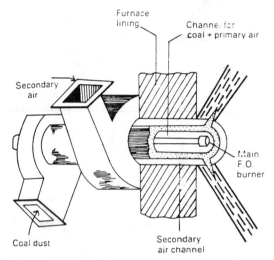

Fig. 12.7 *Vortex burner*

Q. *What is the throughput capacity range of turbulent burners?*

Ans. 1–4 kg/s of fuel.

Q. *What is the range of their heat generation capacity?*

Ans. 25 to 100 MW.

Q. *What is the heat generation capacity of two-scroll and scrollvane burners?*

Ans. It ranges from 75 to 100 MW.

Q. *What is the most important aerodynamic characteristic of vortex burners?*

Ans. It is the vorticity parameter. It ranges from 1.5 to 5—greater the whirling of the secondary air flow greater is its value.

Q. *What are the features of a burner with a high vorticity parameter?*

Ans. Such type of burners produce a wider jet with a larger angle of expansion and with a wider zone of recirculation of the hot furnace gases to the flame root, ensuring quicker fuel preheating and combustion.

Q. *For what kind of solid fuels, are burners with high vorticity parameters used?*

Ans. Low-reactive fuels with a relatively low yield of volatiles.

Q. *What are the axial velocities of primary and secondary air in vortex burners?*

Ans. Primary air velocity varies from 16 to 25 m/s.
Secondary air velocity varies in the range (1.3–1.4) times primary air velocity.

Q. *What is the primary importance of axial velocities of primary and secondary air?*

Ans. To achieve complete burning of the fuel.

Q. *What are combined burners?*

Ans. These are burners that can operate, simultaneously or alternately on a variety of fuels—solids, liquids and gaseous.[Fig. 12.8]

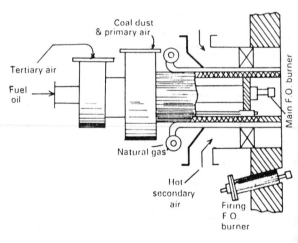

Fig. 12.8 *Combined burner assembly*

Q. *Where are such type of burners installed?*

Ans. In high capacity steam boilers.

Q. *What are the basic factors to be taken into account in the arrangement of burners?*

Ans. Burners are arranged on the furnace walls to
(a) ensure as complete combustion of fuel in the

furnace core as possible

(b) create favourable conditions for slag removal
(c) avoid clinkering of the furnace walls.

Q. *What is embrasure of vortex burners?*

Ans. It is the diameter of their port, D_p.

Q. *How are vortex burners laid on the furnace walls?*

Ans. They are fixed on the furnace walls such that the gap between each other is (2.2–3) D_p while that between the side wall and adjacent burner is (1.6–2) D_p.

Q. *Why is such as arrangement preferred?*

case of high-capacity steam boilers, where the high heat generation rate demands a good many burners that cannot be arranged on a single front wall even in two tiers.

Q. *What are the advantages of the opposite double front arrangement of vortex burners?*

Ans.

1. High heat generation rate.
2. More uniform heat absorption by waterwalls.
3. Less clinkering problem due to the high temperature maintained throughout the furnace forasmuch as the colliding flames are deflected both upwards and downwards
4. Suitable for slag-bottom furnaces.

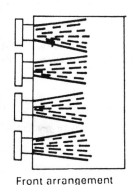
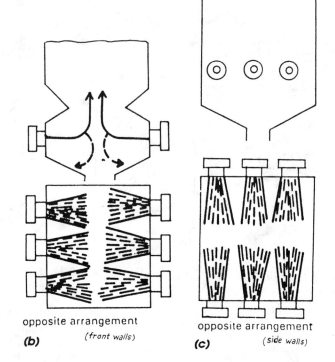

Front arrangement

Fig. 12.9 (a)

opposite arrangement
(b) *(front walls)*

opposite arrangement
(c) *(side walls)*

Ans. To avoid premature interaction between adjacent flames and to prevent flame tongues from licking the walls. (Fig. 12.9a, 12.9b, 12.9c).

Q. *In which case is the opposite double front arrangement made?*

Ans. This arrangement is made particularly in the

Q. *What should be the furnace width in order to ensure efficient interaction of opposite flames propagating from a double-front (opposite) arrangement of turbulent burners?*

Ans. (5–6) D_p.

Q. *In which case are vortex burners laid in an*

opposite arrangement on the side walls?

Ans. Low capacity boilers. The burners are laid in a single tier.

Q. *How are straight flow burners arranged?*

Ans. They are arranged into
   (a) opposite displaced arrangement (Fig. 12.10A)
   (b) corner arrangement (block arrangement) (Fig. 12.10B)
   (c) corner arrangement with tangential jets (tangential arrangement) (Fig. 12.10C)

furnace in the case of a tangential burner arrangement?

Ans. The ratio is 1–1.2, i.e., the furnace is nearly square in cross-section.

Q. *Why is such a selection made?*

Ans. It endows best aerodynamic characteristics in the furnace space.

Q. *What are dry-bottom furnaces?*

Ans. These are furnaces in which slag is removed in the solid state.

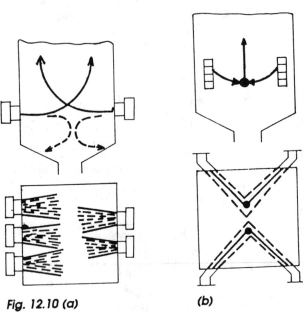

Fig. 12.10 (a)

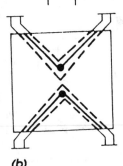

(b)

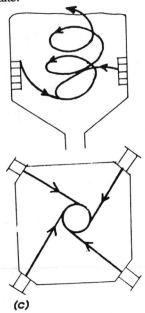

(c)

Q. *Which of these arrangements finds the widest application?*

Ans. Tangential arrangement.

Q. *Why?*

Ans. Because of:
   (a) high heat-generation capacity and hence suitability for high capacity boilers
   (b) uniform heat distribution between the furnace walls
   (c) low clinkering probability.

Q. *What is the ratio of front width to depth of the*

Q. *What are the patterns of flame motion in dry-bottom type furnaces?*

Ans.
   1. Straight flow S-shaped flame (Fig. 12.11A)
   2. Opposite straight flow flame (Fig. 12.11B)
   3. Vertically turbulized flame (Fig. 12.11C)
   4. Combination of straight and horizontally turbulized flame Fig. (12.11D)

Q. *Why are dry-bottom furnaces equipped with a dry-bottom hopper?*

Ans. The dry-bottom hopper intensively cools the products of combustion in the furnace bottom. As a

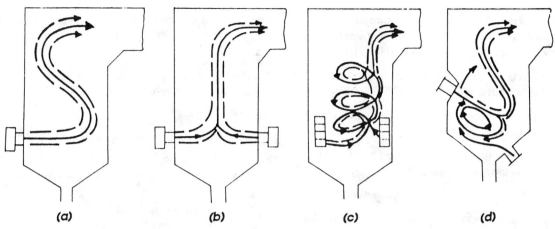

Fig. 12.11

result, the molten slag droplets entering into this zone get quickly cooled, solidified and fall along the sides of the hopper into the slag-collection pit. (Fig. 12.12).

Q. *How much of the total ash content is collected in this way?*

Ans. About 5–10 % of the total ash content of fuel.

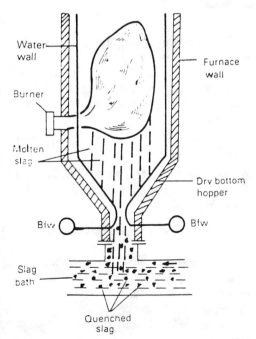

Fig. 12.12 *Dry-bottom furnace fitted with dry-bottom hopper*

Q. *Why are the granulated slag particles collected on the water bath?*

Ans. The water bath acts as a hydraulic seal preventing the entry of cold atmospheric air from beneath into the furnace.

Q. *Why are the heat release rates per m^2 of furnace cross-section and per m^3 of furnace volume not high in the case of dry-bottom type furnaces?*

Ans. The furnace is so designed that its aero-dynamic characteristics will not allow the flue gas to attain ash deformation temperature near the surface of the waterwalls to avoid the problem of clinkering. That is why average heat release rates per m^2 and per m^3 of the furnace are not high (3–4 MW/m^2 and 100–140 MW/m^3 respectively).

Q. *For what kind of pulverized solid fuel, are dry-bottom type furnaces employed?*

Ans. Solid fuels with volatiles exceeding 25%, i.e., coals with moderate or high yield of volatiles.

Q. *How do the different fractions of pulverized fuel burn up in the combined straight flow and horizontally turbulized flame?*

Ans. Burnt up in the straight portion of the flame are the fine fractions of the pulverized solid fuel, whereas the coarser particles are thrown to the bottom, dragged by the jet of secondary air and burnt off completely into vortex motion.

Q. *The excess air ratio, in dry-bottom furnaces, at the furnace exit is 1.15–1.2, despite the fact that the excess air ratio in the burners is lower (1.05–1.1). Why?*

Ans. It is because of the inevitable ingestion of cold air into the furnace from outside.

Q. *What are the slag-bottom type pulverized fuel fired furnaces?*

Ans. These furnaces are designed so as to remove the slag in the molten state.

Q. *What is the fundamental requirement for this?*

Ans. The temperature of the combustion products in the lower portion of the furnace should be higher than the slag fluidity.

Q. *What should be the temperature of the hot gas at the bottom?*

Ans. It should be

$$\Theta_{gas} \geq [\Theta]_{soft}^{Ash} + (50 - 100°C)$$

where, $[\Theta]_{soft}^{Ash}$ = Ash softening temperature

Q. *How can such conditions be provided at the furnace bottom?*

Ans. By shifting the flame closer to the furnace bottom and insulating the waterwall tubes in this zone with carborundum refectories.

Q. *How is this heat-insulation applied?*

Ans. For sound attachment of the refractory to the tubes, pins 10–12 mm in dia and 12–15 mm long are first welded on the waterwall tubes' surface facing the fireside. Then refractory coating is applied to ensure heat insulation.

Q. *What is the nature of the bottom of the slag-bottom type furnace?*

Ans. It is horizontal or slightly inclined towards centre of the furnace.

Q. *What is the profile of the molten slag in the slag-bottom furnace?*

Ans. The slag-bottom type furnace is provided with one or two refractory lined slag holes (500 mm × 800

mm) at the bottom-centre. Molten slag pours off these holes into the slag bath where they solidify on being quenched with water.

Q. *What fraction of the total ash is removed as molten slag in the slag-bottom type furnace?*

Ans. About 20–40 percent.

Q. *How can the combustion of pulverized solid fuel be organized in a slag-bottom type furnace?*

Ans. It can be done in either of the three following ways:
  (a) with straight flow flame
  (b) with intersecting flame jets
  (c) with vortex flame

Q. *How many types of slag-bottom furnace are there with straight flow flame?*

Ans. Two:
  (a) straight-wall furnace [Fig. 12.13(a)]
  (b) constricted-section furnace [Fig. 12.13(b)]

Q. *What are the main disadvantages of the straight-wall (open- shaft) furnace?*

Ans.
  1. Limited combustion control at loads below 70 to 80% of the rated capacity
  2. Slag solidification is encountered on the waterwalls and then at the bottom as a large proportion of generated heat is given up to the upper zone
  3. Only 10–15% of the total ash is removed as liquid.

Q. *What are the advantages of constricting two opposite walls in the slag-bottom furnace with straight flow flame?*

Ans. It isolates the combustion chamber within the furnace
  (a) reducing heat flow substantially to the upper zone
  (b) ensuring high temperature (1600–1800°C) in the combustion chamber
  (c) removing the 20–40% of total ash in molten state
  (d) enabling the boiler to run in a wider range of loads.

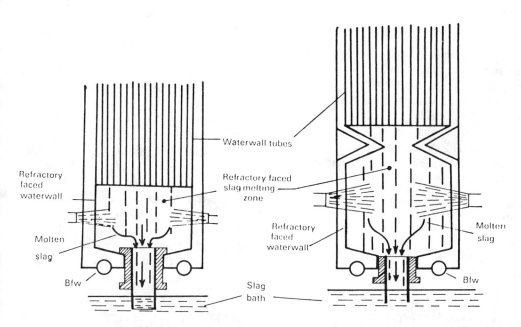

Fig. 12.13 (a)   *Straight-wall furnace*   **(b)**   *Constricted section furnace*

Q.   *How many types of slag-bottom furnaces with intersecting jets are there?*

Ans.   Two types:
(a)   single-wall constricted (Fig. 12.14a)
(b)   both-walls constricted (Fig. 12.14b, c)

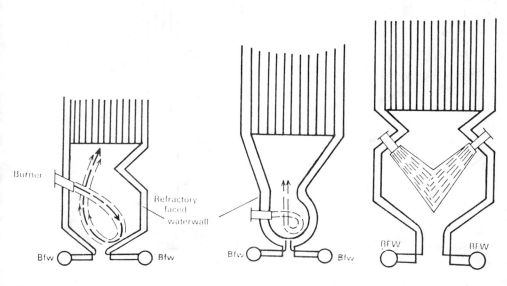

Fig. 12.14 (a)   **(b)**   **(c)**

Q. *What is the flame profile in slag-bottom furnaces with intersecting jets?*

Ans. Such furnaces are fired with straight flow burners which are so arranged as to turbulize the fuel-air feed around the horizontal axis. Since the combustion takes place in the flame, the flame propagates in turbulence around the same horizontal axis. It rebounds from the refractory faced waterwall opposite, makes a full turn and after that hot gases of combustion intersect with the on-coming jets of fresh fuel dust-air mixture and quickly reheat them to ensure sustained combustion.

Q. *How many types of slag-bottom cyclone furnaces are there?*

Ans. Two:

(a) horizontal cyclone-type primary furnace (Fig. 12.15A)

(b) bottom-type primary furnace open at the top (Fig. 12.15B)

Q. *How does combustion in cyclone furnaces proceed?*

Ans. A high degree of turbulence is induced to the coal dust-air mixture in the primary furnace by either tangential high-speed jets of secondary air (80–125 m/s) or tangential fuel dust- air mixture jets from burners. The combustion takes place in the flame.

The fuel particles in the primary furnace are acted upon by two forces: centrifugal and aerodynamic.

The centrifugal force throws the fuel particles onto the walls of the primary furnace while the aerodynamic force tends to lift the fuel particles and gases from the primary furnace. The ratio of these two forces depends on the particle size and that's why the fuel particles get unevenly distributed in the primary furnace. While the coarser particles, upon which the

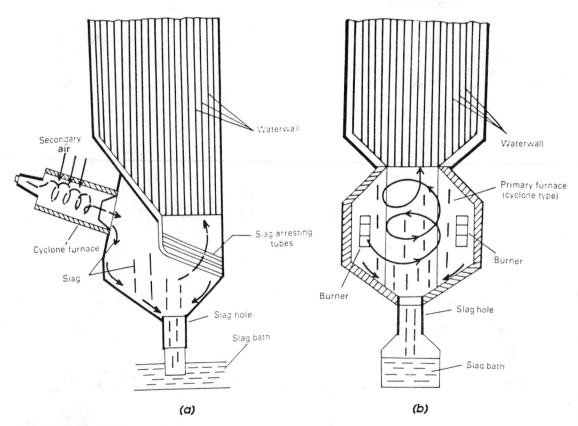

Fig. 12.15 Slag-bottom cyclone furnace

centrifugal force is more prevalent than the aerodynamic force, are thrown against the furnace walls and get involved in the vortex motion until they burn off completely, the finer fractions, under the influence of more aerodynamic force than centrifugal force burn in the centre.

Q. *Briefly describe a slag-bottom horizontal cyclone-type primary furnace.*

Ans. These are usually made to a diameter of 1.8 to 4 m while in length the cyclone is 1.2 to 1.3 times its diameter.

Anything from pulverized fuel to crushed fuel can be used, and this saves the cost of fuel pulverization.

The furnace temperature is as high as 1800–1900°C (2073–2173K) and the heat release rate is high (2–6 MW/ m³).

Large fraction of ash is removed to slag. About 60 to 85% of total ash is removed as slag.

Excess air ratio 1.05–1.10

Heat power of the cyclone 150–400 MW.

Q. *The **Total** heat release rate of horizontal cyclone-type furnace does not exceed 0.2 to 0.3 MW/m³, in spite of your above statement. Why?*

Ans. It is because of the extended shaft that is needed to cool the furnace gases.

Q. *What are the primary advantages of a slag-bottom type furnace over a dry-bottom type furnace?*

Ans.

1. The heat loss with unburnt carbon, when the same kind of fuel is burnt, can be reduced by one-third in a slag-bottom type furnace.
2. The total heat release rate per unit furnace volume of a slag-bottom type furnace comes out to be 20% higher on the average. And that means the slag-bottom type furnace can be made smaller and more compact.
3. Because of better tightening of the bottom portion, there is less infiltration of air into the furnace. As a result, the heat loss with waste gases is somewhat lower.
4. Because of molten slag, the ash disposal system is less expensive.

Q. *What are the drawbacks of a slag-bottom type furnace?*

Ans.

1. Not all fuels are suitable for a slag-bottom type furnace. Fuels with low ash melting point, 1150–1300°C (1423–1573K), can be conveniently burnt while fuels having ash melting point 1350°C (1623 K) present particular difficulties, and the conditions for the formation of molten slag should be accurately calculated.
2. A great amount of heat is lost with the slag since more ash is removed as high temperature slag. And this may be more than the heat saved with the minimization of unburnt carbon.
3. The possibilities of boiler load control are restricted by the conditions of slag removal in the case of single-shaft slag-bottom type furnace.
4. High flame core temperature brings about a higher yield of NO_x than a dry-bottom type furnace.

Q. *What fuels are advantageous for combustion in slag-bottom furnaces?*

Ans.

1. Low-reactive fuels anthracite, semi-anthracite, lean coals
2. Fuels with low ash melting point.

Q. *Why are fuels with low ash melting point not suitable for dry-bottom furnaces?*

Ans. They may cause severe clinkering of the waterwalls.

13

Fuel Oil and Gas Fired Furnaces

Q. *Why can natural gas and fuel oil be burnt in furnaces of the same design?*

Ans. Because of the common combustion characteristics:

1. Both natural gas and fuel oil contain practically no moisture and they give rise to roughly the same volume of combustion products. Thus the blowers of steam boilers run efficiently irrespective of whether fuel oil or natural gas is being burnt in boiler furnace.
2. Combustion of either fuel takes place in the vapour state and the intensity of burning in either case is determined by the conditions of intermixing. Combustion mechanism is the same.
3. Both fuels have nearly the same value of highest allowable heat release rate per unit volume of the furnace. For fuel oil it is 300 kW/m^3 and for natural gas 350 kW/m^3. And as such for the same steam output of a boilers, the furnace dimensions for these two kinds of fuel can be taken to be the same for practical purposes.
4. Both fuels are practically free from ash-formation problem upon combustion. Therefore, clinkering of waterwall tubes does not arise and slag-handling facilities are unnecessary. This is why furnaces for both fuels are designed with a horizontal or slightly inclined bottom.
5. More homogenization of air-fuel mixture is possible as both fuels are in a vapour state prior to ignition. This means virtually complete combustion with less amount of excess air.

6. For both fuels, air can be preheated to the same temperature 250–300°C (523-573K) making it possible to install combined gas-fuel oil burner system.
7. Both fuels produce relatively short flame core-zone near the burners during intensive burning.

Q. *How many type of gas and fuel oil fired furnaces are there?*

Ans. Five types:

1. Open-type furnace with single-front, multi-tier burners layout. (Fig. 13.1a)
2. Double-wall constriction and double-front burners. (Fig. 13.1b)
3. Open-type furnace with opposite double-tier burner arrangement. (Fig. 13.1c)
4. Open-type furnace with opposite cyclone primary furnaces. (Fig. 13.1d)
5. Furnace with straight flow having burners at the bottom. (Fig. 13.1e)

Q. *In how many tiers are the burners laid out in the single-front arrangement?*

Ans. Usually three or four tiers.

Q. *What are the advantages of this arrangement?*

Ans. Less expensive and operation is more convenient.

Q. *What are its drawbacks?*

Ans. The chief disadvantages being:
(a) non-uniform filling of furnace space by the flame

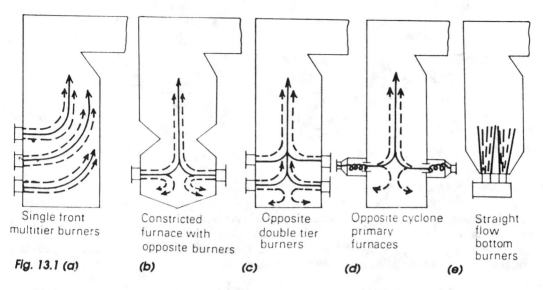

| Single front multitier burners | Constricted furnace with opposite burners | Opposite double tier burners | Opposite cyclone primary furnaces | Straight flow bottom burners |

Fig. 13.1 (a)   (b)   (c)   (d)   (e)

(b) incompatible with furnaces of short depth (< 6m) because of intolerably high heat absorption and, with the consequence, high temperature rise of the rear waterwall.

Q. *What are the primary advantages of opposite arrangement of burners?*

Ans.

1. Higher heat release rate (20–30% higher than single-front burners) in the flame core zone
2. Conditions are favourable for concentrating the flame at the centre resulting in the uniform distribution of flame heat throughout the furnace space.

Q. *When a boiler is switched over from fuel oil to natural gas firing, heat absorption in the furnace space decreases. Why?*

Ans. Flames produced by natural gas have lower emissivity than those produced by fuel oil upon combustion. Since the mode of heat transfer within the furnace space is of radiative type, decrease in flame emissivity results in lowering of heat absorption rate.

Q. *Why are the combustion products at the furnace outlet at a higher temperature in the case of natural gas firing system than fuel oil firing?*

Ans. As heat absorption rate in the furnace space decreases because of lowering of flame emissivity when the furnace is fired with natural gas, the combustion products will exit the furnace at higher temperature.

Q. *What changes are done when a furnace with straight flow having burners at the bottom is switched over from fuel oil firing to natural gas firing? Why?*

Ans. The degree of whirling of secondary air flow is increased. During fuel oil combustion, the degree of swirling of secondary air is decreased to allow the flame to reach up to a greater height of the furnace with the effect that local heat flow to the waterwalls gets noticeably decreased.

And when firing natural gas, the degree of swirling is increased to make the flame wider and shorter.

Q. *Sometimes fuel oil or furnace oil is referred to as an ideal kind of fuel. Why?*

Ans. It is composed almost entirely of hydrocarbons. It contains only traces of oxygen, negligible amounts of moisture and ash. Hence the combustion of F.O. is practically the combustion of hydrocarbons with no solid residue leftover. Hence it is called ideal fuel.

Q. *How is fuel oil burnt in the combustion chamber?*

Ans. Three ways:

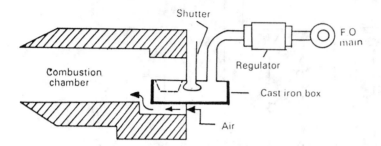

Fig. 13.2 *Simplest and the oldest method for direct firing of fuel oil*

(a) direct combustion of F.O. in combustion chamber, in natural mass
(b) combustion. of pulverized F.O.—pulverization being carried out by atomizers or spray burners.
(c) combustion of vapourized F.O.—vapourization being carried in drip conversion boxes.

Q. *What is the direct method of combustion of F.O.?*

Ans. It is the simplest and oldest method. Oil flows, by gravity, into a cast iron box placed in the combustion chamber. (Fig. 13.2).

Air flow into the combustion chamber is controlled by means of a shutter above the box and channels below the box.

Fuel consumption is 4 kg/h and flame length approximately 1 m.

Q. *For what purpose is it used?*

Ans. Chiefly used in small-capacity room heaters and cooking stoves.

Q. *What do you mean by pulverization of liquid fuel?*

Ans. It means atomization of liquid fuels, i.e., dispersing liquid fuel jet in a fine churn of spray into the dispersing medium.

Q. *Why is it necessary?*

Ans. For efficient combustion of liquid fuels.

Q. *How?*

Ans. As the fuel oil gets pulverized, its surface area increases manyfold and finely dispersed fuel oil par-

ticles become intimately mixed with combustion air. Greater the surface area exposed for combustion and better the intermixing with the air, more complete and rapid will be the combustion process.

Q. *How can F.O. be atomized?*

Ans. By three methods:
(a) **Steam Atomization**—fuel oil, at the exit of the burner, is pulverized by the atomizing action of one or more jets of steam.
(b) **Air Atomization**—fuel oil is atomized in the same way as steam atomization—the only difference is that air instead of steam is used as the atomizing agent.
(c) **Mechanical Atomization**—fuel oil is broken down into finely divided minute particles by means of mechanical devices.

Q. *What are the advantages of steam atomized oil burners?*

Ans.
1. Simplicity in burner design
2. Easy operation of the burner
3. Low preheating temperature (upto 80°C) for fuel oil
4. Very useful in smaller installations.

Q. *What is their main drawback?*

Ans. High consumption of steam for atomization (30–40% of fuel oil consumption).

Q. *How are steam atomized oil burners classified?*

Ans. The classification is based on the nature and

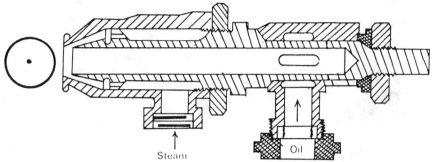

Fig. 13.3 Shukhov burner

shape of the flame produced. Accordingly, they are classified into:

(a) **Round shaped burners**—that produce a round, long or short flame, with slightly expanded central zone.

(b) **Flat burners**—that produce a long tape-like flame or short turbulent flame.

Q. *Apart from the shape of the burner nozzle, what is the other main difference between round-burners and flat-burners?*

Ans. It is the flow paths of the fuel and its atomizer.

In the round burners, the stream of fuel oil is allowed to pass through the inner or outer pipe whereas in the case of flat burners the fuel oil flow is directed either through the same path as the steam or through a channel parallel to the steam channel.

Q. *What is the mechanism of steam atomization of fuel oil?*

Ans. A jet or jets of steam ejected at a pressure of 0.4 to 0.6 MN/m^2 are directed upon a stream of fuel oil. As the steam with high kinetic energy collides with the jet of fuel oil, it, by imparting its kinetic energy to F.O. disintegrates the fuel oil stream into minute droplets, approximately 0.01 mm or even less in diameter. These disintegrated particles are called pulverized fuel oil and that's how steam atomization of fuel oil is effected.

Q. *Briefly describe some steam atomized fuel oil burners.*

Ans.

1. **Shukhov burner:** made of phosphor-bronze.

Consists of two concentric tubes—inner tube for oil and outer tube for steam flow. (Fig. 13.3).

While the cross-sectional area of the oil outlet remains constant, the size of the ring-shaped passage for steam may be increased or decreased by the to-and-fro movement of inner tube with the help of a handwheel.

Fuel oil delivery through the inner tube is controlled by means of a gate valve mounted on the fuel oil main while the steam supply can be doubly controlled by the handwheel as well as by the gate valve installed in the steam main. Consumption of fuel oil is controlled either by change of the velocity of discharge or by changing pressure in the spray burner.

Oil consumption: 3.5–195 kg/h
Oil aperture: 2–16 mm
Steam aperture: 4.5–20 mm

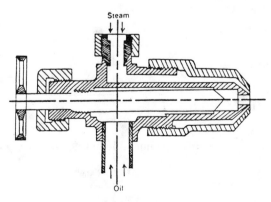

Fig. 13.4 Vagener burner

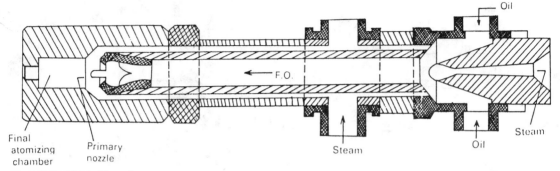

Final
atomizing   Primary
chamber   nozzle

Fig. 13.5 *Varganov burner*

2. **Vagener burner:** Similar to the Shukhov burner. Oil is delivered through the inner tube and steam in the annulus between the inner tube and the nozzle. (Fig. 13.4).

The outer surface of the inner tube is fitted with oblique ribs to produce swirling motion in the steam to provide better intermixing of air with atomized fuel.

Normal capacity = 100 kg F.O. /h

Now this capacity can be increased by 50% without appreciable steam consumption.

3. **Varganov burner:** is fitted with multiple nozzles to effect atomization as well as swirling motion of both oil and steam.

Oil flows through the inner pipe to the inserted nozzle that delivers it into the atomizing chamber in the form of a swirl. At the same time, steam is delivered through the outer pipe to the nozzle. The pattern of threading on the cylindrical portion of the nozzle induces a rotary motion in the same direction

as the F.O. stream. Owing to the peculiar shape of the atomizing chamber, the spiral rotary flow of steam smoothly narrows down before the primary nozzle and strikes the spiral flow of the fuel oil obliquely. This preheats the oil as well as accelerates the flow. (Fig 13.5).

From the atomizing chamber, the steam-fuel oil mixture enters the final atomizing chamber through the primary nozzle. Here, further intermixing takes place.

Ultimately, the ready-mixture issues through the final nozzle into the combustion chamber where a further atomization of fuel takes place under the action of centrifugal force.

4. **Saha's burner:** was designed by A.K. Saha. Oil is distributed to pass through several holes (3mm dia) while atomizing steam is passed through the central pipe to the central nozzle of 4 mm dia. (Fig. 13.6)

Burner capacity 30 – 35 kg F.O./h

Q. *Apart from the classification based on the shape*

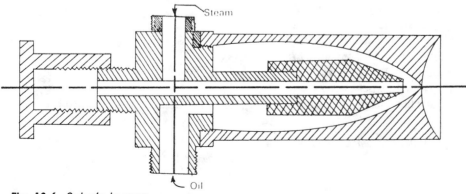

Fig. 13.6 *Saha's burner*

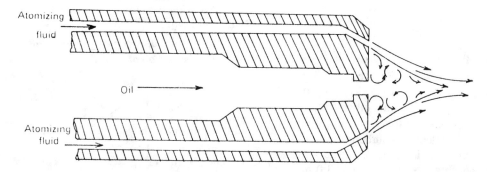

Fig. 13.7 *Internal mixing burner*

of the flame, how can steam and air atomized burners be classified?

Ans. They can be classified as:

  (a) **Internal Mixing Burners**—where atomizing fluid (steam or air) and oil impinge within the burner assembly. (Fig. 13.7).

  (b) **External Mixing Burners**—where steam (or air) is directed into the oil stream or oil is directed into the atomizing fluid at the burner outlet.

Q. *How many kinds of steam (or air) atomizers are there?*

Ans. Three.

  1. **Low pressure atomizer:** using air (0.105–0.115 MN/m²) as the atomizing medium. A larger proportion and sometimes the entire combustion air is used as atomizing fluid. **Turndown Ratio 2 : 1–5 : 1**

  2. **Medium Pressure Atomizer:** using air at 0.23–0.3 MN/m². Less than 10% of the combustion air is required to atomize the fuel. **Turndown Ratio 10 : 1**

  3. **High Pressure Atomizer:** using air or steam at pressure exceeding 0.3 MPa (0.3 MN/m²). Steam requirement : 0.3–0.5 kg/kg of fuel oil.

Q. *What is called turndown ratio?*

Ans. It is the ratio of maximum to minimum load.

Q. *Briefly describe an air atomized burner?*

Ans. A commonly used air atomizer is as shown below. Oil flows through the central pipe fitted with the nozzle at one end and oil regulator at the other end. The oil stream issuing from the nozzle meets at a sharp angle the air stream flowing through the outer cylinder. Ribs are fitted to the path of the air flow to induce vorticity to air stream for better mixing with the pulverized fuel. (Fig. 13.8).

Q. *What amount of air is required for the pulverization of 1 kg of fuel oil?*

Ans. Theoretically 14 kg (approx).

Q. *How is droplet size dependent on the viscosity of F.O.?*

Ans. It is directly proportional to the viscosity of F.O.

Q. *How does droplet size vary with the pressure of fuel oil?*

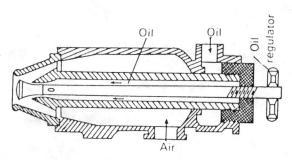

Fig. 13.8 *Air atomized oil burner*

Ans. It varies inversely with the cube root of fuel oil pressure.

Q. *How is mechanical pulverization of fuel oil carried out?*

Ans. This is carried out by delivering fuel oil under excessive pressure (2.5–4.5 MPa, i.e., 2.5–4.5 MN/m^2) into the whirling chamber and forcing it through a narrow hole or nozzle. Fuel oil enters the whirling chamber through a number of tangential channels and it acquires intensively turbulized spiral motion. Depending on the ratio of the tangential and axial velocity components, the whirling fuel oil film expands at the burner outlet and gets disintegrated by the oncoming air flow into numerous fine particles traversing a parabolic path. (Fig. 13.9)

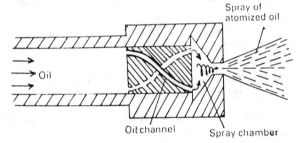

Fig. 13.9 *Mechanical pulverization of oil*

Q. *Upon which factors do the fineness of atomization and the size and shape of the fuel oil particles depend?*

Ans. These depend on the following three factors:

1. **Burner Capacity** which in turn depends on nozzle diameter and fuel oil pressure.

2. **The Spiral Pitch of the Fuel Oil** which is the function of the nozzle diameter

3. **Flow Resistances** as offered by the turbulent flows and frictions.

Q. *How is the throughout capacity of a mechanical burner related to the fuel oil pressure?*

Ans. The relation is:

$$B = b \, \mu \, A \, \sqrt{P \rho}, \text{ kg/s}$$

where B = throughput capacity of the burner, kg/s
b = multiplication factor
μ = flowrate coefficient
A = cross-sectional area of the ejecting nozzle, m^2
P = fuel oil pressure, MPa (MN/m^2)
ρ = fuel oil density, kg/m^3

Q. *What is the throughput capacity of a powerful centrifugal burner?*

Ans. It varies from 3 to 16 t/h for burners with nozzle diameter

$$d_o = 4 - 10 \text{ mm.}$$

Q. *What are the principal characteristics of a mechanical burner?*

Ans. The principal characteristics of a burner are:

1. **Dimensionless Geometric Parameter, Z**

$$Z = R \, r_o / r_i^{\,2}$$

where R = distance between the burner-axis and axis of the tangential channel
r_o = nozzle radius
r_i = radius of the internal gas whirl

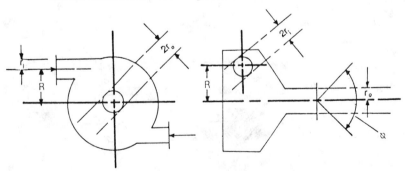

Fig. 13.10

2. **The Coefficient of Free Cross-sectional Area**

$$f = 1 - (r_i/r_o)^2$$

3. **The Flowrate Coefficient, μ**
4. **The Jet Expansion Angle, ϕ**

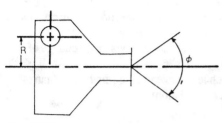

Fig. 13.11

Q. *How do these characteristics vary with the geometric characteristics of a burner?*

Ans. Fig. 13.12.

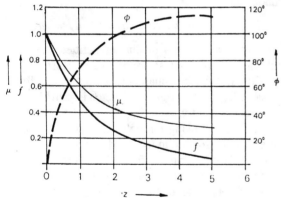

Fig. 13.12

Q. *Describe a mechanical atomizer.*

Ans. **Körting atomizer:** It consists of a pear shaped nozzle into which is concentrically abutted a rod with rectangular threading at its nozzle-end. (Fig. 13.13)

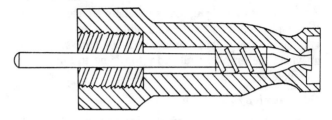

Fig. 13.13 *Körting burner*

Fitting closely to the inner walls of the nozzle, the threading provides high pitched passages. The rod terminates with a pointed end, the outer shape of which conforms exactly to the inner shape of the nozzle.

The outlet aperature of the rod varies from 1 to 2 mm in diameter depending upon the spray capacity of the burner which varies from 50 to 80 kg of fuel oil per hour.

F.O. preheated to 50–60°C is directed to the burners at 5–12 atm. pressure.

Adjustable Mechanical Atomizer: Fitted with gadgets to adjust and alter the nozzle-diameter, thus controlling and adjusting the capacity of the burner. (Fig. 13.14).

Q. *Mechanical burners have a narrow throttling range. Why?*

Ans. Its throughput capacity, i.e., flowrate through the nozzle is directly proportional to the square root of pressure of the fuel oil feed. As a result this type of

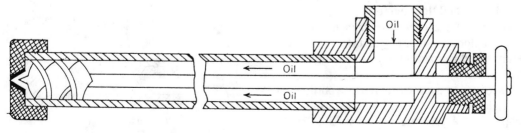

Fig. 13.14 *Adjustable mechanical atomizer*

burner has a narrow throttling range, because doubling the pressure will increase the rate by 40% only.

Q. *What will happen if, during atomization, the fuel oil pressure is gradually decreased?*

Ans. As the fuel oil pressure is gradually decreased, the droplets of atomized fuel oil will grow in size. Ultimately there will be a minimum fuel oil pressure at which the droplets will grow too large to burn completely.

Q. *What is a rotary burner?*

Ans. It is a special type of burner in which atomization of fuel oil is achieved by centrifugal force. (Fig. 13.15).

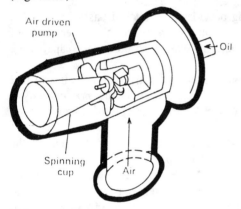

Fig. 13.15 *Rotary fuel oil burner*

The preheated oil streams through the central pipe and ejects upon the inner surface of the spinning cup—a hollow tapered cup revolving round the same horizontal axis of oil flow.

As the oil spreads over in the form of a thin film over the surface of the rotating cup, it is thrown off the periphery of the free end of the cup by centrifugal force. The velocity of oil flow and its thickness over the circumference of the cup-rim remain uniform because of the uniformity of the cup surface.

The result: Good atomization and uniformity in droplet size

The cup is rotated by a turbine driven by a portion (15%) of the atomizing air required for combustion. Air is supplied at a pressure of 0.2–0.35 MPa (MN/m^2).

Q. *What is the turndown ratio of a rotary atomizer?*

Ans. 5 : 1.

Q. *What are its advantages?*

Ans.
1. Good atomization
2. Droplets produced are more uniform in size than those produced by any other method.
3. Much less sensitive to viscosity change of fuel oil.
4. Much less liable to clogging by grit than other mechanical atomizers.

Q. *What is its main drawback?*

Ans. Carbon deposition.

When the burner is taken shutdown, deposition of carbon particles, due to thermal cracking of fuel oil induced by radiation heat from the hot surroundings, takes place on the rotary cup surface, destroying its smoothness.

Q. *Briefly decribe a steam-mechanical burner.*

Ans. It is a common oil burner befitted with an additional capacity of steam atomizing. Instead of centrifugal whirling chamber, an axial whirling device fitted with a conical dissector is provided in the burner assembly.

Oil stream flowing through the central pipe acquires a spiral motion in the whirler and ejects from the conical dissector in the form of jet. (Fig. 13.16).

Steam, supplied at a pressure of 0.2–0.4 MPa, passing through the annular channel at near critical

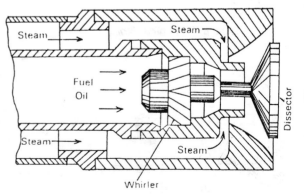

Fig. 13.16 *Steam-mechanical burner*

speed penetrates the fuel oil jet and disintegrates it into fine droplets.

Steam consumption for atomization: 10% of F.O. flowrate

Capacity control: 20–100% of the rated value.

Q. *Why is the quantity of atomizing air requirement much higher (1.5 times higher) than that of atomizing steam when F.O. is atomized?*

Ans. This is due to the higher energy content of steam.

Steam jet with higher thermal and kinetic energy penetrates the fuel oil jet more easily than air jet with lower energy content. Upon collision with the oil jet, steam imparts more thermal as well as kinetic energy to the oil stream which therefore disintegrates into fine droplets. Hence, for a given quantity of fuel oil pulverization, steam requirement is less than air requirement.

Q. *In steam-mechanical burners oil is ejected form the nozzle in the form of a thin hollow cone. What is the range of thickness of this cone?*

Ans. 50–100 μm.

Q. *What is secondary atomization?*

Ans. The oil film ejected from the burner nozzle in the form of a thin hollow cone expands and breaks into fine droplets which proceed at a speed 60–80 m/s. These particles get further atomized by the dynamic pressure of the oncoming stream of the atomizing medium (air or steam). This process is called secondary atomization.

Q. *Upon which factors, does the droplet size of atomized oil depend in the case of steam-mechanical burners?*

Ans.
1. Energy of the atomizing steam
2. Efficiency of steam utilization for fuel oil atomization.

Q. *The size of the largest oil droplet in the case of powerful mechanical burners may be as large as 1.5–2 mm in size. So why should the heat release rate per unit volume of the furnace using these powerful mechanical burners not be more than 200–250 kW/m³?*

Ans. Because of the size, the larger droplets (1.5–2 mm) will take more time to evaporate and burn (roughly equal to 1.5–2 s).

Hence from the correlation between the heat release rate per unit volume and the residence time of the furnace gas (that vertically lifts the larger droplets upwards) we find that heat release rate/m³ of the furnace should not exceed 200–250 kW/m³ so as to ensure complete combustion of the fuel. (Fig. 13.17).

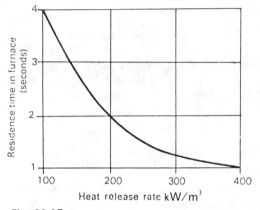

Fig. 13.17

Q. *How many methods are there to control the throughput capacity of a fuel oil burner?*

Ans. By two methods:
(a) by controlling the fuel oil pressure (qualitative method)
(b) by switching some of the burners on and off (quantitative method)

Q. *What difficulties are encountered in the qualitative method of fuel oil control?*

Ans. In order to reduce the load, it requires substantial reduction of initial pressure. For instance, if throughput is to be reduced from 100% to 60% of the rated value, it turns out that fuel oil pressure must be reduced to 0.37 of the initial value forasmuch as throughput capacity is proportional to the square root of fuel oil pressure. On the other hand, a sharp drop in fuel oil pressure in mechanical burners will result in a lower intensity of whirling, producing a thicker film, lower velocity at nozzle outlet and larger droplets. Hence sharp pressure reduction is inadmissible.

Whereas an increase of F.O. pressure to increase the burner throughput will need more intricate and expensive equipment for oil transport and flowrate control. And all these would mean higher operating cost.

Q. *Why is the throughput capacity of the mechanical burners controlled by the combined use of qualitative and quantitative method?*

Ans. As stated above.

Q. *How does quantitative control lend itself to qualitative control?*

Ans. When a boiler furnace is fitted with a good number of oil burners, the load on the boiler can be reduced by switching off some of the burners in a group. This will result in a higher delivery pressure of the F.O. supplied to the rest of the burners in line, thus allowing the boiler load to be further controlled by varying the fuel oil pressure.

Q. *Steam-mechanical burners are very useful for high-capacity steam boilers. Why?*

Ans. Because they ensure throughput control in the whole range of operating loads. At lower loads the fuel oil is atomized by the energy of steam.

Q. *How are gas burners classified?*

Ans.
1. BURNERS IN WHICH GAS-AIR MIXTURE IS PREPARED PRIOR TO COMBUSTION:
   (a) High pressure injection burners with single-stage suction.
   (b) High pressure injection burners with multi-stage suction.
2. BURNERS IN WHICH GAS-AIR MIXTURE IS PRODUCED IN THE COURSE OF COMBUSTION PROCESS:
   (a) Diffusion burners
      (i) cylindrical type
      (ii) flat type
      (iii) combined type
   (b) Irregular turbulent burners:
      (i) parallel tubes (coke-oven burners)
      (ii) ports (open-hearth furnace burners)
      (iii) channels (glass furnace burners)
3. SUBMERGED COMBUSTION BURNERS
   (a) Low pressure injection type of Saha

(b) Kemp and Hammond burner

Q. *What volume of hot air is required to burn 1 m³ of natural gas?*

Ans. Roughly 20 m³.

Q. *Why is the gas supply channel smaller in cross-sectional area than the air duct of gas burner?*

Ans. It is because the volume of combustion air required per m³ of natural gas burned is very large.

Q. *How can the proper intermixing of the gas and air be ensured?*

Ans. By directing the stream of gas in the form of thin jets of high penetrability into the air stream.

Q. *How can the degree of penetrability of natural gas jets in air stream be increased?*

Ans. By raising the gas velocity to as much as 120 m/s.

Q. *What is the air velocity?*

Ans. It is 25–40 m/s.

Q. *How is gas introduced to the air stream?*

Ans.
1. Single jet entering at right angles to airflow. (Fig.13.18)

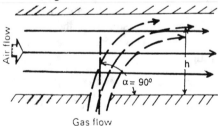

Fig. 13.18

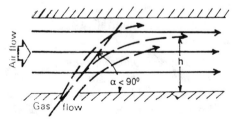

Fig. 13.19

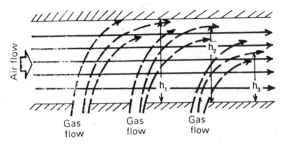

Fig. 13.20

2. Single jet entering at angle $\alpha < 90°$ to airflow. (Fig. 13.19)

3. Multiple jets entering at $\alpha > 90°$ to airflow. (Fig. 13.20)

Q. *What is the depth of penetration?*

Ans. It is the distance *(h)* along the normal from the root of the jet to the point where the direction of the jet becomes coincident with that of the airflow.

Q. *On which factors does the depth of penetration depend?*

Ans. Diameter of the jet and the ratio of gas and air stream velocities.

Q. *What are the principal parameters of a gas-burner?*

Ans.

1. **Relative length (l_m) of the internal mixing zone**

$$l_m = L_m/D$$

where D = diameter of the outlet port of the burner. (Fig. 13.21).

2. **Aerodynamic Parameters** determining the intensity of turbulent mixing

$$n = [\rho_a/\rho_g]\,[v_a/v_g]^2$$

where ρ = density

v = flow velocity

"a" stands for air and "g" stands for fuel gas.

3. **Dimensions, shape and arrangement of gas ports**

Q. *Name two injection burners.*

Ans.

1. Bunsen burner.
2. Teklu burner.

Q. *What are diffusion flames?*

Ans. These are the flames produced when fuel gas and air flow separately into the combustion chamber and undergo intermixing as the burning proceeds.

Q. *What determine the rate of combustion in diffusion flames?*

Ans.

1. Fuel-gas/air ratio
2. Degree of turbulence of gas and air streams.

Q. *What is an important characteristic of such flames?*

Ans. These flames are very luminscent in character which contributes to a high degree of flame emissivity

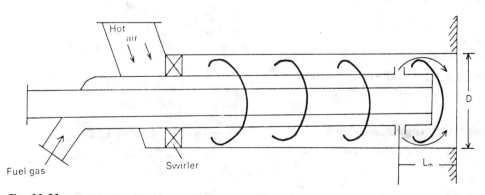

Fig. 13.21

effecting, therefore, a high degree of radiative heat transfer.

Q. *Why are these flames luminscent?*

Ans. Since the intermixing of the gas and air does not take place in the burner proper but in the furnace space, a part of the gas (mainly hydrocarbons) undergoes pyrolysis (thermal dissociation) in the high temperature zone where oxygen is deficient. As a result, carbon particles (soot) are produced. These incandescent carbon particles emit luminous flame.

Q. *What are the advantages of diffusion flame burners?*

Ans.
1. They do not backfire as there is no air within the burner to sustain combustion.
2. Particularly useful where very large volumes of low calorific value gases are to be burned.
3. Very effective when both gas and air are to be preheated, but not to be intermixed because of the dangers of pre-ignition, to produce a high flame temperature.
4. Combustion is silent.
5. Quality and nature of fuel gas can be varied without affecting burner operation.
6. No necessity of accurate adjustment of the gas and air ports.
7. Gas supplied is at low pressure (up to 0.110 kPa) and hence the system is less expensive.

Q. *What flames are called premixed flames?*

Ans. These are flames produced by the burners where gas and air are mixed whilst cold and combustion proceeds as the mixture leaves the burner assembly.

Q. *What are combined gas-fuel oil burners?*

Ans. These burners are provided with the capacity of burning fuel oil and fuel gas simultaneously or separately.

Q. *What is the chief advantage of such burners?*

Ans. Both fuels can be burned under almost optimal conditions.

Q. *What is an air register?*

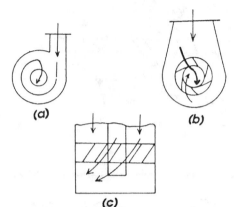

(a)　　*(b)*

(c)

Fig. 13.22 *Different types of air-registers*

Ans. It is an air-guiding device to effect intensive turbulization of airflow for efficient intermixing with the fuel.

Q. *How many types of air-registers are there?*

Ans. Three:
1. Scroll type (Fig. 13.22A)
2. Tangential vane type (Fig. 13.22B)
3. Axial vane type (Fig. 13.22C)

Q. *What is the main drawback of a scroll type register?*

Ans. Too bulky.

Q. *Where is it employed?*

Ans. In low-capacity burners.

Q. *Out of these three air-registers which offers least hydraulic resistance?*

Ans. Air-register with axial vanes.

Q. *How many types of gas-oil combined burners are mainly used in high-capacity boilers?*

Ans. Two types:
1. **Co-axial Gas-Fuel Oil Burner** with central gas supply. (Fig. 13.23)
2. **Gas-Fuel Oil Burner** with peripheral and central gas supply. (Fig. 13.24)

Q. *How does the co-axial gas-fuel oil burner with central gas supply operate?*

Ans. From the central annular header the natural

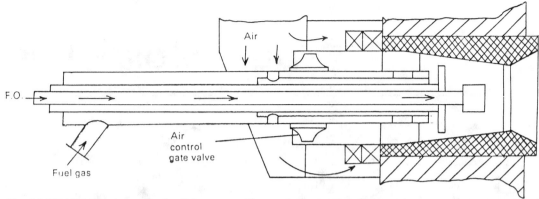

Fig. 13.23 Co-axial gas-fuel oil burner with central gas supply

gas flows through two rows of holes of different diameter, while air is delivered through a tangential vane register imparting vorticity to the air stream. The flowrate of air is controlled by a moving disc valve. So when the boiler load is reduced, air flowrate decreases maintaining the original whirling intensity and the conditions necessary for better fuel-air inter-mixing.

Fuel oil supplied through a pipe symmetrically mounted in the natural gas duct is atomized and burnt in the fuel oil burner (mechanical).

Air velocity = 40 m/s

Gas supply pressure = 2.5–3 kPa

Ignition of the mixture (fuel oil-air or fuel gas-air) is accomplished by an electric ignitor.

Q. *How does the gas-fuel oil burner with peripheral and central gas supply operate?*

Ans. Fuel oil is atomized in a steam-mechanical burner mounted in the central channel to which are charged both air and natural gas. This air is supplied to cool the fuel oil burner. (Fig. 13.24)

The main air supplied through central and peripheral air boxes is passed through axial and tengential vanes. At reduced boiler load, the air flowrate through the peripheral annular channel is diminished by means of a control gate.

Fuel oil is supplied at a pressure of 4.5 MPa.

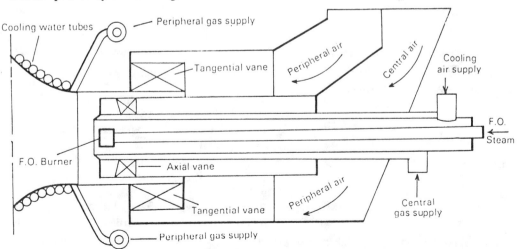

Fig. 13.24 Gas-fuel oil burner with peripheral and central gas supply

Q. *Why is dust collection a necessity?*

Ans. The flue gas exhaust of pulverized coal fired boilers contains flyash which constitutes 80% of the ash in the coal. They are so fine that most of them pass through a 300 mesh screen. They are an environmental hazard from the point of view of human and animal health and crop growth. Hence their concentration in flue gas must be brought down below 0.5 g/Nm^3 before letting it out free in the atmosphere.

Emission Standards for Thermal Power Plants

Rating	Concentration of Particulate Matter in Flue Gas (mg/Nm^3)		
	Protected Area	Other Area	
		After 1979	Other
< 200 MW	150	350	600
> 200 MW	150	150	—

Q. *What is the basic difficulty in eliminating the flyash?*

Ans. It is their fineness that escapes removal. The size of the particles varies from 1 μ to 80 μ.

Q. *What is the ash percentage in Indian coals?*

Ans. 25 to 50%.

Q. *How can dust be removed from the flue gas?*

Ans. Mechanically and electrically.

Q. *What are the basic requirements for a good dust collector?*

Ans.

1. It must be able to remove very fine flyash particles

2. Should have high operating efficiency
3. Flawless operation at all load variations
4. Durable
5. Should occupy minimum floor space
6. Should be wear resistant to erosion problem inflicted by abrasive flyash
7. Low investment cost
8. Low operation and maintenance cost

Q. *How many types of mechanical flyash (dust) collectors are there?*

Ans. Basically two types: Dry Type and Wet Type.
Dry type system incorporates:
Gravitational separators
Cyclone separators.
Wet type system includes:
Spray type scrubbers
Packed bed scrubbers
Impingement type separator.

Q. *How many types of electrical gas cleaning devices are there?*

Ans. Two types: Rod type and Plate type

Q. *How does a gravitational separator work?*

Ans. It works on the basic principle of gravity.
If the dust laden gas is expanded in a duct, the flow velocity will drop and that will settle some dust particles. (Fig. 14.1)
Changing the gas flow direction will bring about some precipitation of the heavier dust particles (Fig. 14.2)
Placing baffles in the path of dust laden gas will bring down some heavier particulate solids due to loss of kinetic energy upon impingement. (Fig.14.3)

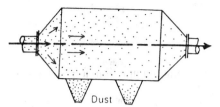

Fig. 14.1

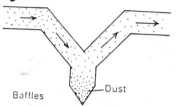

Baffles

Fig. 14.2

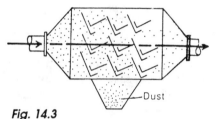

Fig. 14.3

Q. *What are the drawbacks of a gravitational system?*

Ans. It cannot remove fine dust particles. Only large sized and heavier particles can be removed.

It is bulky and requires large space.

Q. *What is a bag house dust collector?*

Ans. It is a flue gas filter made of cloth fabric that collects fine dust particles from the flue gas as it passes through it.

Q. *What is the efficiency of its dust removal?*

Ans. A well designed and properly maintained bag house dust collector can collect 99.9% of the particulate solids of size 1 μ and above. And its efficiency is independent of the amount of dust in the flue gas.

Q. *How does it work?*

Ans. It works reversibly. Dust laden flue gas is allowed to pass through the bags of cloth. These arrest the suspended dust particles and allow the clean gas

to pass out. When the pores of the fabric filters get clogged, it is regenerated by a gentle reverse flow of air.

Q. *What is the limitation of bag house dust collectors?*

Ans. They are used for low sulphur coal (less than 1%).

Q. *Why?*

Ans. The cloth fabric of the bag house filter is sensitive to H_2SO_4 attack, which is formed at 160°C (433K—the dew point temperature of H_2SO_4) and below from sulphur trioxide and water vapour:

$$S \xrightarrow{O_2} SO_2 \xrightarrow{O_2} SO_3 \xrightarrow{H_2O} H_2SO_4$$

SULFUR
FROM
COAL

Q. *Presently, cloth fabric impregnated with teflon is used. Why?*

Ans. It will resist possible deterioration from acid mist should the flue gas temperature drop below DPT of H_2SO_4.

Q. *How does a cyclone separator work?*

Ans. The dust laden flue gas is allowed to enter a conical shell tangentially setting up a swirling motion in the body of gas which casts off heavier particulate solids by imparting to them a centrifugal force. (Fig. 14.4)

Dust particles collected at the bottom are separated out.

Q. *What is the collection efficiency of this system?*

Ans. It ranges from 60 to 99% depending on the size of the dust particles and the pressure drop across the cyclone.

Particle Size	Collection Efficiency		
	P – 2.5 cm of H_2O	P – 6 cm of H_2O	P – 7.5 cm of H_2O
Up to 10 μ	62%	67%	70%
(10 – 20) μ	(95 - 96) %	(97 - 98) %	98%
(20 – 45) μ	(98 - 99) %	99%	99%
45 μ	99%	99.5%	99.6%

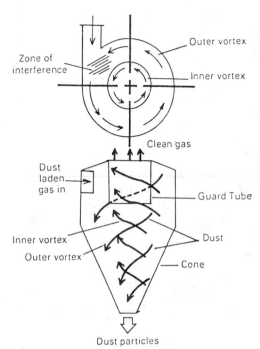

Fig. 14.4 *Cyclone separator*

Q. *What are the advantages of a cyclone separator?*

Ans.
1. Low maintenance cost
2. Higher efficiency for bigger size particles
3. Higher efficiency at higher load.

Q. *What are its disadvantages?*

Ans.
1. Fine dust particles escape separation
2. Non-flexible in terms of volume handled
3. Efficiency declines with the increase of fineness of the particles
4. Considerable loss of pressure
5. More power requirement to produce very high vortex velocity of the flue gas.

Q. *What is an electrostatic precipitator (ESP)?*

Ans. It is a device to precipitate suspended flyash and dust particles from the flue gas by ionizing the particles in an electric field and collecting them subsequently on oppositely charged electric plates (or rods).

Q. *Who first introduced this device?*

Ans. Dr. F.G. Cottell in the year 1906.

Q. *In some cases it is used in combination with a cyclone separator. Why?*

Ans. To upgrade the dust removal efficiency of the system.

Q. *How?*

Ans. Sometimes unburnt coal particles are dragged along with flyash in the flue gas. Large amount of coal particles will adversely affect the collection efficiency of the electrostatic precipitator. Hence, the best part of it is separated out in the cyclone separator before feeding the fluegas to the ESP.

Q. *How does the electrostatic precipitator works?*

Ans. The dust laden flue gas enters the ESP through an inlet channel and passes through a gas distributor that distributes the flue gas evenly over the parallel arrays of discharge electrodes spaced alternately with collection electrodes. The gas flow may be horizontal or vertical. Also, it can be arranged in parallel or series.

The discharge electrodes (usually negative) are in the form of a wire, most commonly, a smooth wire having a small radius of curvature (2.5 mm OD), weighted or supported to retain the right physical tension and location. They are energized at 45–70 kV.

The collecting electrodes come in the shape of plates or tubes of well-reinforced, light-gage metal. They are usually positively charged or grounded.

The high intensity electric field created causes the particulates in the gas stream to acquire negative charges transferred from ionized gas molecules of the same polarity. These charged dust particles accelerate towards the collection electrodes where their charges get neutralized, whereupon they fall into a collection hopper beneath the unit. Waves after waves of charged dust particulates migrate continually to collection electrodes where they transfer their charge and fall into the hoppers by gravity, thus effecting a separation from the gas stream.

The high voltage discharge system is supported by insulators that also isolate it from grounded components.

ESPs operate in the temperature range 95°–450°C (commonly). However, some units operate up to 950°C. The operating pressure is 600 mm of water gauge but operation to 40 atm is possible. The pressure drop through the unit is 15–30 mm water.

Q. *What is the mechanism of dust collection by ESP?*

Ans. Electrostatic precipitation is based on corona discharge that provides a simple and stable means of charging and collecting suspended dust particles in a moving stream of gas.

Step (I): Ionization Gas molecules in the vicinity of high voltage discharge electrode (negative) break down to form a train of plasma—a stream of positively charged ions plus free electrons. This occurs when the applied voltage reaches a critical level.

Step (II): Corona Generation The free electrons being repulsed by the discharge electrode move towards the positive (ground) surface and collide with the gas molecules, forming negative ions in the process.

The negatively charged ions having lower mobility form a negatively charged space-cloud which is called corona. (Fig. 14.5)

The space charge tends to stabilize the corona by reducing the further emission of high-energy electrons.

Step (III): Charging and Collecting Dust Particles Following the establishment of corona, the dust particles entering it become negatively charged by the ions present. Thereafter these ionized particles are attracted by the positively charged electrode called collecting electrode, where their charges are neutralized and they settle on the electrode surface.

Step (IV): Particle Removal is the final step. The collected dust particles are removed, in the dry process, by rapping the collecting surface electromagnetically, pneumatically or mechanically to slough away the particles into a hopper. In the wet process, the deposited material is rinsed with an irrigating liquid.

Q. *What are the advantages of an ESP?*

Ans. It is the most effective method to remove very fine particulates, as fine as 0.01 μ, which escapes removal by mechanical separators.

Very useful for high dust loaded gas. Flue gas containing dust particles as high as 100 g/Nm^3 can be effectively cleaned.

High dust removal efficiency (99 to 99.5%)
Operation is easy and smooth
Minimum draught loss
Least maintenance cost
Both wet and dry dust can be removed.

Q. *What are its disadvantages?*

Ans.

1. High capital cost of equipment.

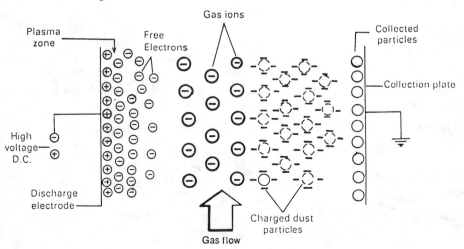

Fig. 14.5 Dust collection mechanism in ESP

2. Power requirement is considerably high and that is why the electricity bill for operation of the unit mounts.
3. Removal efficiency drops with the increase of gas velocity.
4. A good amount of floor-space is occupied.

Q. *What fundamental steps are involved in the electrostatic precipitation of particulate solids?*

Ans. Three steps:
1. charging of the suspended particles
2. collection of the charged particles under the influence of electrostatic field
3. removal of deposited particles from collection plates.

Q. *What factors affect the dust removal by an ESP?*

Ans.
1. Particle size
2. Particle resistivity
3. Field strength
4. Corona characteristics
5. Flue gas velocity
6. Area of collecting surface
7. Rapping.

Q. *How can the collection efficiency of ESP be computed?*

Ans. It can be done with the help of the following equation:

$$\eta = 1 - \exp\left(-A\,V_M/V\right)$$

where V_M = migration velocity of the particle

$$= \frac{r\,\phi_1 \cdot \phi_2}{2\,\pi\,\Theta}$$

r = radius of the particle

ϕ_1 = strength of the particle charging field

ϕ_2 = strength of the particle collecting field

Θ = viscosity or fractional resistant coefficient of the gas

V = gas flow through the precipitator

A = effective collecting area.

Q. *How can the collection efficiency be raised?*

Ans. By increasing the collection area and migration velocity.

Q. *How is it influenced by particle size?*

Ans. Collection efficiency increases with the increase of migration velocity and decreases with the drop of the latter. As migration velocity is directly proportional to particle size, collection efficiency will increase with the increase of particle size and decrease with the fineness of the particle.

Q. *How does gas velocity affect the collection efficiency of an ESP?*

Ans. The efficiency of the collector increases with the decrease of gas velocity and decreases with its increase. (Fig. 14.6)

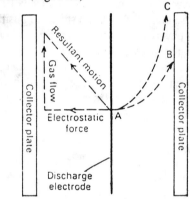

Fig. 14.6

The ionized particulates follow the path of the resultant of two mutually perpendicular forces—electrostatic force and gas flow.

If the electrostatic force > force exerted by gas flow, the particle will traverse the path *AB* and get deposited on the collector plate.

If on the other hand electrostatic force < force exerted by gas flow, the particle will follow the more curved path *AC* and escape precipitation.

Q. *Why should the linear gas velocity be low?*

Ans. This is to avoid turbulence in the interelectrode space and in the hoppers. For turbulence or any eddies in these regions will lead to reentrainment of dust particles.

Also, limitation to higher gas velocities is imposed by the residence time required for particle charging and collection.

Q. *What is the gas velocity range in an ESP?*

Ans. ESPs are known to operate in the gas velocity ranging from 0.6 m/s to 60 m/s. However, its value is greatly dependent on the nature of dust, its density, resistivity and attainable migration velocity.

For stack emissions of boilers burning low sulphur coals, the fly ash velocities of 1.5–2.5 m/s are chosen to avoid reentrainment whereas this value is limited to only 0.6 m/s for particulate emissions from recovery boilers.

Q. *If the inlet velocity of dust laden gas is too high what may happen, other than reentrainment?*

Ans. For the successful operation of an ESP, it is basically required that the proper spacing between the collection and discharge electrodes be maintained. If the inlet velocity of dust laden gas is too high, some of the collection electrodes may buckle and some of the wires (discharge electrodes) may displace slightly to one side. And any such deviations from the designed dimensional spacing generates sparkovers which cause a considerable loss in efficiency and increased wastage of power.

Q. *What is particle resistivity?*

Ans. It is a measure of the ability of a particle to accept a charge, i.e., to become ionized.

Higher resistivity of a particle means greater is its resistance to becoming ionized and so it offers greater resistance to the passage of electric current. Consequently, lower resistivity means it is conductive.

Q. *What factor primarily influences particle resistivity?*

Ans. Particle resistivity depends on the composition and amount of microconstituents present. If the flyash and particulates contain metallic oxides coating their surfaces, their resistivity will be sufficiently low.

Q. *What is the unit of resistivity?*

Ans. Ohm-cm.

Q. *What is the suitable range of resistivity for good collection efficiency?*

Ans. 10^4 to 10^{10} Ohm-cm.

Q. *What is the critical level of resistivity?*

Ans. 10^{10} Ohm-cm.

Q. *If the particle resistivity is higher than this range, how will it affect the ESP's performance?*

Ans. If the particle resistivity is 2×10^{10} Ohm-cm. or higher, ESP efficiency will decrease. This is due to two reasons:

1. Particulates of high resistivity require high current densities and high discharge voltages. ESP operability decreases under these conditions.
2. High resistivity leads to back corona, causing reentrainment of deposited particles back to the gas stream from the collecting electrodes. This means reduction in ESP efficiency as the reentrained particles are to be recollected.

Q. *Why do particulates with low resistivities account for low ESP efficiency?*

Ans. Particulates with low resistivities bring down the ESP efficiency by way of higher reentrainment phenomenon.

If the particles have too little resistivity, usually 10^8 Ohm-cm or less, they will be readily conductive. As soon as they reach the collection electrode, they get neutralized but this happens so fast that the particles still retain some residual momentum that bounces them off the collection plates and causes severe reentrainment.

Q. *What difficulties are encountered with sulphur dust?*

Ans. The sulphur particles have high resistivity. They do not readily sacrifice their negative charge to the collecting electrodes.

Q. *What difficulties are encountered with carbon particulates?*

Ans. Carbon particulates have very low resistivity. They readily give up their negative charge and become positively charged particles. This causes the particulates to be repelled back into the gas stream of negatively charged particles.

Due to their low resistivity they'll be collected and repelled in this way many times and finally escape into the atmosphere.

Hence the presence of large quantities of carbon particulates in the flue gas will adversely affect the collection efficiency of the electrostatic precipitator.

Q. *What is the solution to it?*

Ans. Installation of a cyclone separator upstream of the ESP to separate out as many carbon dust particles as possible to reduce the load on the ESP.

Q. *What is the effect of moisture content on resistivity?*

Ans. Higher the moisture content lower is the resistivity.

Q. *Why does humidity lower resistivity?*

Ans. It increases conductive condensation on the particulate surface.

Q. *What is the effect of temperature on the resistivity of dust particles?*

Ans. Their resistivity sharply increases with the rise of temperature upto a certain range of temperature, reaches a maxima and then falls off with the further rise of temperature (see Fig. 14.7).

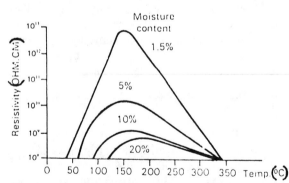

Fig. 14.7 *Dependence of particle resistivity on temperature*

Q. *Does the particle resistivity has any influence upon rapping?*

Ans. Particle resistivity exerts a critical influence on the electrical force causing the particles to stick to the collecting plates. Greater the resistivity of the particles, slower will be their rate of charge transfer, i.e., to get neutralized, and naturally they adhere to the collecting surface with a greater force. And therefore,

higher rapping force is required to dislodge them from the collecting plates.

Q. *Between low-sulphur coal and high-sulphur coal, which one will yield more ash and flue gas per MW of power generation?*

Ans. Low-sulphur coal.

Q. *What will be the problem if we use low-sulphur coal?*

Ans. It will require higher collection efficiency of ESP not only because of higher load of dust and flyash particles but also due to higher resistivity of the dust by several times that of high- sulphur coal.

Q. *How can this problem be coped with?*

Ans. By gas conditioning.

Q. *What is that?*

Ans. Injecting small amounts of sulphur trioxide in the flue gas to reduce the electrical resistivity of the flyash, with the effect that the particulates will become more amenable to collection in the ESP.

To put it simply: It will enhance the collection efficiency of the electrostatic precipitator.

Q. *What is the economy in gas conditioning?*

Ans. It eliminates the need for overdesigning the ESP and reduces the capital cost.

Q. *Humidity lowers the resistivity of the particulate solids in the flue gas. How can this effect be enhanced?*

Ans. By injecting small quantities of conditioning agents. The condensed water droplets on the particulate surface will absorb the conditioning agent (SO_3) having higher conductivity and lower resistivity.

Q. *What are the most commonly used gas conditioning agents?*

Ans. Sulphuric acid and ammonia.

Q. *In which cases sulfuric acid is generally used?*

Ans. Low-sulphur coals.

Q. *In which cases is ammonia used?*

Ans. High-sulphur coals.

Q. *Usually, ash particles carry some sulphate on their surface. Will it enhance the performance of ESP or decrease it?*

Ans. It'll increase the performance of ESP.

Q. *Why?*

Ans. Sulphates have higher electrical conductivity and lower electrical resistivity. Hence deposited sulphates on particulate surface will lend to better electrical performance of the ESP.

Q. *If the sulphate level is lower how will it affect the collection efficiency of the ESP?*

Ans. It will reduce the performance of the ESP.

Q. *What is the critical sulphate level of ash particles?*

Ans. Research works indicate it to be 0.5% of the gas.

Q. *How much SO_3 in the flue gas is sufficient to maintain the resistivity of the flyash below critical level?*

Ans. Usually 10 to 20 ppm.

Q. *High-sulphur (3%) coal needs NH_3 injection in the flue gas for conditioning. Why?*

Ans. High-sulphur coal will yield a greater amount of SO_3 than that required to maintain the resistivity of the dust particles below the critical level. This excess SO_3 is absorbed on dust particles and reduces their resistivity below the desired range. Hence ammonia in combination with steam is injected to the flue gas to take care of this excess SO_3.

Q. *Why is steam also added with ammonia?*

Ans. To complete the following reaction:

$$SO_3 \xrightarrow{\;H_2O\;} H_2SO_4 \xrightarrow{\;2\,NH_3\;} (NH_4)_2\,SO_4$$

Q. *In which state is SO_3 added?*

Ans. Liquid SO_3 and vapour SO_3 (obtained by vapourization of H_2SO_4 and catalytic conversion of SO_2 to SO_3).

Q. *Which method is mostly used?*

Ans. Sulphur burning method.

Molten sulphur is atomized in the combustion chamber with high velocity air and completely burned to SO_2 which is cooled (950° to 350°C) and catalytically converted to SO_3 in a one-stage V_2O_5 bed.

Q. *What is sparkover?*

Ans. The gas pockets between the deposited dust particles behave as insulators. However under the condition of low to medium particle resistivities, the voltage drop across the deposited dust layer may exceed the dielectric strength of the gas in the pores of the layer. As a result of this, a spark will occur in the dust layer and an instantaneous drop in potential across the layer will take place.

Q. *How can a sparkover be avoided?*

Ans. Sparkover can be avoided if the ESP is run at reduced current densities.

Q. *But if the precipitator is to run at reduced current densities, its efficiency will fall. Is it not?*

Ans. Yes; as it will lead to lower migration velocities.

Q. *What is the countermeasure to it?*

Ans. A good solution is to increase the precipitator's collection area.

Q. *What is 'ivy' effect?*

Ans. Ionized dust particles (negatively charged) continually arrive at the collection electrode (ground or positively charged) whereupon their charges get neutralized. However if the rate of charge neutralization is too low, the second wave of dust particles that will arrive at the collection electrode may be so strongly attracted by the electrostatic pull of the electrode that they will sandwich between them and the collection electrode the first group of dust particles, thereby preventing the latter from falling into the hopper.

As this process continues, the dissipation of dust charge becomes increasingly difficult since more and more time is required by the outermost charged particles to bleed through the dust layer to the collection electrode and get neutralized. The particles cannot loose their charge rapidly as their resistivity is too high, so they tend to cling to the surface of the dust layer already deposited on the collection plate. This phenomenon is called 'ivy' effect as it closely resembles an ivy clinging to a wall.

Q. *What is particle reentrainment?*

Ans. It describes the phenomenon of reentry of deposited dust particles from the collector electrode back into the gas stream.

Q. *Why particle reentrainment occurs?*

Ans. It occurs chiefly due to
(a) inadequate surface area of the collection electrodes of ESP
(b) inadequacy in dust removal from the hopper.
Also it occurs during rapping and in cases where particles with low resistivities are encountered. These particles retain some leftover momentum despite their charge-neutralization at the collection electrode. This residual momentum rebounds them back into the gas stream to effect reentrainment.

Q. *How does reentrainment reduce the ESP's performance?*

Ans. Reentrainment reduces the ESP's performance as the reentrained dust particles are again to be removed from the carrier gas.

Reentrainment means more power drain of ESP because of increase of the dust load in the vicinity of charging and collecting electrodes.

Furthermore, if reentrainment occurs in the last field, it may lead to an unacceptable level of particulate emission to the atmosphere.

Q. *What is the effect of 'ivy' effect?*

Ans. As a result of 'ivy' effect (also called fish-net effect), the effective collection potential falls with the increase of thickness of the dust layer adhering to the collection plate. This brings about a steady increase in the power drain of the ESP to maintain the required efficiency. And this fact is clearly unacceptable from the operation point of view. Furthermore, fine particulates, under these circumstances, escape collection.

Q. *How can this problem be overcome?*

Ans. To prevent the 'ivy' effect from happening, it is required that the collection plates and discharge electrodes be periodically vibrated (this process is called rapping) so that the intensity of rapping matches the adhesion characteristics of the deposited particles.

Q. *What is rapping?*

Ans. Rapping is the periodic agitation or vibration of discharge wires and collection electrodes with hammers called rappers.

Rapping can be done mechanically, electromagnetically and pneumatically.

As a result of rapping, dust particles slough away from the electrode surfaces and drop into the collection hoppers.

Q. *By which method can the maximum rapping force be exerted?*

Ans. Rapping induced by mechanical rappers.

Less force is exerted by pneumatic rappers and even lesser by an electromagnetic system.

Q. *What is the spacing of rappers?*

Ans. These can be spaced every 125 mm along the plates.

Q. *In which direction is rapping done?*

Ans. In the direction of gas flow.

Q. *When is rapping done—while the unit is on-line or shutdown?*

Ans. It is carried out when the unit is on-line.

Q. *What is the danger in rapping?*

Ans. Apart from reentrainment loss, rapping may also bring about catastrophic failure of ESP if the support grids, discharge wires and collection electrodes are not designed to withstand the increased rapping force induced to reduce the dust build-up on electrodes. The wires may snap and plates may warp, thus inviting disaster. Also excessive dust puffing may bring about the same tragic failure.

Q. *What is space charge effect?*

Ans. If the rate of ionization of dust particles exceeds substantially the rate of their collection by collecting plates, there occurs a high concentration of dust particles in the vicinity of discharge wires. These negatively charged particles saturate the charging field and effectively reduce the electric field around each discharge electrode.

As this occurs, new batches of charged dust particles leaving the discharge wire are repulsed by the negatively charged dust cloud saturating the charging field.

Therefore, there is a net energy loss in the result and a considerable decline in collection efficiency.

Q. *What is back corona?*

Ans. This is another problem in an ESP particularly those in which collecting electrodes are positively charged. This occurs when electrical breakdown results from high resistivity dust particles at low voltages and currents. As a result, the generation of positively charged ions within the dust layer at the collection electrode and the migration of these ionized dust particles with accelerated speed towards the discharge (negative) electrode takes place. And they tend to neutralize the ions of opposite polarity produced at the discharge wire, thereby reducing ESP's performance.

Q. *What are the corona characteristics of ESP operating at maximum power output?*

Ans. Steeply rising characteristics.

Q. *What is the effect of the temperature of the stack gas on the corona characteristics?*

Ans. It is enhanced with the rise of the temperature of the stack gas.

Q. *Why is a wet-type dust collection method adopted?*

Ans. It is directed solely at reducing the SO_2 emission in the stack gas because of stringent restrictions imposed by the Government to maintain the quality of the environment.

Problems arise due to burning of low-sulphur coal producing ash particles of very low resistivity, making the ESP inefficient.

Hence the control of SO_2 emission in the exhaust gas, when low- sulphur coal is burned, is better ob-

tained by introducing scrubbers—wet-type dust collection system—just upstream of exhaust gas stack.

Q. *What is the benefit of wet-type mechanical dust collectors?*

Ans. Gases like SO_2 and H_2S as well as smokes and fumes can be economically removed.

Q. *How does it work?*

Ans. The dust laden flue gas and washing liquid (water + lime) move counter-currentwise in the scrubbing tower—the gas going up while the liquid raining down.

The SO_2 and H_2S get dissolved while in contact with the alkaline water while dust particles produce a thick liquid with the water collected at the bottom. The clean flue gas escapes from the top.

Q. *What is the mechanism?*

Ans. Liquid (lime-water) sprayed as fine droplets in the scrubber collide with the dust particles, wet their surfaces and become heavy and sticky. When these wetted particles conglomerate, they grow in size until they become heavy enough to drop out of the flue gas stream.

For soluble gases like SO_2 and H_2S, the interfacial mass transfer takes place across the gas-liquid droplet boundary.

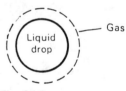

Fig. 14.8

This mass transfer is enhanced by lime which forms compounds with the absorbed SO_2 and H_2S.

$$Ca\,(OH)_2 \;\boxed{SO_2}\!\!> Ca\,(HSO_3)_2$$

$$Ca\,(OH)_2 \;\boxed{H_2S}\!\!> Ca\,(HS)_2 + CaS$$

Q. *How can the efficiency of a scrubber be improved?*

Ans. Finer the water droplets, higher will be the absorption efficiency.

And greater the difference in relative velocities between the dust particles and water droplets, greater will be the collection efficiency of the scrubber.

Q. *How many types of scrubbers are there for the collection of dust and reduction of SO$_2$ in the flue gas?*

Ans. In general there are three:
1. packed type
2. spray type
3. cyclone type

Q. *How does a packed type scrubber function?*

Ans. Dust laden flue gas is fed near the bottom of the scrubber housing a high density polyethylene packing with a 1 to 3m (sometimes 5 to 10 m thick packing layer is used for satisfactory reduction of SO$_2$ and H$_2$S in the stack gas) thick layer.

Lime-water is sprayed through nozzles from the top. (Fig. 14.9)

The flowrate of water is maintained such that it affords good wetting of the packing.

Most of the finer dust particles as well as SO$_2$ and H$_2$S are washed off the flue gas which is then let out to the atmosphere.

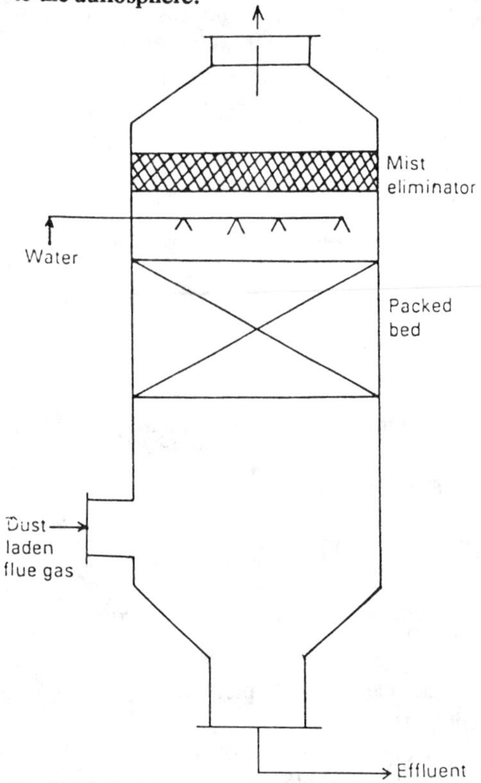

Q. *What is the material of construction of a packed-bed scrubber?*

Ans. Fibre-reinforced plastic.

Q. *Why?*

Ans. It is resistant to corrosion. The scrubber liquor is highly corrosive as it contains dissolved acidic oxides such as SO$_2$, SO$_3$, etc.

Q. *Why is high density polyethylene (HDPE) used as tower packing?*

Ans. It is resistant to corrosion.

Q. *What is the main disadvantage of this system?*

Ans. High pressure drop of the flue gas.

Q. *How does a spray-type scrubber work?*

Ans. It is the simplest kind of all wet type scrubbers.

Flue gas and water droplets move counter-current in the scrubber. The liquid particles wet the dust particles which then conglomerate and drop out.

Water is distributed in a fine spray through successive layers of spray nozzles from the top while the dust laden gas is uniformly distributed across the tower cross-section by a gas distribution plate near the bottom. (Fig. 14.10)

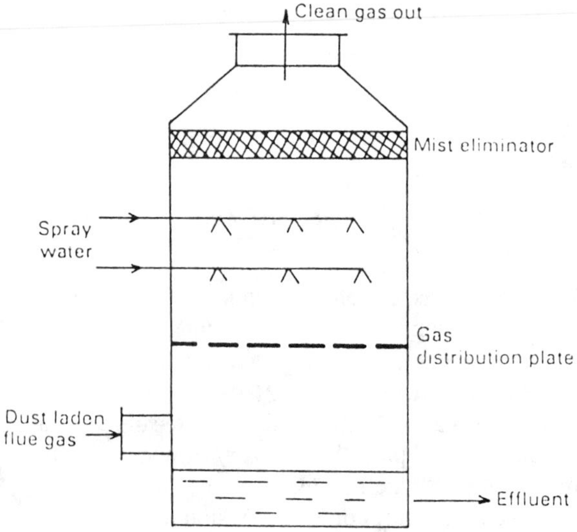

Fig. 14.9 Flue gas scrubber (packed bed type)

Fig. 14.10 Spray-type scrubber

Clean gas escapes from the top via a mist eliminator to avoid entrainment and carryover.

Q. *What is the limitation of spray-type scrubber regarding particle size of the dust?*

Ans. Its application is limited to the particle size 10 μ and higher.

Q. *What are the advantages of spray-type scrubbers?*

Ans.
1. Negligible pressure drop of the flue gas
2. Simplicity in design
3. Low capital cost

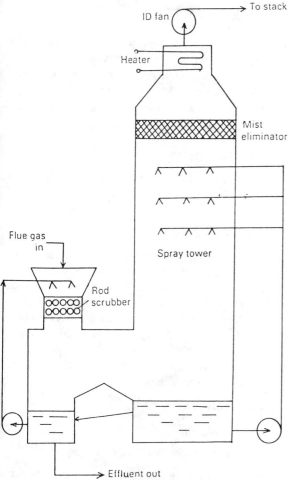

Fig. 14.11 Two-stage spray scrubber

Q. *Why is a two-stage spray scrubber preferred to single-stage spray tower?*

Ans. To enhance the efficiency for SO_2 and particulate solid removal.

Q. *How?*

Ans. It is composed of two systems— rod scrubber and spray tower—hooked up in series.

The particulate solids are removed in the rod scrubber and SO_2 is removed in the spray tower.

Q. *Why is the spray tower reaction effluent fed to the rod scrubber effluent?*

Ans. To supply the seed crystals necessary to provide nucleation sites for non-scale forming homogeneous crystallization of calcium sulphate otherwise it would lead to calcium sulphate scaling due to oxidation in the scrubber. (Fig. 14.11)

Q. *When does scaling occur in the scrubber?*

Ans. In the wet process using lime + water as scrubbing liquid, two type of scales are encountered:

Calcium sulphate scale $CaSO_3 \cdot 0.5H_2O$
Calcium sulphate scale $CaSO_4 \cdot 2H_2O$

These two scales are formed when the solutions are supersaturated to the point where crystallization takes place, with the effect that large, agglomerated crystals, instead of small individual, separate crystals, are formed. The primary disadvantage of microcrystals is that they form on the walls, piping and nozzles of the system causing clogging.

Q. *How can the formation of sulphite scales be avoided?*

Ans. By controlling the pH of the scrubbing liquid (lime + water) and maintaining it in the range 5–7.

Q. *Sulphur dioxide in the flue gas produces bisulphite and sulphite ions in contact with water. So how does sulphate occur?*

Ans. Due to oxidation in the scrubber, reaction tanks and thickners.

Q. *How can this scaling be minimized?*

Ans. By feeding seed crystals of $CaSO_4$ in the scrubbing liquor. They will form isolated nucleation

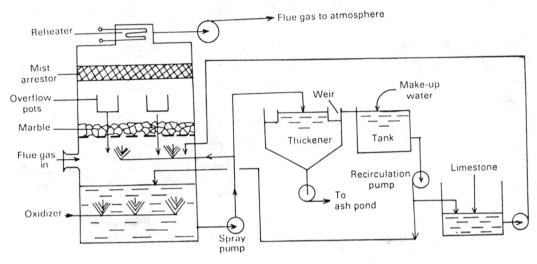

Fig. 14.12

sites for growth of calcium sulphate crystals which will precipitate out, forming no scales.

Q. *Air is bubbled through the reaction tank. Why?*

Ans. Air will induce forced oxidation of sulphite ions to sulphate in the reaction tank:

$$SO_3^{2-} + O_2 + H_2O \longrightarrow SO_4^{2-} + 2OH^- + 2e$$

The best portion of sulphite in the reaction tank is

converted into sulphate and precipitated out as $CaSO_4$. Therefore, when this solution is recycled to the scrubber the concentration of sulphite ions in the solution is so less that there will be very little scaling on the limestone surfaces by $CaSO_4$ formed due to oxidation of $CaSO_3$ by oxygen in the flue gas. (Fig. 14.12)

Q. *What is a cyclone scrubber?*

Ans. It is a kind of wet-type mechanical separator in which the flue gas enters tangentially through a cyclone

Q. *How does it operate?*

Ans. As the flue gas enters tangentially near the base of the scrubber and swirls upwards, it comes in contact with a heavy dose of water droplets sprayed through centrally located spray nozzles. Dust particles absorbed by water droplets get sticky, conglomerate and drop out and collect in the dust settling tank at the bottom. (Fig. 14.13)

The clean overflow liquid from the settling tank is recycled to the spray chamber.

Q. *What about its efficiency?*

Ans. Usually high.

Q. *Any other major advantage?*

Ans. A lion's share (85–90%) of soluble gases like H_2S, SO_2 and SO_3 are absorbed and eliminated from

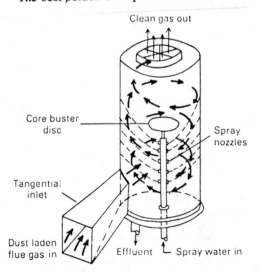

Fig. 14.13 *Cyclone scrubber*

flue gas. SO_2 content (610–860 ppm) is reduced to 120–380 ppm.

Q. *What is the water pressure at the nozzle?*

Ans. It varies. Ranges from 2 to 20 kgf/cm^2.

Q. *What is the draught (draft) loss at full rating?*

Ans. Usually 40 to 50 mm of water column.

Q. *What are the major problems encountered in wet-type scrubbing?*

Ans. Corrosion and erosion problems.

Q. *Why?*

Ans. Sulphur dioxide and sulphur trioxide fumes in the flue gas produce acid mists with water at the saturation temperature (most high efficiency scrubbers quench the flue gas to its saturation temperature).

These acids quickly act upon metals and alloys ; affecting scrubber and auxiliary equipment (like pumps, fans, heat exchangers, etc.) as well.

Q. *What factors enhance corrosive attacks?*

Ans.
1. Humidification of the flue gas
2. High gas inlet temperature
3. High percentage of suspended particulate solids
4. High gas and scrubbing liquid velocities.
5. Acid strength of the recycled liquor

Q. *Sometimes SO_2 and SO_3 laden flue gas is prequenched before entering the main scrubber spray tower. Why?*

Ans. As the SO_2/SO_3 is absorbed by water, an ill-defined junction of wet and dry gases results. In this junction, the scrubbing liquid is sucked into the inlet gas duct by eddy currents creating high localized acid concentrations. To avoid this undesirable phenomenon, the flue gas is prequenched.

Q. *Why is this eddy current created?*

Ans. Just imagine a water droplet surrounded by a flue gas film. As the SO_2/SO_3 inside the gas film is absorbed and dissolved in the droplet within, the contraction of gas film takes place. This gas film

'contraction' of a large multitudes of droplets will create eddy currents.

Obviously, the volume contraction of flue gas also takes place when it is quenched by water at lower temperatures, thus generating eddy current.

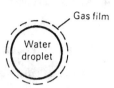

Fig. 14.14

Q. *What is the material of construction of the scrubber to avoid corrosion?*

Ans. Frequently, they are made of stainless steel lined with acid proof bricks. Fibreglass reinforced plastic is also in vogue. Sometimes ceilcote linning is also applied on stainless steel for best results against corrosion and erosion.

Q. *Apart from corrosion resistance, is there any added advantage of acid resistant bricks?*

Ans. Abrasion resistance and resistance to spalling.

Q. *When fireclay acid bricks are used for linning, the flue gas temperature is maintained at 400°C (673 K) or below, Why?*

Ans. Scalling off (i.e. spalling) will occur as the droplets of scrubbing liquor impinge on the brick linning of the scrubber wall.

Q. *Which brick is most suitable for this purpose?*

Ans. Among all acid resistant bricks, silicon carbide brick yields best results.

Q. *Why?*

Ans. Not only is it acid resistant but also resistant to spalling at temperatures as high as 1400°C (1673 K).

Q. *Is it applied widely?*

Ans. High installation cost delimits its use.

Q. *What are the limitations of brick linning?*

Ans.
1. It puts certain limitations on the design of the scrubber.
2. It greatly increases the weight of the vessel.
3. Access to scrubber becomes difficult.

4. Additional cost burden.

Q. *What is ceilcote?*

Ans. It is a thermosetting resin.

Q. *What are its advantages as linning material?*

Ans.
1. Only a 4 to 12 mm thick coating is required.
2. Can be reinforced with glass or cloth to increase its tensile strength and lower its coefficient of thermal expansion
3. Highly resistant to corrosion and thermal shock
4. High impact and abrasion resistance
5. Capable of functioning at high temperature
6. Life span is 10 years or more under severe operating conditions.

Q. *What are the disadvantages of wet-type scrubbers?*

Ans. The quantity of water requirement for scrubbing is very large (20 to 30 litres per 1000 Nm^3 of gas handled).

The system is costly as special material of construction (S.S) and special linning are to be used. For the given gas conditioning, a wet-type scrubber costs 25% more than a double-stage cyclone separator.

Flue gases having a high load of dust particles impair its efficiency.

Its operating cost is about five times that of an ESP.

Q. *Why in many cases is a dry mechanical collector hooked up in series with an electrostatic precipitator?*

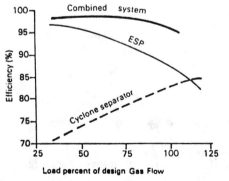

Fig. 14.15 *Efficiency Characteristics of ESP, Cyclone Separator and a combined system*

Ans. To reduce the initial cost of the collector system as well as to maintain an overall high collection-efficiency.

The ESP has a falling efficiency characteristic with increasing load while a cyclone separator (i.e., dry mechanical collector) has a gradual upward efficiency characteristic with increasing load. (Fig. 14.15)

So when these two collector systems are hooked up in series an overall constant high efficiency over a considerable range of load variation results.

Q. *Which one is installed ahead of the other?*

Ans. Mechanical collector is installed before ESP.

Q. *Why?*

Ans. The mechanical collector removes the heavier particles and its efficiency increases with larger particle size. The ESP is best fitted for removing finer particles and exhibits higher performance with reduced dust load. Much of the dust load is reduced by a mechanical collector (cyclone separator) installed upstream of the ESP.

Q. *What is a better location for the ESP—upstream of the air heater or downstream of it?*

Ans. The problem is moot. But arguments favour its location upstream of the air heater.

Q. *Why?*

Ans. ESP upstream of the air heater will keep the heat transfer surface of the air heater clean of dust and soot with the effect that the rate of heat transfer to air will not be impaired over a prolonged period.

Low-sulphur coals (S content 0.4–0.8%) produce, on combustion, ash particles of high electrical resistivity. In such cases the ESP performs poorly until or unless the flue gas temperature is increased (above 270°C, i.e., 543 K) or the gas is conditioned (by adding SO$_3$). If the flue gas temperature after passing through the air heater drops below 543 K, installation of the ESP downstream of the air heater will impair the efficiency of the ESP dealing with low-sulphur coal ash.

Therefore, for efficient performance of ESP dealing with low-sulphur coal ash, it should be allowed to work at flue gas temperature above 543 K and that makes its installation upstream of the air heater imperative.

Finally, whatever be the coal composition, the performance of ESP upstream of the air heater is not affected, provided the flue gas temperature is above 340°C (613 K). It is because at this high temperature, the electrical conductivity of dust particles is substantially higher enabling them to be more responsive to the ESP.

[ESP upstream of air heater is called hot-side ESP].

15

Ash Handling System

Q. *What is ash?*

Ans. Ash is the inc ombustible material that remains when coal is burner .

Q. *How can it be classified?*

Ans. Ash can be classified as either inherent or accidental depending on the source from which it is derived. Accordingly the former is called intrinsic ash and the latter is called extrinsic ash.

Q. *What is intrinsic ash composed of?*

Ans. Oxides of silicon, alkali metals (Na, K) and alkaline earth metals (Ca, Mg).

Q. *How do they come?*

Ans. They come into the body of coal from the original vegetable matter from which coal was formed. Also they might have been ingested into coal during its formation when mineral rich water infiltrated into it.

Q. *Is it separable or inseparable from the coal substance?*

Ans. Inseparable.

Q. *How can it be removed?*

Ans. Through floatation processes.

Q. *What is extrinsic ash?*

Ans. It is dirt, shales, clay, pyrites and ankerites picked up from the adjacent earthy or stony bands in the coal seam.

Q. *How can they be eliminated?*

Ans. Washing and cleaning eliminate most of the extrinsic ash.

Q. *What is the percent of ash content of coal?*

Ans. Ash content in coal may vary from 8 to 10% to as high as 20 to 50% depending upon the grade of coal.

Q. *How does it vary with the quality of coal?*

Ans. Better the quality of coal, lower is the ash content. Hence for inferior grade coal, the ash content is higher than anthracite or semi-anthracite variety of coal.

Q. *Ash is detrimental to the combustion process. Why?*

Ans.

1. It lowers the gross calorific value of coal.
2. It must be disposed of after combustion and as such entails a separate system called ash handling system.
3. Ash with low fusion point forms a deposit on the tube walls, affecting the generation of steam
4. Low fusible ash forms clinkers that contribute to clogging as well as cause severe corrosion of bar grates of the stoker furnace.

Q. *What is the difference between the ash and the mineral matter in coal?*

Ans. Ash left after coal is burned differs in quantity and composition from the original ash present in the coal. For the sake of clarity, the original ash in coal before combustion is usually termed mineral matter.

Composition of Typical Coal Ash (%)

Constituents	I (White)	II (Buff)	III (Red)
SiO_2	57.9	41.1	36.9
Al_2O_3	38.1	33.5	25.7

Fe_2O_3	5.7	17.1	25.1
Na_2O $\big\}$			
K_2O	3.1	2.6	3.3
CaO	0.5	3.8	5.7
MgO	0.7	1.9	3.3

Q. *So coal contains mineral matter that undergoes certain changes into ash. What are these principal changes?*

Ans.

1. DEHYDRATION OF HYDRATES

$$CaSO_4 \bullet 2H_2O \longrightarrow CaSO_4 + 2H_2O \text{ (Loss 2\%)}$$

2. DECOMPOSITION OF CARBONATES

$$CaCO_3 \longrightarrow CaO + CO_2 \quad \text{(Loss 44\%)}$$

$$MgCO_3 \longrightarrow MgO + CO_2$$

3. OXIDATION

$$FeS_2 + \frac{3}{4}O_2 \longrightarrow \frac{1}{2}Fe_2O_3 + 2S \quad \text{(Loss 33\%)}$$

$$S + O_2 \longrightarrow SO_2 \xrightarrow[O_2]{} SO_3$$

4. COMPLEX FORMATION

High temperature combinations of oxides, silicates and free silica

$$Na_2O + SiO_2 \longrightarrow Na_2SiO_3$$

$$Na_2O + Al_2O_3 \longrightarrow 2\,NaAlO_2$$

5. VOLATILIZATION OF ALKALI CHLORIDES

This is usually negligible.

Q. *Why does ash have to be discharged and dumped at a site sufficiently far away from the thermal power plant?*

Ans.

1. It is dusty and irritating. Being easily airborne, it causes inflammation in the eyes.
2. It generates toxic gases and corrosive acids when it comes in contact with water.

Q. *What problems beset the ash handling system?*

Ans.

1. Easily fusible ash forms large clinkers on conglomeration. They must be crushed to a convenient size before discharging onto a conveying equipment.

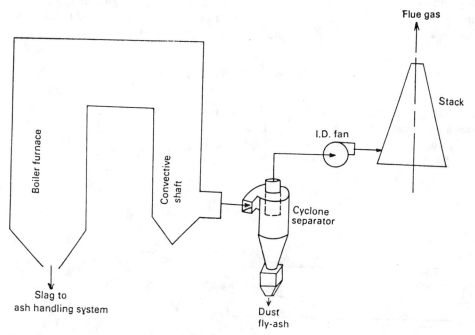

Fig. 15.1 *In a typical coal-fired system, molten ash goes to ash handling system while the flyash is collected in the cyclone separator downstream*

2. Abrasiveness of ash wears out the conveyor parts. Hence a special conveyor must be designed to handle the ash.
3. Transportation of hot ash is difficult as well as hazardous. Hence it must be quenched before transportation. (Fig. 15.1)

Q. *If the silica content of ash is higher, what trouble will originate?*

Ans. Greater the silica content, the greater is the erosion problem.

Q. *Is there any scope of utilization of coal ash?*

Ans. Ash can be utilized for:
1. Land filling
2. Road making
3. Filling abandoned quarries
4. Manufacture of cement.

Q. *Into how many types can the modern ash handling system be classified?*

Ans. Four.

Q. *What are they?*

Ans. Mechanical system
Hydraulic system
Pneumatic system
Steamjet system

Q. *How does a mechanical ash handling system function?*

Ans. The hot ash issuing out of a boiler falls over a conveyor belt through a water seal. (Fig. 15.2)

The hot ash gets cooled on quenching and carried continuously by the conveyor belt to the dumping site or ash-bunker from where it is trucked away.

Q. *What is the main limitation of mechanical system of ash handling?*

Ans. It is generally used for low capacity, coal fired power plants.

Q. *What is the maximum capacity of this system?*

Ans. 5 t/h

Q. *What is the approximate life span of this system?*

Ans. 5 to 10 years.

Q. *What are the major advantages of this system?*

Ans. 1. Low power consumption
2. Continuous ash removal

Q. *Can you mention any power station in India where such an ash handling unit is operating?*

Ans. Khapaskheda power station near Nagpur.

Q. *How does the hydraulic ash handling system differ from a mechanical ash handling system?*

Ans. In the case of a hydraulic ash handling system, water instead of a conveyor belt is used to transport the quenched ash to the ash sump where ash is settled out.

Q. *How many types of hydraulic ash handling system are in vogue?*

Ans. Two types:
Low-velocity system
High-velocity system

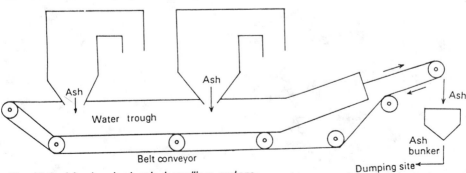

Fig. 15.2 *Mechanical ash handling system*

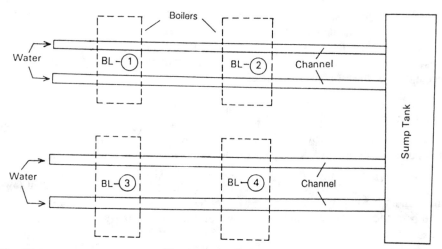

Fig. 15.3 *Low-velocity hydraulic ash handling system*

Q. *How does the low-velocity system operate?*

Ans. Ash from the furnace grate falls into a channel of water possessing low velocity and is carried to sump tank where it is settled, reclaimed and transported while the separated water is recycled. (Fig. 15.3)

One stand-by channel is provided to each boiler.

Q. *What is the velocity of water flow?*

Ans. 3 to 5 m/s.

Q. *Would you prefer a higher limit of velocity or lower limit?*

Ans. Higher limit, of course.

Q. *Why?*

Ans. It expedites the ash transporting.

It reduces the abrasion of the channel (trough) as the ash tends to ride on water instead of scouring the bottom.

Q. *What is the ash carrying capacity of this system?*

Ans. 50 t/h through a distance of 500 m.

Q. *How does the high-velocity system work?*

Ans. High pressure water jetting out from the nozzles fitted at the top (i.e., near the stoker bed) and on the sides strike the hot and molten ash as it rolls off the chain-grate. (Fig. 15.4)

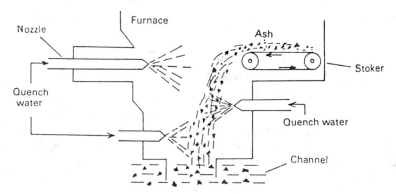

Fig. 15.4 *High-velocity ash handling system*

The water from the top nozzles quenches the ash while that from the side nozzles feeds the necessary driving force to transport the ash down the channel provided at the bottom.

The ash is allowed to settle in the settling pit and the separated water is recycled.

Q. *What is the capacity of this system?*

Ans. Ash as much as 120 t/h can be carried by this system and discharged at a distance of up to 1 km.

Q. *What is the disadvantage of this system?*

Ans. The water channel and the sump meant for conveying the ash should be made of corrosion and wear resistant material as the abrasive ash forms corrosive acids in contact with water.

Q. *Can this system be employed in the case of a pulverized coal fired boiler also?*

Ans. Yes. High pressure water jets directed against the continually flowing molten stream of ash, quench and disintegrate the mass into small particles which drop into a sluice-channel. A second set of nozzles transports them to cast iron downcomers discharging into the main sluice-channel wherefrom it is finally transported to the sump-pond.

Q. *What are the advantages of a hydraulic ash handling system?*

Ans.
1. Clean and dust free operation
2. High ash handling capacity 120 t/h. Hence more suitable for super thermal power stations.
3. Can handle molten ash stream also
4. Ash can be disposed of at a site 1 km from the steam generation plant.
5. There are no working parts in contact with the ash. Hence erosion problem is minimum.

Q. *What are the advantages of a high-velocity hydraulic system over a low-velocity one?*

Ans. It is very difficult to assess.

The water requirement of a low-velocity system is two to three times as much as the high-velocity system, whereas the power requirement to pressurize water through the nozzles in a high-velocity system is two to three times the power requirement for running a low-velocity system.

However despite high power requirement, the high-velocity system handles high-ash-concentrated fluid and minimizes the chocking problem of the sluice.

Because of the higher ash handling capacity of the unit, the high-velocity system allows intermittent operation—20 to 25% of the time required by a continuous low-velocity system. From this point of view, the electricity requirement per ton of ash handled is 1/3 rd to 1/2 of that required by an equivalent low-pressure system.

However the low-velocity system is gaining ground because of:
(a) simplicity of operation
(b) economical in running cost
(c) economical in installation cost.

Q. *How does the pneumatic ash handling system work?*

Ans. A high velocity air stream is directed to collect ash and dust from all discharge points. The ash from boiler units falls through the hopper, passes through crushers and drops into the stream of air.

Air-borne ash and dust particles are separated in cyclone type separators. Finally the air is allowed to pass through a filter and vented to the atmosphere by an exhauster fan. (Fig. 15.5)

The ash collected in the ash separators is trucked away for disposal.

Q. *What quality of ash can be handled by this method?*

Ans. Abrasive ash as well as finely divided ash and soot.

Q. *Air from the final filter bed is discharged into the atmosphere by an exhauster. What kinds of exhauster may you choose?*

Ans. Mechanical exhauster where large tonnage of material is to be handled.

Steamjet exhauster for small and medium size plants.

Waterjet exhauster where a large quantity of water is abundantly available economically.

Q. *What is the ash carrying capacity of a waterjet exhauster?*

Ans. 5 to 30 t/h.

Q. *What are the advantages of a pneumatic ash handling system?*

Ans.

1. Dust free operation is possible as all flyash and dust is eliminated
2. No spillage problem or rehandling trouble

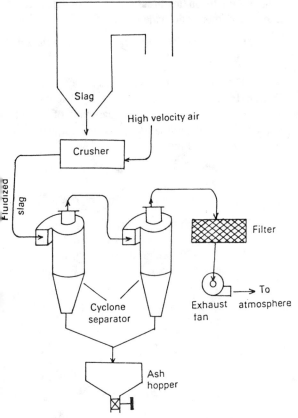

Fig. 15.5 *Pneumatic ash handling system*

33. Dry handling of ash eliminates the chances of ash freezing, picking or sticking in the storage bin
4. Flexibility in application to suit various plant conditions
5. Installation cost per ton of ash handled is less in comparison to other systems.

Q. *What are the disadvantages?*

Ans.

1. Wearing of the crusher as well as pipeline by the abrasive ash inflicting high maintenance charge
2. Noisy operation

Q. *How does a steamjet system function?*

Ans. In this case a high velocity steam instead of air is passed through a pipe to carry off dust and ash particles of considerable size. The ash is separated in the cyclone type separators and deposited in the ash hopper.

Q. *What are the advantages of a steamjet system?*

Ans.

1. Requires less floorspace
2. Can be installed in an inaccessible/awkward position also
3. The self-generated steam can be directed to operate this system. Hence no auxiliary drive is required.
4. The system is economic for ash disposal through a horizontal distance of 200 m and vertical distance of 30m
5. Low installation cost.

Q. *What disadvantages is the steamjet system fraught with?*

Ans.

1. Erosion of pipelines due to abrasive ash
2. Capacity of ash handling is limited to 15 t/h
3. Noisy operation.

16

Carryover, Scale and Sludge

Q. *What are the principal impurities in leakages in condensers?*

Ans. **Chlorides/Sulphates/Bicarbonates** of calcium, magnesium and sodium
Colloidal impurities of organic matter, silicic acid
Suspended matter
Gases (O_2, CO_2, N_2)

Q. *What are the chief impurities in DM water?*

Ans.

1. Sodium compounds
2. Products of metal corrosion
3. Gases like O_2

Q. *Why are these impurities, even in trace amounts, unwanted in BFW?*

Ans. They can precipitate, under favourable conditions, from water and form deposits on the heating surfaces, thus impairing heat transfer and causing high temperature rise of tube walls.

Q. *How is that soluble impurities in water are carried off into steam?*

Ans. Explosion of water bubbles in steam drum disperses fine droplets of water into steam over the water level and this entrained water particles contribute their dissolved impurities to steam.

Q. *Where will these steam-borne impurities deposit?*

Ans. In the superheaters and steam path of turbines.

Q. *Why are deposits in superheaters intolerable?*

Ans. The outlet portion of superheater coils, even at the rated heating loads, operate at the upper admissible temperature limit of the tube metal. Even a slight layer of deposits can, therefore, raise the metal temperature to an inadmissible level and promote creep phenomena and scale formation.

Q. *Why are deposits in the steam path of a turbine extremely undesirable?*

Ans. These increase the roughness of the blades and friction losses and therefore turbine efficiency is reduced. Heavy deposits in the steam path of turbine can cause additional axial pressure requiring a decrease in the turbine power.

Q. *What are the slightly soluble impurities that may be present in boiler water?*

Ans.

1. Calcium and magnesium salts
2. Oxides and Hydroxides of structural materials (Fe, Ni, Cu, Zn and Al) with which the aqueous heat transfer agent may come into contact.

Q. *What are those in readily soluble groups?*

Ans. Sodium salts (Na_3PO_4, Na_2SO_4) and sodium hydroxide.

Q. *What are the principal scale formers in water?*

Ans. $CaSO_4$, $CaCO_3$ and $Mg(OH)_2$

Q. *What is scale?*

Ans. It is the general name ascribed to sufficiently dense deposits appearing on heating or cooling surfaces and capable of reducing the normal heat transfer in a boiler unit or heat exchanger.

Q. *How do scales form?*

Ans. Those dissolved impurities in water which have the lowest solubility product under particular conditions crystallize first. Some of them crystallize on the heating surfaces and form scales. Surface roughness on solid heating surfaces can serve as centres of scale formation.

Q. *What is sludge?*

Ans. These are loose, friable deposits resulting from the settling, accumulation and thickening of suspended substances in the bulk of the solution.

Slimy accumulations in boiler drums, waterwall headers and various tanks are referred to as sludge.

Q. *How does sludge form?*

Ans. Those dissolved substances in water which crystallize in the bulk of the solution form sludge. Disperse and colloidal particles and gas bubbles suspended in water serve as centres of sludge formation.

Q. *How many types of scales are there?*

Ans. Four.
1. **Alkali-earth scales:** $CaCO_3$, $CaSO_4$, $CaSiO_3$, $Ca_3(PO_4)_2$, MgO, $Mg(OH)_2$, $Mg_3(PO_4)_2$
2. **Iron-oxide scales:** $FeSiO_3$, $Fe_3(PO_4)_2$, $NaFePO_4$ (sodium ferrophosphates), Fe_2O_3, Fe_3O_4
3. **Copper scales:** CuO, Cu_2O
4. **Silicate scales:** various compositions

Q. *What is the immediate effect of scale formation?*

Ans. It results in the increased temperature of the tube walls exposed to flue gas.

$$\Theta_{wall} = \Theta_{H_2O} + \frac{\Delta x_{scale}}{k_{scale}} Q + \frac{Q}{h_2}$$

where, Θ_{wall} = wall temperature (°C or K)

Θ_{H_2O} = temperature of water (°C or K)

Δx_{scale} = thickness of scale (m)

k_{scale} = thermal conductivity of scale (W/m K)

Q = rate of heat flow (W/m²)

h_2 = coefficient of heat transfer from the tube wall to water (W/m² K)

Q. *How will you calculate the extent of scale formation over a length of time?*

Ans. It can be determined by the formula:

$M = C_{fw} E G_{fw} t/10^3$, gram

where C_{fw} = concentration of scale- and sludge-forming substances in feed water (mg-equiv./kg.)

E = equivalent weight of scale

G_{fw} = flow rate of feed water (t/h)

t = time in hours

Q. *When will scales form?*

Ans. If the product of the concentrations of anions and cations exceed the solubility product of the scale former

$$[M^{n+}][A^{n-}] = S.P._{scale}$$

where $[M^{n+}]$ and $[A^{n-}]$ represent the concentrations of the cations and anions of the scale forming compound MA which ionizes as

$$MA \rightleftharpoons M^{n+} + A^{n-}$$

and its equilibrium constant being

$$K_{eq} = \frac{[M^{n+}][A^{n-}]}{[MA]} = [M^{n+}][A^{n-}]$$

$$cf. \ [MA] = 1$$

Q. *What is the rate of scale formation of the alkali-earth group?*

Ans. It is

$$\dot{M} = 1.3(10^{-13}) \, CQ^2, \ mg/[(cm^2) \, (h)]$$

where C = concentration of scale forming substances (mg/kg of water)

Q = rate of heat flow, (W/m²)

Q. *Is there any other factor responsible for formation of scale of sparingly soluble substances?*

Ans. Yes. It is the increase in concentration of the substances in the boiler water due to the evaporation effect. The degree of concentration by evaporation is determined by the formula:

$$C_{ev} = \frac{C_{bw}}{C_{fw}}$$

where, C_{bw} = concentration of some compounds in boiler water

C_{fw} = concentration of some compounds in feed-water

Q. *Does it vary? If so, by how much?*

Ans. Yes, it varies from spot to spot in the boiler drum as evaporation differs at different spots.

It varies from 10 to 200 and above.

Q. *Can such spots occur in boiler tubes?*

Ans. Yes. They originate at points where the tube is subjected to a high thermal stress or high rate of heat flow.

Q. *Is there any other factor contributing to scale formation?*

Ans. Substances showing negative temperature-gradient of solubility also contribute to scale formation with rising temperature which brings about the drop in the solubility of these salts.

Q. *Where is the iron-oxide scale formation more prominent?*

Ans. Zones of tubes subject to high rates of heat flow, i.e., in the zone where the burners are fitted, above and below the ignition belt, at the spots where the burner-flame strikes onto the heating surfaces.

Q. *Where are the iron-oxides scales encountered more—high duty or low duty boiler unit?*

Ans. High duty boiler units

Q. *What is the pattern of deposition?*

Ans. They concentrate mainly on the half of the perimeter of a tube facing the boiler furnace.

Q. *What are the basic components of iron-oxide deposition?*

Ans. Iron oxides or iron phosphates (depending on the scale forming conditions).

Q. *How does iron get into the feedwater?*

Ans. As a result of corrosion of structural materials in the water condensate circuit.

Q. *What is the rate of formation of iron-oxide depositions?*

Ans. It is $\dot{M} = 5.7(10^{-14})[C_{Fe}]_{bw}\, Q^2$ mg/[(cm^2) (h)]

Where, $[C_{Fe}]_{bw}$ = concentration of iron in boiler water, mg/kg of water

Q = the rate of heat flow, W/m^2

Q. *When do iron-phosphate depositions occur?*

Ans. It will occur under the conditions when

(a) concentration of phosphates PO_4^{3-} in boiler water is sufficiently high (> 80 mg/kg)

(b) boiler water has low pH values.

Q. *What is the effect of iron-phosphate scaling?*

Ans. It leads to honeycomb corrosion of the metal of the tubes.

Q. *Is the iron-oxide scale uniform throughout?*

Ans. Obviously not. It has two layers:

1. **Internal Layer**—dense and firmly bonded to the metal surface
2. **External Layer**—loose and porous; feebly bonded to the surface of the internal layer.

Q. *How does the internal layer of a ferric oxide scale form?*

Ans. Oxidation of metal at the surface forms the dense and rigid oxide film which is the internal layer of the iron-oxide scale.

Q. *Does it exercise any harmful effect?*

Ans. No. The rigid film of ferric oxide protects the metal underneath from further corrosion.

Q. *How does the external layer form?*

Ans. It forms mainly from the coagulation of the colloidal and disperse particles.

Q. *Does it exercise any harmful effect?*

Ans. Yes. Because of its low thermal conductivity, it impairs the heat removal from the metal surface. (Fig. 16.1)

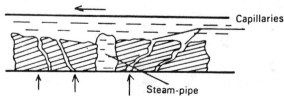

Fig. 16.1

Moreover, it gives rise to the 'wick effect': Water is sucked through numerous capillary pores in the layer and forms steam as it comes in contact with the heating surface. The steam ejects through a wide channel (known as 'steam-pipe'). This local circulation concentrates impurities (including corrosion-active impurities like alkalis and chlorides) at the metal surface and enhances corrosion.

Q. *If the capillary pores get clogged, will the heat transfer rate improve?*

Ans. If the capillaries get clogged due to deposition of other impurities present in water, the heat transfer may be sharply impaired.

Q. *What is the effect of pH on the rate of deposition of the iron-oxide external layer?*

Ans. At subcritical pressures and oxygen concentration of the order 0.4 mg/kg, the rate of deposition is as depicted in the curve. (Fig. 16.2)

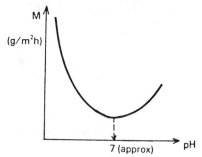

Fig. 16.2

It drops sharply from low pH values, reaches a minimum at pH value approximately 7 and then increases again with the increase of pH.

Q. *Where are the copper scales most likely to be formed?*

Ans. Sections subjected to a high rate of heat flow ($Q \geq 200 \text{ kW/m}^2$) and mostly on the part of the tube circumference facing the furnace.

Q. *What is the main factor responsible for the formation of copper deposits?*

Ans. It is due to the introduction of copper compounds into boiler units with BFW.

Q. *How does it occur in the deposits?*

Ans. Copper is present in both the metallic state and in the form of oxides.

Q. *Is the copper uniformly deposited over the deposition layer?*

Ans. No. The copper content diminishes in the direction from the surface layer toward the tube wall (from 70—90% to 10—15%).

Q. *Why may all the non-volatiles infiltrated into a once-through boiler unit with feedwater remain in the unit as depositions or get entrained by steam into the trubine?*

Ans. It is because the once-through boiler units are operated without blowdown.

Q. *Why does higher steam pressure increase the blade deposits of the turbine?*

Ans. It is due to the solvent action of steam. Higher the steam pressure, higher is the solvent action, which increases the fraction of substances entrained by generated steam and taken away from a boiler.

Q. *Why are the depositions of scale forming compounds of calcium, magnesium and iron-oxides maximum in the region of quasiphase transition?*

Ans. The region of quasiphase transition characterizes itself as the region of maximum heat capacity. Here a substantial reduction in the solubility of many substances(CuO, SiO_3^{2-}, $Mg(OH)_2$, $CaSO_4$ and others) is observed with the growth of enthalpy. (Fig. 16.3)

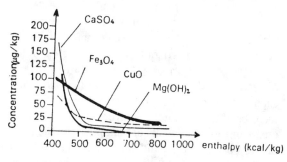

Fig. 16.3

Note: Enthalpy of a substance increases with the increase of its heat capacity.

Q. *What determines the solubility of impurities in superheated steam of subcritical pressure?*

Ans. The properties of superheated steam as well as those of the impurities with which the steam is in contact determine the solubility of impurities in the superheated steam at subcritical pressure.

Q. *Upon which process parameters, are the properties of both superheated steam and impurities dependent?*

Ans. Temperature and pressure.

Q. *How does temperature affect the properties of impurities?*

Ans. It is the temperature upon which the intermolecular forces of attraction and bond strength between the ions of a solid impurity depend substantially. As the temperature is raised the bonds get weakened allowing the solid phase to pass over the steam.

Q. *How does pressure affect the properties of impurities?*

Ans. Though a great pressure can weaken the bond strength between the ions of a solid impurity, in the existing range of operating pressure of boilers it exerts little effect on the behaviour of the solid phase.

Q. *How can temperature influence the capacity of superheated steam to dissolve solid impurities?*

Ans. The pressure remaining constant, the degree of superheat of the steam increases with the rise of temperature and at the same time the density of steam decreases, with the effect that a sharp decline in the dielectric permittivity of water and polarity of water molecules is observed. As a result, the dissolving capacity of superheated steam first decreases due to increase of temperature. However, with a further increase of temperature, at constant pressure, the crystalline bonds in the solid impurities get impaired resulting in the increased solubility of those substances in the superheaded steam. (Fig. 16.4)

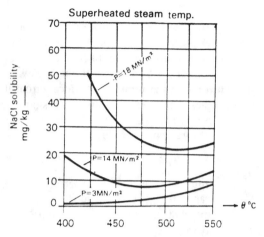

Fig. 16.4 Solubility of NaCl in Superheated steam at subcritical pressures

Q. *How does pressure exert its influence on the dissolving capacity of superheated steam?*

Ans. It exercises a strong influence on the dissolving power of steam. With the increase of pressure, the density of superheated steam increases and consequently the steam gains a higher dissolving power.

Q. *If the concentration of an impurity in steam is lower than its isobaric (constant pressure) solubility, what will happen?*

Ans. Steam will simply dissolve it and carry it off. It will deposit them wherever it suffers an enthalpy drop, i.e., on the turbine blades.

Q. *If, on the other hand, the concentration of an impurity in steam is greater than its isobaric solubility, what will happen?*

Ans. The exces of solid impurity will get deposited in the superheater path while the leftover will deposit on the turbine blades.

Q. *How do the solid impurities pass from water to saturated steam?*

Ans. By two ways:
1. they become steam-borne as the droplets of boiler water contaminate the steam
2. they pass into steam as the latter dissolve them and carry them off.

Q. *How does droplet carryover take place?*

Ans. It results from two processes:
1. breaking of water into fine droplets
2. destruction of steam bubbles

If the steam-water mixture is introduced into the boiler drum above the water level of the drum, the high speed jets of the mixture strike all kinds of baffles and partition plates placed on their path to quench their kinetic energy. After collision with these metal plates and baffles, the water content of the jets breaks into droplets of varying size. The higher the kinetic energy of the steam-water jet entering the drum, the finer the droplets generated. Fine droplets are easily carried off by steam. Along with these droplets the solid impurities pass into the steam.

If the steam-water mixture is introduced into the boiler drum below the level of the boiler water, steam bubbles are formed. These bubbles rise to the disengagement surface, i.e., the water surface of the drum and break through the water film and form droplets. Projected in all directions, the droplets may reach demister (steam separating device) and get entrained by steam thereafter.

Q. *What factor leads to more intensive carryover?*

Ans. It is the high heating load.

Higher the heating loads, higher will be the velocity of exit of steam-water jets into the drum and higher will be their kinetic energy. Higher the kinetic energy of the steam-water jets, greater will be their power of atomization of water mass into finer droplets upon impact with the baffles. As a result a more intensive water carryover takes place.

Q. *Where is the concentration of water droplets highest?*

Ans. Just at the disengagement surface.

Q. *What kind of water droplets will be entrained by steam if its rising velocity is low?*

Ans. Finest water droplets.

Q. *Why does the water content of steam output turn out to be higher at higher heating loads?*

Ans. With the increase of heating loads, the flowrate of steam will increase and that will entrain larger and larger water droplets resulting in higher water content of output steam.

Q. *What is the relationship between the water content of steam and the heating load?*

Ans. It is given by

$$W = AD^x$$

where, W = water content of steam

D = heating load

A and X are functionally dependent on the pressure, concentration, ionic composition of impurities in water and the design of the drum. (Fig. 16.5)

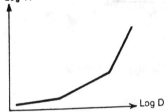

Fig. 16.5 *Effect of heating load on the water content of steam*

Q. *What is the rate of evaporation per m^2 of water surface?*

Ans. It is the average flowrate of steam related per unit square metre of the disengagement surface.

$$R_F = \frac{D}{F}$$

Where F = area of disengagement surface, m^2.

Q. *What is called the rate of evaporation per m^3 of steam space?*

Ans. It is the average velocity of steam related to one cubic metre of steam space

$$R_V = \frac{D}{V}$$

Q. *If at constant rate of evaporation per m^2 of disengagement surface the height of steam space is altered, will it affect the water content of steam?*

Ans. Yes. If the steam-space height is reduced, larger droplets may reach the region of high steam velocities at the inlet to the steam circulating tubes and therefore, the steam will contain more moisture.

If on the other hand, the height of the steam-space is increased, the larger droplets may not have sufficient kinetic energy to reach the steam circulating tubes, with the effect that the water content of the output steam will be considerably lower.

Q. *What is the soaring velocity of water droplets in steam drum?*

Ans. It is the relative velocity of a water droplet at whcih its weight is counterpoised by the force of resistance (↑) offered by the up-flowing steam

$$V_{so} = 1.155 \sqrt{\left(\frac{d}{\varepsilon}\right)\left[\frac{\rho_w}{\rho_s} - 1\right]}$$

where V_{so} = soaring velocity of a water droplet
d = droplet dia
ε = coefficient of resistance
ρ_w = density of water droplet
ρ_s = density of steam

Q. *An increase in pressure will always result in higher water content of steam. Why?*

Ans. With the increase of pressure, the surface tension of water decreases with the effect that the size of the water droplets gets reduced, i.e., more and more fine droplets are produced. These finer water droplets easily become steam-borne increasing the water content of steam.

Q. *But increase in pressure will increase the density of steam and so it will offer greater resistance to rising droplets. Is it not?*

Ans. Yes, of course. But comparatively, pressure has a stronger effect on the transporting ability of steam than on its resistance.

Q. *What is the effect of the salt content of BFW on the water content of steam?*

Ans. This results in a water swell in the steam drum. A large number of water droplets with high concentration of the impurity are ejected and that critically impairs the steam quality.

Q. *What is called the critical concentration of a substance in BFW (Boiler Feed Water)?*

Ans. It is the concentration of the substance at which the water level suddenly swells up and increases the entrainment of moisture in steam (Fig. 16.6)

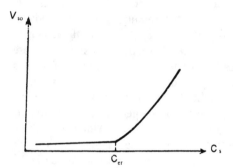

Fig. 16.6

Q. *What determines the quality of steam?*

Ans. It is determined by the quantity of various substances — both volatile and non-volatile, collectively called impurities—present in it.

Q. *What is called the distribution coefficient?*

Ans. The ratio between the quantities of substances distributed in steam and BFW is known as distribution coefficient

$$[K_d]_{total} = \frac{C_s}{C_{bfw}}$$

where C_s and C_{bfw} represent the concentration of salt in steam and boiler feedwater respectively.

Q. *What is carryover in terms of distribution coefficient?*

Ans. It is $C_s = C_{bfw}[K_d]_{total}$

Q. *How many kinds of carryover are there?*

Ans. Two kinds : droplet carryover and selective carryover.

Q. *What is called selective carryover?*

Ans. It is encountered when a substance or a group of substances transfers into the steam phase in quantities comparatively much larger than other substances.

Q. *What is the cause of selective carryover?*

Ans. It originates due to the fact that certain sub-stances (impurities) contained in the boiler water exhibit markedly higher dissolving capacity in high pressure steam in comparison to other substances.

Q. *Can you cite an example?*

Ans. Silicic acid (H_2SiO_3) is a good example. Its solubility in steam increases rapidly with the rise of steam pressure from $7MN/m^2$.

17

Steam Contamination and its Control

Q. *How can the quality of steam be described?*

Ans. It is frequently described in terms of the amounts of Na, SiO_2 and CO_2 dissolved in steam and expressed in mg/kg or microgram/kg.

Q. *Why does contamination occur?*

Ans. Various dissolved substances in the boiler water pass into steam which therefore gets contaminated.

Q. *How can this be expressed?*

Ans. This can be conveniently expressed in terms of the distribution coefficient as the distribution of substances between the liquid and the vapour phase takes place.

$$K_d^{total} = \frac{C_s}{C_{bw}}$$

where K_d^{total} = total distribution coefficient

C_s = concentration of total quantity of any substance (or the sum of substances) in steam, mg/kg

C_{bw} = concentration of total quantity of any substance (or the sum of substances) in boiler water, mg/kg

Q. *How can the true and apparent distribution coefficients be distinguished?*

Ans. The true distribution coefficient (K_t^d) refers to the distribution of a given substance of certain form whereas apparent distribution coefficient (K_a^d) charac-

terizes the distrubution of a given substance in all forms.

Q. *How is K_t^d expressed?*

Ans. It is expressed as $K_t^d = \dfrac{C_s}{C_{bw}} \left[\dfrac{\rho_s}{\rho_{bw}} \right]^n$

ρ_s and ρ_{bw} are the densities of steam and boiler water respectively, kg/m^3.

n = co-ordination number characterizing the nature of substances.

Q. *Does the value of K_t^d vary with temperature and pressure?*

Ans. Yes. The value of K_t^d increases with temperature and pressure, particularly for those substances which have a large value of n

These substances are

CaO	n = 5.5	NaOH	n = 4.1
MgO	n = 3.1	NaCl	n = 1.9
SiO_2	n = 1.9		

Q. *How does the increase of working pressure of the boiler affect the steam quality in once-through boilers?*

Ans. In the once-through boilers, the load of dissolved substances carried by boiler water is partly deposited on the heating surfaces and the rest escapes with the steam as working fluid gets continuously evaporated. As the pressure is increased, the concentration of impurities in steam increases and so the

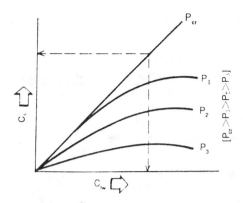

Fig. 17.1

steam quality increasingly deteriorates until it approaches that of feedwater. (Fig. 17.1)

Q. *What factors affect the quality of steam generated in natural circulation boiler units?*

Ans. The quality of saturated steam is affected by
   (a) the design features of a boiler unit
   (b) the operating conditions for drum-type boiler units
   (c) the quality of boiler feedwater.

Q. *What do you mean by "design features of a boiler unit" so far as steam contamination is concerned?*

Ans. It means:
   1. height and volume of steam space in the boiler drum.
   2. the mode of introducing steam-water mixture into the drum
   3. the method of steam extraction from the drum
   4. efficiency of the drum internals, etc.

Q. *How does steam contamination arise from the standpoint of volume of steam space in boiler drum?*

Ans. This arises from the fact that the steam space may be loaded non-uniformly giving rise to local high rate of evaporation in the steam space. This local overloading is produced by a non-uniform steam release from the drum or by concentrated introduction of the steam-water mixture in the drum. This may also arise due to improper design and layout of drum internals or their maloperation.

Q. *What is the rate of evaporation per unit steam space of the drum?*

Ans. It is given by

$$R_d = \frac{G_s v}{V_{ss}} (10^3),\ m^3/(m^3\ h)$$

where G_s = steam generating capacity of the unit, $\frac{t}{h}$

   v = specific volume of steam, m^3/kg
   V_{ss} = steam space of the drum, m^3

Q. *What is the residence time of steam in the drum?*

Ans. It is the inverse of the average rate of evaporation per unit steam space. It is given by

$$\tau = \frac{1}{R_d}$$

Q. *How does steam contamination increase with the increase of the evaporation rate per unit steam space?*

Ans. As the R_d increases, the residence time of steam in the drum steam space decreases and as a result the soaring velocity of steam in the steam space increases, contributing to higher degree of steam contamination.

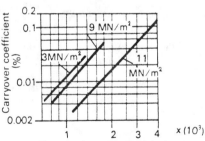

Fig. 17.2

This fact can be best understood from the accompanying figure (Fig. 17.2)

Q. *What factors primarily determine the steam contamination when steam-water mixture is introduced below the water-level of the boiler drum?*

Ans.
   1. Rate of evaporation per unit evaporating surface, R_s, $m^3/(m^2\ h)$
   2. Height of the steam space, h_{ss} (Fig. 17.3)

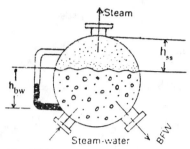

Fig. 17.3

Q. *How can the rate of evaporation per unit evaporating surface be expressed?*

Ans.

$$R_s = \frac{G_s\, v}{A}(10^3),\ m^3/(m^2.h)$$

where, A = evaporating surface area, m^2

Q. *How can steam contamination be reduced by altering the rate of evaporation per unit evaporating surface and the height of steam space?*

Ans. Steam contamination with BFW carrying impurities can be reduced by either decreasing R_s or increasing h_{ss}.

Q. *How does the velocity of steam bubbles moving through boiler water affect the height of steam space in cases of boiler units where steam-water mixture is introduced below the level of water in the drum?*

Ans. Steam bubbles moving through the height of boiler water h_{bw} (see above Figure 17.3) have a retention time equal to

$$\tau_s = \frac{h_{bw}}{v_s}\ \text{where } v_s = \text{steam velocity}$$

So lower the steam velocity, greater is the retention time of the steam-bubbles inside the boiler water and therefore the water becomes more and more saturated with steam and gets converted into a steam-water mixture whose density is considerably lower than that of water. As a result, the volume of the water enlarges, its level in the drum rises and the height of steam space decreases.

Note: This phenomenon is called swelling of boiler water. As the height of steam space decreases, the quality of steam decreases.

Q. *How does steam contamination vary with the height of steam space?*

Ans. Lower the height of steam space greater is the steam wetness. Moreover, for the same steam space height, steam wetness increases with the increase of steam pressure. And for the same degree of wetness, it requires less space height for low pressure steam than for high pressure steam. These can be illustrated with the help of the adjacent figure. (Fig 17.4).

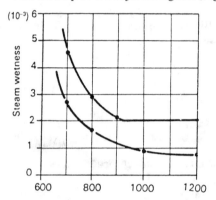

Fig. 17.4 *Steam contamination as a function of height of steam space*

Q. *Why is the concentrated feed of steam-water mixture below the water level in the boiler drum undesirable?*

Ans. It gives rise to high local rates of steam flow through the evaporating surface with the effect that steam contamination intensifies.

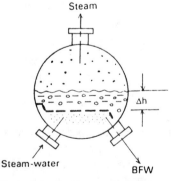

Fig. 17.5 *Baffle resists boiler water swelling and thereby reducing steam contamination*

Q. *How can this be avoided?*

Ans. One good measure is to place a perforated plate (baffle plate) in the body of the boilerwater in the drum. It resists boilerwater swelling and therefore, the evaporating surface is loaded uniformly and contamination of steam is reduced. (Fig. 17.5)

Q. *How does the steam quality vary with the change of the operating parameters of the boiler drum?*

Ans. With the variation of steam demand, the operation of a boiler unit may undergo a change in the level of boiler water and steam pressure. And that results in variation in the quality of steam produced. As the load is increased steam contamination rises. Steam contamination up to point *A* is within admissible limits and beyond that it increases abruptly and becomes inadmissible. (Fig. 17.6).

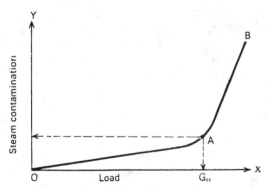

Fig. 17.6 *Steam contamination Vs. Load graph*

Q. *What is critical load?*

Ans. It specifies the boiler load above which steam contamination becomes inadmissible. It is G_{cr}, corresponding to the point *A* in Figure 17.6.

Q. *What is admissible load?*

Ans. All the subcritical loads, i.e., loads in the region to the left of *A* are admissible loads. And the maximum admissible load according to the conventional standard is 70 to 80% of the critical load, i.e.

$$G_{ad}^{max} = (0.7 - 0.8)G_{cr}$$

Note: The design capacity of boiler $= G_{ad}^{max}$

Q. *Is the critical load for a given boiler unit constant?*

Ans. No. It is directly proportional to the height of steam space and efficiency of steam-water separation devices and inversely proportional to the boiler feedwater concentration.

Q. *Why is it undesirable to operate boilers under conditions of sudden and large variation of steam demand?*

Ans. An abrupt increase in load may result in a sharp drop of steam pressure and lead to priming of boiler water. Sudden drop of steam pressure is accompanied by swelling of boiler water reducing the steam space. As a consequence, steam contamination is increased and the steam quality is decreased.

Q. *How does the nature of steam demand influence the quality of steam?*

Ans. As above.

Q. *Due to abrupt change in steam demand, in which case will the quality of steam be affected more — low pressure boiler unit or high pressure boiler unit?*

Ans. High pressure boiler unit.

Q. *For a boiler unit in service, two levels of BFW in the boiler drum are prescribed—one is upper water level and the other is a lower water level. Why?*

Ans. The upper water level is taken so as to ensure high quality steam generation while the lower water level is taken for reliable circulation.

Q. *What is the suggested water level in the boiler drum, as good practice?*

Ans. It is most desirable to operate boiler units with the BFW in the boiler drum maintained at a middle level and without its sharp variations.

Q. *Will the steam generated in a properly designed and well operated boiler unit still have any chance of contamination?*

Ans. Yes.

Q. *How?*

Ans. The quality of steam generated depends upon the quality of boiler feedwater apart from the design features of the boiler unit and its operation.

Q. *How does the steam quality vary with the salt concentration in boiler water?*

Ans. The steam quality deteriorates steadily with the increase of salt content in the BFW (C_{bw}). Upto point A, the carry over of salt to steam due to salt content in the boilerwater is within admissible limits. A further increase in C_{bw} produces a sharp deterioration of steam quality. So point A marks the maximum admissible limit of salt concentration in boiler water. (Fig. 17.7)

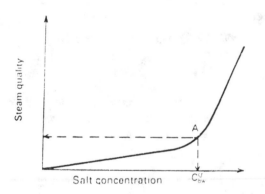

Fig. 17.7 Steam quality as a function of salt concentration in BFW

Q. *What is the critical salt concentration in boiler-water?*

Ans. It characterizes the salt concentration in boiler water above which steam quality becomes inadmissible. It is C_{bw}^{cr} corresponding to the point A in Fig. 17.7.

Q. *What is the admissible salt concentration in the BFW?*

Ans. All the salt concentrations in the BFW in the region to the left of A are admissible. And the maximum admissible salt concentration is taken as 70 to 80 per cent of critical salt content i.e.

$$C_{bw}^{ad} = (0.7 - 0.8)C_{bw}^{cr}$$

Q. *What is foaming?*

Ans. It is the formation of stable foam on the evaporating surface in the boiler drum.

Q. *Why does it occur?*

Ans. Increase in salt concentration in the BFW is the primary factor leading to foaming.

Q. *Does it tell upon the quality of steam?*

Ans. Yes.

Q. *How?*

Ans. Foam swells and disintegrates into minor droplets of water carrying dissolved salts which thus contaminate the steam above the disengagement surface and affect its quality.

Moreover, in the case of intense foaming, flakes of foam may get ingested into the superheater coil imparting a high water content to steam and thereby increasing its wetness and salt carryover.

Q. *What are the major factors contributing to the genesis of foaming?*

Ans.
1. High salt content in boilerwater
2. Presence of oils and grease in boiler water
3. Presence of suspended sludge in a finely dispersed state in boilerwater.
4. High alkali concentration in boilerwater.

Q. *What is priming?*

Ans. Infiltration of BFW into steam lines of the boiler drum.

Q. *Why does it occur?*

Ans. Due to
(a) sharp pressure drop in the drum because of abrupt increase of steam load
(b) penetration of oil or alkali in the BFW in amounts exceeding the allowable standard

Q. *How do oils and grease find their way to the feedwater?*

Ans. This may occur due to
 (a) leakage of fuel oil heater coil
 (b) contamination of steam heating or purging lines by oil from oil burners

 (c) penetration of oil from dirty filters in the hot well.

Q. What is the difference between foaming and priming?

Ans. Foaming characterizes stable foam formation, i.e., continuous formation of bubbles possessing sufficiently high surface tension to prevent them from coalescing and breaking, with the effect that they retain their bubble forms even after their disengagement from the water surface.

Priming is referred to describe the state of unstable foam formation resulting in the discharge of steam contaminated with excessive amount of water droplets because of violent ebullition in the boiler observed periodically.

Q. *What is the cause of foaming?*

Ans. Foaming is chiefly due to contamination of boilerwater with oil, soap, organic matter, suspended matter or other foreign particles.

Q. *What is the cause of priming?*

Ans. Priming is primarily due to surging (swelling) of water in the steam drum and is induced by a high water level in the drum. The violent movement of steam bubbles to the disengagement surface and their explosion causes this swelling and temporarily raises the water level in the steam drum. The exploding bubbles spray water droplets in the steam space and these are carried over by steam.

Q. *What is foamover or foampriming?*

Ans. It is the carryover of foam from the steam drum with the steam.

Q. *In which part of the turbine, are the salts carried over by steam mainly deposited?*

Ans. The salts passed onto steam primarily fall out on turbine blades.

Q. *What is the direct consequence of salt deposition on turbine blades?*

Ans. The smooth blade surface gets rougher, blade-channel profiles get distorted and redistribution of heat drop among turbine occurs due to uneven scaling. As a result turbine efficiency decreases.

Q. *What is the mechanism of salt deposition on turbine blades?*

Ans. The superheated steam fed to the turbine always contains certain non-volatile substances. They may be present in the form of dry, hard particles and in a dissolved state. At steam pressures exceeding 7 MPa (7 MN/m^2) , the superheated steam carries a large fraction of the carried-over salt in the dissolved state. And as the temperature and pressure drops of the steam take place through stages of the turbine, the solubility of these dissolved salts in steam also decreases. So when steam becomes supersaturated with respect to a certain substance at any turbine stage, that very substance will settle on the blades of that turbine stage.

If at the turbine inlet steam is saturated with respect to a substance, the latter starts to deposit onto the blades of even the first turbine stages. However, if the concentration of a particular substance dissolved in steam is less than the concentration at saturation, the fall-out of the substance will occur in the latter stages where a supersaturated solution of that substance in steam is formed.

Dry deposition on the blades also takes place as the dry, hard particles strike and adhere to the blade surfaces.

Q. *What kinds of deposits precipitate on steam turbine blades?*

Ans. The compounds forming deposits on turbine blades can be divided into two classes:
 1. **Water Soluble Compounds:** belonging to this class are sodium silicate (Na_2SiO_3), sodium sulphate (Na_2SO_4), sodium chloride (NaCl) and sodium hydroxide (NaOH).
 2. **Water Insoluble Compounds:** these are iron oxide, aluminium oxide and silicic acid.

Q. *What compounds fall out in the superhigh pressure and high pressure cylinders of supercritical-pressure turbines?*

Ans. These are mainly the oxides of copper, iron and silicon, [CuO, Fe_2O_3 and SiO_2]

Q. *What compounds primarily get deposited in the intermediate pressure and low pressure cylinders of steam turbines?*

Ans. Mostly compounds containing Na_2SiO_3, SiO_2 and to some extent Fe_2O_3.

Q. *How can the quantity of depositon in turbine blades be controlled?*

Ans. The best way of controlling blade deposition of turbines is to generate a high quality of steam. But if deposits form, they are removed by washing the blades with wet steam whereupon water soluble deposits are transferred into the washings giving rise to partial washing-off.

In cases of deposits which are poorly soluble in water, recourse to certain chemicals, such as EDTA, hydrazine and sodium hydrate are taken. These reagents are injected to wet steam to wash out poorly water-soluble deposits from turbine blades.

Q. *How can high quality steam be generated?*

Ans. Methods of producing high quality steam vary depending on the working parameters of the plant. But in general these are:
  (a) Improving the quality of feedwater
  (b) Ensuring effective separation of water droplets from the generated steam
  (c) Harnessing steam washing
  (d) Putting into effect stage-evaporation.

Q. *How can impurities from steam-water cycle be removed to improve the quality of circulating water?*

Ans. The impurities that penetrate into boiler units due to leakage of cooling water in turbine condensers, make-up water and corrosion processes can be brought down to satisfactorily low levels by a blowdown technique. However this method is effective only for drum-type subcritical pressure boiler units.

By blowing down a fraction of the boiler water and its replacement by feedwater make-up, an excessive concentration of undesirable substances in boilerwater is prevented. Since in drum type boilers, the purity of the generated steam is determined by the quality of water from which it is produced, blowdown adequately tackles the problem by ensuring a lower concentration of impurities in the boilerwater. However, blowdown is inapplicable in once-through boilers — both subcritical and supercritical pressure units. In these cases the turbine condensate is treated in a special demineralizing plant to produce polished water which is recycled. And that's how feedwater quality is improved.

Q. *What is blowdown?*

Ans. Blowdown is defined as the removal from a boiler unit or evaporator a fraction of the boilerwater either continuously or intermittently, in order to prevent an excessive concentration of undesirable dissolved substances in boilerwater.

Q. *In which type of boilers can blowdown be put into effect?*

Ans. Drum-type boiler units as well as once-through boiler units befitted with separators and operated at a subcritical pressure.

Q. *How many types of blowdown are known?*

Ans. Two: Continuous and Intermittent type.

Q. *What is the objective of continuous blowdown?*

Ans. It is to maintain a definite concentration of dissolved non-volatile substances in the boilerwater.

Q. *What is the purpose of intermittent blowdown?*

Ans. It is undertaken to remove sludge from a boiler unit.

Q. *What is the general equation of blowdown?*

Ans. Blodown is directed to set up and ensure a salt balance for boiler water, i.e.
   Mass of salts entering a boiler unit
   = Mass of salts removed from the boiler unit
   Therefore, the mass balance (of total dissolved salts) equation is

$$C_{fw}\, G_{fw} = C_s G_s + C_{bw}\, G_{bd} \qquad (17.1)$$

where G_{fw}, G_s and G_{bd} refer to mass flowrate (t/h) of feedwater, steam and blowdown water respectively.

C_{fw}, C_s and C_{bw} correspond to concentration of salt in feedwater, steam and boilerwater respectively.

Q. *What is percent blowdown?*

Ans. It is the ratio of rate of blowdown to the rate of steam generation expressed in percent, i.e., the blowdown rate in relation to steam

$$p_{bd}^s = \frac{G_{bd}}{G_s} \times 100 \qquad (17.2)$$

Again,

$$G_{fw} = G_s + G_{bd} \qquad (17.3)$$

Therefore, from equation 17.1 and 17.3 we get,

$$G_{bd} = \left[\frac{C_{fw} - C_s}{C_{bw} - C_{fw}} \right] G_s \qquad (17.4)$$

$$\therefore \qquad p_{bd}^s = \left[\frac{C_{fw} - C_s}{C_{bw} - C_{fw}} \right] 100 \qquad (17.5)$$

Also percent blowdown can be expressed in relation to feedwater

$$p_{bd}^w = \frac{G_{bd}}{G_{fw}} \times 100 \qquad (17.6)$$

or,

$$p_{bd}^w = \left[\frac{C_{fw} - C_s}{C_{bw} - C_s} \right] 100 \qquad (17.7)$$

Salt content in boiler water

$$C_{bw} = C_{fw} \frac{100}{p_{bd}^w} \qquad (17.8)$$

Salt content in steam

$$C_s = \frac{Y}{100} C_{bw} \qquad (17.9)$$

where Y = coefficient characterizing the carryover of dissolved substances from boiler water to generated steam

$$= \frac{C_s}{C_{bw}}$$

The salt balance equation can also be written as

$$(1 + p_{bd}^s) C_{fw} = (Y + p_{bd}^s) C_{bw} \qquad (17.10)$$

Introducing the carryover coefficient equation 17.5 takes the form

$$p_{bd}^s = \left[\frac{C_{fw} - Y C_{bw}}{C_{bw} - C_{fw}} \right] 100 \qquad (17.11)$$

From this equation we get

$$C_{fw} = \left[\frac{p_{bd}^s + 100Y}{p_{bd}^s + 100} \right] C_{bw} \qquad (17.12)$$

$$C_{bw} = \left[\frac{p_{bd}^s + 100}{p_{bd}^s + 100Y} \right] C_{fw} \qquad (17.13)$$

The percentage of dissolved substances removed by

(a) Steam is

$$\phi_s = \frac{100}{p_{bd}^s + 100} \frac{C_s}{C_{fw}} \times 100 \qquad (17.14)$$

(b) Blowdown is

$$\phi_{bd} = 100 - \phi_s \qquad (17.15)$$

Problem 17.1 For a given drum-type boiler unit operating at subcritical pressure, the salt content of boiler feedwater is 45 mg/kg.

Determine the percent of salt removed by blowdown water and by steam (as carryover) if the blowdown rate in relation to feedwater is 2.2%.

Take carryover coefficient = 0.01%

Solution

We must make use of the formulae 17.14 and 17.15 to determine the percent of salt removed by blowdown water and by steam generated. And for that purpose we must know the value of C_s beforehand.

Step (I) Determination of C_s

Working Formula: $\quad C_s = \dfrac{Y}{100} C_{bw} \qquad (17.9)$

$$= \frac{Y}{100} C_{fw} \frac{100}{p_{bw}^w}$$

Y	C_{fw}	p_{bd}^w	C_s
0.01	45 mg/kg	2.2	$\dfrac{0.01}{100} (45) \dfrac{(100)}{2.2} = 0.2045$ mg/kg

Step (II)   Determination of ϕ_{bd} and ϕ_s

Working Formulae:

$$\phi_s = \frac{100}{p_{bd}^s + 100} \frac{C_s}{C_{fw}} \times 100 \qquad (17.14)$$

$$\phi_{bd} = 100 - \phi_s \qquad (17.15)$$

p_{bd}^w is given but p_{bd}^s is not. So we are to calculate the latter out. This can be done as follows:

Taking $C_s \approx 0$, Eq. (17.5) and (17.7) simplify to

$$p_{bd}^s = \frac{C_{fw}}{C_{bw} - C_{fw}} \times 100 = \frac{\dfrac{p_{bd}^w}{100}}{1 - \dfrac{p_{bd}^w}{100}} \times 100$$

$$= \frac{p_{bd}^w}{100 - p_{bd}^w} \times 100$$

$$= \frac{2.2}{100 - 2.2} \times 100 = 2.249\%$$

C_s	C_{fw}	p_{bd}^s	ϕ_s
0.2045 mg/kg	45 mg/kg	2.249%	$\dfrac{100}{2.249 + 100} \dfrac{0.2045}{45} \times 100$ -0.44% *Ans.*

$$\phi_{bd} = 100 - \phi_s = 100 - 0.44 = 99.56\% \qquad Ans.$$

Problem 17.2 For a drum-type boiler unit, the concentration of chlorine in feedwater and boilerwater is 55 and 245 mg/kg. Determine the blowdown rate for the boiler if 25% of make-up water is used in the feed system of the boiler unit.

Solution   Here we shall use the formula

$$p_{bd}^w = \frac{C_{fw}}{C_{bw}} \times 100$$

(see problem 17.1) to calculate the blowdown rate from the salt concentration in feedwater and boilerwater.

Since 25% of the make-up water is used in the feed system

$$p_{bd}^w = 0.25 \frac{C_{fw}}{C_{bw}} \times 100 = 0.25 \frac{55}{245} (100) = 5.61\% \qquad Ans.$$

Problem 17.3 A drum-type boiler unit (Fig. 17.8) operating at 3.348 MPa (3.348 MN/m²) discharges water to a flash-tank in a continuous-blowdown hookup. The flash tank is maintained at a pressure 0.198 MPa (0.198 MN/m²). Determine the amount of steam separated in the flash-tank per ton of blowdown water if the coefficient accounting for the loss of heat to the surroundings is 0.98.

Solution   **Step (I)** The blowdown flashtank is maintained at a pressure considerably lower than the boiler pressure. And because of this pressure, a fraction of boilerwater, β, gets evaporated in the flashtank such that

$$\beta = \frac{\eta E_{bw} - E_w}{E_s - E_w}$$

where E_{bw} = enthalpy of boilerwater, kJ/kg

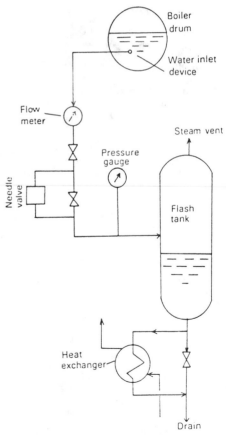

Fig. 17.8

E_w = enthalpy of water in flashtank, kJ/kg

E_s = enthalpy of steam generated in flashtank, kJ/kg

η = coefficient accounting for heat loss to the surroundings

Now

E_{bw} at 3.348 MPa = 1037 kJ/kg

E_w at 0.198 MPa = 503.7 kJ/kg

E_s at 0.198 MPa = 2706 kJ/kg

$$\beta = \frac{(0.98)(1\,037) - 503.7}{2\,706 - 503.7} = \frac{512.56}{2\,202.3} = 0.2327$$

Step (II) Now the quantity of steam generating in the flashtank of the continuous blowdown hookup is

$$G_s^{fl} = \beta G_{bd} = 0.2327(1) = 0.2327 \text{ t/h}$$

Problem 17.4 The double-stage blowdown of a 9.87 MPa boiler unit is as shown below:

Calculate the percent of dry, saturated steam separated from each flashtank if $\eta = 1$.

Solution The fraction of boilerwater that evaporates in the flashtank can be expressed as per cent by the formula:

$$\beta = \frac{\eta E_{bw} - E_w}{E_s - E_w} \times 100$$

1st FLASHTANK

E_{bw} at 9.87 MPa	E_w at 1.55 MPa	E_s at 1.55 MPa	η	β
1 402 kJ/kg	852 kJ/kg	2 793 kJ/kg	1	$\frac{1\,402 - 852}{2\,793 - 852} \times 100$ -28.33% *Ans.*

2nd FLASHTANK

E_{bw} at 1.55 MPa	E_w at 0.198 MPa	E_s at 0.198 MPa	η	β
852 kJ/kg	503.7 kJ/kg	2 706 kJ/kg	1	$\frac{852 - 503.7}{2\,706 - 503.7} \times 100$ -15.81% *Ans.*

Problem 17.5 The total dissolved solids (TDS) in the boilerwater of a drum-type subcritical pressure boiler is 1 175 mg/lt. What should be the maximum admissible content of TDS in softened water so as not to exceed the blowdown rate of 4.75%, if make-up water amounts to 20%?

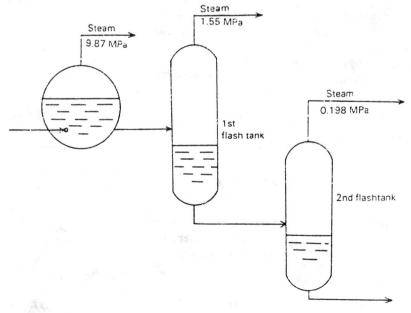

Fig. 17.9

Solution   We shall use the formula 17.5

$$p_{bd}^s = \frac{C_{fw} - C_s}{C_{bw} - C_{fw}} \times 100$$

to determine the maximum admissible TDS in softened water.

If we ignore the salt carryover by steam, the above equation simplifies to

$$p_{bd}^s = \frac{C_{fw}}{C_{bw} - C_{fw}} \times 100$$

In the present case, make-up water amounts to 20% and accordingly the above equation becomes

$$p_{bd}^s = \frac{0.2 C_{fw}}{C_{bw} - 0.2 C_{fw}} \times 100$$

Computing the numerical values,

$$4.75 = \left[\frac{0.2 C_{fw}}{1\,175 - 0.2 C_{fw}} \right] \times 100$$

or

$$1\,175 - 0.2 C_{fw} = 4.21 C_{fw}$$

∴

$$C_{fw} = \frac{1\,175}{4.41} = 266.40 \text{ mg/kg}$$

Ans.

Problem 17.6   The ratio of salt content in boilerwater to that in feedwater for an H.P. boiler fitted with blowdown is 41. Determine the amount of salts getting onto the boiler with feed, removed by blowdown water and as carryover (by steam), expressed as percentage. Take Y = 0.01%

Solution   We shall use the following two fomulae to determine the percentage of salt removed by blowdown (ϕ_{bd}) and that removed as carryover (ϕ_s):

$$\phi_s = \frac{100}{100 + p_{bd}^s} \, Y \frac{C_{bw}}{C_{fw}} (100) \qquad (17.14)$$

[cf. $C_s = Y C_{bw}$; Y = per cent of carryover]

$$\phi_{bd} = 100 - \phi_s \qquad (17.15)$$

Now, $\dfrac{C_{bw}}{C_{fw}}$ is given; Y is also given; p_{bd}^s is unknown.

Step (I) Calculation of p_{bd}^s

Working Formula: $p_{bd}^s = \dfrac{p_{bd}^w}{100 - p_{bd}^w} \times 100$

p_{bd}^w is not known but it is calculated as follows:

$$p_{bd}^w = \frac{C_{fw}}{C_{bw}} 100 = \frac{100}{41} = 2.44$$

$$p_{bd}^s = \frac{2.44}{100 - 2.44} \times 100 = 2.5$$

Step (II) Determination of ϕ_s and ϕ_{bd}

P_{bd}^s	Y	$\dfrac{C_{bw}}{C_{fw}}$	ϕ_s
2.5%	0.01%	41	$\dfrac{100}{100 + 2.5} \dfrac{0.01}{100} (41)(100) = 0.4\%$
			Ans.

$$\therefore \quad \phi_{bd} = 100 - 0.4 = 99.6\% \qquad \textit{Ans.}$$

Problem 17.7   Under identical conditions of the preceeding problem (17.6) determine the amounts of silica (as silicic acid) removed from the boiler as carryover with steam and blowdown water if the concentration of silica in boilerwater is 7.5 mg/kg and in feedwater 0.2 mg/kg.

Take $Y_{SiO_2} = 0.75\%$

Solution

Step (I) Determination of silica carryover by steam

Working Formula $\quad \phi_s = \dfrac{100}{100 + p_{bd}^s} \, Y \dfrac{C_{bw}}{C_{fw}} \times 100$

P_{bd}^s	Y	C_{bw}/C_{fw}	ϕ_s
2.5	0.75/100	7.5/0.2	$\dfrac{100}{102.5} \dfrac{0.75}{100} \dfrac{7.5}{0.2} (100)$
			$= 27.43\%$ *Ans.*

Step (II) Determination of silica removed by blowdown water

$$\phi_{bd} = 100 - \phi_s = 100 - 27.43 = 72.57\%$$

Problem 17.8   The steam load of a boiler is raised from 155 t/h to 175 t/h keeping the rate of blowdown unchanged. If the rate of blowdown in relation to steam is 4%, determine the initial and final salt contents of boilerwater. The salt content of feedwater is 50 mg/kg.

Solution

Step (I) Calculation of initial salt content in boilerwater

Working Formula: $p_{bd}^s = \dfrac{C_{fw} - C_s}{C_{bw} - C_{fw}} \times 100$

$$(17.15)$$

Taking $C_s \approx 0$, it becomes

$$p_{bd}^s = \dfrac{C_{fw}}{C_{bw} - C_{fw}} \times 100$$

Putting the appropriate numerical values,

$$4 = \dfrac{50}{C_{bw} - 50} \times 100$$

$$\therefore \qquad C_{bw} = 50 + \dfrac{50(100)}{4} = 1300 \text{ mg/kg}$$

Ans.

Step (II) Calculation of final salt content in boiler water

Working Formula $\quad G_{bd} = \dfrac{C_{fw} - C_s}{C_{bw} - C_{fw}} G_s \quad (17.4)$

Calculation of G_{bd} at 155 t/h steam load:

Taking $C_s \approx 0$,

$$G_{bd} = \dfrac{50}{1\,300 - 50} (155) = 6.2 \text{ t/h}$$

Calculation of C_{bw} (final) Using the above formula,

$$6.2 = \dfrac{50}{C_{bw} - 50} \times 175;$$

$$\therefore \qquad C_{bw} = 1\,461 \text{ mg/kg}$$

Q. *What are drum internals for ?*

Ans. Their basic objective is to ensure purity of saturated steam generated in the boiler drum.

Q. *How is this done?*

Ans. They quench the kinetic energy of the steam-water mixture jets entering the drum, bring about

coarse separation of steam from water and finally dry the steam, i.e., eliminate water droplets suspended in steam (fine separation).

Q. *What is the driving force effecting the separation of water from steam?*

Ans. It is the density difference between water and saturated steam. Greater the difference, the easier the moisture is separated.

Q. *What factors determine whether a water droplet suspended in steam will be carried-off by steam or settled in the boiler drum?*

Ans. There are two forces that act upon the water droplet suspended in the steam space of the boiler drum. These two are—upward force (i.e., lifting force) due to rising velocity of steam in the drum and downward force due to gravity. The relationship between these two forces and the time of their action determine whether the water droplet will be carried off by steam or settle on the water surface of the boiler drum.

Q. *What is the effect of pressure on steam-water separation?*

Ans. Higher the pressure of the steam, higher will be the rising velocity of steam in the drum and lower will be the density difference between water and saturated steam. And these reduce the ease of separation of water from saturated steam.

Q. *How can this problem be coped with ?*

Ans. This problem can be tackled in a number of ways. One of them is settling separation : Steam is allowed to distribute uniformly over the free steam space of the drum. This will ensure lowest possible rising velocities of steam in the drum creating a favourable condition for the moisture droplets to settle on the evaporating surface. All modern steam-separation circuits involve a large free vapour space to ensure settling separation.

The other method is to deploy deflecting metal baffles and slotted or solid partitions on the path of steam flow. As the wet steam strikes the baffle, its kinetic energy gets quenched and the water droplets adhere to the baffle's surface. The steam changes its direction of motion and the residual water droplets are thrown out of the flow by the action of the centrifugal force.

Q. *What are the simplest and most efficient steam-water separation devices?*

Ans. These are perforated steel plates having perforation dia ranging from 5 to 12 mm.

Q. *Where are they placed?*

Ans. One plate is placed 10–15 cm below the average water level in the drum. This one is called the submerged plate.

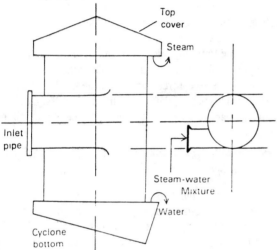

Fig. 17.10 *Cyclone separator*

The other one is placed in the steam space at the top of the boiler drum. This one is called the perforated baffle.

They equalize the distribution of steam over the cross-section of the drum.

Q. *What is the principal steam separating device in high capacity steam boilers?*

Ans. It is the cyclone separator (see Fig. 17.10)

Q. *Where is it arranged?*

Ans. It is mounted inside the boiler drum. Fig. 17.11

Q. *How many kinds of cyclone separators are known?*

Ans. Two types: Internal cyclone separator and external cyclone separator (Fig. 17.12)

Q. *How does a cyclone separator effect steam-water separation?*

Ans. It does so by impactless quenching of the kinetic energy of the steam-water flow. It is a hollow cylinder (dia 300–400 mm) for internal cyclone separators and 350–450 mm for external cyclone separators. And through a special duct, a steam-water mixture enters the cylinder tangentially to the walls so that a rotational motion is imparted to the flow. The centrifugal force produced at the expense of the

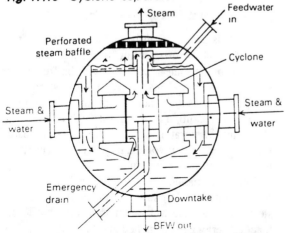

Fig. 17.11 *Cyclone separator mounted inside a boiler drum*

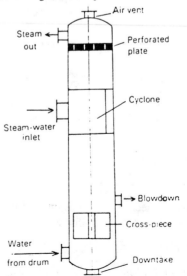

Fig. 17.12 *External cyclone separator*

kinetic energy of the steam throws the water droplets suspended in steam to the walls of the centrifuge while the steam flows uniformly upward along the generating-line in the centre of the cyclone-cylinder.

Q. *What is the important drawback of this steam-separating system?*

Ans. Cyclone steam-water separators increase the hydraulic resistance of the circuit and hence cannot be used where the circulating head is insufficient.

Q. *Are the steam-water separating devices capable of removing substances dissolved in steam?*

Ans. No.

Q. *So what is done for this in addition to mounting separating devices?*

Ans. Boiler drums are fitted with steam washing devices

Q. *What is the basis of the steam washing principle?*

Ans. It is based on the fact that if saturated steam is brought into contact with high quality condensate of feedwater, a redistribution of dissolved solids between the liquid and vapour phase will take place, whereupon a considerable load of dissolved particles in steam will transfer to the wash water and improve the steam quality.

The same conclusion is evident from the equation:

$$K_d = \frac{C_s}{C_{bw}}$$ where C_s and C_{bw} are the concentrations of impurities in steam and boilerwater.

Since the distribution coefficient (K_d) for a particular impurity is constant at a constant pressure, a cleaner steam can be obtained at K_d = constant by decreasing the concentration of impurities in water (C_{bw}), i.e., by washing steam with polished water or condensate.

Q. *In what device is steam washing carried out?*

Ans. Bubble-cap steam washer.

The simplest steam washing device is as shown. It is a perforated plate onto which washing water is poured. (Fig. 17.13)

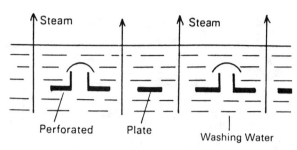

Fig. 17.13 *Bubblecap steam washer*

The steam velocity is so adjusted as not to allow water to pass through the perforations.

Q. *Where is the steam washing carried out?*

Ans. In the boiler drum. Feedwater is used for washing the steam (Fig. 17.11). It is fed to the perforated plate having edges bent up to hold a certain layer of washing water. The flowrate of the feedwater is governed by the steam generation capacity. In modern units, the entire feedwater is charged to a distribution header and the excess of water flows down a slit in the header directly into the water space of the boiler drum without taking part in steam washing.

Q. *What is the basis of multistage evaporation?*

Ans. This can be determined as follows.

Under steady-state operating conditions, the following mass balance equation must hold good in the boiler drum

$$C_{fw} \, G_{fw} = C_s \, G_s + C_{bd} \, G_{bd}$$

Now if $C_s \approx 0$, the above equation transforms to

$$C_{fw} \, G_{fw} = C_{bd} \, G_{bd}$$

$$\therefore \qquad C_{bd} = C_{fw} \frac{G_{fw}}{G_{bd}}$$

Now, writing the distribution coefficient

$$K_d^{total} = \frac{C_s}{C_{bd}}$$

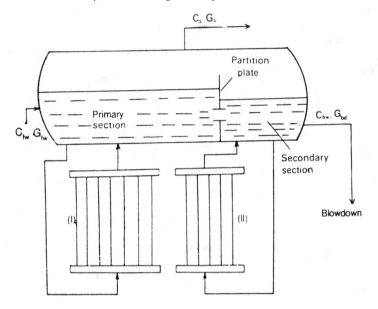

Fig. 17.14 A two-stage evaporation unit

$$\therefore \qquad C_s = C_{bd}\, K_d^{total} = C_{fw}\, \frac{G_{fw}}{G_{bd}}\, K_d^{total}$$

As is evident from this equation, the salt content in steam can be reduced

(a) by reducing G_{fw}. But this is inadmissible as it would mean reducing the boiler load

(b) By decreasing C_{fw}—the salt content in feedwater, which can be done either by demineralizing the make-up water or by reducing the amount of chemically treated water used as make-up for the feed system.

However, demineralization cannot be used in all cases. And a cut in the amount of make-up water will necessitate a greater fraction of condensate or distillate in feedwater. This proves impossible to satisfy in many cases.

(c) by increasing the rate of blowdown. However higher blowdown rate (> 5%) without stage evaporation facility leads to inadmissibly high heat losses. Hence, boiler units are fitted with multistage evaporation systems so that high quality steam can be produced by increasing blowdown without inflicting much heat loss.

Hence salt content in steam can be reduced by taking recourse to stage-evaporation.

Q. *Briefly describe two-stage evaporation.*

Ans. It may be considered as two independent steam boilers sharing the same steam space but having two independent water-flow circuits (No. I and II).

The water space of the boiler drum is separated into two compartments by a partition plate, an opening in which ensures free communication of water between these two compartments. (Fig. 17.14)

Feedwater is charged to the larger compartment called primary section or stage. The blowdown water from the primary section serves as feedwater for the secondary stage of evaporation which is also called the saline section.

Therefore, stage evaporation is based on creating a differential salt concentration in the boiler water by introducing two waterflow circuits forming primary and secondary sections.

A much larger fraction of steam is generated in the primary section having water with salt content considerably less than the blowdown water, whereas only a small fraction of the total steam is generated in the saline section having higher salt content. It follows, therefore, that forasmuch as the steam produced by the primary stage (section) is many times greater than the quantity of steam produced by the secondary section, the overall quality of the steam generated in the boiler unit provided with two-stage evaporation is higher than the quality of steam produced in a single-stage boiler unit operating under otherwise identical conditions.

Q. *How can it be numerically proved that the salt content in steam generated in the primary section is materially less than the salt content in the steam produced in the saline section?*

Ans. The salt content in steam produced in the primary section is

$$C'_s = Y \frac{C'_{fw} G_{fw}}{G'_{bd}} \tag{17.16}$$

The salt content in steam generated in the secondary section is

$$C''_s = Y'' \frac{C'_{bw} G'_{bd}}{G''_{bd}} \tag{17.17}$$

Dividing Equation 17.16 by 17.17 we get

$$\frac{C'_s}{C''_s} = \frac{Y'}{Y''} \cdot \frac{C_{fw}}{C_{bw}} \cdot \frac{G_{fw}}{G'_{bd}} \cdot \frac{G''_{bd}}{G'_{bd}}$$

Now putting, $Y' = Y''$; $G'_{bd} = (p^w_{bd})' G'_{fw}$;

$$G''_{bd} = (p^w_{bd})'' G'_{fw}$$

in the final equation we get

$$\frac{C'_s}{C''_s} = \frac{C_{fw}}{C'_{bw}} \cdot \frac{(p^w_{bd})''}{(p^w_{bd})'} \tag{17.18}$$

Since, $\dfrac{C_{fw}}{C'_{bw}} < 1$ and $\dfrac{(p^w_{bd})''}{(p^w_{bd})'} < 1$, so $C'_s < C''_s$

i.e., the salt concentration in steam produced in the primary stage is less than that produced in the secondary or saline section (stage).

BASIS: STEAM GENERATION CAPACITY OF THE BOILER = 100 t/h

For the boiler unit as a whole

$$(100 + p^s_{bd}) C_{fw} = 100 C_s + p^s_{bd} C_{bw} \tag{17.19}$$

For I stage evaporation

$$(100 + p^s_{bd}) C_{fw} = (100 - n)C'_s + (n + p^s_{bd})C'_{bw}$$

where n = steam generation capacity of II stage (t/h or per cent)

Substituting, $C'_s = Y' C'_{bw}$ the above equation becomes

$$(100 + p^s_{bd}) C_{fw} = (100 - n) Y' C'_{bw} + (n + p^s_{bd}) C'_{bw}$$

or,

$$C'_{bw} = \frac{100 + p^s_{bd}}{(100 - n) Y' + (n + p^s_{bd})} C_{fw}$$

Since, $(100 - n) Y' \ll (n + p^s_{bd})$, so

$$C'_{bw} = \frac{100 + p^s_{bd}}{n + p^s_{bd}} C_{fw} \tag{17.20}$$

For II stage evaporation,

$$(n + p^s_{bd}) C'_{bw} = n C''_s + p^s_{bd} C''_{bw}$$

Putting $C_s'' = Y'' C''_{bw}$, we get

$$(n + p^s_{bd}) C'_{bw} = (nY'' + p^s_{bd}) C''_{bw}$$

or

$$C''_{bw} = \frac{n + p^s_{bd}}{n Y'' + p^s_{bd}} C'_{bw} \tag{17.21}$$

If Y'' and n are small, the product $n Y''$ can be ignored with respect to p^s_{bd}, hence

$$C''_{bw} = \frac{n + p^s_{bd}}{p^s_{bd}} C'_{bw} \tag{17.22}$$

The material balance equation for impurities in steam:

$$100 C_s = (100 - n)C'_s + n C''_s$$

or,

$$C_s = \left(1 - \frac{n}{100}\right) C'_s + \frac{n}{100} C''_s \tag{17.23}$$

If k be the ratio of salt concentrations between the saline section and primary section then

$$k = \frac{C''_{bw}}{C'_{bw}}$$

Hence Equation 17.21 becomes

$$k = \frac{n + p^s_{bd}}{n Y'' + p^s_{bd}} \tag{17.24}$$

or,

$$n = \frac{k - 1}{1 - k Y''} p^s_{bd} \tag{17.25}$$

Problem 17.9 For a given boiler unit it is required to produce steam of the following quality:

Silica concentration in steam = 0.038 mg/kg

Total dissolved solids in steam = 0.22 mg/kg

from the following parameters:

Concentration of TDS in make-up water = 245 mg/kg

Concentration of silicic acid in make-up water = 1.2 mg/kg

Carryover coefficient of silicic acid = 1.2 %

Carryover coefficient of TDS = 0.029 %

Make-up DM water = 25 % of feedwater.

Determine whether it is feasible to generate steam of the above quality without exceeding a blowdown rate of 3%

Solution The steam quality is defined by the expressions:

$$C_s^{TDS} = Y_{TDS} \, C_{bw}^{TDS}$$

and

$$C_s^{H_2SiO_3} = Y_{H_2SiO_3} \, C_{bw}^{H_2SiO_3}$$

where C_s^{TDS} and $C_s^{H_2SiO_3}$ refer to concentrations of TDS and silicic acid in steam.

Y_{TDS} and $Y_{H_2SiO_3}$ refer to the carryover coefficient in relation to total dissolved solids (TDS) and silicic acid respectively. C_{bw}^{TDS} and $C_{bw}^{H_2SiO_3}$ correspond respectively to the concentration of TDS and silicic acid in boilerwater.

So we are to determine C_{bw}^{TDS} and $C_{bw}^{H_2SiO_3}$ to ascertain whether the steam quality will be within the prescribed standard or not.

Now,
$$C_{bw} = \frac{100 + p_{bd}^s}{p_{bd}^s} \, C_{fw}$$

Step (I) Calculation of C_{fw}^{TDS}

$$C_{fw}^{TDS} = \alpha_{TDS} \, C_{ch}^{TDS}$$

where α_{TDS} = fraction of chemically treated water with salt content C_{ch}^{TDS}

$$C_{fw}^{TDS} = \frac{25}{100} (245) = 61.25 \text{ mg/kg}$$

Step (II) Calculation of $C_{fw}^{H_2SiO_3}$

$$C_{fw}^{H_2SiO_3} = \alpha_{H_2SiO_3} \, C_{ch}^{H_2SiO_3} = \frac{25}{100} (1.2) = 0.3 \text{ mg/kg}$$

Step (III) Calculation of C_{bw}^{TDS}

$$C_{bw}^{TDS} = \frac{100 + 3}{3} (61.25) = 2\,102.91 \text{ mg/kg}$$

Step (IV) Calculation of $C_{fw}^{H_2SiO_3}$

$$C_{bw}^{H_2SiO_3} = \frac{100 + 3}{3} (0.3) = 10.3 \text{ mg/kg}$$

Step (V) Determination of Steam Quality

Now,

$$C_s^{TDS} = Y_{TDS} C_{bw}^{TDS} = \frac{0.029}{100} (2\,102.91) = 0.6098$$
$$\text{mg/kg}$$

$$C_s^{H_2SiO_3} = Y_{H_2SiO_3} C_{bw}^{H_2SiO_3} = \frac{1.2}{100}(10.3) = 0.1236$$
$$\text{mg/kg}$$

Conclusion: It is evident from the calculated result that at blowdown rate 3%, the desired steam quality will not be achieved.

Problem 17.10 Now to raise the steam quality as specified in Problem 17.9, the boiler unit is fitted with means of multistage evaporation, keeping the blowdown rate to 3%.

Determine the steam generating capacity of the secondary section.

Also determine the salt and silicic acid contents of steam in each section and in the whole boiler.

Solution Since the value of $k (= C''_{bw}/C'_{bw})$ usually falls within the range 3 – 10, let's assume $k = 5$

Step (I) Calculation of steam generation capacity of secondary section

Working Formula: $n = \dfrac{k-1}{1 - kY''} p_{bd}^s$ (17.25)

Ignoring kY'', the above equation becomes

$$n = (k - 1) p_{bd}^s = (5 - 1)(3) = 12\% \text{ percent}$$

Ans.

Step (II) Calculation of salt content in boilerwater in the primary section

Working Formula: $(C'_{bw})^{TDS} = \dfrac{100 + p_{bd}^s}{n + p_{bd}^s} C_{fw}^{TDS}$

(17.26)

p_{bd}^s	n	C_{fw}^{TDS}	$(C'_{bw})^{TDS}$
3%	12%	61.25 mg/kg	$\dfrac{100+3}{12+3}(61.25)$ = 420.58 mg/kg

Step (III) *Calculation of silica content in boiler-water in the primary section*
Working Formula:

$$(C'_{bw})^{H_2SiO_3} = \frac{100 + p_{bd}^s}{n + p_{bd}^s} C_{fw}^{H_2SiO_3}$$

p_{bd}^s	n	$C_{fw}^{H_2SiO_3}$	$(C'_{bw})^{H_2SiO_3}$
3%	12%	0.3 mg/kg	$\dfrac{100+3}{12+3}(0.3)$ = 2.06 mg/kg

Step (IV) Steam quality in primary section
Salt content in the steam (I-stage)

$$(C'_s)^{TDS} = Y'_{TDS}(C'_{bw})^{TDS}$$

$$= \frac{0.029}{100}(420.58)$$

$$= 0.12 \text{ mg/kg}$$
Ans.

Silicic acid content in the steam (I-stage)

$$(C'_s)^{H_2SiO_3} = Y'_{H_2SiO_3}(C'_{bw})^{H_2SiO_3}$$

$$= \frac{1.2}{100}(2.06) = 0.0247 \text{ mg/kg}$$
Ans.

Step (V) Calculation of salt content in boiler-water in the secondary section
Working Formula:

$$(C''_{bw})^{TDS} = \frac{n + p_{bd}^s}{p_{bd}^s}(C'_{bw})^{TDS}$$

P_{bd}^s	n	$(C'_{bw})^{TDS}$	$(C''_{bw})^{TDS}$
3%	12%	420.58 mg/kg	$\dfrac{12+3}{3}(420.58)$ = 2102.9 mg/kg

Step (VI) Calculation of silicic acid content in boilerwater in the secondary section
Working Formula:

$$(C''_{bw})^{H_2SiO_3} = \frac{n + p_{bd}^s}{p_{bd}^s}(C'_{bw})^{H_2SiO_3}$$

p_{bd}^s	n	$(C'_{bw})^{H_2SiO_3}$	$(C''_{bw})^{H_2SiO_3}$
3%	12%	2.06 mg/kg	$\dfrac{12+3}{3}(2.06)$ = 10.3 mg/kg

Step (VII) Steam quality in the secondary section
Salt content in the steam (II - stage)

$$(C''_s)^{TDS} = Y''_{TDS}(C''_{bw})^{TDS}$$

$$= \frac{0.029}{100}(2\,102.9) = 0.6098 \text{ mg/kg} \qquad Ans.$$

Silicic acid content in the steam (II - stage)

$$(C_s'')^{H_2SiO_3} = Y''_{H_2SiO_3}(C''_{bw})^{H_2SiO_3}$$

$$= \frac{1.2}{100}(10.3) = 0.1236 \text{ mg/kg}$$
Ans.

Step (VIII) Steam quality in the boiler as a whole
Working Formula:

$$C_s = \left(1 - \frac{n}{100}\right)C'_s + \frac{n}{100}C''_s \quad (17.23)$$

Average salt content,

$$C_s^{TDS} = \left(1 - \frac{12}{100}\right)(0.12) + \frac{12}{100}(0.6098)$$

$$= 0.1787 \text{ mg/kg}$$
Ans.

Average silicic acid content:

$$C_s^{H_2SiO_3} = \left(1 - \frac{12}{100}\right)(0.0247) + \frac{12}{100}(0.1236)$$

$$= 0.036 \text{ mg/kg}$$
Ans.

Problem 17.11 A high pressure boiler has the following feedwater quality:

Salt content = 45 mg/kg

Silicic acid = 0.23 mg/kg

It is outfitted with 2-stage evaporation such that carryover coefficient for silica in primary and secondary saline sections is 0.75 and 0.55 % respectively.

If the rate of boiler blowdown is 2.75 % determine:

(a) concentration of total dissolved solids in the boilerwater in both primary and secondary sections

(b) concentration of silicic acid in the boilerwater in both primary and secondary sections

(c) silicic acid concentration in the overall steam produced

Given $C_{bw}'' / C_{bw}' = 4$

Solution **Step (I):** Let's first calculate the percentage of steam generated in the primary section.

Working Formula:

$$n = \frac{k-1}{1 - k\,Y''}\,(p_{bd}^s) \tag{17.25}$$

Since, $kY'' \ll 1$, so $n = (k-1)\,p_{bd}^s = (4-1)$

$$(2.75)$$

$$= 8.25\%$$

Step (II) TDS in boilerwater of primary section

Working Formula

$$(C_{bw}')^{TDS} = \frac{100 + p_{bd}^s}{n + p_{bd}^s}\,(C_{fw})^{TDS} \tag{17.20}$$

p_{bd}^s	n	C_{fw}^{TDS}	$(C_{bw}')^{TDS}$
2.75%	8.25%	45 mg/kg	$\dfrac{100 + 2.75}{8.25 + 2.75}(45)$
			$= 420.34$ mg/kg *Ans.*

Step (III) Silicic acid content in boilerwater of primary section

p_{bd}^s	n	$C_{fw}^{H_2SiO_3}$	$(C_{bw}')^{H_2SiO_3}$
2.75%	8.25%	0.23 mg/kg	$\dfrac{100 + 2.75}{8.25 + 2.75}(0.23)$
			$= 2.148$ mg/kg *Ans.*

Step (IV) TDS in boilerwater of secondary section

$(C_{bw}')^{TDS}$	n	p_{bd}^s	$(C_{bw}'')^{TDS}$
420.34 mg/kg	8.25%	2.75%	$\dfrac{8.25 + 2.75}{2.75}(420.34)$
			$= 1681.36$ mg/kg *Ans.*

Step (V) Silicic acid content in boilerwater in the secondary section.

$(C_{bw}')^{H_2SiO_3}$	n	p_{bd}^s	$(C_{bw}'')^{H_2SiO_3}$
2.148 mg/kg	8.25%	2.75%	$\dfrac{8.25 + 2.75}{2.75}(2.148)$
			$= 8.59$ mg/kg *Ans.*

Step (VI) Average silicic acid content in the overall steam

Working Formula:

$$C_s^{H_2SiO_3} = \left(1 - \frac{n}{100}\right)(C_s')^{H_2SiO_3} + \frac{n}{100}(C_s'')^{H_2SiO_3} \tag{17.23}$$

$$(C_s')^{H_2SiO_3} = Y'_{H_2SiO_3}\,(C_{bw}')^{H_2SiO_3}$$

$$= \frac{0.75}{100}(2.148)\,\text{mg/kg} = 0.0161\,\text{mg/kg}$$

$$(C_s'')^{H_2SiO_3} = Y''_{H_2SiO_3}\,(C_{bw}'')^{H_2SiO_3}$$

$$= \frac{0.55}{100}(8.59)\,\text{mg/kg} = 0.0472\,\text{mg/kg}$$

$$C_s^{H_2SiO_3} = \left(\frac{1 - 8.25}{100}\right)(0.0161) + \frac{8.25}{100}(0.0472)$$

$$= 0.0186\,\text{mg/kg}$$

     Ans.

Problem 17.12 For a high pressure boiler fitted with multistage evaporation, a project envisages that the total dissolved solid in the boilerwater in the primary section should not exceed 420 mg/kg and the upper limit of silicic acid content in the secondary section should be 8.5 mg/kg.

The make-up water amounts to 15% of feedwater. The make-up water quality:

TDS content = 400 mg/kg

Silicic acid content = 2 mg/kg

Determine:

(a) the limiting quality standard for boilerwater

(b) the blowdown rate prescribed by the standard

(c) the concentration ratio for boilerwater filling I and II-stage

(d) the content of total dissolved solid in the I and II stage evaporating section.

Solution

Step (I) Calculation of C_{fw}^{TDS}

$C_{fw}^{TDS} = \alpha_{TDS}\, C_{ch}^{TDS}$ (see Problem 17.9)

$= \dfrac{15}{100}(400) = 60\ \text{mg/kg}$

Step (II) Calculation of $C_{fw}^{H_2SiO_3}$

$C_{fw}^{H_2SiO_3} = \alpha_{H_2SiO_3}\, C_{ch}^{H_2SiO_3} = \dfrac{15}{100}(2) = 0.3\ \text{mg/kg}$

Step (III) Determination of blowdown rate

Now, $(C'_{bw})^{TDS} = 420\ \text{mg/kg}$ (given)

$(C''_{bw})^{H_2SiO_3} = 8.5\ \text{mg/kg}$ (given)

But, $(C'_{bw})^{TDS} = \dfrac{100 + p_{bd}^s}{n + p_{bd}^s}\, C_{fw}^{TDS}$

or, $420 = \left[\dfrac{100 + p_{bd}^s}{n + p_{bd}^s}\right] 60$

or, $7 = \dfrac{100 + p_{bd}^s}{n + p_{bd}^s}$ (I)

Also, $(C''_{bw})^{H_2SiO_3} = \dfrac{n + p_{bd}^s}{p_{bd}^s}\,(C'_{bw})^{H_2SiO_3}$ (II)

and $(C'_{bw})^{H_2SiO_3} = \dfrac{100 + p_{bd}^s}{n + p_{bd}^s}\, C_{fw}^{H_2SiO_3}$ (III)

From equations (II) and (III) we get

$(C''_{bw})^{H_2SiO_3} = \dfrac{100 + p_{bd}^s}{p_{bd}^s}\, C_{fw}^{H_2SiO_3}$

or, $8.5 = [(100 + p_{bd}^s)/p_{bd}^s]\, 0.3$

or, $p_{bd}^s = 3.658\% \approx 3.66\%$

                        Ans.

Step (IV) Determination of fraction of total steam generated in the secondary section

Computing the value of p_{bd}^s in equation (I) we have,

$7 = \dfrac{100 + 3.66}{n + 3.66}$ $\therefore\ n = 14.808 - 3.66 = 11.14\ \%$

Step (V) Calculation of $(C''_{bw})^{TDS}$

$(C''_{bw})^{TDS} = \dfrac{n + p_{bd}^s}{p_{bd}^s}\,(C'_{bw})^{TDS}$

$= \dfrac{11.14 + 3.66}{3.66}(420)\ \text{mg/kg}$

$= 1698.36\ \text{mg/kg}$

                        Ans.

Step (VI) Concentration ratio for boilerwater filling I and II stages

$k = \dfrac{(C''_{bw})^{TDS}}{(C'_{bw})^{TDS}} = \dfrac{1\,698.36}{420} = 4$ *Ans.*

Problem 17.13 If the carryover coefficient for silicic acid in the primary and secondary sections in Problem 17.12 are 0.8 and 0.55% respectively, determine the quality of overall steam in relation to the silicic acid content.

Solution The quality of overall steam can be determined with the aid of the formula:

$C_s = \left(1 - \dfrac{n}{100}\right)C'_s + \dfrac{n}{100}\,(C''_s)$ (17.23)

Hence the silicic acid concentration in the overall steam is

$C_s^{H_2SiO_3} = \left(1 - \dfrac{n}{100}\right) Y'_{H_2SiO_3}\,(C'_{bw})^{H_2SiO_3}$

$\qquad\qquad + \dfrac{n}{100}\, Y''_{H_2SiO_3}\,(C''_{bw})^{H_2SiO_3}$

Now, $(C'_{bw})^{H_2SiO_3} = \dfrac{100 + p_{bd}^s}{n + p_{bd}^s}\, C_{fw}^{H_2SiO_3}$

$= \dfrac{100 + 3.66}{11.14 + 3.66}(0.3) = 2.10\ \text{mg/kg}$

n	11.14%
$Y'_{H_2SiO_3}$	0.8%
$Y''_{H_2SiO_3}$	0.55%
$(C'_{bw})_{H_2SiO_3}$	2.10 mg/kg
$(C''_{bw})^{H_2SiO_3}$	8.5 mg/kg
$C_s^{H_2SiO_3}$	$(1 - 11.14/100) (0.8/100) (2.10) +$
	$\dfrac{11.14}{100}\left(\dfrac{0.55}{100}\right)(8.5) = 0.020$ mg/kg *Ans.*

Problem 17.14 Taking data from Problems 17.9 and 17.10, determine the percentage of silicic acid carried over by steam from the primary and secondary section. Also compare these carryovers of silicic acid with that carried over by steam before the boiler was fitted with 2-stage evaporation.

Solution

Step (I) In the primary section of the two stage evaporator, the following material balance equation can be drawn:

$$(100 + p_{bd}^s)\, C_{fw} = (100 - n)\, C'_s + (n + p_{bd}^s)\, C'_{bw}$$

If C_{fw}, C'_s and C'_{bw} are taken as the silicic acid concentration in feedwater, in the steam of the primary section and in the boilerwater of the primary section, then

$(100 - n)C'_s =$ carryover of silicic acid by steam in the primary section

$(n + p_{bd}^s)\, C'_{bw} =$ carryover of silicic acid by boilerwater in the primary section.

Therefore, the relative percentage of silicic acid carried over by steam from the total amount of silicic acid entering with the feedwater is

$$R_1^{H_2SiO_3} = \frac{(100 - n)C'_s}{(100 + p_{bd}^s)\, C_{fw}}\,(100)$$

$$= \frac{100 - 12}{100 + 3}\left(\frac{0.0247}{0.3}\right)(100) = 7\%$$

Step (II) In the secondary section, the amount of silicic acid carried over by steam is $n\, C''_s$.

$$R_2^{H_2SiO_3} = \frac{n\, C''_s}{(100 + p_{bd}^s)\, C_{fw}}\,(100)$$

$$= \frac{12\,(0.1236)}{(103)\,(0.3)}\,(100) = 4.8\ \%$$

Ans.

Step (III) Before the boiler was provided with 2-stage evaporation, the silicic acid carryover by steam amounted to

$$\frac{100\,(0.1236)}{(100 + 3)\,(0.3)}\,(100) = 40\%$$

Ans.

Problem 17.15 If in Problem 17.10, the flow of boilerwater from the 2nd-stage to 1st-stage is at a rate of 4% of the steam generation capacity of the boiler, determine, with the help of the data in Problem 17.9 and 17.10, the amount of silicic acid carryover by steam from the boiler.

Solution   If "r" be the water flow from the II-stage to the I-stage, expressed as a percentage of G_s (the steam generating capacity of the boiler unit), then Equations 17.20 and 17.22 modify to

$$C'_{bw} = \frac{(100 + p_{bd}^s)\, C_{fw} + r\, C''_{bw}}{r + n + p_{bd}^s} \qquad (17.26)$$

$$C''_{bw} = \frac{r + n + p_{bd}^s}{r + p_{bd}^s}\, C'_{bw} \qquad (17.27)$$

Putting the appropriate numerical values we get from equation 17.26,

$$C'_{bw} = \frac{(100 + 3)\,(0.3) + 4C''_{bw}}{4 + 12 + 3} = 1.62 + 0.21 C''_{bw}$$

From equation 17.27 we get,

$$C''_{bw} = \frac{4 + 12 + 3}{4 + 3}\, C'_{bw} = 2.71 C'_{bw}$$

Solving these two equations we get,

$$C'_{bw} = 1.62 + 0.21\,(2.71)C'_{bw}$$

$$\therefore \quad C'_{bw} = 3.75 \text{ mg/kg}$$

$$C''_{bw} = 2.71(3.75) \text{ mg/kg}$$

$$= 10.16 \text{ mg/kg}$$

Therefore, the concentration of silicic acid in steam

(a) **Primary section:** $C'_s = Y'_{H_2SiO_3} C'_{bw}$

$$= \left(\frac{1.2}{100}\right)(3.75) = 0.045 \text{ mg/kg} \qquad Ans.$$

(b) **Secondary section:**

$$C''_s = \left(\frac{1.2}{100}\right)(10.16) = 0.1219 \text{ mg/kg} \qquad Ans.$$

Hence the amount of silicic acid carryover by the overall steam

$$C_s = \left(1 - \frac{n}{100}\right)C'_s + \frac{n}{100} C''_s$$

$$= \left(1 - \frac{12}{100}\right)(0.045) + \frac{12}{100}(0.1219)$$

$$= 0.054 \text{ mg/kg}$$

$$Ans.$$

18

Prevention of Deposit Formation in Boiler Units

Q. *How can the formation of depositions in drum-type boiler units prevented?*

Ans. This can be achieved by adopting the following measures:
1. Intensive softening of make-up water
2. Boiler water treatment
3. Preventing BFW from contamination with the products of corrosion of structural material
4. Elimination of leakage of cooling water in the turbine condenser
5. Prevention of local high heat flow rates through steam generating tubes.

Q. *What is the aim of boiler water treatment?*

Ans. Basically it is designed to render the scale formers, whatever may penetrate into the boiler unit, harmless and to eliminate their deposition on heat transfer surfaces.

Q. *How is this done?*

Ans. By introducing orthophosphates or salts of EDTA (ethylene diamine tetracetic acid) into boiler units.

Q. *What is the object of phosphate treatment of BFW?*

Ans. To create conditions that'll favour the formation of sludge instead of scale, i.e., scale formers $CaSO_4$; $CaSiO_3$ will precipitate as loose sludge along with other sludges, viz. $CaCO_3$; $3Ca_3(PO_4)_2 \cdot Ca(OH)_2$ which are removed from the boiler unit by blowdown. To be more particular, phosphate treatment is directed to prevent the formation of calcium scales rather than scales in general.

Q. *Why?*

Ans. Phosphate treatment leads to the formation of scale of magnesium triphosphate $Mg_3(PO_4)_2$.

Q. *What are phosphate-dosers?*

Ans. Trisodium phosphate, Na_3PO_4
Tripolyphosphate, $Na_5P_3O_{10}$
Hexametaphosphate, $(NaPO_3)_6$

Q. *When trisodium phosphate is dissolved in water what principal reaction takes place?*

Ans. TSP dissolves in water to liberate phosphate ions (PO_4^{3-}) which undergoes hydrolysis to render the solution alkaline

$$PO_4^{3-} + H_2O \rightleftharpoons HPO_4^{2-} + OH^-$$

Q. Why is this reaction particularly useful for boilers operating at higher pressure (>7 MN/m^2)?

Ans. For boiler units operating at higher pressure, the above reversible reaction forms the very basis of the method of purely phosphate alkalinity. As the solution is concentrated on evaporation, the equilibria shifts from right to left eliminating the aggressive free sodium hydrate which would otherwise give rise to intergranular corrosion.

Q. *However there is the nitrate treatment method to control the aggressivity of boiler water. Why can this system not be adopted in the case of boilers operating at higher pressures (>7 MN/m^2)?*

Ans. At higher pressures, nitrates are not stable and sodium nitrate decomposes, rendering the treatment useless. However, the phosphate ion (PO_4^{3-}) is of sufficient strength and it does not disintegrate or dissociate in boiler water at higher steam pressure and rising temperature and permits prevention of calcium scale formation.

Q. *Why is the BFW made sufficiently alkaline when polyphosphates are used?*

Ans. It is because they undergo hydrolysis in aqueous solution to produce mono and disodium phosphates which are potentially acidic:

$$NaH_2PO_4 \Leftrightarrow Na^+ + HPO_4^{2-} + H^+$$

$$Na_2HPO_4 \Leftrightarrow 2Na^+ + H^+ + PO_4^{3-}$$

That is why the boiler water must be sufficiently alkaline to ensure their neutralization.

Q. *Why does the method of purely phosphate alkalinity commonly use a mixture of trisodium phosphate and one of the acid phosphates, say* $(NaPO_3)_6$?

Ans. It is because the BFW may entail some carbonate or bicarbonate ions that must be eliminated to avoid scaling.

Hence recourse to acid phosphate is taken so that the mixture of Na_3PO_4 and $(NaPO_3)_6$ conforms to the composition Na_2HPO_4 that removes HCO_3^- and CO_3^{2-} ions as follows:

$$HPO_4^{2-} \Leftrightarrow H^+ + PO_4^{3-}$$

$$H^+ + HCO_3^- \rightarrow H_2O + CO_2 \, (\uparrow)$$

$$2H^+ + CO_3^{2-} \rightarrow H_2CO_3 \rightarrow H_2O + CO_2 (\uparrow)$$

Q. *However prolonged treatment of water with the introduction of acid phosphates has in some cases resulted in scaling iron phosphates in sections of water-wall tubes exposed to high heat flowrate. The result is honey-comb corrosion. How can this be eliminated?*

Ans. This can be eliminated by maintaining in the boiler water a ratio $A_{ph}/A_{tot} = 0.5 - 0.6$
where A_{tot} = total alkalinity
A_{ph} = alkalinity due to PO_4^{3-} ions

Q. *Why during phosphate treatment, is the pH of boiler water maintained above 8.5?*

Ans. The phosphate treatment only prevents the formation of calcium scale rather than scale in general and hence creates conditions so that all Ca^{2+} ions are eliminated as loose sludge of hydroxylapatite $3Ca_3(PO_4)_2 . Ca(OH)_2$. Since this compound contains two OH^- ions care is to be taken to have the BFW with a sufficient concentration of hydroxyl ions, i.e., the boiler water with pH in excess of 8.5.

Q. *If this condition is not obeyed what will happen?*

Ans. It will lead to the formation of iron phosphate deposits.

Q. *What should be the lower limit of phosphate ion concentration in boiler water?*

Ans. This can be specified from the following formula

$$i_{ph} = N + 0.1C_{SiO_3^{2-}} + 0.01C_{SO_4^{2-}}$$

where N = constant = 5 mg/kg for boilers without multistage evaporation and 3 mg/kg for boiler with multistage evaporation

$C_{SiO_3^{2-}}$ and $C_{SO_4^{2-}}$ are the concentrations of SiO_3^{2-} and SO_4^{2-} ions in feedwater, mg/kg.

Q. *How can the amount of trisodium phosphate introduced continuously into the boiler in service be determined?*

Ans. This can be done with the help of the formula:

$$P_1 = \frac{100}{C_{ph}} \left[28.5H_t(1 + p_s) + p_s i_{ph} \right]$$

where

28.5 = equivalent of PO_4^{3-} ion in hydroxylapatite

C_{ph} = concentration of PO_4^{3-} in the commercial product introduced, %

H_t = total hardness of feedwater, mg-equiv/kg

p_s = blowdown in fractions of boiler capacity

P_1 = amount of phosphate chemical, kg/h

Also, it can be determined in terms of feedwater rate and steam-generating capacity of boiler as follows:

$$P_2 = \frac{G_{fw}}{10C_{ph}}\left[28.5H_t + P_{fw}\, i_{ph}\right]$$

$$P_3 = \frac{G_s}{10C_{ph}}\left[28.5H_t(1 + p_s) + P_s\, i_{ph}\right]$$

G_{fw} = feedwater rate, t/h

G_s = steam-generating capacity of the boiler, t/h

p_{fw} = blowdown in fraction of feedwater.

Q. *How can the rate of phosphate feed be determined in terms of volume?*

Ans. It can be determined by using the formula:

$$V_{ph} = \frac{G_s}{10C_s\rho n}\left[28.5H_t(1 + p_s) + P_s\, i_{ph}\right]$$

where C_s = concentration of TSP in solution, %

ρ = density of TSP in solution, kg/m^3

$n = C_{PO_4}^{3-}/C_{Na_3PO_4} = 0.579$

Problem 18.1 The feedwater of a boiler unit operating without multistage evaporation upon analysis shows that the concentrations of silicate and sulphate ions are equal to 0.015 gm/kg and 0.055 gm/kg respectively.

Determine the excessive amount of phosphates.

Solution The PO_4^{3-} concentration can be determined by using the formula:

$$i_{ph} = N + 0.1C_{SiO_3}^{2-} + 0.01C_{SO_4}^{2-}$$

N	$C_{SiO_3^{2-}}$	$C_{SO_4^{2-}}$	i_{ph}
5 for boiler operating without multistage evaporation	0.015 gm/kg − 15 mg/kg	0.055 gm/kg − 55 mg/kg	5 + 0.1(15) + 0.01(55) − 7.05 mg/kg *Ans.*

Problem 18.2 A boiler fitted with multistage evaporation is to be started with BFW quality $C_{SiO_3}^{2-}$ = 10 mg/kg; $C_{SO_4}^{2-}$ = 120 mg/kg.

Determine the quantity of trisodium phosphate ($Na_3PO_4 \cdot 12H_2O$) required, if the water space of the

boiler is 80 m^3. Take the content of PO_4^{3-} in commercial TSP equal to 25%.

Solution

Step (I) Calculation of excess phosphate required

Working Formula:

$$i_{ph} = N + 0.1C_{SiO_3}^{2-} + 0.01C_{SO_4}^{2-}$$

N	$C_{SiO_3^{2-}}$	$C_{SO_4^{2-}}$	i_{ph}
3	10 mg/kg	120 mg/kg	3 +0.1(10) + 0.01 (120) − 5.2 mg/kg

Step (II) Quantity of trisodium phosphate required

This can be determined by the formula:

$$P_o = V_b\, \frac{i_{ph}}{10C_{ph}}$$

V_b	i_{ph}	C_{ph}	P_O
80 m^3	5.2 mg/kg	25%	80(5.2)/10(25) − 1.664 kg *Ans.*

Problem 18.3 It is found that the requirement of phosphate ion concentration in BFW for a boiler to be lined up is 18 mg/kg. The boiler water space measures 60 m^3. Calculate the amount of trisodium phosphate ($Na_3PO_4 \cdot 12H_2O$) required to be fed in the feedwater.

Also determine the percentage fraction of TSP that will be bound by the calcium content present in BFW that fills the boiler.

Calcium hardness in feedwater = 10 μg-equiv./kg

Express the result as tricalcium phosphate.

Solution

Step (I) Amount of trisodium phosphate required

Working Formula:

$$P_O = \frac{V_b\, i_{ph}}{10C_{ph}}$$

V_b	i_{ph}	C_{ph}	P_O
60m^3	18 mg/kg	25% (see above)	60(18)/10(25) − 4.32 kg Ans.

Step (II) Quantity of Ca^{2+} ions in 60 m^3 of BfW

60m^3 water = 60(1000) kg of water

Ca^{2+} hardness = 10 μg-equiv./kg = $10(10^{-3})$ mg-equiv./kg

∴ Quantity of Ca^{2+} ions present in 60 m^3 water
= 60 (1000) (10×10^{-3}) mg-equiv. = 600 mg-equiv.

∴ 600 mg-equiv. of Ca^{2+} ions will be bound by 600 mg-equiv. of total TSP added.

Step (III)

Mol. Wt. of $Na_3PO_4 \cdot 12H_2O$	Eq. Wt. of $Na_3PO_4 \cdot 12H_2O$
3(23) + 31 + 4(16) + 12(18) = 380	380/3 = 126.666

Hence the amount of TSP present (in terms of mg-equiv.)

$$= \frac{4.32(1000)10^3}{126.666} = 34.10 \times 10^3 \text{ mg-equiv.}$$

Hence the fraction of TSP that will be bound by the calcium present $= \dfrac{600}{34.105 \times 10^3} = 0.01759$

i.e. 1.759%   Ans.

Problem 18.4 A steam generation plant equipped with three boilers each with a steam generation capacity 120 t/h receives boiler feedwater with a hardness of 7.5 μg-equiv./kg.

Estimate the daily consumption of trisodium phosphate in the three boilers and also its percentage removed from the boilers with blowdown water.

The silicate and sulphate ion concentrations in feedwater are respectively 45 mg/kg and 325 mg/kg.

Rate of blowdown = 3.2 %

Take concentration of PO_4^{3-} in commercial trisodium phosphate = 25%

Assume no multistage evaporation.

Solution

Step (I) Determination of the concentration of excess PO_4^{3-} ions

Working formula:

$$i_{ph} = N + 0.1C_{SiO_3^{2-}} + 0.01C_{SO_4^{2-}}$$

N	$C_{SiO_3^{2-}}$	$C_{SiO_4^{2-}}$	i_{ph}
5	45 mg/kg	325 mg/kg	5 + 0.1(45) + 0.01 (325) = 12.75 mg/kg

Step (II) Daily consumption of TSP in three boilers

Working Formula:

$$P_3 = \frac{G_s}{10C_{ph}} [28.5H_t(1 + p_s) + p_s i_{ph}]$$

G_s	3 × 120 × 24 t/day
C_{ph}	25%
H_t	7.5 μg-equiv./kg = 7.5×10^{-3} mg-equiv/kg
p_s	3.2% = 0.032
i_{ph}	12.75 mg/kg
P_3	$\dfrac{3(120)(24)}{10(25)} [28.5(7.5 \times 10^{-3})1.032 + 0.032(12.75)]$ = 21.724 kg/day Ans. (0.905 kg/h)

Step (III) TSP removed with blowdown

Rate of TSP consumed = 0.905 kg/h

Out of this TSP spent in water conditioning

= 0.032(12.75) kg/h

= 0.408 kg/h

∴ TSP wasted with blowdown water

= 0.905 – 0.408 = 0.497 kg/h

∴ % of TSP lost with blowdown water

$$= \frac{0.497}{0.905} 100 = 54.917$$

Ans.

Problem 18.5 If purely phosphate alkalinity is maintained in boilers, obtain the formula to determine the excess amount of PO_4^{3-} in boiler water, i_{ph} (mg/kg), required to neutralize feedwater alkalinity ΔA_{fw} (mg-equiv./kg).

Solution The alkalinity in feedwater is neutralized by di- and monosodium phosphate in accordance with the following reaction:

$$HPO_4^{2-} \rightarrow PO_4^{3-} + H^+ \xrightarrow{OH^-} PO_4^{3-} + H_2O$$

$$H_2PO_4^{2-} \rightarrow PO_4^{3-} + 2H^+ \xrightarrow{2OH^-} PO_4^{3-} + 2H_2O$$

And the PO_4^{3-} concentration in the feedwater required to neutralize alkalinity ΔA is

$$i_{fw} = \frac{95}{n}(\Delta A)$$

where n = neutralization capacity for acid phosphates. It is 2 for monosodium phosphate and 1 for disodium phosphate.

Again, $\Delta A = A_{fw} - 0.4H_t$

where H_t = total hardness in feedwater

0.4 = factor accounting for the quantity of alkali and phosphates consumed in the formation of $3Ca_3(PO_4)_2 \cdot Ca(OH)_2$ [Hydroxylapatite]

Hence the excess PO_4^{3-} (mg/kg) in boiler water

$$i_{ph} = \frac{100(95)}{n\,p_s}(A_{fw} - 0.4H_t)$$

Problem 18.6 Purely phosphate alkalinity is maintained in boilers. The alkalinity and hardness of feedwater are respectively 14 and 4.5 µg-equiv./kg. Blowdown rate is 2.5 % and a standard excess of PO_4^{3-} concentration equal to 35 mg/kg is maintained.

Determine if there is any excess phosphate in the boiler over the prescribed standard.

If so, what mixture (molar ratio) of mono and disodium phosphate is required to ensure the prescribed standard of PO_4^{3-} alkalinity.

Solution

Step (I) Determination of excess phosphate (if any)

The neutralization of excess alkali is carried out as follows:

$$H_2PO_4^- + OH^- \rightarrow PO_4^{3-} + H_2O$$

Here

n	A_{fw}	H_t	P_s	Excess PO_4^{3-} ion concentration
1	14 µg-equiv/kg	4.5 µg-equiv/kg	2.5	$i = \dfrac{100(95)}{1(2.5)}[14 - 0.4 \times 4.5]$ $\times 10^{-3} = 46.36$ mg/kg

The calculated value of i_{ph} shows that the PO_4^{3-} ion concentration in the boiler water is higher than prescribed by the standards.

Step (II) Therefore, it is necessary to make the use of a mixture of monosodium and disodium phosphate to ensure the concentration of the phosphate, $\left[C_{PO_4^{3-}}\right]_{st}$ as prescribed by the standard.

Therefore, we are to calculate the value of n—the neutralizing capacity of the phosphate mixture

$$n = \frac{100(95)}{P_s\left[C_{PO_4^{3-}}\right]_{st}}(A_{fw} - 0.4H_t)$$

$$= \frac{100(95)}{2.5(35)}[14 - 0.4(4.5)] \times 10^{-3} = 1.32$$

Step (III) Proportion of mono and disodium phosphate

Let x be the fraction of monosodium phosphate

$$1 + 2x = 1.32 \; ; \quad \therefore \quad x = 0.16$$

Therefore, the mixture should be $Na_2HPO_4 + 0.16NaH_2PO_4$ Ans.

Q. *Can the phosphate treatment ensure scale-free operation?*

Ans. No. It fails to prevent the formation of iron-oxide and cuprous salt depositions. Frequently phosphate chemicals themselves become the source of depositions, viz. iron phosphates.

Q. *Then how can the boiler water be conditioned for scale-free, sludge-free operation?*

Ans. This can be done by treating the boiler water with complexing agents that ensure not only scale-free, sludge-free operation but also corrosion-free operation under certain circumstances.

Q. *What are these agents?*

Ans. EDTA (ethylene-diamine tetracetic acid) Trylon - B (disodium salt of above)

$$CH_2 = CH_2$$

ethylene

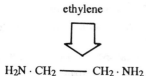

$$H_2N \cdot CH_2 \longrightarrow CH_2 \cdot NH_2$$

ethylene diamine

Trylon – B

A quadridendate compound

Q. *How is the treatment with Trylon-B carried out?*

Ans. A 1.5% or weaker solution of Trylon-B alkalized by NaOH or NH₄OH to pH 8.5 is introduced into the feedwater of a drum-type boiler.

Q. *At which point?*

Ans. At the suction of BFW pumps.

Q. *How does it act?*

Ans. This complexing agent traps the cations of scale formers into strong, well soluble chelates (complexonates). As a result no sludge and scale forming compounds form in the boiler units.

Trylon - B treatment also enables binding iron into a complex and that reduces its chance of carryover with steam.

Q. *What are sequestering agents?*

Ans. Sequestration means masking of metal ions by formation of soluble chelate compounds. Hence sequestering agents come to mean those specific chelating agents that react with certain metal ion or ions to produce soluble chelates.

Q. *What are chelates?*

Ans. Cyclic co-ordination complexes in which the ring is closed by two or more points of attachment of the co-ordinating group with the metal ion

Cupric ethylene diamine cation

Q. *How is the EDTA or Trylon-B treatment more efficient than phosphate treatment in preventing carryover of iron with steam?*

Ans. Compared with phosphate treatment, it can reduce the carryover of iron with steam by 50% or more.

Q. *How can the quantitative estimation of the consumption of complexing agent for boiler water treatment be done?*

Ans. This can be done by the formula

$$\dot{M} = [186H_{fw} + 6.7C_{Fe} + 6C_{Cu}]G_s(10^{-2}), \text{ g/h}$$

where

H_{fw} = feedwater hardness, microgram-equiv./kg

C_{Fe} = iron concentration in feedwater, µg-equiv/kg

C_{Cu} = copper concentration in feedwater, µg-equiv/kg

G_s = steam generation capacity of the boiler, t/h

Q. *What is the main objective difference in boiler water treatments with complexing agent between medium pressure (4 MPa = 4 MN/m²) boiler unit and superhigh pressure (15 MPa = 15 MN/m²) boiler unit?*

Ans. The main object of boiler water treatment, in the case of a medium pressure boiler is to prevent the formation of alkali-earth and iron oxide depositions, as the BFW in this case contains much higher concentrations of Ca^{2+}, Mg^{2+} and Fe^{2+} ions than the BFW in the case of superhigh pressure boiler units.

The feedwater for superhigh pressure boilers is deeply deionized (softened). So the hardness is minimum. Hence the complexing agent treatment is directed mainly to preventing the formation of iron-oxide scale, reduce the carryover of iron-oxides with steam and protect the units against steam-water corrosion.

Q. *Despite the effectiveness of complexing agents such as EDTA and Trylon -B, the feedwwater of intermediate pressure boiler units is subjected to phosphate treatment preferentially. Why?*

Ans. This is solely due to the economic point of view. The phosphate treatment is cheaper. And its use is justified so long as PO_4^{3-} ions are sufficient to ensure scale-free as well as reliable operation of tube metal subjected to high rate of heat flow.

Q. *But EDTA will also decompose at higher temperature. Will it continue to preserve its complexing capacity beyond that?*

Ans. EDTA and its salts start to decompose at about 423K and decomposition proceeds at a higher rate with the increase of temperature and ends up at about 573K. But still the gaseous products of decomposition of primary complexonates preserve their complexing capacity. When the steam condenses and the gaseous decomposition products redissolve, the complexing capacity revives.

Q. *Why does it sometimes become necessary to make the boiler water more alkaline?*

Ans. It will make the scale-forming substances penetrating into a boiler unit precipitate as $CaCO_3$ and $Mg(OH)_2$ in the form of loose sludge not adhering to surfaces while in a non-alkaline medium (e.g., in turbine condensers) the calcium carbonate will deposit on the heating or cooling surfaces as a dense scale.

Moreover, alkalization of boiler feedwater is applied when it is necessary to neutralize the ammonium salts of strong acids

$$[NH_4]_2SO_4 \Leftrightarrow 2NH_4^+ + SO_4^{2-}$$

$$NH_4^+ + H_2O \xrightarrow{\text{hydrolysis}} NH_3 + H_3O^+$$

$$H_3O^+ + OH^- \xrightarrow{\text{neutralization}} 2H_2O$$

or when it is necessary to raise the alkalinity of BFW to neutralize the acid phosphates.

Q. *How can the quantity of alkaline reagents introduced into a boiler unit to maintain a certain alkalinity of BFW be determined?*

Ans. It can be determined by the formula on the basis of 100% pure chemical reagent

$$G_A = G_s(10^{-3})[H_{nc}(1 + p_s) + p_s A_{bw}]E, \text{ kg/h}$$

where H_{nc} = non-carbonate hardness, mg-equiv./kg

G_s = steam generating capacity of the boiler, t/h

p_s = blowdown in fraction of boiler capacity

A_{bw} = required/admissible alkalinity of BFW, mg-equiv/kg

E = equivalent of the given alkaline reagent.

Q. *If the calcium and magneisum hardness in BFW be H_{Ca} and H_{Mg}, then how can the amount of scale-forming substances penetrating into the boiler unit be derived from the above formula?*

Ans. The rate of sludge formation in feedwater is

$$G_{sl}^{fw} = G_s(10^{-3})[(1+p_s)(50H_{Ca}) + (1+p_s)(29H_{Mg})],$$
$$\text{kg/h}$$

$$= G_s(10^{-3})(1+p_s)[50H_{Ca} + 29H_{Mg}], \text{kg/h}$$

Q. *Which reagent is usually used to make the soft BFW more alkaline?*

Ans. Sodium hydroxide.

Q. *Why not soda ash or sodium bicarbonate which are cheaper than caustic soda?*

Ans. The use of these alkalizing reagents is restricted only in boiler units operated at a pressure of up to 1.6 MPa (1.6 MN/m^2), because of the thermal and hydrolytic decomposition of these compounds:

$$2NaHCO_3 \xrightarrow[\Delta]{} Na_2CO_3 + H_2O + CO_2$$

$$CO_3^{2-} + H_2O \longrightarrow CO_2(\uparrow) + 2OH^-$$

Since CO_3^{2-} is unstable at higher temperatures, as a result, the effectiveness of alkaline treatment decreases and the generated steam gets contaminated with carbon dioxide. (Fig. 18.1)

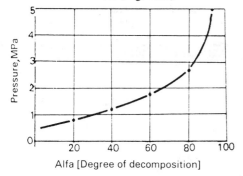

Fig. 18.1 *Instability Curve of* CO_3^{2-} *ions*

Q. *What happens if steam is contaminated with carbon dioxide?*

Ans. It is undesirable since it will give rise to carbon dioxide corrosion of metal in the condensate circuit.

Q. *How is the quantity of carbon dioxide passing into the generated steam found out?*

Ans. This can be determined by the equation

$$\dot{M}_{CO_2} = 22[A_b(1 + \alpha) + \alpha A_c] = 22[A_b + \alpha A_t], \text{mg/kg}$$

where
α = degree of decomposition of caustic soda
$A_t = A_c + A_b$ = total alkalinity, mg-equiv./kg
A_c = carbonate alkalinity, mg-equiv./kg
Factor 22 is the equivalent weight of CO_2.

Problem 18.7 A steam generation plant runs with three boilers each of 120 t/h steam generation capacity. If the actual alkalinity of feedwater is 225 microgram-equiv./kg, determine the daily cnsumption of NaOH per ton of feedwater to maintain the boiler water alkalinity at 10 mg-equiv./kg. Take the blowdown rate as 3%. Also determine the daily NaOH requirement.

Solution: This can be done by using the formula:

$$G_A = \left[\frac{p_s}{1+p_s}(A_{bw}) - A_{fw}\right]E, \text{mg/kg of feedwater.}$$

where p_s = blowdown rate
A_{bw} = boilerwater alkalinity, mg-equiv./kg
A_{fw} = feedwater alkalinity, mg-equiv./kg
E = equivalent of the alkalizing reagent.

p_s	A_{bw}	A_{fw}	E_{NaOH}	G_A
3%	10 mg-equiv/kg	225 micro-equiv/kg – 0.225 mg-equiv/kg	40	$\left[\frac{0.03}{1.03}(10) - 0.225\right]$ 40 – 2.65 mg/kg – 2.65 g/t

Therefore daily NaOH requirement

$$= 3 \times 120 \times 24 \times 2.65 = 22900 \text{ gm} = 22.9 \text{ kg.}$$
Ans.

Problem 18.8 Three boiler units each having steam generation capacity of 150 t/h requires internal water treatment of their feedwater to maintain alkalinity of BFW at 10 mg- equiv./kg. If soda ash is used as the alkalizing reagent, determine the amount of annual comsumption of soda ash on the basis of

7920 h running hours of each boiler with a blowdown rate of 5%.

Take non-carbonate hardness of feedwater equal to 2.5 mg- equiv/kg.

Solution On the basis of 100% pure Na_2CO_3, the annual consumption of soda ash can be determined by the formula

$$G_{Na_2CO_3} = G_s(10^{-3})\left[H_{nc}(1+p_s) + p_s A_{bw}\right] E_{Na_2CO_3}$$

Step (I) One boiler unit (150 t/h) is considered

G_s	H_{nc}	p_s	A_{bw}	$E_{Na_2CO_3}$
150 t/h	2.5 mg-equiv/kg	5%	10 mg-equiv/kg	$150(10^{-3})[2.5(1 + 0.05) + 0.05 (10)] 53$ $- 24.8437$ kg/h

Step (II) Annual consumption of Na_2CO_3

$G_{Na_2CO_3}$	No. of boiler units	Total running hours	Total consumption of Na_2CO_3 per Annum
24.8437 kg/h per boiler	3	7920 h	24.8437(3)(7920) kg $- 590287.5$ kg $- 590.287$ t

Problem 18.9 A boiler house is fitted with six low-pressure boilers each of steam-generating capacity 5 t/h. The feedwater is a mixture of condensate and raw water the total hardness in which is 1.8 mg-equiv./kg and carbonate hardness 1.2 mg-equiv./kg. Internal treatment of BFW is applied by using Na_2CO_3 to maintain the alkalinity of boiler water to 15 mg-equiv./kg. If the rate of blowdown is 5%, calculate the rate of consumption of Na_2CO_3.

Solution Consumption rate of soda ash for internal treatment of boiler water with known hardness can be carried out by using the formula:

$$G_{Na_2CO_3} = G_s(10^{-3})[H_{nc}(1 + p_s) + p_s A_{bw}]E_{Na_2CO_3}$$

Step (I) Calculation of non-carbonate hardness (H_{nc})

Working Formula: $H_t = H_c + H_{nc}$

where

H_t = total hardness, mg-equiv./kg

H_c = carbonate hardness, mg-equiv./kg

H_{nc} = non-carbonate hardness, mg- equiv./kg

H_t	H_c	H_{nc}
1.8 mg-equiv./kg	1.2 mg-equiv./kg	(1.8–1.2) $- 0.6$ mg-equiv./kg

Step (II) Consumption rate of soda ash

G_s	H_{nc}	p_s	A_{bw}	No. of boilers	$G_{Na_2CO_3}$
5 t/h	0.6 mg-equiv/kg	5%	15 mg-equiv/kg	6	$6(5 \times 10^{-3})$ $[0.6(1 + 0.05) + 0.05(15)] 53$ $- 2.194$ kg/h *Ans.*

Problem 18.10 Determine the volume of the storage tank, in the preceding problem, required for daily consumption of soda ash taken in the form of 4% solution with density nearly equal to 1000 kg/m^3. Assume the tank is 75% filled with the solution.

Solution

Step (I) Volumetric rate of flow of soda ash solution

Solution strength	Mass flow rate of soda ash	Volume flowrate
4% – 4 kg Na_2CO_3/100 kg soln.	2.194 kg/h	$\dfrac{2.194}{40} \dfrac{kg}{h} \times \dfrac{m^3}{kg}$ $- 0.05485$ m^3/h

$$- \frac{4 \text{ kg } Na_2CO_3}{100 \times \dfrac{1}{1000} m^3 \text{ solution}}$$

$- 40$ kg Na_2CO_3/1 m^3 of solution

Step (II)

Vol. flowrate of soda ash	Daily requirement of soda ash soln.	Required capacity of tank
0.05485 m^3/h	0.05485 (24) m^3	0.05485 (24)/0.75 $- 1.755$ m^3 *Ans.*

Problem 18.11 A boiler of an electric power station is filled with 65 m^3 of water. This BFW is to be made alkaline with a solution of 35% of NaOH solution available at the end of the battery limit so as to ensure a 1% solution of NaOH in it. Determine the volume of reagent solution required.

Density of 35% NaOH solution = 1 380 kg/m^3

Density of 1% NaOH solution = 1 000 kg/m^3

Solution The feedwater must be made alkaline so as to obtain 65 m^3 of BFW of 1% NaOH solution.

Step (I)

The mass of BFW

= 65(1 000)

= 65 000 kg solution (1% NaOH)

Mass of NaOH present in it

= 65 000(1/100)

= 650 Kg

Step (II) Volume of 35% NaOH solution required

35 kg NaOH contain in 100 kg solution (35% NaOH)

$$= \frac{100}{1380} \text{ kg (m}^3\text{/kg)} = \frac{100}{1380} \text{ m}^3 \text{ solution}$$

650 kg NaOH will contain in $\frac{100}{1380} \frac{650}{35}$ m^3 solution

= 1.345 m^3 solution   *Ans.*

Problem 8.12 Three boilers, 100 t/h of capacity each, are generating steam at 3 MPa (3 MN/m^2) from the BFW having bicarbonate alkalinity 0.10 mg-equiv./kg and carbonate alkalinity 0.25 mg-equiv./kg. Determine the amount of CO$_2$ contained in the steam.

Take the degree of decomposition of sodium carbonate at 3 MPa pressure = 0.9

Solution: The quantity of CO$_2$ passing into steam can be determined with the aid of the formula

$$\dot{M}_{CO_2} = 22[A_b (1 + \alpha) + \alpha A_c'], \text{ mg/kg}$$

A_b	A_c	α	$\dot{M}_{CO_2}$
0.10 mg-equiv/kg	0.25 mg-equiv/kg	0.9	22 [0.10 (1 + 0.9) + 0.9 (0.25)] = 9.13 mg/kg *Ans.*

Problem 18.13 A steam generation plant is fitted with three boilers each of steam generation capacity 10 t/h. Feedwater contains total hardness 2.5 mg-equiv./kg and carbonate hardness 1.2 mg-equiv./kg. Soda ash is used as the alkalizing agent to maintain boilerwater alkalinity equal to 15 mg-equiv./kg.

Determine

(a) the hourly consumption (kg/h) of soda ash if blowdown rate is 5%
(b) the rate of sludge formation, kg/h
(c) the concentration of sludge in boilerwater, mg/kg. Take average equiv. of Ca^{2+} and Mg^{2+} ions present = 43.

Solution

Step (I) Consumption rate (kg/h) of soda ash

(a) The concentration of soda ash required to maintain the prescribed alkalinity in BFW is given by

$$C_{Na_2CO_3} = (H_t - A_i) (1 + p_s) + p_s A_{bw}, \text{ mg-equiv./kg}$$

where H_t = total hardness in feedwater, mg-equiv./kg

A_i = initial alkalinity of feedwater, mg-equiv./kg

p_s = blowdown rate

A_{bw} = boilerwater alkalinity, mg-equiv./kg

H_t	A_i	p_s	A_{bw}	$C_{Na_2CO_3}$
2.5 mg-equiv/kg	1.2 mg-equiv kg	5%	15 mg-equiv kg	(2.5 – 1.2) (1 + 0.05) + 0.05 (15) = 2.115 mg-equiv./kg

(b) *Soda ash consumption*

Feedwater requirement of each boiler = 10 t/h

Overall feedwater requirement of SGP

= 3(10)

= 30 t/h

Hence the rate of consumption of soda ash

$$= 30 \times 10^3 \frac{\text{kg}}{\text{h}} \times 2.115 \frac{\text{mg-equiv.}}{\text{kg}}$$

$$= 30 \times 10^3 \times 2.115 \times 53 \text{ mg/h}$$

$$= 30 \times 10^{-3} \times 2.115 \times 53 \text{ g/h}$$

= 3.362 kg/h

                                        Ans.

Step (II) Rate of sludge formation (kg/h)

The rate of sludge forming in BFW can be conveniently expressed as

$$Q_{sl} = G_s (1 + p_s) H_t E_a, \ g/h$$

where G_s = rate of steam generation, t/h

p_s = blowdown rate

H_t = total hardness in feedwater, mg-equiv./kg

E_a = average equivalent of the sum of Ca^{2+} and Mg^{2+} ions present in BFW.

G_s	p_s	H_t	E_a	Q_{sl}
30 t/h in three boilers	0.05	2.5 mg-equiv/kg	43	30(1 + 0.05)(2.5)(43) − 3386 g/h − 3.38 kg/h *Ans.*

Step (III) Concentration of sludge in BFW

The sludge concentration in BFW is given by

$$C_{sl} = \frac{1 + p_s}{p_s}(H_t) E_a, \ mg/kg$$

p_s	H_t	E_a	C_{sl}
0.05	2.5 mg-equiv./kg	43	$\frac{1.05}{0.05}$ (2.5) (43) − 2257.5 mg/kg *Ans.*

Problem 18.14 An industrial boiler house with total steam generating capacity 75 t/h at 1 MPa (1 MN/m^2) utilizes a mixture of raw water (60%) and condensate. The raw water is of the following quality:

Total hardness = 3.9 mg-equiv./kg

Carbonate hardness = 2.2 mg-equiv./kg

Calcium hardness = 2.6 mg-equiv./kg

Soda ash is used as the alkalizing reagent to maintain boiler water alkalinity at 12 mg-equiv./kg.

Blowdown rate is 5%

Determine:

(a) the concentration of CO_2 in steam

(b) the rate of sludge formation in the boiler

(c) the concentration of the sludge.

Take $E_a = 43$.

Solution

Step (I) Since the feedwater is a mixture of 60% raw water containing carbonates and bicarbonates,

the CO_2 carryover to steam produced is entirely due to raw water.

Hence the concentration of CO_2 in generated steam

$$C_{CO_2} = 22[A_b(1 + \alpha) + \alpha A_c] \times (60\%), \ mg/kg$$

where A_b = bicarbonate hardness in raw water, mg-equiv./kg

$$= H_t - H_c$$

A_c = carbonate hardness in raw water, mg-equiv./kg

α = degree of dissociation of Na_2CO_3 at the working pressure of the boiler

$\alpha = 0.34$ cf. PRESSURE vs. ALFA cruve at 1 MPa

Factor 22 = equiv.wt. of CO_2

A_b	α	A_c	C_{CO_2}
3.9 − 2.2 − 1.7 mg-equiv./kg	0.34	2.2 mg-equiv./kg	22 [1.7 (1.34) + 0.34 (2.2)] (0.6) − 39.94 mg/kg

Step (II) Rate of sludge formation in boilers

Working Formula:

$$G_{sl} = G_s(1 + p_s)[50H_{Ca} + 29H_{Mg}], \ g/h$$

G_s	p_s	H_{Ca}	$H_{Mg} = H_t - H_{Ca}$	G_{sl}
75 t/h	0.05	2.6 mg−equiv. kg	3.9 − 2.6 − 1.3 mg−equiv. kg	75 (1 + 0.05) [50(2.6) + 29(1.3)] × (0.6) − 7923.6 g/h − 7.92 kg/h *Ans.*

Step (III) Concentration of sludge in boiler

Working Formula:

$$C_{sl} = \frac{1 + p_s}{p_s}(H_t) E_a, \ mg/kg$$

p_s	H_t	E_a	C_{sl}
0.05	3.9 mg-equiv./kg	43	$\frac{1.05}{0.05}$(3.9)(43)(0.6) − 2113 mg/kg *Ans.*

Q. *How can the scale formation on the waterside heating surfaces of once-through boiler units be prevented?*

Ans. This can be done by eliminating the sources of feedwater contamination.

Q. *What are the primary sources of feedwater contamination?*

Ans. 1. Presence of dissolved oxygen in BFW from deaerator to boiler drum. It acts as a depolarizer and contributes to corrosion

$$O_2 + 4e + 2H_2O \longrightarrow 4OH^-$$

2. Presence of dissolved CO_2 in the condensate leading to corrosion of heaters and evaporators

$$CO_2 + H_2O \longrightarrow HCO_3^- + H^+$$

$$2H^+ + Fe \rightarrow H_2 + Fe^{2+} \xrightarrow{\;HCO_3^-\;} Fe(HCO_3)_2$$

3. Leakage of cooling water in turbine condenser—a tube and shell—type heat exchanger.

Q. *How does the infiltration of cooling water into the intertubular space of a turbine-condenser occur?*

Ans. This occurs through the expansion joints between the condenser tubes and the tube plate as well as through the hair-cracks in the walls of the tubes.

Q. *How can the leakage of cooling water in the turbine-condenser be measured?*

Ans. This can be measured in terms of leakage which is given by the formula:

$$(\phi)_1 = \frac{C_{cond} - C_s}{C_{cw} - C_s} \times 100$$

where, C refers to concentration. Subscripts "cond", "s" and "cw" stand for condensate, steam and cooling water respectively.

Q. *So how can this problem be solved?*

Ans. By incorporating a demineralizing unit downstream of the turbine condenser. This will eliminate from the condensate the ionic and dispersed impurities and bring down the water condition within the prescribed limit.

Q. *What type of corrosion takes place in the feedwater circuit in once-through boiler units?*

Ans. Oxygen corrosion due to dissolved oxygen.

$$O_2 + 4e + 2H_2O \rightarrow 4OH^-$$

$$2Fe \rightarrow 2Fe^{2+} + 4e$$

$$\text{(Adding)} \; 2Fe + O_2 + 2H_2O \rightarrow 2Fe(OH)_2$$

If sufficient amount of oxygen is present, ferrous hydroxide gets oxidized to ferric hydroxide which falls out on the tube wall

$$2Fe(OH)_2 + \frac{1}{2} O_2 + H_2O \rightarrow 2Fe(OH)_2(\downarrow)$$

$$\xrightarrow[\Delta]{} \; Fe_2O_3 \cdot 3H_2O$$

Besides, due to oxygen corrosion, formation and deposition of copper oxides also takes place.

Q. *At which section of the heating surface, does the iron oxide mainly deposit?*

Ans. In the region of intense heat flow, i.e., at the bottom radiant heat-transfer part.

Q. *In which pattern?*

Ans. On the semi-circumference of the heating tubes facing the furnace. (Fig. 18.2)

Q. *How can this scaling be avoided in once-through boiler units?*

Ans. By EDTA treatment **Fig. 18.2** of feedwater.

This forms a complex with Fe^{2+} and Fe^{3+} ions and eliminates the accumulation of iron oxides at the bottom radiant zone of heat-absorbing surfaces.

Q. *However the iron-complex of EDTA undergoes thermolysis leading to precipitation of magnetite on tube walls. So how can this method be effective?*

Ans. Most of the magnetite precipitates before the water reaches the bottom radiant heat-absorbing parts. Nearly 70–80% of the magnetite falls out in the economizer. That's how the effectiveness of EDTA treatment is retained.

Q. *How does the formation of solid carbonate deposits take place in the tubes of turbine condensers?*

Ans. It takes place due to decomposition of soluble bicarbonate:

$$Ca(HCO_3)_2 \rightarrow CaCO_3 + CO_2(\uparrow) + H_2O$$
$$\text{(scale)}$$

Q. *How is the formation of calcium carbonate deposit in turbine condensers prevented?*

Ans. By adopting any one of the following processes:
- (a) acidification
- (b) lime treatment
- (c) phosphate treatment
- (d) recarbonation process

Q. *How does the acidification eliminate carbonate scaling?*

Ans. Acidification carried out by introducing sulphuric acid to the makeup water adds free H^+ ions that destroy the HCO_3^- ions as follows:

$$HCO_3^- + H^+ \rightarrow H_2O + CO_2(\uparrow)$$

Q. *Why is it important to meter out the correct amount of acid introduced into the make-up water?*

Ans. This is because excess acidity will increase the corrosivity of water while less acidity will fail to ensure scale-free operation.

Q. *How can lime treatment serve the purpose, i.e., prevent carbonate scaling?*

Ans. Lime $Ca(OH)_2$ converts all HCO_3^- ions into CO_3^{2-} without the formation of CO_2

$$HCO_3^- + OH^- \rightarrow CO_3^{2-} + H_2O$$

Q. *Between acidification and lime treatment, which one is more efficient?*

Ans. Lime treatment. By acidification, only a small fraction of HCO_3^- ions are destroyed, whereas by lime treatment entire bicarbonate ions are removed from the water.

Q. *So the lime treatment system in the return water circuit is preferable. Is it not?*

Ans. Though lime $Ca(OH)_2$ treatment destroys all the bicarbonate ions, it is seldom applied to return water circuit since this involves the installation of an expensive plant.

Q. *What is the recarbonation process?*

Ans. It is a process by which CO_2 concentration in water containing HCO_3^- ions is increased by blowing the furnace gas through the layer of water or mixing it with water by means of a water-operated ejector.

Q. *How does the dissolved free carbon dioxide so introduced prevent scaling?*

Ans. It shifts the following equilibra from right to left:

$$HCO_3^- + H^+ \Leftrightarrow H_2O + CO_2$$

and therefore bicarbonate decomposition to carbonate ceases

$$2HCO_3^- \xrightarrow{\Delta} H_2O + CO_2 + CO_3^{2-}$$

Q. *How many methods are known to clean boiler units and heat exchanger off deposits?*

Ans. Two methods—mechanical and chemical.

Q. *How is mechanical cleaning of deposits carried out?*

Ans. It involves the use of various type of scrappers, wire brushes, etc.

Q. *In which case is this method suitable?*

Ans. For small boiler units and simple heat exchangers.

Q. *How is chemical cleaning performed?*

Ans. It is carried out by circulating cleaning agents —acids, alkalies and complexing agents—through the boiler or heat exchanger circuit.

Q. *What is the difference between prestarting cleaning and planned or scheduled cleaning?*

Ans. The prestarting cleaning is carried out in the case of newly installed boilers before they're put into operation. This is directed primarily to removing rust and welding rag, iron-scale, burrs and other impurities from the steam-water circuit.

Whereas scheduled cleaning is the periodic cleaning in the course of service of the boilers, aimed at removing all kinds of deposits formed during operation.

(*Note:* Both are chemical cleaning)

Q. *What reagents are used for chemical cleaning?*

Ans. Hydrochloric acid, caustic soda, ammonium hydroxide, citric acid, maleic acid and other organic acids as well as the complexing agent EDTA.

Q. *Why is hydrochloric acid preferred?*

Ans. It is abundantly available and is the cheapest of all cleaning chemicals.

Q. *In which cases is it particularly used?*

Ans. For cleaning low-capacity boiler units, heat exchangers and surface condensers of turbines.

Q. *Why is it not favoured for cleaning high-capacity units, particularly supercritical-pressure boiler units?*

Ans. Due to the aggressiveness of the acid as well as the intricate layout of cleaning.

Q. *What is the basic advantage of organic acids?*

Ans. They are less aggressive than HCl.

Q. *What are the disadvantages of using organic acids as chemical cleaning agents?*

Ans. Firstly, these chemicals are expensive. Secondly, they present difficulties in wastewater treatment.

Q. *What is the most successful reagent?*

Ans. It has been found by experience that a mixture of complexing agent (EDTA) and organic acid (say, citric acid) taken in a 1:1 proportion the most suitable reagent.

Q. *How is scheduled washing of boiler units performed?*

Ans. At first, the boiler unit is washed with a mixture of EDTA and organic acid solution to dissolve away the scales.

When all the deposits have been removed, the solution is displaced from the boiler unit by polished water. Boiler surfaces are flushed with polished water to dislodge traces of leftover impurities.

It is then followed by a neutralization operation with an alkaline solution.

Finally, the cleaning is terminated by passivating the metallic surfaces by nitrite-ammonium or hydrazine-ammonium or complexon passivation treatment.

19

Characteristics of Steam-water Flow

Q. *What is circulation velocity?*

Ans. It is the velocity of water at the saturation temperature

$$v_{cir} = \frac{G_m}{A\,\rho}$$

where

G_m = mass flowrate of steam-water mixture through the tube system, kg/s.

A = free cross-sectional area for the passage of steam-water mixture, m^2

ρ = density of water at saturation temperature, kg/m^3

Q. *What is circulation ratio?*

Ans. It is the inverse of mass steam content

$$K = \frac{1}{x} = \frac{G_m}{G_s}$$

where, G_s = mass flowrate of steam through the tube system

Q. *What is mass steam content?*

Ans. It is the ratio of the quantity of steam produced to the quantity of circulating water for the same interval of time:

$$x = \frac{G_s}{G_m}$$

Q. *What is volume steam content?*

Ans. It is the volume fraction of steam in a steam-water mixture flow having identical velocities of water and steam

$$\beta = \frac{V_s}{V_s + V_w}$$

Q. *What is actual steam content?*

Ans. It is the fraction of the tube cross-section occupied by steam

$$\varphi = \frac{A_s}{A}$$

Q. *What is the relation between actual steam content and volume steam content?*

Ans. $\dfrac{\varphi}{\beta} = \dfrac{v_m}{v_s}$

i.e., the ratio of actual steam content to volume steam content is equal to the ratio of velocity of steam-water mixture to the velocity of steam

$$\beta = \frac{V_s}{V_s + V_w} = \frac{v_s A_s}{v_m A} = \frac{v_s}{v_m}\,\varphi$$

Q. *When does the actual steam content equalize with the volume steam content?*

Ans. φ will become equal to β, if $v_s \approx v_m$. This condition will prevail if the relative velocity between steam and water reduces to zero

$$v_r = v_s - v_w = 0$$

i.e.,

$$v_s = v_w = v_m$$

Since with the increase of pressure, the relative velocity of steam decreases, as the pressure approaches the critical value φ tends to β.

Q. *Will the regime of flow pattern of steam-water mixture through heating tubes will vary with the spatial arrangement of tubes?*

Ans. Yes; the structure and flow-mode, under identical conditions, will vary on the spatial arrangement of the tubes, i.e., whether horizontally laid or vertically laid.

Q. *How many kinds of flow-modes of steam- water mixture in vertical tubes are known?*

Ans. Five types:
1. Bubble Flow
2. Emulsion Flow (Frothy Flow)
3. Slug Flow
4. Dispersed annular flow (Wet-wall Flow)
5. Inverse Disperse-Annular Flow (Dry-Wall Flow)

Q. *What is bubble flow?*

Ans. It is a low-velocity adiabatic flow of steam-water mixture characterized by rare fine vapour bubbles distributed in it. It contains a very insignificant amount of steam. (Fig. 19.1)

Fig. 19.1 Bubble flow

Q. *What is emulsion flow?*

Ans. With the increase of steam content and higher mass velocity of steam-water mixture, the bubble flow is converted into emulsion flow. It is characterized by a relatively large number of fine vapour bubbles (Fig. 19.2)

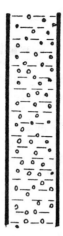

Fig. 19.2 Emulsion flow

Q. *What is slug flow?*

Ans. If mass velocity of steam-water mixture is low but steam content increases with higher heat transfer, the fine bubbles of steam will coalesce to form large cylindrical bubbles (called slugs) with a diameter comparable to that of the tube and length many times the diameter. This flow regime is called slug flow. Each 'slug' is followed by a trail of fine vapour bubbles suspended in the thin liquid bridge separating two consecutive slugs. (Fig. 19.3)

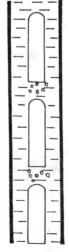

Fig. 19.3 Slug flow

Q. *What is wet-wall flow?*

Ans. It is the next stage of slug flow when the steam content further increases.

If the steam/water ratio increases, the water bridges separating the slugs will get narrowed down and ultimately disappear with the slugs merging into a long, continuous vapour column trapping atomized water droplets within.

This column moves along the core of the tube and is separated from the tube wall by a narrow film of water layer moving along the wall. This flow regime is called disperse-annular flow or wet-wall flow (Fig. 19.4)

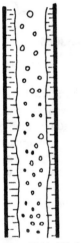

Fig. 19.4 Wet-wall flow

Q. *What is dry-wall flow?*

Ans. The water film separating the vapour column from the tube wall, in wet-wall flow, gradually loses its water mass at higher steam pressure and higher steam velocity. With the increase of steam pressure and steam velocity, more and more of water from the film is carried off as droplets, decreasing the thickness of the water-film on the tube surface. This reduced film quickly evaporates forming a vapour film while the water droplets removed from the water film coalesce to form a continuous stream of water in the core. Thus the liquid fills the core of the tube while a vapour film separates it from the heat-exchange surface. This type of flow is called dry-wall flow. (Fig. 19.5)

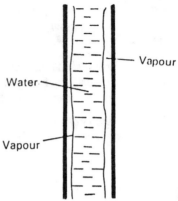

Fig. 19.5 Dry-wall flow

Q. *What is the nature of flow of the steam-water mixture in horizontal tubes?*

Ans. The steam-water flow in a horizontal tube is characterized by a non-uniform distribution of steam and water over the tube cross-section. This is because of difference in density between steam and water at any temperature lower than critical temperature. At temperatures below critical temperature, the density of steam is lower than that of water. So steam will flow primarily along the upper tube wall while the main mass of water will flow along the lower tube wall. However, the flow-mode depends also on the velocity of flow and diameter of the tube (see below).

Q. *How many kinds of flow-modes of steam- water mixture in horizontal tubes are known?*

Ans. Two types:
1. Nearly symmetrical flow of water and steam
2. Asymmetrical flow of steam-water mixture.

Q. *How is symmetrical flow of steam-water mixture in a horizontal tube observed despite a density difference between steam and water?*

Ans. Such flow is observed in horizontal tubes with diameter 30–40 mm and inlet velocity of water greater than 1 m/s.

The vapour bubbles formed initially on the tube wall detach themselves, under these circumstances, move along the mainstream of water and coalesce to form large bubbles which increase in number with the increase of quantity of steam produced. Eventually, there results a combined flow of two phases, the

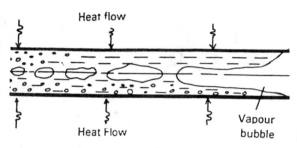

Fig. 19.6 *Symmetrical flow of steam-water mixture in horizontal tube*

distribution of which becomes nearly symmetrical about horizontal axis and resembles wet-wall flow in vertical tubes. (Fig. 19.6)

Q. *What happens, i.e. how will the flow-mode change if the inlet velocity of water is less than 0.5 m/s in horizontal steam-generating tubes?*

Ans. At such a low inlet velocity of water into horizontal steamgenerating tubes, the two-phase flow is separated resulting in asymmetrical motion of water and steam along the flow path. (Fig. 19.7)

Steam bubbles formed will merge into larger formations which will attach themselves to the upper wall of the tube, as steam has lower density than water. These steam bubbles will flow along the upper wall of the tube while water will flow predominantly along the lower wall of the steam generating tube.

Q. *How do the pressure, temperature, density and dielectric permittivity change in the water-steam path of subcritical and supercritical boilers?*

Ans. For subcritical pressure boilers abrupt change in thermophysical characteristics takes place within the boiler furnace where conversion of water to steam takes place. As the water to steam transition takes place, the density of water abruptly changes to that of steam. Similarly dielectric permittivity of water changes abruptly to that of steam but temperature remains constant in this zone. (Fig. 19.8)

For supercritical pressure boilers, the transition of water to steam does not take place, unlike subcritical pressure boilers, at constant temperature. There is a smooth, gradual rise in the temperature in this zone (Zone-II). (Fig. 19.9)

The change in thermophysical properties (density, dielectric permittivity) of the working fluid in the zone of high heat capacity is rather smooth, not abrupt as with subcritical pressure boilers.

Q. *How do the density and dielectric permittivity of water vary with pressure?*

Ans. The variations have been shown graphically. (Fig. 19.10)

As is evident from the figure the density of water is substantially different from that of steam at any particular pressure. But this difference becomes narrower and narrower as the pressure of the working fluid is increased. Eventually the density of both steam and water coincide at critical pressure. Analogous statements can be made in the case of dielectric permittivity of the working fluid.

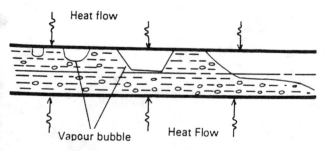

Fig. 19.7 *Asymmetrical flow of steam-water mixturre in a horizontal steam-generating tube*

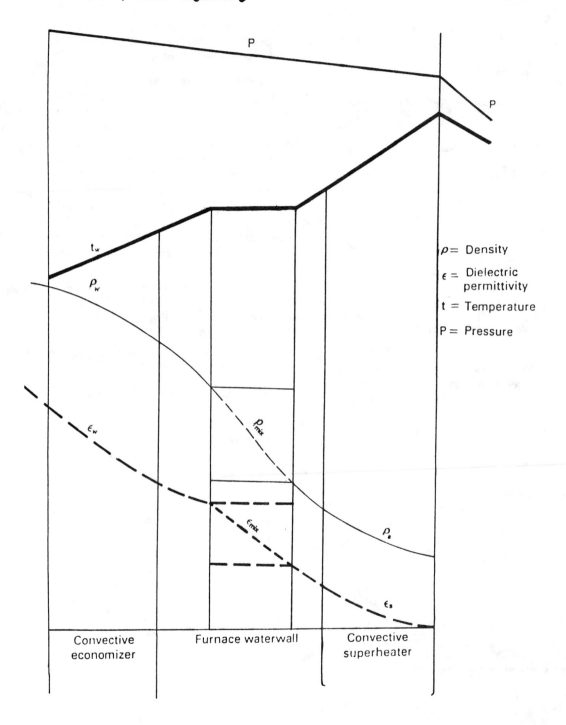

$\rho=$ Density

$\epsilon =$ Dielectric permittivity

$t =$ Temperature

$P =$ Pressure

Fig. 19.8

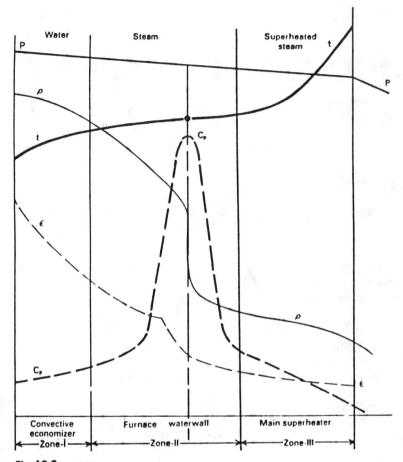

Fig 19.9

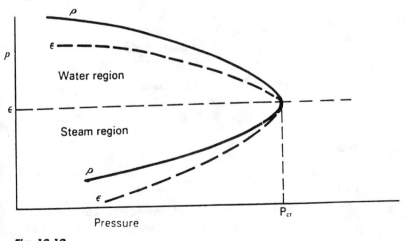

Fig. 19.10

20

Temperature Conditions and Heat Transfer

Q. *Into how many zones can a boiler be divided on the basis of mode of heat transfer?*

Ans. Three zones.

ZONE-I consisting of waterwalls, i.e., heating surfaces arranged within the furnace, receives heat mainly by radiation from the hot (1200–1400°C) gaseous products of combustion in the furnace core.

ZONE-II comprising mainly of platens and waterwalls, receives heat partly by radiation and partly by convection. In this zone gas temperature is quite high (800–1200°C) and so a substantial portion of heat transfer takes place by radiation.

ZONE-III composed of economizer, air heater and some superheater banks receives heat primarily by convection as this zone has a relatively low temperature (800–900°C) of combustion products contributing a very negligible fraction of radiant heat transfer.

Q. *In order to ensure stable operation of a boiler furnace, it is basically required to remove actively, through heating surface, the heat released in the furnace. How can this be ensured?*

Ans. By virtue of motion of the working fluid at a definite mass velocity.

Q. *How does the boiler load variation in free-circulation plants affect the mass velocity of the steam-water mixture through the evaporating tubes?*

Ans. For steam generating plants operating on free-circulation, i.e., gravity circulation, the working fluid—steam-water mixture—moves under the free circulation head set up when the tubes are heated up. (Fig. 20.1)

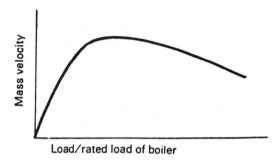

Fig. 20.1 *Circulation velocity of steam-water mixture in SGP changes with the load of its boiler*

When the boiler load is increased, the mass flowrate at the inlet to the evaporating tubes rises sharply at first and on reaching a maximum value gets almost stabilized or even decreases somewhat.

Q. *Why does this decrease in the mass flowrate of the working fluid take place in a free-circulation boiler even after its maximum value has been reached with the increase of boiler load?*

Ans. It is because with the increase of steam generation rate, the resistance to the flow of working fluid through the evaporating tubes increases due to higher unit volume of steam generation.

Q. *What is the effect of boiler load on the mass velocity of the working fluid in a multiple forced circulation boiler?*

Ans. In such types of boilers, circulation of water and steam-water mixture through evaporating tubes is effected by a circulation pump, with the effect that

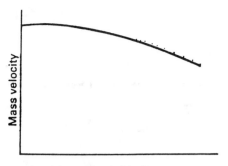

Fig.20.2 *Mass velocity of working fluid in a multiple forced circulation boiler remains practically constant with the change of boiler load*

mass flowrate of working fluid remains practically constant irrespective of boiler load. (Fig. 20.2)

That's why the mass velocity vs. boiler load curve is roughly parallel to the load axis except for a slight declination at higher load which is due to the higher resistance to flow offered by higher steam generation.

Q. *What is the effect of boiler load variation on the mass flowrate of the working fluid in the case of a once-through boiler and forced circulation boiler?*

Ans. In the case of oncethrough boiler units, the mass flowrate of working fluid is directly proportion-

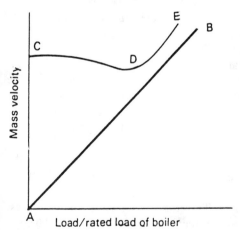

Fig. 20.3 *Mass velocity of working fluid in once-through boiler increases with boiler load while that in the case of combined circulation changes on the principle of multiple forced circulation and once-through operation*

al to the boiler load. Hence mass velocity vs. boiler load diagram is a straight line (*AB*). (Fig. 20.3)

For combined circulation boilers working on the principle of multiple forced circulation at low loads and on the once-through principle at high load, the mass velocity vs. boiler load curve is as shown (*CDE*). Naturally, *CDE* is the resultant of two curves as depicted in the earlier two diagrams (see above). The line *CD* corresponds to operation on the principle of forced circulation while *DE* corresponds to that on once-through operation.

Q. *Why is load-shedding restricted to no less than one-third of the rated load in the case of once-through boiler plants?*

Ans. Since in the case of such boilers, the mass flowrate of water and steam-water mixture is proportional to the boiler load, load-shedding of more than 30% of the rated load would reduce the mass flowrate to an inadmissibly low value causing tube burnouts.

Q. *On which factors do the temperature conditions in evaporating tubes, for a given heat release rate in a furnace, depend?*

Ans. On three parameters:
1. mass flowrate of working fluid
2. pressure of the flow
3. steam content of water-steam mixture.

Q. *How does the boiling crisis (i.e., burnout conditions) originate in the evaporating tubes with bubble flow of the steam-water mixture?*

Ans. So long as the temperature of the tubes is maintained at a substantially low level *AB*, the steam content in the steam-water mixture with bubble flow is moderate and a thick liquid film adheres to the tube wall. (Fig. 20.4)

As the working fluid flows and is heated up increasingly along the tube length, more and more liquid mass of the water film gets converted into vapour bubbles which stick to the inner wall of the tube. These vapour bubbles coalesce, grow in size and eventually form a continuous vapour film separating the flowing liquid from the heated tube wall. Presently, the main mass of the liquid flows through the central core while a film of vapour surrounds it. Since the heat

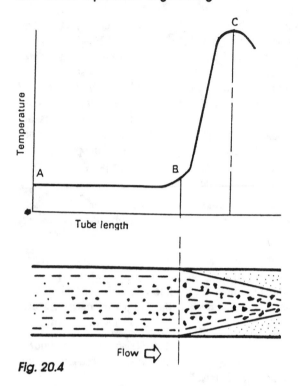

Fig. 20.4

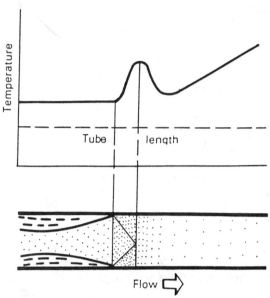

Fig. 20.5 *Boiling crisis occuring with steam—water mixture in disperse-annular flow*

transfer coefficient of water vapour is much less than liquid water, the prevailing condition sharply impairs the conditions of heat transfer. The effect of a sharp fall of heat transfer due to change from bubble boiling to film boiling is called boiling crisis.

Q. *What is critical load?*

Ans. It is the heating load at burnout conditions.

Q. *What happens to the tube walls under boiling crisis?*

Ans. Under these conditions, the layer of super-heated steam adhering the inner tube wall grows thicker and the overall heat transfer coefficient plummets sharply, with the effect that there is greater intensity of incident heat on tube wall surface than the rate of heat rmoval by the working fluid. This brings about a sharp rise in the wall temperature and often in the burnout of the tube wall.

Q. *Will the boiling crisis arise in the case of evaporating tubes with steam-water mixture in disperse-annular flow?*

Ans. Yes.

Q. *How?*

Ans. The disperse-annular flow is wet-wall flow in which a water film flows along the tube walls and hence wets it while steam saturated with water droplets moves in the core of the tube. (Fig. 20.5)

When heated, the water film gets thinner due to evaporation and breaks off, leaving a microfilm on the wall. This microfilm also gets destroyed under a particular heating load and disintegrated into individual islets and streaks whose number and size diminish in the course of further evaporation. As a result, a continuous layer of steam adheres to the wall while the diminutive number of water droplets carried down the mainstream of flow in the core gradually evaporate along the direction of flow. Since the vapourization is more intense at the walls, the water droplets are repulsed from the tube walls into the steam flow. Therefore, under these circumstances, heat transfer occurs through a continuous layer of steam instead of a continuous layer of water before heating. This sharply decreases the heat transfer coefficient and the wall temperature increases resulting in boiling crisis.

Q. *What is the important difference between the boiling crisis in bubble flow and that in disperse-annular flow?*

Ans. When a boiling crisis occurs, the temperature of the tube wall rises much more substantially in the case of bubble flow than in the case of disperse-annular flow.

Q. *Why do the burnout conditions in tubes heated unevenly occur at much higher values of critical load than in tubes heated uniformly from all sides?*

Ans. For unevenly heated tubes, the front side facing the furnace is most intensively heated while the rear side is heated least intensively. This results in transverse circulation of the working fluid from the less heated tube zone to the more heated tube zone, i.e., from the rear side of the tube wall to the front side. (Fig. 20.6) Due to this turbulence, some quantity of heat is transferred by conduction to the tube material in the direction opposite to the heat flow from the furnace. This brings about a cooling effect of the working fluid, thus retarding the boiling crisis.

Q. *Why in modern powerful steam boilers, do evaporating tubes have their inner surface screw threaded?*

Ans. This will enhance the turbulization of the working fluid and deter a boiling crisis, by introducing a cooling effect.

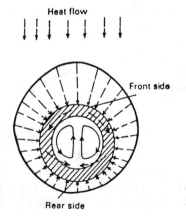

Fig. 20.6 Unevenly heated boiler tube sets up transverse circulation of the working fluid from the less heated zone to the more heated tube zone

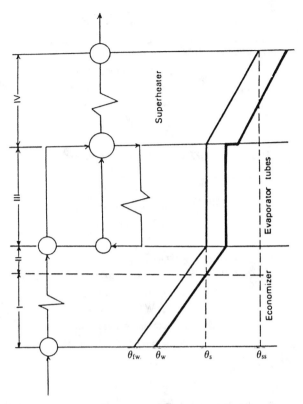

Fig. 20.7 Temperature variation in the steam—water path of a drum type subcritical pressure boiler

Q. *How do the temperature and mode of heat transfer vary in the steam-water path of a drum type subcritical pressure boiler?*

Ans. Temperature variations in the water-steam path of a drum-type subcritical pressure boiler can be best described by Figure 20.7.

The temperature of the feedwater (Θ_{fw}) is raised to the saturation point (Θ_s) in the economizer. And therefore, in the evaporating tubes the entire heat absorbed goes to evaporate the steam-water mixture at a fixed temperature (Θ_s); while in the superheater, the temperature of the steam is further raised to the value Θ_{ss}—the temperature of the superheated steam.

Q. *What is the mass steam content at the boundary between sections III and IV?*

Sec-tion	Equipment	Temperature Condition	Mode of Heat Transfer	Remarks
(I)	ECONOMIZER	Feedwater temparature (θ_{fw}) as well as wall temperature (θ_w) gradually increase but both remain below the saturation temperature $\theta_{fw} < \theta_w < \theta_s$	Convection. Heat transfer occurs due to temperature gradient $\Delta\theta$ $= \theta_w - \theta_{fw}$	Heating intensity is low (5–10 kW/m^2) but coefficient of heat transfer is high (2–5 kW/m^{2}K) and that's why the wall temperature is only a few degrees higher than the feedwater temperature.
(II)	ECONOMIZER	Feedwater temperature (θ_{fw}) gradually approaches saturation temperature (θ_s) while wall temperature (θ_w) gradually exceeds θ_s $\theta_{fw} \rightarrow \theta_s < \theta_w$	Unstable bubble boiling	Surface boiling has already set in on the tube walls. As the vapour bubbles are formed, they get detached from the tube walls, pass into the bulk of water flow and condense there, raising the temperature of the flow.
(III)	EVAPORATOR TUBES	The temperature of a steam-water mixture θ_{fw} becomes equal to boiling temperature θ_s. Wall temperature θ_w remains higher than θ_s throughout $\theta_{fw} = \theta_s < \theta_w$	Stable bubble boiling	Heat transfer coefficient is substantially large (50–100 kW/m^2 K) and therefore, the temperature difference between tube wall and steam-water flow is marginal. Heat transfer takes place independent of flow velocity but it is dependent on the heating intensity and pressure of the flow.
(IV)	SUPERHEATER	Steam is superheated to the rated superheat temperature, (θ_{ss}). Along with it increases the wall temperature.	Convective heat transfer through single phase medium, i.e., steam	Since steam has a lower coefficient of heat transfer than water film, heat transfer to superheated steam worsens and the temperature of the tube wall increases.

Ans. Unity, i.e., the entire water content in the steam-water mixture is evaporated to steam.

Q. *What is the temperature profile in the steam-water path of once-through subcritical pressure boilers?*

Ans. Analogous to drum type, the temperature of feedwater in a once-through subcritical-pressure boiler is raised to the saturation point θ_s in the economizer, with the effect that the entire heat absorbed by the evaporating tubes goes to evaporate the water and the temperature of the steam-water mixture remains nearly constant at θ_s throughout the evaporating tubes. Finally, the steam at θ_s is superheated to superheating temperature θ_{ss} in the superheater. (Fig. 20.8)

In the economizer section I, both the feedwater and wall temperature lie below the saturation temperature θ_s

i.e. $\theta_{fw} < \theta_s$; $\theta_w < \theta_s$

Heat transfer takes place by convection through water-film adhering to the tube wall. Because of a high transfer coefficient (5–10 kW/m^{2}K) the wall temperature is only marginally greater than temperature of the flow.

In section II, the temperature of the flow θ_{fw} is lower than θ_s but the wall temperature is higher than

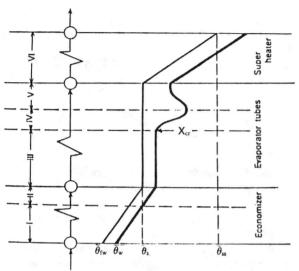

Fig. 20.8 *Temperature profile in the steam-water path of once-through subcritical pressure boiler*

θ_s all through. Heat transfer takes place by unsteady surface boiling whereupon vapour bubbles formed on the tube wall pass into the bulk of the water flow, condense and raise the flow temperature.

In the evaporator sections III, IV and V, the temperature of the water-steam mixture remains essentially constant at θ_s throughout, as the water mass is gradually converted into steam. Heat transfer takes place through stable bubble boiling.

But the steam content varies along the length of the evaporating tubes in the range $0 < x < 1$. And at steam content corresponding to x_{cr}, at the boundary between sections III and IV, boiling crisis occurs due to deterioration of heat exchange conditions as the water film dries up. This results in a sharp increase of wall temperature.

However, after attaining a maxima, the wall temperature of the evaporating tubes sharply decreases as the increase of steam velocity brings about an increase in the heat transfer coefficient, with the effect that more of the heat deposited to the tube wall is transferred to the steam. At the boundary between sections IV and V the entire liquid water is

converted into steam. And any heat added thereafter goes to superheat the steam.

Q. *What is the effect of flow pressure on boiling crisis?*

Ans. As the pressure of the flow is increased, it goes to reduce the surface tension, so boiling crisis will occur at a much lower value of x_{cr} (critical dryness fraction, i.e., mass steam content).

Q. *What is the effect of higher heating load on boiling crisis?*

Ans. Higher heating load means a quicker evaporation rate of water from the steam-water flow and that means boiling crisis will arise at a lower value of x_{cr}.

Q. *What is the principal effect of boiling crisis on the evaporating tube?*

Ans. Since heat transfer conditions are deteriorated in the zone of boiling crisis (which can move along the length of the evaporating tube under certain circumstances) the tube metal in this zone is subjected to abrupt temperature changes that will induce thermal fatigue in the affected region.

Q. *How can this be reduced?*

Ans. The high temperature gradient between the internal surface of the tube and the flow in the transition zone of boiling crisis can be reduced to within 80°C—a tolerable limit by maintaining a sufficiently high flow velocity.

Q. *What is the effect of tube bends of the evaporating surface on the temperature conditions of the working fluid in a steam generation plant?*

Ans. The tube bends induce a centrifugal force on the working fluid towards the outside of the bend. The fluid in the core of the flow being less intensively heated than that near the wall, it is much denser than the fluid adjacent to the wall. And therefore, the fluid in the core of the flow possesses higher kinetic energy than that near the tube wall. As a result, the fluid in the core of the flow is thrown, under the influence of centrifugal force, towards the outside of the bend where it displaces the fluid of lower density towards the centre of the tube. Thus two symmetrical vortices

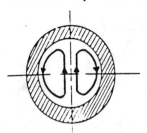

Fig. 20.9 *Two symmetrical vortices are developed in the working fluid at the tube bend*

are developed at the bend, Fig. 20.9 over and above the main axial flow.

The narrow bend left between these two opposing vortices on the inside wall cools less intensively than it would otherwise, resulting in deterioration of heat transfer in this local zone.

Whereas, on the outside of the bend, the heat transfer conditions are better because of higher heat removal by transverse motion of two opposing symmetrical vortices and the wall temperature remains lower.

Q. *What is the coefficient of heat spread?*

Ans. The coefficient of heat spread for a particular tube point subjected to uneven heating is defined as the ratio of the actual temperature to the temperature that would appear at that point should the heating be uniform with the maximum heat flowrate q_{max}.

Q. *Why is the range of deteriorated heat transfer (boiling crisis) larger in horizontal tubes (15–20mm dia) than in vertical tubes of the same dia though heating is uniform over the periphery?*

Ans. It is because the asymmetry of the steam-water flow, in the horizontal tubes, under the influence of gravitational forces worsens the internal heat transfer in the upper region more than in the lower region. Whereas, for a uniformly heated vertical tube this a symmetry in the flow of the steam-water mixture does not arise.

Q. *What problems arise in horizontal evaporating tubes having separated flow of the working fluid at subcritical pressure?*

Ans. Because of separated flow, heat transfer is unsymmetrical resulting in higher temperature on the upper surface and lower temperature on the lower surface of the horizontal tubes. The separation of flow

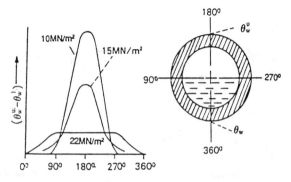

Fig. 20.10 *Unsymmetrical heat transfer results from separated flow*

between two phases—water and steam—is effected by gravitational forces. (Fig. 20.10)

Waves may form on the separating surface with their tops periodically splashing onto the overheated wall of the upper surface. Repeated water splashing cause a sharp cooling of the tube wall which will be subjected to thermal shock due to periodic heating and cooling. This can result in fatigue in the tube metal.

Q. *Why are horizontal tubes in free- circulation boilers not heated?*

Ans. Free-circulation boilers are characterised by low flow velocities in evaporating tubes which, if laid horizontal, will give rise to metal fatigue in the tube due to periodic heating and cooling resulting from separation of flow, as stated above.

Q. *Why does this problem not arise in the case of once-through boilers over a wide range of heating loads?*

Ans. In this case, flow separation does not arise because of high flow velocities.

Q. *What will be the effect on horizontal evaporating tube heating, for separated flow of the working fluid, if the tube wall thickness is increased or the tube is replaced by a similar tube of metal with higher thermal conductivity?*

Ans. In either case, the heat deposited on the tube surface will be spread through the tube metal and thereby the elevation of temperature of the upper portion of the tube with respect to its lower portion will get somewhat diminished, reducing thermal fatigue arising from flow separation.

Q. *What is the effect of deposits on the temperature conditions on the heating surface?*

Ans. Impurities present in the BFW in a dissolved or suspended state will form deposits on the inner wall of the evaporating tubes. And these grow in thickness with the operation of the boiler. Since the deposits have thermal conductivities of their own, the inner surface temperature of the tube will be

$$\theta_w^{in} = \theta_f + q_{in}\left(\frac{1}{\alpha} + \frac{\Delta x}{\lambda}\right)$$

where,

θ_f = flow temperature

q_{in} = heating intensity on the inside surface

α = coefficient of heat transfer from the inner wall to the working fluid.

Δx = thickness of deposit

λ = thermal conductivity of the deposit

For the clean tube, the above equation reduces to

$$\theta_w^{in} = \theta_f + \frac{q_{in}}{\alpha}$$

Depending on chemical composition, the thermal conductivity of the deposits may vary:

from 0.1 to 0.5 W/m.°K for mineral deposits
from 3 to 5W/m.°K for ferric oxides

As such, with larger thickness (Δx) of the deposited layer and with a lower thermal conductivity (λ), the thermal resistance across the wall to the working fluid is higher and that results in higher wall temperature (θ_w^{in}) — even a layer of deposit of a few fractions of a millmetre is enough to raise the wall temperature to a value inadmissible for reliable operation of the tube metal. With the increase of running hours of boilers, the deposits grow thicker and thermal resistance of the deposited tube well goes higher, with the effect that tubes of the waterwall get intensively heated raising the wall surface temperature to a level that delimits the continuous operation of the steam generation plant.

21

Hydrodynamics of
Closed Hydraulic System

Q. *What feature ensures reliable and lasting operation of the free-circulation circuit of boilers?*

Ans. For boilers with free-circulation circuit, the vertical evaporating tubes (i.e., uptake tubes) are arranged inside the boiler furnace and downtake tubes outside it resulting in a density difference between the water columns in the downtake tubes and the columns of steam-water mixture in the uptake tubes. (Fig. 21.1)

Because of this density difference, a continuous circulation of water and steam-water mixture is developed in the circuit and that ensures a continuous and efficient heat removal from the heating surfaces. Therefore, the tube metal is maintained within a

Steam & water
drum

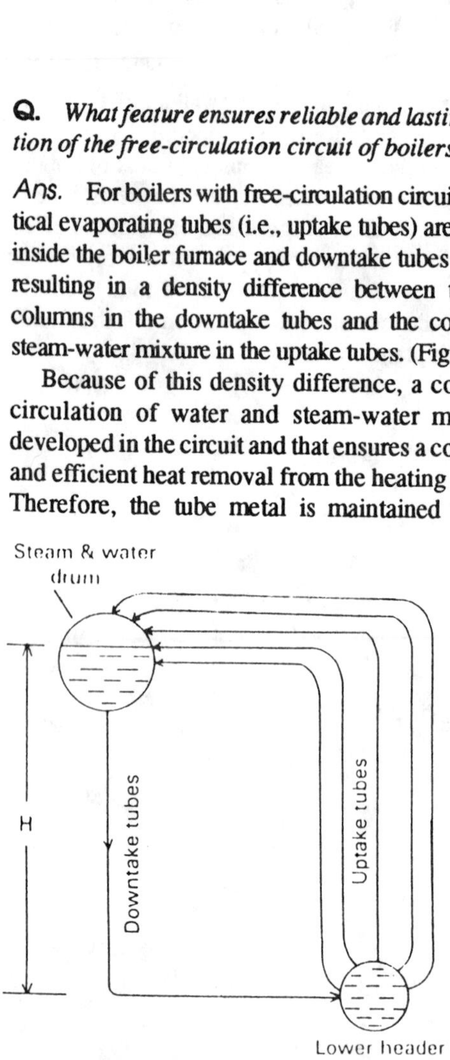

Fig. 21.1 *Free-circulation circuit of boilers puts uptake tubes inside the furnace and downtake tubes outside it.*

tolerable temperature level ensuring prolong and reliable operation of the free-circulation circuit.

Q. *What is the driving head of the free- circulation circuit?*

Ans. It is given by

$$S_{drv} = H\,(\rho_{H_2O} - \rho_{mix}\,)\,g = \Sigma\,\Delta\,P_{Hydr.}$$

where H = height of water level in the boiler drum from the base of the evaporating tubes

ρ_{H_2O} = density of water

ρ_{mix} = density of steam-water mixture

$\Sigma\,\Delta\,P_{Hydr}$ = sum of all hydraulic resistances in the circuit

$$= \Delta\,P_{up} + \Delta\,P_{dwn}$$

$\Delta\,P_{up}$ = resistance in the uptake tubes

$\Delta\,P_{dwn}$ = resistance in the downtake tubes

Q. *What is the useful circulation head?*

Ans. It is the difference between the driving circulating head and resistance of the uptake tubes

$$S_{usf} = S_{drv} - \Delta\,P_{up} = \Delta\,P_{dwn}$$

i.e., useful circulating head is expended to overcome the resistance of the downtake tubes.

Q. *What is the effect of circulation velocity on useful head?*

Ans. This can be determined with the help of S_{usf} vs. v_{cir} curves obtained at various values of resolved steam velocity $v_{s^{res}}$ (Fig. 21.2)

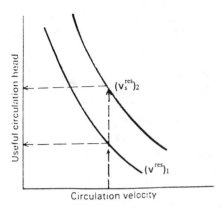

Fig. 21.2

$$v_s^{res} = G_s/(A\,\rho_s)$$

where

G_s = mass flowrate of steam, kg/s

A = area of cross-section of steam generating tube, m²

ρ_s = steam density, kg/m³

From the curves it is evident that for the same circulation velocity, useful circulation head increases with the increase of resolved steam velocity, i.e., with the increase of heating rate.

Resolved Steam Velocity: is the velocity that steam would have in a particular cross-section of a steam-generating tube if it is completely occupied by steam.

Q. *What is the effect of pressure on the useful circulating head?*

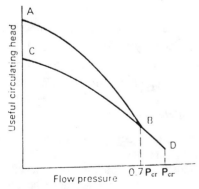

Fig. 21.3 *Effect of flow pressures on the useful circulating head*

Ans. This can be determined by plotting the useful circulating head against different flow pressures under identical conditions. (Fig. 21.3)

From the curves it is evident that the useful circulating head decreases with the increase of working pressure. It is because as the pressure is increased, the density of the steam-water mixture in the uptake tubes increases and as a result, ΔP_{up} increases and S_{usf} decreases.

$$S_{usf} = S_{dry} - \Delta P_{up}$$

(*AB* has been drawn by ignoring the relative velocity of steam and water while *CD* has been drawn by taking this relative velocity into account.)

As the pressure approaches the critical value, S_{usf} gradually reduces to a low value whereupon free circulation becomes practically non-existent.

Q. *What is the ultimate pressure at which reliable free-circulation is still possible?*

Ans. 18–19 MPa (18–19 MN/m²)

Q. *Why on plotting S_{usf} vs. P_{flow}, are two curves are obtained — one when the relative velocity of steam and water is considered and the other when the relative velocity of steam and water is ignored?*

Ans. The relative velocity of steam with respect to water is functionally dependent on the flow pressure which in turn determines the value of the useful circulating head. When the pressure is low, the steam formation is large and so the relative velocity of steam is rather high. Therefore, for a particular pressure (below critical pressure), there will be two useful circulating heads — one due to consideration of the relative velocity of steam and the other due to ignoring it.

Q. *How does the density of steam-water mixture vary with the circulation velocity in the uptake and downtake tubes?*

Ans. For uptake tubes, the density of steam-water mixture will increase with the increase of the circulation velocity while for downtake tubes it is just the reverse. (Fig. 21.4)

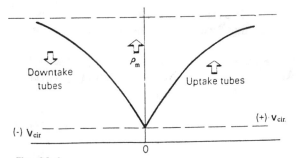

Fig. 21.4

Q. *What is the effect of steam-slippage and friction upon the useful circulating head in both uptake and downtake tubes?*

Ans. In uptake tubes, the useful circulating head decreases due to steam-slippage and friction with the increase of circulating velocity. (Fig. 21.5)

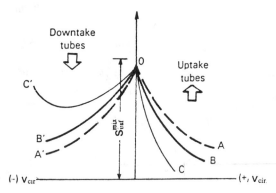

Fig. 21.5

Whereas in the case of downtake tubes these two factors tend to increase the useful circulating head with the increase of circulation velocity.

OA and *OA'* correspond to theoretical S_{usf} vs · v_{cir} curves when friction factor and steam slippage are ignored.

OB and *OB'* correspond to S_{usf} vs · v_{cir} curves when steam slippage has been considered.

OC and *OC'* correspond to S_{usf} vs · v_{cir} when both steam slippage and friction factors are taken into account.

Q. *Why does the useful circulating head decrease with the increase of circulation velocity in uptake tubes of closed hydraulic systems?*

Ans. For upward motion of the steam- water mixture through the evaporating tubes, the increase of circulation velocity increases the relative velocity between steam and water which in turn decreases the actual steam content but increases the density of the mixture resulting, in a decline of S_{usf}.

$$cf.\ S_{usf} = S_{drv} - \Delta P_{up} = H\,(\rho_{H_2O} - \rho_{mix}) - \Delta P_{up}$$

Q. *Why does the useful circulating head increase with the increase of circulation velocity in the downtake tube of closed hydraulic system?*

Ans. For downward motion, the increase of the relative velocity of steam and water (with the increase of circulation velocity), increases the actual steam content and decreases the density of the steam-water mixture, resulting in an increase of S_{usf}.

Q. *What is called the make-up velocity?*

Ans. It corresponds to the velocity of flow of make-up water supplied to the evaporating tubes such that the rate of steam generated is compensated by the rate of water supply when there is no circulation through the evaporating tubes.

Q. *What does the term make-up velocity* ($v_{m'u}$) *imply?*

Ans. It implies two things?
1. the quantity of water supplied to the evaporating tubes is low
2. the entry of water to the evaporating tubes is possible either from the bottom or from the top.

Q. *What is the effect of pressure on the total hydraulic characteristic of a steam-generating tube both at lower and higher pressure?*

Ans. This can be better understood graphically.

The curve to the right of the ordinate represents the hydraulic characteristic corresponding to the ascending motion of the working fluid while that to the left to the descending motion. (Fig. 21.6)

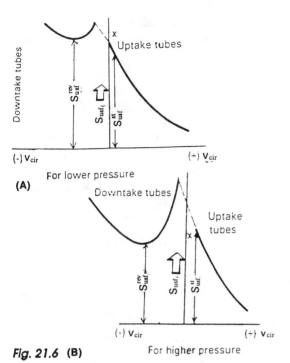

For lower pressure

(A)

Fig. 21.6 **(B)**   For higher pressure

At point X on the right-hand branch, the circulation velocity is practically nil and circulation virtually ceases. The velocity of feedwater at this point is called make-up velocity.

Q. *Why does the hydraulic characteristic corresponding to descending motion lie always higher than that corresponding to ascending motion for lower pressure evaporating tubes, whereas in the case of higher pressure systems a considerable portion of it lies lower than the portion of the hydraulic characteristic corresponding to ascending motion?*

Ans. It is because, at lower pressure, the effect of relative steam velocity is high and therefore S_{usf} decreases in ascending tubes and increases in the descending tubes with the increase of circulation velocity. Therefore, the hydraulic characteristic of descending tubes lies higher than that of ascending tubes.

Whereas at higher pressure, the steam velocity and hence hydraulic resistance are low and therefore a sizeable portion of the hydraulic characteristic of the

descending tube lies below the portion of the hydraulic characteristic of ascending tubes.

Q. *What is known as circulation stagnation?*

Ans. It marks the phenomenon of zero circulation, i.e., the point at which circulation virtually ceases in a natural circulation system. This point is reached when the circulation velocity equals the make-up velocity.

Q. *What is the useful stagnation head?*

Ans. It is the useful head corresponding to zero circulation. (The ordinate of the point X on the hydraulic characteristic of ascending tubes.)

Q. *In which of the following cases will the circulation stagnation arise—evaporating tubes connected into the water space of the steam-drum or evaporating tubes connected into the steam space of the steam-drum?*

Ans. In both cases circulation stagnation may arise.

Q. *Why?*

Ans. It is because, in the first case (when the evaporating tubes are connected to the water space) with circulation stagnation, the water is at liberty to enter the evaporating tubes from the bottom as well as from the top and it moves very slowly upwards or downwards in the evaporating tubes while steam moves only upwards bubbling through the column of water in the ascending tubes. (Fig. 21.7)

Whereas in the latter case, the very slowly moving useful head of water is insufficient to overcome the hydraulic resistance in the descending tubes and to lift

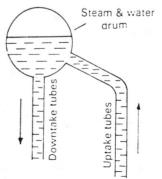

Fig. 21.7

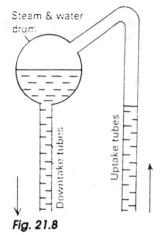

Steam & water drum

Downtake tubes

Uptake tubes

Fig. 21.8

the working fluid to the topmost level in the ascending tubes of the circuit. (Fig 21.8)

Q. *What is circulation reversal?*

Ans. In the hydraulic characteristic of descending tubes (for both lower pressure and higher pressure system), there is a dip, i.e., a minima beyond which the useful circulation head increases with the decrease in the circulation velocity, i.e., ascending motion occurs in the downtake tubes. That is, it is at this point that the change from ascending to descending motion occurs in the uptake tubes. This transition is called circulation reversal.

Q. *What do you mean by useful head of reversal, S_{usf}^{rev}?*

Ans. It is the useful circulating head at which the change from ascending to descending motion takes place in an evaporating tube.

Q. *What is the effect of clinkering on the hydraulic characteristics of steam-generating tubes?*

Ans. Since clinker has a low thermal conductivity, so a clinkered tube will receive substantially less heat than an unclinkered (i.e., clean) tube. And therefore, differently heated tubes will produce different quantities of steam and therefore will develop different driving circulating heads— clean tubes will develop the highest driving head. Differently slagged tubes will develop different driving heads, always lower than that in the clean tubes.

Now that all the ascending tubes are connected to common headers and operate under the same useful circulating head, S_{usf}

$$S_{usf} = \Delta P_{dwn}$$

the useful circulating head will be the same for all the tubes in the circuit. However due to non-uniform heating this useful head, common for all tubes, will correspond to different flowrates of water circulating in these tubes with the effect that little water will pass through some of the tubes. As a result, such circulation disturbances as circulation starvation or even circulation reversal may occur.

Q. *What is circulation reliability?*

Ans. Absence of circulation disturbances.

Q. *How can it be ensured?*

Ans. By taking recourse to reliability criteria as cited below:

1. Criterion of Stagnation

$$\left[S_{usf}^{st} / S_{usf}^{cir} \right] > 1.1$$

2. Criterion of Circulation Reversal

$$\left[S_{usf}^{rev} / S_{usf}^{cir} \right] > 1.1$$

3. Criterion of Free Water Level (Fig. 21.9)

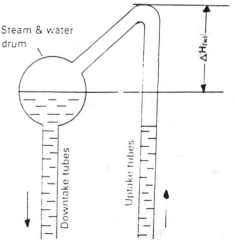

Steam & water drum

Downtake tubes

Uptake tubes

ΔH_{fwl}

Fig. 21.9

$$\left[\left(S_{usf}^{st} - \Delta P_{fwl}\right) / S_{usf}^{cir}\right] > 1.1$$

where, ΔP_{fwl} = pressure loss in raising the steam-water mixture above the free water level of the boiler drum.

Q. *How may steam bubbles form in the downtake tubes?*

Ans. If the pressure drop at the inlet of the downtake tubes becomes greater than the water level in the boiler drum, the pressure of water at the inlet to the downtake tubes will drop to such a low level that the water will boil in the zone of reduced pressure. And thus steam bubbles will form and be dragged down by the circulating water into the downtake tubes.

Q. *What is the effect of this?*

Ans. Since steam bubbles take some time to get condensed, these bubbles are entrained by the water-flow over an appreciable length of the downtake tubes. Therefore, the downtake tubes get filled with a steam-water mixture instead of water alone. Since steam offers higher hydraulic resistance to flow than water, downtake tubes filled with substantial quantity of steam bubbles will suffer from loss of driving circulating head.

Q. *What are the main causes for the appearance of steam in downtake tubes?*

Ans.
1. Sudden pressure drop in the boiler
2. Entrainment of steam bubbles from the steam space of the drum
3. Entrainment of water bubbles from the water space of the drum

Q. *How does the entrainment of steam bubbles from the steam space of the drum to downtake tubes takes place?*

Ans. As the water enters the downtake tubes, its hydrostatic pressure head decreases substantially due to appearance of velocity head as well as due to unavoidable energy losses on local resistances.

If the water pressure at the entry of downtake tubes is less than its pressure at the water-level in the drum, steam will form in the zone of reduced pressure at the entry of the downtake tubes. Under these circumstan-

ces, steam cones may form at the disengagement surface (i.e., water-steam separating surface) so that a large amount of steam will ingress from the steam space into the system of downtake tubes giving rise to disturbances in the water circulation of the circuit.

Q. *Does the formation of steam cone take place in the case of boilers with free circulation only?*

Ans. No. It may occur in boilers with both free and forced circulation.

Q. *How can this be prevented?*

Ans. By maintaining the water level in the drum at a safe height above the inlets to the downtake tubes, so that the hydrostatic pressure head in the drum is always greater than the pressure head at the inlet to the downtake tubes.

Q. *Does the diameter of the downtake tubes have any function regarding the formation of steam cone?*

Ans. Yes; the probability of formation of steam cone increases with the increase of the diameter of the downtake tubes.

Q. *How does the entrainment of steam from the water-space of the drum into the downtake tubes take place?*

Ans. This phenomenon can occur if the circulation velocity of water in the circuit is substantially high, whereupon steam bubbles formed below the disengagement surface will not get any chance to separate and are thus entrained by the downward flow of water into the downtake tubes.

Q. *When may such a case arise?*

Ans. If the outlets of the evaporating tubes in the drum are placed close to the inlet of the downtake tubes.

Q. *How can the entrainment of steam bubbles from the water-space of the boiler drum to downtake tubes be prevented?*

Ans. This can be reduced by:
(a) proper arrangement of partitions
(b) separating the steam-water mixture in cyclones.

22

Deaeration and Deoxygenation

Q. *What is deaeration?*

Ans. Deaeration is the process of removing dissolved corrosive gases (O_2 and CO_2) from water.
This process is also called degasification.

Q. *Why are these gases corrosive?*

Ans. Dissolved oxygen acts as a depolarizer and contributes to the corrosion of metal

$$O_2 + 4e + 2H_2O \longrightarrow 4OH^-$$

Carbon dioxide dissolved in water liberates H^+ ions that attack metal

$$CO_2 + H_2O \Leftrightarrow H_2CO_3 \Leftrightarrow H^+ + HCO_3^-$$
$$\text{CARBONIC ACID}$$

Q. *What is deoxygenation?*

Ans. It is the process of removal of only oxygen from water.

Q. *How can deaeration be distinguished from deoxygenation?*

Ans. Deaeration is the removal of both O_2 and CO_2 from water while deoxygenation involves the removal of O_2 only.
Deaeration is a thermal method. Water is heated to bring about degasification, whereas deoxygenation is carried out mainly by chemical techniques.

Q. *What is the basis of thermal deaeration?*

Ans. Upon heating the water, the solubility of gases dissolved in it decreases because their partial pressures in the gas phase above the water level drop off.*

So when water at a given pressure is heated to boiling point degasification is effected:

$$x_{H_2O} = n_{H_2O}/\Sigma n \;;\; x_{O_2} = n_{O_2}/\Sigma n \;;\; x_{CO_2} = n_{CO_2}/\Sigma n$$

Now

$$x_{H_2O} + x_{O_2} + x_{CO_2} = 1$$

where x stands for the mole fraction of a component.
As the mole fraction of O_2 and CO_2 decreases in the vapour phase so does their partial pressure

$$p_{O_2} = x_{O_2} \cdot (\text{Total pressure})$$

$$p_{C\bar{O}_2} = x_{CO_2} \cdot (\text{Total pressure})$$

(Since the solubility of gases in a liquid is proportional to their partial pressure above the liquid phase, the decrease in the partial pressures of oxygen and carbon dioxide as a result of heating of water lowers the solubility of O_2 and CO_2 in water.)

Q. *What is the effect of BFW pressure on deaeration?*

Ans. No effect, in practice.

Q. *Can deaeration be effected at a pressure higher or lower than the atmospheric pressure?*

---

* (As the water is heated, more and more water molecules transfer from the liquid phase to vapour phase. This brings about an increase in the mole fraction of water in the vapour phase, consequently reducing the mole fraction of O_2 and CO_2 in the vapour phase.

Ans. Yes; this can be done as the feedwater pressure exerts practically no effect on deaeration.

Q. *What is a deaerator?*

Ans. It is the apparatus wherein deaeration is carried out.

Q. *How are deaerators classified?*

Ans. Deaerators are classified according to the working pressure under which they operate:

Type Of Deaerator	Working Pressure
1. **Vacuum Deaerator**	$0.116 \, MN/m^2$
2. **Atmospheric Deaerator**	$0.12 - 0.17 \, MN/m^2$
3. **High Pressure Deaerator**	$0.17 - 0.7 \, MN/m^2$

Atmospheric deaerators can be subdivided into direct-contact (mixing) deaerators and overheated-water deaerators.

Also deaerators can be classified in accordance with the mode of steam-water distribution in them:

1. atomizing
2. tray-type
3. film-type

Q. *How do direct-contact deaerators function?*

Ans. They consist of a deaerator column fitted on the top of the storage tank.

The deaerator column is a hollow cyclindrical vessel provided with

(a) perforated, horizontal trays arranged one above the other.
(b) water distribution device at the top.
(c) Steam distributor at the bottom. (Fig. 22.1)

Feedwater enters the deaerator from the top and is evenly distributed, by means of the distributing device, into the perforated trays fitted at the bottom. Water fills the perforations (dia 5–7 mm) and rains

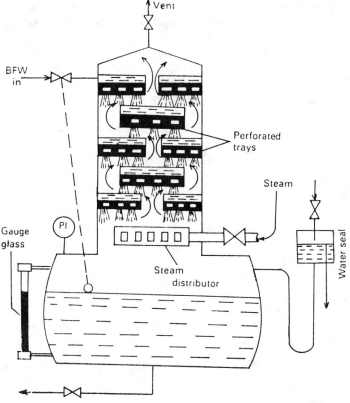

Fig. 22.1   A direct-contact deaerator

down and comes in contact with heating steam delivered into the lower portion of the deaerator column through the stream distributor. As a result of heat exchange between the steam going up and the feedwater stream flowing down, the water gets heated up to its boiling point and the gases (O_2, CO_2) dissolved in it are transferred to the gas phase. These gases together with non-condensing vapour are vented into the atmosphere through a vent valve or pass into a vent condenser where the steam is condensed and the O_2 and CO_2 are vented.

The deaerated water is collected in the storage tank which is fitted with gauge glass, pressure gauge and hydraulic seal to avoid the formation of high pressure or vacuum in the deaerator.

Q. *Does the storage tank of a deaerator function as a storage for water only?*

Ans. No. Here the remaining dissolved gases are discharged from water. Over and above, desorption takes place here.

Q. *Why does the desorption of dissolved gases from water take place in the storage tank, since degasification takes place in the deaerator column?*

Ans. It is due to the fact that some individual streams of water may bypass or not get fully degasified in the deaerator column, thereby carrying some dissolved oxygen and carbon dioxide into the storage tank. Therefore, removal of these dissolved gases takes place in the storage tank that affords enough space above the water level for desorption to take place.

Q. *What is steam-jet deaeration?*

Ans. It is a modern process that injects heating steam into the body of water in the storage tank.

Q. *Why is steam-jet deaeration more effective than direct-contact deaeration?*

Ans. When jets of steam at increased pressure are introduced in the body of feedwater, it agitates the water more intensively over and above heating it. The rate of heat transfer is much higher and it keeps the water in a state of boiling. These ensure more complete removal of dissolved gases than direct-contact deaeration which may suffer from partial degasification in some individual streams of water due to lack of adequate contact with the heating steam.

Q. *What is the demerit of steam-jet deaeration?*

Ans. It utilizes steam of increased pressure.

Q. *What conditions must be fulfilled for successful operation of the deaerator?*

Ans.
1. The temperature of deaerated water must be equal to the saturation temperature of water corresponding to the pressure at which the deaerator operates.
2. Sufficient heating steam must be delivered to the deaerator to ensure continuous boiling of water undergoing deaeration.
3. The feedwater charged to the deaerator must be disintegrated into fine droplets to ensure better heat transfer between the heating steam and feedwater. Hence perforated trays find widest application.
4. Deaerator must be provided with sufficient venting to purge all the non-condensing gases out of the system and to ensure minimum partial pressures of these gases in the upper part of the deaerator. Reduction of partial pressure will enhance desorption of dissolved gases.
5. If two or more deaerators are operated in parallel, they must all be operated at same pressure.

Q. *Why is it necessary to maintain the deaerated temperature equal to the saturation temperature of water corresponding to the pressure in the deaerator?*

Ans. It'll keep the deaerated water boiling and that will ensure the removal of leftover dissolved gases from the water in the storage tank.

Q. *What will happen if the heating steam supplied to deaerator is insufficient?*

Ans. Insufficient steam supply will cause a drop in the deaerator pressure and will fail to provide enough heat to maintain the water in continuous boiling condition—a must for deaeration. As a result, there will be only partial deaeration.

Q. *If the venting of non-condensing gases from the deaerator is not sufficient what harm will it cause?*

Ans. If the non-condensing gases (O_2, CO_2) collected at the top of the deaerator column are not adequately vented, it'll increase the partial pressures of oxygen and carbon dioxide (p_{O_2}, p_{CO_2}). Increase of partial pressures of O_2 and CO_2 will set up the reverse process of degasification, i.e., cause transfer of O_2 and CO_2 from the gas phase to liquid (water) phase and that'll impair deaeration.

Q. *Why are several deaerators working in parallel maintained at same operating pressure?*

Ans. To avoid water flow from one storage tank to the other.

Q. *What are the limits for an atmospheric-type deaerator to operate without steam-jet deaeration?*

Ans.
1. The carbon dioxide content in heating steam should not exceed 3–5 mg/lt.
2. The bicarbonate alkalinity in deaerated water should not be less than 0.3 mg-equiv./lt.
3. Residence time of deaerated water in storage tank should not be more than 15 minutes.

Q. *What are the causes that contribute to increased concentration of oxygen and carbon dioxide in deaerated water in the storage tank?*

Ans.
1. Inadequate deaerator vent leading to accumulation of non-condensing gases and increase in their partial pressures
2. High feedwater flowrate inadmissible at the given water temperature causing hydraulic impacts
3. Delivery of O_2-containing condensate directly into the storage tank and to the deaerator
4. Frequent pressure drop in the deaerator
5. Faulty deaerator internals
6. Delivery of overheated condensate on to the second tray
7. Delivery of relatively "cold" flows with higher O_2 - content to 2nd/3rd tray of deaerator.

Q. *What is vacuum deaeration?*

Ans. Deaeration is carried out in vacuum (rarefaction), i.e., at a pressure less than atmospheric pressure.

Q. *In which cases is it carried out?*

Ans. Low-capacity, low-pressure boilers, requiring water at a temperature less than 373 K.

Q. *Why is vacuum deaeration needed for water deaeration at a temperature less than 373 K?*

Ans. Since the saturation temperature of water at 1 atm. pressure is 373K, water at a temperature less than 373K will boil if the pressure is less than atmospheric. Hence vacuum is needed to effect deaeration of water at temperature less than 373 K.

Q. *Does the amount of vacuum depend on the temperature of the water deaerated?*

Ans. Yes. The lower the temperature, the higher must be the vacuum while higher the temperature, the lower the vacuum.

Q. *What methods are practised to carry out chemical deoxygenation of water?*

Ans. Two methods:
1. Sulphite treatment of water
2. Hydrazine treatment of water

Q. *What is the basis of chemical deoxygenation of water?*

Ans. This is based on the principle that there are certain water soluble substances that are capable of binding water-dissolved oxygen by virtue of a chemical reaction with it in aqueous solution.

Q. *What are these substances that are used for chemical deoxygenation of water in SGP?*

Ans. Sodium sulphite (Na_2SO_3) and hydrazine (N_2H_4)

Q. *What is the basis of sodium sulphite treatment?*

Ans. It is based on the reducing property of sodium sulphite as a result of which it itself gets oxidized to sodium sulphate by the following chemical reaction with oxygen dissolved in water:

$$2Na_2SO_3 + O_2 \longrightarrow 2Na_2SO_4$$

Q. *What is basically required from this reaction when it is applied in practice?*

Ans. The fundamental requirement is that this reaction must go to completion in the time interval during which water flows from the sulphite dosing point to the inlet of the economizer of the boiler unit.

Q. *What is done for this purpose?*

Ans. Water temperature is maintained at 353 K and an excess sulphite concentration (at 2mg/1) is ensured.

Q. *How is sodium sulphite dosed?*

Ans. It is added in the form of aqueous solution (3 to 6%).

Q. *How is the quantity of sodium sulphite to added determined?*

Ans. It is determined with the help of the formula:

$$C_{Na_2 SO_3} = 8\, C_{O_2} + 2,\ mg/kg$$

where $C_{Na_2 SO_3}$ = concentration of sodium sulphite solution, mg/kg

C_{O_2} = concentration of dissolved oxygen, mg/kg

Q. *When sodium sulphite solution is prepared and stored, what important precaution should be taken?*

Ans. Care should be taken to seal the Na_2SO_3— solution from contact with atmospheric air.

Q. *Why?*

Ans. Otherwise, the sodium sulphite solution will absorb atmospheric oxygen and get oxidized to inert sodium sulphate, thereby decreasing the deoxygenation capacity of the sulphite solution.

Q. *Why is sodium sulphite unsuitable for deoxygenation treatment in high-pressure boiler units?*

Ans. At high-pressure and high temperature, sodium sulphite becomes unstable and undergoes decomposition rendering sulphite treatment for deoxygenation meaningless:

$$Na_2SO_3 + H_2O \longrightarrow 2NaOH + SO_2$$

Q. *What is the basis of chemical deoxygenation using hydrazine?*

Ans. Hydrazine is a strong reducing agent and in aqueous solution (it is highly soluble in water) it reacts with dissolved oxygen

$$N_2H_4 + O_2 \longrightarrow N_2 + 2H_2O$$

and thus lends a convenient system to remove oxygen dissolved in water.

Q. *How is it commercially available?*

Ans. It is available as:

Hydrazine hydrate: $N_2H_4 \cdot H_2O$
Hydrazine hydrochloride: $N_2H_4 \cdot HCl$
Hydrazine sulphate: $N_2H_4 \cdot H_2SO_4$

Q. *What is the difference between hydrazine hydrate and hydrazine sulphate?*

Ans. Hydrazine hydrate is a colourless liquid showing weak alkalinity in aqueous solution. It readily absorbs O_2, CO_2 and moisture from air.

Hydrazine sulphate is a solid, poorly soluble in cold water. Its aqueous solution exhibits acidic properties.

Q. *What are the principal factors that affect the deoxygenation (reducing) capacity of hydrazine?*

Ans.

1. TEMPERATURE—hydrazine will remove dissolved oxygen at an appreciable rate only at water temperature $\geq 373K$
2. pH VALUE—at pH value < 7, hydrazine fails to prevent corrosion and even intensifies it.

However, at pH 9—9.5, temperature 373—380 K, hydrazine in excess of 20 microgram/kg eliminates all dissolved oxygen in a matter of just 2 to 3 seconds.

Q. *Hydrazine hydrate but not hydrazine sulphate can be used as a suitable deoxygenation reagent in once-through boiler units. Why?*

Ans. It is because hydrazine hydrate leaves no solid residue on decomposition

$$N_2H_4 + O_2 \longrightarrow N_2 + 2H_2O$$

Therefore, the total dissolved solid in feedwater is not increased.

Moreover, the decomposition products (N_2, H_2O) being harmless, no corrosion takes place to add the products of corrosion to the boiler feedwater. Since, in once-through boiler units, water is continuously evaporated, lower the salt content of feedwater, lower is the risk of deposit formation. That's why hydrazine hydrate is most suitable in these type of boiler units.

On the other hand, hydrazine sulphate adds sulphate radical to the feedwater. Moreover, it is acidic in nature. Hence it may contribute to the formation of ferrous sulphate that will fall out on the heating surfaces in once-through boiler units impairing the overall heat transfer rate.

Q. *For which type of boiler units, is hydrazine sulphate suitable?*

Ans. Only fit for drum-type boiler units.

Q. *Why?*

Ans. The drum-type boiler units are fitted with blowdown facility and that'll reduce the TDS of boiler feedwater within tolerable limits if its value is increased by ferrous sulphate formed when hydrazine sulphate is used as the deoxygenation reagent.

Q. *Does the hydrazine only eliminate dissolved oxygen when treated with boiler feedwater?*

Ans. Apart from binding dissolved free oxygen in water, hydrazine also reduces oxides of iron and copper formed due to corrosion of metal:

$$6Fe_2O_3 + N_2H_4 \longrightarrow 4Fe_3O_4 + N_2 + 2H_2O$$

$$2CuO + N_2H_4 \longrightarrow 2Cu + N_2 + 2H_2O$$

It also reduces ferric hydroxide to ferrous hydroxide

$$4Fe\,(OH)_3 + N_2H_4 \longrightarrow 4Fe\,(OH)_2 + N_2 + 4H_2O$$

and curbs the nitrite corrosion in high-pressure boiler units.

23

Water Treatment and Demineralization

Q. *What is meant by clarification of water?*

Ans. It means the removal of coarse-dispersed suspended solids as well as colloidal impurities from water.

Q. *How is it achieved?*

Ans. Coarse-dispersed suspended solids are separated from water by a sedimentation and filtration processes.

Q. *How is water clarified by sedimentation?*

Ans. It is a process carried out in large settling tanks wherein a large bulk of water is kept stationary to allow coarse-dispersed suspended solids to settle by gravity on the bottom of the settling tank and the sediment is removed periodically by the slow motion of horizontal scrappers.

Q. *What are the disadvantages of sedimentation?*

Ans.
1. It is a very slow process taking a long time to remove appreciable quantities of suspended solids
2. It never ensures complete elimination of suspended impurities from water
3. It requires large-capacity settling tanks.

Q. *What is filtration?*

Ans. It is the process of clarification of water by passing it through a filter bed composed of a porous material that retains coarse suspended solids on its surface and in the pores.

Q. *What is the driving force of filtration?*

Ans. It is the pressure drop through the filter bed.

Q. *On which factors does the pressure drop through the filtering bed depend?*

Ans.
1. Height of filtering bed
2. Rate of filtration
3. Particle size (i.e., grain diameter) of the filtering material
4. Degree of contamination of the filter bed by impurities trapped

Q. *What is meant by coagulation of water?*

Ans. It is the process of removal of colloidal impurities from water by conglomeration of small colloidal particles into bigger particles having enough mass to settle by the action of gravity.

Q. *What is the process opposite to coagulation?*

Ans. It is called peptization, i.e., transition of a solid into a colloidal-state.

Q. *What is the mechanism of coagulation?*

Ans. The colloidal particles owe their existence to the electric charges developed on the surfaces by the metal cations from which they form, or by the absorption of cations or anions from the solution (in the case of organic sols). Since all the colloidal particles of a given colloidal solution carry like charges, they repel each other and as a result they cannot spontaneously conglomerate into larger particles.

Coagulation is effected by dosing of stimulants that neutralize the charges of the colloid particles which thereby coagulate on collision with similar other neutralized particles.

On standing, the coagulated particles grow in size and become too large to outweigh gravity by the buoyant force exerted upon them by water. And they settle on the bottom as gelatinous and porous flakes.

Q. *What are these stimulants called?*

Ans. Coagulants.

Q. *What chemicals are used as coagulants in water-treatment plants?*

1. Alum, K_2SO_4, $Al_2 (SO_4)_3 \cdot 24H_2O$
2. Aluminium sulphate, $Al_2 (SO_4)_3 \cdot 18H_2O$
3. Ferrous sulphate, $FeSO_4 \cdot 7H_2O$
4. Ferric chloride, $FeCl_3$

Q. *In which state are they introduced into water?*

Ans. They are dosed in the form of a solution of 5–10% concentration.

Q. *Why is it necessary to maintain some degree of alkalinity in the water being treated with a coagulant?*

Ans. It is because the H^+ ions liberated due to hydrolysis of metal ions (Fe^{2+}, Al^{3+}, etc.) of the coagulants:

$$Fe^{2+} + 2H_2O \Leftrightarrow Fe (OH)_2 + 2H^+$$

$$Al^{3+} + 3H_2O \Leftrightarrow Al (OH)_3 + 3H^+$$

reduce the water alkalinity:

$$HCO_3^- + H_2O \rightleftharpoons H_2O + CO_2 + OH^-$$

$$H^+ + OH^- \longrightarrow H_2O$$

and since the process of coagulation is very sensitive to pH (It should be 5–7 when an Al-coagulant is used; 8–10 when Fe-coagulant is used), some amount of alkaline substances are added to adjust the pH of water.

Q. *What alkaline agents are used?*

Ans. $NaOH$, Na_2CO_3, $Ca(OH)_2$.

Q. *How can their amounts be determined?*

Ans. The amount of alkali to be added for correcting the pH is determined by the formula:

$$A_{cg} = \overset{\bullet}{M}_{cg} + 0.4 - A_t$$

where

A_t = total alkalinity, mg-equiv./kg

A_{cg} = alkalinity of coagulated water, mg-equiv./kg.

$\overset{\bullet}{M}_{cg}$ = dose of coagulant, mg-equiv./lt.

Q. *Why is pH the important factor of consideration during the process of coagulation?*

Ans. As coagulation proceeds, it is expedient to create conditions in water so as to precipitate out most of the coagulants, otherwise these added salts will increase the load on the ion-exchanger in the demineralizing plat as well as contaminate the ion-exchange resin to an inadmissible degree.

This entails the risk of penetration of Fe^{2+} and Al^{3+} ions into the boiler units.

Therefore, it is essential to precipitate out these metal ions as less soluble hydroxides produced due to hydrolysis:

$$Al^{3+} + 3H_2O \Leftrightarrow Al (OH)_3 \downarrow + 3H^+$$

$$Fe^{2+} + 2H_2O \Leftrightarrow Fe (OH)_2 + 2H^+$$

$$2Fe (OH)_2 + \frac{1}{2} O_2 + H_2O \longrightarrow Fe (OH)_3 \downarrow$$

However, the solubility of these hydroxides $Fe(OH)_3$ and $Al(OH)_3$ is a function of the pH. Lower the pH, higher is the acidity of the medium, so more of the precipitated hydroxide will pass into the solution. Again for aluminium hydroxide precipitate, if pH is maintained at ≥ 8, it will dissolve by forming aluminate ions:

$$Al (OH)_3 + OH^- \longrightarrow AlO_2^- + 2H_2O$$

Hence, when aluminium compounds are used as coagulants, the water pH is maintained in the range 5–7. For iron compounds it is 8–10.

Q. *What basic steps are involved during the process of softening of water?*

Ans.

1. Estimation of the amounts of chemical required on the basis of chemical analysis of raw water

2. Dosing of the reagents in calculated amounts
3. Rapid and thorough mixing of the chemicals with the raw water to effect quick flocculation
4. Settling the flocs and separation of precipitated solids
5. Removal of clarified water.

Q. *What is the reason for water hardness?*

Ans. Hardness of water is due to the presence of calcium and magnesium salts: bicarbonates, carbonates, chlorides, sulphates and nitrates. Presence of iron, aluminium and manganese salts also contributes to water hardness, but these are not normally present in appreciable quantities. Normal carbonates are also found, but only occasionally and in highly alkaline water. Nitrates are usually present in minor quantities while it has been found that sulphates on the average exceed chlorides.

Q. *How can hardness be classified?*

Ans. It can be classified as carbonate (more accurately, bicarbonate) and non-carbonate hardness.

Classification	Carbonate Hardness	Non-Carbonate Hardness
Calcium hardness	Calcium bicarbonate, $Ca(HCO_3)_2$ Calcium carbonate, $CaCO_3$	Calcium sulphate, $CaSO_4$ Calcium chloride, $CaCl_2$
Magnesium hardness	Magnesium bicarbonate, $Mg(HCO_3)_2$ Magnesium carbonate, $MgCO_3$	Magnesium sulphate, $MgSO4$ Magnesium chloride, $MgCl_2$

Q. *What is the basis of water softening by the precipitation method?*

Ans. The scale forming cations Ca^{2+} and Mg^{2+} are precipitated from water by binding them with anions like CO_3^{2-} and OH^- to produce less soluble compounds, viz. $CaCO_3$, $MgCO_3$ and $Mg(OH)_2$ so that most of these harmful cations are eliminated from the system.

Q. *How are these CO_3^{2-}, OH^- ions added to the water?*

Ans. By adding the following reagents individually or in combinations:
1. slaked lime, $Ca(OH)_2$
2. caustic soda, $NaOH$
3. soda ash, Na_2CO_3

Q. *What is lime treatment?*

Ans. It is treating raw water with a solution of slaked lime.

Q. *What chemical processes occur during lime treatment?*

Ans. The bicarbonate hardness is eliminated:

$$HCO_3^- + OH^- \rightarrow H_2O + CO_3^{2-} \xrightarrow{Ca^{2+}} CaCO_3 \downarrow$$

Magnesium hardness is removed partly:

$$Mg^{2+} + 2OH^- \longrightarrow Mg(OH)_2 \downarrow$$

Alkalinity of initial water is decreased somewhat:

$$CO_2 + 2OH^- \longrightarrow CO_3^{2-} + H_2O$$

Q. *What are the effects of lime treatment?*

Ans. The lime treatment:
(a) affects carbonate hardness ; it is decreased to 0.7–1 mg-equiv./lt.
(b) removes magnesium ions from water if there are OH^- ions present in sufficient excess
(c) removes water-dissolved CO_2
(d) reduces total dissolved solids (TDS)
(e) decreases the total hardness in lime-treated water.

Q. *What is the main drawback of this treatment?*

Ans. It does not ensure deep water softening. Magnesium ions are removed only partially.

Q. *What is lime-soda treatment?*

Ans. In this process, water is treated with a mixture of soda ash and lime and the water is heated to 360–370 K, whereupon water can be softened to a residual hardness of 200 microgram-equiv./kg.

Q. *What processes develop during lime-soda treatment?*

Ans. Permanent hardness is mostly eliminated:

$$CaCl_2 + Na_2CO_3 \longrightarrow CaCO_3 \downarrow + 2NaCl$$

$$MgSO_4 + Na_2CO_3 \longrightarrow MgCO_3 \downarrow + Na_2SO_4$$

Q. *What is the basic advantage of lime-soda treatment over lime treatment?*

Ans. The non-carbonate hardness that remains after lime treatment is eliminated by lime-soda treatment. This residual hardness is mainly due to Ca^{2+} which is precipitated by soda-ash (Na_2CO_3)

$$Ca^{2+} + CO_3^{2-} \longrightarrow CaCO_3 \downarrow$$

Q. *What is soda-sodium hydrate treatment?*

Ans. In this process water is treated with an NaOH + Na_2CO_3 mixture.

Q. *Why is the solubility of calcium carbonate in water given fundamental importance in the softening of water?*

Ans. Primarily because it is the chief scale forming agent and also contributes to corrosion. Hence knowledge of the solubility data of this compound is necessary to avoid scale formation and control corrosion as well as to predict the results of many of the softening reactions occurring in this process.

Q. *What is the solubility of calcium carbonate in water?*

Ans. Pure calcium carbonate added to distilled water in the absence of carbon dioxide will go into solution to the extent of 15–20 ppm at room temperature.

Q. *What factors affect the solubility of calcium carbonate?*

Ans.

1. Temperature
2. Presence of Ca^{2+} or CO_3^{2-} ions (Common Ion Effect)
3. pH
4. Inhibitors.

Q. *What is the effect of temperature on the solubility of calcium carbonate in water?*

Ans. Calcium carbonate has a negative temperature gradient of solubility, i.e., its solubility decreases with the increase of temperature:

$$\left[S_{CaCO_3}\right]^{15°C} = 1.12 \left[S_{CaCO_3}\right]^{20°C}$$

$$\left[S_{CaCO_3}\right]^{25°C} = 0.88 \left[S_{CaCO_3}\right]^{20°C}$$

where $\left[S_{CaCO_3}\right]^{15°C}$, $\left[S_{CaCO_3}\right]^{20°C}$ and $\left[S_{CaCO_3}\right]^{25°C}$ represent the solubility of calcium carbonate at 15°, 20°C and 25°C respectively.

Q. *How does the common-ion-effect exert an influence upon the solubility of $CaCO_3$?*

Ans. When raw water is treated with lime, $Ca(OH)_2$, to convert all the calcium bicarbonate to the normal carbonate, the Ca^{2+} ions liberated by the lime

$$Ca(OH)_2 \Leftrightarrow Ca^{2+} + 2OH^-$$

increase the concentration of Ca^{2+} to such a level that solubility product of calcium carbonate is reached

$$[S.P.]_{CaCO_3} = \left[Ca^{2+}\right]\left[CO_3^{2-}\right] = 4.8 \times 10^{-9} \text{ at } 20°C$$

whereupon calcium carbonate is precipitated, with the effect that calcium hardness can be reduced to about 35 ppm at 15° to 20°C.

However if the raw water contains calcium bicarbonate as well as magnesium bicarbonate, the added lime will convert the latter into soluble magnesium carbonate

$$Ma(HCO_3)_2 + Ca(OH)_2 \longrightarrow$$

$$MgCO_3 + CaCO_3 \downarrow + 2H_2O$$

which will increase the concentration of CO_3^{2-} ions in water thereby depressing the solubility of calcium carbonate even below 35 ppm by the same principle of common-ion-effect.

Q. *What is the effect of pH on the solubility of $CaCO_3$?*

Ans. The solubility of calcium carbonate in water is minimum at pH 9.4 but rises steeply thereafter with the increase of pH.

Q. *Why does the solubility of calcium carbonate increases sharply at pH above 9.4?*

Ans. This is due to the formation of soluble calcium hydroxide

$$CaCO_3 + 2OH^- \Leftrightarrow Ca(OH)_2 + CO_3^{2-}$$

The solubility product of $Ca(OH)_2$ is much higher than $CaCO_3$.

Q. *What is the effect of inhibitors on the solubility of $CaCO_3$?*

Ans. The inhibitors present in the water in the form of tannins, organic wastes, sewage, etc., retard the precipitation of $CaCO_3$ or distort its crystal formation and thereby the apparent solubility of calcium carbonate gets increased.

Q. *Why has it become the standard procedure to express all types of hardness as calcium carbonate?*

Ans. The compound calcium carbonate ($CaCO_3$) has a molecular weight of 100 and that renders calculations greatly simplified if all types of hardness are expressed as calcium carbonate. That's why bicarbonate, carbonate, hydroxide alkalinity as well as sulphate, chloride, nitrate and phosphate ions present are reported as calcium carbonate.

Conversion Factors for Converting PPM* of Ions into PPM of Calcium Carbonate

	For ppm, as $CaCO_3$	
	Divide by	Multiply by
Cations		
Calcium, Ca^{2+}	40.08	2.495
Magnesium, Mg^{2+}	24.305	4.114
Sodium, Na^+	45.978	2.175
Potassium, K^+	78.196	1.279
Anions		
Chloride, Cl^-	70.906	1.410
Sulphate, SO_4^{2-}	96.057	1.041
Nitrate, NO_3^-	124.008	0.806
Phosphate, PO_4^{3-}	63.314	1.579
Bicarbonate, HCO_3^-	122.034	0.819
Carbonate, CO_3^{2-}	60.009	1.666
Hydroxide, OH^-	34.0146	2.940

* PPM – Parts Per Million

Q. *What are the analytical data required for the raw water for softening calculations?*

Ans.
1. Free carbon dioxide which is equal to 2.27 (ppm of CO_2)
2. Bicarbonate
3. Calcium bicarbonate
4. Magnesium bicarbonate
5. Sodium bicarbonate
6. Bicarbonate hardness excess of calcium
7. Non-carbonate hardness
8. Calcium non-carbonate hardness.

Q. *How can bicarbonate hardness be estimated?*

Ans. It is equal to M-alkalinity, M stands for methyl orange. Titrating a measured volume of raw water with the standard mineral acid using methyl orange as the indicator

$$HCO_3^- + H^+ \longrightarrow H_2O + CO_2$$

Normal carbonates are rarely present in the raw water. If these are present then there will be phenolphthalein alkalinity,
    i.e., carbonate = 2 [P-alkalinity].
    Therefore, the carbonate concentration

    = M-alkalinity—2 (P-alkalinity)

Q. *How is the bicarbonate hardness excess of calcium estimated?*

Ans. It is obviously magnesium bicarbonate or sodium bicarbonate or their sum.

Q. *How is non-carbonate hardness computed?*

Ans. It is equal to the excess of total hardness over M-alkalinity.

Q. *How is calcium non-carbonate hardness computed?*

Ans. It is equal to the calcium in excess of calcium bicarbonate.

Q. *What is the basis of silica removal from raw water?*

Ans. Silica is removed from the raw water by the precipitation method but it is not carried down as a true precipitate.

Silica is found in raw water as coarse-dispersed solids and as colloids. While the former can be readily separated by a settling process the later is removed by the precipitation method. In colloidal state silica exists as silicic acid (H_2SiO_3) produced due to hydrolysis of silicates:

$$SiO_3^{2-} + 2H_2O \Leftrightarrow H_2SiO_3 + 2OH^-$$

The presence of mineral acid hastens the process. The colloidal silica goes down only when it is absorbed by the hydroxide of iron, magnesium and aluminium

These metal hydroxides result from the hydrolysis of coagulants.

Q. *What factors should be taken into consideration to effect silica removal by ferric hydroxide or magnesium hydroxide?*

Ans.

1. Silica removal occurs by absorption
2. There is an optimum pH range for each process
3. The efficiency of silica removal is affected by temperature
4. As the silica concentration is reduced, the efficiency of the process gradually falls off and thereby increasingly larger quantities of $Fe(OH)_3$ or $Mg(OH)_2$ are required for each additional ppm of silica removed
5. The effectiveness of silica removal is basically controlled by the degree of hydrolysis of ferric oxide or magnesium oxide added.

Ferric hydroxide forms hydrous ferric oxide which then dehydrates in the absence of water:

$$Fe(OH)_3 \Leftrightarrow Fe_2O_3 \cdot nH_2O \Leftrightarrow Fe_2O_3 + nH_2O$$

Magnesium oxide undergoes hydrolysis as

$$MgO + H_2O \Leftrightarrow Mg(OH)_2$$

It has been found that hydroxides precipitated in situ are far more effective than externally precipitated oxides.

6. The reactions are rapid and take less time than that provided for settling process.

Q. *How is ferric hydroxide introduced in raw water?*

Ans. It is introduced in the form of commercial ferric sulphate marketed as coagulant which, upon hydrolysis by caustic soda or lime produces ferric hydroxide in raw water.

Q. *What is the optimum pH value for silica removal by ferric hydroxide?*

Ans. It is about 9.

Q. *What is the efficiency of the process?*

Ans. Silica content in raw water can be brought down to as low as 2 ppm but it varies with the quality of the water.

Q. *How much quantity of ferric sulphate is required for each ppm of silica removed?*

Ans. For raw water containing silica to the extent less than 10 ppm, the ferric sulphate required is 15–20 ppm for each ppm of silica removed. At higher silica concentrations, the quantity of ferric sulphate required for each ppm of silica removed may be as low as 10 ppm.

Q. *Why does this method suit the cold-softening process best?*

Ans. It has been found that efficiency of silica removal by ferric hydroxide decreases with the increase of temperature and vice versa.

Q. *What is the chief disadvantage of this process?*

Ans. This method increases the non-carbonate hardness of water in the form of calcium sulphate.

Q. *How is magnesium hydroxide introduced into raw water for silica removal?*

Ans. Sometimes the source of magnesium hydroxide is the natural magnesium hardness present in the water, which produces $Mg(OH)_2$ when raw water is treated with excess lime.

$$Mg^{2+} + Ca(OH)_2 \longrightarrow Mg(OH)_2 + Ca^{2+}$$

Over and above, magnesium hydroxide can be supplied from various sources: Magnesium sulphate, magnesium oxide and dolomitic lime.

Q. *What is the efficiency of this process?*

Ans. At 20°C, about 1ppm of silica is removed for about 6–7 ppm magnesium hardness precipitated. And the efficiency of silica removal increases rapidly as the temperature is raised further.

Q. *Magnesium oxide prepared from calcined magnesite can be added directly to raw water in the dissolver whereas dolomitic lime cannot be so applied. Why?*

Ans. Magnesium oxide normally reacts with water (20°C) in 20–30 minutes and goes into solution. This rate of dissolution is a function of the pH—lower the pH more rapid is the rate of dissolution.

Dolomitic lime, usually available as hydrated dolomitic lime [32% MgO, 62% $Ca(OH)_2$] is a source of magnesium oxide which, however, will not dissolve in cold water in an environment of high pH. Since the calcium hydroxide present in the dolomitic lime provides a high pH rapidly, the associated MgO remains inert.

Q. *Then how can this magnesium oxide of dolomitic lime be brought into solution?*

Ans. For this purpose, it is basically necessary that the pH of the water should be lowered below the pH range that prevails for the precipitation of magnesium hydroxide. This requires a separate dissolution tank where the bicarbonate alkalinity or the free carbon dioxide of raw water will dissolve magnesium oxide.

If the natural water fails to dissolve sufficient MgO to ensure necessary silica removal, carbon dioxide must be added from an external source to the dissolving tank:

1. $HCO_3^- \Leftrightarrow H^+ + CO_3^{2-}$

$$MgO + 2H^+ \longrightarrow Mg^{2+} + H_2O$$

2. $CO_2 + H_2O \Leftrightarrow H_2CO_3$

$$MgO + H_2CO_3 \longrightarrow MgCO_3 + H_2O$$

Important: It is well nigh impossible to predict precisely what amount of Mg^{2+} ions must be precipitated from any water supply to ensure a specific silica removal.

Q. *Cooling water is subjected to many treatments?*

Ans.
1. Biocide treatment
2. Chlorine treatment
3. Quat-2C treatment
4. Alkali/Acid treatment
5. Phosphate treatment
6. Chromate treatment
7. Betz Polynodic 620BE treatment, etc.

Q. *What is biocide treatment?*

Ans. Dosing sodium penta and trichlorophenate to cooling water.

Q. *What is the mode of dosing?*

Ans. Shock dosing.

Q. *How is this shock dosing carried out?*

Ans. 200 kg sodium pentachlorophenate is dosed to 2400 m^3 of cooling water in 1 hour. (Interval : Fortnight)

In another typical cooling water system, 80 kg of sodium pentachlorophenate and 40 kg of sodium trichlorophenate are added to 2000 m^3 of cooling water within 1 hr.

Q. *Are they added directly?*

Ans. Sodium trichlorophenate is dosed directly while a 5% solution of sodium pentachlorophenate is injected.

Q. *At what interval is this dosing made?*

Ans. Usually 15 days.

Q. *What is its objective?*

Ans. To kill the algae and rupture the algal mat.

Q. *Why is killing of bacteria an imperative?*

Ans. Bacteria (nitrosifying) converts dissolved ammonia to NO which is converted to NO_2 by nitrifying bacteria. The latter is acidic in an aqueous layer causing corrosion.

Q. *Why is chlorine treatment made?*

Ans. To kill bacteria. Chlorine is a bactricide.

Q. *At what pH is this made?*

Ans. 7.5–8.0

Q. *Why is this range preferred?*

Ans. Chlorine reacts with water in this range of pH which is slightly alkaline to produce hypochlorite that liberates nascent oxygen to burst out bacteria cells.

$$Cl_2 + OH^- \longrightarrow Cl^- + H^+ + OCl^- \longrightarrow \overset{\bullet}{O} + Cl^-$$

Q. *What quantity of chlorine is added?*

Ans. One typical installation handling 2400 m^3 of cooling water doses 60 kg Cl_2 per day.

Q. *What is the mode of chlorine dosing?*

Ans. It may be dosed continuously from a chlorinator at the varied rate as desirable.

Shock dosing of chlorine is also made. Dosing 60 kg of chlorine in just 2–3 hrs, i.e., 20–30 kg/h rate is one typical example.

Q. *Is chlorine injected directly?*

Ans. It is not injected directly from a 1 ton cylinder. It is dissolved in water and then injected.

Q. *Why is chlorine not injected directly from the cylinder?*

Ans. To avoid:
  (a) uneven distribution.
  (b) loss of chlorine.

Q. *What is QUAT-2C?*

Ans. It stands for Quaternery Ammonium Compound.

Q. *Is it a biocide?*

Ans. Yes.

Q. *How is it added?*

Ans. Shock dosing.

Q. *At what rate is it dosed?*

Ans. It varies with the total bulk of cooling water handled. One installation (2000 m^3 C/W capacity) doses 120 kg of this compound in just 10–15 minutes.

Q. *At what interval is it dosed?*

Ans. Usually a week.

Q. *Is chlorination continued during biocide dosing?*

Ans. No. 48 hours after.

Q. *What is Betz Polynodic 620BE?*

Ans. It is a phosphate doser.

Q. *Why is it dosed?*

Ans. To eliminate the formation of scales of calcium and magnesium carbonates.

Q. *How does it eliminate these?*

Ans. It forms gelatinous, or colloidal calcium/magnesium phosphate with the scale former, Ca^{2+} and Mg^{2+}, thereby inhibiting their precipitation as carbonates.

Q. *In the recent trend toward more alkaline cooling waters, bromine replaces chlorine in cooling water treatment. What are the advantages of bormination over chlorination in cooling water treatment?*

Ans. Depending on plant conditions, the potential advantages of bromine treatment are:
  1. Lower chemical costs
  2. Reduced corrosion, i.e., under-deposit corrosion of condenser tubes (copper alloy) as initiated by Cl_2 can be avoided thereby eliminating the need for condenser retubing

3. Reduced requirement for cleaning of main surface condensers
4. Reduced environmental hazard
5. Reduction of lost output at plants fitted with condensers having marginal design surface area.

Q. *What are the environmental benefits from the use of bromine for cooling water treatment?*

Ans.
1. Bromine is more effective a biocide than chlorine. Therefore it allows major dose reductions.
2. Bromine and its compounds (bromamines) undergo rapid decay thereby minimizing their impact on aquatic life in the effluent receiving water mass.

Q. *Why does bromine treatment of cooling water have an edge over chlorine treatment regarding corrosion control?*

Ans. Bromination leads to improved biofouling control than chlorination of cooling water. The immediate impact of this are reduced under-deposit corrosion and microbiologically induced corrosion (MIC) of stainless steels.

Q. *In what kind of water—acidic or alkaline or neutral — are the benefits of bromination most apparent?*

Ans. Alkaline waters and those waters that contain ammonia.

Q. *Why alkaline waters are most suitable for the efficacy of bromine treatment?*

Ans. Bromine acts as a biocide in the form of HOBr (hypobromous acid) which acts as an oxidizing agent.

$$Br_2 + H_2O \Leftrightarrow HOBr + HBr$$

The stability of HOBr depends on the pH of water. HOBr is primarily available as an effective biocide in the alkaline range—higher alkalinity than required for chlorine.

Q. *What is the principal mechanism underlying bromine's biocidal activity?*

Ans. The biocidal activity of bromine is explained in three distinct steps:
1. FORMATION OF HYPOBROMOUS ACID Bromine reacts with water forming hypobromous acid

$$Br_2 + H_2O \Leftrightarrow HBr + HOBr$$

2. CHEMICAL EQUILIBRIUM BETWEEN HOBr AND HYPOBROMITE IONS

$$HOBr \Leftrightarrow H^+ + OBr^-$$

Higher the pH of the solution, more the equilibria shifts towards the right bringing about greater concentration of hypobromite ions (OBr^-)

3. DISSOCIATION OF HYPOBROMITE IONS (OBr^-)
Hypobromite ions dissociate to liberate nascent oxygen that ruptures the cells of algae

$$OBr^- \longrightarrow O + Br^-$$

Q. *Why does hypobromous acid remain available as a more effective biocide than hypochlorous acid at a significantly higher pH?*

Ans. HOBr and HOCl both act as biocides.
The dissociation curves of these two show that the efficacy of bromine is retained at a significantly higher pH than chlorine. (Fig. 23.1)
This explains why biocidal activity of bromine is superior to chlorine in alkaline cooling waters. It renders, for example, much lower survival rate of

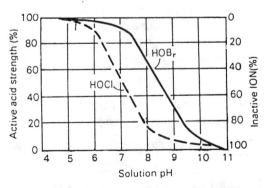

Fig. 23.1 *Dissociation curves of hypohalous acids. HOBr remains available as an effective biocide at a pH significantly higher than HOCl*

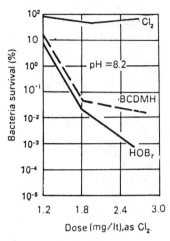

Fig. 23.2 *Bactricidal Power. Both HOBr and BCDMH are far stronger biocidal to pseudomona Bacteria in alkaline water*

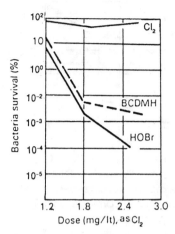

Fig. 23.3 *Bactricidal power. Both HOBr and BCDMH are far stronger biocidal to pseudomona bacteria in alkaline water*

slime-forming bacteria *Pseudomona* under equivalent dosages at pH 8.2. (Fig. 23.2)

Q. *Is there any other effective brominating agent?*

Ans. Yes; bromochlorodimethylhydantoin (BCDMH).

Q. *When Cl_2 or Br_2 or BCDMH is dosed to the cooling waters containing ammonia or nitrogen-based compounds what do they form?*

Ans. They form haloamines.

Q. *What are the basic differences between bromamines and chloramines?*

Ans. The biocidal activity and the mode of decay are the two significant features that distinguish bromaines from chloramines.

Bromine as both HOBr and BCDMH is a far more effective biocidal than chlorine for PSEUDOMONA bacteria in cooling waters containing ammonia.

In contrast to chloramines, bromamine reactions are reversible by the pH effect.

Bromamines decay more rapidly than do chloramines.

Because of their higher biocidal activity and rapid decay, bromamines are far more beneficial to the environment as compared to the more persistent chloramines. (Fig. 23.3)

Q. *How does the problem of corrosion in cooling water circuit arise from chlorination or bromination?*

Ans. Chlorine and bromine are the oxidizing biocides. Overfeeding of these biocides, it has been reported, stimulates corrosion of some metals and alloys. Copper alloys are particularly sensitive to these. Hence these biocides should be applied prudently.

Free oxidant (Cl_2 or Br_2) attacks the metal which passes into the solution

$$Cu \longrightarrow Cu^{2+} + 2e$$

$$Cl_2 + 2e \longrightarrow 2Cl^-$$

and the degree of corrosion largely depends on the oxidation-reduction potential and pH at the metal-solution interface. For copper alloys, the corrosion is an oxidation process which is much more energetically favoured by the higher oxidation potential of chlorine than bromine. Moreover this copper corrosion is lower in a slightly alkaline condition when the concentration of oxidant is less but it accelerates with rapidity when oxidants are introduced at higher pH.

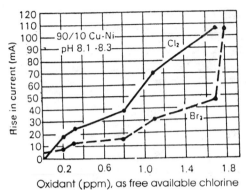

Fig. 23.4 *Electrochemical measurements indicate chlorine to be more corrosive on 90/10 copper-nickel alloy than bromine*

Laboratory tests carried out to study the comparative effects of chlorine and bromine on 90/10 Cu/Ni alloy using natural cooling water having TDS (total dissolved solids) value 1400 ppm, pH 8.2 ± 0.1 at 298 K revealed that chlorine exerts two to three times the corrosive influence of bromine at equivalent oxidant levels (see Fig. 23.4). Potentiostatic measurements were carried out to observe the rise and decay of corrosion effect while oxidant levels (ppm) were measured continuously as free available chlorine (FAC). In separate test runs, various dosage levels of each oxidant were used and the current readings were recorded at the maximum sustained rise upon treatment. The rise in corrosion current plotted against oxidant concentration (ppm) generated the above curves indicating the corrosive influence of chlorine on copper alloy to be considerably higher than bromine.

However, similar experiments carried out at pH 6–7 did not produce a significant increase in corrosion current even at higher concentrations of free Cl_2 or Br_2.

Q. *Why at pH levels slightly in the acidic range (6–7), is the corrosion of copper or copper alloys by free chlorine or bromine not significant even at higher oxidant levels?*

Ans. At lower pH, i.e., neutral to acidic range, copper needs higher oxidation potential for its oxidation to copper ions.

Q. *What is the comparative effect of chlorination and bromination of cooling water on iron?*

Ans. Treatment of cooling water by chlorine or bromine yields halide ions that attack iron at low pH. The effect of bromide ions is similar to that of chloride ions. So far as stress-corrosion cracking of austenitic stainless steels is concerned, the aggressivity of bromide ions is considerably lower than chloride ions.

Q. *What is the effect of free available chlorine or bromine on other metals in the cooling water circuit?*

Ans. Manganese fouling (i.e., manganese corrosion) has been reported. Manganese requires, like copper, much lower potential in alkaline water for its oxidation.

Q. *What is the remedy to this problem?*

Ans. Prudent use of biocide (Cl_2 / Br_2), lower oxidant dosage and reduced oxidation potential are the controlling factors to combat manganese corrosion.

Q. *To what extent does the change over from chlorine to bromine for cooling water treatment reduce corrosion?*

Ans.

Chemical Dosed	Treatment Rate (kg/day)	Duration of Dosing	Corrosion Rate (mil/yr) During Dosing	Daily Average
Cl_2 (only)	680	60 min	6–8	0.8
Cl_2/Br_2	340	30 min	4–8	0.3
Cl_2/Br_2	91	30 min	0.3	0.1

Q. *Texas Utilities Electric in its two gas-fired supercritical units of 565 MW and 775 MW capacity uses lake water for cooling. However, the lake water being typically highly alkaline (pH 8.8–9.1), its auxiliary cooling water system had experienced severe corrosion problem with poor biofouling control while very high rates of chlorination were applied. However upon switch over to bromination, corrosion rates have been reduced substantially.*

What do you infer from this practical case study?

Ans. The efficacy of bromine in reducing substantially the microbiologically influenced corrosion associated with poor biofouling control is much higher than chlorine, though chlorine is potentially a more powerful oxidizing agent than bromine at any given pH. Bromine exercises better biofouling control than chlorine.

Q. *Severe manganese fouling and pitting corrosion of 304-stainless steel tubings of condensers had been reported from a 3-unit (total capacity 1900 MW) lignite-fired plant of Monticello using lake water for cooling while continuous chlorination was being practised for biofouling control. The pH of lake-water varied in the range 8–8.8 in summer.*

However upon adoption of bromine treatment for more than 3 years, the condenser-tube failures have been drastically reduced.

What do you infer from this case study?

Ans. Chlorine having higher oxidation potential than bromine had induced more rapid rate of corrosion of manganese/iron at elevated pH > 8.

Moreover, its efficacy as a biofoulant is less than that of bromine in substantially high alkaline cooling water. Therefore, the switchover to bromine chemistry has enhanced the life of the condensers tubes and reduced biofoulant induced corrosion.

Q. *What do the initials AVT stand for?*

Ans. All-Volatile boilerwater Treatment.

Q. *Why the sodium limit of 3 ppb for both drum and once-through boilers on AVT is the most stringent requirement as per EPRI (Electric Power Research Institute) of USA?*

Ans. It is because the sodium compounds (NaOH, NaCl) are the major boiler-tube and turbine blade corrodants.

Q. *Why is adherence to an even lower limit of sodium, where possible, sometimes suggested?*

Ans. It is primarily due to the fact that sodium limit of 3 ppb will not always ensure the prevention of sodium hydroxide deposition in low-pressure turbines.

Q. *But why is the sodium level (as per EPRI recommendation) for steam 5 ppb for drum-type boilers on phosphate treatment?*

Ans. It is because a fraction of sodium will be tied up in the form of sodium phosphate compounds which are not corrosive.

Q. *What are the target values of sodium, as per EPRI, for boiler water?*

Ans. For boilers operating at a pressure of 60 atm. the maximum permissible value for sodium concentration is about 6 ppm, for boilers with working pressure of 140–145 atm it is about 2.8 ppm and is as low as 0.55 ppm for 195–197 atm boilers.

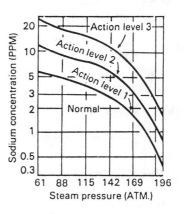

Fig. 23.5 *Sodium target values. Thes are based on those for reheat/superheated steam*

Q. *Why are the values (Na$^+$ concentration) set for higher pressure boilers lower than usually observed?*

Ans. This is required to maintain high standard of turbine-steam purity and to minimize phosphate hideout during load changes.

Q. *What action is to be taken, if for a 60 atm boiler, the Na$^+$ limit exceeds 6 ppm?*

Ans. If it is greater than 6 ppm but less than 12 ppm, action level (1) is to be taken, i.e., to return to normal levels within one week.

Q. *What are the values of Na$^+$ concentration for Action Level (1) for 115 atm, 145 atm and 195 atm boiler units?*

Ans.

Boiler Operating Press	Na$^+$ Concentration Range for Action Level (1)
115 atm	4 < [Na$^+$] < 8 ppm
145 atm	2.8 < [Na$^+$] < 5 ppm
195 atm	0.55 < [Na$^+$] < 0.95 ppm

Q. *What are the values of Na$^+$ concentration for Action Level (2) for 115 atm, 145 atm and 195 atm boilers?*

Ans.

Boiler Operating Press	Na$^+$ Concentration Range for Action Level 2
115 atm	8 < [Na$^+$] < 15 ppm
145 atm	5 < [Na$^+$] < 12 ppm
195 atm	0.95 < [Na$^+$] < 2 ppm (approx)

Q. *What do you mean by Action Level (2)?*

Ans. It means return to normal level must be effected within 24 hours.

Q. *If for example sodium concentration exceeds 12 ppm in a 145 atm boiler or 2 ppm in a 195 atm boiler what measure should be taken?*

Ans. Switchover to Action Level (3), i.e., shutdown to be taken within four hours.

Q. *Why?*

Ans. To avoid rapid corrosion of boiler-tubes as effected by highly acidic boilerwater pH.

Q. *Plants using AVT assume all the sodium to be present as NaOH. Why?*

Ans. It is due to the fact that sodium hydroxide is more volatile than other sodium compounds (NaCl, Na$_2$SO$_4$) available in boilerwater. (Fig. 23.6)

Q. *How does sodium get into the condensate?*

Ans. It finds its way to the condensate via make-up water treatment system and condenser leakage.

Q. *Why is chloride monitoring required?*

Ans. It exercises a deleterious effect on steam turbines, boilers and condensate-polisher performance.

Q. *What types of deleterious effects does chloride have on steam turbines?*

Ans. It initiates corrosion fatigue, stress-corrosion cracking and pitting type corrosion on bladings.

Q. *What are the deleterious effects chloride exercises on boilers?*

Ans. Corrosion, hydrogen damage and pitting.

Q. *What is the limit of chloride in steam?*

Ans.

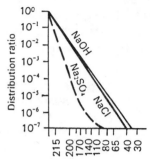

Fig. 23.6 *Distribution ratios of various sodium salts in steam. Sodium Hydroxide has the highest distribution ratio. Hence it is more easily carried over by steam than any other sodium salt found in boiler water*

3 ppb for reheat units ⎫
⎬ according to EPRI
5 ppb for non–reheat units ⎭

Q. *What is the basis of these values?*

Ans. These are based on the solubility of sodium chloride in superheated steam.

Q. *What is the basis of boilerwater chloride limits?*

Ans. Boilerwater chloride limits are derived on the basis of mechanical and vapourous carryover.

Q. *In which form of compound, is chloride normally carried over in AVT (All-Volatile boilerwater Treatment) plants?*

Ans. Ammonium chloride.

Q. *Why in plants using AVT, are target values of chloride ions assumed on the ammonium chloride rather than sodium chloride?*

Ans. Ammonium chloride is more volatile than sodium chloride and so measurement on the basis of ammonium chloride rather than sodium chloride

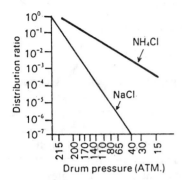

Fig. 23.7 *Distribution ratios of NaCl & NH₄Cl in steam. Ammonium chloride has higher carryover coefficient than sodium chloride*

gives more conservative boilerwater chloride values. (Fig. 23.7)

Q. *What is the chloride limit in make-up water for boilers using phosphate treatment?*

Ans. It is 3 ppb (as per EPRI).

Q. *Why is sulphate carryover in steam undesirable?*

Ans. This is because of its corrosive action to turbines during operation and to boilers both on- and off-load.

Q. *What are the sulphate limits in superheated steam?*

Ans.

3 ppb for reheat units

5 ppb for non–reheat units

⎫ according to EPRI
⎭

Q. *Why are these identical to those of chlorine?*

Ans. Sulphate limits, as prescribed above for superheated steam, are identical to those of chlorides, as the solubility of sodium sulphate is similar to that of sodium chloride.

Q. *What is the basis of these sulphate values?*

Ans. These are based on mechanical and vapourous carryover with steam.

Q. *In which form/forms silica may occur in boiler-water?*

Ans. Silica may occur either in dissolved state or colloidal state.

Q. *Which 'form' is called reactive silica?*

Ans. Silica in dissolved form is called reactive silica.

Q. *Which 'form' is called unreactive silica?*

Ans. Silica in colloidal form is called unreactive silica

Q. *Why is dissolved silica called reactive silica?*

Ans. Silica in dissolved form is volatile and easily carried off by steam. Its solubility in steam commensurates with steam-temperature and pressure. This dissolved silica carried over by steam precipitates where steam pressure (and enthalpy) drops, i.e., in the turbine blades. These are very difficult to dislodge from sites with the effect that the efficiency and capacity of the turbine get reduced.

Q. *Is it true that quite a significant amount of colloidal silica passes into condensate/feedwater undetected with make-up water?*

Ans. Yes.

Q. *What are the silica limits for steam?*

Ans.

Steam Generation Unit	Silica Limit
Reheat type	10 ppb
Non-reheat type	20 ppb

Q. *Which one of the two — vapourous carryover and mechanical carryover — is the major source of silica in steam?*

Ans. Vapourous carryover.

Q. *Why is a silica target-value of maximum 10 ppb for condensate and make-up water selected?*

Ans. It will ensure maximum blowdown control and flexibility in operation during excursions.

Q. *Why is phosphate controlled to ensure a sodium to phosphate molar ratio in the range 2.3 to 2.8 with excursions limited to 2.9?*

Ans. To prevent the presence of free NaOH.

Low sodium-to-phosphate ratios are maintained also to avoid acid attack.

Q. *Why are excursions limited to a minimum ratio of 2.1?*

Ans. To avoid the formation of phosphoric acid

Q. *What is the effect of dissolved oxygen in boiler-water?*

Ans. Dissolved oxygen in boiler water is unwanted because of its corrosive effect throughout the cycle and the carryover and deposition of corrosion products as a result.

Q. *How can its presence in boilerwater be minimized?*

Ans. By using chemical scavengers like hydrazine in high pressure units and mechanically with deaerators.

Q. *What is the target value, according to EPRI, of dissolved oxygen in the deaerator effluent?*

Ans. 7 ppb.

Q. *Why is the dissolved oxygen monitoring at the inlet to deaerator required?*

Ans. It serves as a check on the effectiveness of deaeration and chemical scavenging.

Q. *If a chemical scavenger (say hydrazine) is added just upstream of the deaerator what will be the effect?*

Ans. This will supplement mechanical scavenging of dissolved oxygen from the BFW and improve the deaerator's efficiency.

Q. *What decides whether the chemical scavenger (N_2H_4) feedrate is adequate or needs modification?*

Ans. The hydrazine residual at the deaerator inlet compared to the dissolved oxygen concentration serves as the basis for control of the feedrate of hydrazine.

Q. *Can you identify an inexpensive as well as reliable on-line method for monitoring the overall level of contamination?*

Ans. Specific conductivity measurement.

Q. *How can the specific conductivity measurement lead to the determination of contaminants and their form?*

Ans. The boilerwater impurities that lend themselves to specific conductivity measurement are all ionic in nature. And the specific conductivity (i.e., ionic mobility) of such an ionic impurity depends on the form of ion—salt, acid or alkaline form. Therefore, the specfic conductivity measurement monitors the overall level of the contaminants that are present and their form.

Q. *What does degassed cation conductivity reflect?*

Ans. It reflects only the total nonvolatile anionic impurity level.

Q. *What do economizer-inlet cation conductivity target values reflect?*

Ans. They reflect the sum of the cation conductivities contributed by individual anions and infiltration of carbon dioxide to air in leakage.

Q. *What is the targeted value of condensate and steam cation conductivity in power-plants?*

Ans. Below 0.1 µS/cm (micro Siemens/cm)

Q. *What is the level of sodium and chloride impurities in condensate that power-plant operators generally strive to maintain?*

Ans. Less than 3 ppb, i.e., less than 3 µg/lt.

Q. *Why is a reliable water-quality monitoring system required?*

Ans.

1. To measure accurately and consistently the concentrations of impurities in the boilerwater to ensure high water and steam quality
2. To record data
3. To alert the operators when the prescribed limits are exceeded

Q. *What is water demineralization?*

Ans. It means the removal of cations and anions of substances dissolved in water.

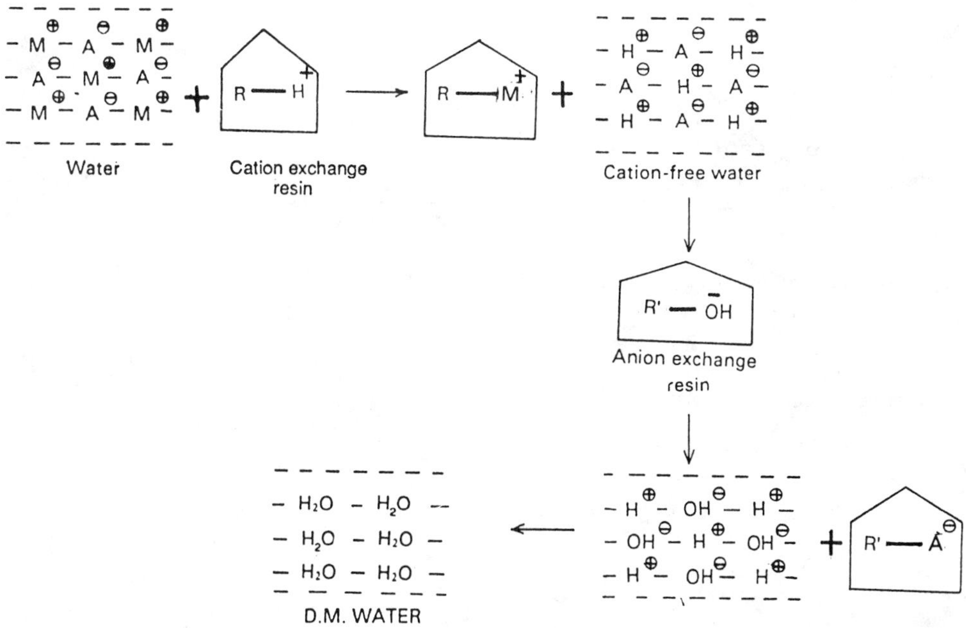

Fig. 23.8 Demineralization process

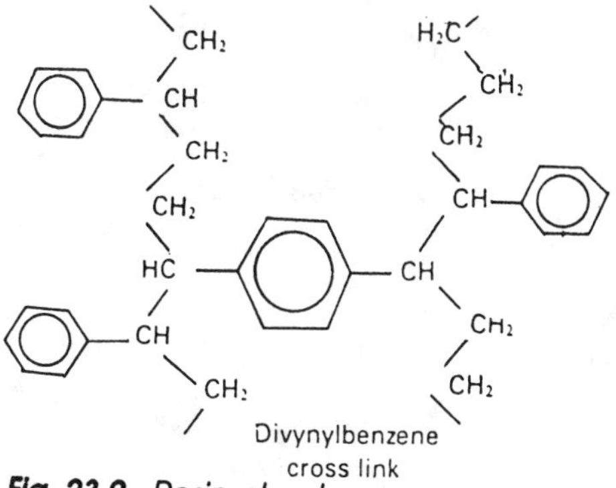

Fig. 23.9 Resin structure

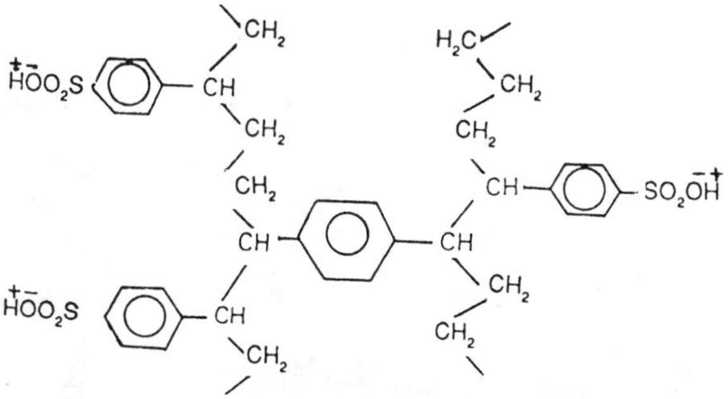

Fig. 23.10 Sulphonated derivative of styrene-divinylbenzene polymer is a cation exchange resin

Q. *How is demineralization carried out?*

Ans. Pretreated water free from coarse-dispersed and colloidal impurities is passed through a series of cation exchangers where cations like Ca^{2+}, Mg^{2+}, Na^+, NH_4^+ and other metal ions are substituted by H^+ ions. Then it is passed through a series of anion exchanger where anions like SO_4^{2-}, NO_3^-, NO_2^-, HCO_3^-, CO_3^{2-} and Cl^- are substituted by OH^- ions. The resulting water is demineralized water. (Fig. 23.8)

Q. *What is ion-exchange resin?*

Ans. Resins are cross-linked polymers of styrene and divinylbenzene. They are hard, insoluble substances formed into beads. (Fig. 23.9)

They attain ion-exchange properties when an acid group (say SO_3H) or a basic group [say $- \overset{+}{N}(CH_3)_3$] is introduced to its pendant phenyl groups as a result of electrophilic substitution.

Q. *How can a cation-exchange resin be produced?*

Ans. When such a polymer of divinylbenzene is sulphonated, an insoluble resin with sulphonic acid groups appended to its surface as a result of electrophilic substitution on the pendant phenyl groups (from styrene units) is obtained. (Fig. 23.10)

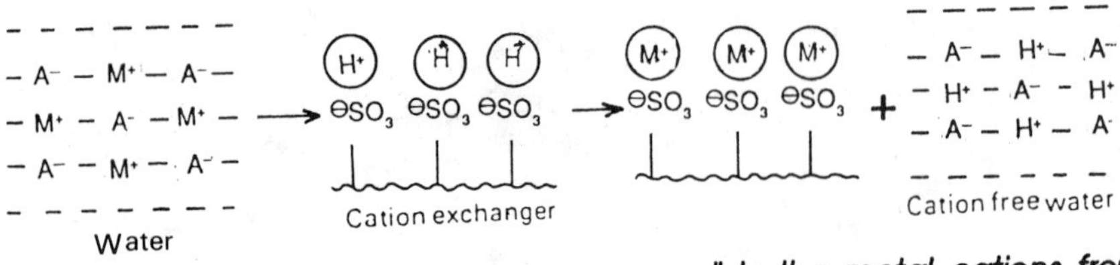

Fig. 23.11 *Cation exchanger removes reversibly the metal cations from the treated water*

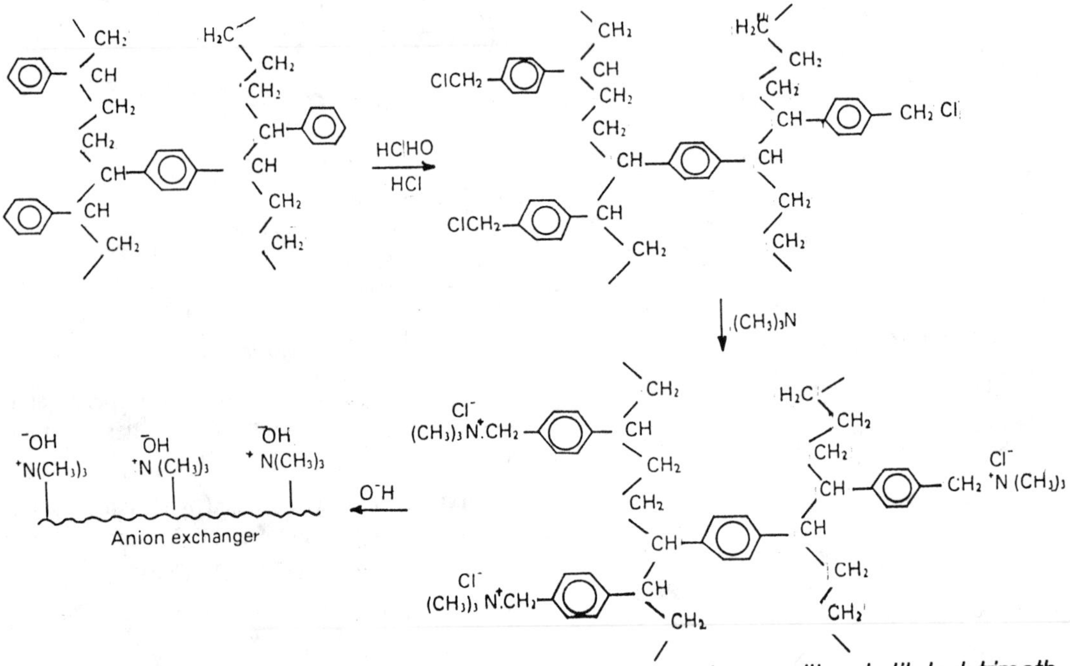

Fig. 23.12 *Anion exchanger is a styrene-divinylbenzene polymer with substituted trimeth-lamine cations*

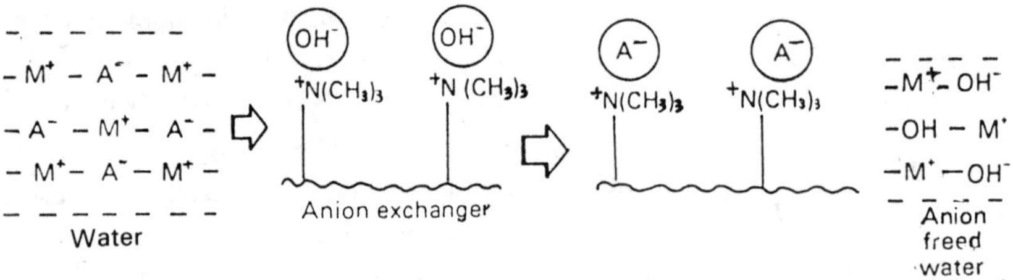

Fig. 23.13 *Anion exchanger removes reversibly the negative radical or ion from treated water*

This sulphonic acid substituted resin is called cation-exchange resin or cation exchanger.

Q. *Why does the sulphonated resin serve as a cation exchanger?*

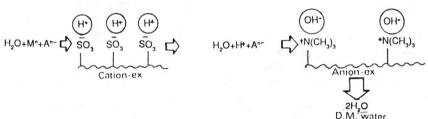

Fig. 23.14 *The process of demineralization is realized through the combined action of cation and anion exchangers*

Ans. The protons of substituted SO_3H (sulphonic acid) groups are only held electrostatically and are very labile. They easily migrate away in an adjacent water phase as long as another positive ion from the aqueous solution can move in to take its place and balance the negative sulphonate groups. (Fig. 23.11)

Q. *How can an anion exchange resin be produced?*

Ans. The polymerized divinylbenzene upon chloromethylation followed by trimethyl amination yields another resin known as anion exchange resin. (Fig. 23.12)

Q. *How does this resin serve as an anion exchanger?*

Ans. The hydroxyl group of substituted trimethyl amine $(- \overset{+}{N} (CH_3)_3\overline{O}H)$ is held by a weak electrostatic force on the surface of the base molecule of the resin and is capable of being replaced by another negative radical or ion from the aqueous phase. That is how it lends the resin anion exchange properties. (Fig. 23.13)

Q. *How can the overall process of demineralization be described schematically?*

Ans. It involves two basic steps:
1. removal of metal cations (Ca^{2+}, Mg^{2+}, Fe^{2+}, Na^+, K^+, Al^{3+} represented by M^{n+}) in the cation exchanger
2. removal of the anions (like CO_3^{2-}, HCO_3^-, SO_4^{2-}, Cl^-, etc., represented by A^{n-}) by the anion exchanger. (Fig. 23.14)

Q. *Why is softened water and not untreated water fed to the DM-unit?*

Ans. The water treatment reduces the quantum of total dissolved solids (TDS) in water and therefore, softened water contains less amount of TDS, organic impurities, coarse-dispersed and colloidal impurities than untreated water. Therefore, the load on the DM - unit becomes much less when softened water instead of untreated water is used.

Q. *Why should the water charged to a DM- Unit be free from coarse dispersed and colloidal impurities?*

Ans. These may precipitate on the resin surface affecting the overall performance of the demineralizing unit.

Q. *Why is the water first passed through a cation-exchanger and then anion-exchanger but not the other way around?*

Ans. When water is treated in the sequence:

WATER $\longrightarrow$ Cation-Exchanger $\longrightarrow$

Anion-Exchanger $\longrightarrow$ DM WATER

the water at the exit of the cation-exchanger becomes acidic due to substitution of metal cations by the H^+ ions of resin

$$R\overset{+}{—H} + M^+A^- \longrightarrow R—M^+ + H^+ + A^-$$
$$(TDS)$$

These freed H^+ ions go to liberate CO_2 from HCO_3^- ions present

$$HCO_3^- + H^+ \longrightarrow H_2CO_3 \longrightarrow H_2O + CO_2$$

in the aqueous phase and thereby reduce the load on the anion- exchanger as well as lower the pH so that insoluble sulphates are not precipitated

$$Ca^{2+} + SO_4^{2-} \longrightarrow CaSO_4 \downarrow$$

to impair the activity of the resin bed.

If, on the other hand, treated water is passed first through an anion-exchanger and then a cation-exchanger, the water at the exit of the anion-exchanger will be alkaline

$$\underset{\text{ANEX}}{R-OH} + A^- + M^+ \longrightarrow R-A + OH^- + M^+$$

as a result of which carbonates and sulphates of certain metal cations will be precipitated over the resin bed impairing the exchange-efficiency of the anion exchanger.

Q. *How is the degree of demineralization measured?*

Ans. It is measured conveniently by the conductivity test. As the water gets demineralized its conductivity λ decreases or resistivity ρ increases

$$\lambda = \lambda^+ + \lambda^- \ \text{Ohm}^{-1} \ \text{cm}^{-1}$$
$$= (0.1 - 0.2) \times 10^{-6} \ \text{Ohm}^{-1} \ \text{cm}^{-1} \text{ for deep}$$

demineralized water

where λ^+ = conductivity of cations

λ^- = conductivity of anions

$\rho = (5 - 10) \times 10^6$ Ohm cm. for deep demineralized water

Q. *Why is quantitative estimation of TDS in DM water not preferred?*

Ans. It is a laborious and time consuming process.

Q. *Apart from ion-exchange method what are the other methods of demineralizing water?*

Ans.
   1. Electrodialytic method
   2. Hyperfiltering (reverse osmosis) method
   3. Thermal method
   4. Magnetic method.

Q. *What is the drawback of the ion-exchange method?*

Ans. When the resin beds are exhausted, i.e., when all the H^+ ions of the cation-exchanger and all the OH^- ions of the anion-exchanger have been replaced by cations and anions respectively, the resin beds are regenerated by using concentrated sulphuric acid (cation-exchanger) and caustic alkali (anion-exchanger) respectively. And as this takes up great quantities of these chemicals the expenditure mounts.

Q. *Why is sulphuric acid and not nitric acid used during regeneration of the cation exchanger?*

Ans. Nitric acid is an oxidizing agent which will oxidize resin molecules which are polymeric organic compounds.

Q. *Why is not hydrochloric acid commonly used for regeneration of cation-exchangers?*

Ans. It is costly.

Q. *Why are organic acids not used for regeneration of cation-exchangers?*

Ans. These are weak acids and will accomplish only partial regeneration of exhausted cation exchangers. Moreover, they're costly.

Q. *Why is a mixed bed used in a DM-Unit?*

Ans. Mixed bed containing both cation- and anion-exchange resins is used to arrest silica slippage

DM WATER → $\boxed{\text{MIXED BED}}$ → POLISHED WATER

[Silica : 0.7 ppm]   [Silica : 0.01 ppm]

(Fig. 23.15)

Fig. 23.15

Q. *What is sodium slippage?*

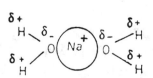

Fig. 23.16 *Hydrated sodium ion*

Ans. Due to them relatively smaller size, sodium ions have higher charge density on their surface. That is why they get highly hydrated, rendering the Na^+ ions to slip past the resin (cation-exchanger) bed and contribute to the increase in electrical conductivity of DM water output.

Q. *How much will the conductivity of DM water be increased if Na^+ slippage is 1 ppm?*

Ans. Usually 5 micro MHO/cm.

Q. *What is the difference between a Strong Acid Cation Exchanger (SAC) and a Weak Acid Cation Exchanger (WAC)?*

Ans. The only distinguishing feature is that the former contains strong acid groups such as— SO_3H (sulfonic acid group) and the latter contains weak carboxylic acid group (—COOH) substituted on the appended phenyl nucleus of the resin—a porous and permeable, inert, polymeric organic substance. That is, the former has — SO_3H groups and the latter —COOH groups in the cation exchange sites.

Q. *What is the advantage of an SAC over a WAC ?*

Ans. SAC can function well at all pH values while WAC fails to function effectively at pH below 5.

SAC can remove all cations associated with both strong and weak acids whereas WAC can remove cations of weak acids only, i.e, cations associated with water alkalinity, silicic acids and organic acids.

Q. *What is the chief advantage of Weak Acid Cation Exchangers (WAC) over Strong Acid Cation Exchangers (SAC)?*

Ans. Weak Acid Catex (WAC) have comparatively very high regeneration efficiency as a result of which they require only 110–120% excess acid over theoretical quantity of acid for regeneration, whereas SAC require as much as 150% to 300% more acid than the theoretical quantity.

Another advantage of WACs is that they require very dilute acid, as low as 0.5%, for their efficient regeneration. This dilute acid can be readily available from waste acid regenerant of SAC.

Q. *What is the difference between Strong Base Anion Exchangers (SBA) and Weak Base Anion Exchangers (WBA)?*

Ans. The Strong Base Anex (SBA) have quarternary amine (—$\overset{+}{N} \cdot R_4$) groups while Weak Base Anex (WBA) have primary amine (R $\cdot \overset{+}{N} H_{3-}$), secondary amine ($R_2 \cdot \overset{+}{N} H_{2-}$) and tertiary amine ($R_3 \overset{+}{N} H_-$) groups substituted on the appended phenyl nucleous of the resin for exchange of anions.

Q. *What is the advantage of a Strong Base Anex (SBA) over a Weak Base Anex (WBA) ?*

Ans. SBA can remove all anions—the conjugate base of all acids right from sulphuric, hydrochloric and nitric to weak acids like silicic acid and carboxylic acids whereas WBA can readily remove SO_4^{2-}, Cl^-, NO_3^-—the conjugate bases of strong acids H_2SO_4 HCl and HNO_3 but not the anions of weak acids, viz, silicic acids, carboxylic acids, etc.

Unlike WBA, SBA can function well in the entire range of pH.

Q. *What is the advantage of WBA over SBA?*

Ans. Weak Base Anex have higher regeneration efficiency than Strong Base Anex. As such WBA require only 120–140% excess of theoretical regenerant whereas SBA require as much as 150–300% excess over theoretical regenerant.

SBA must be regenerated with strong alkali, i.e., caustic soda (NaOH) solution alone. WBA, in contrast, require less expensive alkali, viz, soda ash (Na_2CO_3) solution or ammonia solution for regeneration. They can be regenerated even with the spent caustic regenerant of SBA.

Q. *What parameters characterize the ion exchangers?*

Ans. A number of parameters are involved. The most important among them being:
1. Operating exchange capacity
2. Granulometric composition (grain size)
3. Bulk mass
4. Swelling factor
5. Thermal stability.

Q. *What is operating exchange capacity of an ion exchanger?*

Ans. It is the gram-equivalents of ions absorbed by 1 m^3 of ion exchanger. Hence it is expressed in terms of g-equiv./m^3. So when this value is reached during servic cycle, the ion exchanger is said to have been exhausted.

Q. *Why cannot the initial water having high salt content ($\geq$ 800 ppm) be softened sufficiently in one softener?*

Ans. This is because of the counterion effect—a phenomenon which occurs as follows:

As the water with high salt content is passed through the bed of the hydrogen cation exchanger, the concentration of H$^+$ ions grows in the water. As it flows downward, more and more H$^+$ of cation exchangers are replaced by cations like Ca^{2+}, Mg^{2+}, N$^+$H$_4$, etc. The concentration of H$^+$ ions becomes maximum at the bottom part of the bed. This increased concentration of H$^+$ ions counteracts the dissociation of cation exchangers, obeying the La Chatelier Brown principle,

$$R—H \rightleftharpoons R^- + H^+$$

H-exchanger

and thereby reducing the rate of ion-exchange process near the bottom zone of the bed. Therefore, a good amount of cationic load remains in the effluent.

Q. *It is said that a decrease in HCO$_3^-$ ion concentration in the water being subjected to hydrogen cation exchange treatment intensifies the counterion effect. How does this happen?*

Ans. The presence of HCO$_3^-$ ions has some favourable effect in counteracting the counterion effect as induced by liberated H$^+$ ions by rich load of cations in water passing through a bed of H-cation exchanger. These HCO$_3^-$ ions bind the H$^+$ ions

$$HCO_3^- + H^+ \longrightarrow H_2CO_3 \Leftrightarrow H_2O + CO_2$$

and thereby reduce the concentration of H$^+$ ions accumulated in water during the course of cationic exchange. So H$^+$ ion-enrichment cannot take place (near the bottom of the bed) and therefore counterion effect is much reduced.

Therefore it is evident that with the decrease of bicarbonate concentration in the initial water, the counterion effect intensifies.

Q. *What is the swelling factor?*

Ans. Ion exchange resins swell in water. The degree by which they swell in volume is defined by swelling factor, ϕ, which is given by

$$\phi = A\,H\,M_b/Q_c$$

where A = Area of cross-section of cation-exchange bed, m^2

H = Height of the bed, m

M_b = Bulk mass of cation exchange resin, t/m^3

Q_c = Quantity of cation exchange resin, t

Q. *What is its significance in the practical field?*

Ans. Swelling factor must be taken into consideration during charging of ion exchange resin in the vessel. Otherwise, the bed thickness may be rather high and entail the risk of resin loss during backwashing.

Q. *What do you mean by the term breakthrough?*

Ans. It refers to the amount of contaminating cations (Ca^{2+}, Mg^{2+}, Na$^+$, for example) and anions Cl$^-$, SO$_4^{2-}$, etc.) entrained by the water softened by H- cation exchange and anion exchange bed respectively.

During regeneration of hydrogen cation exchanger, all the absorbed cations, viz, Ca^{2+}, Mg^{2+} and Na$^+$ are not removed. Particularly those saturating the cation exchange sites of the resin at the bottom layers of the bed cannot be removed by the common straight-flow regeneration technique.

Now during the service cycle these retained cations are liberated by the H$^+$ ions transferred to the water from the top layers of the bed. These cations will then deteriorate the quality of softened water. Hence the magnitude is determined in terms of breakthrough, i.e., the milligram of cations entrained per kg of effluent water.

$$C_{br} = \frac{C}{1 + \dfrac{k\,\eta_r}{(100 - \eta_r)}}, \text{mg/kg}$$

where

C = Concentration of removed ions from treated water, mg/kg

k = Distribution coefficient of ions between ion-exchange bed and water

η_r = Degree of regeneration, %

A similar explanation is valid for an anion exchanger.

Q. *To what type of ion-exchange material, does the term silicon capacity refer to?*

Ans. Strongly Basic Anion Exchangers (SBA).

Q. *What does it mean?*

Ans. It is an important characteristic of SBA defining the quantity of silicate anions (SiO_3^{2-}), in gram-equivalents, absorbed by 1 m^3 of the resin.

Q. *On which factors does the silicon capacity of the anion exchanger depend?*

Ans.
1. pH value of the water
2. Initial concentration of silicate ions $HSiO_3^-$ in water
3. Specific alkali consumption for regeneration
4. Filtering rate, etc.

Q. *How can penetration of large amounts of carbon dioxide into the water-condensate circuit of a boiler-turbine unit be avoided?*

Ans. This can be done by taking recourse to a greater dosage of ammonia into the feedwater.

Q. *Why?*

Ans. This ammonia will pass into solution and maintain the desired pH level (9–9.2) in the boiler feedwater. This will transform all dissolved CO_2 into carbonate anions which will be conveniently removed by the anion exchanger

$$NH_3 + H_2O \Leftrightarrow NH_4OH \Leftrightarrow \overset{+}{N}H_4 + \overset{-}{O}H$$

$$CO_2 + H_2O \Leftrightarrow H^+ + HCO_3^- \Leftrightarrow CO_3^{2-} + 2H^+$$

$$H^+ + OH^- \longrightarrow H_2O$$

Q. *What is the normal flowrate of water in the Catex and Anex during a service cycle?*

Ans.

FLOWRATE:	WAC	SAC	Anion Exchangers
	6 – 9	18 – 24	9–24

(m^3/h per m^2 of resin bed)

Q. *Why is the flowrate of water, during a service cycle, through a WAC lower than that through an SAC?*

Ans. The rate of cation exchange in a WAC is much slower than that in an SAC. Hence lower flowrate is to be maintained in a WAC to avoid slippage of cations.

Q. *When is the regeneration cycle to be started?*

Ans. It is to be started as soon as the resin beds are exhausted, i.e., when all the exchange sites of catex and anex have been saturated with undesirable ions picked up from the water in the course of its demineralization.

Q. *When can it be said that the SAC bed is on the verge of exhaustion and ready for regeneration?*

Ans. It is indicated by a sharp drop in hydrogen ion (H^+) concentration and/or fall of conductivity of water after the SAC bed.

Q. *Why?*

Ans. As the cation exchanger bed nears its exhaustion, a very small number of active sites will be available on the resin surface for the exchange of their H^+ ions by the metal cations present in the water. Therefore, there will be an abrupt decline in the H^+ ion concentration in water at the outlet of the SAC.

Moreover, as the H^+ ions are more labile than any other cation, these have higher ionic conductivity than any other cation and therefore an abrupt decline in H^+ ion concentration as a consequence of exhaustion of the resin bed means a sharp drop in conductivity of water at the exit of the SAC bed.

ion concentration as a consequence of exhaustion of the resin bed means a sharp drop in conductivity of water at the exit of the SAC bed.

Q. *When can it be said that the WAC bed is on the verge of exhaustion and needs regeneration?*

Ans. It is marked by a sharp increase in hardness and/or alkalinity in the water just after the WAC bed.

Q. *Why?*

Ans. Weak Acid Cation Exchangers (WAC) can remove only those cations which are associated with weak acids, viz., carbonic acid, silicic acid and organic acids, so the exhaustion of WAC bed means an increase in these cations in the exit water. Had these cations been removed by the WAC bed, there would have been sufficient concentration of H^+ ions (transferred from resin) in water which would combine with the anions of these weak acids and prevent their hydrolysis to cause alkalinity in water.

When WAC is active $\begin{cases} H^+ + HSiO_3^{1-} \rightleftharpoons H_2SiO_3 \\ H^+ + CO_3^{2-} \rightleftharpoons H_2CO_3 \end{cases}$

When WAC is exhausted

$\begin{cases} HSiO_3^- + H_2O \xrightarrow{hydrolysis} H_2SiO_3 + OH^- \\ CO_3^{2-} + H_2O \xrightarrow{hydrolysis} HCO_3^- + OH^- \end{cases}$

Accounts for alkalinity

Q. *What indicates the exhaustion of an anion exchanger?*

Ans. It is marked by a sharp increase in the silicate concentration or abrupt fall of conductivity followed by its rise again in the water at the outlet of anion exchanger.

Q. *What principal steps are involved in the regeneration cycle?*

Ans. Four:
1. Backwashing
2. Acid injection to Catex bed
   Alkali injection to Anex bed
3. Slow rinse
4. Fast rinse.

Q. *What is backwashing?*

Ans. It is the upward flow of softened water through the resin bed. The water is forced through the under-

drain of the bed. It flushes up the bed and agitates each bead in a turbulent motion and finally discharges out of the service water distributor and runs into waste.

Q. *What is the purpose of backwashing?*

Ans. Backwashing dislodges the resin fines and filtered particulates from the interspaces of resin beads and therefore cleans the resin surface to ensure efficient regeneration by acid (for Catex regeneration) or alkali (for Anex regeneration) treatment.

Q. *What is the normal duration of backwashing?*

Ans. It is carried out till a clear backwash effluent results. Normally it takes 10 to 15 minutes to complete backwashing.

Q. *What is the flowrate of backwash water?*

Ans. It usually ranges from 12 to 24 m^3/h per m^2 of Catex bed and 3 to 9 m^3/h per m^2 of Anex bed.

Q. *On which factors backwash flowrate depend?*

Ans.
1. Type of resin
2. Backwash-water temperature.

Q. *What is the strength of acid taken to regenerate a cation exchanger?*

Ans. Strong Acid Cation Exchangers are regenerated with 2–6% H_2SO_4

Weak Acid Cation Exchangers are regenerated with 0.5–0.7 % H_2SO_4

Q. *What is the flowrate of acid for regeneration?*

Ans. SAC beds are regenerated with an acid flowrate of about 5-10 m^3/h per m^3 of resin bed.

WAC beds are regenerated with an acid flowrate of about 10 m^3/h per m^3 of resin bed.

Q. *What is the strength of alkali taken to regenerate anion exchangers?*

Ans. 4 – 5% NaOH solution.

Q. *What is the flowrate of regenerant alkali?*

Ans. 5 – 10 m^3/h per m^3 of resin.

Q. *Is the regenerant fed to the exhausted resin bed in upward flow or downward flow?*

Ans. The regenerant is charged to the resin bed from the top and distributed evenly through the regenerant distributor. The regenerant flows down the bed by gravity and regenerates the beads in the process, and discharges out through the bottom drain into waste.

Q. *Why are some demineralizing plants befitted with sorption filters charged with activated charcoal and set upstream of anion exchangers?*

Ans. This is a measure to trap the organic substances present in the treated water. Since these substances cannot be completely removed by conventional pretreatment, they find their way to the anion-exchange grains and get sorbed as ions. Macro molecules of organic substances infiltrating into the pores of anion-exchangers, particularly SBA, are very difficult to dislodge from the exchange sites during the course of regeneration and therefore poison the anion exchange resins and reduce their exchange capacity. However, activated charcoal filters mounted upstream of anion exchanger will trap these unwanted organic molecules and safeguard the exchanger.

Q. *Why is alkali or acid injection followed by slow rinsing?*

Ans. After alkali (Anex bed) or acid (Catex bed) treatment, the rejuvenated resin beds are subjected to a low downward flow of rinse water containing an insignificant amount of ions. This is carried out to remove any alkali or acid remaining.

Q. *Why is slow rinse to be followed by fast rinse?*

Ans. Fast rinsing of resin bed, at a rate of 15–20 m^3/h per m^3 of resin, follows slow rinsing to dislodge any local regenerant trapped in the interspace of resin beads. The flow is downward.

Q. *How is a mixed bed demineralizer regenerated?*

Ans. Mixed bed demineralizer is an infinite series of cation exchanger and anion exchangers packed in a single column. When exhausted as determined by conductivity test and/or silica measurement—the bed can be regenerated through the following successive steps:

1. BACKWASHING: the bed thoroughly to separate completely the anion exchange resin

from cation exchange resin into two beds. Anion exchangers being lighter than the more compact cation exchangers form the top bed.
2. ACID INJECTION: Regenerating acid (H_2SO_4) is charged to the column through the bottom distributor. It moves up through the catex bed which gets regenerated in the process and the acid flows out of the interface collector to the drain.
3. ALKALI INJECTION: The caustic is introduced from the top via the upper distributor and moves down the anex bed which whereupon gets regenerated. The alkali then flows out of the interface collector to the drain.
4. RINSING: The bed is then subjected to a short rinse to purge out any excess acid or alkali remaining.
5. MIXING: The anion and cation exchange resins are then thoroughly agitated by air introduced from the bottom to mix them uniformly.
6. FINAL RINSING: The column is filled up with polish water and rinsed till the effluent is of acceptable quality to declare the M.B. ready for line-up.

Q. *What combination of cation exchanger and anion exchanger should be most economic where silica and CO_2 removal from feedwater is not essential?*

Ans. Strong Acid Cation Exchanger (SAC) is to be hooked up in series upstream of a Weak Base Anion Exchanger (WBA)

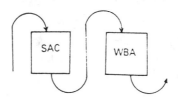

Fig. 23.17

While the metal cations are absorbed in the SAC bed, silicate and carbonate anions slip through the anion exchanger downstream as it is a weak base type.

Q. *What is the advantage of such a system?*

Ans. Equipment as well as regeneration costs are low.

Q. *However, if silica removal is not required but CO_2 elimination is necessary, what demineralization system would serve the purpose most economically?*

Ans. One degasifier (DG) should be lined up in series downstream of the above system. (Fig. 23.18)

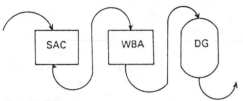

Fig. 23.18

The CO_3^{2-} anions, escaping arrest by WBA, form gaseous carbon dioxide with the protons H^+ liberated in the SAC bed.

$$CO_3^{2-} + H^+ \rightleftharpoons HCO_3^- \xrightarrow{H^+} H_2CO_3$$
$$\longrightarrow H_2O + CO_2$$

The carbon dioxide is eliminated in the degasifier downstream of WBA.

Q. *What is the conductivity of effluent water after degasifier above?*

Ans.

$\lambda = 10\text{–}20 \ \mu\text{S/cm}.$

cf. [1 μ S/cm = 1 microsimens/cm

= 10^{-6} Ohm^{-1} cm^{-1}/cm]

Q. *Which demineralizing system will be technically suitable to remove silica (along with other ions) from treated water where alkalinity is low?*

Ans. A Strong Acid Cation Exchanger (SAC) and Strong Base Anion Exchanger (SBA) joined in series will serve the purpose. (Fig. 23.19)

SAC bed eliminates all cations associated with both strong and weak acids and SBA bed removes all anions associated with both strong and weak acids, viz., silicic acid.

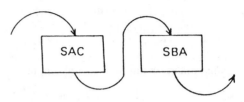

Fig. 23.19

Q. *What will be the silica level in the effluent in this case?*

Ans. It will be in the range 0.02 — 0.1 ppm.

Q. *What will be the conductivity of this effluent?*

Ans. 5 — 10 μS/cm, i.e., $(5 - 10) \times 10^{-6}$ Ohm^{-1} cm^{-1}/cm

Q. *What is the chief disadvantage of this set-up?*

Ans. High chemical costs.

Q. *Which demineralizing system will ensure efficient removal of silica from treated water whose alkalinity is high?*

Ans. One degasifier inserted between SAC and SBA beds in the above combination will fulfill the task.

Much of the carbonate alkalinity will be eliminated in the degasifier. (Fig. 23.20)

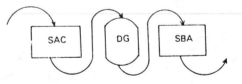

Fig. 23.20

$$2H^+ + CO_3^{2-} \longrightarrow H_2O + CO_2$$

(effluent of SAC bed)
and thereby reducing the load on SBA.

Q. *What will be the silica level in the effluent?*

Ans. 0.02 — 0.1 ppm.

Q. *What is the advantage of this system?*

Ans. Low chemical cost.

Q. *Why?*

Ans. Degasifier eliminates the CO_3^{2-} ions and thereby minimizes the load on the Strong Base Anion Exchanger downstream and this will ensure longer service period of SBA before it is being subjected to regeneration. This buys economy in regenerant expenditure.

Q. *If silica removal (along with other ions) is required from water having high alkalinity, high chloride and sulphate concentrations, what demineralization system will be both technically and economically viable?*

Ans. One each of Strong Acid Cation Exchanger, Weak Base Anion Exchanger, Degasifier and Mixed Bed hooked up in series will fulfill the requirement.

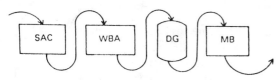

Fig. 23.21

While SAC bed eliminates all the cations belonging to both mineral acids (HCl, H_2SO_4, etc.), and weak acids (silicic acid, carbonic acid), the WBA bed removes all anions of strong acids (HCl, H_2SO_4). As the effluent of WBA bed passes through the degasifier, the CO_3^{2-} ions are eliminated as CO_2. Silica that is not arrested in the WBA bed is eliminated in the M.B.

Q. *What is the silica content of the final effluent of this system?*

Ans. 0.02 — 0.1 ppm.

Q. *What is the conductivity of the final effluent?*

Ans. 5 — 15 μS/cm

Q. *What demineralizing system will effectively remove silica (together with other ions) from water with high alkalinity, high hardness and high chloride and sulphate?*

Ans. A WAC–SAC system linked to a WBA–SBA system via a degasifier will ensure demineralization plus silica removal. (Fig. 23.20)

Fig. 23.22

The WAC bed reduces the load on the SAC bed by removing all cations associated with weak acids like silicic acid.

The SAC bed removes all cations.

The degasifier eliminates CO_3^{2-} ions and thereby reduces the load on WBA bed which further reduces the load on the SBA bed by scavenging out SO_4^{2-}, Cl^- anions readily.

Q. *What is the silica level in the final effluent of this system?*

Ans. It ranges from 0.02 to 0.1 ppm.

Q. *What is the advantage of this system?*

Ans. Lowest chemical costs.

Q. *What are the disadvantages of this system?*

Ans. 1. Equipment cost is high
2. Repumping is required

Q. *What convenient DM system will ensure polish water output economically from the treated water which has high alkalinity and high sodium content?*

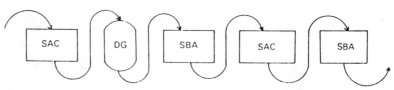

Fig. 23.23

Ans. The following arrangement (Fig. 23.23) of exchangers and degasifier hooked up in series may ensure this.

Nearly all cations associated with both strong and weak acids are eliminated in the 1st SAC bed.

Followed by this, the degasifier removes all carbonate ions from the acidic effluent of SAC and thereby reduces the load on SBA that

$$CO_3^{2-} + 2H^+ \longrightarrow H_2O + CO_2$$

removes all anions—those of both strong and weak acids.

Any leftover cations and anions slipping out will be arrested in the final SAC and SBA beds respectively.

Q. *What is the silica content in the final effluent?*

Ans. As low as 0.01 to 0.05 ppm.

Q. *What is the conductivity of the polish water obtained this way?*

Ans. $0.1 - 5\,\mu S/cm$, i.e., $(0.1 - 5) \times 10^{-6}\ ohm^{-1}\ cm^{-1}/cm$

Q. *What is the advantage of this system?*

Ans. Low chemical costs.

Q. *What are the disadvantages of this combination?*

Ans. It requires repumping and the equipment cost is high.

Q. *If polish water (i.e., highly pure DM water) is required from treated water which has high alkalinity and high dissolved solids, what convenient DM system will ensure this?*

Ans. A combination of SAC, degasifier and SBA will ensure DM water (Fig. 23.24)

which when passed through a mixed bed which is an infinite series of cation and anion exchange systems will produce polish water.

Q. *What is the silica content of the polish water output of this system?*

Ans. As low as 0.01 to 0.05 ppm.

Q. *What is the conductivity of the final effluent?*

Ans. $1\,\mu S/cm$.

Q. *What is the advantage of this system?*

Ans. Low chemical costs.

Q. *And disadvantages?*

Ans. It requires repumping and the equipment cost is high.

Q. *What factors are responsible for the malfunctioning of the demineralization system?*

Ans. There are basically four principal factors:
1. Mechanical failures
2. Resin degradation and/or fouling
3. Improper operational controls
4. Abrupt changes in the characteristics of feedwater.

Q. *What types of mechanical failures are responsible for the malfunctioning of a DM system?*

Ans. This means physical defects in the equipment of DM system and may come in the form of broken or plugged distribution laterals, clogged underdrain screens and malfunctioning of multiport valves.

Q. *What do you mean by resin degradation?*

Ans. It refers to the breakdown of ion exchange properties of resin inflicted by some external agencies or processes.

Q. *How many types of resin degradation are there?*

Ans.
1. Physical degradation of resin
2. Chemical degradation of resin
3. Resin degradation by fouling.

Q. *How may physical degradation of resin occur?*

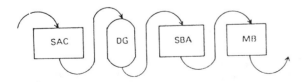

Fig. 23.24

Ans. It may be due to
  (a) osmotic shock
  (b) thermal shock
  (c) mechanical shock.

Q. *What is osmotic shock?*

Ans. Periodic expansion of resin beads during service cycle and contraction during regeneration cycle induce in the resin structure a shock that is called osmotic shock.

Q. *Why does this expansion and contraction of resin beads occur?*

Ans. Hydrogen ion (H^+) is the smallest cation while the metal cations Na^+, Ca^{2+}, Mg^{2+}, etc. are much larger—their size increases with their atomic number. Hence when fully regenerated cation exchange resin is treated with water containing metal cations, the resin beads expand as their H^+ ions, at the exchange sites, are increasingly being replaced by these cations. Larger the size of the cations replacing the H^+ ions of the resin, greater will be the expansion of resin beads.

Similarly when these resin beads fully loaded with cations are regenerated, the large cations are replaced by much smaller size H^+ ions and as a consequence the cation exchange resin beads contract.

In an analogous way the expansion of anion exchange resin beads occurs when their OH^- ions replaced by much larger anions, CO_3^{2-}, SO_4^{2-} and Cl^- during the service cycle and contraction occurs when these resin beads saturated with large anions are replaced by much smaller OH^- ions during regeneration cycle.

Q. *What is thermal shock?*

Ans. It is the strain induced to the resin structure due to rapid variations of temperature.

Q. *How does it occur?*

Ans. Thermal strain is inflicted to resin when it is subjected to a cold regeneration cycle after a hot service cycle and then repeating the total cycle all over again.

Also the resin may suffer from thermal degradation if it is subjected to operations at temperatures higher than those for which it is designed.

Q. *How is mechanical strain induced to a resin bed?*

Ans. This is due to the load exerted by the resin fines upon the rest of the bed. The resin fines shift to the top of the bed and increase the pressure drop over the bed. As this pressure drop mounts, more resin beads break down to fines which migrate to the top of the bed and inflict more mechanical strain on the resin bed.

Q. *What is the effect of mechanical strain on resin bed?*

Ans.
  1. Production of more fines in a cumulative way
  2. Increase of pressure drop in the resin bed
  3. Increased channeling
  4. More resin loss during backwashing.

Q. *What is chemical degradation of resin?*

Ans. It refers to breakdown of chemical structure of resin resulting from fission of divinylbenzene cross-link of the resin polymer.

Q. *Why does this occur?*

Ans. This occurs due to oxidation of C—C bonds of the cross-link by dissolved oxygen or chlorine

$$Cl_2 + H_2O \longrightarrow 2HCl + O$$

This nascent oxygen produced by dissolved chlorine is powerful enough to rupture the divinyl cross-link.

Q. *What is the effect of chemical degradation of resin?*

Ans. This brings about
  (a) hydration of resin and swelling of resin beads
  (b) greater pressure drop across the bed
  (c) more channeling
  (d) more loss in ion-exchange capacity.

Q. *What is resin degradation by fouling?*

Ans. Resin may degrade due to fouling arising from surface coating of resin beads by
  (a) metal oxides and inorganic salts
  (b) organic substances
  (c) oil
  (d) microbiological organisms.

Q. *What metal oxides and inorganic salts are known to foul resin beads?*

Ans. These are:

Metal Oxides	Inorganic Salts
Fe_2O_3	$CaCO_3$
Fe_3O_4	$Mg(OH)_2$
CuO	$CaSO_4$
ZnO	Silicate salts
Al_2O_3	

Q. *How does iron oxide fouling occur?*

Ans. Iron may find its way to the resin bed either as insoluble form (Fe^{3+}) or soluble form (Fe^{2+}). Its sources being the corrosion products of water distribution pipeline, contaminated regenerants, organically sequestered iron suspended in the raw water, contamination of turbine condensate by cooling water due to leakage in the turbine condenser, etc. Insoluble ferric oxide (Fe_2O_3) is retained on the surface of the resin beads by virtue of filtration action through the bed. The impervious layer of ferric oxide on resin surface inhibits efficient resin-water contact and thereby reducing the exchange capacity of the resin.

The soluble ferrous ions, on the other hand, find their way to exchange sites by being exchanged with H^+ ions of the hydrogen cation exchanger. These absorbed Fe^{2+} ions are difficult to dislodge during the regeneration cycle, from the exchange site particularly after aging.

Q. *What is the source of copper oxide fouling?*

Ans. The copper and brass tubes of the turbine condenser are the source of copper oxide fouling.

The corrosion and erosion of such tubes transfer CuO and Cu_2O in the condensate which carries them down and deposits them on the resin bed and thereby causes copper oxide fouling.

Q. *How does alumina fouling occur?*

Ans. If the feedwater brings in alumina or aluminium salts because of malfunctioning of alum clarifier in the water treatment plant, flocs of alumina (Al_2O_3) or soluble Al^{3+} ions will appear in the resin bed. While insoluble Al_2O_3 will deposit itself on the resin surface in the same way as does Fe_2O_3, soluble aluminium ions will be absorbed at the cation exchange sites of the resin by replacing hydrogen ions (H^+). Since these Al^{3+} ions are held so tightly to the resin, they cannot be easily removed by straight-flow regeneration techniques, thereby decreasing the number of available sites for cation exchange.

The insoluble Al_2O_3 layer on the resin beads is too impervious to allow efficient resin-water contact, with the effect that ion-exchange capacity of H-ion exchanger is reduced.

Q. *Since supercritical pressure once-through boiler units require highly pure feedwater, how can the concentration of these dispersed and ionic impurities in BFW be brought down to the prescribed level?*

Ans. Such removal can be ensured by a special demineralizing system befitted with mechanical filters and mixed bed units lined up with a series of cation exchanger, degasifier and anion exchanger.

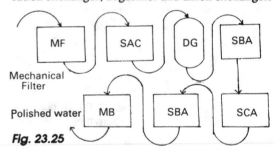

Fig. 23.25

Mechanical filters will remove the dispersed impurities, mainly the corrosion products while the mixed bed will almost completely eliminate any cations or anions escaping absorption in the ion exchangers upstream.

Q. *What are mechanical filters?*

Ans. These may be
   (a) cellulose tubular-element precoat-type filters
   (b) sulphonated-coal filters
   (c) electromagnetic filters

Q. *What are the advantages of cellulose tubular-element precoat-type filters?*

Ans. They
   (a) possess high unit capacity 500 m^3/h

(b) ensure reliable filtration off suspended impurities in water.

Q. *What are its drawbacks?*

Ans. It is beset with certain basic operational difficulties which are:
(a) removal and precoating of cellulose layer
(b) the necessity of fitting cellulose traps downstream to arrest drained cellulose.

Q. *What is the chief advantage of sulphonated coal over the cellulose variant of mechanical fliters?*

Ans. Better operational reliability. Besides, it also removes, unlike cellulose filters, the NH_4^+ ions from water.

Q. *What is the usual bed thickness of sulphonated coal type mechanical filters?*

Ans. 0.8 to 1 m.

Q. *What is the basic limitation of electromagnetic filters?*

Ans. They remove from water only the magnetic substances leaving all other dispersed impurities free to pass out. That is only the particles of Fe_3O_4 are retained by electromagnetic filters.

Q. *Then why is the use of electromagnetic filters, instead of cellulose tubular-element precoat-type filters or sulphonated coal filters justified?*

Ans. Since the corrosion products consist predominantly of Fe_3O_4, the use of electromagnetic filters is justified.

Q. *Why does calcium carbonate and magnesium hydroxide fouling occur?*

Ans. This is due to inadequate performance of lime softener and/or filtration operation in the raw water treatment plant. Minute particles of calcium carbonate or magnesium hydroxide may be carried over along with feedwater into the demineralizing system or they may also result from their after-precipitation downstream of pretreatment due to malfunctioning of the softener.

Q. *How does calcium sulphate fouling occurs?*

Ans. The possibility of calcium sulphate fouling arises during the regeneration of the cation exchanger bed with sulphuric acid. If the regenerant acid (H_2SO_4) concentration is high and its flowrate is low, the calcium ions replaced by H^+ ions of acid from the saturated resin will form sparingly soluble calcium sulphate which will settle as insoluble solid on the surface of the resin beads and thereby cause Ca-sulphate fouling of the resin.

Q. *What is the minimum concentration of sulphuric acid that will entail the risk of $CaSO_4$ fouling of resin beads?*

Ans. The possibility of calcium sulphate fouling becomes significant if the H_2SO_4 concentration exceeds 5% strength.

Q. *Higher the strength of H_2SO_4, greater is the risk of calcium sulphate fouling. Why?*

Ans. The calcium sulphate precipitates from the solution as soon as the ionic product of Ca^{2+} and SO_4^{2-} ions reaches the solubility product of $CaSO_4$.

$$\text{IONIC PRODUCT OF} \atop \overline{Ca^{2+} \ \& \ SO_4^{2-} \ \text{IONS}}$$

$$S.P._{CaSO_4} = [Ca^{2+}][SO_4^{2-}] = 2.3 \times 10^{-4}$$

i.e. $[Ca^{2+}][SO_4^{2-}]$ represents the ionic product of Ca^{2+} and SO_4^{2-} ions.

The presence of sulphuric acid of higher strength (> 5%) increases the concentration of SO_4^{2-} ions in the aqueous phase and with that the ionic product of Ca^{2+} and SO_4^{2-} ions reaches the value of solubility product of $CaSO_4$ early, causing the precipitation of calcium sulphate. This is due to the common ion effect induced by the sulphate ions of the sulphuric acid added as regenerant.

Q. *If the flowrate of regenerant acid (H_2SO_4) is below 10 m^3/h per m^3 of resin bed, calcium sulphate fouling also occurs. Why?*

Ans. Low flowrate of regenerant acid solution will endow ample opportunity to facilitate the nucleation and growth of the $CaSO_4$ crystals in the aqueous phase and will allow them to settle on the surface of the resin beads. On the other hand, higher flowrate will drift the crystals away before they grow appreciably in size to settle by gravity, besides scavenging them off the resin beads if they chance to settle on them (resin).

Q. *Why does silica fouling occur?*

Ans. This occurs due to improper regeneration of both strong and weak base anion exchangers. This is due to the formation of colloidal silicic acid within the resin beads as follows: When anion exchangers are regenerated by alkali, the silicate anions $HSiO_3^-$ together with other anions are replaced from the exchange sites of the resin by the hydroxyl ions OH^- liberated by alkali. Since silicate ion is the salt of a weak acid, it undergoes hydrolysis to produce soluble silicic acid:

$$HSiO_3^- + H_2O \Leftrightarrow H_2SiO_3 + OH^-$$

This soluble silicic acid eventually polymerizes into a colloidal form of silicic acid which causes silica fouling.

Q. *How can fouling silica be minimized?*

Ans. Leaching out the colloidal silica entrapped in the resin beads by warm (50°C) water before subjecting the anion exchangers to warm alkali (NaOH solution) for regeneration.

Q. *How does organic fouling occur?*

Ans. If the raw water source is surface water, there is the likelihood that it may get contaminated with sewage effluent and decaying organic bodies which may add some heavy organic acids, viz., humic acid, fulvic acid and tannic acid to the raw water. These acids, mainly in the colloidal form, escape separation during pretreatment of water and get absorbed at the exchange sites of the anion exchanger. Upon being absorbed by the resin, they close the resin pores and as a consequence the capacity of the ion exchanger is reduced and regeneration becomes incomplete.

Q. *Why do they attack the anion exchangers?*

Ans. Most of these organic foulants, being basically acidic in nature, will naturally seek basic exchange sites and that's why anion exchangers are mostly affected.

Q. *Apart from extrinsic organic foulants, can the organic fouling of resin take place?*

Ans. Yes. The breakdown of divinylbenzene crosslink in the cation exchangers because of oxidation and transfer of the reaction products to the anion exchangers downstream my also irreversibly foul the anion exchange resins.

Q. *Between a weak base anion exchanger and a strong base anion exchanger which one is more resistant to organic fouling?*

Ans. Weak Base Anion Exchanger (WBA).

Q. *Why?*

Ans. The Weak Base Anion Exchangers actually exchange their hydroxyl ions for the anions of organic acids and thus scavenge them out of water.

Note: Because of this scavenging action WBA protects costlier SBA downstream.

Q. *How does oil fouling affect the ion exchange process of resin?*

Ans. Oil infiltrating the resin bed will coat the beads and agglomerate them into larger particles, with the effect that there will be an increased pressure drop across the bed, besides channeling and premature breakthrough.

Q. *How does microbial fouling affect the ion exchange resin?*

Ans. Microbiological organisms have been found to anchor themselves on the surface of resin and they grow and proliferate by deriving their food from such compounds as nitrates, phosphates, ammonia, oils and other organic compounds. As these microbes, particularly in slimy form, grow, increased pressure drop across the resin bed occurs and water–resin contact is reduced, causing highly contaminated effluent.

Q. *What do you mean by improper operational controls that contribute to malfunctioning of DM systems?*

Ans. Improper operational controls mean deviation from the designed values of:

(a) service flowrate
(b) backwash flowrate
(c) slow-rinse flowrate
(d) fast-rinse flowrate
(e) regenerant flowrate
(f) regenerant concentration
(g) regenerant dosage.
(h) water temperature.

Q. *If the service flowrate is higher than the designed values, what difficulties may be encountered in a demineralizing system?*

Ans. It may cause:

(a) bed compaction and higher pressure drop across the bed as a consequence
(b) channeling
(c) premature breakthrough.

Q. *If, on the contrary, the service flowrate is below the designed value, what difficulties may crop up?*

Ans.

1. Ion slippage
2. Reduced capacity of the DM system.

Q. *If the concentration of regenerant is higher than the prescribed value what will happen?*

Ans.

1. It will lead to osmotic shock
2. It will foul the H-cation exchanger with $CaSO_4$.

Q. *If the concentration of the regenerant is very low, i.e., below the prescribed range of strength what will be the effect?*

Ans. Unsatisfactory regeneration.

Q. *If the regenerant flowrate is very high what will be the effect?*

Ans. Unsatisfactory regeneration because of reduced contact-time.

Q. *If from the record it is found that there is a gradual increase of rinsing time during each regeneration cycle, what conclusion may you draw,* provided *the same rinse flowrate is maintained everytime?*

Ans. This simply indicates that the resin bed is getting fouled.

Q. *If there is permanent deviation of influent water characteristics turbidity, total hardness, silica, organic impurities, heavy metallic compounds and the presence of dissolved oxidizing agents—what will be the effect on the demineralization of water?*

Ans. It may lead to:

(a) production of water having inadmissible degrees of impurities
(b) capacity loss
(c) degradation and/or fouling of resin.

Q. *What factors are responsible for a high pressure drop across the resin bed?*

Ans.

1. Bed compaction
2. Too much accumulation of resin fines in the bed
3. Higher service flowrate
4. Slimy microbial growth
5. Channeling
6. Increased turbidity
7. Broken underdrain
8. Clogging of internal distributor with resin fines.

Also, a defective pressure gauge may show a high pressure drop.

Q. *If bed compaction be the cause of high ΔP across the resin bed what is the remedy to it?*

Ans. Thorough agitation of the bed either pneumatically (i.e., by blowing air through the bed) or mechanically (i.e., by stirring) after backwash.

Q. *What is the remedy to the high pressure drop across the bed due to the increased load of turbidity in the influent water?*

Ans. It needs adequate pretreatment of the feedwater and a mid-cycle backwash.

Q. *How can microbial growth in the form of slime be inhibited?*

Ans. Proper pretreatment of feedwater to kill the micro-organisms in it.

Q. *How can channeling be reduced?*

Ans. By adjusting the service flowrate and back-wash.

Q. *How the problem of high ΔP across the resin bed due to accumulation of resin fines be eliminated?*

Ans. These resin fines being lighter than beads themselves collect on the top of the resin bed. They must be removed by thorough backwashing over an extended period while keeping the manhole open and scrapping off the fines from the bed-top.

Q. *If the internal distributor is clogged by resin fines causing high ΔP across the bed, what should be done to eliminate this?*

Ans. Backwashing and cleaning the distributor.
It is to be repaired if found damaged.

Q. *What are the reasons for a low pressure drop across the resin bed?*

Ans.
1. Low flowrate
2. Reduced bed depth due to resin loss.

Q. *What should be done in the first case?*

Ans. Service flowrate should be adjusted to the proper value.

Q. *What factors are responsible for resin loss?*

Ans. Resin loss may be due to
  (a) backwash
  (b) underdrain failure
  (c) wearing out of resin due to ageing.

Q. *What are the causes of the loss of capacity of an ion exchanger?*

Ans.
1. Higher load of exchangeable ions in the in-fluent water.
2. Inadequate regeneration
3. Resin degradation by oxidants
4. Resin fouling
5. Resin ageing
6. Resin loss taking place

7. Excessive rinsing
8. Channeling and improper distribution of feed-water
9. Capacity overtaxed in previous run.

Also a defective flow recorder may show a false indication of loss of capacity.

Q. *If the increase in the load of exchangeable ions in the feedwater is responsible for capacity loss, what is the remedy for it?*

Ans. Ionic load in water should be reduced. And therefore adequate pretreatment of the raw water to reduce its removable ions as much as possible is necessary so that the exchangeable ionic load of the influent water entering the DM unit is within the desirable range for satisfactory functioning of the ion exchangers.

Q. *If inadequate regeneration is the cause of "loss of capacity", what should be done to counter it?*

Ans. Proper regeneration. And that needs correct concentration and correct flowrate of the regenerant as well as the correct temperature of regeneration. For this purpose it may be necessary to check the regenerant concentration over and above its dilution system.

Q. *If resin deterioration by oxidation has been detected what should be done?*

Ans. Carryover of dissolved chlorine in influent water should be checked and complete dechlorination should be ensured.

Q. *How can resin fouling, causing loss in capacity, be avoided?*

Ans. The following measures should be undertaken to counteract resin fouling:
1. Proper pretreatment of water
2. Proper backwash
3. Checking the regenerants for foulants
4. Checking influent water for foulants
5. Ensuring proper regeneration of resin

Q. *What is the remedy to the problem of resin ageing?*

Ans. Resin analysis should be checked to determine the useful resin life. Normally cation exchangers suffer 5% yearly loss in capacity and anion exchangers 20%. If the checking shows more loss, the resin is better replaced.

Q. *If resin loss is detected, what should be done?*

Ans.

1. Carryover of resin with backwash water is to be checked.
2. Excessive backwash is to be avoided.
3. Designed backwash rate is to be maintained.

Q. *What are the possible factors responsible for the poor quality of the output water of a DM Plant?*

Ans.

1. Catex bed exhausted
2. Anex bed exhausted
3. Channeling
4. Leaky valves causing leakage of exchangeable ions
5. Overrunning capacity
6. Poor regeneration
7. Fouling of resin
8. Resin deterioration due to ageing and/or oxidation
9. Excessive service flowrate
10. Low service flowrates
11. Low temperature
12. High pH in treated water due to regenerant (Caustic alkali) valve leakage
13. Increase of Na^+ ions in influent water
14. Higher hardness of influent water
15. High silica in the feedwater
16. Low pH in treated water due to overrunning of anion exchanger capacity

Q. *What are the reasons for excess leakage of Na^+ ions in treated water?*

Ans. It may be due to
(a) increase in Na^+ load in influent water
(b) inadequate rinse
(c) oxidation of resin
(d) capacity overrun.

Q. *How can this problem be overcome?*

Ans.

1. Quality of influent water is to be analyzed
2. Analysis of resin is to be done to check the presence of cation exchange resin on anion exchange resin
3. The quality as well as concentration of the acid (regenerant) are to be checked
4. Regeneration of the cation exchanger with higher concentration of acid is to be performed.

Q. *Why does hardness in the effluent water of a DM plant occur?*

Ans. It may be due to
(a) slippage of a cation through the cation exchanger
(b) hardness in water used during the regeneration cycle.

Q. *What is the remedy for the high pH of the treated water of a DM plant?*

Ans.

1. Checking any leakage of regenerant (caustic alkali) valve. Zero-leaking operation of the valve is to be ensured.
2. pH monitoring equipment is also to be checked and rectified if found malfunctioning, as a faulty pH meter will show wrong pH values.

Q. *What are the reasons of low pH of the treated water of a DM plant?*

Ans.

1. Valve leakage
2. Overrunning of anion exchanger capacity
3. Organic fouling of anion exchangers.

Q. *What measures should be adopted to battle this problem?*

Ans.

1. Any valve leakage of regenerant (acid) is to be checked and leakage rectified
2. Proper treatment of influent water in the water treatment plant upstream to destroy its organics before it is delivered to the battery limit of the DM unit
3. Anion exchanger bed is to be checked for its performance
4. Also it is necessary to check the pH monitoring instrument

Q. *Why does silica slippage in the treated water of a DM plant occur?*

Ans.

1. Higher silica load of influent water
2. Organic fouling of anion exchanger
3. Anion exchanger capacity overrun
4. Higher temperature of influent water
5. Silica deposition on anion exchange resin.

Q. *How can silica slippage be contained?*

Ans.

1. Pretreatment of influent water is necessary to destroy the organics
2. Regeneration of anion exchanger at higher regeneration level is to be ensured
3. The resin bed is to be preheated with warm water (50°C) to dislodge the absorbed silica and then it is to be regenerated by warm (50°C) alkali

Q. *If overrunning capacity be the cause of loss of capacity of the ion exchange process, what should be done?*

Ans.

1. Throughput rate is to be reduced
2. End of run determination is to be evaluated.

Q. *In which forms does the malfunctioning of the mixed bed of a DM unit appear?*

Ans.

1. Overall poor quality of output water in general
2. Higher silica slippage
3. Low pH of effluent
4. High pH of effluent
5. High conductivity of effluent.

Q. *Why does poor quality of treated water from MB occur?*

Ans. It may be due to one or more of the following causes:

1. Inadequate regeneration of ion exchangers
2. Valve leakage
3. Inadequate bed depth
4. Poor mixing after regeneration
5. Improper resin separation during backwash for regeneration.

Q. *What are the causes of silica slippage from a Mixed Bed?*

Ans.

1. Resins have got exhausted
2. Improper anion exchanger regeneration because of inadequate separation during backwash
3. Fouling of anion exchange resin by organic substances
4. Not maintaining the correct quality and concentration of regenerant.

Q. *Why does low pH of mixed bed effluent occur?*

Ans. It may be due to
 (a) premature exhaustion of anion exchange resin of MB due to silica or organic fouling.
 (b) Improper mixing of anion and cation exchanger resin beads.

Q. *What is the remedy to this problem?*

Ans.

1. Chemical cleaning of the fouled component of the mixed bed is required.
2. Regeneration should be supplemented by thorough mixing of the cation and anion exchange resins.

Q. *What are the causes of high pH of the MB effluent?*

Ans.

1. Load variation influent to mixed bed.
2. Overexhaustion of the cation exchanging component.

Q. *What are the reasons for high conductivity of water at a mixed bed outlet?*

Ans.

1. Channeling and poor distribution
2. Unsatisfactory regeneration due to improper separation of cation and anion exchange resins of the MB during backwash
3. Fouling of anion exchange component by organics
4. Overexhaustion.

24

Scaling of Fireside of Heating Surfaces

Q. *What are the harmful effects of ash deposition on boiler tubes?*

Ans. Uneven ash deposition on fireside steam generating surfaces of drum-type boiler may lead to alarming disturbances of circulation. The situation aggravates with higher steam generating parameters (temperature and pressure). For once-through super-critical pressure boilers, ash deposits on waterwalls of the furnace bring about a large amount of temperature maldistribution.

Under the aggressive slag blanket, corrosion may occur and result in tube rupture.

Large aggregate of slags detached from the furnace waterwalls can cause mechanical breakage of water-wall tubes at the furnace bottom upon impaction.

Q. *What are the adverse effects of flyash deposition?*

Ans. Flyash clogs convective gas duct with the effect that greater resistance is imparted to the motion of the flue gas. And this may result in a loss of steam-generating capacity as high as 50 to 70% from the rated value. Over and above, this may bring down non-stop boiler operation to 20 — 30 days.

Not only this, flyash deposits on convective heat transfer surfaces build up additional thermal resistance and lead to a loss of boiler efficiency.

Q. *What factors should be considered and what measures should be taken in order that heating surface of boilers is maintained in clean state?*

Ans. Modern furnace engineering lays crucial importance on maintaining boiler heating surfaces

clean. And that necessitates the following factors to be considered:
1. boiler design
2. boiler operation
3. economic efficiency of SGP

The measures that are imperative in this regard is to study
1. chemical and mineral composition of fuels fired
2. physical properties of fuel
3. heating rates
4. flue gas composition
5. effect of temperature

Q. *What are the sources of scale that form on the outer surface of waterwalls in the boiler furnace?*

Ans. The genesis of scale forming on the fireside surface of waterwalls is attributed to the various mineral impurities present in the organic mass of solid fuels which are burnt in boiler furnaces.

Q. *How do these mineral impurities in solid fuels deposit as scale on the outer wall of the boiler tubes?*

Ans. When pulverized solid fuels are burnt, a part of the mineral impurities in the fuel are melted and combined into larger aggregates which precipitate as slag at the furnace bottom while the majority, in the form of fine particles, are carried off as ash from the furnace. Depending on their melting point, viscosity and thermal conductivity these ash particles adhere to the external surface of the steam generating tubes and form scale.

Initially, the low fusible compounds, such as NaCl, Na_2SO_4, $CaCl_2$, $MgCl_2$ and $Al_2(SO_4)_3$ with m.p. 970–1 125 K are vapourized in the high temperature zone and first condensed from the flue gas stream onto the clean surface of boiler tubes which is at a temperature of always less than 970 K. Upon condensing, these low melting compounds form a primary sticky layer onto which are deposited high-melting ash particles. Together they form a strong bond and firmly adhere to the tube surface as scale. This process is also called slagging.

Q. *How can the ash-slags be classified?*

Ans. This classification is based on the ash composition:

1. LOW FUSIBLE ASH containing compounds like NaCl , Na_2SO_4 , $CaCl_2$, $MgCl_2$ and $Al_2(SO_4)_3$ with melting point in the range 700 – 850°C (973–1 123 K). They vapourize in the high temperature zone of the flame core and condense on the surface of relatively cooler waterwall tubes.

2. MEDIUM FUSIBLE ASH containing compounds like Na_2SiO_3, K_2SO_4 and FeS having melting point in the range 900–1 100°C (1 173–1 373 K). They form a primary sticky layer on the waterwall and platen tubes if the high temperature zone is very close to the tubes.

3. HIGH MELTING ASH containing oxides of iron, calcium, magnesium silicon (SiO_2, CaO, MgO, Al_2O_3, Fe_2O_3, etc.) with their melting point 1600–1800°C (1873–2073 K) exceeding the highest temperature of the flame core. They remain solid throughout

Besides, there is another convenient way to classify slag and ash deposits encountered in steam generation plants. Depending on the bonding characteristic of slag/ash particles and the mechanical strength of deposited layer, slag and ash deposits are classified into:

1. loose deposit
2. bonded deposit, also called sintered deposit.

Q. *What are the loose deposits?*

Ans. These are the deposits that form without chemical reactions, and exhibit no capillary attraction and contain no sticky component in the matrix.

Q. *What are the effects of loose deposits of slag/ash?*

Ans.
1. Reduction of thermal efficiency of heating surfaces
2. Clogging (partial) of clear space of gas ducts.

Q. *What is the effect of built-in ash collectors on coarse particles in flue gas?*

Ans. They arrest the coarse particles suspended in flue gas and thereby diminish their concentration in it.

Q. *Does the built-in ash collectors exert any influence on the formation of loose deposits?*

Ans. They intensify the formation of loose deposits.

Q. *What is bonded deposit?*

Ans. It is a dense layer of slag firmly bonded to the heat transfer surfaces of the boiler. At higher flue gas temperatures, most of the ash particles suspended in the flue gas remain in the plastic stage and may form a sticky mass on the outer surface of waterwall and superheater tubes. It quickly grows in thickness by arresting flyash particles from flue gas and subsequently such physico-chemical processes as caking, crystallization, sulfation etc. may occur to strengthen the bond.

Q. *What factors govern the formation of bonded deposits?*

Ans. Bonded deposits are mainly encountered in solid fuel fired boilers. They are also known to occur in fuel oil fired boilers if the fuel oil contains such mineral components as sodium, vanadium, calcium, magnesium etc.

They form on high-temperature heating surfaces. Aerodynamic factors coupled with chemical reactions involved in the deposited layer of ash govern the formation of bonded deposits. The deposited ash must contain chemically active as well as high binding (i.e. low fusible) components.

Q. *What factors are responsible for the formation of bonded ash deposits?*

Ans. The principal factors responsible for the formation of bonded ash deposits are:

1. the mineral matter present in the fuel
2. the behavioral pattern of mineral constituents of fuel during the process of combustion.
3. furnace temperature
4. heating rate
5. length of exposure of mineral matter of fuel to high temperature
6. flue gas composition
7. temperatures of flue gas, flyash and heating surfaces in those zones where formation of ash deposits is more likely.
8. physico-chemical processes taking place inside the deposited mass of ash.

Q. *What is the nature of scale deposited on the fireside surfaces of the waterwall in boiler furnace?*

Ans. It is a double layer deposit—the internal layer is dense, rigidly adhering to the tube surface (bonded deposits) while the external layer consists of fine solid particles (loose deposits) of high-melting ash loosely held to the surface of the inner layer.

Q. *How does this double layer deposit form?*

Ans. The inner layer is formed due to the condensation of low-melting ash from the flue gases on the tube surface as a thick sticky film that arrests the high-melting fine ash particles falling on it. Saturated with fine ash particles, the molten slag on the tube surface solidifies to form the primary dense layer of slag firmly bonded to the tube surface. The temperature of the outer surface of the primary layer then gradually increases and that inhibits further condensation of low-melting ash from the flue gas. Therefore, further deposits of high melting ash particles on the rough outer surface of the primary layer become

loosely held and these fine particles form loose specks of secondary layer. (Fig. 24.1)

Q. *What is the reason for an ash being low-melting or high-melting?*

Ans. The fusibility of ash is determined by the amount of different metal oxides and silica present in the ash.

These metal oxides (CaO, MgO, FeO and Fe_2O_3) form eutectics with silica (SiO_2) or alumina (Al_2O_3). Usually ash produced by most solid fuels contains metal oxides in the range 5–40%. Higher percentage of metal oxides lowers the melting point of the eutectic giving rise to low-melting ash while $Al_2O_3 + SiO_2$ content higher than 80% tends to increase the melting-point of the ash which, therefore, becomes high-melting.

Q. *Does the mechanical strength of primary ash deposits have any effect on slagging of heat exchange surfaces in the boiler?*

Ans. Yes. It is one of the principal characteristics that determine the slagging properties of fuel.

Q. *On which factors does the mechanical strength of primary ash particles depend?*

Ans.
1. Mineral fraction of fuel
2. Method of fuel combustion

Q. *How can the primary ash deposits be classified on the basis of their mechanical strength?*

Ans. The mechanical strength of ash particles is determined by their compressive strength, σ. On the

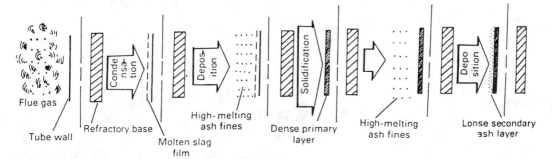

Fig. 24.1 Mechanism of double-layer deposit formation

basis of their different compressive strengths (σ), the primary ash deposits can be divided into

1. LOOSE or FRIABLE: These are easily crushed in fingers and have $\sigma \leq 1$ MN/m^2
2. DENSE: These are weakly sintered bonded. They are hardly crushed in fingers. σ in the range 1—3 MN/m^2
3. STRONG: Particles having σ in the range 3—7 MN/m^2
4. VERY STRONG: Particles having $\sigma > 7$ MN/m^2

Q. *Some primary ash deposits contain predominantly (up to 80%) hematite* Fe$_2$O$_3$. *How are ash deposits of ferric oxide produced?*

Ans. Such deposits are frequently encountered in those boilers that burn high-sulfur coals, e.g. brown coals. These coals contain iron pyrites which upon incomplete oxidation form low-fusible (melting point 1175—1225K) compounds viz. FeO-FeS eutectic. So when such fuels are burnt, these low-fusible eutectics get vaporized and deposited in relatively cooler part (convective duct) of the boiler.

Q. *What is sulfation?*

Ans. It is a kind of sintering that may occur in the furnace if there is an unfavourable mineral composition in the fuel. That is, if the calcium oxide content in the ash is more than 40% and if SO$_2$ is present in the flue gas, the process of sintering may start in the friable deposited layer formed on the heating surfaces. It results in the growth of a dense, strongly bonded slag deposit on the tube surfaces of water-walls, platens and superheaters.

$$CaO + SO_2 + \frac{1}{2} O_2 \longrightarrow CaSO_4$$

Q. *What is the upper temperature limit of sulfation?*

Ans. Depending on the proportions of acidic (SiO$_2$, Al$_2$O$_3$, TiO$_2$) as well as basic (CaO, MnO, Fe$_2$O$_3$) oxides present in the ash, the upper temperature limit of sulfation is 1375—1575 K.

Q. *Does the intensity of slagging depend on the geometry of tube layout?*

Ans. Yes. Horizontal and slightly inclined tubes suffer from more intensive slagging than vertical tubes.

Q. *In which zone of the boiler furnace are the loose deposits likely to form with no dense underlayer on the tube surfaces?*

Ans. It occurs in the zone where the flue gas temperature is as low as 600–700°C (873–973 K), i.e., the convective shaft of the furnace. (Fig. 24.2)

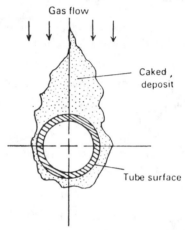

Fig. 24.2 *Formation of loose deposits with no dense underlayer*

Q. *Why is the subsurface dense layer absent?*

Ans. It is because the low-melting compounds that form the dense, primary layer on the tube surfaces due to condensation from flue gases have already been exhausted before this low temperature region (600–700°C).

Q. *Does the direction of flue gas flow exert any influence upon the magnitude of friable ash deposition?*

Ans Yes. The back side of the tubes relative to the gas flow direction collect more deposit of loose ash particles than the front sides which collect loose deposit only at low flue gas velocity. (Fig. 24.3)

Q. *Why?*

Ans. The deposits tend to collect on that part of the tube surface which is in the turbulent zone as per the profile of fluid flow. (Fig. 24.4)

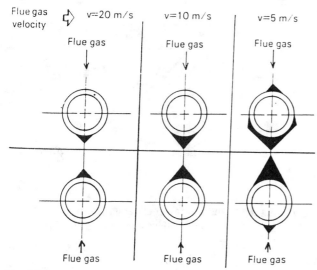

Fig. 24.3

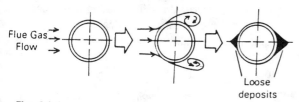

Fig. 24.4

Q. *In which case will the loose deposits of ash particles form on the front-side of the tubes?*

Ans. Only at low flue gas velocity (< 5 to 6 m/s) or if the flue gas carries a load of very fine flyash.

Q. *What is the size of the ash particles loosely deposited on the tube surfaces?*

Ans. Their sizes vary predominantly in the range of 10–30 μ m.

Q. *What about the ash particles that lie outside this range?*

Ans. Ash particles having a size of less than 10 μm are so fine that they are called inertialess particles. They move along the flue gas profile and therefore, the probability of their settling on the tube surface is very low.

Ash particles having a size greater than 30 μm are called coarse particles. Because of their larger mass and higher kinetic energy they're less likely to adhere

to the tube surface. On collision they kick off loose deposits from the tube surface.

Q. *Does the loose deposit of ash particles affect the heat transfer?*

Ans. Yes; they impair the heat transfer across the tube wall by reducing the overall heat transfer coefficient.

Q. *How can this be estimated?*

Ans. This can be estimated by the factor called fouling coefficient which is given by

$$\varepsilon = \delta_{slg}/k_{slg}, \, m^2 K/W$$

where

δ_{slg} = average thickness of the slag, m

k_{slg} = thermal conductivity of the slag, W/m K

It determines the thermal resistance of slag layers.

Q. *On which factors does the degree of tube fouling depend?*

Ans. It depends on:
(a) time of exposure of the tube surface to ash laden flue gases (Fig. 24.5)
(b) size composition of the ash particles—finer fractions causing more intensive fouling than coarse fractions

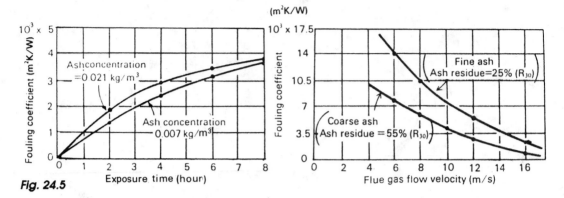

Fig. 24.5

(c) flue gas flow velocity—lower the gas flow velocity, thicker is the deposition. (Fig. 24.5)
(d) mode of tube layout in the boiler furnace—staggered or in-line as well as the longitudinal pitch or the staggered tube bunches.

Q. *Why does the degree of fouling decrease as the gas velocity increases?*

Ans. The degree of fouling is determined by the thickness of the loose deposits comprising chiefly of medium fractions (10–30 μm). Ash particles in this range continuously settle on the tube surface while ash particles in the coarse range (> 30 μm) continually destroy the layer on collision. Ultimately the final thickness of the deposited loose layer is governed by the dynamic equilibrium between these two processes.

As the flow velocity of the flue gases increases, the kinetic energy of the ash particles increases and with that rises the destructive effect of coarse particles in proportion to the third power of the velocity. Therefore, the net result is that the thickness of the loose deposits on the tube surface decreases.

Q. *How does the degree of tube fouling depend on the arrangement of tube in the boiler furnace?*

Ans. The tube fouling depends not only on the type of tube arrangement—staggered or in-line formation—but also on the longitudinal pitch p_2 of the staggered bunches. This has been demonstrated in the following figure. It is clear from the graph that under the conditions of identical gas velocity and tube

diameter, the fouling coefficient for in-line tubes is about 2 to 3.5 times that for staggered tubes. (Fig. 24.6)

Q. *Why is it not advisable to operate pulverized coal fired water-tube boilers at gas velocity as low as 3–4 m/s?*

Ans. It is due to inadmissible fouling which occurs in this gas velocity range. The degree of fouling of fireside heating multiplies with the fall of flue gas velocity and it increases significantly if the flue gas velocity is reduced to 3 to 4 m/s.

Q. *What should be the nominal gas velocity at the rated load?*

Ans. 5–6 m/s.

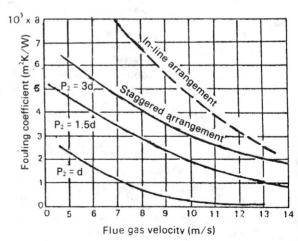

Fig. 24.6

Q. *What kind of deposits form on the heating surfaces in boilers fired on high-sulphur fuel oil?*

Ans. Two kinds of deposits:
  (a) sticky deposits
  (b) dense glassy deposits.

Q. *In which temperature zone do they fall out on the tube surfaces?*

Ans. In the zone with gas temperature $\leq 600°C$ (873 K)

Q. *What is the composition of sticky deposits?*

Ans. It consists primarily of vanadium compounds (mainly vanadium pentoxide) and sulphates.

Q. *What is the composition of dense deposits?*

Ans. It consists mainly of iron sulphate and the oxides of calcium and sodium.

Q. *What is the particularity of the deposits that form during fuel oil burning?*

Ans. These deposits have a tendency to develop quickly impairing greatly the rate of heat transfer.

Q. *How are such depositions occur in fuel oil fired boilers?*

Ans. Fuel oils burnt in boilers are characterized by low ash content (0.07—0.15%), whereas some of them may be rich in sulfur, as high as 3—3.5%. Besides the ash contains an assortment of mineral fractions—mainly the compounds of calcium, magnesium, sodium, silicon, vanadium and iron. These specific mineral impurities together with a high content of sulfur lead to ash depositions.

When fuel oil is burned, the combustion products may contain all sorts of vanadium oxides: VO, V_2O_3, V_2O_4 and V_2O_5—the higher valent oxides have lower melting point and hence particularly more harmful from the standpoint of fouling of the heat exchange surfaces. Major portion (60—80%) of ash of fuel oils constitutes mostly of low-melting components of flyash. Represented symbolically,

$$\Sigma\, Na_2O + V_2O_5 + SO_3$$

these low-melting compounds of flyash plays a decisive role in forming ash deposits.

Fouling of waterwall tubes in the form of sticky deposits begins at temperatures above 815K. This sticky film of sodium vanadates ($5Na_2O \cdot V_2O_4 \cdot 11V_2O_5$), endows itself as the substrate to arrest flyash particles. The temperature of the primary deposited layer gradually increases because of impaired heat transfer characteristics through this deposited sticky mass to the tubewall. With the increase of its surface temperature, ash layer become more plastic and capable of retaining higher load of fusible fractions of flyash. As a result, the overall process leads to the formation of strong bonded deposits.

Q. *Does the Na/V ratio plays any role in determining the characteristics of deposited ash in fuel oil fired boilers?*

Ans. Yes; with the increase of sodium isto vanadium ratio in ash, its sticky property increases. It is found to be maximum with Na/V ratio in the range of 3.7—9.5.

Q. *It has also been reported that the process of fouling of heating surfaces also occur at temperatures below or slightly above the dew point of flue gases (320—500K). Why does this occur and what is the mechanism of such an attack?*

Ans. This is due to the condensation of sulfuric acid and water vapour from the flue gases. They form an adhesive film sticking to the tube surface and arresting and retaining higher quantities of flyash particles which dissolve in the viscous acidic film producing a solution of mineral components of fuel, hydrocarbons and corrosion products of sulfuric acid. They constitute of 70—80% water-soluble salts, mainly sulfates.

Q. *Below dew point, fouling of heating surfaces by flyash is more intense. Which factors are principally responsible for this?*

Ans. The rate of fouling increases with
  1. the increase of sulfur content of fuel
  2. the increase of excess air/fuel ratio
  3. the deep shedding of boiler load causing a substantial drop of temperature of heating surfaces.

Q. *How can this temperature drop be estimated?*

Ans. This can be estimated from the relationship

$$\theta_w = \left[\theta_w\right]_r \left[\frac{G}{G_r}\right]^{1/3}$$

where,

θ_w = wall temperature at current load

$[\theta_w]_r$ = wall temperature at rated load

G = current steam load

G_r = rated steam load

Q. *Does the ash-deposition on low-temperature heating surfaces also occur in solid-fuel fired boilers?*

Ans. Yes. But they differ substantially from those formed in fuel oil fired boilers. Flue gas, in solid-fuel fired boilers contains much higher load of flyash containing substantial quantity of coarse particles which under high gas velocity (5—6 m/s) produce abrasive effect on the primary ash deposits.

Q. *How can the fouling ability of flyash be characterized?*

Ans. This can be conveniently expressed in terms of the ratio of the sums of basic and acidic components in the ash i.e. by

$\Sigma B / \Sigma A$

where,

$\Sigma B = CaO + MgO + Fe_2O_3 + Na_2O + K_2O$

$\Sigma A = SiO_2 + Al_2O_3 + TiO_2$

Q. *What are the known methods of cleaning the heating surfaces from deposits?*

Ans. There are several methods:
1. steam jet blasting
2. water jet blowing
3. air blasting
4. shot blasting
5. vibration cleaning
6. heat-wave cleaning
7. thermal cleaning.

Q. *What is steam jet blasting?*

Ans. Saturated/superheated steam with pressure at about 4 MPa(4 MN/m^2) is injected with the help of multi-jet retractable blowers to clean radiant heating surfaces and convective superheaters.

Q. *What is water-jet blowing?*

Ans. Jets of water issuing at high velocity are directed on the fouled surfaces of heating tubes. Upon collision with the mass of deposit, the water particles transfer their momentum to the deposits which therefore get dislodged.

Q. *What is the advantage of a water jet over steam and air jets in cleaning deposits from the heating surfaces of boiler tubes?*

Ans. The effectiveness of these three types of jets depend on the kinetic energy they carry. Greater the kinetic energy of the projectile, greater will be its range and consequently, greater will be its capacity to remove deposits.

Water jets possess the highest kinetic energy and so have the highest range. This coupled with the higher thermal conductivity of water than steam or air makes water jets superior to steam and air jets in cleaning waterwalls.

Q. *What is the demerit of a water blowing system?*

Ans. Because of its high thermal conductivity water jets have the highest thermal effect on dense deposits. As such it may bring about sharp quenching of the tube metal unless its amount is carefully calculated.

Q. *What is vibration cleaning?*

Ans. High frequency vibration is transmitted to the heating tubes by means of vibrators attached to rods cooled by water. The transmitted sonic waves disturb the bond between the deposits and tube metals and therefore dislodge the scales from the tubes' surface. By this process platens and in-line tube banks are cleaned.

Q. *What is shot-blasting?*

Ans. It is an efficient method of cleaning deposits from convective heating surfaces (fireside) by utilizing the kinetic energy of flying cast iron shots of 3 to 5 mm dia. Shots are projected on the deposited surfaces by high speed jets of air. Airborne, they're carried upwards and distributed over the entire cross-section of the downtake shaft of the boiler furnace. The process lasts 20 to 60 s and its intensity varies from 150–200 kg/m^2 of the convective shaft cross-section.

Q. *What is heat-wave cleaning?*

Ans. This method is based on the application of low-frequency sound waves generated as a result of an explosion perfected in a special explosive-combustion impulse chamber. The acoustic waves set the tubes vibrating which weakens the bond between the deposits and the tube metal.

Q. *What is thermal cleaning?*

Ans. It is based on raising the temperature of the regenerative air heaters to $250 - 300°C$ by cutting the air supply to the apparatus. The sticky deposits get dried up while the sulphuric acid is vapourized.

Q. *What will be the effect of improper selection of flue gas velocity, apart from tube fouling in the boiler furnace?*

Ans. It may lead to intensive erosion of tube metal in certain places, resulting in thinning of the walls and even break-through of the tubes.

Q. *What is this process called?*

Ans. Abrasive erosion.

Q. *What is the mechanism of abrasive erosion (or wear)?*

Ans. Coarse ash particles, because of their mineral composition, are very hard and they frequently possess sharp edges.

Flue gas-borne coarse ash particles possess sufficient kinetic energy and when these hard and sharp particles strike the tube surfaces they chip off microscopic layers of the tube metal. Since this is a continuous process, it leads to gradual thinning of the tube wall at critical places.

Besides ash, particles of unburnt fuel (coal), particularly those of a harder variety (anthracite or semi-anthracite) are capable of causing abrasion of the tube surfaces.

Q. *On which factors does the abrasion wear of boiler tubes by ash particles depend?*

Ans. It depends on several factors:
1. abrasiveness of ash particles, i.e., SiO_2 content in ash
2. total ash content of fuel
3. flue gas velocity
4. angle of attack

Q. *What is the angle of attack of gas flow for greatest abrasion wear of staggered tubes in the crossflow of the gas path?*

Ans. $30° - 60°$

Q. *Which factors determine the intensity of wear of heating surfaces?*

Ans.
1. Kinetic energy of ash and unburnt fuel particles
2. The flux of ash particles, i.e., the number of ash particles passing through unit surface area per unit time
3. Non-uniformity of ash concentrations in the gas flow
4. The time of exposure.

25

Corrosion of Waterside Heating Surfaces

Q. *What is corrosion?*

Ans. Corrosion is the destruction of materials—metals, non-metals and refractories—under the action of aggressive surrounding media (e.g., steam, water, furnace gas) as a result of chemical and or electrochemical processes taking place on the surface of the materials.

(*Note:* We'll limit our discussion to the corrosion of metals and alloys only)

Q. *What is erosion?*

Ans. Erosion is purely a physical phenomenon which describes the wearing of surface material by such mechanical processes as friction, impact, etc.

Q. *In how many forms does corrosion manifest itself?*

Ans. Two forms: Uniform and non-uniform corrosion.

Non-uniform corrosion is also called local corrosion.

Q. *How can uniform corrosion be distinguished from non-uniform corrosion?*

Ans. Uniform corrosion describes the destruction of metal over the entire surface whereas in non-uniform corrosion, the destruction of metal is limited to a local spot or point and may penetrate deeply into the body of the metal.

Q. *How can non-uniform corrosion be classified?*

Ans. It may be classified as:
  (a) spot corrosion

  (b) point corrosion
  (c) honeycomb corrosion
  (d) intergranular (intercrystalline) corrosion
  (e) selective corrosion

Q. *What is selective corrosion?*

Ans. It is the selective destruction of one of the component metals of an alloy system, e.g., the dissolution of only zinc from a zinc-copper alloy (brass) system.

Q. *Where is this type of corrosion encountered in power plants?*

Ans. In the turbine condensers. The brass tubes of turbine condensers are exposed to river or seawater for cooling purposes and as a result they undergo dezincification, i.e., selective dissolution of zinc.

Q. *Does the dezincification tell upon the structure and performance of turbine condensers?*

Ans. Yes. The elimination of zinc makes the brass brittle and leaves a dendritic copper structure. It leads to leakages in the tubes and deteriorates the heat transfer efficiency of the condenser.

Q. *How can chemical corrosion be distinguished from electrochemical corrosion?*

Ans. Chemical corrosion involves purely chemical processes that characterize the direct interaction between the metal and the surrounding media without the generation of electric current. For example, iron is attacked by superheated steam at 650–775 K and higher:

$$3Fe + 4H_2O \longrightarrow Fe_3O_4 + 4H_2$$

This is purely a chemical interaction between iron and steam and does not involve any electric current.

Electrochemical corrosion describes the destruction of metal by electrochemical attack whereupon one of the two metal components preferentially dissolves in the presence of electrolytic solution and a flow of electric current is set up between these two different metal components otherwise electrically connected.

Corrosion, i.e., oxidation takes place at the anode wherefrom metal passes into solution

$$(Zn \longrightarrow Zn^{2+} + 2e)$$

And deposition of metal (or liberation of hydrogen gas), i.e., reduction takes place at the cathode

$$(Cu^{2+} + 2e \longrightarrow Cu)$$

The free electrons liberated at the anode go to the cathode by the external conductor and set up an electric current between the cathode and anode.

Hence electrochemical corrosion involves both chemical processes (oxidation and reduction) as well as the flow of electric current.

Q. *What is polarization?*

Ans. It describes the electrode process leading to a diminishing rate of corrosion because of the drop in the potential difference between the anode and cathode.

The corrosion rate is proportional to the current strength which is directly proportional to the potential difference between the cathode and anode

$$I = (V_c - V_a)/R$$

where R = resistance of the overall circuit.

As the corrosion proceeds, the p.d. $(V_c - V_a)$ gets smaller and smaller until it reaches such a small value that the corrosion process practically ceases. This phenomenon is called polarization.

Q. *What leads to polarization?*

Ans. Polarization is an electrode phenomenon effecting the cessation of the corrosion process.

The rate of electron transfer from the anode to cathode is initially equal to the rate of reduction of

cations (take for example H^+ ions) in the cathode. However with the lapse of time, some of the hydrogen molecules liberated

$$H^+ + e \longrightarrow H$$

$$H + H \longrightarrow H_2$$

get deposited on the surface of the cathode and form a layer of H_2-gas bubbles. This hydrogen-gas-layer on the cathode surface deters the penetration of the oncoming H^+ ions to reach the cathode surface

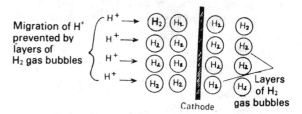

Fig. 25.1 *Migration of H^+ ions prevented by layers of H_2 gas bubbles*

But due to the electrostatic pull of the cathode these H^+ ions get adhered to the surface of the hydrogen gas layer and form between them and the anode a back e.m.f. (electromotive force) that reduces the original potential difference between the primary anode and cathode. As the p.d. between the cathode and anode decreases, the flow of electric current nearly stops leading to termination of corrosion.

Q. *What is depolarization?*

Ans. The electrode process that reduces polarization (i.e., intensifies electrochemical corrosion) is called depolarization.

Q. *What are depolarizers?*

Ans. These are the substances that contribute to depolarization.

Q. *What do they do?*

Ans. They bind the free electrons.

Q. *What are the fundamental requirements for an electrochemical corrosion to proceed?*

Ans.

1. A galvanic couple, i.e., two metals having different electrode potential
2. An electrolyte in contact with the galvanic couple
3. Depolarizers.

Q. *How is the galvanic cell produced on the inner surface of steel tubes carrying boiler water?*

Ans. The material of construction of boiler tubes is not uniform throughout and the galvanic cell is set up because of electrochemical heterogeneity of the various sections of the surface and the presence of water as electrolyte. (Fig. 25.2)

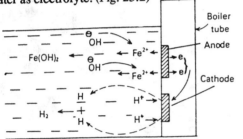

Fig. 25.2 *Galvanic cell is set up on boiler tube surface because of inhomogeneity of tube material*

If one section be electrically more negative than the other section on the surface, the electrochemical attack takes place and the destruction of iron from the anode occurs.

Q. *What substances act as depolarizers in a steam generation unit?*

Ans. Depolarizers are those which bind the free electrons. Such substances are:

1. OXYGEN: $O_2 + 2H_2O + 4e \longrightarrow 4OH^-$

   This process is called oxygen depolarization.

2. HYDROGEN: $2H^+ + 2e \longrightarrow 2H \longrightarrow H_2$

   This process is called hydrogen depolarization.

3. SODIUM NITRITE:

   $NO_2^- + H_2O + e \longrightarrow NO + 2OH^-$

Q. *How do nitrites appear in boiler water?*

Ans. It appears because of the oxidation of ammonium ions by nitrifying bacteria

$$NH_4^+ \longrightarrow NO_2^- + NO_3^-$$

These bacteria anchor at the surface of the pipelines and mechanical filters.

Q. *How can these bacteria be eliminated?*

Ans. By periodical chlorination of bacteria-infected pipelines.

Q. *What factors lead to the development of electrochemical corrosion?*

Ans. Two types of factors: Internal and external. Among the internal factors are
   (a) the type of metal and its grade
   (b) the cleanliness of the metal, i.e., whether having metallic or non-metallic inclusions
   (c) the presence of stress due to manufacturing processes or heat treatment.

External factors are:
   (a) the nature of the surrounding medium
   (b) pH of the electrolyte
   (c) presence of depolarizers.

Q. *Under normal working conditions what factor is most decisive in determining the intensity of corrosion of the heat transfer equipment on the water-side?*

Ans. It is the dissolved oxygen in water.

Q. *How does it intensify the corrosion?*

Ans. It acts as a depolarizer, and readily accepts the free electrons liberated by the iron of the walls of the boiler tubes or equipments

$$Fe - 2e \longrightarrow Fe^{2+}$$

$$O_2 + 2H_2O + 4e \longrightarrow 4OH^-$$

$$Fe^{2+} + 2OH^- \longrightarrow Fe(OH)_2$$

So iron will continue to dissolve from the surface of the tubes or vessels in contact with the boiler water so long as dissolved oxygen remains in the latter.

Q. *What happens as a result of this oxygen corrosion?*

Ans. It causes pronounced local destruction of the metal surface leading to the formation of pits. This type of oxygen corrosion occurs at pH ≥ 7

At pH < 7, i.e., in acidic media, the presence of dissolved oxygen leads to uniform corrosion.

Q. *How does oxygen come into the boiler water?*

Ans. It gets into the water from air.

Q. *Why is deaerated water or condensate not allowed to come in contact with the atmosphere?*

Ans. To avoid contamination by O_2 (as well as CO_2) from air, that may lead to corrosion.

Q. *Why is contamination by CO_2 to be avoided?*

Ans. Dissolved CO_2 in water liberates H^+ ions

$$CO_2 + H_2O \Leftrightarrow H_2CO_3 \Leftrightarrow H^+ + HCO_3^-$$

that act as a depolarizer intensifying the corrosion of metal surfaces

$$H^+ + e \longrightarrow H$$

$$H + H \longrightarrow H_2 \uparrow$$

Q. *Why is carbon dioxide corrosion more likely to occur in equipment and pipelines in contact with condensate than those in contact with softened water or natural water?*

Ans. Softened water or natural water contains to some extent HCO_3^- ions which establish the following equilibria with H^+ ions

$$HCO_3^- + H^+ \Leftrightarrow H_2CO_3 \Leftrightarrow H_2O + CO_2$$

Therefore, the presence of HCO_3^- ions (e.g., in softened or natural water) will press the equilibria to the right and deter the dissolution of CO_2 in water. Hence CO_2– corrosion does not occur in pipelines/equipment in contact with softened or natural water.

However, the metal surface in contact with the condensate is very prone to CO_2 attack, as carbon dioxide quickly dissolves in the condensate because of the absence of bicarbonate ions in the condensate water.

Q. *Why is a CO_2 solution in water more aggressive than an HCl solution in water at the same pH value?*

Ans. Hydrochloric acid is a strong acid and so the entire HCl is completely dissociated into H^+ and Cl^- ions. Therefore, as the corrosion process develops, the neutralization of H^+ ions on the cathode continues. However, this decrease in H^+ ions is not compensated for unlike the case of CO_2 dissolved in water

$$CO_2 + H_2O \Leftrightarrow H_2CO_3 \Leftrightarrow H^+ + HCO_3^-$$

in which case as the concentration of available H^+ ions is depleted a fresh supply of H^+ ions is made by the dissociation of H_2CO_3.

That's why a CO_2 solution is more aggressive than a HCl solution at the same pH value.

Q. *How does the presence of bicarbonate ions in the course of steel corrosion lead to further corrosion?*

Ans. In the course of steel corrosion ferrous ions are liberated, which combine with the bicarbonate ions if present in the solution.

$$Fe^{2+} + 2HCO_3^- \rightleftharpoons Fe(HCO_3)_2$$

However, under certain circumstances ferrous bicarbonate undergoes decomposition liberating free CO_2

$$Fe(HCO_3)_2 \rightleftharpoons Fe(OH)_2 + 2CO_2$$

This free CO_2 again dissolves in water to give rise to carbon dioxide corrosion

$$CO_2 + H_2O \rightleftharpoons H^+ + HCO_3^-$$

$$Fe - 2e \longrightarrow Fe^{2+}$$

$$2H^+ + 2e \longrightarrow 2H \longrightarrow H_2$$

However, this corrosion is possible if water contains sufficient dissolved O_2 to oxidize ferrous hydroxide and throw it out of equilibria.

$$4Fe(OH)_2 + O_2 + 2H_2O \longrightarrow 4Fe(OH)_3 \downarrow$$

Q. *How can corrosion products deter further corrosion?*

Ans. Since corrosion is a surface phenomenon, the corrosion products are formed on the surface of the metal.

Now if the corrosion products firmly adhere to the metal surface and form a layer sufficiently dense, they will then effectively isolate the base metal from the surrounding aggressive media and also prevent the penetration of depolarizers to the metal surface, with the effect that the corrosion process terminates.

Q. *How does the rate of oxygen-corrosion vary with the temperature of the medium?*

Ans. The rate of oxygen-corrosion increases steadily and continuously, in the case of a closed system (i.e., oxygen-contaminated water is heated in a closed vessel), with the increase of temperature whereas in the case of an open system (i.e., in a tank exposed to atmosphere), the corrosion-rate increases first with the rise of temperature, reaches a maximum at 350–360 K and then drops. (Fig. 25.3)

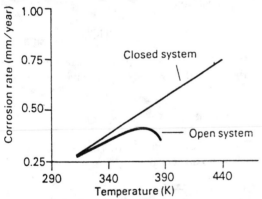

Fig. 25.3 *Dependence of oxygen-corrosion on temperature*

Q. *How does carbon dioxide make its way up into the steam-water circuit of a thermal power plant?*

Ans. It appears as bicarbonate (HCO_3^-) and carbonate CO_3^{2-} ions in the boiler feedwater makeup as well as in the condensate of turbine condensers (because of leakage of cooling water). In the boiler unit these ions decompose and liberate CO_2.

Q. *What is the most harmful effect of CO_2-corrosion?*

Ans. It is the contamination of feedwater with the products of corrosion (oxides of iron).

Q. *What is the effect of feedwater on CO_2-corrosion?*

Ans. It proliferates with the increase of feedwater temperature.

Q. *How can carbon dioxide corrosion best be prevented?*

Ans. This can be done by:
(a) Appropriate methods of makeup water treatment to reduce adequately the amount of free CO_2 in steam
(b) Eliminating the condensate of heating steam from regenerative feedwater heaters
(c) Treating feedwater with ammonium ions
(d) Treating steam with film forming amines
(e) Complete venting of CO_2, O_2 from the steam space of heaters.

Q. *How will ammonium ions introduced into the feedwater makeup control carbon dioxide corrosion?*

Ans. Ammonium ions bind the carbon dioxide in aqueous phase:

$$NH_3 + H_2O \Leftrightarrow NH_4^+ + OH^-$$

$$CO_2 + H_2O \Leftrightarrow H^+ + HCO_3^-$$

$$NH_4^+ + HCO_3^- \longrightarrow (NH_4)HCO_3$$

$$NH_4^+ + (NH_4)HCO_3 \longrightarrow (NH_4)_2CO_3 + H^+$$

$$H^+ + OH^- \longrightarrow H_2O$$

Q. *But ammonia escapes from the solution phase to vapour phase on heating in a boiler unit. How will it curb CO_2-corrosion in the condensate?*

Ans. Ammonia escapes into the steam in the boiler unit. And when the steam condenses in the turbine condenser subsequently, the vapour ammonia again gets dissolved in the condensate to produce ammonium ions. And carbon dioxide contained in the condensate as CO_3^{2-} and HCO_3^- ions is completely bound by ammonium ions.

Q. *If ammonia dissolves in the condensate, will it change the pH of the latter?*

Ans. Yes. It will make the condensate alkaline and the pH value of the condensate will be ≥ 8.3.

Q. *In which form is ammonia introduced into the steam-water circuit?*

Ans. Ammonia is added into the feedwater as the:
  (a) aqueous solution NH_4OH
  (b) aqueous solution of NH_4Cl or $(NH_4)_2SO_4$.

Q. *What is the difference between using NH_4OH and NH_4Cl or $(NH_4)_2SO_4$ as the ammoniating agent ?*

Ans. NH_4OH is distinctly basic

$$NH_4OH \Leftrightarrow NH_4^+ + OH^-$$

However, NH_4Cl or $(NH_4)_2SO_4$ being salts of a strong acid and weak base undergo hydrolysis to liberate H^+ that makes the solution of these reagents potentially acidic

$$NH_4Cl \Leftrightarrow NH_4^+ + Cl^-$$

$$\text{(aq.)} \qquad \text{(aq.)} \quad \text{(aq.)}$$

$$NH_4^+ + H_2O \rightleftharpoons NH_4OH + H^+$$

Hence, when these latter reagents are used to introduce ammonia in the steam-water circuit, care must be taken to ensure that the boiler water is sufficiently alkaline.

Q. *What is the effect of oxygen corrosion in a steel economizer?*

Ans. Poorly deaerated feedwater circulating in the steel economizers results in oxygen corrosion (pitting) with the effect that flaws in the tubes appear. Through these flaws water gushes out with tremendous velocity and erodes the adjacent tubes producing through-holes in them. Since the economizer tubes are closely packed, tubes fail one after another, inflicting great damage to the overall economizer.

Q. *Do the iron economizers suffer from oxygen corrosion as do the steel economizers?*

Ans. No.

Q. *Why?*

Ans. The oxygen present in the feedwater produces on the tube surface a hard, passive film of Fe_2O_3 that prevents further oxidation of the base layer.

$$Fe \longrightarrow Fe^{2+} + 2e$$

$$Fe^{2+} + 2OH^- \longrightarrow Fe(OH)_2$$

$$4Fe(OH)_2 + O_2 + 2H_2O \longrightarrow 4Fe(OH)_3$$

$$4Fe(OH)_3 \longrightarrow 2Fe_2O_3 + 6H_2O$$

Q. *In which section of the economizer, is oxygen corrosion most prevalent?*

Ans. In the inlet sections.

Q. *In the boiler drum, the riser tubes suffer from a lesser degree of oxygen-corrosion than downcomer tubes. Why?*

Ans. It is due to the deaerating effect of steam bubbles in the riser tubes.

Q. *What is outage corrosion?*

Ans. It is the out-of-service corrosion. It is a type of electrochemical corrosion affecting idle boiler units.

Q. *How does this corrosion arise?*

Ans. When a boiler is taken out-of-service and left shutdown for a long period for maintenance jobs, the boiler pressure gradually diminishes and vacuum develops in the drum. This sucks in air, with the effect that contamination of boiler water with oxygen takes place. Even if the boiler water is completely drained from the boiler drum, its internal surface remains wet and gets contaminated with infiltrated oxygen. And this initiates oxygen corrosion.

Q. *Why are the superheaters thoroughly washed before laying for a long shutdown?*

Ans. The deposits adhering to the superheater tube surface are capable of dissolving in the moisture that forces its way into the boiler unit alongwith infiltrated air during long outage time. This becomes enriched with oxygen and contributes to the development of electrochemical corrosion.

Q. *How can outage corrosion be avoided in the boiler unit?*

Ans. This can be prevented by adopting the following methods:

1. Dry storage of boiler units
2. Wet storage of boiler units
3. Protecting the boiler unit by coating it with a 10% $NaNO_2$ solution film
4. Filling the boiler unit with hydrazine solution
5. Treating the internal surfaces of boiler unit with complexing agents
6. Preserving the boiler unit in a nitrogen atmosphere.

Q. *How is dry storage of boiler units put into effect?*

Ans. After cooling down the boiler unit, all the mains and pipelines are isolated, boiler water is drained, the deposits are removed and the surfaces are dried keeping the air vents open.

Following the completion of drying, water absorbents (anhydrous $CaCl_2$, CaO and silica gel) filled in trays are placed in the boiler drum in about $2kg/M^3$ of the drum space. All manholes are tightly closed. The absorbents are periodically checked and replaced if necessary.

Q. *How is wet storing of the boiler unit performed?*

Ans. After the shutdown, the boiler unit is cooled, the drum is drained to remove all sludge, it is isolated, filled with deaerated water under a deaerator pressure and boxed up.

Q. *Why does the coating of a nitrite film on the internal surfaces of the boiler unit prevent oxygen corrosion?*

Ans. The nitrite ions passivate the metal surface by producing a thin film of ferric hydroxide on the surface

$$3NaNO_2 + Fe + 3H_2O \rightarrow 3NO + Fe(OH)_3 + 3NaOH$$

and this film prevents the penetration of dissolved oxygen, which acts as a depolarizer, to the metal underneath the film.

Q. *How does filling the boiler unit with hydrazine solution protect the internal surfaces of the unit from oxygen corrosion?*

Ans. Hydrazine itself reacts with dissolved oxygen

$$N_2H_4 + O_2 \longrightarrow N_2 + 2H_2O$$

to remove any traces of local oxygen that may be present in the deaerated water filling the boiler unit during shutdown.

Q. *At which points of the boiler units, does intercrystalline (i.e., intergranular) corrosion appear?*

Ans. In riveted and expanded joints exposed to boiler water.

Q. *How does it appear?*

Ans. First, it appears in the form of a very fine crack invisible to the naked eye. Gradually it develops with time and eventually becomes quite detectable to the eye.

Q. *Why is it called intergranular corrosion?*

Ans. It originates at the boundary between the adjacent crystal grains under stress and propagates along the grain boundary to relieve the internal stress. That's why it is called intergranular corrosion.

Q. *What factors lead to the initiation of intercrystalline corrosion?*

Ans. The following three factors must be present and act simultaneously to give rise to intercrystalline corrosion:

1. presence of internal stress (tensile stress close to yield point) in the body
2. leakage in riveted or expanded joints
3. aggressiveness of boiler water.

Q. *If any of these three factors is absent, will intercrystalline corrosion occur?*

Ans. No.

Q. *Among these which one is a primary cause for the genesis of intercrystalline corrosion in boiler units?*

Ans. It is the thermal strain in the boiler elements associated with high internal stresses in the metal.

Q. *How does it originate?*

Ans. It arises due to non-uniform heating or cooling of the boiler components, abrupt load-shedding in the boiler and sudden shutdown of the boiler unit for technical reasons. The rapid and non-uniform cooling produces a thermal strain in the riveted and expansion joints and causes circular cracks to appear in tubes at the joints.

Q. *Is it true that intercrystalline corrosion also appears at the spots under local cooling in the boiler drum?*

Ans. Yes.

Q. *What are these spots?*

Ans. These are where phosphate solution feed pipe, water-gauge pipes, feedwater lines and intermittent-blowdown pipes are connected to the boiler drum.

Q. *What about the aggressivity of boiler water to prevent fracturing due to intercrystalline corrosion?*

Ans. Regarding the aggressivity of boiler water it has been found that sodium hydrate concentration of as much as 5 to 10% is sufficient to react with metal. That's why the relative alkalinity of boiler water should not go higher than 20%.

Q. *What is relative alkalinity?*

Ans. It is an important indicator of the boiler water's aggressiveness. Expressed in percent, it is determined by the relationship

$$A_{rel} = \frac{40\,(\,100A_{bw} - 2C_{PO_4^{3-}}\,)}{C_{bw}} - \times 100$$

where

A_{bw} = boiler water alkalinity, mg/1

$C_{PO_4^{3-}}$ = phosphate ion concentration in boiler water, mg/l

C_{bw} = total dissolved salt content in boiler water, mg/1.

Q. *What methods are applied to control the aggressivity of boiler water?*

Ans. Two methods are applied — one is nitrate treatment and the other is phosphate treatment.

NITRATE TREATMENT: makes use of dissolved sodium nitrate whose concentration in boiler water must be as

$$C_{NO_3^-} \geq (\,0.35 — 0.4\,)\,40\,A_{bw}$$

However, this method, though an effective process of controlling free sodium hydrate, is applicable to boiler units operating at a pressure ≤ 7 MPa.

PHOSPHATE TREATMENT: makes use of trisodium phosphate in solution to eliminate the aggressive free sodium hydrate from boiler water. This method, also known as the method of purely phosphate alkalinity, is applied to boiler units operated at pressures higher than 7 MPa.

Q. *Why is nitrate treatment inapplicable to boiler units operating at pressures higher than 7 MPa?*

Ans. At higher pressure and consequently higher temperature the nitrate ions become unstable and decompose rendering the treatment futile

$$2NaNO_3 \xrightarrow[\Delta]{} 2NaNO_2 + O_2$$

Q. *What is vapour-water corrosion?*

Ans. It is the destruction of metal by the chemical attack of water vapour at elevated temperature.

Water vapour attacks carbon steel at temperatures 673K and higher producing magnetite and hydrogen

$$3Fe + 4H_2O \longrightarrow Fe_3O_4 + 4H_2$$

Whereas it attacks alloy-steels above 800–850K.

Q. *What factors contribute to vapour-corrosion of metals?*

Ans. These are:
(a) high rates of heat flow (≥ 230 kW/m^2)
(b) disturbed circulation in steam-water circuit
(c) film-type boiling leading to metal overheating.

Q. *In which sections of boiler units, is vapour (steam) corrosion most prevalent?*

Ans. Superheater coils and waterwall tubes.

Q. *Why?*

Ans. These are the zones of high heat flowrates and temperature reaches critical values.

Q. *How can vapour-water corrosion be controlled and prevented?*

Ans. By eliminating those factors that lead to tube overheating.

Q. *Can this be prevented by changing or improving the water control of boiler units?*

Ans. No.

Q. *Why?*

Ans. It is because vapour-water corrosion depends on the operating conditions of the boiler units as well as the hydrodynamic processes involved.

Q. *What is hide-out corrosion?*

Ans. This refers to the destruction of metal underneath a layer of sludge settling on the waterside surfaces of boiler tubes. Since it remains hidden under the layer of sludge, it is called hide-out corrosion.

Q. *Why does it occur?*

Ans. It occurs due to insufficient treatment of boiler water containing suspended substances which fall out on the walls of the boiler tubes forming a solid sediment.

Also it occurs due to the destruction of iron underneath the deposits of iron-oxide that act as a depolarizer.

$$Fe + 4Fe_2O_3 \longrightarrow 3Fe_3O_4$$

Q. *A layer of Fe_2O_3 on metal (Fe) surface passivates the base metal by sealing it from dissolved oxygen. Then how can it contribute to corrosion?*

Ans. It is because, Fe_2O_3 itself acts as a source of oxygen. It supplies oxygen that functions as a depolarizer to hasten the process of corrosion.

$$3Fe_2O_3 \xrightarrow[\Delta]{} 2Fe_3O_4 + \frac{1}{2}O_2$$

$$Fe \longrightarrow Fe^{2+} + 2e$$

$$2Fe(OH)_2 + \frac{1}{2}O_2 + H_2O \longrightarrow 2Fe(OH)_3$$

$$\xrightarrow[\Delta]{} Fe_2O_3 + 3H_2O$$

Q. *How does it appear?*

Ans. It appears on the metal surface as pits or cavities filled with Fe_2O_3 at the top and Fe_3O_4 at the bottom.

Q. *How can this be prevented?*

Ans. By reducing the penetration of oxide of iron Fe_2O_3 (≤ 100 microgram/kg) and the oxides of copper (CuO, Cu_2O) into the boiler units.

Q. *What methods are adopted to prevent intensive corrosion in the water-condensate circuits of the supercritical-pressure boiler-turbine units?*

Ans. The following methods are adopted to slow down the corrosion process.

   1. Alkaline method   2. High Alkaline method
   3. Neutral method   4. Complexonate method

Q. *At what pH, is the alkaline (correction) method carried out?*

Ans. At pH range 9.0–9.2

Q. *Why?*

Ans. It has been found that the feedwater of supercritical pressure boiler units becomes least aggressive at a pH value equal to 9.

Q. *What alkalizing agent is used?*

Ans. Ammonia (in an amount so that its concentration does not exceed 800 microgram/kg in feedwater).

Q. *How does it maintain the pH within the prescribed range?*

Ans. Ammonia dissolves in water to produce a weak base NH_4OH which dissociates to liberate hydroxyl ions:

$$NH_3 + H_2O \Leftrightarrow NH_4OH \Leftrightarrow NH_4^+ + OH^-$$

It is these OH^- ions that ensure the prescribed pH value.

Q. *Why in addition to ammonia, is hydrazine also added?*

Ans. Hydrazine is a strong reducing agent. Its use is directed to arrest the water-dissolved oxygen and reduce it to water

$$N_2H_4 + O_2 \longrightarrow N_2 + 2H_2O$$

and thereby prevent corrosion processes.

Q. *At which points in the water-condensate circuits of boiler-turbine unit, are the ammonia and hydrazine dosed?*

Ans. Ammonia is injected downstream of the deaerator or L.P. Regenerative feedwater heater.

Hydrazine is introduced downstream of the DM (demineralizing) plant.

Q. *What would happen if ammonia were injected downstream of the DM plant in place of hydrazine?*

Ans. The pH value of water downstream of the DM plant becomes less than 9. And if ammonia was introduced to correct the pH of DM water, it would increase the concentration of copper in the BFW upstream of the deaerator above the prescribed limit (5 microgram/kg) and in the case of low-pressure regenerative BFW heaters it would intensify the corrosion of the heating surfaces made of brass.

Q. *Does the pH value set up by ammonia remain constant over the entire section of the steam-condensate circuit in the boiler-turbine unit?*

Ans. No.

Q. *Why?*

Ans. The pH of ammonia solution is functionally dependent on the solubility of ammonia in water, which is directly proportional to the temperature of water. Higher the water temperature, lesser will be the solubility of ammonia and accordingly lesser will be the free hydroxyl ions

$$NH_3 + H_2O \Leftrightarrow NH_4OH \Leftrightarrow NH_4^+ + OH^-$$

and consequently the pH of the solution will be lower.

Q. *Will there be any harm if the pH in the water-condensate circuit of the boiler-turbine unit is raised to 9.5–9.6 by adding more ammonia?*

Ans. Yes. A pH value in this range is inadmissible since it will give rise to corrosion of the brass tubes of the turbine condenser.

Q. *Why is it difficult to ensure anticorrosive conditions in all sections of the water-condensate circuit when ammonia is used as the alkalizing reagent?*

Ans. It is because the pH of the circuit varies due to variation of the solubility of ammonia in feedwater at different temperatures in different sections of the water-condensate circuit.

At a temperature of 453 K, the amount of ammonia needed to maintain the pH of feedwater at 9 is 600 microgram/kg while at 523 K, the amount of ammonia needed becomes as high as 1400 microgram/kg to ensure the same pH. However at the turbine condenser (303–305 K) such a concentration of ammonia will raise the pH to a level 9.5–9.6 which is dangerous as it will lead to caustic embrittlement of the brass tubes of the condenser.

Q. *Then how can this difficulty be overcome?*

Ans. This problem can be tackled by using such chemicals such as morpholine and cyclohexylamine instead of ammonia. These are alkaline amines and required to a considerably lesser extent than ammonia.

Q. *How is an anticorrosive environment created by the high-alkalinity method?*

Ans. It is done by dosing more ammonia to the water-condensate circuit to raise the value equal to 9.5–9.6.

Q. *Why is this method adopted?*

Ans. It has been found by experience that this method is highly effective. It has proved itself extremely reliable in protecting the pipelines and the equipment against corrosion.

Q. *What is the main disadvantage of this method?*

Ans. It expedites the corrosion of brass. Since all the tubes of turbine condensers are generally made of brass, they must be replaced by steel tubes (perlitic grade) in order to fit this environment.

Q. *What is the basis of corrosion prevention by the neutral method?*

Ans. Using no chemical reagents, this method relies entirely on the protective ferric oxide film produced on the steel surface by the dissolved oxygen at higher pressures and elevated temperatures. It is this protective film that inhibits further corrosion.

Q. *What conditions must be fulfilled to adopt this technique successfully?*

Ans. The following conditions must be rigidly observed for the successful application of the neutral method:

1. The water must be deeply deionized, i.e., it should be polished water with conductivity $\leq 0.15 \, \mu$ S/cm at 298 K.
2. pH range 7–7.5
3. The quantity of dissolved oxygen
   20–200 microgram/kg at 473 K
   10–20 microgram/kg above 473 K

Q. *What are the additional advantages of the neutral method of preventing corrosion in the water-condensate circuit?*

Ans.

1. It lends to simplified chemical control
2. It affords great savings in reagents such as ammonia, hydrazine, sulphuric acid and caustic alkali
3. The time gap between two successive regenerations of ion-exchange resin-beds in the DM plant is increased
4. It decreases the concentration of iron in the water-condensate circuit.

Problem 25.1 The oxygen corrosion of iron is an electrochemical process that involves the following reactions:

$$2Fe + O_2 + 2H_2O \longrightarrow 2Fe(OH)_2$$

$$4Fe(OH)_2 + O_2 \longrightarrow 4Fe(OH)_3 \downarrow$$

Calculate the amount of dissolved oxygen required to inflict damage by corrosion on a layer of metal 0.15 mm thick over an area 100 m^2 surface accompanied by the formation of ferric hydroxide.

Also determine the amount of magnetite formed. (Density of iron = 7.8 gm/c.c.)

Solution:

Step (I) Overall Corrosion Reaction
Multiplying the 1st reaction by two and adding the result with 2nd reaction we get the overall reaction

$$4Fe + 3O_2 + 6H_2O \longrightarrow 4Fe(OH)_3$$

$$4(55.8) \quad 3\,(16 \times 2)$$

Step (II) Oxygen Consumption in Corrosion
By stoichiometry, the amount of oxygen involved in corrosion of iron

$$M_{O_2} = 3O_2/4Fe = 3\,(16 \times 2)/4\,(55.8)$$

$$= 0.4301 \text{ gm/gm of Fe}$$

Mass of iron in the layer 0.15mm thick and 100 m^2 surface area

$$M_{Fe} = (0.15/10)(100 \times 10^4)(7.8) \text{ cm (cm}^2)(gm/c.c.)$$

$$= 1.17 \times 10^5 \text{ gm}$$

Hence the amount of oxygen consumed in the given corrosion

$$= 0.4301 \times 1.17 \times 10^5 = 50321.7 \text{ gm} = 50.321 \text{ kg}$$
Ans.

Step (III) Amount of Magnetite Formed

$$2Fe + \frac{3}{2}O_2 \longrightarrow Fe_2O_3$$

$$2(55.8) \qquad (2 \times 55.8 + 3 \times 16) = 159.6$$

Mass of magnetite formed

$$M_{Fe_2O_3} = \frac{159.6}{2\,(55.8)} \times 1.17 \times 10^5 \text{ gm} = 167.322 \text{ kg}$$
Ans.

Problem 25.2 Determine the amount of hydrogen liberated in the corrosion process involved in Problem 25.1 assuming H$^+$ ions instead of oxygen involved as depolarizer.

Solution:

Step (I) Basic Equation of Corrosion with H_2-Depolarization

$$Fe \longrightarrow Fe^{2+} + 2e$$

$$2H^+ + 2e \longrightarrow 2H \longrightarrow H_2$$

Adding these two we get the overall corrosion-reaction

$$Fe + 2H^+ \longrightarrow Fe^{2+} + H_2$$
$$(55.8) \qquad\qquad (2)$$

Step (II) Quantity of H_2 Liberated

$$M_{H_2} = (2/55.8)\,(1.17 \times 10^5)\ gm$$

$$= 4193.54\ gm$$

$$= 4.193\ kg$$

Ans.

Problem 25.3 Calculate the number of pits that will form as a result of pitting corrosion in the feedwater circuit of a boiler unit during its one year service. Base your calculations on the following data:

boiler capacity = 120 t/h

blowdown rate = 4%

operating hours of the boiler in 1 year = 7920 hr.

dimension of pits (cone-shaped):

base diameter = 9 mm

depth = 3 mm

oxygen content in BFW = 0.025 mg/kg

(assume the entire dissolved oxygen take part in pitting corrosion)

density of steel = 7.8 gm/c.c.

Solution:

Step (I) Blowdown (t / h)

Working Formula: $p_{bd}^s = \dfrac{G_{bd}}{G_s} \times 100$

where p_{bd}^s = per cent blowdown

G_{bd} = blowdown rate, t/h

G_s = boiler's steam generation capacity, t/h

p_{bd}^s	G_s	G_{bd}
4	120 t/h	$4 \times 120/100 = 4.8$ t/h

Step (II) Feedwater Flowrate (t/h)

By mass balance:

$$G_{fw} = G_{bd} + G_s$$

$$G_{fw} = 4.8 + 120 = 124.8\ t/h$$

Step (III) Quantity of Dissolved Oxygen involved in entire Corrosion Process during 1-yr service of the boiler

Total mass of feedwater,

$$M_{fw} = 124.8 \times 7920t$$

$$= 124.8 \times 7920 \times 10^3\ kg$$

Total mass of dissolved oxygen,

$$M_{O_2} = (124.8 \times 7920 \times 10^3)\,(0.025)\ mg$$

$$= 24.7104\ kg.$$

Step (IV) Mass of Metal involved per Pit

Volume of metal involved per pit = $\dfrac{1}{3}\pi r^2 h$

(Fig. 25.4)

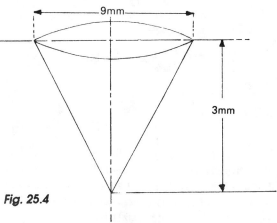

Fig. 25.4

$$= (1/3)\,(\pi)(9/2)^2\,(3)\ mm^3$$

$$= 63.617\ mm^3 = 0.0636\ cm^3$$

Mass of metal involved per pit,

$$m_{Fe} = 0.0636 \times 7.8\ gm$$

$$= 0.4962\ gm$$

Step (V) Mass of Oxygen involved per Pit

The overall reaction of corrosion of iron involving oxygen as depolarizer

$$4Fe + 3O_2 + 6H_2O \longrightarrow 4Fe(OH)_3$$
$$4(55.8) \quad 3(32)$$

Mass of iron involved per pit = 0.4962 gm

∴   Mass of oxygen involved per pit

$$= \frac{3\,(32)}{4\,(55.8)} \times 0.4962 = 0.2134 \text{ gm}$$

Step (VI)   Total Number of Pits

The total number of pits formed during 1-yr service of the boiler

$$N = \frac{M_{O_2}}{m_{O_2}} = \frac{24.7104 \times 10^3}{0.2134} = 115794 \qquad Ans.$$

Problem 25.4 Determine the amount of Fe_2O_3 that forms in the boiler during 7920 h of service under the conditions of Problem 25.3.

Solution: The amount of ferric oxide formed can be calculated with the help of the following reaction:

$$2Fe + \frac{3}{2}O_2 \longrightarrow Fe_2O_3$$
$$2(55.8) \qquad\qquad 159.6$$

$$M_{Fe_2O_3} = \frac{159.6}{2\,(55.8)} \times (0.4962 \times 115\,794)$$

$$= 82169.66 \text{ gm}$$

$$= 82.169 \text{ kg}$$

Ans.

Problem 25.5 Assuming that the entire amount of magnetite forming in the boiler under conditions as described in Problem 25.4 remains in the boiler as deposit, calculate its thickness falling out of 10% of the total heating surface (1200 m^2).

Take the density of magnetite = 5.25 gm/c.c.

Solution: Mass of the deposit = 82.169 kg
   Density of deposit = 5.25 gm/c.c.
   Volume of deposit = 82169/5.25 cm^3
   $\qquad\qquad\qquad = 15651.238 \text{ cm}^3$
Area of heating surface covered by deposit

$$= 1200\,(10/100) = 120 \text{ m}^2 = 120 \times 10^4 \text{ cm}^2$$

Therefore, the depth of deposit

$$= 15651.238/(120 \times 10^4)$$

$$= 0.013 \text{ cm} = 0.13 \text{ mm} \qquad Ans.$$

Problem 25.6 The degree of vapour-water corrosion in superheater coils is measured in terms of the quantity of hydrogen entrained by the superheater steam.

In a superheater of a boiler unit, the concentration of hydrogen in superheated steam amounts to 0.12 mg/kg. If the steam generating capacity of the boiler unit is 200 t/h, estimate the amount of metal affected by vapour-water corrosion.

Solution: The amount of iron damaged by vapour-water corrosion can be calculated with the help of the reaction:

$$3Fe + 4H_2O \text{ (steam)} \longrightarrow Fe_3O_4 + 4H_2$$
$$3(55.8) \qquad\qquad\qquad\qquad 4(2)$$

Now the rate of hydrogen passing to superheated steam

$$= 200 \times 10^3 \times 0.12 \text{ mg/h} = 24 \text{ g/h}$$

Therefore, the amount of metal attacked by steam

$$= \frac{3\,(55.8)}{4 \times 2} \times 24 = 502.2 \text{ g/h}$$

Ans.

Problem 25.7 A superheater coil has a total heating surface of 800 m^2. If 10% of this surface area gets affected by vapour-water (steam) corrosion, determine the time required to reduce the superheater tube thickness by 0.1 cm.

Steam contains hydrogen 0.1 mg/kg
Average steam load of the boiler = 200 t/h

Solution:
Step (I)   Mass of Iron involved in Vapour-Water Corrosion

Total heating surface area of superheater = 800 m^2

Heating surface affected by vapour-water corrosion = 800 (10/100)

$$= 80 \text{ m}^2.$$

Volume of iron attacked by vapor-water corrosion

$$= (80 \times 10^4)(0.1) = 80\,000 \text{ cm}^3$$

Mass of iron involved in corrosion

$$M_{Fe} = \text{Volume of Iron} \times \text{Density}$$

$$= 80\,000 \times 7.8 = 624\,000 \text{ g} = 624 \text{ kg}.$$

Step (II) Mass of Hydrogen evolved in the vapour-water Corrosion

$$3Fe + 4H_2O \longrightarrow Fe_3O_4 + 4H_2$$

$$3(55.8) \qquad\qquad 4(2)$$

Mass of hydrogen liberated when 624 kg of iron suffers from vap-water corrosion is

$$M_{H_2} = \frac{4\,(2)}{3\,(55.8)} \times 624 = 29.82 \text{ kg}$$

Step (III) Time Interval for the above Corrosion

Rate of hydrogen evolution

$$= (200 \times 10^3)\,(0.1) \text{ mg/h} = 20 \text{ g/h}$$

Required time $= 29.82 \times 10^3/20 = 1491 \; h$

Ans.

Q. *What are the gases other than carbon dioxide and oxygen that may find their way to the feedwater system and contribute to corrosion?*

Ans. These are hydrogen sulphide, ammonia, sulphur dioxide, chlorine, etc.

Q. *What is the source of hydrogen sulphide?*

Ans. It comes chiefly from dissolved sulphides which are occasionally found in the surface water as a result of bacterial decomposition of organic substances or from miscellaneous pollution. However, for well-waters the only source of sulphides is the soluble mineral sulphides.

Q. *Why is hydrogen sulphide in a dissolved state corrosive?*

Ans. Hydrogen sulphide dissolves in water to produce weak acid:

$$H_2S \Leftrightarrow H^+ + SH^-$$

$$\text{(aq.)}$$

These liberated protons (H^+) act as a depolarizer and contribute to corrosion

$$Fe \longrightarrow Fe^{2+} + 2e$$

$$2H^+ + 2e \longrightarrow H_2$$

Hydrogen sulphide, even in small quantities, reacts acidic and may be responsible for active corrosion of iron even in the presence of oxygen. Moreover, it has been observed that turbines having nickel alloy parts undergo damage due to the action of hydrogen sulphide (carried over by steam) upon nickel.

Q. *How can the hydrogen sulphide from water be removed?*

Ans. This can be done satisfactorily by the reduction of the pH value of the water followed by good aeration.

Q. *What does aeration in driving off hydrogen sulfide?*

Ans. It removes dissolved sulphides by the scrubbing action of air and by oxidation.

Q. *Why should pH be lowered to eliminate hydrogen sulphide?*

Ans. Hydrogen sulphide, upon dissolving in water, undergoes the following ionic equilibria:

$$H_2S \Leftrightarrow H^+ + SH^-$$

$$\text{(aq.)}$$

which is shifted to the left if more H^+ ions are added, i.e., the pH is lowered. So at low pH, lesser amount of H_2S will dissolve in water, facilitating H_2S removal from the feedwater system.

Q. *Why is the presence of sulphide ions in feedwater unwanted?*

Ans. Many metals are rapidly corroded by relatively high concentrations of sulphides. Stainless steel and brass which can effectively withstand attack in oxidizing medium undergo aggressive attack by water containing sulphides.

Q. *Why is hot water more corrosive than cold water?*

Ans. With the increase of the temperature of water, the convection current increases and viscosity decreases and these two factors accelerate the diffusion of oxygen through the metal surface film.

Moreover, with the increase of water temperature, its pH value decreases which in turn accelerates the deposition of atomic hydrogen (depolarizer) on cathodic areas.

Q. *What is the effect of water alkalinity on the corrosion of iron?*

Ans. Though it has been established by experience and tests that the corrosion rate of iron by alkalies diminishes to a certain minimum as the concentration of the base increases, high concentrations of alkalies to the point of supersaturation as are encountered under the conditions of film-boiling, greatly accelerate rather than retard corrosion of steel.

Q. *What is the mechanism of attack?*

Ans. It proceeds via the mechanism of protective film destruction. The upper surface of tubes which are only slightly inclined from the horizontal as is the case in dry- bottom furnaces and in the floors of slag-bottom furnaces, may be blanketed with steam causing high local temperatures of the metal surfaces. Intermittent splashing of water frequently wets these partially insulated surfaces leading to higher concentration of NaOH in these areas. High caustic alkalinity tends to rupture the protective ferric-oxide (Fe_2O_3) film. The combined effect of higher temperatures and destruction of this oxide film causes grooving of tubes and ultimate failure.

Q. *What is the remedy?*

Ans.
1. Cut in caustic alkalinity of boiler feedwater
2. Controlling heat input
3. Improving circulation
4. Increasing the inclination of the tubes.

Q. *In many cases severe corrosion has been reported beneath the layer of protective ferric oxide film resulting in the loss of metal at the corroded area and even accompanied by embrittlement of steel. Why?*

Ans. This is due to alkaline attack. Though a dilute solution of sodium hydroxide apparently assists in developing an impervious oxide film, higher con-

centration of alkali tends to open it up while a very concentrated solution actually destroys it by forming soluble iron complexes. Therefore, local high concentrations of sodium hydroxide impair the protective film, enhance the generation of hydrogen and increase the rate of hydrogen embrittlement of metal.

Q. *What are the mineral acids that may contaminate the boiler feedwater supplies?*

Ans. These are: Hydrochloric, nitric, sulphuric acids.

Q. *Where do they come from?*

Ans. They come from industrial wastes discharged into surface or underground supplies. Many mine waters have a high acid content. Hence mine drainage is frequently responsible for the mineral acids in the feedwater.

Q. *What is the effect of these mineral acids in the BFW if these are not neutralized?*

Ans. They will cause rapid destruction of metals of boilers and auxiliary equipment in the form of pitting corrosion that may lead to tube failure or equipment-breakdown.

Q. *It has been observed that iron salt present in the form of ferric sulphate increases the corrosivity of water whereas the opposite effect has been noted when iron is present in the form of ferrous sulphate, when the dissolved oxygen concentration, the velocity and temperature of water and other factors are constant. Why?*

Ans. Ferric ions act as a depolarizer:

$$Fe^{3+} + e \longrightarrow Fe^{2+}$$

So iron present in the feedwater in the ferric state increases the aggressivity of water.

Whereas, ferrous ions present in water react with dissolved oxygen and inhibit corrosion.

Q. *Why does the presence of magnesium chloride in water increase the corrosiveness of the latter?*

Ans. Dissolved magnesium chloride undergoes hydrolysis to produce hydrochloric acid

$$MgCl_2 + 2H_2O \Leftrightarrow Mg(OH)_2 + 2HCl$$

Since magnesium hydroxide separates out as insoluble precipitate, the above equilibria continues to shift to the right to generate more and more hydrochloric acid that accounts for increasing corrosivity of the feedwater.

Q. *A relatively small amount of dissolved magnesium chloride may lead to active corrosion. Why?*

Ans. Dissolved magnesium chloride undergoes hydrolysis liberating hydrochloric acid that attacks iron as:

$$MgCl_2 + 2H_2O \Leftrightarrow Mg(OH)_2 + 2HCl$$

$$Fe + 2HCl \longrightarrow FeCl_2 + H_2$$

The ferrous chloride resulting as a preliminary corrosion product reacts further with magnesium hydroxide to produce back magnesium chloride:

$$FeCl_2 + Mg(OH)_2 \longrightarrow Fe(OH)_2 + MgCl_2$$

which again undergoes hydrolysis and initiates the cycle again.

This explains why a small amount of dissolved magnesium chloride may lead to active corrosion that may go on indefinitely.

Q. *Sodium chloride is mildly corrosive; so its presence in the feedwater does little harm. But in the presence of magnesium sulphate, it becomes actively corrosive. Why?*

Ans. Sodium chloride is a neutral salt. So its aqueous solution is not corrosive. But in the presence of magnesium sulphate, it reacts with the latter to form magnesium chloride:

$$2NaCl + MgSO_4 \longrightarrow Na_2SO_4 + MgCl_2$$

which undergoes hydrolysis, liberating hydrochloric acid that actively corrodes the metal surface

$$MgCl_2 + 2H_2O \longrightarrow Mg(OH)_2 + 2HCl$$

Q. *How can this corrosion by magnesium salt be inhibited?*

Ans. This can be done by neutralizing the magnesium chloride with an alkali:

$$MgCl_2 + 2NaOH \longrightarrow Mg(OH)_2 + 2NaCl$$

Q. *Are there any magnesium salts other than magnesium sulphate or chloride that may lead to corrosion in the boiler?*

Ans. Magnesium carbonate and magnesium nitrate in the presence of sodium chloride give rise to active corrosion:

$$MgCO_3 + 2NaCl \longrightarrow MgCl_2 + Na_2CO_3$$

$$MgCl_2 + 2H_2O \longrightarrow Mg(OH)_2 + 2HCl$$

$$Mg(NO_3)_2 + 2NaCl \longrightarrow MgCl_2 + 2NaNO_3$$

$$MgCl_2 + 2H_2O \longrightarrow Mg(OH)_2 + 2HCl$$

$$NaNO_3 + HCl \longrightarrow NaCl + HNO_3$$

Q. *Does corrosion occur under the scales?*

Ans. As a general rule, scale formation inhibits corrosion since the deposited salts mechanically protect the metal and prevent direct contact between the corrosive agent and the metal.

Q. *But still corrosion beneath the scale or even under the protective coating applied to protect the metal surfaces have been reported. Why?*

Ans. Corrosion results when the protective coating gets cracked, permitting the water, containing a fairly large concentration of corrosive salts, to concentrate between the laminations produced by a hair crack of the protective layer. These salts dissociate at the metal surfaces and corrode the latter.

Q. *How many forms of pitting corrosion arise in boiler deterioration?*

Ans. Three types:
1. Air-bubble pitting
2. Scab pitting
3. Soft scab pitting.

Q. *What does air-bubble pitting looks like?*

Ans. It appears in the form of isolated hemispherical pits often capped by loose, soft corrosion products. These, when dried, are black iron oxide covered with a layer of red dehydrated ferric oxide.

Q. *Where can this type of pitting corrosion be located?*

Ans. Likely to be found on any surfaces in contact with water where air bubbles may collect, but mainly on the roof of steam and water drums.

Q. *What is the principal cause of this corrosion?*

Ans. Air bubbles liberated attach themselves to metal surfaces, thereby triggering anodic attack due to localized electrolytic action, and oxygen acts as a depolarizer.

Q. *What does scab pitting looks like?*

Ans. It appears as deep localized pitting capped by a scab of hard scale. This is the scale of black magnetic iron oxide which cannot be dislodged by wire brushes.

Q. *Where does this pitting occur?*

Ans. Usually in hotter zones of heating surfaces. Here the corrosion products, as soon as they form, get quickly consolidated in a hard scab by heat.

Q. *What is the cause of scab pitting?*

Ans. Use of partially treated or untreated impure boiler water containing dissolved salts as well as gases, rendering water pH ranging from 4.8 to 9.4.

Q. *What does soft scab pitting look like?*

Ans. They appear as broad, shallow pitting covered with fairly hard corrosion products composed of black magnetic iron oxide capped by a layer of red dehydrated ferric hydroxide.

Q. *Where does this soft scab pitting appear?*

Ans. Usually in superheater tubes and headers and in main steam lines.

Q. *What is the cause of soft scab pitting?*

Ans. This is chiefly due to deposition of salts by priming in superheaters, steam main followed by condensation when the boiler is shutdown giving rise to formation of droplets of electrolyte and the initiation of a pit by differential aeration.

Q. *General wastage is a type of internal deterioration of the boiler. In how many varied forms does it appear?*

Ans. It appears in the form of:
1. Tube-wall thinning
2. Necking and Grooving
3. Galvanic attack

Q. *What does tube-wall thinning look like?*

Ans. It appears as a general reduction in the thickness of the tube-wall over a considerable area with or without the presence of the blankets of corrosion products.

Q. *Where can this be located?*

Ans. More or less evenly distributed over the water wall surfaces. In hotter areas most likely.

Q. *What is the cause of it?*

Ans. Acidity in feedwater.

Q. *What do necking and grooving look like?*

Ans. They appear as a circumferential band of corrosion.

Q. *Where can these be found?*

Ans. On the waterside of the fire tubes (where they enter the end plates or combustion chamber), stay tubes and stay bolts.

Q. *Why do they occur?*

Ans. They result from stress. Expansion and contraction of adjacent endplates result in stresses that cause repeated localized flaking of protective films.

Q. *What signals the appearance of galvanic attack?*

Ans. Reduction in metal thickness in localized areas.

Q. *What is the usual location of this attack?*

Ans. In the areas where steel is not homogeneous in composition and also near places adjacent to non-ferrous boiler mountings.

Q. *What is the cause of galvanic attack?*

Ans. Ferrous and non-ferrous metals in contact with an electrolyte or steel areas of non-homogeneous composition in contact with the electrolyte set up microunits of the galvanic cell wherein a potential differential is set up between the anode and cathode. The metal dissolves at the anode.

Q. *How can corrosion-fatigue responsible for boiler deterioration classified?*

Ans. They may be classified as:
  (a) corrosion-fatigue in tubes
  (b) corrosion-fatigue in steam and water drums
  (c) corrosion-fatigue in steam mains.

Q. *What does the corrosion-fatigue in tubes look like?*

Ans. It appears as a network or series of cracks over the tube surfaces in the waterside.

Q. *What is the usual location of corrosion-fatigue in tubes?*

Ans. In the hottest zone of boiler and superheater tubes and also frequently in inherently curved tube or tubes which got buckled due to prolonged operation.

Q. *What is the cause of this type of corrosion-fatigue?*

Ans. Non-uniform circulation of working fluid in tubes causing alternate overheating and cooling inducing thermal fatigue stress in the walls. Presence of residual stresses and a corrosive environment also accelerate this type of cracking.

Q. *What does corrosion-fatigue in steam and water drums look like?*

Ans. It appears as fissures covered under the layers of corrosion products. That's why they're difficult to locate. It is usually associated with deep pitting.

Q. *What is the usual location of such types of attack?*

Ans. It starts at the arboring in tube plates, extends axially, eventually joining two or more holes.

Q. *What is the cause of fatigue-corrosion in steam and water drums?*

Ans. Bending stresses induced due to the lack of a common neutral axis between the wrapper plate and the tube coupled with excessive cold working due to initial rough machining operation are responsible for such a type of attack. It is accelerated by thermal stress and the presence of an aggressive environment.

Q. *What is the appearance of fatigue-corrosion in steam mains?*

Ans. It appears as lines of cracks on the bores of pipes.

Q. *What is the usual location of these type of cracks?*

Ans. Bores of steam pipelines particularly in the base of corrugations.

Q. *What is the cause of corrosion-fatigue in steam mains?*

Ans. Bending stresses due to movement during operation. This is accelerated by the presence of a corrosive environment.

Q. *How can boiler deterioration at elevated temperatures be classified?*

Ans. This can be classified as:
  (a) high temperature oxidation
  (b) bursts
  (c) creep-cracking at high temperature.

Q. *What is the appearance of high-temperature oxidation?*

Ans. It appears as a continuous layer of hard black scale of magnetic iron oxide.

Q. *Where is the high temperature oxidation most prevalent?*

Ans. Internally it occurs in fire row and superheater tubes. It also occurs in drums and headers.
    Externally it occurs on surfaces in direct contact with furnace gases.

Q. *What is the cause of it?*

Ans. For internal attack, the poor circulation of the working fluid is responsible. Improper circulation results in overheating in a steam atmosphere.

For external attack, local flame impingement, poor circulation and scaling on the waterside are responsible.

Q. *What is the appearance of boiler deterioration in the form of bursts occurring at elevated temperatures?*

Ans. They appear as splits in the form of longitudinal gaps. These may or may not be associated with balloning.

Q. *Where do they usually appear?*

Ans. Hottest part of the evaporator and superheater tubes.

Q. *Why do they occur?*

Ans. Water starvation in the evaporator and steam starvation in the superheater tubes are chiefly responsible for tube bursts. They lead to overheating which deteriorates the mechanical properties of the tube material with the effect that tubes burst even under working pressure.

Q. *What does the creep-cracking occurring at elevated temperatures look like?*

Ans. It appears as cracks and splits associated with layers of oxide scale on tube surfaces.

Q. *What is the usual location of such a type of attack?*

Ans. In superheater tubes.

Q. *What is the cause of such an attack?*

Ans. Overheating at around 550°C coupled with restricted expansion induce thermal stresses in metal and they are accountable for creep-cracking at elevated temperatures.

Q. *It has been found in practice that boilers that had been in service for many years under normal operating conditions experienced no difficulty. But when they were greatly overloaded under increased demand, tube failures occurred. Why?*

Ans. These tube failures are due to creep-cracks at high temperature. At the higher ratings defective circulation of the working fluid develops in certain tubes, resulting in overheating of metal and that brings about creep, causing tube failure.

Q. *What morphological changes occur during this type of tube failure?*

Ans. Photomicrographs of the metal at points of greatest heating revealed the complete spheroidization of carbides. And also the masses of spheroids representing the original pearlitic grains were elongated in the direction of greatest stress. At temperature 650°C and above, low-carbon steels or completely speroidized steel will recrystallize, in the presence of a critical strain.

Q. *Intergranular disintegration on the waterside of steam generating tubes has been reported in high pressure boiler units. What is the appearance of this intergranular corrosion?*

Ans. In the affected areas, layers of metallic copper, oxides of copper and black oxide of iron are found deposited. Beneath these deposits large corrosion pits are formed and beneath these pits severe deterioration of the metal in the form of intergranular disintegration takes place.

Q. *Wherefrom do metallic copper and its oxides appear?*

Ans. These are the corrosion products of the condenser and other heat exchanger tubes which are made of copper-alloy in general.

Q. *What agents accelerate intergranular corrosion?*

Ans. Dissolved gases like ammonia and carbon dioxide in the condensate accelerate this type of attack.

Q. *What role does copper play in stimulating intergranular corrosion?*

Ans. Copper deposits create an anodic condition beneath the black iron oxide resulting in corrosion of the metal and production of atomic hydrogen. This atomic hydrogen penetrates the steel, causing decarburization and intergranular corrosion.

Q. *Boiler-tube failures have been reported in certain cases where corrosion of the metal is found under an adherent formation of black iron oxide accom-*

panied by embrittlement of the adjacent steel. What is the most apt designation of this failure?

Ans. Corrosion embrittlement.

Q. *What is the mechanism of corrosion embrittlement?*

Ans. This type of corrosion is initiated by excessive heat or improper circulation of the working fluid resulting in the rise of metal temperature—not high enough for overheating but sufficient to cause film-boiling. Under this condition, water vapour dissociates and the metal reacts with the oxygen producing a magnetic oxide film on the metal surface and liberating hydrogen:

$$H_2O \longrightarrow 2H + O$$

$$2Fe + 3O \longleftarrow\!\!\!\longrightarrow Fe_2O_3$$

As the corrosion propagates, the corrosion product deposited on the tube surface gets washed away and the tube cross-section gets thinner and thinner.

The hydrogen formed now penetrates the steel along the crystallographic planes or grain boundaries and reacts with the oxides, nitrides and carbides of steel to form gaseous products within the metal. As a result tremendous pressures are developed causing rupture of metal at the grain boundaries.

Moreover, if this corrosion proceeds sufficiently, the thinned tube portion becomes weakened enough not to withstand the high operating pressures of the boilers and may be blown out of the tube walls by the internal steam and water pressure.

Q. *It has been found that magnetic oxides, when allowed to accumulate, compact and adhere to the protective oxide films, can damage the latter. Why?*

Ans. Once the heavy magnetic oxides formation upon the protective oxide film has been established, the protective film gets penetrated with the effect that the resultant deposit, not being homogeneous, no longer protects the steel. This is followed by the attack of the metal by steam or water beneath the black iron oxide layer. The protective film is destroyed substantially and the corrosion rate is enhanced.

26

Corrosion of Fireside Heating Surfaces

Q. *How can the fireside corrosion of heating surfaces in boiler units can be classified?*

Ans. It can be classified as:
(a) high temperature corrosion (600°C and above)
(b) low temperature corrosion (<600°C)

Q. *How does the high temperature corrosion of heating surfaces on the fireside of the boiler unit take place?*

Ans. Corrosion of the waterwall tubes in the boiler furnace takes place in the zone of the flame core because of the attack of the tube-metal by sulphurous gases present in the combustion products.

Superheater tubes and their fastening elements experience damage of metal by the vanadium oxides present in the flue gases.

Q. *What kind of fuels give rise to corrosion on the fireside of waterwalls in the boilers?*

Ans. Pulverized coal (anthracites, semianthracite, lean coals) having very low yield of volatiles and fuel-oil with high sulphur content.

Q. *Which section of the waterwall tubes is most afflicted by corrosion of combustion gases?*

Ans. It is most predominant in the tube section at the burner level or immediately above it.

Q. *What is the rate of corrosion of the metal surface in this zone of the tube surface?*

Ans. It may be as high as 2.5–4mm per annum rendering waterwall tubes, usually 5 to 6mm thick, out of service in less than a year.

Q. *What is the main corrosion-active component in the furnace gases?*

Ans It is hydrogen sulphide.

Q. *What is the mechanism of the hydrogen sulphide corrosion of tube metal?*

Ans. It proceeds via following steps:
1. **Chemical reaction of the tube metal with H_2S**

$$Fe + H_2S \longrightarrow FeS + H_2$$

2. **Oxidation of ferrous sulphide layer**

$$FeS + 2O_2 \longrightarrow FeSO_4$$

3. **Flaking-off of ferrous sulphate:** The ferrous sulphate layer produced on the tubewall surface flakes off due to erosion exposing the fresh tube metal to further corrosion. (Fig. 26.1)

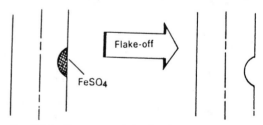

Flake-off

$FeSO_4$

Fig. 26.1 *Ferrous sulphate flakes off as a result of hydrogen sulphide attack*

Q. *How can damage due to corrosion of the external surface of waterwall tubes by hydrogen sulphide be avoided?*

Ans. The effective way is to mix and distribute evenly the fuel and air between the burners so that excess air ratio in each burner is always greater than unity. This will reduce the chances of formation of zones deficient in oxygen where hydrogen sulphide attack is likely to take place.

Burners should be aligned to keep off the flame from direct heating of the waterwall.

Q. *What is vanadium corrosion?*

Ans. It refers to the attack of superheater tubes by the vapours of vanadium pentoxide formed in the furnace gases.

If the fuel contains sodium, the furnace gases will contain, among others, sodium vanadate $(Na_2 O \cdot 5V_2O_5 \cdot V_2O_4)$ — a low fusing compound (m.p. $\approx 600°C$) forming a thin film of corrosive liquid on the superheater tube surface at $610 - 620°C$. This film attacks carbon, low-alloyed and austenitic steels.

Q. *Do the oxides of sulphur have any influence on vanadium corrosion?*

Ans. Yes. It is augmented by the presence of oxides of sulphur in the flue gases. Particularly dangerous is sodium pyrosulfate $(Na_2S_2O_7)$. The corrosivity of V_2O_5 increases manifold in combination with the oxides of sulphur.

Q. *How can vanadium corrosion be reduced?*

Ans. Two ways:
(a) keeping the superheater temperature below 600°C (873 K)
(b) dosing alkaline additives to fuel oil.

Q. *What are these alkaline additives?*

Ans. For instance, $MgCl_2$ added in amount 0.6 to 0.8 kg per ton of fuel oil.

Q. *Where does the low temperature corrosion occur?*

Ans. It occurs on the heating surfaces of the air heaters only.

Q. *Why does this corrosion take place?*

Ans. This corrosion takes place due to the attack of the metal by sulphuric acid vapour in the flue gases.

Q. *How does sulphuric acid vapour form in the flue gas flow?*

Ans. The sulphur present in the fuel is oxidized to sulphurous anhydride during the combustion of fuel

$$S + O_2 \longrightarrow SO_2$$

The sulphur dioxide so produced gets further oxidized to sulphur trioxide (sulphuric anhydride) by atomic oxygen

$$SO_2 + O \Leftrightarrow SO_3$$

In the zones where temperature drops below 500°C, the sulphur trioxide reacts with water vapour and produces sulphuric acid vapour which

$$SO_3 + H_2O \longrightarrow H_2SO_4$$

is carried off by the gas stream.

Q. *How is the atomic oxygen available in the boiler-furnace to oxidize SO_2 to SO_3?*

Ans. It is produced by the chain reactions of combustion and thermal dissociation taking place in the high-temperature zone of the flame.

Q. *But the formation of SO_3 from SO_2 is an exothermic reaction. So how will it be formed in the high temperature zone where atomic oxygen is available?*

Ans. Because of the exothermic nature of the reaction

$$SO_2 + \frac{1}{2}O_2 \Leftrightarrow SO_3$$

the sulphur trioxide that appears noticeably on the boundry of the flamecore dissociates completely on completion of combustion.

However, its concentration is regained as the temperature in the gas path decreases gradually so that at gas temperatures 1200–1250°C (1473–1523K), the dissociation of SO_3 ceases almost totally.

Apart from atomic oxygen, the presence of soot particles deposited on the heating surfaces may serve as a catalyst to oxidize SO_2 to SO_3 by oxygen in excess air.

Q. *What is dew point?*

Ans. It refers to the temperature at which moisture condenses on a solid surface. It is also called thermodynamic dew temperature.

Q. *Has it any significance regarding corrosion on the fireside of heating surfaces?*

Ans. Yes. If the dew point of sulphuric acid vapour in its partial pressure in the flue gases is reached, a film of highly corrosive sulphuric acid will settle on the tube surfaces and corrode the metal.

Q. *What is the dew point of sulphuric acid vapour?*

Ans. It is roughly in the range 140–160°C (413–433K).

Q. *What is the effect of temperature on the rate of acid corrosion, on the heating surfaces in contact with flue gases produced by combustion of high-sulphur fuel oil?*

Ans. This can be best understood with the help of the rate of corrosion (g/m^2.h) vs. temperature curve. (Fig. 26.2)

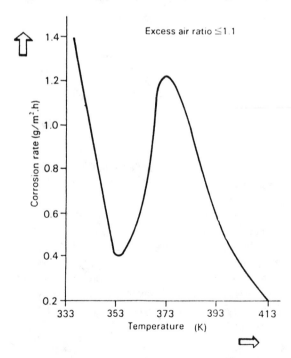

Fig. 26.2 *Acid corrosion as a function of temperature of heat transfer surface*

As the curve shows, the intensity of corrosion is very pronounced in two temperature ranges — one at 60–80°C (333–353K) and the other at 90–120°C (363–393K).

As the wall temperature decreases from 145°C (418K, dew point) the corrosion rate increases sharply, reaches a maxima in the temperature range 100–110°C (373–383K) and then drops steeply, reaches a minima at 80°C (353K) and again rises sharply at wall temperature below 800°C (353K).

Q. *Why does the intensification of corrosion take place in the 60–80°C (333–353K) range?*

Ans. It is due to the dissolution of metal by sulphurous acid condensing on heating surfaces in this temperature range

$$H_2O + SO_2 \longrightarrow H_2SO_3$$

Sulphurous acid is an unstable acid dissociating at temperature above 80°C (353K).

Q. *Why does intensive corrosion take place in the temperature range 90–120°C (363–393K) ?*

Ans. It is mainly due to the attack of H_2SO_4 condensing on the metal surface. The rate of corrosion increases with the increase of the rate of condensation of H_2SO_4 vapours.

Q. *It has been observed that the rate of corrosion initiated by the oxides of sulphur at 80–120°C (353–393K) is much less if lowest allowable excess air is maintained. Why?*

Ans. When lowest allowable excess air is supplied to ignite the high sulphur fuel oil, the combustion gases contain a much higher precentage of SO_2 than SO_3.

Since sulphur dioxide-corrosion is prevalent at temperatures less than 80°C (353K), because of the thermal instability of sulphurous acid

$$H_2SO_3 \longrightarrow H_2O + SO_2$$

flue gases containing a lesser percentage of SO_3 and consequently lesser amount of H_2SO_4 that would otherwise initiate intensive corrosion in the temperature range 80–120°C (353–393K), become less aggressive in this temperature range. (Fig. 26.3)

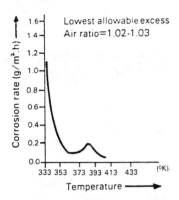

Lowest allowable excess
Air ratio=1.02-1.03

Fig. 26.3

Q. *Can the flyash reduce SO_2-SO_3 corrosion?*

Ans. Yes; provided it contains certain basic metals like calcium, magnesium, sodium, etc. These are basic in nature, and will react with acidic SO_2/SO_3 to form chemical compounds and reduce their harmful effect.

Q. *What should be the wall temperature to avoid low temperature corrosion?*

Ans. The wall temperature should be always 10–15°C higher than the sulphuric acid dew point. (Fig. 26.3)

27

Evaporators

Q. *What is an evaporator?*

Ans. It is a boiler auxiliary where generation of steam takes place.

Q. *Where is it located?*

Ans. The evaporating surfaces, also called steam-generating surfaces, are always laid out for the most part within the shaft of the furnace.

Q. *What kind of heat does it absorb?*

Ans. Radiant heat (mainly).

Q. *How much of the total heat released in the furnace is absorbed by the waterwalls of the evaporator?*

Ans. 30–60% of the total heat released in the furnace.

Superheated Steam		BFW Temp.	Heat Distribution		
Pressure	Temp.		Evaporator	Super-heater	Econo-mizer
4 MN/m^2	713°K	418°K	60%	20%	20%
10 MN/m^2	813°K	488°K	49%	30%	21%
14 MN/m^2	843°K	503°K	32–39%	36–46%	22–25%

Q. *Why is the economizer of medium pressure (4 MN/m^2) generally of the boiling type?*

Ans. It is because the total heat demand (62%) for generating steam is too high to be covered by the radiant heat absorbed by the evaporator's heating surfaces. Hence a part of the heat spent on water

evaporation is transferred to the economizer wherein water is, therefore, not only heated to the saturation temperature but also partially vapourized.

Q. *Why can the economizer of high-pressure (14 MN/m^2 or more) drum-type boilers be of the non-boiling type?*

Ans. In these cases the fraction of the heat utilized for steam generation in the evaporator is considerably less. As a result the total radiant heat absorbed by the evaporating surfaces in the furnace shaft is enough to produce the required quantity of steam and so the economizer can be of the non-boiling type.

Q. *What forms the evaporating surfaces of evaporators?*

Ans. Boiler-water tubes arranged in the furnace shaft in the form of waterwalls constitute the evaporating surfaces. They absorb raidant heat.

Q. *How many types of waterwalls are used in practice?*

Ans. They may be:
 (a) smooth or bare tubes in design (Fig. 27.1A)
 (b) gas-tight with finned tubes (Fig.27.1B)
 (c) gas-tight with welded-on-rectangular finned tubes. (Fig. 27.1C)
 (d) smooth tubes with intertubular space filled with building-up metal (Fig. 27.1D)

Q. *Where are the bare-tube waterwalls employed?*

Ans. In all boiler systems fitted with balanced draft furnaces.

Q. *How are the waterwalls arranged in natural circulation boilers?*

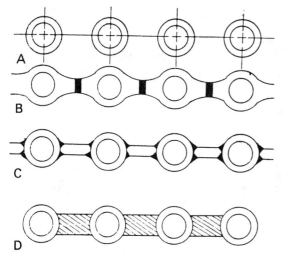

Fig. 27.1 *Different types of waterwall tubes*

Ans. They are laid out in the furnace shaft almost exclusively in vertical pattern or at a steep angle.

Q. *How are the waterwalls arranged in once-through and multiple forced circulation boilers?*

Ans. They can be oriented within the furnace space at any angle, i.e., the waterwalls may be horizontal, vertical or ascending-descending.

Q. *What are the causes of non-uniform heating of the waterwalls of evaporators?*

Ans. The principal cause of non-uniform heating is slagging of tubes. Since slagging is never uniform over the entire evaporating surface of waterwalls, heavily slagged and consequently poorly heated tubes receive less heat than do clean tubes. That is why the slagged tubes develop a lower driving circulating head than clean tubes with the effect that less cooling (circulating) BFW passes through them. Being cooled inadequately, such slagged tubes may become overheated in the tube portions free from slag.

Secondly, the uneven heating across the width of the waterwalls of the circulation circuit may be due to an improperly designed circulation circuit or due to clinkering or misalignment of some tubes. Tubes in the middle of a furnace wall generally receive about twice as much radiant heat as absorbed by the corner tubes. Besides, tubes around the burner port and those

at a certain height from the port are not directly exposed to the radiant heat of flame, i.e., they turn out to be shaded.

Sometimes some tubes become misaligned when tube fastenings get loosened. Protruded from tube rows, these off-rank tubes receive more heat than the adjacent tubes.

Q. *How can the circulation ratio be increased to improve the operating conditions of waterwalls?*

Ans. This can be done by increasing the cross-section of the downtake tubes. This will reduce the hydraulic resistance in the downtake (descending) tubes and ensure a higher flowrate of water through the uptake tubes as well as through the slagged (poorly heated) tubes—uptake and downtake as well. Therefore, a more or less uniform circulation head is established.

Q. *What is waterwall sectionalizing?*

Ans. It is a system that combines a group of water-wall tubes heated similarly into a section which is fed separately with BFW.

Q. *Why is this done?*

Ans. To avoid uneven heating of evaporating tubes. (Fig. 27.2)

The corner waterwall tubes cannot develop the same useful head $S_{usf} = \Delta P_{des}$ (i.e., the pressure drop in descending tubes) as do the middle waterwall tubes because the former receive substantially lesser heat than the latter ones. Therefore, circulation stagnation or circulation reversal is likely to appear in these corner tubes. However, if this bunch of corner tubes can be sectionalized into a separate circulation circuit by placing partitions in the upper and lower headers, the circulation velocities in the two circuits—one comprising the middle waterwall tubes and the other, corner tubes—can be roughly equalized. And for the same non-uniformity of heating across the width of waterwall, the useful circulation head in the separated section will decrease somewhat but the margin to circulation stagnation and circulation reversal will enlarge considerably.

Q. *What kind of waterwall is applied in gravity circulation boilers?*

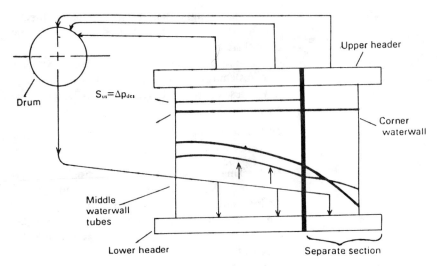

Fig. 27.2 Waterwall sectionalizing is intended for avoiding uneven heating of evaporating tubes

Ans. Usually bare-tube waterwalls.

Q. What are the dimension of waterwall tubes in high pressure boilers?

Ans. ID 40 – 50 mm of uptake tubes.

ID 60 – 160 mm of downtake tubes.

Downtake tubes may be much larger (ID: 600–800 mm) for high capacity boilers.

Q. What is the arrangement of waterwalls in the furnace of a high pressure boiler?

Ans. Front-, rear- and side-waterwalls comprise of a system of parallel vertical tubes. All the upper headers are mounted approximately on the same height and suspended from the boiler structure, i.e., the boiler structure carries the suspended waterwall system. (Fig. 27.3)

The entire waterwall tube system is attached with movable rigidity belts of steel, which on heating, can move freely with the tubes.

Q. How are the waterwalls arranged in once-through boiler units?

Ans. They're arranged either horizontally or vertically.

Q. What is the pattern of layout of horizontal waterwall tubes in once-through boilers for enhancing the reliability of operation?

Ans. For enhancing circulation reliability, the waterwall tubes must be more or less uniformly heated. In a horizontal layout, the evaporating tubes are wound around the periphery of the furnace so that if one of the furnace sides is heated more intensively than the others this non-uniformity of heating will have no ultimate effect. Since all tubes are wound on all the sides of the furnace, any degree of non-uniform heating across the width or depth of the furnace will be distributed equally in all parallel sides.

For high capacity boilers, the evaporating tubes are divided into a number of parallel tube bands which are arranged in the furnace, i.e., the mainstream of boiler feedwater is divided into a number of parallel flows. The parallel bands of two side waterwalls are slightly inclined (12°–15°) while the front and rear waterwalls are laid horizontally. (Fig. 27.4)

Q. Why for high-capacity high-pressure once-through boilers with horizontally arranged waterwall tubes, are the tubes arranged in a number of parallel tube bands?

Ans. For high-capacity, high-pressure boiler units, the waterwall tubes are made of small diameter (ID: 25 – 40 mm) to economize on the tube mass.

But because of high-capacity, the number of parallel tubes must be increased to maintain the same mass flowrate. The greater number of tubes and smaller

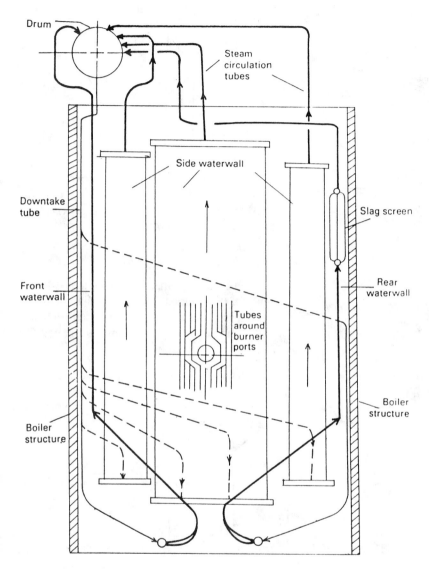

Drum

Steam circulation tubes

Side waterwall

Downtake tube

Slag screen

Front waterwall

Rear waterwall

Tubes around burner ports

Boiler structure

Boiler structure

Fig. 27.3 *Layout of waterwalls in the furnace of a high pressure boiler*

tube diameter make it necessary that the width of the tube band should be increased and with wider tube band the uneven heating along the height of the furnace in the parallel tubes of a band becomes increasingly prominent. For this reason, the waterwall tubes are divided into a number of parallel tube bands and the mainstream of the feedwater flow is divided into a number of parallel flows which are fed to the system of parallel tube bands.

Q. *How are the vertical waterwalls in once-through boiler units arranged?*

Ans. They are arranged similar to those in gravity circulation boilers. They are laid up on the whole surface of the furnace walls and are sectionalized across the furnace width for more uniform heating. These individual sections are joined in series through downtake tubes to result a multipass system. (Fig. 27.5)

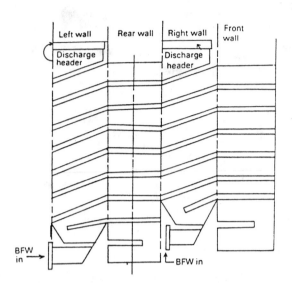

Fig. 27.4 *Layout pattern of horizontal water-wall in once-through boiler units*

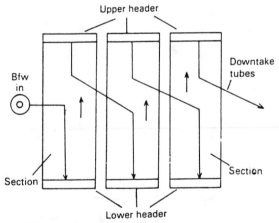

Fig. 27.5 *Layout pattern of vertical waterwall in once-through boiler units*

For high-capacity, once-through vertical boiler units, the narrow vertical waterwall tubes are sectionalized into parallel bands. These parallel bands are connected with one another in series. (Fig. 27.6)

Q. *How are the waterwalls in the gas-tight boiler arranged?*

Ans. They are arranged in the form of vertical sections.

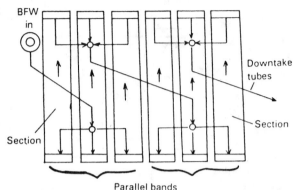

Fig. 27.6 *Sectionalizing of vertical waterwall tubes for high-capacity, once-through boiler units*

Q. *What is the most important requirement for gas-tight waterwalls for reliable operation?*

Ans. The basic requirement is that the temperature difference of the working fluid in the adjacent tubes (for single-pass scheme) or between adjacent sections $(\theta'_2 - \theta'_1)$ and $(\theta_2 - \theta_1)$ (for two-pass scheme) should not exceed 50–100°C. (Fig. 27.8)

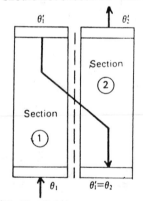

Fig. 27.7 *All-welded 2-Pass waterwall*

Measures are taken to keep the temperature difference ($\Delta \theta$) between welded sections as low as possible. (Fig. 27.8)

Q. *If these measures are not taken what will happen?*

Ans. The welded sections will develop excessive temperature stresses that may even lead to the breakthrough of tubes.

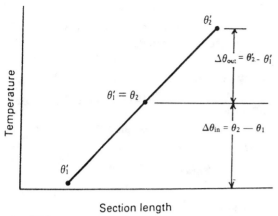

Fig. 27.8

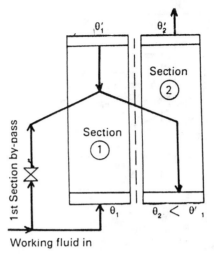

Fig. 27.9 Cold flow bypassing

Q. *What measures are taken to reduce the temperature difference between adjacent sections to below 50–100°C?*

Ans. The following methods are adopted in practice:

1. intermixing of the working fluid along its path
2. cold flow by-passing
3. recirculating the combustion products
4. recirculating the working fluid.

Q. *How does cold flow by-passing ensure the reduction of $\Delta\theta$ between welded sections?*

Ans. This is done by by-passing a portion of the cold flow, i.e., a portion of the working fluid is by-passed around the 1st-section of the heated waterwall tubes. (Fig. 27.9) This results in a higher outlet temperature of the working fluid from the 1st-section because of the increased heat absorption per unit flowrate. Therefore, the temperature difference between the welded sections $[\Delta\theta_{out} = \theta_2' - \theta_1']$ is thus effectively reduced. (Fig. 27.10)

Q. *How much quantity of working fluid is by-passed?*

Ans. It is determined by the ratio of the mass velocities of the working fluid and the heat release rate in the furnace. It is roughly equal to 20% of the rated load.

Q. *How can the $\Delta\theta$ between the welded sections be reduced to an allowable limit by recirculating the working fluid?*

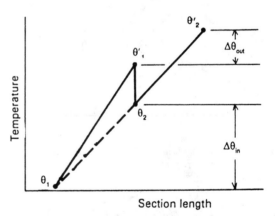

Fig. 27.10 Cold flow bypassing reduces effectively the temperature difference between the welded sections

Ans. This is done by drawing a portion of the hot working fluid from the exit of the 2nd-section and feeding it to the stream at the inlet to the 1st-section. (Fig. 27.11)

Therefore, the temperature of the working fluid at the inlet to section (1) is raised and that reduces the gain in the enthalpy of the working fluid. Therefore, the margin between θ_1 and θ_2 is reduced, i.e., $\Delta\theta_{IN}$ is decreased to a safe value. (Fig. 27.12)

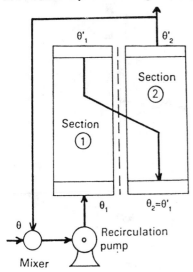

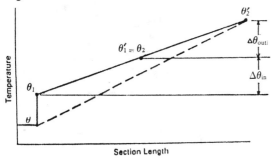

Fig. 27.11

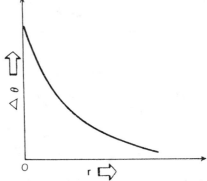

Fig. 27.12 *Working fluid recirculation reduces effectively the temperature difference between the welded sections*

Q. *On which factor does the temperature conditions of adjacent sections of waterwalls depend?*

Ans. These depend on the recirculation ratio.

Q. *What is understood by the recirculation ratio?*

Ans. It is the ratio of the mass flowrate of the working fluid with recirculation (G_r) to the mass flowrate of the working fluid without recirculation (G) i.e.

$$r = \frac{G_r}{G}$$

Q. *How does it influence the temperature difference between adjacent welded sections?*

Ans. As the recirculation ratio increases, i.e., G_r increases, the temperature at the inlet of section (1), i.e., θ_1 increases. Therefore, for the same heating intensity, the $\Delta\theta_{IN}$, i.e., $(\theta_1' - \theta_1)$ between two adjacent welded section decreases. So higher the value of r, lower will be the temperature difference between the adjacent sections. This can be better understood from the $\Delta\theta$ vs. r curve. (Fig. 27.13)

Fig. 27.13 *Temperature difference between the adjacent welded sections decreases with the increase of circulation ratio*

Q. *What is the additional advantage of the method of recirculation of the working fluid?*

Ans. It makes start-up possible at low loads.

Q. *At how much low-load does it ensures start-up?*

Ans. As low as 15% of rated load with recirculation.

Q. *Why?*

Ans. Because this system ensures reliable cooling of the waterwalls at low-loads during start-up when the furnace is lighted up with fuel oil having a high heat release rate.

Q. *Why are the vertically laid waterwalls of high-capacity boilers divided vertically into stages by parting joints and connected with one another by mixing headers?*

Ans. To avoid the risk of non-uniform heating which may be dangerous.

For high-capacity boilers, waterwall tubes are intensively heated in the furnace and so with the increase in the tube length the rate of heat absorption increases and with that the probability of maldistribution of heat increases. For this reason the waterwalls of high-capacity boilers are divided vertically into stages.

As a result of this separation, the increase in enthalpy of the working fluid in each stage decreases and that brings down the highest temperature of the tube walls.

Q. *How does the recirculation of the combustion products ensure reliable operation of waterwalls?*

Ans. Recirculation of the gaseous products of combustion taken off the zone at a lower temperature (350–380°C) downstream of the economizer and injected to the furnace zone of high température dilutes the oxidant in the zone of high heat release rate and prolongs the combustion process. As a result, the heat release rate reduces and furnace temperature decreases and therefore, the probability of maldistribution of heat in the tubewalls reduces.

Q. *What are the plus points of gas-tight all-welded waterwall tubes over bare-tube waterwalls?*

Ans. Gas-tight all-welded waterwall tubes require less mass per unit area of radiant heat-absorbing surface than that of bare-tube waterwalls.

The fins of the welded waterwalls lend themselves as active heating surfaces and they transfer heat to the working fluid, increasing the rate of heat absorption.

The fins welded to the gas-tight waterwalls behave like good clamps holding the tubes perfectly in-line. Therefore, the probability of misalignment of tubes, which may occur in the bare-tube waterwalls is practically nil. Since misalignment is absent no tube can shadow the other.

Gas-tight waterwalls compared to bare-tube waterwalls have relatively a thin layer of refractory layer and therefore can be waterwashed without risk of mostening the heat-insulating surface. Hence, the corrosion that may take place in the inaccessible places is minimized.

Q. *Why is uniform heating of all-welded gas-tight waterwall tubes more essential than that of bare-tube waterwall?*

Ans. It is because non-uniform heat distribution between the adjacent welded tubes will develop substantial thermal stresses on the welded fins which may fail.

Q. *If the uniformity in heat distribution between the all-welded gas-tight waterwalls of high-capacity boilers is sought by increasing the number of separate flows what complicacies may arise?*

Ans. It will
(a) complicate the automatic control.
(b) increase the number of fittings '
(c) tell upon the reliability of operation of waterwalls increasing the temperature difference between welded tubes
(d) increase the number and mass of unheated tubes, causing a sharp rise in the hydraulic resistance of the circuit.

Q. *So how can this uniformity in the heating of gas-tight waterwalls of high capacity boilers be ensured?*

Ans. By increasing the number of passes connected in series.

Q. *Where are the refractory-faced waterwalls mounted?*

Ans. These are placed in the zones of intensive combustion, such as slagbottom furnaces, cyclone furnaces and in boiler furnaces that burn low volatile fuels.

Q. *How many types of refractory-faced waterwalls are used in practice?*

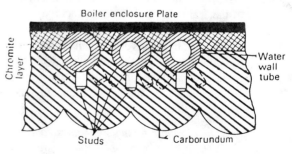

Fig. 27.14 Smooth-tube (vertical waterwall)

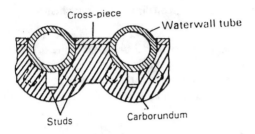

Fig. 27.15 *Membrance-type (vertical water-wall)*

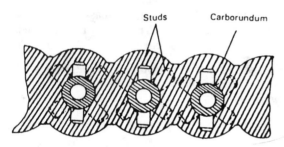

Fig. 27.16 *Platen wall*

Ans. Three types:
1. smooth-tube (vertical waterwall) (Fig. 27.14)
2. membrane-type (vertical waterwall) (Fig. 27.15)
3. platen wall (Fig. 27.16)

Q. *How are refractory-faced waterwalls made?*

Ans. These are made from metal tubes welded with studs (10–12 mm diameter and 15–25 mm in length) that form the frame for holding a refractory paste made of carborundum or chromite. The refractory facing is compacted by means of a pneumatic hammer to ensure better contact of the refractory with the surface of the studs and the tubes. Heat is transferred from the refractory face to the watercooled tube directly as well as through the studs.

Q. *Why is the chromite facing sometimes covered on the fireside by a layer of carborundum paste?*

Ans. To impart a higher resistance to slag attack.

Q. *Does the amount of heat absorption by refractory-faced waterwalls depend on the number of studs?*

Ans. Yes; it increases with the increase of the density of the stud arrangement.

Q. *What do you mean by the term "stud density"?*

Ans. It is the ratio of the total cross-sectional area of all studs per unit surface area of the waterwall tubes.

Q. *What is the stud density for high-capacity boilers?*

Ans. It varies from 0.15 to 0.25.

Q. *What factor delimits the stud density?*

Ans. It is the technological feasibility of stud welding.

Q. *How does the temperature vary across the refractory-faced waterwall?*

Ans. The temperature distribution across the refractory-faced waterwall is illustrated by the accompanying figure. (Fig. 27.17)

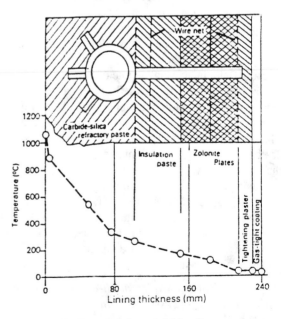

Fig. 27.17 *Temperature distribution across the refractory-faced waterwall*

Q. *What problem is commonly encountered with refractory-faced waterwalls?*

Ans. It is the burn-off the refractory facing and the studs as they are subjected to high operating temperature conditions.

Q. *On what factors does the life of a refractory-faced waterwall depend?*

Ans. There are several factors:

1. temperature conditions of the furnace
2. thermal conductivities of the refractories
3. refractoriness of the refractory material
4. contact resistance between the refractory facing and the metal
5. size and shape of the studs
6. material of the studs
7. nature of the slag.

Q. *What is a steam superheater?*

Ans. It is a boiler auxiliary designed to superheat saturated steam produced in the evaporator to a specified temperature.

Q. *How are superheaters classified?*

Ans. Depending on the mode of heat transfer from the furnace to superheater surfaces, the superheaters are divided into:
- (a) CONVECTIVE SUPERHEATERS which absorb heat mainly by convection
- (b) RADIANT SUPERHEATERS which absorb radiant heat from the flame in the furnace core
- (c) PLATEN SUPERHEATERS that absorb both radiant and convective heat.

Q. *Where are these located?*

Ans. Convective superheaters are placed in the convective gas duct.

Radiant superheaters are arranged on the furnace walls.

Platen superheaters are mounted on the top of the furnace and partially in the flue gas duct between the convective and radiant heating surfaces.

Q. *How many kinds of superheaters are deployed?*

Ans. Two kinds:
1. Primary superheater
2. Intermediate superheater.

Q. *What is the function of primary superheaters?*

Ans. To superheat the saturated steam just produced in the evaporators.

Q. *What is the function of intermediate superheaters?*

Ans. It is to reheat the turbine-exhaust steam to a superheat temperature once more.

Intermediate superheaters are also called reheaters.

Q. *What is called superheating?*

Ans. Production of steam at a temperature higher than saturation temperature is called superheating.

Q. *What is degree of superheat?*

Ans. It is the temperature imparted to steam above its saturation temperature at a given pressure.

Q. *How can the condition of superheat be expressed?*

Ans. This can be done in two ways:
1. in terms of the degree of superheat for a given pressure
2. the actual temperature and pressure of the steam.

Q. *When does the significance of the term 'superheat' fail?*

Ans. At steam pressure above critical point. Because, beyond this point no saturation steam exists, so the term superheat is no longer applicable

Q. *Why is superheated steam less erosive and corrosive than wet, saturated steam?*

Ans. Wet, saturated steam carries water droplets but superheated steam has no entrained moisture and as such the latter is less erosive and corrosive than the former.

Q. *Why are desuperheaters also installed with superheaters?*

Ans. Mainly due to two reasons:

1. to control the superheated steam temperature by injecting DM water to superheated steam
2. to produce saturated steam to run the boiler auxiliaries.

Q. *If saturated steam is charged to the boiler auxiliaries for their operation, then why is steam not directly taken from the boiler drum?*

Ans. If the wet, saturated steam used for running the boiler auxiliaries is directly taken from the steam drum, then the superheater may not receive as much steam as is required for the necessary cooling of its coils. This phenomenon is called superheater starvation and may lead to superheater tube failure. To avoid this problem, the entire saturated steam separated in the boiler drum is charged to the superheater and then a part of the output steam is desuperheated, according to requirements, for running the boiler auxiliaries.

Q. *Why is saturated steam, not superheated steam used to run small boiler auxiliaries?*

Ans. The use of superheated steam requires more stringent measures to be adopted regarding close clearances, special lubrication and expansion control, etc., which mean higher construction costs.

Q. *On which factors does the ratio of the superheating surface to boiler convection surface depend?*

Ans.

1. Fuel burned
2. Steam temperature as required.

Q. *Why is the installation of convective superheater alone not favoured in large steam generation units with high superheat demands?*

Ans. For generating steam with a high degree of superheat, in a large steam generation plant, the use of convective type superheater alone is rather impractical because of two fundamental reasons:

1. Convective superheaters are not befitted for high degree of heat absorption as are radiant superheaters. As a result, much of the generated heat in the boiler furnace is lost to flue gases escaping at higher temperature. This means loss of overall performance of the steam generation plant.

2. Due to the problem of slagging, the flue gas temperature at the furnace exit is fixed and furnace parameters should be controlled so as not to allow the flue gas temperature to exceed this limit. The use of convective superheater alone in a large steam generation unit won't allow us to bring down the flue gas exit temperature below the prescribed limit.

Q. *What factor(s) cause the vulnerability of a radiant superheater?*

Ans. Overheating.

Q. *What factors contribute to the overheating of radiant superheaters?*

Ans.

1. Vulnerability during start-up.
2. Deposition of scale on the steam-side of superheater coils impairing heat transfer.
3. Flame impingement directly upon superheater coils.

Q. *How can flame impingement be avoided?*

Ans. By:

1. burner adjustment
2. tilting the burner
3. increasing the distance between the burner and the radiant superheater surface
4. proper flame control.

Q. *How can the failure of radiant superheater coils due to scale formation on the steam-side be prevented?*

Ans. The only solution is to enhance steam purification before it is charged to the superheater.

Q. *When may the possibility of superheater tube burnout arise?*

Ans. It will occur under conditions where there is insufficient steam flow to cool the superheater tubes adequately and such conditions may arise:

1. DURING START-UP: until or unless steam generation begins and steam flows in to protect the superheater tubes, the heat absorbing surface of the radiant superheater is cooled by radiation to the waterwall tubes. Now if steam flow fails

to be established in superheater tubes or a series of tubes while the furnace attains the operating temperature, there is a chance of tube burnout in the superheater.

2. NON-DRAINABLE SUPERHEATER: if a portion of the superheater is not completely drainable, then it will hold some quantity of steam condensate when the superheater is under shutdown. Now when the superheater is taken in line, there will be no flow of steam through the superheater circuit until this stored condensate evaporates or is blown out. Because of stoppage of steam flow, the superheater will starve, causing tube burnout.

3. IMPROPER DESIGN: if the areas adjacent to burner ports are not properly designed, it may give rise to deflection of the flame which in turn may strike the superheater tubes causing burnouts.

Q. *In which cases are separately fired superheaters installed?*

Ans. Separately fired superheater is installed in cases where:
   (a) absolute control of the temperature of process steam is required
   (b) high degree of superheat is to be maintained in the process steam at a distance considerably far away from the SGP battery limit
   (c) process plants or turbo-alternators require a higher steam superheat temperature than can be maintained in the existing steam generation plant
   (d) process requires steam of high temperature (870 K and above) without regard to pressure.

Q. *What factors influence the design of convective superheaters?*

Ans.
1. Flue gas velocity
2. Steam velocity
3. Temperature difference between the flue gas and steam
4. Direction of gas flow—parallel or cross-flow.
5. Cleanliness of the tubes.

Q. *Which factors primarily influence the design of radiant superheaters?*

Ans.
1. Temperature of the furnace
2. Temperature of the steam
3. Amount of heat absorbing surface exposed to radiant heat
4. Cleanliness of the tube surface.

Q. *What is the tube dia of convective superheaters?*

Ans. The ID of convective superheater tubes varies from 20 to 30 mm.

Q. *What is the material of construction of convective and radiant superheaters?*

Ans. They are made from high-alloyed Cr-Ni austeni[?] class steel. A few typical compositions being:

	Cr	Ni	W	Mo	Ti	Nb	B	Fe	*Working Temperature*
1.	13–15	13–15	2–3	0.45 – 0.6	—	—	—	rest	up to 925°K
2.	17–19	11–13	—	—	0.65	—	—	rest	-do-
3.	15–17	15–17	2–3	0.5 – 1	—	0.5 – 1	0.005	rest	up to 980°K
4.	15–17	12–15	2–3	—	—	1–1.3	0.005	rest	-do-

Q. *Why are these made from the austenitic-class of steel?*

Ans. It is because the austenitic-class of steels have high heat and corrosion resistance.

Q. *How does it gain higher corrosion resistance?*

Ans. It is due to the presence of high content of chromium. Chromium metal produces a hard and impervasive layer of Cr_2O_3 on the tube surface exposed to flue gas. This prevents further penetration of oxygen or corrosive gases to the base metal and thus resists high temperature corrosion.

Q. *Why are titanium and niobium added?*

Ans. To prevent austenitic-steels from inter-granular corrosion.

Q. *How?*

Ans. These metals (Ti, Nb) are strong carbide-forming elements. They can combine with the entire available carbon and bind it into their carbides and thus prevent the formation of chromium carbides precipitates at the boundaries of the austenitic grains.

Q. *Why are molybdenum and tungsten added?*

Ans. These elements when alloyed with austenitic steels impart to the latters, higher heat resistance.

Q. *Why?*

Ans. They form strong compounds, Fe_2Mo and Fe_2W, which are highly dispersed throughout the matrix of the austenitic steel, thus imparting to the latter higher heat resistance.

Q. *Finned tubes absorb more heat than smooth tubes. Yet superheaters are usually made from smooth tubes. Why?*

Ans. It is because smooth tubes are:
 (a) much less expensive than finned tubes.
 (b) much simpler to manufacture
 (c) easier to clean
 (d) less prone to fouling by ash.

Q. *What is the main drawback of smooth tube superheaters?*

Ans. Since the reheater steam pressure (3–4 MN/m^2) is not high, it is required that the hydraulic resistance of the reheater should not be high. That's why reheater tubes are made of substantially larger diameter.

Q. *Why is the reheater arranged in a zone of moderate heating?*

Ans. It is because if the reheater is placed in a high temperature zone, the intensive heating of the external surfaces of the reheater tubes can lead to an inadmissible rise of temperature due to a low value of the coefficient of heat transfer from the inner wall to the steam. This may turn out to be too high for the pearlitic class of steel from which reheater tubes are made.

Q. *Why will it be too high a temperature for the pearlitic class of steel from which reheater tubes are made?*

Ans. Pearlitic class of steel is low-alloyed steel. They do not contain tungsten or a sufficient quantity of molybdenum to gain higher heat resistance property. Since they also lack titanium and niobium—the stabilizing elements—higher heating temperature may lead to intercrystalline corrosion due to chromium carbide precipitation.

Q. *What is the material of construction of reheater tubes?*

Ans. These are made of low-alloyed pearlitic-class steels.

	Cr	Mo	V	Si	Nb	B	Fe	Operating Temperature
1.	1–1.25	1–1.2	0.15–0.3	–	–	–	rest	up to 855°K
2.	1.5–2	0.5 – 0.7	0.2–0.4	0.4 – 0.7	–	0.005	rest	-do-
3.	2–2.5	0.5 – 0.7	0.2–0.4	–	0.5 – 0.8	–	rest	-do-

Ans. They suffer from lower heat absorption capacity at moderate flue gas velocities.

Q. *What is the ID of the reheater tubes?*

Ans. It is larger than that of convective superheater tubes and may be upto 50 mm.

Q. *Why are the reheater tubes made substantially larger in diameter?*

Q. *However placing the reheater tubes in the moderate temperature zone will decrease the overall* *heat transfer to the steam requiring higher surface area of the tubes, which may be economically unfavourable. So what can be done to overcome this problem?*

Ans. The overall heat transfer to the steam in the reheater can be substantially raised even in the moderate temperature zone by designing the reheater tubes into internally finned or rifled tubes. (Fig. 28.1B). This will extend the internal surface area of the tubes and will intensify the rate of heat transfer. (Fig. 28.1)

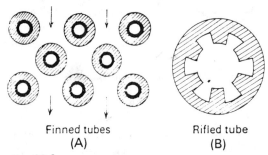

Finned tubes (A)   Rifled tube (B)

Fig. 28.1

Q. How many types of superheater coils are used?

Ans. These may be single pass or multi-pass. Multi-pass tubes may again be divided into:
(a) 2–pass (Fig. 28.2C)
(b) 3–pass (Fig. 28.2D)
(c) 4–pass (Fig. 28.2E)
(d) glove–type multi-pass tubes. (Fig. 28.2A)

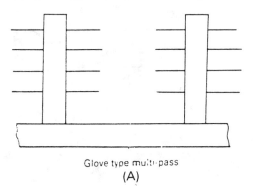

Glove type multi-pass (A)

Q. How many types of convective superheaters are known?

Ans. Depending upon the direction of motion of

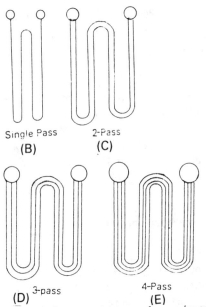

Single Pass (B)   2-Pass (C)

(D) 3-pass   4-Pass (E)

Fig. 28.2 *Different types of superheater coils*

steam and combustion products in a convective superheater, the superheaters may be divided into:
(a) counterflow type (Fig. 28.3A)
(b) parallel-flow type (Fig. 28.3B)
(c) mixed-flow type (Fig. 28.3C & 28.3D)

Q. Out of these three types of convective superheaters, which one develops the highest temperature gradient between the combustion products and steam?

Ans. The counterflow type superheater.

Q. What is the main drawback of this type of convective superheater?

Ans. It is the problem of burn-through of the last coils at elevated temperatures.

Q. Why does this occur?

Ans. This occurs due to the fact that both the working fluid (steam) and the combustion products are at the highest temperature here, subjecting the tube metal to a very demanding temperature condition.

Q. Out of these three types of convective superheaters, which one has the lowest temperature gradient between the combustion products and steam?

Ans. Parallel-flow superheater.

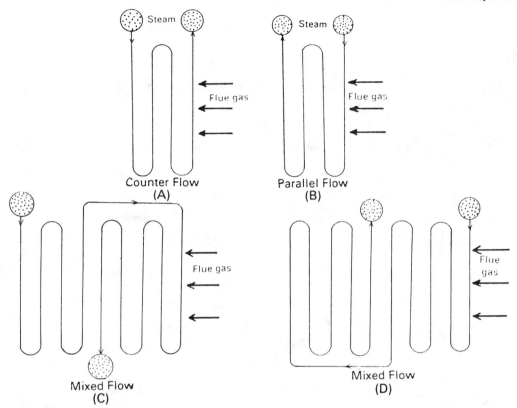

Fig. 28.3 _Different types of convective superheaters_

Q. *Why?*

Ans. Here the superheater tube metal is exposed to the most favourable temperature conditions because the tube coils at the outlet which have the steam at the highest temperature are in contact with the flue gases which have already been partially cooled at the inlet section of the superheater.

Q. *What about the mixed-flow superheater?*

Ans. It strikes a good balance between counterflow and parallel-flow super heaters, i.e., temperature gradient is not too high, nor is the cost economically unfavourable.

Q. *How can the superheater coils be arranged?*

Ans. Either vertically or horizontally. (Fig. 28.4)

Q. *What are the advantages of vertical superheaters over horizontal superheaters?*

Ans. Vertical superheaters are much less intricate in design, as regards their fastening, than horizontal superheaters.

They arrest less slag than their horizontal counterparts.

Q. *What is the chief disadvantage of vertical superheaters?*

Ans. The main drawback is that these are undrainable, i.e., the condensate cannot be completely drained off the circuit. And this may lead to internal corrosion of tubes and create certain technical difficulties during boiler firing.

Q. *Can the condensate from a horizontal superheater be completely drained?*

Ans. Yes.

Q. *Where is the radiant superheater located in once-through boiler units?*

Ans. It is usually located on the furnace roof and the walls of the horizontal flue duct.

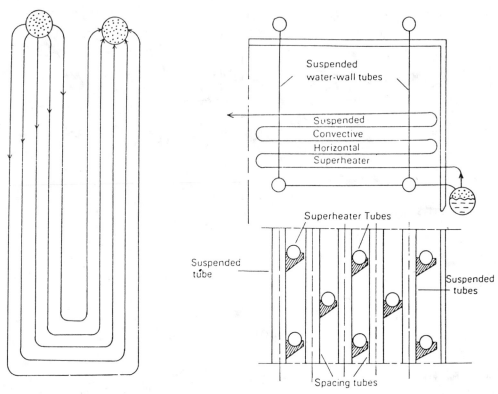

Fig. 28.4 (A) *Superheater coils laid out vertically*   (B) *Superheater coils arranged horizontally*

Q. *Where is the radiant superheater of a drum-type boiler unit located?*

Ans. On the furnace roof, if the heating surface of superheater is not too large. Otherwise it is placed partly on the vertical walls of the furnace.

Q. *What are the advantages of radiant super-heaters over convective superheaters?*

Ans.

1. Fewer elements are required than in convective superheaters
2. Better access to tubes
3. They have low hydraulic resistance in the steam-path and low resistance on the gas side
4. Ensure better furnace cooling and that reduces furnace temperature and maintenance
5. Over the entire load range 25–100%, the steam temperature characteristics remain substantially horizontal.

Q. *Why do radiant superheaters offer very much lower resistance on the gas side than do convective superheaters?*

Ans. Convective superheaters are placed directly in the path of flue gases which must come in contact with convective superheater tubes to transfer heat by the process of convection. Whereas radiant superheater coils are arranged on the furnace walls. They receive heat by radiation from the flame. They do not obstruct the flow of combustion products and hence offer less resistance to the flue gas stream.

Q. *What is the reason behind the high thermal efficiency of platen superheaters?*

Ans. It is because they absorb both radiant and convective heat.

Q. *How much of the total heat of superheating is absorbed by platens?*

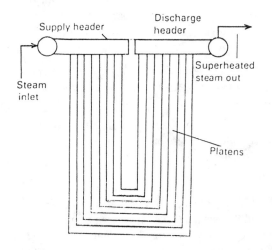

Fig. 28.5 *Platen*

Ans. As much as 50% of the total heat of superheating.

Q. *What is a platen?*

Ans. A platen is an assembly of smooth tubes* set into the pattern of a flat-gastight band with inlet and outlet headers. (Fig. 28.5)

Q. *How are platens arranged?*

Ans. They can be arranged either horizontally or vertically with a spacing of 500–1000 mm between them.

Q. *Since platens operate under lower temperature conditions, they may suffer from slagging problems. How is this problem tackled?*

Ans. The ash deposition problem is taken care of by platens themselves. Softened ash particles that deposit and solidify on the surface of the platens form loose deposits which are self-cleaned due to vibration. Therefore, ash deposits cannot grow to an appreciable thickness.

However for boilers fired with high-slagging fuel grades, dense deposits can be a problem on platen surfaces.

Q. *Are all the tubes of the platens uniformly heated?*

Ans. No. Front tubes are much more strongly heated than other tubes.

Q. *What may be the outcome of this?*

Ans. The heavily heated external tubes may turn out to operate under critical conditions.

Q. *How can the reliability of operation of platens therefore, be enhanced?*

Ans. These can be done by either of the three methods:
1. shadowing the external tubes from intensive heating by a system of tubes of some other heating surfaces carrying working fluid at much lower temperature. (Fig. 28.6)

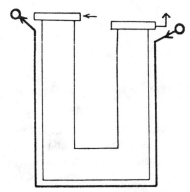

Fig. 28.6

2. shortening the length of the external tubes. (Fig. 28.7)

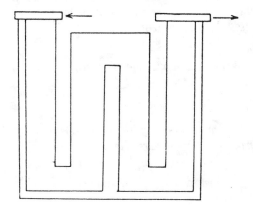

Fig. 28.7

---

* (finned tubes have been tested)

3. enlarging the diameter of the tubes operating under the heaviest conditions. (Fig. 28.8)

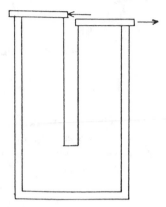

Fig. 28.8

Q. *Why is the superheater temperature not allowed to rise substantially over the upper limit of the rated value?*

Ans. To avoid superheater tube failure due to creep and scaling. Creep is a physical phenomenon characterized by plastic deformation under the action of constant load at an elevated temperature. As the temperature is increased, the rate of creep is increased precipitating the premature rupture of superheater tubes.

Scaling leads to the thinning of metal tubes and is aggravated by the intensive heating loads or high stresses due to pressure in tubes.

Q. *What is the service life of superheater tubes for reliable operation?*

Ans. About 15 years.

Q. *How much is the service life of superheaters affected with the rise of temperature above the rated value?*

Ans. A 15 to 20°C rise in operating temperature over the rated value of the superheater can cut short the service life of a superheater by half.

Q. *In which cases is the risk of rise of superheater temperature over the rated (average) value maximum?*

Ans. In the cases of superheaters fitted to high-capacity boilers because of higher heat absorption of the superheaters.

Q. *Why?*

Ans. The superheaters of high-capacity boilers are placed in the gas ducts having a large area of cross-section. And therefore, for horizontal ducts, superheater tubes are heated extremely unevenly over the width and height of the ducts while for vertical ducts they are unevenly heated over the width and depth of the ducts. That is, non-uniform heating of superheater tubes occurs along the length of the duct for horizontal superheaters and along the height of the gas duct for vertical superheaters.

Q. *How can this non-uniform heating of superheater tubes across the width of the gas duct be diminished?*

Ans. This can be done by sectionalizing the superheater across its width as well as along the length of the duct. This is achieved by transferring steam in a cross-over manner, as shown. (Fig. 28.9A & 28.9B)

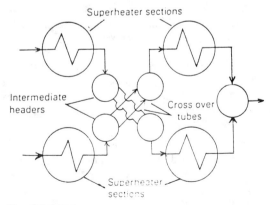

Fig. 28.9(A)

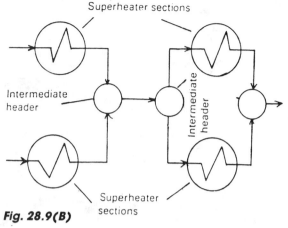

Superheater sections

Intermediate
header

Intermediate
header

Superheater
sections

Fig. 28.9(B)

Q. *What factors determine the location of super-heaters in the flue gas duct?*

Ans. There are two important parameters that chiefly determine the position of the superheater in the gas duct. These are: arrangement of evaporator and steam parameters.

Q. *How does the evaporator arrangement determine the position of superheater in the case of medium-pressure boilers?*

Ans. For medium-pressure boilers with operating pressure 4MPa (4 MN/m^2) and superheated steam temperature about 710–715°K, the evaporating surfaces occupy the lion's share of the furnace space since they absorb 60–62% of the total heat absorbed in the boilers. Therefore, the superheater which absorbs 20% of the total heat absorption is of convective type and occupies a position immediately behind the membrane wall. (Fig. 28.10)

Q. *Where is the superheater of a high- pressure (P = 10 MPa i.e. 10 MN/m^2 and θ_{ss} = 810–815°K) boiler located?*

Ans. In this case the superheater is composed of: roof superheater which is partly radiant partly convective: a radiant platen superheater and a convective superheater.

The platen section is arranged in the upper section of the furnace and in front of the rear waterwall tubes. (Fig. 28.11)

The convective section is located in the horizontal portion of the gas duct.

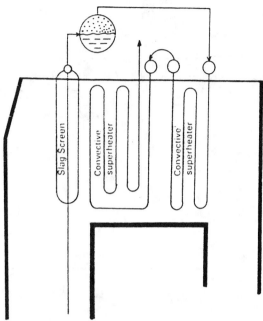

Fig. 28.10 *Layout of superheaters in the case of medium pressure boilers. They are of convective type*

All the three sections are connected in series. And steam from the boiler drum is fed to the superheater in the following sequence:

Steam ⟶ ROOF SUPERHEATER ⟶ PLATEN
SECTION

CONVECTIVE SECTION ↵

Q. *How are the superheaters positioned in the case of high-pressure (P = 14 MPa i.e. 14 MN/m^2 and θ_{ss} = 820°K) boilers?*

Ans. The superheater layout is the same as in the case of 10 MPa/θ_{ss} = 810°K superheater; besides one reheat superheater (called reheater) is added in the convective shaft of the flue gas duct (Temp 1120°K) (Fig. 28.12)

Q. *Why is the radiant section of the superheater connected to the steam path first in the sequence, i.e.*

Steam Drum — steam → Radiant Superheater

Convective Superheater ↵

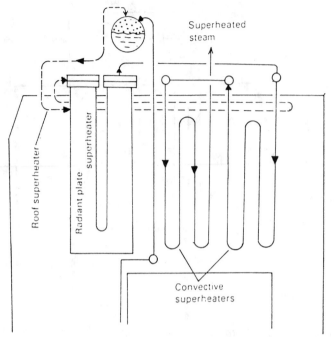

Fig. 28.11 *Convective superheaters are located in the horizontal portion of the flue gas duct*

Ans. The radiant section of the superheater operates under the most heavily heated conditions. So it runs the greater risk of tube surface temperature exceeding the rated value. Hence it must be adequately cooled

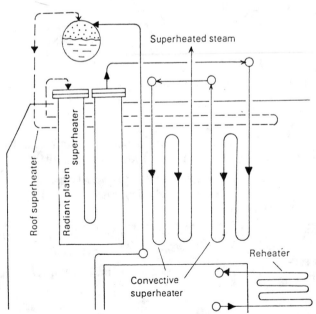

Fig. 28.12 *A reheater is introduced into the convective shaft of the flue gas duct over and above the existing convective superheater*

to keep tube metal temperature within the upper limit of the admissible temperature. That's why saturated (relatively much colder) steam is fed to the radiant superheater straight from the steam drum rather than the superheated steam from the convective section of the superheater. By this method, the reliability of operation of superheater tubes is enhanced.

Q. *What is necessary to regulate the temperature of superheated steam?*

Ans.
1. To perform load changes
2. To alter operating variables.

Q. *Why is the term "control" applied to mean the regulation of superheated steam temperature?*

Ans. When it is required to put into effect load changes and carry-out corrections for the operating variables without changing the equipment to provide the desired superheated temperature, it is called control.

Q. *How are load changes controlled?*

Ans. By
(a) changing the number of boilers in operation
(b) varying the number of burners in line
(c) varying the rate of fuel burning.

Q. *How the operating variables can be controlled?*

Ans. By the method of:
(a) excess air control
(b) gas recirculation
(c) gas by-pass control
(d) adjustable burner control
(e) control of combined radiant-convective superheaters
(f) control of superheat with sootblowers
(g) control of differently fired twin furnaces
(h) control of separately fired superheaters.

Q. *How does the method of excess air control provide the desired temperature to the superheated steam?*

Ans. If the excess air supply is increased, the superheated steam temperature at the exit of the convective superheater, operating at partial load, will increase. This is because, as more excess air is introduced the

furnace heat absorption decreases, for the same steam generation rate, while the gas mass flowrate increases with its increased total heat content that goes to raise the degree of superheat. (Fig. 28.13)

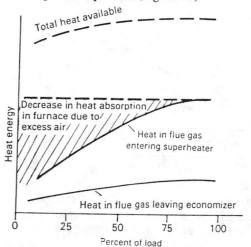

Fig. 28.13 *Ecess air suply raises the superheated steam temperature at the outlet of convective superheater*

However, the radiation superheater responds conversely when excess air flow is increased because of the dilution of flame emissivities of the combustion products in the furnace.

Q. *How does the gas recirculation method serve to control the superheated steam temperature?*

Ans. Recirculation of a portion of the combustion products to the furnace increases the mass flowrate of the flue gases through the convective superheater and that increases the superheated steam temperature at the outlet of the convective superheater. Therefore, by regulating the percentage of recirculating combustion gases, the desired superheat steam temperature can be maintained. (Fig. 28.14)

Q. *At which point of the furnace are the recirculating gases introduced?*

Ans. The hot gases are introduced below the combustion zone.

Q. *What are the advantages of the gas recirculation method of superheat control?*

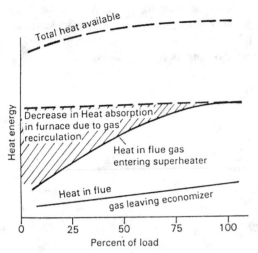

Fig. 28.14 *Flue gas recirculation increases the superheated steam temperature at the outlet of convective superheater*

Ans.

1. The efficiency of the process is relatively high
2. Ensures close regulation over a wide operating range

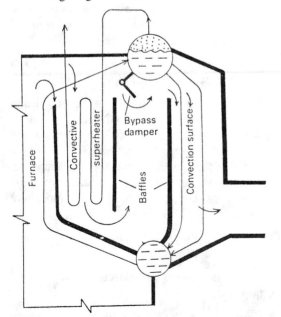

Fig. 28.15 *Gas bypass method to control superheated steam temperature*

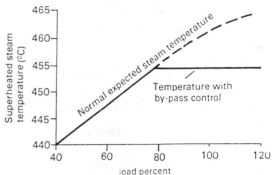

Fig. 28.16 *Superheated steam temperature control by gas bypass method*

3. Minimizes slagging that helps to longivate plant operation
4. Befits well with high-capacity modern units.

Q. *What are its drawbacks?*

Ans.

1. High initial investment
2. Relatively higher operation cost.

Q. *How does the method of gas-bypass control ensure superheated steam temperature control?*

Ans. In this case the superheater is designed oversize so that at partial load conditions, say 70–80%, it is capable of producing the required degree of superheat. (Fig. 28.16) And the boiler tube banks are disposed in such a manner that a portion of the combustion gas can be by-passed around the superheater elements with the aid of by-pass dampers. These dampers can be controlled manually or automatically to ensure the desired superheat temperature of the steam at the superheater exit. (Fig. 28.15)

Q. *What are the drawbacks of a gas bypass control system?*

Ans.

1. Though made of alloy steel, the regulating dampers suffer from high temperature corrosion and fatigue
2. Variable draft loss
3. Control is better and more sensitive provided dampers are not opened more than 50%
4. Hunting of dampers occurs with abrupt load change
5. Control is not too prompt.

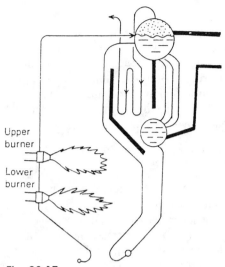

Fig. 28.17

Q. *If this system is associated with so many drawbacks, why it is so popular?*

Ans. Because of low investment cost.

Q. *How does the adjustable burner control method lend itself to superheat temperature control?*

Ans. Satisfactory superheat temperature can be maintained by adjustable burner control which chan-

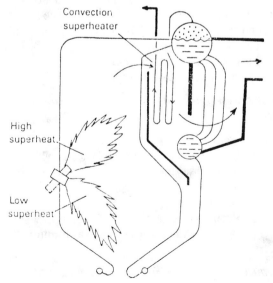

Fig. 28.18

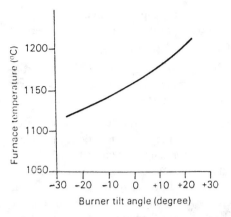

Fig. 28.19

ges the radiant heat absorbing capacity of the waterwalls and ensures the required flue gas temperature at the furnace outlet.

For low loads, it is preferable to maintain the flame in the upper zones of the furnace and so the upper burners are taken in line. And for heavy loads, it is best to light up both the upper and lower burners so that the entire furnace volume may be utilized to maintain the desired outlet temperature of the combustion products. Thus by shifting the principal combustion zone, the furnace outlet temperature can be regulated and that in turn ensures effective control of superheat temperature over a wide load range. (Fig. 28.17)

The furnace outlet temperature can also be regulated (Fig. 28.19) by tiltable burners (Fig. 28.18) which can be tilted in the vertical plane to shift the position of the combustion zone.

Q. *Why is the adjustable burner control system used in conjunction with interstage desuperheating?*

Ans. This is done to ensure:
   (a) quicker response to load change
   (b) effective constant superheat temperature over a wide load range
   (c) no hunting
   (d) wider operating range
   (e) close control ($\pm 3°C$)
   (f) operation simplicity.

Q. *Frequently adjustable burner control is used in conjunction with gas bypass control. Why?*

1. To increase the capacity range
2. To ensure a very close control of superheat temperature ($\pm 1.5°C$)

Q. *How does the control of combined radiant-convective superheaters ensure the regulation of the superheat temperature of steam?*

Ans. This system works relatively simply and on the basis of compensating characteristics. An increase in excess air (or flame length) reduces the radiant heat transfer but increases the convective heat transfer. A reduction in excess air (or flame length) imparts the opposite effect. Variation in the heat absorption in the two superheaters alters the temperature of the superheated steam passing through them. That's how a radiant-convective superheater combination functions over the entire control range and regulates the final superheat temperature.

Q. *How can the superheat temperature be controlled in differentially fired twin furnaces?*

Ans. Such a system of generator may have:
 (a) a radiant superheater in one furnace while convective superheater common to both furnaces (Fig. 28.20)
 or
 (b) a superheater in one furnace section only

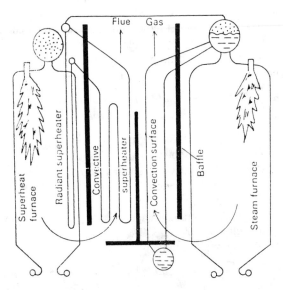

Fig. 28.20 *Twin-fired furnace*

or
 (c) more superheating surface exposed to one furnace than to the other.
  So by varying the firing rate (i.e., by introducing a differential firing rate) in each furnace section, the rate of heat absorption in the radiant and convective superheaters can be varied on compensating basis and that lends an element of superheat control.

Q. *How sootblowers can be deployed for the control of superheat?*

Ans. This is made possible by employing the sootblowers to clean the
 (a) radiant superheater tubes' surface, when the superheat is low, to increase the net heat absorbed by the superheater and thereby raising the superheat temperature
 (b) other furnace surfaces, when the superheat is high, to enhance the effectiveness of furnace cooling, i.e., to increase the rate of heat absorption by waterwall with the ultimate effect that the percentage of heat absorbed by the superheater is reduced and superheat temperature drops.

Q. *How does the separately fired superheater ensure superheat temperature control?*

Ans. These superheaters operate independent of the operation of the saturated steam boiler. The degree of superheat control imparted by them is achieved by:
 (a) adjustable burner control
 (b) gas recirculation method
 (c) control of excess air supply.

Q. *Are there any suitable methods to control superheat?*

Ans.
 1. Blending the superheated steam with the returned steam
 2. Raising the steam velocity through the superheater
 3. Recirculating a portion of the superheated steam back to the steam and water drums.

Q. *Where is the recirculation of a part of the superheated steam to steam and water drums carried out?*

Ans. In the Loeffler positive circulation boilers.

Q. *What is called the control characteristic of a superheater?*

Ans. It is the relationship between the boiler load and the superheated steam temperature.

Q. *Is it same for all superheaters?*

Ans. No.

Q. *How does it vary for radiant, convective and combined superheaters?*

Ans. The control characteristics of these three superheaters are represented by θ_{ss} vs. G_s (the steam load) curves. (Fig. 28.21)

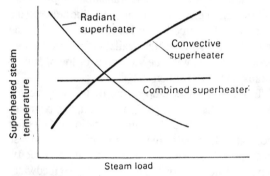

Fig. 28.21

For radiant superheaters it falls off with the increase of load while it increases with temperature for convective superheaters.

With combined superheaters having properly selected dimensions of the radiant and convective sections it remains constant.

Q. *Why does the control characteristic of a radiant superheater fall off with the increase of steam load?*

Ans. As the boiler load is increased, the quantity of fuel burning is increased. However this increased rate of fuel burning influences very little those parameters that govern the quantity of heat absorbed by the radiant heating surfaces. Therefore, the heat absorption in a radiant superheater does not increase in the same scale as the flowrate of steam through it. As a result, the heat absorption per unit steam flowrate decreases.

Q. *What are the parameters that govern the quantity of heat absorbed by radiant superheaters?*

Ans. Combustion temperature, flame emissivity and the thermal efficiency of the waterwalls.

Q. *Why does the control characteristic of a convective superheater increase with the increase of steam load?*

Ans. In this case as the rate of fuel burning increases with the increase of steam load, the quantity of flue gases produced increases and that raises the rate of convective heat transfer almost in the same proportion as the boiler load.

With the higher rate of fuel burning the flue gas velocity increases resulting in an increased rate of convective heat transfer in proportion to the power of 0.6–0.65 of gas velocity.

Thus as a result of increased rate of fuel combustion, the radiant heat absorption in the furnace decreases bringing about higher exit temperature of flue gases from the furnace. Therefore, the temperature gradient in the region of the convective superheater becomes higher.

The increased gas velocity and the higher furnace-exit temperature of flue gases with the increase of boiler load are responsible for the superheated steam-temperature increasing at a higher rate than does the boiler load.

Q. *Is it possible to obtain the constant-temperature superheated steam despite the increase of steam load?*

Ans. Yes it is possible theoretically by properly selecting the dimensions of radiant and convective superheaters.

Q. *Why can this not be met under practical conditions?*

Ans. The temperature of the superheated steam is affected by certain operating conditions such as:
1. temperature of the BFW
2. excess air ratio in the furnace
3. the degree of cleanliness (regarding slagging) of waterwalls and the superheater
4. moisture content of the fuel.

Q. *How does the boiler feedwater temperature affect the superheated steam temperature?*

Ans. For drum-type boilers, the decrease of feed-water temperature results in a higher superheated steam temperature. It is because, for drum-type boilers, the heating surface of the superheater is fixed and as the as feedwater temperature decreases the steam generation capacity of waterwall decreases with the effect that for constant fuel consumption the quantity of heat absorbed by the superheater from the flue gases increases per unit flowrate of steam.

For once-through boilers lower feedwater temperature leads to reduced superheated steam temperature. It is because, as the feedwater temperature decreases, the temperature of the generated steam decreases and as the rate of heat absorption, in the superheater, per unit flowrate of steam remains constant so this results in a drop in superheated steam temperature.

Q. *How does the excess air ratio influence the superheated steam temperature?*

Ans. An increase in the excess air ratio in the furnace results in greater quantity of combustion products passing through the convective shaft resulting in a higher rate of heat transfer in the zone of the convective superheater. So as a result of more intensive heat exchange in this zone the temperature of the superheated steam increases.

Q. *How does the moisture content in fuel influence the superheated steam temperature?*

Ans. As the moisture content in the fuel increases, the superheated steam temperature increases.

Q. *Why?*

Ans. It is because, with the increase of moisture content of the fuel, the mass of combustion products passing through the superheater increases. This results in higher convective heat transfer and consequently the superheated steam temperature rises.

Moreover, higher the moisture content of fuel, greater is the number of triatomic molecules in the combustion products. Since these have an elevated flame emissivity, they increase the heat transfer coefficient on the fireside resulting in higher temperature of superheated steam.

Q. *How does cleanliness of superheater tubes affect the superheated steam temperature?*

Ans. Slagging fouls the superheater tubes and destroys the cleanliness of the tube's surface.

Since slags have a lower heat transfer coefficient than clean metal tubes, slagging impairs the rate of overall heat transfer in the superheater and thus decreases the temperature of the superheated steam.

Q. *Why is it difficult in the case of once-through boilers to maintain a steady temperature of superheated steam with the increase of steam load?*

Ans. In the case of once-through boiler units there is no definite zone of phase transition (liquid water $\longrightarrow$ steam), i.e., a distinct boundary line between the evaporator and superheater is absent and as such the actual heating surface area available for superheating the steam varies with the variations of the operating parameters of the boilers.

In this case, the temperature of the superheated steam, with the increase of boiler load, can be maintained constant by controlling the ratio of the boiler feedwater flowrate to fuel consumption. Since these kinds of boilers have a low accumulating capacity, they are rather sensitive to the variations of feedwater flowrate and fuel consumption. This culminates in variations in the temperature of superheated steam.

Q. *What are the reasons for abnormally high superheated steam temperature with constant load?*

Ans.
1. Too high excess-air
2. Too low feedwater temperature
3. Slagged up furnace
4. Secondary combustion
5. Malfunctioning of steam temperature indicating and controlling equipment.

Q. *What are the reasons for abnormally low superheated steam temperature with constant load?*

Ans.
1. Too low excess-air
2. Too high feedwater temperature
3. Excessive slagging of superheater tubes (external surface)

4. Excessive moisture carryover from the boiler drum
5. Malfunctioning of steam temperature indicating and controlling equipment.

Q. *Will the superheated steam temperature get affected when the boiler is priming?*

Ans. Yes. It will drop suddenly.

Q. *How far is the temperature of the superheated steam allowed to vary from the rated value?*

Ans. It is allowed to vary from (+) 5° to (–) 10°C of the rated value.

Q. *Is it possible to maintain the superheated steam temperature within such a narrow range despite the variations of steam load even in the case of the combined system of the radiant-convective superheaters?*

Ans. No.

Q. *Then what measures can be undertaken so as not to allow the superheated steam temperature to vary beyond this admissible range?*

Ans. There are two methods for controlling the superheated steam temperature. These are: steam control and gas control.

Q. *How far these controls should be effective?*

Ans. These methods should be effective enough to maintain the superheated steam temperature within the specified limit in the load range from 30% to 100% while in the case of reheated steam from 60% to 100% of the rated value.

Q. *What is the basis of steam control?*

Ans. This method is based on curtailing the enthalpy of the superheated steam by transferring a part of its heat to the feedwater or by injecting polish water (deeply deionized water) to the superheating steam. As a result, the temperature of the superheated steam decreases.

Q. *What is the basis of gas control?*

Ans. This method is based on the fact that if, by any suitable means, the rate of heat absorption on the

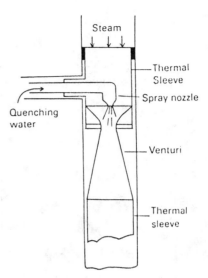

Fig. 28.22 *Spray attemperator*

fireside of the heating surfaces can be reduced or regulated then as a consequence the temperature of the superheated steam can be reduced or regulated.

Q. *What are these "suitable means" to control the rate of heat absorption on the fireside of heating surfaces?*
Ans.
1. Changing the flame position in the furnace
2. Recirculating the combustion products
3. By-passing a portion of combustion products around the superheater.

Q. *How is steam control accomplished?*

Ans. This is carried out in surface-type attemperators as well as in spray-type attemperators.

Q. *What is an attemperator?*

Ans. The attemperator (desuperheater) is basically a heat exchanger in which the superheated steam is cooled either by a direct heat exchange with cold condensate sprayed into the flow of the superheated steam (SPRAY-TYPE) (Fig. 28.22) & (Fig. 28.23) or indirect heat exchange with feedwater in tube-and-shell type heat exchanger (SURFACE TYPE).

Q. *Why is it not economically favourable to inject a condensate to the flow of reheated steam as a method of controlling temperature?*

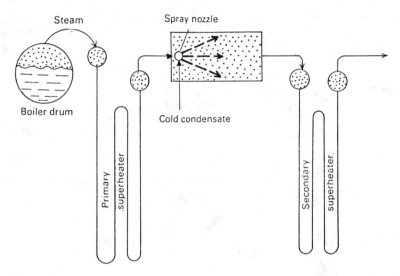

Fig. 28.23 Spray-type attemperator

Ans. From the reheater, the superheated steam goes to the medium pressure cylinder of the steam turbine. As such the additional quantity of MP steam produced due to the injection of condensate to the flow of superheated steam enters along with the main steam into the turbine by-passing the high pressure cylinder. This excess steam, if it exceeds the requirement of the turbine under operating load, must be vented to avoid pressure build up in the MP steam header. That's why the injection of condensate to reheated steam is economically unfavourable.

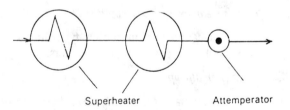

Fig. 28.24 Attemperator located downstream of superheater

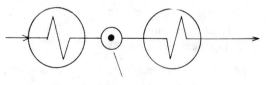

Fig. 28.25 Attemperator located in-between two superheaters

Q. *Where is the attemperator mounted?*

Ans. It is located either downstream of the superheater (Fig. 28.24)
    or placed in-between two superheaters (Fig. 28.25)
    or placed upstream of the superheater (Fig. 28.26)

Q. *Is the method of locating attemperator downstream of the superheater used for controlling the temperature of the main superheated steam?*

Ans. No.

Q. *Why?*

Ans. It does not safeguard the superheater against excessive rise of steam temperature.

Q. *Are the other two types of attemperator arrangements used in practice?*

Ans. Yes. They adequately safeguard the superheater and the turbine.

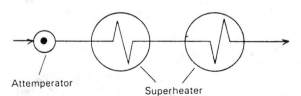

Fig. 28.26 Attemperator located upstream of the superheater

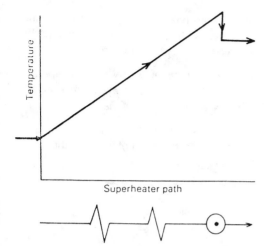

Fig. 28.27 *Attemperator mounted downstream of superheaters*

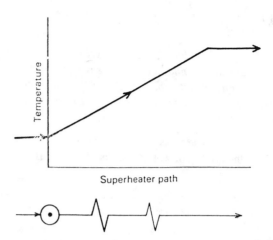

Fig. 28.29 *Attemperator provided upstream of the superheaters*

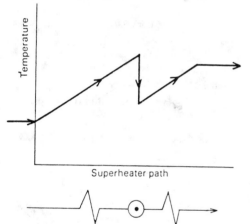

Fig. 28.28 *Attemperator mounted in-between two superheaters*

Q. *How does superheated steam temperature vary with these three versions of attemperator arrangement?*

Ans. (Fig. 28.27) (Fig. 28.28) and (Fig. 28.29)

Q. *Briefly describe the spray-type attemperator?*

Ans. It consists of an atomizing nozzle inserted into the superheated steam header through a pipe connection. The nozzle holes are 3–6 mm in diameter and the nozzle is covered by a metallic jacket having a gap of 6–10 mm from the header wall. This protective jacket is fitted to avoid direct contact of the cold condensate

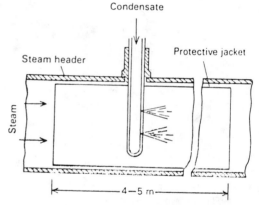

Fig. 28.30 *Spray-type attemperator in action*

with the header wall which is heated to the temperature of superheated steam.

Cold condensate is sprayed through the atomizing nozzle into the flow of superheated steam and the protective jacket ensures the complete evaporation of the droplets as they travel along the length of the jacket which is 4–5 m long. (Fig. 28.30)

Q. *Why is spray-type attemperator arranged close to the superheater outlet?*

Ans. To quench the superheated steam before its temperature before the spray nozzle runs too high.

If this is not done the local temperature of the steam header upstream of the attemperator will be so high that it may impair the operating conditions of the header metal in that place.

Q. *Why is the condensate used for spraying through the attemperator is obtained by condensing saturated steam directly from the boiler instead of using feedwater as the quenching fluid in some cases of drumtype boilers?*

Ans. It is because spray-type attemperators are sensitive to the quality of water used for spraying. If the feedwater of drum-type boilers are mineralized, then the spray water is obtained by condensing a part of the steam taken directly from the drum.

Q. *How is more than one spray-type attemperator positioned in the circuit to prevent an excessive temperature rise downstream of the superheaters?*

Ans. Here, two arrangements of two and three spray-type attemperators are shown schematically for superheat control. (Fig. 28.31)

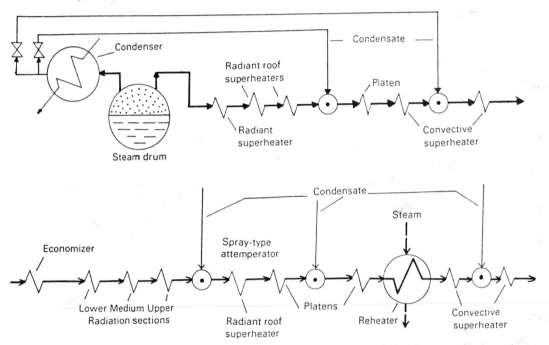

Fig. 28.31 *More than one spray-type attemperator inducted in the circuit prevents excessive temperature buildup downstream of the superheaters*

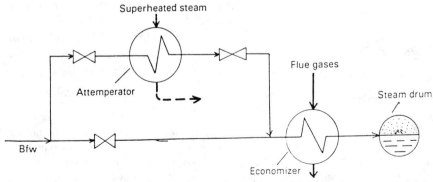

Fig. 28.32 *BFW is shunted through a surface-type attemperator*

Q. *How is a surface-type attemperator arranged in the circuit?*

Ans. It is joined in series with the economizer. (Fig. 28.32)

Q. *In which cases are the surface-type attemperators used?*

Ans. They are employed in low-capacity boilers.

Q. *Why?*

Ans. The surface-type attemperator with indirect cooling mechanism is not as efficient as the spray-type attemperator in controlling the high superheat temperature of the large quantity of steam generated in high-capacity boiler units.

Q. *Why is a live-steam reheater employed?*

Ans. It is used to equalize the temperature between the superheated steam (live steam) produced in the radiant superheater and that (reheated steam) produced in the reheater.

A decrease in boiler load is accompanied by a sharp rise in the temperature of the live steam in the radiant superheater and a substantial drop in the reheated steam temperature in the convective reheater. (Fig. 28.33)

Hence to equalize the temperature in both, it is justifiable to abstract some of the heat from the live steam (main steam) and transfer the same to the

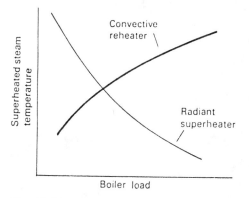

Fig. 28.33

reheated steam. This is achieved in a heat-exchanger that is called live-steam reheater.

Q. *Briefly describe a live-steam reheater.*

Ans. It is a U-shaped shell-and-tube heat exchanger comprising of 10–20 U-tubes of 25–30 mm dia mounted in a 300–400 mm header.

Live steam (superheated) flows through the tubes and the reheated steam through the shell-side. A part of the flow of the live steam is by-passed past the heat exchanger to control the temperature.

Q. *Why is the live-steam reheater made U shaped?*

Ans. To ensure:
(a) better compensation of thermal expansion of the system of tubes
(b) more compactness.

Q. *What is the range of steam temperature control of the live-steam reheater?*

Ans. 30–45°C

Q. *How does the variation in the boiler load affect the performance of a live-steam reheater?*

Ans. As the boiler load is decreased, the rate of heat absorption of the live-steam superheater increases ; if the boiler load is increased the heat absorption in the live-steam reheater decreases.

Q. *Why?*

Ans. Since the live-steam reheater is connected in series downstream of the radiant superheaters wherein superheated steam temperature increases with the decrease of boiler load, the rate of heat absorption of live-steam reheater increases with the higher outlet temperature of superheated steam from the radiant superheater.

If the boiler load is increased, it'll bring about a sharp drop in the temperature of the live-steam outlet of the radiant superheater and consequently the heat absorption in the live-steam reheater will decrease.

Q. *Show the arrangement of live-steam reheater connected in the circuit.*

Ans. In the following arrangement (Fig. 28.34) the entire high pressure superheated steam having a high degree of superheat passes through the live-steam reheater that reheats the medium pressure steam. A

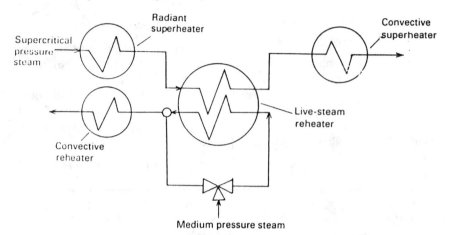

Fig. 28.34 *A typical live-steam reheater circuit. Entire H.P. superheated steam passes through the reheater to heat up the M.P. steam*

portion of the M.P. steam is by-passed around the live-steam reheater for necessary control of temperature of the reheated steam.

In the following arrangement (Fig. 28.35) a by-pass valve has been inserted in the high pressure superheated steam path so that only about 5% of this steam is allowed to enter the live-steam reheater and the rest passes through the by-pass valve when the boiler is running at its rated load.

At reduced load, the flow rate of MP steam decreases but the temperature of the HP superheated steam exit of the radiant superheater increases, so under this condition a greater portion of HP steam is allowed to pass through the live-steam reheater to ensure a constant reheated steam temperature.

Q. *What is the drawback of the system that sends the entire HP superheated steam through the live-steam reheater?*

Ans. Loss of high power due to substantial hydraulic resistance in the superheated steam path.

Q. *What is the drawback of the system that introduces a by-pass valve to the HP superheated steam path?*

Ans. This by-pass valve (control valve) must be capable of functioning at a high temperature and pressure and high flowrate. It must have a large cross-section.

Q. *How is gas control effected?*

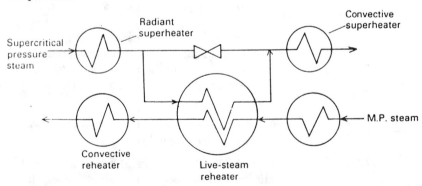

Fig. 28.35 *Another typical live-steam reheater circuit. Only about 5% of the H.P. superheated steam is allowed to pass through the reheater for heating the M.P. steam while the rest is bypassed*

Ans. It is effected by:
   (a) recirculating or by-passing the combustion products

                  or

   (b) changing the flame position in the furnace

                  or

   (c) switching on and off the burner tiers.

Q. *What are the drawbacks of the gas control method?*

Ans. It complicates the boiler operation by influencing the temperature of the superheated steam. It involves a loss of heat with waste gases. Additional power is required for draft.

Q. *Why is gas control employed only in combination with steam control in modern boilers?*

Ans. As it alone cannot satisfactorily control the temperature of the superheated steam in the specified range.

Q. *Why does the recirculation of combustion products taken off at a lower temperature and delivered into the furnace affect the temperature of the superheated steam generated?*

Ans. Combustion products tapped at the lower temperature (520–650 K) zone of the convective shaft and blown to the furnace back makes redistribution of heat even between the heating surfaces. The extent of heat redistribution dependes on the recirculation ratio, i.e., the fraction of the total combustion products recycled. Therefore, it affects the rate of steam generation, the temperature of steam and the temperature of the flue gases. Consequently, the temperature of the superheated steam alters.

Q. *How does the recycling of the combustion products affect the temperature of the superheated steam with the change in boiler load?*

Ans. This can be best understood with the help of superheated steam temperature vs. boiler load graph.

   With the increase of recirculation ratio, the direct heat absorption in the furnace decreases and the temperature of the combustion products at the outlet of the furnace increases. Moreover, the flowrate of gases through the superheater increases. This increased flowrate and

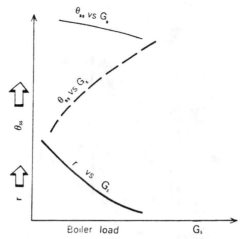

Fig. 28.36

higher outlet temperature of the combustion products increase the convective heat transfer and therefore, the temperature of the superheated steam also increases. Since the increase in recirculation ratio is accompanied by a fall of the boiler load, it can be said that with the drop of boiler load the temperature of the superheated steam increases. (Fig. 28.36)

   Similarly with the increase of boiler load, the recirculation ratio is decreased and that results in higher furnace temperature, and consequently the temperature of the superheated steam in the radiant superheater increases. Therefore, with the change in boiler load the temperature of the superheated steam changes very little, i.e., the fluctuation of superheated steam temperature is minimized.

Q. *At which point in the furnace are the recirculated combustion products introduced?*

Ans. The cold (520–620°K) combustion products taken off the convective shaft can be admitted into the upper or lower portion of the furnace. (Fig. 28.37)

Q. *How is the furnace operation affected if the cold recirculated combustion products are introduced into the lower portion of the furnace?*

Ans. When the cold recirculated flue gases are introduced into the lower portion of the furnace, it decreases the furnace temperature and consequently decreases the rate of radiant heat absorbed by the

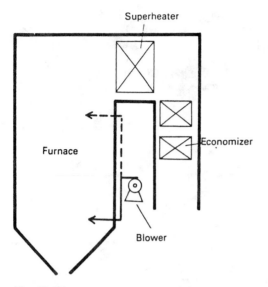

Fig. 28.37

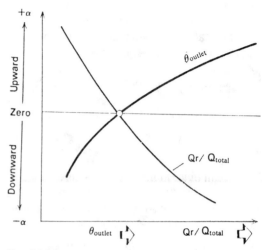

Fig. 28.38

waterwalls. Therefore, the temperature of the combustion products at the outlet of the furnace increases. Since, as the recirculation increases the total mass flowrate of gases through the superheater, the increased flowrate and higher outlet temperature of the combustion products increase the superheated steam temperature in the convective superheater.

Q. *How is the furnace operation affected if the cold recirculated combustion products are introduced into the upper portion of the furnace?*

Ans. It affects the furnace operation little but prevents the superheater tubes from slagging by lowering the inlet gas temperature to the superheater by a substantial degree. Moreover, the lowering of the temperature of the flue gases in the superheater somewhat decreases the rate of heat absorption by steam.

Q. *How does the flame position affect the temperature of the superheated steam?*

Ans. The furnace temperature and the pattern of temperature distribution within the furnace determine the rate of heat absorption by the evaporator waterwalls which, as a consequence, determines the furnace outlet temperature of the flue gases entering the convective gas duct. Therefore, by varying the position of the flame, the

heat absorption in the furnace and the furnace outlet temperature of the flue gases can be varied. This as a consequence brings about a change in the rate of heat absorption by the reheaters and convective superheaters which are arranged in the convective shaft.

Q. *How will the furnace outlet temperature of flue gases be affected if the burners are tilted down?*

Ans. If the burners are tilted downwards, the temperature of the furnace core increases and that increases the radiant heat absorption by the waterwalls in the furnace, i.e., the fraction of the total absorbed heat shared by radiant heat (Q_r / Q_{total}) increases. Therefore, the temperature of the flue gases at the furnace outlet decreases. (Fig. 28.38)

Q. *How will the flue gas temperature at the burner outlet be affected if the burners are tilted up?*

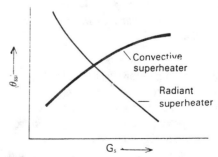

Fig. 28.39

Ans. It will decrease the rate of heat absorption by the waterwalls and therefore, the furnace outlet temperature of the flue gases in the convective shaft increases.

Q. *How will you orient the burners, when the boiler load drops, so as to maintain a steady temperature of the reheated steam?*

Ans. As the boiler load drops, the temperature of the superheated steam in the convective superheater diminishes. In order to maintain a constant temperature of the superheated steam in the reheater, the temperature of the flue gases at the furnace outlet is increased and this is done by tilting the burners upwards. (Fig. 28.39)

Q. *How far is the gas control method by tilting burners capable of covering boiler load variation to maintain a constant temperature of the reheated steam?*

Ans. It can cover boiler load range of 100% to 70% to maintain a constant temperature of the reheated steam.

Q. *How can the position of the flame in the furnace be changed other than by tilting the burners?*

Ans. This can be done by turning on and off some of the burner tiers.

Q. *Suppose a boiler furnace has three tiers of burners such that operation with any two can ensure 100% steam generating load. Depending on load variation which tier should be turned on?*

Ans. At low loads, the rate of steam superheating decreases. Therefore the burners in the upper tiers are turned on.

At high loads, the steam superheating increases ·and consequently the burners of the lower tiers are turned on.

Q. *How can the temperature of the reheated steam be controlled by by-passing of the combustion products?*

Ans. This can be done by

(a) providing a free gas duct between the reheater sections for passing the combustion products through it. The gas flowrate is controlled by the gate valve. (Fig. 28. 40)

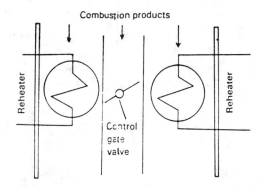

Fig. 28.40

(b) distributing the combustion products between the reheater sections and economizer mounted in parallel gas ducts (split duct) by means of control gate valves. (Fig.28. 41)

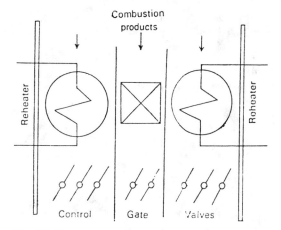

Fig. 28.41

(c) distributing the combustion products between the reheater sections and economizer mounted in parallel gas ducts by means of a control fan. (Fig. 28.42)

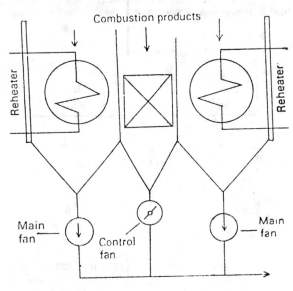

Fig. 28.42

Q. *What is done in the case of tube failure?*

Ans. What should be done in the case of one or more tube failures of a natural circulation boiler is dictated by the size of the failure.

If the failure is such that the water level can be maintained in the gauge glass by using the normal make-up system, the boiler can be kept in service until it is more convenient to take it off the line and make the necessary repairs.

If the failure is such that the water level cannot be maintained in the gauge glass, then the boiler should be shut down immediately.

A leak in the superheater tube should be investigated at the earliest possible time and the unit should be taken for normal shutdown.

An economizer tube leak can be detected by sound and/or increased make-up water requirement. The leak should be detected at the earliest possible time.

29

Economizers and Air Heaters

Q. *What are the economizer and air heater?*

Ans. They are heat recovery equipment that pick up heat from the combustion products of a boiler furnace to heat up the boiler feedwater (ECONOMIZER) and combustion air (AIR HEATER).

Q. *Why are they installed?*

Ans. Forming essential parts of a complete system of steam generating unit they are installed to conserve as much heat of fuel combustion in the boiler furnace as is possible and to increase the capacity of a boiler or decrease the size of a prospective boiler for a stipulated output. That is, the fundamental basis of their installation is to reduce the steam output cost.

Q. *If instead of installing these heat recovery equipment, in order to conserve as much combustion-gas-heat as possible, the entire temperature drop of the flue gases is allowed to take place over the boiler surfaces what would happen?*

Ans. This would be uneconomical and would increase steam generation cost and reduce boiler efficiency.

In the boiler proper, the flue gas temperature is allowed to drop to a level at which it becomes more economical to discontinue heat absorption through boiler surfaces (waterwalls) at saturation temperature and to initiate this heat absorption again at a later point downstream, i.e., in an economizer or air heater which operates under the law of diminishing returns. Since there always exists a most economical flue gas temperature above and below which steam production costs mount, the excessive reduction in flue gas temperature within the boiler will increase the capital investment as well as fixed charges that will eat away any gain in efficiency.

Q. *How far do these heat recovery equipment (economizer, air heater) bring about fuel economy and boost the overall efficiency of the boiler?*

Ans. It has been seen in practice that fuel saving to the extent of 1% is achieved for each 16°C reduction in the flue gas temperature in the heat recovery equipment (calculation is based on coal with calorific value 7 222 kcal/kg and CO_2 yield 13%).

Moreover, when these are installed, the overall efficiency of the boiler increases by 3–5% .

It is only by incorporating economizers and air heaters in the SGP has it been possible to bring down the flue gas temperature at the stack outlet to as low as 150°C (or even lower) and thereby to attain boiler efficiency as much as 88–90% .

Q. *When may the installation of these heat recovery systems not stand justified?*

Ans. In the following cases:
1. if fuel is available abundantly at low cost
2. if the steam generation unit is taken in line only occasionally
3. if the unit is meant for standby service.

Q. *What factors determine the temperature to whcih the flue gases, passing through the economizer and air heater, are to be cooled?*

Ans. This temperature is determined by:
(a) the amount of heat that can be extracted from the flue gases.

(b) inlet temperature of the BFW to the econo-mizer and combustion air to the air heater

(c) dew point of the flue gases

(d) optimum exit temperature below which any gain in efficiency is offset by higher capital investment and fixed charges.

Q. *Are these heat recovery equipment associated with any disadvantages?*

Ans. These equipment:

(a) need sound operation and good maintenance

(b) are sensitive to dust and dirt-laden flue gases which may clog the heat exchangers

(c) bring about draft loss and hence draft fans are necessary to overcome the resistance imposed by these heat recovery equipments

(d) require that a part or all of the flue gases, at low load, is to be by-passed to avoid cold-end corrosion

(e) entail design complications if their service is required at low loads.

Q. *How is the heat recovery system (ECONO-MIZER and AIR HEATER) arranged?*

Ans. It must be accommodated within the limited space of the plant and the arrangement should ensure good accessibility for operation and maintenance, over and above harnessing design economy.

Usually the air heater is placed behind the economizer. Sometimes it is located in-between two economizer sections but it is never placed above the economizer. Into this circuit is incorporated a flyash collector which is placed either ahead of the economizer or in-between the economizer and air heater or behind the air heater.

An ID fan is fitted to the economizer for flue gas circulation while the air heater is provided with both ID fan (flue gas circulation) and FD fan (combustion air circulation). If the unit size is small, a by-pass around the heat recovery equipment and ID fan is embodied in the unit to ensure natural draught (draft) chimney operation during emergencies.

The heat exchange tubes of the economizer and air heater may be disposed horizontally or vertically in the gas duct for single or multiple passes and cross-flow of flue gases. The combined heat recovery sys-tem may be located above, alongside, adjacent or behind the boiler.

Q. *What factors are considered in designing the economizer and air heater of a boiler?*

Ans. They must be adequately designed so as to:

(a) maximize the rate of heat transfer from the flue gases

(b) buy economy in metal use

(c) minimize wear of tube surface due to abrasive ash particles

(d) minimize fouling and corrosion.

Q. *From the standpoint of heat transfer how does the air heater differ from the economizer?*

Ans. It differs from an economizer in that the temperature gradient between the hot combustion products and air is much lower than that between the combustion products and water (BFW). And the over-all heat-transfer coefficient in the air heater is lower than that in the economizer.

Q. *Why is the surface area of air heater very large (larger than the entire surface-area of all elements of the steam-water path)?*

Ans. It is due to a very low heat transfer coefficient between the combustion products and air. Therefore, the air heater needs a very large surface area for heat exchange so as to heat up the combustion air to the desired temperature level.

Q. *How much heating surface area does an air heater require?*

Ans. It depends on the capacity of the boiler. It may range from 30 m^2 to $300\,000 \text{ m}^2$.

Q. *Why is such a large surface area required?*

Ans. Because of a very low value of the heat-trans-fer coefficient between the combustion products and air.

Q. *How does the temperature gradient vary during air heating?*

Ans. As the figure shows, the temperature of air varies more sharply than that of the combustion products. (Fig. 29.1)

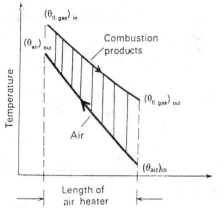

Fig. 29.1 Temperature profile in the air heater

Q. Why is the air, in the air heater, heated more quickly than the combustion products are cooled?

Ans. The air that is heated in the air heater contains less moisture (triatomic molecule) than the combustion products which have higher water vapour concentration (depending on the moisture and hydrogen content of the fuel) as well as such triatomic gases as CO_2, SO_2, etc. Triatomic molecules have higher heat capacities. Therefore, the volume and heat capacities of the combustion products are much greater than those of air composed predominantly of diatomic molecules such as N_2 and O_2. Hence the ratio of water equivalents of air and combustion products:

(Sp. Heat × Volume) of AIR/(Sp. Heat × Volume) of COMBUSTION PRODUCTS.

is always less than unity which means that air is heated more quickly than the combustion products are cooled.

Q. Why does the temperature gradients, as the air is being heated, decrease to a minimum at the hot end of the air heater?

Ans. It is because the temperature of combustion air in the air heater rises more quickly than the combustion products are cooled. Therefore, the margin between the heating curve of air and the cooling curve of combustion products narrows down at the hot end (i.e., at the inlet of hot combustion products or the outlet of combustion air). Therefore, the temperature gradient between the combustion products and the air is at its lowest in the hot end of the air heater.

Q. What is the temperature difference between the combustion products and the air in the hot end of the air heater?

Ans. Usually it lies in the range of 30–50°C.

Q. Why is it not desirable to allow the temperature difference between the combustion products and air at the hot end of the air heater to drop below 30–40°C?

Ans. If Δθ at the hot end is allowed to drop below 30–40°C, it means an increase in temperature of the hot air. As the temperature difference in the hot portion becomes too low, the rate of heat exchange in this zone will be poor. Therefore, it will bring about a higher temperature of combustion products at the outlet of the air heater. Or it will need a higher heat transfer surface area of the air heater to bring down the temperature of the waste gas (combustion products) at the outlet of the air heater to the desired level.

Q. How does preheating the air influence the temperature of the waste gas (combustion products at the outlet of the air heater) at various ratios of water equivalents?

Ans. This effect of air heating on waste gas temperature is depicted in the adjacent figure (Fig. 29.2). Computation has been done on the basis: C_a, C_{cp} are the specific heats of air and combustion products respectively. V_a, V_{cp} are the volume of air and combustion products respectively.

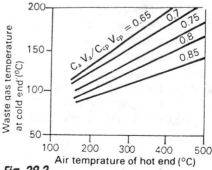

Fig. 29.2

Air inlet temp. = 30°C
Δθ at hot end = 40°C

As can be seen from the figure, to preheat air to 400°C at W.Eq. ratio = 0.8, the temperature of the waste gases at the outlet of the air heater is about 150°C. Whereas for fuels with higher moisture content (W.Eq. ratio = 0.65), the same degree of air preheating requires that the waste gas temperature outlet of the air preheater should be about 200°C.

Q. *What conclusion can you draw from it on the efficiency of the air heater regarding the moisture content of fuel?*

Ans. As the moisture content in the fuel increases, the ratio of water equivalents between air and combustion products decreases and with that the efficiency of the air heater decreases.

Q. *What are the advantages of air preheating?*

Ans.

1. Improved combustion and efficient use of fuel.
2. Stabilized fuel ignition which improves low load combustion.
3. Increased steam generation capacity (kg/m² h) and higher boiler efficiency which increases by 1% for every 20°C rise of the preheated air temperature
4. Better utilization of low grade, high ash fuel
5. Higher thermal efficiency as most of the heat from the combustion products is recovered
6. Greater load flexibility
7. Reduction in boiler heating surface due to increase in overall boiler efficiency because of improved combustion
8. Greater the preheat, lower the SO_3 in the flue gas and therefore the stack outlet temperature can be lowered resulting in greater heat utilization.
9. Air preheating ensures complete burning of fuel resulting in less slagging and cleaner flue gas and this reduces boiler outage for cleaning.

Q. *What are the disadvantages of air preheating?*

Ans.

1. Increases maintenance cost of stoker and furnace refractory
2. Entails the risk of serious damage due to deposition and ignition of combustibles

3. Operation of entire steam generation unit may be critically hampered due to clogging
4. Requires an F.D. fan and ducts for air passage and that needs careful design calculation for draft loss
5. Any air leakage necessitates higher fan output, i.e., taxes F.D. fan more electric power unit
6. Leakage is difficult to detect until the corrosion is in an advanced stage requiring a major repairing or total replacement and greater outage time

Q. *How does the boiler load variation affect the efficiency of the air preheater?*

Ans. As the boiler load is increased, it simultaneously increases the volume and temperature of the flue gases. This results in higher convective heat transfer through the air heater and improves the efficiency of the latter.

Q. *Where is the air heater installed?*

Ans. It is installed in the convective shaft of the furnace and downstream of the economizer.

Q. *Why?*

Ans. Its purpose is air preheating by utilizing the waste heat of the combustion products downstream of the economizer, where the temperature of the combustion products is sufficiently high, 625–725K. By abstracting the waste heat from the combustion products, the flue gas temperature is brought down to 395–435K at the outlet of the air preheater, before letting them escape to atmosphere.

Q. *How does high-sulphur fuel combustion affect the air preheater?*

Ans. Air preheater operates in the zone of lowest temperature of combustion products, therefore a part of its surface near the cold end (where the waste gas temperature may be 120–160°C) will be at a temperature equal to the dew point of sulphuric acid (140°C) produced on reaction of SO_3 with the moisture content in the combustion products. This corrosive acid film deposited on the surface of the air heater at the cold end will contribute to severe corrosion of the tube metal surface.

Q. *How many kinds of air heaters exist?*

Ans. Two:
1. recuperative air heaters
2. regenerative air heaters

Q. *What is a recuperative air heater?*

Ans. These are mostly tubular heat exchangers in which heat from combustion products or some other heat source is transferred continuously to the air as the air flows through shell-side and combustion products through the tube side.

Q. *What is a regenerative air heater?*

Ans. These are a kind of heat exchanger in which the heat transfer surface is alternately heated and cooled by charging it alternately with combustion products and air.

Q. *How many methods are used in practice to heat up air?*

Ans. There are three types of air heating:
1. direct heat transfer from combustion products (as in a tubular air heater) (Fig. 29.3)
2. indirect heating through an intermediate heat-transfer agent (solid packing) (Fig. 29.4A, B)
3. combined heating in which air is heated up by low-temperature waste-steam or water as well as by combustion gas. (Fig. 29.5)

Q. *In which apparatus is direct heating of air by combustion products carried out?*

Ans. Recuperative air heater.

Q. *Briefly describe it?*

Ans. These are predominantly shell-and- tube heat exchangers in which combustion products flow

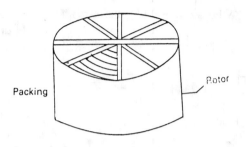

Fig. 29.4(A)

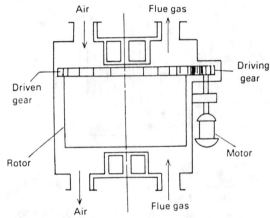

Fig. 29.4(B)

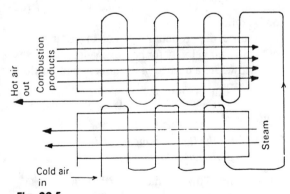

Fig. 29.5

through the tubes and air moves between the tubes, i.e., in the shell-side in cross-current flow with the combustion products. (Fig. 29.6)

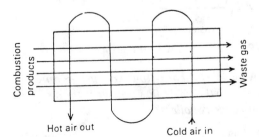

Fig. 29.3

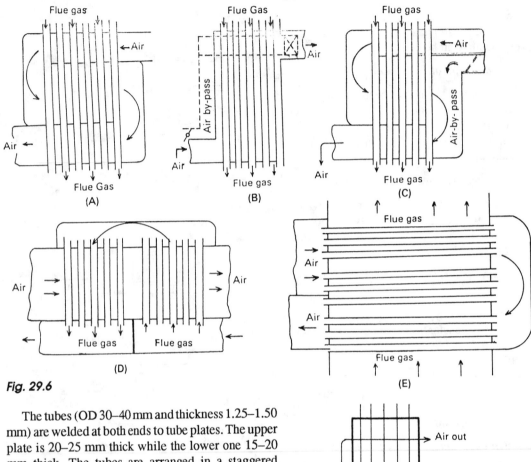

Fig. 29.6

The tubes (OD 30–40 mm and thickness 1.25–1.50 mm) are welded at both ends to tube plates. The upper plate is 20–25 mm thick while the lower one 15–20 mm thick. The tubes are arranged in a staggered configuration.

A number of partitions are placed in the air path to get optimal air velocity for heat exchange. Air boxes are fitted where a change in the direction of air flow is necessary.

The tube system is capable of expanding upwards when heated.

Q. *What is the flow scheme of air in the air heater of boilers of moderate capacity?*

Ans. Single-flow scheme (Fig. 29.7)

Q. *What is the flow scheme of air in the air heater of high- capacity boilers?*

Ans. Split-flow scheme in which air is distributed in the heater into two or more separate flows. (Fig. 29.8)

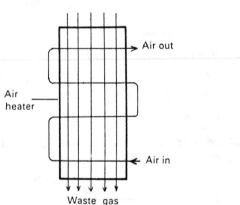

Fig. 29.7 *Air heater with single-flow scheme*

Q. *Why is the split-flow scheme adopted in preference to the single-flow scheme in the air heater of high-capacity boilers?*

Ans. High-capacity boilers need a larger quantity of combustion air for preheating. Therefore, the air

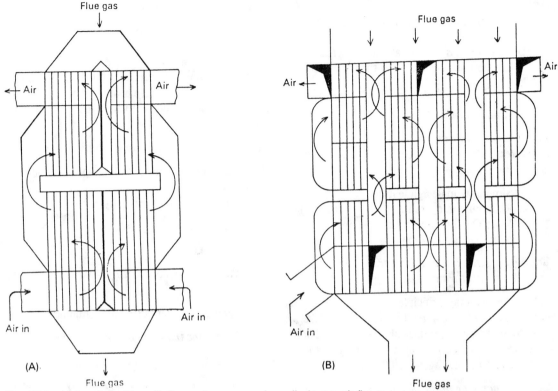

Fig. 29.8 *Air heater with split-flow scheme can handle large air flowr*

preheater with single-flow air path must be of greater height to accommodate larger air flowrate. This uneconomically large height can be substantially cut short by increasing the number of passes for air by adopting the split-flow scheme for air dividing the air in the heater into two or more separate flows.

Q. *Is the heat-transfer coefficient at the gas side greater or the coefficient of heat-transfer at the air side, in the air heater?*

Ans. The heat-transfer coefficient at the air side is greater than the heat-transfer coefficient at the gas side.

Q. *How can the heat-transfer coefficient at the gas side of the air heater be increased?*

Ans. By replacing straight-tubes with corrugated-tubes having a sinusoidal shape of constant radius of curvature.

Q. *But corrugated-tubes will get clogged by ash particles when the combustion products flow through them impairing the reliability of operation of the air heater. Is it not?*

Ans. Yes.

Q. *How to overcome this problem?*

Ans. By reversing the flow scheme, i.e., by taking air through the tubes and combustion products through the shell.

Q. *Between the straight-tube air heater and corrugated-tube air heater, which one is more efficient?*

Ans. The corrugated-tube air heater is more efficient than the straight-tube air heater.

Q. *What are the chief advantages of corrugated-tube air heaters?*

Ans.

1. Higher efficiency than a straight-tube air heater
2. Require less metal (about 50% less) than straight-tube air heaters of the same capacity
3. About $\frac{1}{3}$ rd less costly than conventional type straight-tube air heaters.

Q. *What are the basic advantages of straight-tube air heaters?*

Ans.

1. Simplicity in design feature
2. Reliability in operation
3. More gas-tight operation than any other type of air heater.

Q. *What is the basis of heat transfer through an intermediate heat-transfer agent?*

Ans. The working principle of this system is that heat instead of being directly transferred to the air from the combustion products through the wall of the heat transfer surface is transmitted first to an inter-mediate heat-transfer agent and then to air.

Q. *In which type of air heater is this principle put into practice?*

Ans. Regenerative air heaters.

Q. *What material is used as intermediate heat-transfer agent?*

Ans. Usually steel sheets used as solid packing act as intermediate heat-transfer agents.

Q. *Are the sheets plain or corrugated?*

Ans. Both kinds are employed.

Q. *What is the most commonly used regenerative air heater?*

Ans. Rotary type regenerative air heater.

Q. *Where is the packing placed?*

Ans. It is placed into a hollow cylindrical drum called rotor which slowly rotates about its vertical axis by a gear and motor drive mechanism. The rotor is divided into several sectors by radial partitions.

Q. *How is the air preheating accomplished in the regenerative air heaters?*

Ans. As the rotor rotates slowly (0.5–0.35 rpm), the gas and air paths are intersected alternately by the rotating packing. Therefore, when the packing in some of the sectors receive heat from the combustion products, heat is given out to air by the packing in the other sectors.

Q. *What is the diameter of the rotor?*

Ans. It usually comes with a diameter of 10 m. In some cases rotor dia as large as 15 to 18 m are used for regenerators in high-capacity plants.

Q. *What is the heating surface area of packing?*

Ans. 300–350 m^2 per cubic metre of rotor packing.

Q. *What is the contact time of gas/air flow with the packing?*

Ans. It is usually less than half-a-minute.

Q. *What is the thickness of packing sheets?*

Ans. 0.6–1.25 mm.

Q. *Do the packings in a regenerative air heater come in the form of sheets only?*

Ans. No. In some cases packing is made up of balls of metal, ceramic or glass.

Q. *What are the basic drawbacks of corrugated packing?*

Ans.

1. Rapid destruction of thin sheets of metal due to erosion and corrosion
2. Low heat-transfer coefficient in longitudinal flow.

Q. *What is the relation between the flows of com-bustion products and air in the rotary regenerative air heater?*

Ans. These two flows are in counter-current rela-tions.

Q. *How is combined air heating accomplished?*

Ans. Air is preheated by low-temperature steam (or hot water) in a steam-air heater (or low-pressure economizer-air heater) and then finally heated in the tubular air heater by the combustion products

Q. *What are the advantages of this combined system of air heating?*

Ans. It fits well in combating low-temperature corrosion when high-sulphur fuel is used.

Secondly, the waste-heat of the turbine exhaust steam is utlized in air preheating, thus decreasing the heat loss in the circuit.

Q. *What is the steam air heater?*

Ans. It is a tubular heat exchanger in which low-temperature steam flows in the tubes while air flows in the shell-side in cross-current flow. (Fig. 29.9)

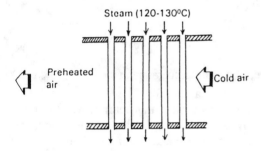

Steam (120-130ºC)

Preheated air

Cold air

Fig. 29.9 *Steam-air heater*

Q. *What is the temperature of this heating steam?*

Ans. About 120–130°C.

Q. *Where does this steam come from?*

Ans. It is the worked-off steam from the steam turbine.

Q. *At which point in the circuit is this steam-air heater introduced?*

Ans. Downstream of the air blower. [Fig. 29.10]

Q. *How does the combined scheme of air preheating with economizer and air heater operate?*

Ans. In this scheme air is preheated in a closed cycle low-pressure economizer-air heater system.

The preheated air is finally heated in a tubular air heater by hot combustion products. [Fig. 29.11]

Q. *What is the advantage of this circuit?*

Ans. Air inleakage is substantially reduced.

Q. *Why is the system made closed?*

Ans. The system operates by closed cycle so that contamination of the condensate by corrosion products is prevented.

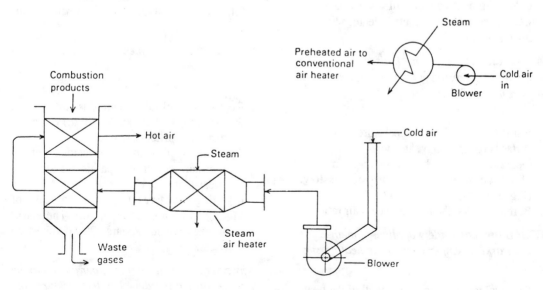

Fig. 29.10 *Steam-air heater is provided downstream of the air blower*

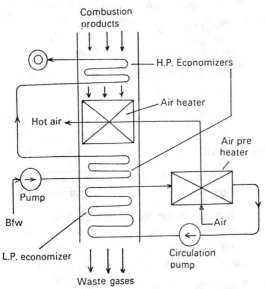

Fig. 29.11 Combined system of air heating

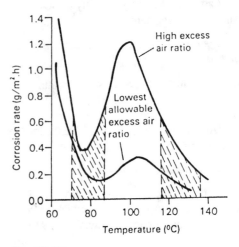

Fig. 29.12

Q. *What is the temperature of deaerated water in the closed loop?*

Ans. It is kept slightly greater than 140°C—the dew point of H_2SO_4.

Q. *Why?*

Ans. Otherwise acid vapour will·condense on the heat exchange surface and contribute to acid corrosion and loss of metal.

Q. *Among all the known methods of combating low-temperature corrosion of heating surfaces which ones are the most efficient?*

Ans.

1. Raising the working temperature of heating surfaces (metal) above the dew point.
2. Operating the air heater in those regions of the curve where the corrosion is low. (Fig. 29.12)
3. Burning fuel at the lowest excess air ratio.

Q. *What is the most widely applicable method for raising the working temperature of the heating surfaces?*

Ans. The local temperature of the wall of the heating surface is given by

$$\theta_w = (\theta_a)_i + \left(\frac{q}{\alpha}\right)_i$$

where, $(\theta_a)_i$ = air temperature at the inlet

q = rate of heat transfer

α = coefficient of heat transfer from wall to the air.

So by increasing the inlet air temperature, the working surface temperature can be raised.

Therefore, the most widely applied method is to preheat air in the steam-air preheater and then subject the preheated air to conventional air heaters.

Q. *What are the other methods in fighting low-temperature corrosion in air heaters?*

Ans.

1. Replacing steel tubes with glass tubes (OD: 30 – 40 mm; thickness: 3.5 – 4 mm) where the corrosion is particularly intensive
2. Application of corrosion-resistant paints to heating surfaces. These paints are acid-and heat-resistant enamels
3. Injecting additives (e.g., aq. solution of $MgCl_2$) to high-sulphur fuel oil. These will neutralize the acid vapours of the combustion products.

Q. *Why can the burning high-sulphur fuel oil at the lowest excess air ratio reduce low-temperature corrosion?*

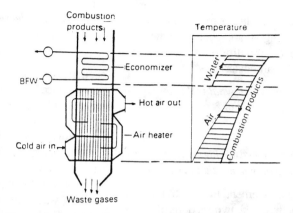

Fig. 29.13

Ans. Low-temperature corrosion due to condensed film of sulphuric acid produced on combustion of high-sulphur fuel oil can be minimized by using minimal excess air ratio. With low excess air ratio, the excess of O_2 supplied to fuel burning is low. Therefore, combustion products will contain less SO_3 and therefore, the sulphuric acid corrosion will be lower.

Q. *Graphically describe the distribution of temperature gradients in a single-stage arrangement of economizer and air heater.*

Ans. Figure 29.13

Q. *What is the difference between the temperature distribution curves of the economizer and air heater hooked up in a single-stage arrangement?*

Ans. In the economizer, the temperature gradient at the hot end increases while that at the hot end of the air heater decreases.

Q. *Why does the temperature gradient at the hot end of the economizer increase?*

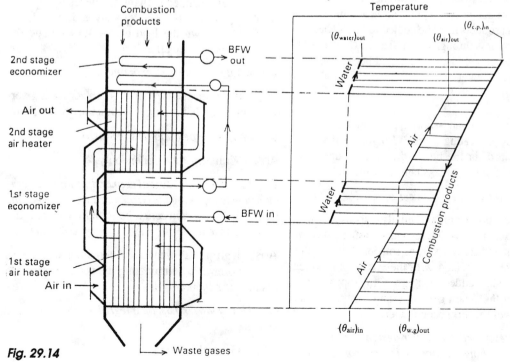

Fig. 29.14

Ans. It is because water has a much higher heat capacity than air causing an appreciably higher rate of cooling of the combustion products in the economizer than in the air heater.

Q. *In the single-stage circuit what is the limit of air preheating for efficient operation?*

Ans. 250–350°C.

Q. *If the air is to be preheated to a higher temperature, say 350–450°C, what arrangement should be done?*

Ans. Two stage arrangement in which the economizer is placed between the two sections of the air heater. (Fig. 29.14)

Q. *Why is a hot section of the economizer placed before the second stage of the air heater?*

Ans. The tubes of air heaters are made of carbon steel that limits the allowable surface temperature to 500°C. Therefore, this means that for air preheating to 400–425°C, the temperature of combustion products should not exceed 570–580°C. Since the temperature of the combustion products downstream of the superheater is appreciably higher (600–650°C), in order to protect the metal of the air heater tubes, the combustion products must be cooled to 580°C before allowing it to pass through the air heater. Hence a hot section of the economizer is mounted before the 2nd stage of the air heater.

Q. *What is an economizer?*

Ans. It is a tube-and-shell type heat exchanger designed for preheating the BFW before it is fed to the evaporator via the boiler drum. The tubes are either straight or given the shape of coils.

Q. *Why is the term economizer applied?*

Ans. An economizer is essentially a shell-and-tube type heat exchanger which recovers waste heat from the flue gas and transfers it, as sensible heat, to the BFW before it is mixed with the circulating boiler feedwater. This added heat goes to improve the economy of the steam generator and herein lies the implication of the term economizer.

Q. *The present trend in boiler design is to increase the heat absorbing surface of the economizer and*

proportionally decrease that of the steam generator, i.e., waterwall. Why?

Ans. It is due to a greater heat transfer in the economizer than in the convection surfaces of the boiler. Since BFW enters the economizer at a temperature much lower that of the boiler steam, more effective heat transfer takes place at this point than in the convection waterwalls of the boiler.

The non-steaming economizers with air heater included occupy 25–30% of the heating surface as compared to the boiler heating surface. This figure doubles when the air heater is excluded from the system. For steaming economizers this figure comes out to be 25–50% .

Q. *Compared with the air preheater, what distinct advantages does an economizer possess?*

Ans.
1. It befits small capacity plants and those operating at low pressure well because of low initial cost
2. Lower requirement of auxiliary power
3. Endows quicker steam generation capacity when operated with on-off cycles
4. Requires less space for a given duty
5. Cuts down the load on the furnace and the burners more effectively.

Q. *What is an integral economizer?*

Ans. It is a special type of economizer having vertical banks of tubes arranged within the boiler casing.

Q. *In which type of boilers is it used?*

Ans. With bent-tube boilers only.

Q. *Sometimes an integral economizer may be of the intertube type having the elements placed between the tubes of a boiler convection bank. Why is this arrangement chosen?*

Ans. It affords a compact and economical design of the economizer by eliminating the need for separate casing and supports.

Q. *How many kinds of integral economizers are known?*

Ans. Generally three:

1. Single-drum type
2. Double-drum type
3. Steaming economizer type.

Q. *How does a double-drum type economizer function?*

Ans. It comprises of two drums—one upper and the other lower, interconnected by banks of vertical tubes. Into the compartmented upper drum is introduced the boiler feedwater which then moves down one bank of tubes to the lower drum from which it flows back again (completing a 2-pass unit) before being discharged into the boiler drum.

The flue gas path is baffled vertically to ensure counterflow in the multipass (usually 2-or 3-pass) arrangements; also it may be in a single or multipass-crossflow arrangement with the water circuit in some cases.

Q. *In which cases are steaming economizers installed?*

Ans. These are meant for generating steam at higher output rates, i.e. these are provided for high-capacity steam generating units.

Q. *Why is a steaming economizer incompatible with high-feed makeup boiler units?*

Ans. It is because of the inherent problems associated with suitable treatment of boiler feedwater at high rate.

Q. *What are the dimensions of economizer tubes.*

Ans. ID : 20–30 mm and thickness : 2.5–3.5 mm.

Q. *Why are small dia tubes chosen?*

Ans. To maximize the rate of heat transfer and minimize fouling.

Q. *What is the extent of draft loss in an economizer?*

Ans. It usually ranges from 60–75 mm of water column. It rarely exceeds 100 mm of water gauge.

Q. *Why is the BFW delivered to the economizer at a pressure 10–20% over the boiler pressure?*

Ans. This will make the necessary allowance for the pressure drop of BFW as it passes through the

economizer tubes, BFW regulator (if any), piping and valves.

Q. *At which point is the BFW regulator located in the economizer?*

Ans. At the discharge end of the economizer.

Q. *Why is it placed there?*

Ans. To increase the feedwater pressure as well as to minimize the possibility of steam formation.

Q. *Is it at all possible to operate the economizer at a pressure lower than boiler pressure?*

Ans. Yes.

Q. *In which case is it possible?*

Ans. Occasionally, some cast-iron economizers are installed to operate at a pressure lower than boiler pressure.

Q. *Then how does this system function in conjugation with the boiler working at higher operating pressure?*

Ans. A booster pump is fitted into the water circuit downstream of the economizer. This pump boosts up the hot BFW at the outlet of the economizer to a pressure necessary for delivering it to the boiler drum.

Q. *Does the inlet temperature of the BFW to the economizer affect the efficiency of the latter?*

Ans. Yes; as the inlet feedwater temperature is raised, the efficiency of the economizer decreases.

Q. *What is the maximum temperature limit of the BFW normally before its entry into the economizer?*

Ans. Normally it is not allowed to exceed 105°C, however in some high-pressure plants the BFW is preheated to 200°C before its entry to the economizer.

Q. *Why is the boiler water often recirculated to the economizer?*

Ans.
1. To increase the pH of feedwater by 2–4% which is adequate.
2. To raise the temperature to avoid cold end–condensation.

Q. *What is the pH value of water in steel tube economizers?*

Ans. 8 to 9.

Q. *Why is the final temperature of BFW at the economizer outlet usually kept at a temperature about 10°C less than the saturation temperature?*

Ans. Steam formation will occur within the economizer tubes if the boiler feed is restricted or cut off.

Q. *What harm will it cause?*

Ans. This will cause water hammer and thermal stress in the economizer tubes. When the boiler feed is restricted or cut off, a considerable portion of water in the economizer will convert to steam which will escape rapidly as soon as the boiler feed is resumed, leaving the economizer partially empty. When water surges in to fill this void space, it will give birth to water hammer as well as temperature strains that may finally result in economizer tube failure.

Q. *In which case does the above limitation on the BFW temperature at the outlet of economizer not exist?*

Ans. In the case of steaming economizer where steam formation takes place within the economizer itself.

Q. *Earlier steam generating units would instal economizer by-pass flues. What are the utilities of such by-passing?*

Ans. This system
   (a) prevents cold-end condensation during start-up
   (b) harnesses the natural draft when economizer or ID fan goes out of service
   (c) prevents the formation of steam in the economizer.

Q. *What are the usual materials of construction of economizer tubings?*

Ans. These are made of
   (a) steel for high pressure steam generators
   (b) cast iron for lower pressure steam generators.

Q. *What kind of steel?*

Ans. Low alloyed pearlitic-class steels: Mn: 0.9–1.25% ; Si: 0.7–1.0%

Q. *What are the advantages of steel tubes used in the economizer?*

Ans.
   1. Ensure a more compact economizer design.
   2. Permit steam generation at elevated pressure and temperature and at higher rate.
   3. Ensure a higher rate of heat transfer for a given weight
   4. Cost is relatively low.

Q. *Are these steel tubes bare or provided with extended surfaces?*

Ans. Both kinds are known. Fins, rings and studs are welded to the top and bottom of steel tubes of the economizer.

Q. *What modification is done to increase the rate of heat transfer from the flue gas side and to increase the compactness of the apparatus?*

Ans. Fins are welded on the gas-side tube surface of the economizer. These increase the effective heating surface area on the gas-side and hence the heat absorption.

Q. *What are the advantages of finned tubes over plain tubes used in the economizer?*

Ans.
   1. This system increases the heating surface resulting an overall gain in heat absorption on the gas-side and hence higher steam output rate
   2. Prevents the tube from sagging
   3. Minimizes slag deposition on the tube top
   4. Finned tubes with rectangular fins occupy a volume which is 1/3–1/4th less than the volume occupied by plain tubes for the same draft and same amount of metal.
   5. The economizer becomes more compact. The dimensions of the economizer are curtailed by 40 – 50%

Q. *Other than rectangular fins what fins are used in finned-tube economizers?*

1. Trapezoidal fins
2. Helical fins
3. Circular fins.

Q. *In which cases can economizer tubes be made of cast-iron?*

Ans. Cast iron materials are best suited for the purpose of low pressure operation. Also these are found suitable when:

1. BFW is not fully treated
2. BFW is not totally deaerated
3. Operational attention and maintenance are hampered.

Q. *How many types of economizers are used in practice?*

Ans. Two types: Boiling and non-boiling.

Q. *What is a boiling type economizer?*

Ans. This type of economizer is characterized by the transformation of a part of the boiler feedwater (BFW) into steam.

Q. *How much of BFW is converted into steam in a boiling type economizer?*

Ans. Maximum 25% .

Q. *What is a non-boiling type economizer?*

Ans. In this type of economizer, the BFW is heated below the saturation point.

Q. *What are membrane type economizers?*

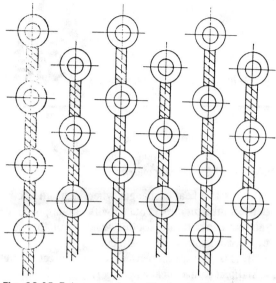

Fig. 29.15 *Tubes welded with steel membranes enhance the heat transfer characteristics of membrane type economizer*

Ans. This is made up of plain tubes with steel-membranes (steel sheets 2–3 mm thick) welded in between the straight portion of the tubes. (Fig. 29.15)

Q. *What are the advantages of membrane type economizers over plain tube type economizers?*

Ans.

1. Higher efficiency
2. More reliable in operation
3. Requires less metal for the same amount of heat absorption.

30

Steam Condensers

Q. *What is a steam condenser?*

Ans. It is a heat exchanger wherein steam is condensed either in direct contact with cooling water or indirect contact with cooling water through a heat transfer medium separating them.

That is, a steam condenser is either a direct contact or indirect contact heat exchanger.

Q. *How many types of steam condenser are known?*

Ans. Two main types:
1. JET CONDENSER (direct contact heat exchanger)
2. SURFACE CONDENSER (indirect contact heat exchanger)

Q. *What is a jet condenser?*

Ans. It is a direct contact heat exchanger in which steam to be condensed comes into direct contact with the cooling water (cold condensate) which is usually introduced in the form of a spray from a jet. (Fig. 30.1)

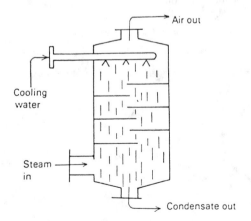

Fig. 30.1 *Jet condenser*

Upon contact with the cooling water, the steam gives up its enthalpy and gets cooled and ultimately settles as condensate.

Q. *What is a surface condenser?*

Ans. It is a shell-and-tube heat exchanger in which steam is condensed on the shell-side while cooling water flows through the tubes. The condensate and cooling water leave the system separately.

Q. *How many types of jet condensers are known?*

Ans. Three types:
1. Parallel flow type
2. Contra flow type
3. Ejector type.

Q. *What is a parallel flow jet condenser?*

Ans. It is a kind of jet condenser in which both exhaust steam and cooling water enter the condenser at the top, both flow downward and the steam condensate discharges out from the bottom of the condenser.

Q. *How many types of parallel flow jet condenser exist?*

Ans. Two types:
1. Low level parallel flow jet condenser
2. High level parallel flow jet condenser.

Q. *How does the low level parallel flow jet condenser function?*

Ans. Exhaust steam is introduced from the top of the condenser while cold condensate is sprayed upon it through multiple jets. As steam comes in contact with cold condensate particles, its enthalpy drops and it settles at the bottom as condensate. (Fig. 30.2)

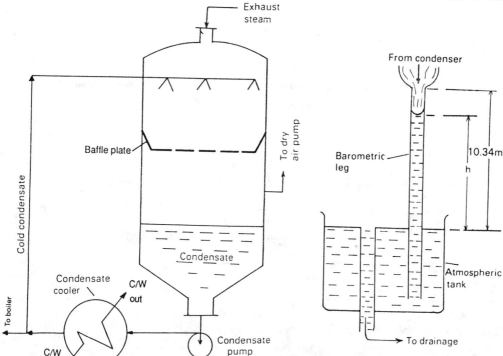

Fig. 30.2 *Operating principle of low-level-parallel-flow jet condenser*

Fig. 30.3 *Operating principle of high-level parallel-flow jet condenser*

Vacuum is maintained in the condenser with the aid of a dry air pump which also extracts the air present in the condenser.

The condensate collected at the bottom is cooled in the condensate cooler by cooling water and is partly recycled and partly sent as boiler feedwater.

Q. *What is the principle of operation of a high-level-parallel-flow jet condenser?*

Ans. This condenser, also called barometric condenser, works as follows:

The condenser is mounted on a long pipe (at least 10.34m) called barometric leg which acts in a way identical to a barometer.

Now if water is used in a barometer then the barometric height would be 10.34 m. If some vacuum exists in the condenser, the height of water column (*h*) will be less than 10.34 m. Now it is possible, by using this condenser leg, to drain away the condensate from the condenser.

This vacuum is maintained by constant high level discharge from the atmospheric tank through the atmospheric leg that dips deeply into the water in the atmospheric tank. (Fig. 30.3)

Q. *What is a contra-flow jet condenser?*

Ans. In this type of jet condenser the cooling fluid (cold condensate) and exhaust steam flow in a counter-current direction—steam goes up and cold condensate rains down.

Q. *How many types of contra-flow jet condenser exist?*

Ans. Two types:
1. LOW LEVEL CONTRA-FLOW JET CONDENSER
2. HIGH LEVEL CONTRA-FLOW JET CONDENSER

Q. *How does the low-level contra-flow jet condenser operate?*

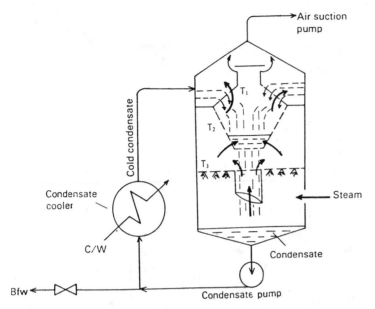

Fig. 30.4 *Operating principle of low-level contra-flow jet condenser*

Ans. Exhaust steam enters the condenser near the bottom and moves up through a pipe P. As it goes up, it comes in direct contact with cold condensate distributed into the condenser through perforated trays T_1, T_2 and T_3. (Fig. 30.4)

Vacuum is maintained through an air suction pump.

Condensate collected at the bottom is partly sent for as BFW and the rest is recycled via the condensate cooler.

Q. *Between the low-level contra-flow jet condenser and low-level parallel-flow jet condenser which one is more efficient?*

Ans. Low-level contra-flow jet condenser is more efficient.

Q. *Why?*

Ans. It is because the hottest steam comes in contact with the hottest water. So as steam goes more and more up, it comes into contact with more and more colder condensate and that enhances the steam condensation.

Q. *How does the high-level contra-flow jet condenser work?*

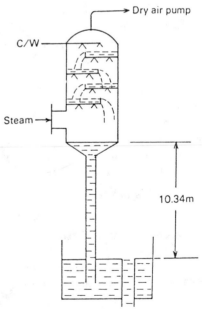

Fig. 30.5 *Operating principle of high-level contra-flow jet condenser*

Ans. The condenser is placed 10.34 m above the hot-well. The condenser is operated at a pressure somewhat less than atmospheric pressure such that

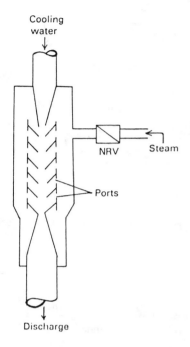

Fig. 30.6 Ejector condenser

the water level in the tail pipe remains always below 10.34 m. And therefore, any steam condensing to the condensate is spontaneously drained off. (Fig. 30.5)

Q. *What is an ejector condenser?*

Ans. It is one kind of jet condenser in which the mixing of cooling water and steam takes place in a series of combining cones and the kinetic energy of the steam is expended to drain off the condensate and cooling water from the condenser.

Q. *How does it work?*

Ans. Cooling water is forced through a series of cones and gets mixed with steam coming through ports.

As the cooling water 1'ows through the series of nozzles, it suffers more and more pressure drop and at the same time its velocity gradually increases. (Fig. 30.6)

Due to this pressure drop, more and more steam is drawn through the ports, gets intimately mixed with the cooling water jet and condenses thereafter.

A NRV (non-return valve) is provided on the steam inlet line to prevent the backflow of condensate in case of failure of the cooling water supply.

Q. *How many types of surface condensers are known?*

Ans. Four types:
1. Down flow type
2. Central flow type
3. Inverted flow type
4. Evaporative condenser type.

Q. *How does the down flow type surface condenser act?*

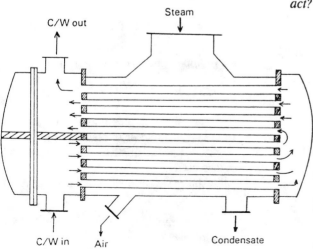

Fig. 30.7 Downflow type surface condenser

Ans. Exhaust steam is admitted to the top of the condenser which is a tube-and-shell type crossflow heat exchanger. Cooling water flows through the tubes and extracts heat from the steam which is on the shell-side. After having been condensed on the surface of the water tubes, steam is converted into condensate which is discharged from the condenser bottom. (Fig. 30.7)

Q. *How does the central flow type surface condenser works?*

It is also a shell-and-tube type crossflow heat exchanger at the centre of which is located the suction of an air extraction pump so that the entire steam moves radially inward and comes in better contact with the outer surface of the nest of tubes through which the cooling water flows. The steam condensate is extracted from the bottom by the condensate-extraction pump (Fig. 30.8)

Q. *How does the inverted type surface condenser work?*

Ans. In this type of condenser, steam is admitted at the bottom and flows upwards in cross-flow with the cooling water flowing in the tubes. The air extraction pump draws its suction from the top of the condenser, maintaining a steady upward current of steam which after having been condensed on the outer surface of

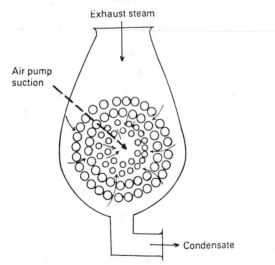

Fig. 30.8 Working principle of central flow type surface condenser

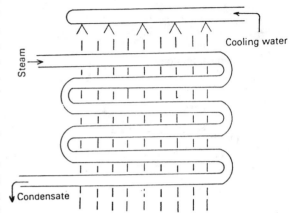

Fig. 30.9 Working principle of evaporative condenser

water tubes is removed by the condensate extraction pump.

Q. *How does the evaporative condenser function?*

Ans. Exhaust steam from the turbine is condensed inside the finned tubes as cooling water rains down from the top through the nozzles. A part of the cooling water in contact with the tube surface evaporates by drawing enthalpy from the steam which upon losing its latent heat condenses and discharges out as condensate. (Fig. 30.9)

Q. *Where is the evaporative condenser used in practice?*

Ans. In those cases where the shortage of cooling water is acute.

Q. *What are the prime functions of a condenser?*

Ans. There are two important functions of a condenser:

1. It reduces the back pressure upon the turbine by a considerable degree and therefore, the work done per kg of steam during expansion is increased

2. The exhaust steam condensate can be recycled as boiler feedwater

Q. *What are the auxiliary equipment required for operating a steam condenser?*

Ans.

1. Cooling water (which may be cold condensate) circulation pump. Generally, it is a centrifugal one
2. Arrangement for cooling the condensate (i.e., a heat exchanger) in case the condensate is recycled to extract heat from the exhaust steam
3. An air pump or steam ejector to remove air and other non-condensing gases from the condenser
4. An extraction pump (usually centrifugal) to remove the condensate from the condenser.

Q. *What should be the basic criteria for an efficient steam condenser?*

Ans.

1. Maximum amount of steam condensed per unit area of available heat transfer surface.
2. Minimum quantity of circulating coolant required
3. Minimum heat transfer surface required per kW capacity
4. Minimum power drawn by the auxiliaries.

Q. *Why is vacuum maintained in the steam condenser?*

Ans. By maintaining a vacuum in the steam condenser, the efficiency of the steam-power plant can be increased as greater the vacuum in the system, greater will be the enthalpy drop of steam. Therefore, more work will be available per kg of steam condensing.

Secondly, the non-condensate (air) can be removed from the condensate-steam circuit by pulling and maintaining a vacuum in the steam side. Therefore, the condensate can be used as boiler feed.

Q. *What are the advantages of a jet condenser over a surface condenser?*

Ans.

1. Simplicity in design
2. Lower in manufacturing cost
3. Lower maintenance cost
4. Occupies lesser floor space
5. Requires lesser amount of cooling water.

Q. *What are the advantages of a surface condenser over a jet condenser?*

Ans.

1. It imparts to power generation plant a higher thermal efficiency
2. The condensate can be reused as boiler feedwater
3. Auxiliary power requirement is less than that of a jet condenser
4. Less amount of air is carried to the boiler.

Q. *What are the limitations of a surface condenser?*

Ans.

1. It is very bulky and as such requires more floor space
2. Its manufacturing, running and maintenance costs are high.

Q. *What should be the requirements of an ideal surface condenser used for steam power plants?*

Ans.

1. Uniform distribution of exhaust steam throughout the heat transfer surface of the condenser
2. Absence of condensate subcooling
3. There should not be any leakage of air into the condenser
4. There should not be any tube leakage
5. The heat transfer surface in contact with cooling water must be free from any deposit as scaling reduces the efficiency of heat exchangers.

Q. *What do you mean by vacuum?*

Ans. Vacuum means any pressure below atmospheric pressure.

Q. *How is vacuum in a condenser usually measured?*

Ans. It is measured by means of a Bourdon pressure gauge which is calibrated to read the pressure in mm of mercury below atmospheric pressure.

Q. *If the gauge pressure of a condenser is 630 mm of Hg, what will be the absolute pressure in the condenser?*

Ans. It means the pressure in the condenser is 630 mm below atmospheric pressure.

The atmospheric pressure is 760 mm of Hg, the absolute pressure in the condenser

$= 760 - 630 = 130$ mm Hg $= 130 \, (0.1334) = 17.34$ kN/m^2

Q. *On what factors does the degree of vacuum in a condenser depend?*

Ans. It depends on the partial pressure of steam and the partial pressure of air in the condenser.

Q. *What will be the absolute pressure in the condenser?*

Ans. $P_{abs} = P_{steam} + P_{air}$

where

P_{abs} = Absolute pressure in the condenser

P_{steam} = Partial pressure of steam

P_{air} = Partial pressure of air in the condenser.

Q. *What is the vacuum efficiency of a condenser?*

Ans. It is the ratio of the actual vacuum at the steam inlet to the maximum obtainable vacuum in a perfect condensing plant, i.e., it is the ratio of actual vacuum to ideal vacuum.

$$\text{Vacuum Efficiency} = \frac{\text{Actual Vacuum}}{\text{Ideal Vacuum}}$$

$$= \frac{P_{atm} - (P_{sat} + P_{air})}{P_{atm} - P_{sat}}$$

where

P_{atm} = Atmospheric pressure

P_{sat} = Saturation pressure of steam at condenser inlet

P_{air} = Partial pressure of air in the condenser.

Problem 30.1 A surface condenser operates at a vacuum of 635 mm Hg and exhaust steam enters the condenser at 49°C. Calculate the vacuum efficiency of the condenser if the barometric pressure is 760 mm Hg.

Solution Vacuum efficiency of the surface condenser

$$= \frac{\text{Actual Vacuum}}{\text{Ideal Vacuum}} = \frac{\text{Actual Vacuum}}{P_{atm} - P_{sat}}$$

Now, actual vacuum = 63.5 cm Hg

P_{sat} at 49°C = 0.12 kgf/cm^2

$= 0.12/0.01359$ cm of Hg

$= 8.83$ cm Hg

Ideal Vacuum $= 76 - 8.83 = 67.17$ cm Hg

Vacuum Efficiency $= (63.5/67.17)(100)$

$= 94.53\%$

Ans.

Problem 30.2 The vacuum in a condenser is 685 mm Hg when the barometer is recording 765 mm Hg. Correct this reading to the standard barometer of 760 mm Hg.

Solution

Absolute pressure in the condenser = [Actual barometer reading] − [Actual Vacuum]

$= 765 - 685 = 80$ mm of Hg

Corrected Vacuum $= 760 - 80 = 680$ mm Hg

Q. *What is condenser efficiency?*

Ans. The widely accepted definition is

Condenser Efficiency

$$= \frac{\text{Rise of Temperature of Cooling Water}}{\left[\begin{array}{c}\text{Temp. corresponding} \\ \text{to the Vacuum in} \\ \text{the condenser}\end{array}\right] - \left[\begin{array}{c}\text{Inlet temperature} \\ \text{of Cooling Water}\end{array}\right]}$$

Note: There is actually no standard and universal definition of condenser efficiency.

Problem 30.3 A surface condenser operates under a vacuum of 680 mm Hg. The cooling water inlet and outlet temperatures being 33°C and 40°C respectively.

Determine the efficiency of the surface condenser if the barometer reading is 760 mm Hg.

Solution

Step (I) Absolute Pressure in the Condenser

Condenser Efficiency

$$= \frac{\Delta \theta \text{ of Cooling Water}}{\left[\begin{array}{c}\theta_{sat} \text{ corresponding} \\ \text{to the abs. press. in} \\ \text{the condenser}\end{array}\right] - \left[\theta_{in} \text{ of Cooling Water}\right]}$$

Now, absolute pressure in the condenser

$= 760 - 680 = 80$ mm Hg

$= 80 \, (1.0322/760)$ kgf/cm^2

$= 0.10865$ kgf/cm^2

Step (II) Saturation Temperature Corresponding to 0.10865 kgf/cm²

Saturation Pressure	Saturation Temperature
0.12 kgf/cm²	49°C
(–) 0.10 "	(–) 45.5°C
0.02 "	3.5°C

Difference 0.02 kgf/cm² correspond to sat. temp. difference 3.5°C

Difference 0.00865 kgf/cm² correspond to sat. temp 3.5 (0.00865)/0.02 = 1.5°C

Hence the saturation corresponding to 0.10865 kgf/cm²

$$= 45.5 + 1.5 = 47°C$$

Step (III) Condenser Efficiency

$$\eta_{cond} = \frac{40 - 33}{47 - 33} = 0.5 \text{ i.e. } 50\%$$

Ans.

Problem 30.4 The vacuum produced in a surface condenser is 685 mm of Hg when the barometer reading is 760 mm of Hg. The condensing steam temperature is 36°C. The net volume of steam space is 25 m³.

Determine:
(a) the partial pressure of air
(b) the mass of air present in the condenser.

Solution

Step (I) Partial Pressure of Air

$$P_{abs} = P_{sat} + P_{air}; \quad \therefore \quad P_{air} = P_{abs} - P_{sat}$$

Now, $P_{abs} = (760 - 685)$ mm of Hg

$$= 75 \text{ mm Hg} = 75 \ (0.001359) \text{ kgf/cm}^2$$

$$= 0.10192 \text{ kgf/cm}^2$$

Partial pressure of air $= 0.10192 - p_{sat}$ at 36°C

$$= 0.10192 - 0.06 = 0.04192 \text{ kgf/cm}^2$$

Ans.

Step (II) Mass of Air in the Condenser

Applying the characteristic gas equation

$$PV = mRT \text{ we get,}$$

$$(0.04192 \times 10^4) \ (25) = m(29.27) \ (273 + 36)$$

$$\therefore \ m = 1.1588 \text{ kg} \approx 1.16 \text{ kg}$$

Ans.

Problem 30.5 A certain surface condenser condenses 20 t/h of exhaust steam (49°C) from a turbo-alternator. The dryness fraction of the steam is 0.85.

Determine the flowrate of cooling water.
Cooling water inlet temperature = 33°C
Cooling water outlet temperature = 42°C
Condensate temperature at the exit = 49°C.

Solution

Step (I) The flowrate of cooling water can be computed from the heat balance equation of the surface condenser, where

$$\begin{array}{c} \text{Total heat rejected by} \\ \text{condensing steam} \end{array} = \begin{array}{c} \text{Net heat gained by} \\ \text{cooling water} \end{array}$$

Now, the latent heat of the condensing steam at 49°C = 568.6 kcal/kg

Since it is 0.85 dry so, the total heat rejected by condensing steam

$$= (20 \times 10^3/3600) \ (568.6 \times 0.85) \text{ (kg/s) (kcal/kg)}$$

If the flowrate of cooling water be Ṁ, then

$$\dot{M} \ (1)(42 - 33) = (20 \times 10^3/3600)(568.6 \times 0.85)$$
$$\text{(kcal/kg.°C) (°C)} \qquad \text{(kcal/s)}$$

$$\therefore \ \dot{M} = (20 \times 10^3/3600) \ (568.6 \times 0.85)/9$$

$$= 298.339 \text{ kg/s} = 1074 \text{ t/h}$$

Ans.

Problem 30.6 Designed to handle 15 t/h of steam, a surface condenser receives steam at 0.10 × 10⁴ kgf/m² and 0.95 dry. The cooling water inlet and outlet temperatures being 27°C and 36°C respectively.

Determine the flowrate of cooling water if the condensate leaves at saturation temperature.

Solution

Step (I) Heat Load of Condensing Steam

STEAM at 0.10×10^4 kgf/m²:

Total heat of the water, $H_w = 45.4$ kcal/kg

Latent heat, $L = 570.5$ kcal/kg

Total heat of the condensate, $H_{w_c} = 45.4$ kcal/kg

Heat extracted from condensing steam per unit time

$$= [15 \times 10^3/3600][H_w + xL - H_{w_c}](kg/s)(kcal/kg)$$

$$= (15 \times 10^3/3600)(0.95)(570.5) \ kcal/s$$

$$= 2258.229 \ kcal/s$$

The cooling water takes care of this rate of heat extraction.

Step (II) Flowrate of Cooling Water

If $\dot{M}_{cw}$ be the flowrate of cooling water and its specific heat = 1 kcal/kg. °C , then

$$\dot{M}_{cw} \ (1) \ (36 - 27) = 2258.229$$

$$\therefore \ \dot{M}_{cw} = 2258.229/9 = 250.914 \ kg/s = 903.29 \ t/h$$

Ans.

Problem 30.7 A surface condenser operates at a vacuum of 699.2 mm of Hg when the barometric head is 760 mm of Hg.

It receives steam at a rate 10 t/h which is 0.9 dry. Calculate the
(a) condenser efficiency
(b) cooling water flowrate.
Cooling water inlet temperature = 33°C
Cooling water outlet temperature = 40°C.
The condensate leaves the condenser at saturation temperature.

Solution
Step (I) Absolute Pressure in the Condenser

$$P_{abs} = 760 - 699.2 = 60.8 \ mm \ of \ Hg$$

$$= 60.8 \times 0.001359 \ kgf/cm^2$$

$$= 0.0826 \ kgf/cm^2$$

Step (II) Condenser Efficiency

Saturation temperature corresponding to 0.0826 kgf/cm^2 = 42°C

Cooling Water Temperature		Temperature Corresponding To Vacuum in the Condenser	Condenser Efficiency
Inlet	Outlet		(η_{cond})
33°C	40°C	42°C	$\dfrac{40 - 33}{42 - 33}(100)$ = 77.77% Ans.

Step (III) Cooling Water Flowrate

$$\boxed{\text{STEAM, at } 0.0826 \ kgf/cm^2} \to \text{Latent heat}$$

$$= 572.6 \ kcal/kg$$

Heat load of condensing steam

$$= \frac{10 \times 10^3}{3600} (H_w + xL - H_{w_c});$$

$$= \frac{10^2}{36} (xL) \ kcal/s$$

$$= (10^2/36) \ (0.9) \ (572.6)$$

$$= 1431.5 \ kcal/s$$

$H_w =$ total heat of water at 42°C

$L =$ latent heat of steam at 42°C

$H_{w_c} =$ total heat of condensate at 42°C

$$(cf. \ H_w = H_{w_c})$$

$x =$ dryness fraction of steam

Let the mass flowrate of cooling water be $\dot{M}_{cw}$

$$\therefore \ \dot{M}_{cw} \ (C_p)_{cw} \ (\Delta \ \theta)_{cw} = 1431.5 \ \ or,$$

$$\dot{M}_{cw} = 204.5 \ kg/s = 736.2 \ t/h$$

Ans.

Problem 30.8 A surface condenser condenses steam at the rate of 20 t/h which leaves the condenser at the corresponding saturation temperature. The condenser operates at a vacuum of 680 mm Hg. Determine:

(a) condenser efficiency
(b) flowrate of cooling water
(c) number of condenser tubes.
Given:

Cooling water inlet temperature = 33°C
Cooling water outlet temperature = 43°C
Mean velocity of cooling water = 2.5 m/s
Internal diameter of condenser tubes = 19.6 mm
Barometer pressure = 760 mm Hg
Steam is 0.95 dry.

Solution
Step (I)

Vacuum in the Condenser	Barometric Pressure	Absolute Pressure in the Condenser
680 mm Hg	760 mm Hg	760 – 680 = 80 mm Hg = 80 (0.001359) = 0.10872 kgf/cm^2

Step (II)

Dry Satd. Steam Pressure	Satd. Steam Temperature	Total Heat of the Condensate	Latent Heat
0.12 kgf/cm^2 (–) 0.10 kgf/cm^2	49°C (–) 45.4°C	49 kcal/kg (–) 45.4 kcal/kg	568.6 kcal/kg (–) 570.5 kcal/kg
0.02 kgf/cm^2	3.6°C	3.6 kcal/kg	(–) 1.9 kcal/kg
0.00872 kgf/cm^2	1.56°C	1.5 kcal/kg	(–) 0.8284 kcal/kg
0.10872 kgf/cm^2	45.4 + 1.56 = 46.96°C	46.96 kcal/kg	570.5 – 0.8284 = 569.671 kcal/kg

Step (III) Condenser Efficiency

$$\eta_{cond} = \frac{\text{Rise in Temperature of Cooling Water}}{\left[\begin{array}{c}\text{Temp. corresponding}\\ \text{to vacuum in}\\ \text{the condenser}\end{array}\right] - \left[\begin{array}{c}\text{Inlet temperature}\\ \text{of cooling water}\end{array}\right]}$$

$$= \frac{43 - 33}{46.96 - 33} = 0.716 \text{ i.e. } 71.6\%$$

Ans.

Step (IV) Cooling Water Flowrate

Heat rejected by condensing steam = Heat absorbed by cooling water

or, $(20 \times 10^3/3600)(0.95)(569.671)$

$= \dot{M}_{cw}(C_p)_{cw}(\Delta \theta)_{cw} = \dot{M}_{cw}(1)(43 - 33)$

$\therefore M_{cw} = 300.7$ kg/s

Ans.

$= 300.7/1000$

$= 0.3$ m^3/s

Ans.

Step (V) Number of Condenser Tubes

Volumetric flowrate through a single tube

$= (\pi/4) d^2 v$

where d = ID of condenser tubes

v = velocity of flow of C/water

n be the total number of condenser tubes, then

$(\pi/4) d^2 v n = 0.3$

or $(\pi/4)(0.0196)^2 (2.5)(n) = 0.3$

or, $n = 398$

Ans.

Problem 30.9 The following parameters were recorded for a surface condenser:

Condenser vacuum = 690 mm Hg
Barometric pressure = 760 mm Hg
Mean condenser temperature = 37°C
Condenser exit temperature = 30°C
Rise in temperature of cooling water = 10°C
Rate of steam condensation = 25 t/h
Rate of cooling water flow = 1200 t/h
Determine:
(a) the mass of air present per unit volume of the condenser
(b) the dryness fraction of steam at the inlet to condenser
(c) the vacuum efficiency of the condenser.

Solution
Step (I) Absolute Pressure in the Condenser

Condenser Vacuum	Barometric Pressure	Absolute Pressure in the Condenser
690 mm Hg	760 mm Hg	760 – 690 = 70 mm Hg = 70 (0.001359) = 0.09513 kgf/cm^2

Step (II) Mass of Air/M^3 of Condenser Volume

Partial Pressure of Steam at 37°C	Abs. Press. in the Condenser $P_{abs} = p_s + p_{air}$	Partial Pressure of Air in the Condenser
0.064 kgf/cm^2	0.09513 kgf/cm^2	0.09513 – 0.064 = 0.03113 kgf/cm^2

Applying the characteristic gas equation,

$PV = mRT$, we get

$$\frac{m}{V} = \frac{P}{RT} = \frac{0.03113 \times 10^4}{(29.27) \cdot (273 + 37)}$$

$$= 0.0343 \ \text{kg/m}^3$$

Ans.

Step (III) Dryness Fraction of Steam

| STEAM, 37°C | → Sensible heat, $H_w = 36.99 \ \text{kcal/kg}$

Latent heat, $L = 576.4 \ \text{kcal/kg}$

The dryness fraction of steam can be determined from the heat balance equation:

$$\dot{M}_s \, (H_w + xL - H_{cond}) = (\dot{M} \, C_p \, \Delta \, \theta)_{cw}$$

or $25 \, (36.99 + x \, 576.4 - 30) = (1200) \, (1) \, (10)$

$$\therefore x = 0.82$$

Ans.

Step (IV) Vacuum Efficiency

$$\eta_{cond} = \frac{\text{Condenser Vacuum}}{\text{Ideal Vacuum}}$$

$$= \frac{690}{760 - (0.064/0.001359)} \, (100) = 96.77\%$$

Ans.

Problem 30.10 A surface condenser operating at a mean temperature of 30°C and vacuum 700 mm Hg receives exhaust steam from a steam turbine at the rate of 10 t/h.

The air leakage is 0.35 kg per ton of steam admitted.

Barometric pressure is 760 mm of Hg.

Determine the capacity of a suitable dry air pump required for the condenser.

Take the volumetric efficiency of the pump as 85%.

Solution

Step (I) Absolute Pressure in the Condenser

Barometric Pressure	Condenser Vacuum	Absolute Press. in the Condenser
760 mm Hg	700 mm Hg	760 − 700
		= 60 mm Hg
		60 (0.001359)
		= 0.08154 kgf/cm^2

Step (II) Partial Pressure of Air

Partial Press. of Steam at 30°C	Abs. Press. in the Condenser	Partial Press. of Air in the Condenser $p_{air} = P_{abs} - p_s$
0.0432 kgf/cm^2	0.08154 kgf/cm^2	0.08154 − 0.0432 = 0.03834 kgf/cm^2

Step (III) Capacity of Air Pump

Leakage of air = 0.35 kg/t of steam admitted

Therefore, the rate of air leakage = 0.35 (10) = 3.5 kg/h

Applying the characteristic gas equation, $PV = mRT$ we get,

$$(0.03834 \times 10^4) \, (\dot{V}) = 3.5 (29.27) \, (273 + 30)$$

$$\therefore \dot{V} = 80.96 \ \text{m}^3/\text{h}$$

$$\therefore \ \text{Capacity of air pump} = 80.96/(85/100)$$

$$= 95.249 \ \text{m}^3/\text{h}$$

Ans.

Q. *What are the sources of air in the condenser?*

Ans. Air may find its way into the condenser through

(a) boiler feedwater as dissolved gases. From boiler it is carried off by steam and to the turbine and finally to the condenser

(b) flange leakage

(c) cooling water (for jet condenser) containing a certain amount of dissolved air in it.

Q. *What are the effects of air leakage in the condenser?*

Ans.

1. It increases the back pressure on the turbine with the effect that there is less heat drop and low thermal efficiency of the plant.

2. The pressure of air in the condenser lowers the partial pressure of steam which means steam will condense at a lower temperature and that will require greater amount of cooling water.

3. It reduces the rate of condensation of steam, because air having poor thermal conductivity impairs the overall heat transfer from the steam-air mixture.

Problem 30.11 A jet condenser is on duty to maintain a vacuum of 700 mm of Hg while condensing dry steam at a rate of 2.5 t/h.

Determine the mass flowrate of cooling water.

The inlet temperature of cooling water = 25°C

The cooling water plus condensate discharges at 40°C.

Solution

Step (I) Absolute Pressure in the Condenser

Vacuum in the Condenser	Barometer Pressure	Absolute Pressure in the Condenser
700 mm Hg	760 mm Hg	760 − 700 = 60 mm Hg = 60 × 0.001359 = 0.08154 kgf/cm²

Step (II) Steam at 0.08154 kgf/cm²

Steam Pressure (kgf/cm²)	Steam Temperature (°C)	Sensible Heat (kcal/kg)	Latent Heat (kcal/kg)
0.10	45.4	45.4	570.5
(−) 0.08	(−) 41.1	(−) 41.1	(−) 572.9
Diff. 0.02	4.3	4.3	(−) 2.4
0.00154	$\frac{4.3(0.00154)}{0.02}$	$\frac{4.3(0.00154)}{0.02}$	$\frac{(-)(2.4)(0.00154)}{0.02}$
	= 0.33°C	= 0.33	= (−) 0.1848
0.08154	41.43	41.43	572.71

Step (III) Heat Extracted for Condensing Steam

Heat released by condensing steam

$$= 2.5 \times 10^3 (572.71 + 41.43 - 40)$$

$$= 2.5 \times 10^3 \times 574.14 \text{ kcal/h}$$

Step (IV) Cooling Water Flowrate ($\dot{M}_{cw}$)

$$\dot{M}_{cw} (C_p)(\Delta \theta)_{cw} = 2.5 \times 10^3 \times 574.14$$

$$\dot{M}_{cw} = 2.5 \times 10^3 \times 574.14/[(1)(40-25)]$$

$$= 95690.33 \text{ kg/h}$$

$$= 95.690 \text{ t/h}$$

Ans.

Problem 30.12 A surface condenser receives steam (0.09 kgf/cm² and 0.9 dry) at a rate of 15 t/h. The temperature of the condensate is 37°C which is same for the air extraction pipe.

The amount of air leakage is 5.25 kg/h.

Determine the heat transfer surface required if the average heat transmission rate is 25000 kcal/m² h.

Also determine the cylinder diameter for the dry air pump, if it is single acting at 72 rpm with stroke/bore ratio 1.25 and volumetric efficiency 80%

Solution

Step (I) Rate of Heat Extraction

STEAM, 0.09 kgf/cm²	Sensible heat = 43.25 kcal/kg

Latent heat = 571.7 kcal/kg

Therefore, heat extraction rate

$$= 15 \times 10^3 (43.25 + 571.7 - 37) \text{ kcal/h}$$

$$= 8669.25 \times 10^3 \text{ kcal/h.}$$

Step (II) Heat Transfer Surface of Condenser

Average heat transmission rate = 25000 kcal/m².h

$$= 25 \times 10^3 \text{ kcal/m}^2 \text{ h}$$

Heat extraction rate = 8669.25 × 10³ kcal/h

Therefore, heat transfer surface of the condenser

$$= 8669.25 \times 10^3/(25 \times 10^3) \text{ m}^2 = 346.7 \text{ m}^2$$

Ans.

Step (III) Volume of Air Handled by the Pump

Partial Press. of Steam at 37°C	Condenser Pressure	Partial Pressure of Air
0.064 kgf/cm²	0.09 kgf/cm²	0.09 − 0.064 = 0.026 kgf/cm²

Applying the characteristic gas equation $PV = mRT$ we get,

$$(0.026 \times 10^4)(V) = (5.25/60)(29.27)(273 + 37);$$

or, $V = 3.053 \text{ m}^3/\text{min}$

Step (IV) Pump Cylinder Diameter

Volumetric Flowrate of Air	Volumetric Efficiency of the Pump	Capacity of the Pump
3.053 m³/min	80%	3.053/0.8 = 3.817 m³/min

$$\therefore \frac{\pi}{4} D^2 L N = 3.817$$

where D = cylinder dia

L = stroke length

N = rpm

or $(\pi/4)(D^2)(1.25D)(72) = 3.817$

$\therefore D = 0.3779$ m $= 37.79$ cm

Ans.

Problem 30.13 A surface condenser operates at a pressure of 0.14 kgf/cm² absolute. Dry, saturated steam is condensing at a rate 20 t/h. From the given data calculate the heat transfer area required for the surface condensation of steam and also determine the number of tubes required.

Cooling water inlet temperature = 33°C
Cooling water outlet temperature = 40°C
Condensate temperature at exit = 50°C
ID/OD of surface condenser tubes = 19/26 (mm)
Length of each tube = 3m
Number of passes = 2
Overall heat transfer coefficient for surface condenser, $U_o = 0.85$ kcal/s/m²/°C

Solution
Step (I) Steam Parameters

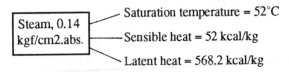

Steam, 0.14 kgf/cm2.abs.
— Saturation temperature = 52°C
— Sensible heat = 52 kcal/kg
— Latent heat = 568.2 kcal/kg

Step (II) Nature of cooling
Steam condenses at 52°C
Condensate leaves at 50°C.
Therefore, there is undercooling.

Step (III) Heat Load of Condenser

$q = \dot{M}_s \, (H_w + L - H_{cond})$

$= (20 \times 10^3/3600)\,(52 + 568.2 - 50)$ kcal/s

$= 3167.78$ kcal/s

Step (IV) Log-Mean – Temperature – Difference of Surface Condenser

	Inlet	Outlet
Steam	52°C	52°C
C/W	33°C	40°C
Diff.	19°C	12°C

$\therefore \Delta\Theta_{lm} = \dfrac{19-12}{\ln\left(\dfrac{19}{12}\right)} = 15.23°C$

Step (V) Heat Transfer Surface
Working Formula: $q = U_o \, A_o \, \Delta\theta_{lm}$

where q = rate of heat transfer, kcal/s

U_o = overall heat transfer coefficient, kcal/s/m²/°C

A_0 = external surface area of all tubes, m²

Computing the known values,

$3167.78 = 0.85\,(A_o)\,(15.23);$

$\therefore A_o = 245$ m²

Ans.

Step (VI) Number of Condenser Tubes
If n be the number of tubes required per pass, then

$2\,(n)\,(\pi D_o L) = A_o$ where D_o = OD of tubes

or $2\,(n)\,(\pi)\,(0.026)\,(3) = 245$

$\therefore n = 500$

L = length of single tube

Factor (2) for double pass.

Hence the total number of tubes required = 1000

Ans.

31

Steam Turbines

Q. *What is the principle of steam turbine?*

Ans. If high-velocity steam is allowed to blow on to a curved blade, the steam will suffer a change in direction as it passes across the blade, and leaves it as shown. (Fig. 31.1)

As a result of its change in direction across the blade, the steam will impart a force to the blade. This force will act in the direction shown. (Fig. 31.1)

Now if the blade were free, it would move in the direction of force as depicted. If, therefore, a number of blades were fixed on the circumference of a disc which is free to rotate on a shaft, then steam blown across the blades in the way described, would cause the disc to rotate. This is the working principle of a steam turbine.

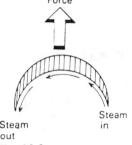

Fig. 31.1

Q. *What is the definition of a steam turbine?*

Ans. A steam turbine is a prime mover that derives its energy of rotation due to conversion of the heat energy of steam into kinetic energy as it expands through a series of nozzles mounted on the casing or produced by the fixed blades.

Neilson defines it: The turbine is a machine in which a rotary motion is obtained by the gradual change of the momentum of the fluid.

Graham's definition: The turbine is a prime mover in which a rotary motion is obtained by the centrifugal force brought into action by changing the direction of a jet of a fluid (steam) escaping from the nozzle at high velocity.

Q. *What is the essential distinguishing feature between a steam turbine and reciprocating steam engine?*

Ans. In a steam turbine, the heat energy of steam is converted into kinetic energy by allowing it to expand through a series of nozzles and this kinetic energy of steam is then imparted to the turbine blades mounted on a shaft free to rotate to drive this prime mover.

In a reciprocating steam engine, the pressure energy of steam is directly utilized to overcome the external resistance. Here, the utilization of the K.E. of input steam is negligibly small.

Q. *What are the advantages of steam turbines over reciprocating steam engines?*

Ans.

1. Steam turbine has higher thermal efficiency than reciprocating steam engines
2. The brake horse power of steam turbines can range from a few HP to several hundred thousand HP in single units. Hence they are quite suitable for large thermal power stations
3. Unlike reciprocating engines, the turbines do not need any flywheel as the power delivered by the turbine is uniform
4. Steam turbines are perfectly balanced and hence present minimum vibrational problem
5. High rpm 18000–24000 can be developed in steam turbines but such a high speed generation is not possible in the case of reciprocating steam engines
6. Some amount of input energy of steam is lost as the reciprocating motion of the piston is converted to circular motion

This loss does not arise in the case of steam turbines.

7. Unlike reciprocating steam engines, no internal lubrication is required for steam turbines due to the absence of rubbing parts.
8. Steam turbines, if well designed and properly maintained, are more reliable and durable prime-movers than steam engines.

Q. *How can steam turbines be classified?*

Ans. They can be classified with respect to:
1. The action of steam:
   (a) **Impulse** (b) **Reaction** (c) **Impulse and reaction combined**
2. the number of step reductions involved:
   (a) **Single stage** (b) **Multi-stage**
3. whether there is one or more revolving vanes separated by stationary reversing vanes.
4. the direction of steam flow:
   (a) **Axial** (b) **Radial** (c) **Mixed** (d) **Tangential** (e) **Helical** (f) **Reentry**
5. the inlet steam pressure:
   (a) **High pressure** (b) **Medium pressure** (c) **Low pressure**
6. the final pressure:
   (a) **Condensing** (b) **Non-condensing**
7. the source of steam:
   (a) **Extraction** (b) **Accumulator**

Q. *What are the differences between impulse and reaction turbines?*

Ans. The impulse turbine is characterized by the fact that it requires nozzles and that the pressure drop of steam takes place in the nozzles (Fig. 31.2)

The reaction turbine, unlike the impulse turbines has no nozzles, as such. It consists of a row of blades mounted on a drum. The drum blades are separated

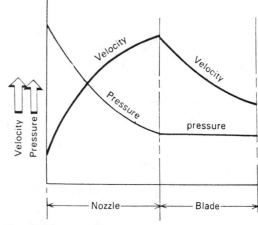

Fig. 31.2

by rows of fixed blades mounted in the turbine casing. These fixed blades serve as nozzles as well as the means of correcting the direction of steam onto the moving blades (Fig. 31.3)

In the case of reaction turbines, the pressure drop of steam takes place over the blades. This pressure drop produces a reaction and hence cause the motion of the rotor (Fig 31.4)

Q. *Why are simple impulse turbines not so common?*

Ans.
1. Since the whole pressure drop from boiler to condenser pressure takes place in a single row of nozzles, the velocity of the steam entering the turbine is very high. If some of this velocity is used up in a single row of turbine blading, as in the de Laval turbine, the speed of the rotation of the wheel will be too high to

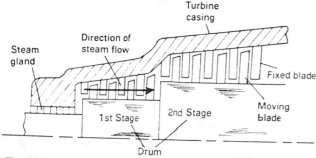

Fig. 31.3

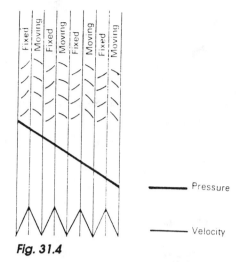

Fig. 31.4

fixed blades are mounted on the casing while the moving blades are keyed in series on a common shaft. The function of the fixed blades is to correct the direction of entry of steam to the next row of moving blades.

The high velocity steam leaving the nozzles passes on to the 1st row of moving blades where it suffers a partial velocity drop. (Fig. 31.5)

Its direction is then corrected by the next row of fixed blades and then it enters the 2nd row of moving blades. Here the steam velocity is again partially reduced. Since only part of the velocity of the steam is used up in each row of the moving blades, a slower turbine results. This is how velocity compounding works.

Q. *In which turbine is velocity compounding utilized?*

Ans. Curtis turbine.

Q. *How is pressure compounding accomplished?*

Ans. This is accomplished by an arrangement with alternate rows of nozzles and moving blades.

Steam enters the 1st row of nozzles where it suffers a partial drop of pressure and in lieu of that its velocity gets increased. The high velocity steam passes on to the 1st row of moving blades where its velocity is reduced. (Fig. 31.6)

The steam then passes into the 2nd row of nozzles where its pressure is again partially reduced and velocity is again increased. This high velocity steam

be useful for practical purposes, as there is the danger of structural failures due to excessive centrifugal stresses.

2. Steam exits from the turbine with a sufficiently high velocity, meaning a considerable loss of kinetic energy.

Q. *How can these disadvantages be overcome?*

Ans. By compounding:
 1. Velocity compounding
 2. Pressure compounding
 3. Pressure-Velocity compounding.

Q. *How is velocity compounding accomplished?*

Ans. This is accomplished by an arrangement with alternate rows of fixed blades and moving blades. The

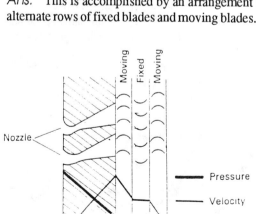

Fig. 31.5

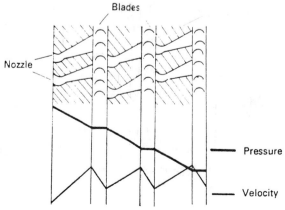

Fig. 31.6

passes from the nozzles to the 2nd row of blades where its velocity is again reduced.

Thus pressure drop takes place in successive stages. Since a partial pressure drop takes place in each stage, the steam velocities will not be so high with the effect that the turbine will run slower.

Q. *In which turbine is this pressure compounding used?*

Ans. Rateau and Zoelly turbines.

Q. *How is pressure-velocity compounding accomplished?*

Ans. It is a combination of pressure compounding and velocity compounding.

Steam is expanded partially in a row of nozzles whereupon its velocity gets increased. This high velocity steam then enters a few rows of velocity-compounding whereupon its velocity gets successively reduced. (Fig. 31.7)

The velocity of the steam is again increased in the subsequent row of nozzles and then again it is allowed to pass onto another set of velocity compounding that brings about a stagewise reduction of velocity of the steam.

This system is continued.

Q. *In which turbines, is this pressure-velocity compounding principle employed?*

Ans. Moore turbines, Curtis turbines.

Q. *What are the advantages of velocity compounding?*

Ans.
1. The velocity compounding system is easy to operate and operation is more reliable.
2. Only two or three stages are required. Therefore, first cost is less.
3. Since the total pressure drop takes place only in nozzles and not in the blades, the turbine casing need not be heavily built. Hence the economy in material and money.
4. Less floor space is required.

Q. *What are the disadvantages of velocity compounding?*

Ans.
1. Steam velocity is too high and that is responsible for appreciable friction losses.
2. Blade efficiency decreases with the increase of the number of stages. (Fig. 31.8)
3. With the increase of the number of rows, the power developed in successive rows of blades decreases. Forasmuch as, the same space and material are required for each stage, it means,

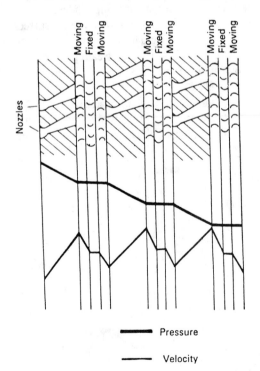

Fig. 31.7

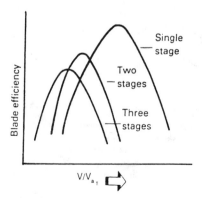

Fig. 31.8

therefore, that all stages are not economically efficient.

Q. *Where are velocity compounded steam turbines mostly employed?*

Ans. They are chiefly used as the prime mover for:
- (a) centrifugal pumps
- (b) centrifugal compressors
- (c) low capacity turbo-generators
- (d) feed pumps of high capacity power plants.

Q. *What are the losses in steam turbines?*

Ans.

1. **Residual Velocity Loss:** This is equal to the $V_o^2/2gJ$ kcal/kg where V_o is the absolute velocity of the steam at the blade exit.
2. **Loss due to Friction:** Friction loss occurs in the nozzles, turbine blades and between the steam and rotating discs.
   This loss is about 10%.
3. **Leakage Loss**
4. **Loss due to Mechanical Friction:** Accounts for the loss due to friction between the shaft and bearing.
5. **Radiation Loss:** Though this loss is negligible, as turbine casings are insulated, it occurs due to heat leakage from turbine to ambient air which is at a much lower temperature than the turbine.
6. **Loss due to Moisture:** In the lower stages of the turbine, the steam may become wet as the velocity of water particles is lower than that of steam. So a part of the kinetic energy of steam is lost to drag the water particles along with it.

Q. *Describe the velocity diagram of an impulse turbine.*

Ans. Refer to Fig. 31.9, V_1 = absolute velocity of steam entering the moving blade, m/s.

V_2 = absolute velocity of steam at outlet from the blade, m/s.

V_b = linear velocity of moving blade, m/s

V_{r_1} = relative velocity of steam to moving blade, at inlet, m/s.

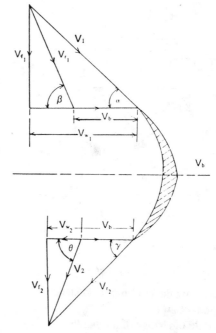

Fig. 31.9

V_{r_2} = relative velocity of steam to moving blade at outlet, m/s.

V_{f_1} = velocity of flow at inlet, m/s. It is the axial component of V_1.

V_{f_2} = velocity of flow at outlet, m/s. It is the axial component of V_2

V_{w_1} = velocity of whirl at inlet, m/s. It is the horizontal component of V_1

V_{w_2} = velocity of whirl at outlet, m/s. It is the horizontal component of V_2

α = nozzle angle
β = blade angle at inlet
γ = blade angle at outlet
θ = angle of discharge of the exit steam with the tangent to the wheel.

Since both the inlet and the exit velocity triangles have the common vector, V_b, these two can be combined into a single diagram, for the sake of convenience in solving the problems. (Fig. 31.10)

Assuming there is no frictional loss at the blades,

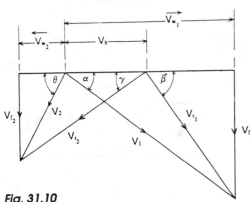

Fig. 31.10

$$V_{r_1} = V_{r_2}$$

Q. *What is the work done on the blades per second?*

Ans. Now, work done = Force × Distance

By Newton's Laws,

Force = Rate of change of momentum

$$= \text{(Mass)} \times \text{(Change of Velocity)}.$$

The change in velocity of the whirl produces the force on the blades. Therefore, change in the velocity of whirl $= -V_{w_2} - V_{w_1}$

i.e, change in the velocity of whirl

$$= -(V_{w_1} + V_{w_2})$$

Workdone on the blade/s

$$= \frac{W}{g}(V_{w_1} + V_{w_2})(V_b)\ \text{kgf}\cdot\text{m/s}$$

The negative sign shows that this force acts in a direction opposite to the rotation.

Q. *What is the rim horse power of the turbine wheel?*

Ans. It is the horse power developed by the wheel

$$= \frac{W}{g}(V_{w_1} + V_{w_2})(V_b/75)$$

$$(\text{1HP} = 75\ \text{kgf.m/s})$$

Q. *What is the blade efficiency?*

Ans. Also called diagram efficiency, the blade efficiency is the ratio of the work done on the blades to the energy supplied to the blades.

$$\eta_b = \frac{\text{Work done on the blade/kg of steam}}{\text{Energy supplied to the blade/kg of steam}}$$

$$= \frac{(V_{w_1} + V_{w_2})}{\dfrac{V_1^2}{2}}(V_b) = 2V_b\frac{(V_{w_1} + V_{w_2})}{V_1^2}$$

Q. *What is the stage efficiency?*

Ans. If ΔH be the total heat drop in the nozzle ring, then the total energy supplied per stage is $J\,\Delta H$ kg · m/kg of steam.

$$\therefore \quad \text{Stage Efficiency} = \frac{V_b\,(V_{w_1} + V_{w_2})}{W\,J\,\Delta H}\left(\frac{W}{g}\right)$$

$$= \left(\frac{V_b}{g}\right)\frac{(V_{w_1} + V_{w_2})}{J\,\Delta H}$$

Problem 31.1 Steam issues from nozzles, inclined at an angle 24° to the direction of motion, with a velocity 1000 m/s on to the blades of a single stage impulse turbine.

The linear velocity of the blades = 400 m/s

Blade angles at inlet = Blade angles at outlet.

Steam enters the blade without shock.

Frictional loss over the blades is nil.

Steam flowrate = 4000 kg/h

Calculate:

(a) inlet angle of the blades

(b) force exerted on the blades in the direction of their motion.

(c) power developed by the rotor.

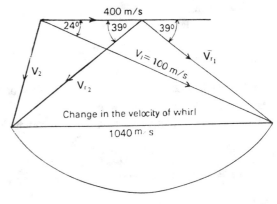

Fig. 31.11

Solution From the combined velocity diagram, inlet angle of the blades = 39°. (Fig. 31.11)

Force exerted on the blade

= [Mass flowrate of steam] × [Change in the velocity of whirl]

= (4000/3600) (1040)

= 1.155 kN

Power developed

= Force × Distance/time

= 1.155 (V_b) = 1.155 (400) = 462 kW

Ans.

Problem 31.2 Steam with a velocity of 600 m/s is directed upon the blades of a single stage impulse turbine at an angle 22° to the direction of motion. The average diameter of the blade ring is 1250 mm. The blade angles at inlet and exit are the same and equal to 37°.

Determine:
(a) the speed of the turbine
(b) the absolute velocity of steam leaving the blades
(c) torque on the turbine rotor when the steam flowrate is 5 t/h

Base your calculation on shockless entry of steam and its frictionless flow over the blades.

Solution
Step (I) Velocity Diagram

Horizontal *AE* is drawn. Drawn subtended 22 to *AE*, *AC* represents 600 m/s in the scale 1 cm ≡ 100 m/s. (Fig. 31.12)

Since it is known that inlet angle ($\angle EBC$) will be 37°, so $\angle ABC$

= 180° − 37° = 143°.

∴ $\angle ACB$ = 180° − (22° + 143°) = 15°.

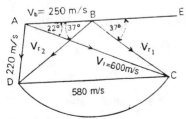

Fig. 31.12

Hence from *C*, *CB* is drawn making 15° angle with *AC*.

Step (II) Turbine Speed

From the velocity diagram,

AB = blade speed = V_b = 250 m/s

∴ πdN = 250 m/s; or $N = \dfrac{250}{(\pi)1.25}$

= 63.6619 rps

≈ 3820 rpm

Ans.

where *N* = turbine speed.

Step (III) Absolute Velocity of Steam at Outlet

Now, absolute velocity of steam leaving the blades, V_2 is represented by *AD*.

V_2 = 200 m/s from the velocity diagram

Ans.

Step (IV) Torque

Force exerted on the turbine rotor
= Steam flowrate × Change in the velocity of whirl

= (5000/3600)/(580)

= 805.555 N

Therefore, torque on the turbine rotor

= $F \cdot (d/2)$ = 805.555 (1.25)/2 = 503.47 Nm

Ans.

Problem 31.3 The nozzle angle for a simple impulse turbine is 20° and the steam leaves the nozzle at 400 m/s. The blade is 180 m/s. What should be the inlet and outlet angles for the blades so that the blades experience no axial thrust?

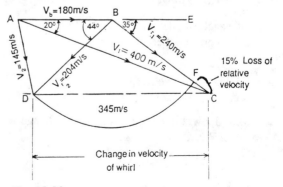

Fig. 31.13

Due to friction the velocity of the steam as it passes over the blades is reduced by 15%.

Also determine the power developed if the steam flow is 10 kg/s and the kinetic energy of the steam finally leaving the turbine.

Solution

Step (I) Velocity Diagram

AB is drawn to scale (1 cm ≡ 50 m/s) to represent the blade speed V_b = 180 m/s. (Fig. 31.13)

From A, and at 20° to AB, a line AC, to the same scale is laid off such that AC represents the absolute velocity of the steam as it leaves the nozzle.

$$V_1 = AC = 400 \text{ m/s}$$

BC is joined to complete the triangle ABC. Now BC represents the relative velocity of steam at the inlet to the blade.

From the diagram, $BC = V_{r_1} = 240$ m/s.

Now, in its passages over the blade, 15% of this velocity is lost. Length FC corresponds to 15% of BC.

[15% BC = (15/100) (240) = 36 m/s]

Taking B as centre, arc FD is drawn—its radius being BF. Now in order that there will be no axial thrust on the blades, there must be no changes in the velocity of flow, i.e., $V_{f_1} = V_{f_2}$.

This means that peaks of both the inlet and outlet velocity triangles must be at the same level. Hence a horizontal is drawn from C and it intersects the arc at D. AD and BD are joined.

Therefore, AD = velocity of flow at the outlet = 145 m/s from the figure.

Step (II) Blade Angles

.Blade angle at inlet, $\beta = \angle EBC = 35°$
Blade angle at outlet, $\gamma = \angle ABD = 44°$ }

Ans.

Step (III) Power Developed

Power developed for steam flowrate 10 kg/s

= Mass Flowrate × Blade Velocity × Change in Velocity of Whirl

$$= 10 \ (180) \ (345) \ \frac{\text{kg}}{\text{s}} \ (\text{m/s}) \ (\text{m/s})$$

$$= 621000 \ (\text{kg m/s}^2) \ \text{m/s} = 621 \times 10^3 \ \text{N m/s}$$

$$= 621 \times 10^3 \ \text{W} = 621 \ \text{kW}$$

Ans.

Step (IV) K.E. of Steam

The kinetic energy of steam finally leaving the wheel

$$= \frac{1}{2} \ (\text{Mass Flowrate}) \ (\text{Relative Velocity of steam at outlet})^2$$

$$= \frac{1}{2} \ (10) \ (204)^2 \ \text{J/s} = 208 \ \text{kJ/s}$$

Ans.

Problem 31.4 The rotor of a single stage impulse turbine is 500 mm in dia. It runs at 150 rps. Nozzles are at 20° to the plane of the wheel. Steam leaves the nozzles at 750 m/s. If the friction factor is 0.8 determine:

(a) the horse power developed per kg steam per second

(b) the blade efficiency.

Take blade outlet angle = 30°

Solution

Step (I) Blade Velocity

Blade speed, N = 150 rps.

Blade velocity, $V_b = 2 \pi Nr$

$$= 2 \ (\pi) \ (150) \left(\frac{500}{1000} \ \frac{1}{2} \right) = 235.6 \ \text{m/s}$$

Step (II) Velocity Diagram

$\alpha = 20°$; $\gamma = 30°$; $V_1 = 750$ m/s. (Fig. 31.14)

From the horizontal EF, AB is cut off so that

$AB = V_b = 235.6$ m/s.

With the known parameters, ABC is completed.

From the diagram:

$BC = V_{r_1}$ = relative velocity of steam to moving blade at inlet

= 540 m/s (from the diagram)

$$V_{r_2} = 0.8 \ (V_{r_1}) = 0.8 \ (540) \ \text{m/s}$$

= 432 m/s

From point B, BD is drawn making an angle 30° to AB. BD is cut-off so that BD = 432 m/s. AD is joined.

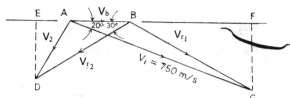

Fig. 31.14

Now AD represents the absolute velocity of steam at the blade outlet.

From the diagram, $AD = 260$ m/s.

From points C and D two perpendiculars CF and DE are dropped on EF.

EF corresponds to a change in the velocity of whirl. From the diagram, $EF = 834$ m/s.

Step (III) Power Developed

Therefore, the horse power developed by the wheel per kg

$$= \frac{1}{9.81} (834) (235.6)/75 = 267$$

Ans.

Step (IV) Blade Efficiency

Blade efficiency $= 2 V_b [V_{w_1} + V_{w_2}]/V_1^2$

$= 2(235.6) (834)/(750)^2$

$= 0.6986$ i.e. 69.86%

Ans.

Problem 31.5 A De Laval type steam turbine receives dry steam at 16 kgf/cm².abs. and discharges at 1.8 kgf/cm². abs. The mean diameter of the blade is 1500 mm and the rotational speed of the blade is 4074 rpm. If 10% of the available enthalpy drop is lost to friction in the nozzle, calculate:

(a) the absolute velocity of steam at outlet from the blade
(b) the diagram efficiency.

Take nozzle angle = 22°
blade friction factor = 0.8

Solution

Step (I) Dryness Fraction of Exhaust Steam

Since the expansion of steam in the nozzle is isentropic (cf. the process is adiabatic), so

Entropy of steam before expansion = Entropy of steam after expansion

or, $1.535 = 0.355 + x (1.712 - 0.355)$

$x = 0.869$.

Therefore, steam is 0.869 dry at the turbine outlet.

Step (II) Adiabatic Enthalpy Drop

Initial Enthalpy, kcal/kg H_1	Final Enthalpy, kcal/kg H_2	Enthalpy Drop, kcal/kg $\Delta H = H_1 - H_2$
$204 + 463$ -667	$116.6 + 0.869 (528.5)$ $= 575.86$	$667 - 575.86$ -91.14

Step (III) Velocity of Steam at Inlet

10% of this enthalpy drop is lost to friction, so the rest 90% goes to increase the kinetic energy of steam. Therefore, the absolute velocity of steam at the inlet to the blade

$$V_1 = \sqrt{2 g J (0.9) (\Delta H)}$$

(cf. $J = 427$ kgf.m/kcal)

$= \sqrt{2 (9.81) (427) (0.9) (91.14)} = 828.97$ m/s

≈ 829 m/s

Step (IV) Velocity Diagram

Blade rpm = 4074
Blade velocity, $V_b = (2 \pi) (1.5/2) (4074/60)$ m/s
$= 320$ m/s (cf. $V_b = 2 \pi r N$)

From the diagram, $BC = 550$ m/s $= V_{r_1}$ [Fig. 31.15]

$\angle CBE = 42°$ = blade angle at inlet.

As the blade is symmetrical, the blade angle at the outlet will be also 42°.

So $\angle ABD$ is drawn equal to 42°.

D is such a point on BD as that

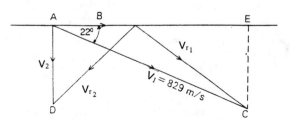

Fig. 31.15

$BD = V_{r_2} = 0.8V_{r_1} = 0.8(550) = 440$ m/s

AD is joined to complete the outlet triangle.

Step (V) Steam Velocity and Diagram Efficiency

From the diagram, the absolute velocity of steam at outlet

$V_2 = AD = 330$ m/s

Diagram efficiency $= 2V_b (V_{w_1} + V_{w_2})/V_1^2$

From the diagram,

$V_{w_1} + V_{w_2} = AE = 770$ m/s

Diagram efficiency $= 2(320) (770)/(829)^2 = 0.717$ i.e. 72%

Ans.

Problem 31.6 Steam at 10 kgf/cm². abs. and dry is supplied to the nozzles of a simple impulse turbine at a rate 10 t/h. As the steam passes over the blades it suffers a velocity drop of 15% due to friction. The blade speed is 5560 rpm. Blade dia = 1030 mm.

Determine the
- (a) net horse power developed at the shaft coupling
- (b) steam condition when it
    - (i) leaves the nozzle
    - (ii) leaves the blade
    - (iii) flows through the exhaust pipe.

Given Exhaust steam pressure = 0.8 kgf/cm² abs.
Nozzle angle = 22°
Blade angle at outlet = 30°
Different losses sum up to 2.5 h.p.
Nozzle efficiency = 0.82

Solution

Step (I) Steam Condition After Expansion

Since the expansion of the steam in the nozzle is adiabatic, i.e., isentropic, so

Entropy of steam before expansion = Entropy of steam after expansion

or $1.575 = 0.293 + x(1.777 - 0.293)$

$\therefore x = 0.863$

Step (II) Enthalpy Drop

Enthalpy of steam (10 kgf/cm² abs.) at nozzle inlet = 663.3 kcal/kg

Enthalpy of steam (0.8 kgf/cm² abs.) at nozzle outlet

= 93 + 0.863 (543.4) = 561.95 kcal/kg

Therefore, enthalpy drop of steam in the nozzle

= 663.3 − 561.95 = 101.35 kcal/kg

Useful enthalpy drop = 0.82 (101.35) = 83.107 kcal/kg

Step (III) Steam Velocity at Nozzle Exit

Steam velocity at the exit of the nozzle

$= \sqrt{2 g J \Delta H}$

$= \sqrt{2 (9.81) (427) (83.107)}$

= 834 m/s

(cf. $J = 427$ kgf.m/kcal)

Step (IV) Velocity Diagram

Blade speed,

$V_b = 2 \pi r N = 2(\pi) (0.1030/2) (5560) = 300$ m/s

Taking 1cm ≡ 100 m/s, AB is drawn to represent V_b equal to 300 m/s. (Fig. 31.16)

BAC is drawn equal to 22°.

In the same scale AC is drawn equal to 834 m/s, i.e.

$AC = V_1 = 834$ m/s = velocity of steam inlet to blade.

BC is joined. From the diagram,

$BC = V_{r_1} = 570$ m/s.

Now the relative velocity of steam at the exit

$V_{r_2} = V_{r_1} (1 - 15/100) = 570 (0.85) = 484.5$ m/s.

$\angle ABD$ is drawn equal to 30°. Point D is such that $BD = 484.5$ m/s.

AD is joined.

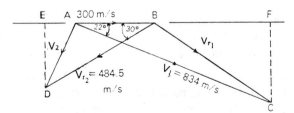

Fig. 31.16

$AD = 272$ m/s (from the diagram)

From points, D and C, perpendiculars DE and CF are dropped on AB extended.

$EF = \Delta$ Velocity of whirl $= 890$ m/s (from the diagram)

Step (V) Blade H.P.

Blade h.p.

$= \dfrac{W}{g}$ [Δ Velocity of Whirl] [Blade Velocity]/75

$= \dfrac{(10 \times 1000/3600)\,(890)\,(300)/75}{9.81} = 1008$

Allowing different losses, we get net h.p. at the shaft coupling

$= 1008 - 2.5 = 1005.5$

Ans.

Step (VI) Dryness Fraction of Steam at Nozzle Exit

Let the dryness fraction of steam at nozzle exit be x_e.

Useful enthalpy drop of steam $= 83.107$ kcal/kg
Initial enthalpy of steam $= 663.3$ kcal/kg
Final enthalpy of steam $= 663.3 - 83.107$

$= 580.193$ kcal/kg

$\therefore 580.193 = 93 + x_e\,(543.4);$ or $x_e = 0.896$

Ans.

Step (VII) Dryness Fraction of Steam at Blade Outlet

Energy available at the shaft coupling

$= 1005.5\,(75)/427 = 176.61$ kcal/s

Steam flowrate $= 10(1000)/3600$ kg/s $= 2.778$ kg/s
Energy drop (available for workdone) per kg of steam

$= 176.61/2.778 = 63.574$ kcal.

Now, from the diagram, the velocity of steam at blade exit is

$AD = V_2 = 272$ m/s.

Therefore, kinetic energy of steam

$= \dfrac{1}{2}\,(W/g)\,(V_2)^2$ kgf.m

$= \dfrac{1}{2}\,(W/g)\,(V_2)^2/J$ kcal

$\therefore$ K.E./kg of steam $= V_2^2/(2gJ)$

$= (272)^2/(2 \times 9.81 \times 427) = 8.831$ kcal

Since the loss of K.E. due to friction reappears as heat, so enthalpy of steam at the blade exit (at 0.8 kgf/cm². abs.)

$H_{exit} = H_{in} -$ Energy Drop for W/Done $-$ K.E. loss due to friction

$= 663.3 - 63.574 - 8.831 = 590.895$ kcal/kg

If x_o be the dryness fraction of steam at the blade exit, then

$H_{exit} = 93 + x_o\,(543.4)$

or $590.895 = 93 + x_o\,(543.4);$ or, $x_o = 0.916$

Ans.

Step (VIII) Steam Condition at the Exhaust Pipe

Assuming the entire K.E. of steam is converted into heat in the exhaust pipe, then enthalpy of steam at exhaust pipe

$H_{exh} = h_{in} -$ Energy Drop for W/done

$= 663.3 - 63.574 = 599.726$ kcal/kg.

If x_p be the dryness fraction of steam at the exhaust pipe then,

$93 + x_p\,(543.4) = 599.726;$ $\therefore x_p = 0.932$

Ans.

Problem 31.7 The following data are available for a single stage impulse turbine.

Steam pressure inlet to nozzles $= 15$ kgf/cm². abs.
Steam pressure at nozzle outlet $= 1.5$ kgf/cm².abs.
Steam condition at entry to the nozzles $= 0.98$ dry
Velocity carryover from previous stage $= 120$ m/s
Energy lost to friction $= 5\%$ of the theoretical available energy
Nozzle angle $= 20°$
Blade angle at outlet $= 27°$
Mean blade speed $= 40\%$ of velocity of steam at the nozzle exit
Turbine output $= 495$ h.p.
Steam flowrate $= 5$ t/h.
Determine
(a) the blade velocity coefficient
(b) the energy lost in the blade per second

(c) the diagram efficiency
(d) the overall stage efficiency.

Solution

Step (I) Steam Condition at Nozzle Outlet

Due to adiabatic (isentropic) expansion in the nozzles,

Entropy of steam before expansion = Entropy of steam after expansion

or, $0.551 + (1.541 - 0.551)(0.98)$

$= 0.341 + (1.727 - 0.341)(x)$

$\therefore x = 0.85$

Step (II) Enthalpy Drop

Initial enthalpy of steam (0.98 dry, 15 kgf/cm². abs.)

$= 200.7 + (466)(0.98) = 657.38 \text{ kcal/kg}$

Final enthalpy of steam (0.85 dry, 1.5 kgf/cm². abs.)

$= 111 + 532 (0.85) = 563.20 \text{ kcal/kg}$

Therefore, adiabatic enthalpy drop in the nozzle

$= 657.38 - 563.20 = 94.18 \text{ kcal/kg}$

Step (III) Useful Energy at the Nozzle Exit

Theoretical enthalpy drop of steam in the nozzle = 94.18 kcal/kg

$$\text{K.E. of steam at nozzle inlet} = \frac{1}{2}\frac{(120)^2}{9.81(427)} = 1.718$$

kcal

$$[\text{cf. K.E.} = \frac{1}{2}\frac{W}{g}\frac{V^2}{J} \text{ kcal}]$$

Therefore, theoretical energy available at the nozzle exit = 94.18 + 1.718 = 95.898 kcal/kg

Energy lost to friction in the nozzle = 5% of the theoretical enthalpy drop in the nozzle.

Therefore, useful kinetic energy available = 0.95 (95.898) kcal/kg steam = 91.10 kcal/kg steam

Step (IV) Steam Velocity at Nozzle Exit

This is also the absolute velocity of steam at the blade inlet, i.e.

$$V_1 = \sqrt{2g\,J\,\Delta H} = \sqrt{2\,(9.81)\,(427)\,(91.10)}$$

$= 873.6 \text{ m/s}$

Step (V) Mean Blade Velocity

$V_b = 40\%$ of $V_1 = 0.4\,(873.6) = 349.44 \text{ m/s}.$

Step (VI) Change in Velocity of Whirl

H.P.

$= (W/g)$ (Change in velocity of whirl) $(V_b/75)$

or, $495 = \dfrac{(5 \times 10^3/3600)}{9.81}$ (Change in velocity of whirl) (349.44/75)

Change in the velocity of whirl = 750.40 m/s

Step (VII) Velocity Diagram

AB is drawn to represent blade velocity, V_b = 349.44 m/s in suitable scale 1 cm ≡ 100 m/s.

From points A and B, the angles BAC and ABD are drawn equal to 20° (nozzle angle) and 27° (blade angle at outlet) respectively. (Fig. 31.17)

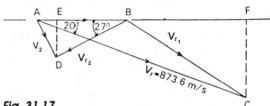

Fig. 31.17

AC is cut off to represent absolute velocity of steam at turbine inlet

$V_1 = 873.6 \text{ m/s}$ in the same scale.

CF is the perpendicular dropped on AB extended.

EF is measured out equal to 750.40 m/s to represent the change in the velocity of whirl.

At point E, $DE \perp AB$ is drawn. It meets BD at D.

Then BD = relative velocity of steam to moving blade at outlet. From the diagram: $BC = V_{r_1} = 557 \text{ m/s}$

$BD = V_{r_2} = 309 \text{ m/s}$

$AD = V_2 = 158 \text{ m/s}$

Step (VIII) Blade Velocity Coefficient

Blade velocity coefficient

= Ratio of relative velocities of steam relative to blade at exit and at inlet

$= 309/557 = 0.5547$

Ans.

Step (IX) Energy Lost

The energy lost in the blades per second

$$= \frac{1}{2} \frac{W}{g} \left(V_{r_1}^2 - V_{r_2}^2 \right)$$

$$= \frac{1}{2} \frac{(5000/3600)}{9.81} \left[(557)^2 - (309)^2 \right] \text{ kgf m}$$

$$= 15203 \text{ kgf m}$$

Ans.

Step (X) Diagram Efficiency

Workdone per kg steam

$$= \frac{\text{H.P.}(75)}{\dot{M}(427)} = \frac{495(75)}{(5000/3600)(427)} = 62.60 \text{ kcal}$$

where $\dot{M}$ = Steam flowrate, kg/s

Diagram efficiency = Workdone/Useful K.E. available at nozzle exit

$$= 62.60/91.10 = 0.6871 \text{ i.e. } 68.71\%$$

Ans.

Step (XI) Overall Stage Efficiency

Overall stage efficiency = Workdone/Theoretical K.E. available at the nozzle exit

$$= 62.60/95.898 = 0.6527 \text{ i.e. } 65.27\%$$

Ans.

Problem 31.8 A reaction turbine has speed 2650 rpm and drum dia 2000 mm. At a particular ring, the blade height is 175 mm and discharge angle is 30°. The exhaust steam from the ring is 0.94 dry and has pressure 0.8 kgf/cm² abs. Find the power developed at that particular ring if the steam consumption is 54 t/h.

Receiving angle is 50°.

Solution

Step (I) Area of Flow

The area of flow of steam at the ring with drum dia 2000 mm and blade height 175 mm is

$$= \pi(D + h)h = \pi(2 + 0.175)(0.175) \text{ m}^2$$

$$= 1.1957 \text{ m}^2$$

where D = drum dia ; h = blade height

Step (II) Blade Speed

$$V_b = \pi(D + h)N/60 = \pi(2 + 0.175)(2650/60)$$

$$= 301.78 \text{ m/s}$$

Step (III) Velocity Diagram

In the case of reaction turbine V_1 is made equal to V_{r_2} and $V_2 = V_{r_1}$

$$\alpha = \gamma$$

Here, $\alpha = \gamma = 30°$ (Fig. 31.18)

$\beta = \theta$, the blade angle for shockless entry = 50°

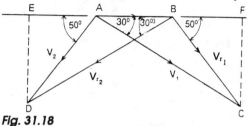

Fig. 31.18

(here)

Step (IV) Power Developed at the Particular Ring

Change in the velocity of whirl = EF = 840 m/s

Velocity of blade = $V_b = AB$ = 301.78 m/s

Mass flowrate of steam = $54 \times 10^3/3600$ kg/s = 15 kg/s

H.P. developed at the rim

$$= (W/g)(\Delta \text{ Velocity of Whirl})(V_b/75)$$

$$= (15/9.81)(840)(301.78)/75$$

$$= 5168.09 \text{ i.e. } 5168$$

Ans.

Problem 31.9 At a stage in a reaction turbine, the mean drum diameter is 1.5 m and the mean blade height is 150 mm. The blade tip angles of fixed and moving blades at inlet and exit are 30° and 20° respectively.

The turbine is supplied with steam at the rate 720000 kg/h and the stage efficiency is 85%.

Determine:

(a) the axial velocity of steam flow
(b) the blade speed
(c) the power output of the stage
(d) the specific enthalpy drop in this stage
(e) the percentage of increase in relative velocity in the moving blades due to expansion in these blades.

The steam is dry, saturated at 15 kgf/cm² abs.

Solution

Step (I) Axial Velocity

Axial velocity of flow of steam

V_a = Volumetric Flowrate of steam/Annular area of flow

Now, at 15 kgf/cm^2. abs., the sp. vol. of steam

= 0.134 m^3/kg

Mass flowrate of steam = 720000 kg/h

= 720000/3600 kg/s = 200 kg/s

Volume flowrate of steam

= Mass flowrate × Sp. vol. of steam

= 200 (0.134) = 26.8 m^3/kg

Annular area of flow = π $(D + h)$ h

= π (1.5 + 0.15) (0.15) = 0.7775 m^2.

$$\therefore \ V_a = \frac{26.8}{0.7775} = 34.469 \quad \text{i.e.} \quad 34.47 \text{ m/s}$$

Ans.

Step (II) Velocity Diagram

On a horizontal EF, a perpendicular CF is drawn at F such that CF = 34.47 m/s which is the axial velocity of flow of steam. Now, A is another point on EF such that

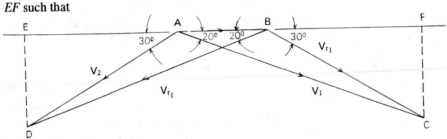

Fig. 31.19

$AF = CF \cot 20° = 34.47 \ (2.747) = 94.69$ m/s

Similarly B is a point such that $BF = CF \cot 30°$

= 34.47 (1.732) m/s

= 59.70 m/s

From point A, AE is cut off such that AE = 59.70 m/s

Now, $AB = AF - BF = 94.69 - 59.70 = 34.99$ m/s

Step (III) Blade Speed

Blade speed is represented by AB. From the velocity diagram, (Fig. 31.19)

AB = 34.99 m/s = V_b

i.e. blade speed = 34.99 m/s

Ans.

Step (IV) Power Output

Now the power developed at this stage is

H.P. = (W/g) (Δ Velocity of Whirl) $(V_b/75)$

Now the change in the velocity of whirl

= $EF = AE + AB + BF$

= 59.70 + 34.99 + 59.70 m/s = 154.39 m/s.

V_b = 34.99 m/s; W = 720000 kg/h = 200 kg/s

$\therefore$ H.P. = (200/9.81) (154.39) (34.99)/75

= 1468.46

Ans.

Step (V) Specific Enthalpy Drop

Let the specific enthalpy drop in this stage be ΔH.

$\Delta H = V_b \ (V_{w_1} + V_{w_2})/\eta$ where,

η = stage efficiency.

= 34.99 (154.39)/0.85 = 6355 kJ/kg

Ans.

Step (VI) Per cent Increase in Relative Velocity

From the diagram,

relative velocity at inlet, $V_{r_1} = BC = CF \ \text{cosec} 30°$

= 34.99 (2) = 69.98 m/s

relative velocity at outlet,

$V_{r_2} = BD = CF \ \text{cosec} 20°$

= 34.99 (2.92) = 102.30 m/s

Increase in relative velocity = 102.30 − 69.98

= 32.32 m/s.

% increase in relative velocity

= (32.32/69.98) (100) = 46.18%

Ans.

Problem 31.10 The following data are available for a reaction turbine:

Steam pressure at inlet = 20 kgf/cm^2 abs.
Steam temperature at inlet = 300°C
Steam pressure in the condenser = 0.2 kgf/cm^2 abs.
Speed of the turbine = 5600 rpm
Power developed = 25000 h.p.
Reheat factor = 1.05
Stage efficiency = 80%
Axial velocity of steam = 75% of the blade velocity
Blade height = 0.10 times of mean blade diameter
Calculate the
(a) steam consumption per horse power hour
(b) drum dia if the exhaust steam (0.2 kgf/cm^2, abs.) is 0.94 dry.

Solution

Step (I) Adiabatic Specific Enthalpy Drop

As the steam expands in the turbine adiabatically (assumed), from 20 kgf/cm^2 abs. to exhaust pressure 0.2 kgf/cm^2 abs., its specific enthalpy drops from

720 kcal/kg (from Mollier Chart)
⇓
530 kcal/kg (-do-)

∴ $\Delta H = 720 - 530 = 190$ kcal/kg

Step (II) Work done

Now, workdone during expansion = $\Delta H \times \eta_{int}$

where η_{int} = internal efficiency

= Stage efficiency × Reheat Factor

= 0.80 × 1.05

= 0.84

∴ Work done = 190(0.84) = 159.6 kcal/kg of steam

Step (III) Steam Consumption Per H.P. Hour

Now 1h.p. = 75 kgf. m/s

1 h.p.h = 75(3600) kgf m = 75(3600)/427 kcal

Work done/kg of steam = 159.6 kcal

Steam consumption/ h.p.h

= 75(3600)/[(427) (159.6)] kg = 3.96 kg

Step (IV) Steam Consumption

Horse power developed at the turbine = 25000

Steam consumption/h.p.h = 3.96 kg

∴ Steam consumption/h = 25000(3.96) kg

= 99000 kg

∴ Steam consumption/s = 99000/3600 kg

= 27.5 kg

Step (V) Blade and Axial Velocities

Blade velocity, $V_b = \pi DN/60$

speed in rpm = $\pi D(5600)/60$ where, N = blade
= 293.215D

D = average blade dia

Axial velocity, $V_a = 0.75 V_b = 219.911D$

Step (VI) Drum Dia

Now the specific volume of exhaust steam (0.2 kgf/cm^2 abs. and 0.94 dry) = 0.94 (7.79) m^3/kg = 7.3226 m^3/kg

Now, Area × Axial velocity of flow = Mass flowrate of steam × Sp. Volume of steam

∴ $\pi Dh (V_a) = 27.5 (7.3226)$

where h = blade height = 0.1D (as given)

or $D(0.1D)(219.911D) = 201.3715$

∴ $D = 1.428$ m

∴ $h = 0.1 (1.428) = 0.1428$ m

Therefore, drum dia

= $D - h = 1.428 - 0.1428 = 1.285$ m

Ans.

Q. *What is the cause of turbine deposits?*

Ans. The turbine deposits are steam-borne foreign matter settled on turbine blades. Substances dissolved in the BFW transfer partly from the water to the steam, during the process of evaporation. They get dissolved in the steam and are carried into the turbine.

Q. *What type of deposits are formed?*

Ans. These deposits are broadly divided into two main types:
1. Water-Soluble deposits
2. Water-Insoluble deposits

Q. *What are the water-soluble deposits?*

Ans. $NaCl$, Na_2SO_4, $NaOH$ and Na_3PO_4.

Q. *What are the water-insoluble deposits?*

Ans. SiO_2 (mainly).

Q. *Where do water-soluble deposits prevail?*

Ans. In the high-and intermediate-pressure sections of steam turbines.

Q. *Where do silica-sediments prevail?*

Ans. At the end of the intermediate-pressure section as well as in the low-pressure section of the turbine.

Q. *How to prevent turbine deposition?*

Ans. By upgrading the quality of steam. That is by ensuring proper
  (a) boiler feedwater quality
  (b) steam boiler model
  (c) boiler design
  (d) boiler operation.

Q. *What are the consequences of turbine deposi-tions?*

Ans. It has three kinds of effects:
  1. **Economic Effect**
     (a) reduction in turbine output
     (b) decrease in efficiency requiring higher steam consumption
  2. **Effect of Overloading and Decreasing Reliability in Operation**
     (a) Pressure characteristic in the turbine gets disturbed with the effect that thrust and overloading of thrust bearing increase
     (b) Blades are subjected to higher bending stresses
     (c) Natural vibrations of the blading are affected
     (d) Vibration due to uneven deposition on turbine blading
     (e) Valve jamming due to deposits on valve stems
  3. **Corrosion Effect**
     (a) Fatigue corrosion
     (b) Pitting corrosion
     (c) Stress corrosion.

Q. *How does deposit formation on turbine blades affect turbine efficiency?*

Ans. About 500 g of deposits distributed more or less evenly all over the blading section can bring down turbine efficiency by 1%.

Q. *How can the detection of deposits in a turbine be made during operation?*

Ans.
  1. Pressure Monitoring
  2. Internal Efficiency Monitoring
  3. Monitoring Exhaust Steam Temperature
  4. Monitoring Specific Steam Consumption.

Q. *How does pressure monitoring ensure detection of turbine deposits?*

Ans. Pressure of steam expanding in the turbine is measured at characteristic points, i.e., at the wheel chamber, points of pass-out, inlet/outlet of H.P., I.P and L.P. stages of the turbine.

The turbine manufacturer provides the pressure characteristics in the form of graphs.

At 1st commissioning, the user supplements these theoretical curves with those derived from actual measurements. These are actual pressure characteristics for a clean turbine. Now these pressure characteristics are compared with those obtained during operation in the later period.

Under identical conditions, an increase in pressure shows the formation of deposits.

For a steam throughput in the range 70 — 100%, an increase in wheel chamber pressure of more than 10% indicates severe blade depositions.

Q. *How does the internal efficiency monitoring lend to the detection of turbine deposits?*

Ans. The internal efficiency of the turbine is given by

$$\eta_i = \frac{[\Delta H]_{pro}}{[\Delta H]_{ad}} \times 100$$

where $[\Delta H]_{pro}$ = process heat drop

$[\Delta H]_{ad}$ = adiabatic heat drop

In the economic rating range, the η_i varies negligibly with steam flow. However, blade depositions restrict the passage of steam flow, i.e., change the inner condition of the steam turbine with the effect that η_i alters in comparison to original values.

The process heat drop and adiabatic heat drop are obtained from a Mollier-chart for the corresponding values of steam parameters—pressure and temperature—at initial and final conditions. And η_i is calculated by computing these values in the above equation.

Q. *By monitoring the exhaust steam temperature, how can the blade deposition be predicted?*

Ans. Immediately after the 1st commissioning, the different values of exhaust temperature for different steam flowrates are precisely determined and plotted against steam flow. This will produce the first actual graph. This is for a clean turbine.

Similar graphs are to be drawn at later periods for comparing with the initial graph.

A rise in exhaust steam temperature under the same conditions refers to deposit formation.

An increase of exhaust steam temperature by more than 10% in the range of 70–100% steam flow indicates inadmissible blade depositions. Shutdown is to be taken and blades are to be washed off deposits.

Q. *How can monitoring specific steam consumption tell you that there has been blade deposition?*

Ans. The specific steam consumption in a turbine is given by

$$d = D/E$$

where D = steam flowrate, kg/h

E = electricity output at terminals, in kWh.

D and E are measured as precisely as possible and corrected over the specific volume, after the 1st commissioning. And the value of d is determined. This ascertains the specific steam consumption for a clean turbine.

This value is compared with the other values of d determined in the same way after some given periods of operation. Any deviation from the initial value is due to blade deposits in most cases.

Q. *How can the deposits be removed?*

Ans. Water - soluble deposits may be washed off with condensate or wet steam.

Water-insoluble deposits are removed mechanically after dismantling the turbine.

Experience shows that water soluble deposits are embedded in layers of water-insoluble deposits. And when the washing process is carried out, water soluble parts of the deposit dissolve away leaving a loose, friable skeleton of water-insoluble deposits which then break loose and wash away.

Q. *How is the washing of turbine blades carried out with the condensate?*

Ans. The washing is carried out with the condensate at 100°C.

The turbine is cooled or heated up to 100°C and filled with the condensate via a turbine drain.

The rotor is turned or barred by hand and the condensate is drained after 2–4 hours.

It is then again filled with the condensate at 100°C (but upto the rotor centre-level), the rotor is rotated and the condensate is drained after sometime. This process is repeated several times.

Q. *How is turbine blade washing with wet steam carried out?*

Ans. Wet steam produced usually by injecting cold condensate into the superheated steam, is introduced to the turbine which is kept on running at about 20% of nominal speed.

For back-pressure turbine the exhaust steam is let out into the open air through a gate valve. For a condensing turbine, the vacuum pump is kept out of service while cooling water is running, with the effect that the entering cooling steam is condensed. The condensate is drained off.

The washing steam condition is gradually adjusted to a final wetness of 0.90–0.95.

It is important:
(a) not to change washing steam temperature by 1°C/min
(b) to keep all turbine cylinder drains open.

Q. *What are the possible causes of excessive vibration or noise in a steam turbine?*

Ans. These are:

(a) Misalignment
(b) Worn bearings
(c) Worn coupling to driven machine
(d) Unbalanced coupling to driven machine
(e) Unbalanced wheel
(f) Piping strain
(g) Bent shaft

Q. *How will you detect that misalignment is the probable cause of excessive vibration?*

Ans. Coupling to the driven machine is to be disconnected.

The turbine is to be run alone.

If the turbine runs smoothly, either misalignment, worn coupling or the driven equipment is the cause of the trouble

Q. *How can the misalignment be rectified?*

Ans. The bolts holding the flanges together are to be tightened. The coupling is to be checked for squareness between the bore and the face. At the same time axial clearance is to be checked. Using gauge block and feeler gauges, the gap between coupling faces 180° apart is to be measured. After rotating the coupling-half 180°, the gap at the same points is to be measured. After this, the other coupling is to be rotated 180° and the gap at the same points is to be remeasured. These measures should come within a few thousands of an inch. Dividing the coupling faces into four 90° intervals, the distance between the coupling faces at this intervals is to be measured with the aid of a gauge block and feeler gauges. These gap measurements should come within 0.005 inch for proper angular shaft alignment. After proper alignment at room temperature, the two halves of the coupling are to be connected.

Q. *What is the remedy for worn bearings causing excessive vibration?*

Ans. The bearings should be replaced.

Q. *What should be done in case unbalanced coupling to the driven machine is suspected to be the cause of excessive vibration?*

Ans. The coupling halves are to be removed.
Checking for unbalance is to be carried out.

Q. *What should be done if excessive vibration is due to an unbalanced wheel?*

Ans. The turbine wheel is to be checked if it has got unbalanced duc to overspeeding.

The turbine wheel must be rebalanced or replaced.

Q. *Why does such 'unbalancing' occur?*

Ans. If the turbine is kept idle for a long spell without complete drainage of exhaust casing, the solid matter can deposit in the lower half of the wheel causing unbalance.

Q. *How can the problem of 'excessive vibration or noise' due to piping strain be avoided?*

Ans. The inlet as well as exhaust steam lines should be firmly supported to avoid strains from being imposed on the turbine.

Adequate allowance should be made for expansion of steam pipes due to heat.

Q. *What is the remedy for a bent shaft causing excessive vibration?*

Ans. The runout of the shaft near the centre as well as the shaft extension should be checked.

If the runout is excessive, the shaft is to be replaced.

Q. *What are the factors that contribute to bearing failure in a steam turbine?*

Ans. The possible causes of bearing failure are:
(a) not using correct type bearing and inadequate mounting
(b) improper lubrication
(c) inadequate water cooling
(d) misalignment
(e) bearing fit
(f) excessive thrust
(g) unbalance
(h) rusting of bearing.

Q. *What should be done in the case of bearing failure due to improper bearing?*

Ans. Correct type of ball bearings should be used.
The worn out ball bearings should be replaced.

Q. *How can bearing failure due to improper lubrication be avoided?*

Ans. Only the recommended lubricant should be used.

Q. *How can bearing failure due to inadequate water cooling be prevented?*

Ans. The jacket temperature should be maintained in the range of 37°–60°C

The flow of cooling water should be adjusted accordingly.

Q. *How does bearing fit brings about bearing failure?*

Ans. It is desirable that ball bearings should fit on the turbine shaft with a light press fit. If the fitting is too tight, it'll cause cramping. And on the other hand if the fitting is too loose it'll cause the inner race to turn on the shaft. Both conditions are undesirable. They result in wear, excessive vibration and overheating. And bearing failure becomes the ultimate result.

Q. *How to avoid excessive thrust to save the bearings?*

Ans. To avoid excessive thrust, the coupling should be kept clean and installed such that no excessive thrust is imparted to the turbine from the driven equipment.

Q. *What causes the rusting of the bearing?*

Ans. Rusting on bearing surface may occur due to improper storage.

Q. *What factors cause excessive steam leakage under carbon rings?*

Ans.
1. Dirt under rings
2. Shaft scored
3. Worn or broken carbon rings.

Q. *What causes dirt deposition under the carbon rings?*

Ans. Steam borne scale or dirt foul up the rings if steam is leaking under the carbon rings.

Q. *What is the remedy for worn carbon rings?*

Ans. These should be replaced with a new set of carbon rings. The complete ring is to be replaced.

Q. *What are the possible causes for the turbine not running at rated speed?*

Ans. The possible causes are:
(a) too many hand valves closed
(b) oil relay governor set too low
(c) inlet steam pressure too low or exhaust pressure too high
(d) load higher than turbine rating
(e) throttle valve not opening fully
(f) safety trip valve not opening properly
(g) nozzles plugged
(h) steam strainer chocked.

Q. *What are the possible causes of the speed of the turbine rotor increasing excessively as the load is decreased?*

Ans.
1. Throttle valve not closing fully
2. Wearing of throttle valve seats.

Q. *What factors contribute to excessive speed variation of the turbine?*

Ans.
1. Improper governor droop adjustment
2. Improper governor lubrication
3. Throttle assembly friction
4. Friction in stuffing box
5. High inlet steam pressure and light load
6. Rapidly varying load.

Q. *How can the speed variation be reduced by making a governor droop adjustment?*

Ans. If the internal droop setting is increased, the speed variation will reduce.

Q. *How does improper governor lubrication arise?*

Ans. In the event of low governor oil level or if the oil is dirty or foamy, it will cause improper governor lubrication.

Q. *What is the remedy to it?*

Ans. The dirty or foamy lube oil should be drained off, governor should be flushed and refilled with a fresh charge of proper oil.

In the event of low level, the level should be **built** up by make- up lube oil.

Q. *How can the problem of excessive speed variation due to throttle assembly friction be overcome?*

Ans. The throttle should be dismantled. All the moving parts should be checked for free and smooth movement. Using very fine grained emery paper, the throttle valve, seats and valve steam should be polished.

Q. *What is the possible cause of slow start up of a steam turbine?*

Ans. This may be due to high starting torque required by the driven equipment.

Q. *What are the possible causes of a governor not operating?*

Ans.
1. Restriction of throttle valve reflex
2. Failure of governor control on start-up.

Q. *What may be the cause of the failure of governor control on start-up?*

Ans. If it is found that after start-up, the speed increases continuously and the governor is not closing the throttle valve, it may be that the governor pump has been installed in the wrong direction.

Q. *What may be the possible causes for the SAFETY TRIP tripping during load variation?*

Ans.
1. Light load and high inlet steam pressure
2. SAFETY TRIP set very close to the operating speed of turbine.

Q. *What is the remedy for the 2nd problem?*

Ans. The setting of SAFETY TRIP should be made at 25% (approx) of the rated speed.

Q. *What may be the possible causes for the SAFETY TRIP to trip at normal speed?*

Ans.
1. Excessive vibration
2. Leakage in the pilot valve
3. Deposition of dirt in the safety trip valve

Q. *How does the dirty SAFETY TRIP valve trip the safety trip at normal speed?*

Ans. Dirt may find its way to the safety trip valve and get deposited around the spring end cap end. This will block the clearance between the safety trip valve and the spring end cap. As a result the steam pressure in the spring cap gets lowered allowing the valve to close.

Q. *What is the remedy to it?*

Ans. The spring end cap as well as safety trip valve should be cleaned.

Q. *Of all the factors that contribute to the unreliability of steam turbines, which one is the most prominent?*

Ans. It is the problem of turbine blade failures that chiefly contribute to the unreliability of steam turbines.

Q. *What factors are responsible for turbine-blade failures?*

Ans. In the high pressure cylinder, the turbine blades are mostly affected by
  (a) Solid-particle erosion (SPE)
  (b) High-cycle fatigue.
Whereas in the last few stages of the low-pressure cylinder, the blade damage is mainly afflicted by
  (a) erosion
  (b) corrosion
  (c) stress/fatigue damage mechanism.
According to EPRI (Electric Power Research Institute, USA) data stress-corrosion cracking and fatigue are the chief exponents for turbine-blade failures in utility industries.

FACTORS	BLADE FAILURES
Unknown	26%
Stress-Corrosion Cracking	22%
High-Cycle Fatigue	20%
Corrosion-Fatigue Cracking	7%
Temperature-Creep Rupture	6%
Low-Cycle Fatigue	5%
Corrosion	4%
Other causes	10%
	TOTAL 100%

Besides, many damage mechanisms operate in combination of

(a) poor steam/water chemistry
(b) certain blade-design factors that vary from one turbine manufacture to other
(c) system operating parameters.

Q. *How does the damage to turbine-blades tell upon the efficiency of the unit?*

Ans. The damage to blade profiles changes the geometry of steam flow path and thereby reducing the efficiency of the unit.

Q. *What is the remedy to the damage to blade profiles?*

Ans. Upgrading the turbine.

And depending on the extent of damage, upgrading may involve
1. weld repair of affected zones of the blade
2. replacement of damaged blades by new ones and of new design
3. replacement of base material
4. application of protective coatings to guard against corrosion and erosion damage.

Q. *What does "upgrading" generally means in the context of steam turbines?*

Ans. Upgrading is a most widely used term. It encompasses a variety of meanings viz. life extension, modernization and uprating of steam turbines.

Q. *In which case does upgrading imply life extension of steam turbines?*

Ans. For a capital-short electric utility plant, upgrading comes to mean extending the life of that plant scheduled for retirement.

Q. *In which cases does upgrading mean uprating the turbine capacity?*

Ans. For a electric utility system facing uncertain load growth, upgrading is chiefly uprating. It is an inexpensive way to add capacity in small increments.

Q. *When does upgrading mean modernization of utility industry?*

Ans. Upgrading is really modernization to all those units other than those facing uncertain load growth and low-capital utility system. It involves replace-ment of damaged parts/components by state-of-the-art components without scrapping the entire machine.

Q. *Which factors affect the extent of an upgrading programme?*

Ans.
1. Age of the unit
2. How it has been operated.

Note: Turbines less than quarter of a century old can simply be upgraded to their original design conditions.

Q. *Can the replacement blades for achieving higher reliability improve the efficiency of the unit?*

Ans. Yes.

Q. *How?*

Ans. When the damaged blades of low-pressure cylinder of steam turbines are replaced by longer and new blades, the efficiency of LP section improves. Efficiency is further boosted by smoother surfaces—when old blades are replaced by new ones or coated or remachined—by virtue of reduced friction loss.

Q. *In which part of the steam turbine does stress corrosion cracking (SCC) occur?*

Ans. In the wet stages of the low-pressure turbine.

Q. *In which part of the steam turbine does corrosion fatigue occur?*

Ans. Also in the wet stages of the LP cylinder.

Q. *Where is pitting corrosion mostly prevalent?*

Ans. Upstream of LP stages as well as wet stages of LP cylinder.

Q. *What are the points of SCC attack?*

Ans. SCC attack predominates where corrodents deposit and build up i.e. in those blading areas where flowing steam cannot provide a washing effect.

Q. *What are these points in particular?*

Ans.
1. Tie wires
2. Tie-wire holes
3. Brazings

4. Blade covers
5. Tenon holes.

Q. *At what points corrosion fatigue does show up?*

Ans. It attacks trailing edges, near the base of the foil and also the blade-root serrations.

Q. *What is done when cracks due to SCC or corrosion-fatigue are found?*

Ans. The damaged blade is usually replaced as repairing is difficult.

Q. *What is the solution to the problem of SCC/corrosion fatigue of steam turbine blades?*

Ans. It involves changing the blade material as well as minimizing the presence of corrodents in steam to a permissible level.

Q. *How many types of particle-impact damage occur in turbine blades?*

Ans.
1. Erosion/corrosion
2. Foreign-particle impacts
3. Solid-particle erosion
4. Water damage.

Q. *In which cases does erosion/corrosion damage appear?*

Ans. It is commonly encountered in nuclear steam turbines and old fossil-fuel-fired units that employ lower steam temperatures and pressures.

Q. *In which cases does moisture- impingement and washing erosion occur?*

Ans. These are encountered in the wet sections of the steam turbine.

For nuclear power plants, these wet sections can involve parts of high-pressure cylinder.

Q. *What steps are taken to minimize damage from moisture?*

Ans. The following measures are employed at the design stage:
1. Stellite inserts
2. Hardening of the base metal
3. Moisture-removal devices

to combat impingement corrosion due to moisture.

Q. *But despite these preventive measures, damage due to moisture impingement has been found, in certain cases, in the shield and beyond. Why?*

Ans. Shields are designed and fabricated on the basis of predicted range of steam/water quantities impacting the blades at specific angles.

Now if the operating conditions deviate significantly from design parameters then the erosion damage will occur. And in some cases it may go beyond normal erosion wear and warrant repair.

Also the corrosion of casing can occur due to blockage/clogging of water drains or extraction thereby forcing the water back into the casing. If this condensate water is carried over to steam path and impacts the blade, thermal-fatigue failure can occur within a short period.

Q. *How does solid-particle erosion occur?*

Ans. Solid-particle erosion, i.e. SPE occurs in the high-pressure blades. And it takes place when hard particles of iron exfoliated by steam from superheater tubes, reheater tubes, steam headers and steam leads strike on the surface of turbine blades.

Q. *When does SPE damage usually occur?*

Ans. It occurs usually during startup or abrupt load change.

Q. *How does the foreign-particle damage of turbine blades arise?*

Ans. It occurs due to impact on blades by foreign particles (debris) left in the system following outages and become steam-borne later.

Q. *Is there any factor other than corrodents and erodents that contributes to turbine blade failure?*

Ans. Yes; turbine blade damage and failures can be effected by vibration and fatigue.

Q. *How do the problems of vibration and fatigue arise?*

Ans. These arise due to
1. flow irregularities introduced because of manufacturing defects, e.g. lack of control over tolerances.

2. system operating parameter, e.g. low flow may excite various modes of vibration in the blades.

Q. *Why are some groups of blades, particularly the first or control stages more prone to fatigue failures than others?*

Ans. Blades in the first or control stages are under partial-arc admission that forces the blades to move into and out of the steam flow causing alternating high-and low-impact forces. This periodic change of impact forces imparts fatigue stress that makes such groups of blades susceptible to fatigue failure.

Q. *In which zone of steam turbines has temperature-creep rupture been observed?*

Ans. Damage due to creep is encountered in high temperature (exceeding 455°C) zones. That is, it has been found to occur in the control stages of the high-pressure and intermediate-pressure turbines where steam temperature sometimes exceed 540°C. In the reheat stage, it has been observed that creep has caused complete lifting of the blade shroud bands.

Q. *When "stall flutter" occurs?*

Ans. This problem is encountered when operating limits are exceeded i.e. when turbine exhaust pressure exceeds the value what has been recommended by the manufacturer.

Q. *What is its effect?*

Ans. It induces damaging stresses in the blades.

Q. *What is the harm if the rotor is oversped?*

Ans. Oversped rotor grows radially causing heavy rub in the casing and the seal system. As a result, considerable amount of shroud-band and tenon-rivet head damage occurs. (Fig. 31.20)

Q. *Why it has been that failures become too common at lacing or tie-wire holes for blades having this type of stiffening?*

Ans. These areas have peculiar susceptible characteristics to damage mechanism. But the reason is not clear.

It may be that the following causes may act singly or in combination to bring about blade failure:

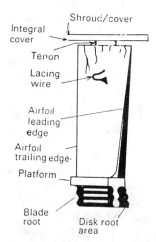

Fig. 31.20 BLADE COMPONENTS

(a) inadequate manufacturing techniques
(b) poor surface finish at the hole
(c) failure to properly align the wire holes
(d) genesis of very complex stress patterns due to shear stress in the brazed material during operation.

Q. *In how many patterns are tie wires used?*

Ans. In one design, tie wire is passed through the blade vane.

In another design, an integral stub is jointed by welding/brazing.

Q. *Which pattern experiences less problem?*

Ans. The integral stub form.

Q. *What other parts of the steam turbine blades suffer from damage?*

Ans.
1. Blade roots
2. Shroud band

Q. *Why do blade roots suffer from damage?*

Ans. Fatigue is the common cause to the effect of blade root damage. Also this factor is often assisted by a generic type of fault in design or manufacturer.

Moreover, the root-fillet radii are subjected to a high degree of stress concentration with the effect that they crack relatively easily.

Q. *Why do shroud bands suffer from damage?*

Ans. Steam borne solid particles and moisture strike the shroud band continually and in that process they remove material from rivet heads until the rivet becomes too weak to exercise its clamping effect whereupon it fails to hold the band in place.

Q. *What is to be done for erosion-induced damage on high-and low-pressure stage blading?*

Ans. In such cases welding repair can be a good solution and this can be carried out during a normal maintenance outage without removing the blade. Using oxyacetylene torch, stellite is generally deposited onto the damaged site. Following this, the weld is subjected to stress-relieving and reprofiling.

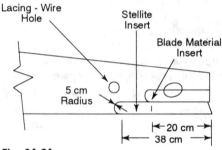

Fig. 31.21

In case of erosion penetrating the erosion shield and extending to the base material, a filler material of consistent or identical composition of blade material is used.

In some cases use is made of Inconel alloy to build up the metal base. Therefore, using welding or brazing technique, a new shield can be attached to the blade. If brazing technique is followed, the rebuilt section is stress-relieved prior to the attachment of shield to it. If, on the other hand, the shield is attached by welding, then they are stress-relieved together.

Q. *In some weld-repair cases, it has been found that the stellite survived while the filler material eroded away. Why?*

Ans. If Inconel is used as the filler material, it has the inferior resistance to erosion in comparison to the stellite insert. So filler material erodes away underneath.

Q. *What is to be done in case of cracks originating at the lacing-wire holes?*

Ans. These are to be weld-repaired. However the following factors must be considered:
  (a) the length of the crack that appears on the pressure and/or suction face
  (b) whether the cracks propagate towards inlet end, discharge end and or both.

Q. *How can damaged tenons be repaired?*

Ans. By adopting modern welding techniques, tenons can be rebuilt. This in some cases results in extended blade life.

Q. *What steps/modifications should be implemented to curtail the damage from moisture impingement?*

Ans.
  1. The drainage system should be redesigned. Larger drains are to be provided.
  2. More effective water-catchers are to be installed.
  3. Radial seals are to be eliminated to remove water before it can chance upon the blades.
  4. Nozzle trailing edges are to be thinned to promote the formation of smaller and less harmful droplets.

Q. *How can the fatigue damage on high-pressure blades be corrected?*

Ans. Fatigue-damage on high-pressure blades arises due to vibration induced by partial-arc admission. This can be corrected by switching over to full arc admission technique.

Q. *Is there any adverse effect of full-arc admission operation?*

Ans. At low loads, this results in a heat-rate penalty, due to throttling over the admission valves.

Q. *So can you recommend this technique as a permanent measure?*

Ans. No.
This can be recommended in extreme cases or at best temporarily.

Q. *So what should be the more sound approach?*

Ans. The more reasonable and better approach is to replace the damaged blades with new ones that are stiffened by

1. serrating the interface surface of individual blades so they interlock, or
2. welding the blades together.

In some cases, a single monolithic block is machined out to manufacture the blades in a group.

In some other cases, blades themselves are directly welded into the rotor.

Q. *What design modification is adopted to reduce susceptibility of last low-pressure stages to fatigue failure?*

Ans. One modification is to join the blade segments together at the shroud band.

Q. *How does this modification reduce the vibration-fatigue damage?*

Ans. Joining the blade segments together at the shroud band increases the length of the arc—to a maximum of 360°—that alters the natural frequency of the blade grouping from the operating vibration mode.

This design has gained considerable success in commercial service.

Q. *Why are free-standing blades in the last low-pressure stage favoured more, in some cases, than those that are coupled and shrouded together?*

Ans. These free-standing blades are known to provide good and adequate protection against stresses and aggressive environment.

They eliminate all areas viz. shroud/tenon interface and tie-wire/hole area where corrodents can collect.

Q. *Is there any other alternative to improve low-pressure turbine reliability?*

Ans. One sound approach is titanium replacement blades. This will improve turbine reliability if particularly corrosion fatigue is the principal mode of failure of the final two stages.

Tests results, conducted at a Texas Utility, have confirmed that titanium alloys possess superior fatigue strength than the conventional blading alloys in the presence of corrodents and stresses commonly encountered in the last-stage environment of the turbines.

NOTE: Three US utilities have replaced existing low-pressure blades with titanium alloys as far back as 1972 and since then no service related failures have been recorded.

Q. *Titanium replacement blades are a costly approach. Is there any method that can provide cost-effective protection?*

Ans. Yes. It is the technique of blade coating.

Extensive laboratory test program followed by field verification trials have confirmed that certain sacrificial coatings can be applied to the existing blade designs to impart increased resistance to corrosion and erosion. Even the wholesale blade replacement or design changes can be avoided, in certain cases, by use of coatings.

Q. *What are these coatings?*

Ans.

1. Ion deposited aluminium
2. Diffused nickel/cadmium.

Q. *How effective are they?*

Ans. Test results showed that the aluminium coating remained fully protective even at erosion affected areas after one full year of operation of the steam turbine. It is expected that the aluminium coating will withstand chemically aggressive environment except the extremes in feedwater/steam chemistry.

Nickel/cadmium system has also proved to be a very effective coating against corrosion, and it is more protective than the deposited aluminium coating against erosion.

Q. *Is there any drawback of nickel/cadmium coating system?*

Ans. Difficulties are encountered in applying this coating uniformly to complex blade geometries.

Q. *Is there any solution?*

Ans. Modification of coating application process seems to be the only viable solution.

Q. *Do you think that turbine blade failure is the only cause of unreliability of steam turbines? Does* *upgrading of turbine means replacement of blades and/or improvement of blade design?*

Ans. Not exactly.

Like the blades, the steam-turbine rotors are highly stressed components. They are subject to cracking by a variety of failure mechanisms. Rotor failures do occur. And when they occur the result is catastropic with the complete destruction of the unit and the total loss of generating capacity.

Therefore, special attention should be given to rotor upgrading and repairing techniques.

Q. *In which section of the steam-turbine rotors is the problem of rotor failure mostly prevalent?*

Ans. Rotor failures occur mostly on the large low-pressure rotors.

Q. *What are the basic causes of the problem of rotor failure?*

Ans.
1. Normal wear.
2. Fatigue failure due to high stress.
3. Design deficiency.
4. Aggressive operating environment.

Q. *What types of cracking occur in the LP rotor/ shaft?*

Ans.
1. Radial axial-bore cracks
2. Circumferential cracks

Q. *What is the cause of axial-bore cracks?*

Ans. Inadequate toughness of rotor steel and transient thermal stresses.

Q. *What should be the remedial action?*

Ans. For new rotors, control cleanliness of the steel. i.e. inclusion and segregates free and more homogeneous steel shaft is required.

For current rotors, replace the rotor, grind and overbore.

Q. *What is the cause of circumferential cracks?*

Ans. High-cycle fatigue with or without corrosion.

Q. *What is the remedy to this failure?*

Ans. For existing rotor, weld repair may be a choice; otherwise retire it.

For new rotors, materials with improved pitting resistance should be used.

Q. *Do the radial axial-bore cracks occur in the LP rotor/shaft alone?*

Ans. No. These are also known to occur in the HP as well as IP rotors.

Q. *What are the causes of radial axial-bore cracks on HP/IP rotors/shafts?*

Ans. The predominant cause is creep which may act with or without low cycle fatigue.

Also the cracks result due to poor creep ductility due to faulty heat treatment process.

Q. *Is there any other type of cracking occurring in HP/IP rotors and causing rotor failures?*

Ans.
1. Blade-groove-wall cracking
2. Rotor-surface cracking

Q. *Why do these two types of cracking take place?*

Ans. The cause to the effect of blade-groove-wall cracking is creep with/without low cycle fatigue.

Faulty heat treatment procedure results in poor creep ductility that may also contribute to this type of faults.

Whereas thermal fatigue have been identified as the single cause to rotor-surface cracking.

Q. *What remedial measures you can suggest to cope with radial axial-bore cracks?*

Ans. For new rotors, modified heat treatment process is recommended while for existing rotors derating the turbine or replacement of the rotor may be a solution.

Q. *What should be the remedial actions for blade-groove-wall cracking?*

Ans. Modified heat treatment of new rotors is a sound and lasting remedy.

For current rotors, cracks should be machined off and lighter blades should be installed. Better, retire the cracked shaft.

Q. *What is the remedy for rotor-surface cracking?*

Ans. Current rotor/shaft should be machined off (skin-peeling).

Q. *Failures of IP rotors of shrunk-on-disc design due to stress-corrosion-cracking (SCC) has been observed at alarming frequency in the nuclear powerplants of USA.*

Why does SCC occurs at intermediate pressure stage?

Ans. Steam turbines of nuclear powerplants usually operate on more wetter steam than those of thermal powerplants. So even at the intermediate pressure stage, steam becomes wet and it precipitates the impurities i.e. corrodents dissolved in it. These corrodents deposit and build up on rotor shaft causing stress-corrosion-cracking.

Q. *What is shrunk-on-disc rotor?*

Ans. These are built by heat expanding the discs, so that upon cooling they shrink on the main rotor forging.

Q. *Is there any alternative to the shrunk-on-disc design?*

Ans. Yes. Two designs are available at present.
1. Welded rotor in which each individual discs are welded, instead of shrunk, onto the main shaft.
2. Monobloc rotor in which the entire shaft and blade assembly is manufactured from a single forging.

Q. *What are the stresses to which a steam turbine rotor is subjected during its service life?*

Ans.
1. Mechanical stress
2. Thermal stress
3. Electrically induced stress.

Q. *Why does mechanical stress occur in turbine rotors?*

Ans. The factors that contribute to mechanical stress in the shaft are the centrifugal forces and torques generated due to revolving motion of the shaft as well as bending arising during steady-state operation.

Q. *Why do thermal stress occur in the turbine rotors?*

Ans. Transient operating phases i.e. startup and shutdown are the genesis of thermal stress induced to the turbine shaft.

Q. *Why do electrically induced stresses occur in turbine rotors occur?*

Ans. They originate due to short circuits and faulty synchronization.

Q. *Is it true that a rotor with an axial-through bore experiences more stress than a solid rotor of same configuration? If so, why?*

Ans. Yes; rotors with an axial-through bore suffer from twice as much tangential elastic stresses at the centre as the solid rotor of identical configuration.

This is because of uneven stress distribution in the area of the wheels and blade roots and rotor central bore. Stress in the wheels and the blade root areas is different from that in the rotor central bore.

Q. *So bored rotors have higher stress levels. Still they are favoured in preference to solid rotors. Why?*

Ans.
1. Bored shaft is more free from inhomogeneities than solid shaft. A bore removes inclusions and segregates from the centre of the rotor.
2. Direct evaluation of the mechanical properties of the core-bar samples is possible.
3. Non-destructive tests viz. ultrasonic testing and magnetic particle inspection testing of the rotor from the bore are possible.
4. A bore eliminates entrapped gases (which increase shaft porosity) that gravitate to the centre.

Note: Advanced steel-making techniques today make it possible to manufacture "clean" solid rotor/shaft. Hence the foregoing reasons, according to some experts, why a bored shaft should be opted in preference to a solid shaft stands obsolete.

Q. *Why were cracks at the bore common for high-pressure and intermediate-pressure rotors of the early sixties?*

Ans. These rotors were manufactured from forgings which were not "clean" steel and that's why cracks were initiated at the sites of inclusions and segregation bands in the steel. This coupled with low inherent toughness of rotor materials resulted in bore cracks.

Q. *What is the potential problem of shrunk-on-disc type rotor?*

Ans. It is the failure due to circumferential cracks which are not limited to old rotors of early models (1960) but they also take place on present-day rotors.

Q. *What is the nature of circumferential cracking in shrunk-on-disc rotors?*

Ans. Regions of high stress concentration give birth to this type of cracking. It begins in corrosion pits and propagates towards the bore by high-cycle fatigue. It may culminate in a catastrophe, if it penetrates the bore (happily this usually does not occur).

Q. *Rim cracking continues to be a problem of shrunk-on-disc type rotors in utility steam turbines. Where does it occur?*

Ans. Rim cracking has been located at the outer corners of the grooves where the blade root attaches to the rotor.

Q. *Why does rim cracking occur?*

Ans. Poor low-term creep rupture strength or low ductility of the nickel/cadmium/molybdenum/vanadium (NiCdMoV) is the cause of rim cracking.

Q. *What is the nature of rotor surface cracks?*

Ans. They are shallow in depth and have been located in heat grooves and other small radii at labyrinth-seal areas along the rotor.

Q. *Usually it has been found that SCC (Stress Corrosion Cracking) attack takes place particularly at keyways of shrunk-on-disc rotors of low-pressure turbines. Why are keyways prone to SCC attack?*

Ans. Keyways shrunk-fit each disc onto the rotor shaft. They improve the rigidity of the connection between the disc and the central shaft. However, keyways are subjected to abnormal centrifugal forces due to high overspeed, that reduce the amount of shrink. Tangential stresses tend to gravitate at the keyway connection and steam tends to condense there. As a result corrodent impurities like chlorides concentrate at keyways. This factor coupled with high stress concentration lead to SCC attack on keyway areas.

Q. *It has been suggested that a monobloc rotor is a good substitute of shrunk-on-disc rotor. What are the plus points of monobloc design?*

Ans.
1. It is a one-piece-construction, and thus has inherent rigidity.
2. Advanced steelmaking techniques enable building of monobloc rotors almost free from non-metallic inclusions and gas bubbles. Even large monobloc rotors of clean steel are being manufactured today.
3. It exhibits lower inherent stresses.
4. The chance of disc-loosening during operation is eliminated.
5. Highly stressed keyway is elminated.

Q. *Does the monobloc rotor have any drawback or disadvantage?*

Ans. There are two limitations:
1. It is difficult to test and inspect large solid rotors.
2. Manufacturing is difficult. In fact only limited manufacturing capabilities are available to produce the forgings.

Q. *Is there any other alternative rotor design for upgrading steam turbines?*

Ans. Yes. It is the welded rotor which has been claimed by some manufacturer as being the good alternative for upgrading the steam turbines with damaged shrunk- on-disc rotors.

Q. *What are the advantages of welded rotors?*

Ans.
1. Welded rotor is a composed body built up by welding the individual segments. So the limitations on forgings capacity do not apply.
2. Welding discs together results in a lower stress level. Therefore, more ductile materials can be chosen to resist SCC attack.
3. There is no keyways. So regions of high stress concentrations are eliminated.

Q. *Suppose yours is an old utility steam turbine. The rotor is of shrunk-on-disc type. You want to replace it and retrofit welded rotors designed by another manufacturer. What primary factors will you consider and what are fundamental steps you will follow?*

Ans. First step required is to ensure the compatibility with the balance on the components.

Rotor critical speeds are to be checked very carefully.

Precisely to be measured are the geometrical factors at the interfaces.

New vibration behaviour is to be characterized for the whole machine.

Care must be taken not to disturb the overall steam-circuit parameters. And therefore, the thermodynamic design is to be adopted to existing machine conditions.

32

Cycles for Steam Power Plants

Q. *What is the most ideal steam power plant cycle?*

Ans. Carnot cycle.

Q. *Why is it so?*

Ans. Carnot cycle offers the greatest efficiency possible between any two given limits of temperature.

Q. *What processes would be involved if a Carnot cycle were to be operated in a steam power plant?*

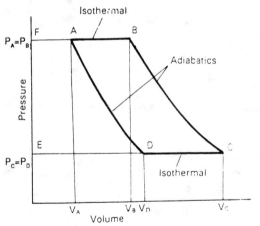

Fig. 32.1 *Carnot cycle*

Ans. **AF** BFW is fed to boiler at boiler pressure P_B and volume V_A. (Fig. 32.1)

AB BFW is converted into steam, in the boiler, at pressure P_B and volume of the steam is V_B, temperature T_B.

BC Frictionless expansion (adiabatic) of steam in the turbine.

CD After expansion, the exhaust steam is condensed isothermally (T_D) in a condenser. Volume of steam drops from V_C to V_D.

DE Partially condensed steam at D and pressure $P_C = P_D$ is fed from the condenser to the feed pump.

DA The residual steam is compressed, in the feed pump, frictionless adiabatically to boiler pressure P_A. This step converts wet steam at condenser pressure to water at boiler pressure.

Q. *How can the efficiency of this cycle be computed from a temperature-entropy diagram?*

Ans. The foregoing processes involved in a steam plant operating in a Carnot cycle can be represented in a T–φ diagram, as shown. (Fig. 32.2)

The network output from the plant is the area *ABCD*.

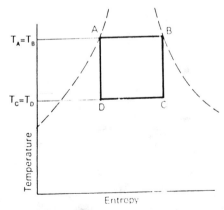

Fig. 32.2 *Carnot cycle on temperature-entropy diagram*

(work output = work done by turbine–work done by feed pump

= *FABCDE–FADE*

= area *ABCD*)

The area *ABCD* is enclosed by two isothermal processes and two adiabatic processes. Hence this is a Carnot cycle whose efficiency

$$\eta = \frac{T_B - T_C}{T_B}$$

Q. *How far is this Carnot cycle practical for operation in a steam power plant?*

Ans. This cycle, applied to a steam power plant, is practical up to a point.

The isothermal generation of steam in the boiler (process *AB*) with volume expansion from V_A to V_B is reasonable.

Also reasonable is the adiabatic expansion of the steam in the turbine (process *BC*).

Q. *What is the impractical part of this cycle?*

Ans. The impractical part of this cycle is in the handling of steam in the condenser and feed pump.

Steam is only partially condensed in the condenser and the condensation must cease at *D*.

Furthermore, the feed pump must be capable of handling both wet steam and water.

Q. *How can this cycle be made more practicable for its application in steam power plants?*

Ans. A slight modification to this cycle is introduced: Instead of stopping the condensation in the condenser at some intermediate condition, the condensation is allowed to continue until it is complete at point *D*. At this point there is all water. This condensate can be successfully dealt with in the feed pump pumping it to the boiler at boiler pressure P_B. (Fig. 32.3 & 32.4)

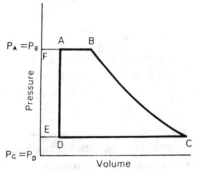

Fig. 32.3

Q. *What is the effect of adopting this cycle?*

Ans. Though this cycle is more practical, it gives rise to lesser thermal efficiency than the Carnot cycle.

Q. *What processes are involved when a steam power plant is working on the Rankine cycle?*

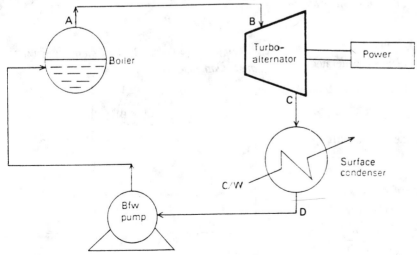

Fig. 32.4

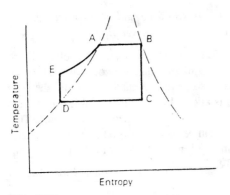

Fig. 32.5

Ans. **FA** BFW at boiler pressure $P_B = P_A$ is delivered to the boiler. Its volume is V_A. (Fig. 32.5)

AB In the boiler, BFW is converted to steam at pressure P_B. The volume of the steam generated is V_B. An isothermal process.

BC Adiabatic frictionless expansion of steam in the turbo-alternator generating electricity.

CD Total condensation of the exhaust steam in the condenser. An isothermal process.

DE Pumping of condensate from condensate pressure $P_C = P_D$ to boiler pressure P_B.

EA Heating the BFW at constant pressure till it attains the saturation temperature corresponding to boiler pressure $P_B = P_A$.

Q. *How can the efficiency of the Rankine cycle be computed?*

Ans. Total heat supplied by boiler

$= H_B - (h_{w_D} + W)$ per Kg of steam generated.

where, H_B = Enthalpy of steam at B, kcal/kg

h_{w_D} = Liquid enthalpy at D, kcal/kg

W = Mechanical work done by the pump per kg water.

Availble work output $= H_B - H_C$ per kg of steam.

Net work available per kg of water

$= (H_B - H_C) - W$

$$\text{Rankine efficiency} = \frac{(H_B - H_C) - W}{H_B - (h_{w_D} + W)}$$

Compared with other heat quantities W is very small and hence ignored for all practical purposes.

Hence, Rankine efficiency $= \dfrac{(H_B - H_C)}{(H_B - h_{w_D})}$

Q. *How can the work done during the Rankine cycle be determined from a P–V diagram?*

Ans. The work done by the feed pump is neglected and if the steam expands in the turbine polytropically $(PV^n = C)$.

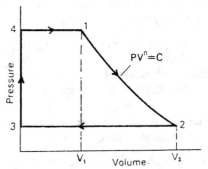

Fig. 32.6

Workdone = Area 4123 (Fig. 32.6)

= Area under $4 - 1$ + Area under $1 - 2$ − Area under $2 - 3$

$$= P_1 V_1 + \frac{P_1 V_1 - P_2 V_2}{n - 1} - P_2 V_2$$

$$= \frac{n}{n - 1}(P_1 V_1 - P_2 V_2)$$

Q. *What will be the Rankine efficiency when super-heated steam is used?*

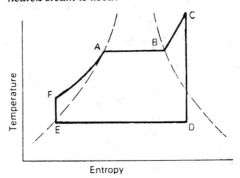

Fig. 32.7

Ans. Rankine efficiency for the complete cycle generating superheated steam and expanding it through the turbine is (Fig. 32.7)

$$\eta = \frac{H_C - H_D}{H_C - h_{w_E}}$$

Problem 32.1 A power plant using steam as the working fluid operates on a Rankine cycle. The turbine receives steam at 3500 kN/m^2 and 623K and exhausts to the condenser at 10 kN/m^2. The condensate is recycled to the boiler by a BFW pump.

Determine:

(a) the energy supplied in the boiler (including the superheater) per kg of steam generated
(b) the dryness fraction of steam entering the condenser
(c) the Rankine efficiency.

Solution
Step (I) Energy Supplied to Steam

$$H_C = 3139 - \left(\frac{1.5}{2}\right)(3139 - 3095) = 3106 \text{ kJ/kg}$$

$$h_{w_E} = 192 \text{ kJ/kg}$$

Energy supplied in the boiler (including the superheater)

$$H_C - h_{w_E} = 3106 - 192 = 2914 \text{ kJ/kg}$$

Ans.

Step (II) Dryness Fraction of Steam Entering the Condenser

Since the expansion of steam in the turbine takes place adiabatically, i.e., isentropic, so

$$\varphi_C = \varphi_D$$

$$\varphi_C = 6.961 - (1.5/2)(6.961 - 6.588)$$

$$= 6.681 \text{ kJ/kg. K}$$

Now, $\varphi_D = 0.649 + x(8.152 - 0.649)$

$$\therefore 6.681 = 0.649 + x(8.152 - 0.649)$$

or, $x = 0.80$

Therefore, the steam entering the condenser is 0.8 dry

Ans.

Step (III) Rankine Efficiency of the Cycle

$$\eta = \frac{H_C - H_D}{H_C - h_{w_E}} = \frac{3106 - [192 + 0.8(2393)]}{3106 - 192} = 0.34$$

i.e. Rankine efficiency is 34%

Ans.

Problem 32.2 A thermal power plant using steam as a working fluid operates on a Rankine cycle. Steam (0.65 MN/m^2 and 0.85 dry) is fed to the turbine wherein steam is allowed to expand following the law $PV^{1.136}$ = constant. If the condenser pressure is 15 kN/m^2, determine the specific work done.

Solution
Step (I) Final Volume

Pressure (kN/m^2)		Specific Volume (m^3/kg)	
Initial	Final	Initial	Final
650	15	0.293(0.85) – 0.249	V_2

For expansion $P_1 V_1^{1.136} = P_2 V_2^{1.136}$

$$V_2 = (P_1/P_2)^{1/1.136}(V_1)$$

$$= (650/15)^{1/1.136}(0.249) \text{ m}^3/\text{kg}$$

$$= 6.871 \text{ m}^3/\text{kg}.$$

Therefore, specific work done

$$= \frac{n}{n-1}(P_1 V_1 - P_2 V_2)$$

$$= \frac{1.136}{0.136}(650 \times 0.249 - 15 \times 6.871)$$

$$= 491.02 \text{ kJ/kg}$$

Ans.

Problem 32.3 A steam turbine, assumed to be working on the Rankine cycle, receives steam at 15 kgf/cm^2 abs. and temperature 240°C and exhausts at 0.36 kgf/cm^2 abs.

Sketch the temperature–entropy diagram for this cycle. Describe the Carnot cycle on the same sketch for the same limits of temperature.

Determine the efficiency of each cycle.

Solution
Operating cycle *ABCDE*
Carnot cycle 1234 (Fig. 32.8)

Step (I) Dryness Fraction of steam Entering the Condenser

$\varphi_C = \varphi_D$ as the expansion is isentropic.

$\varphi_C = 1.593$ kcal/kg K from steam table.

$\varphi_D = 0.236 + x(1.606)$ kcal/kg K from steam table.

$\therefore \ 1.593 = 0.236 + x(1.606)$

or, $x = \dfrac{1.357}{1.606} = 0.8449$

Step (II) Total Energy Available for Work Done

$H_C = 692.5$ kcal/kg of steam.

$H_D = 72.9 + 0.8449 \ (555.6)$

$= 542.357$ kcal/kg of steam

Total energy available for work output

$= 692.5 - 542.357 = 150.143$ kcal/kg of steam

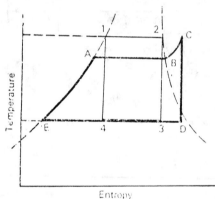

Fig. 32.8

Step (III) Efficiency of the Cycle

Rankine efficiency

$= \dfrac{H_C - H_D}{H_C - h_{w_E}} = \dfrac{150.143}{692.5 - 72.9} = 0.2423$

i.e. 24.23%

Ans.

Step (IV) Carnot Efficiency

Temperature of the superheated steam = 240°C

Temperature of the exhaust steam = 72.9°C from steam tables

∴ Carnot efficiency operating in this temperature range

$= \dfrac{(240 + 273) - (72.9 + 273)}{(240 + 273)}$

$= 0.3257$

i.e. 32.57%

Ans.

Problem 32.4 A steam power plant operates on a Rankine cycle. The steam (28 kgf/cm² abs.) is dry saturated at the inlet. The exhaust steam pressure is 0.90 kgf/cm². abs.

Determine the thermal efficiency of the cycle
(a) ignoring the feed pump work
(b) considering the feed pump work.

Solution

Step (I) Steam Parameters

STEAM, 28 kgf/cm² abs.	Total enthalpy – 669.5 kcal/kg Saturation temp. – 229°C Enthalpy of evaporation – 434.1 kcal/kg

STEAM, 0.90 kgf/cm² abs.	Saturation temp. – 96.2°C Enthalpy of evaporation – 541.3 kcal/kg Liquid enthalpy – 96.2 kcal/kg

Step (II) Dryness Fraction of Exhaust Steam

Since the expansion of the steam in the turbine is isentropic,

$\varphi_B = \varphi_C$

$\therefore \ \ln\left[\dfrac{T_B}{T_O}\right] + \left[\dfrac{L_B}{T_B}\right] = \ln\left[\dfrac{T_C}{T_O}\right] + \left[\dfrac{x \, L_C}{T_C}\right]$

Computing the known values,

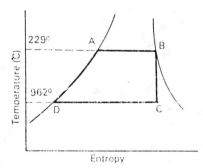

Fig. 32.9

$$\ln \frac{229 + 273}{273} + \frac{434.1}{(229 + 273)}$$

$$= \ln \frac{96.2 + 273}{273} + \frac{x(541.3)}{96.2 + 273}$$

$$\therefore \quad x = 0.799 \approx 0.8$$

Step (III) Rankine Efficiency Ignoring Feed Pump Work

$$H_B = 669.5 \text{ kcal/kg};$$

$$H_C = 96.2 + 0.8(541.3)$$

$$= 529.24 \text{ kcal/kg}$$

$$h_{w_D} = 96.2 \text{ kcal/kg}$$

$$\therefore \quad \eta = \frac{H_B - H_C}{H_B - h_{w_D}} = \frac{669.5 - 529.24}{669.5 - 96.2}$$

$$= 0.2446 \quad \text{i.e. } 24.46\%$$

Ans.

Step (IV) Pump Work

$$W = (\Delta P) \text{ (Specific volume of exhaust steam)}$$

$$= \frac{(P_B - P_C)(v_c)}{427} \text{ kcal/kg (cf. 1 kcal} \equiv 427 \text{ kgf.m)}$$

$$= \frac{(28 - 0.90)(0.00104)(10^4)}{427} = 0.66 \text{ kcal/kg}$$

Step (V) Rankine Efficiency Taking into Consideration Feed Pump Work

$$\eta = \frac{H_B - (H_C + W)}{H_B - (h_{w_D} + W)} \qquad \text{(Fig. 32.10)}$$

$$= \frac{669.5 - (529.24 + 0.66)}{669.5 - (96.2 + 0.66)}$$

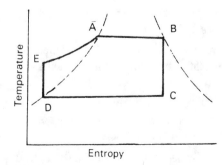

Fig. 32.10

$$= 0.2437 \quad \text{i.e. } 24.37\%$$

Ans.

Problem 32.5 Steam at 1100 kN/m^2 and 523K is delivered to the turbo-alternator where it expands isentropically to a pressure of 280 kN/m^2. Then its pressure drops to 35 kN/m^2 at constant volume and it is exhausted to a surface condenser where the steam is completely condensed with no undercooling. The condesate is then pumped back to the boiler.

Estimate:
(a) the efficiency of the cycle
(b) the specific steam consumption per MWH
(c) the Carnot efficiency for the same temperature limits of the cycle.

Solution The power plant works on the Rankine cycle. It can be diagramatically represented as shown (*P–V* diagram), ignoring the feed pump's work. (Fig. 32.11)

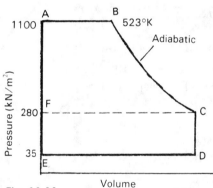

Fig. 32.11

Step (I) Sp. Enthalpy of Steam at B.

$$H_B = 2942.9 - 0.1(2942.1 - 2901.8)$$

$$= 2938.87 \text{ kJ/kg}$$

Step (II) Dryness Fraction of Steam at C.

Expansion B $\longrightarrow$ C is isentropic, so

$$\varphi_B = \varphi_C$$

Now, $\varphi_B = 6.925 - 0.1(6.925 - 6.544)$

$$= 6.887 \text{ kJ/kg K}$$

$$\varphi_C = 1.646 + x\,(7.015 - 1.646) = 1.646 + 5.369x$$

$\therefore 6.887 = 1.646 + 5.369 \, x$

or, $x = 0.976$.

Step (III) Sp. Enthalpy of Steam at C

$H_C = 551.5 + 0.976(2170) = 2669.42 \text{ kJ/kg}$

Step (IV) Sp. Volume of Steam at C

$v_C = x \, (0.645) = 0.976 \, (0.645) = 0.6295 \text{ m}^3/\text{kg}$

Step (V) Work Done in the Cycle

$W = \text{Area } ABCDE = \text{Area } ABCF + \text{Area } CDEF$

$= (H_B - H_C) + (P_D - P_C) \, v_C \text{ (Fig. 32.11)}$

$= (2938.87 - 2669.42) + (280 - 35)(0.6295) \text{ kJ/kg}$

$= 423.677 \text{ kJ/kg}$

Step (VI) Cycle Efficiency

$$\eta = \frac{\text{Network output}}{\text{Net energy supplied}} = \frac{W}{H_B - h_{w_E}}$$

$$= \frac{423.677}{2938.87 - 304} = 0.160 \quad \text{i.e.} \quad 16\%$$

Ans.

Step (VII) Specific Steam Consumption

$1 \text{ MWH} = 10^6 \, (3600) \text{ J} = 10^3 \times 3600 \text{ kJ}$

Available work output per kg of steam

$= 423.677 \text{ kJ}$

Specific steam consumption

$= 10^3 \times 3600/423.677$

$= 8.497 \times 10^3 \text{ kg of steam}$

$= 8.497 \text{ ton of steam}$

Ans.

Step (VIII) Carnot Efficiency

$$\eta = \frac{T_B - T_D}{T_B} = \frac{523 - 346}{523}$$

$$= 0.338 \quad \text{i.e.} \quad 33.8\%$$

Ans.

Q. *Apart from the Rankine cycle, what other cycles are adopted in steam power plants?*

Ans.

1. Reheat cycle
2. Regenerative cycle
3. Reheat - Regenerative combined cycle
4. Binary cycle.

5. Superposed cycle.

Q. *Why is the reheat cycle adopted?*

Ans. In modern thermal power plants, the pressure drop of steam due to its expansion in the turbine is considerable. As a result of this, any superheated steam supplied to the turbine may become wet after partial expansion. This wet steam passing over the turbine blades for a prolonged period will corrode and erode the turbine blades. Just to avoid these problems, the superheated steam after partial expansion in the turbine is reheated at constant pressure in the reheater—a kind of superheater—wherefrom the superheated steam is again delivered to the turbine so that the expansion that will follow will be again in the dry and superheated region. (Fig. 32.12A & 12B)

These are the reasons why the reheat cycle is adopted.

Q. *What is the main purpose of the reheat cycle?*

Ans. Its chief purpose is to increase the dryness fraction of steam passing through the lower stages of the steam turbine.

Q. *Up to what dryness fraction is steam allowed to fall during its expansion in the turbine?*

Ans. It is allowed to fall upto 0.88 and not below it.

Q. *Apart from avoiding corrosion and erosion problem, what are the other gains of the reheat cycle?*

Ans. It increases the thermal efficiency and decreases the specific steam consumption.

Because of reduced specific steam consumption, the size of the boiler and its auxiliaries for the same output is reduced.

Q. *The reheat pressure is generally kept within 20% of the initial pressure of the steam. Why?*

Ans. If the reheat cycle is operated at low pressure, i,e., less than 1/5th of the initial superheated steam pressure, the overall cycle efficiency may fall below the Rankine efficiency.

Q. *Why is the reheat cycle preferred only for high capacity plants, 50MW and above?*

Ans. For high capacity plants, superheated steam pressure may be as high as 100 kgf/cm^2 abs. and

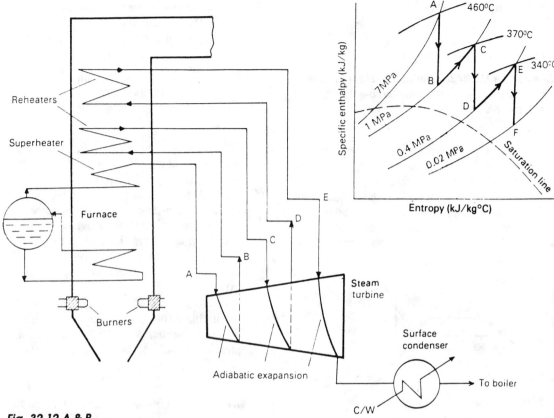

Fig. 32.12 A & B

higher. And at such high pressures, the reheat cycle can be adopted successfully with cycle efficiency higher than Rankine efficiency.

Higher the steam pressure, greater the pressure drop that can be allowed in the turbine before reheating it to the superheated temperature for re-expansion in the turbine and that means higher work output and greater thermal efficiency.

Q. *Why is the reheat cycle not preferred for low capacity plants?*

Ans. It'll not justify the cost of reheater installation.

Q. *How can the efficiency of the reheat cycle be determined from the temperature–entropy diagram?*

Ans. Reheat cycle is an improvement of the Rankine cycle by increasing the temperature and pressure of the steam to be delivered to the turbine.

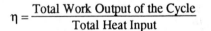

$$\eta = \frac{\text{Total Work Output of the Cycle}}{\text{Total Heat Input}}$$

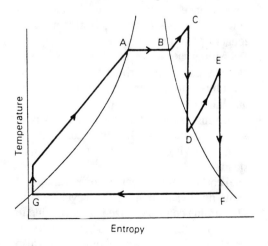

Fig. 32.13

$$= \frac{[H_C - H_D] + [H_E - H_F]}{[H_C - h_{w_H}] + [H_E - H_D]}$$ (Fig. 32.13)

(ignoring the pump work)

Q. *What are the advantages of the reheat cycle over the Rankine cycle using the same turbo-generator and same steam conditions at the inlet to the turbine?*

Ans.

1. Modern power plants use higher pressure steam and due to isentropic expansion, the saturation line is reached earlier, with the effect that most of the turbine stages operate in the saturated steam region. This is highly undesirable as wet steam causes heavy blade erosion due to the impact of water particles on turbine blades. (Fig. 32.14)

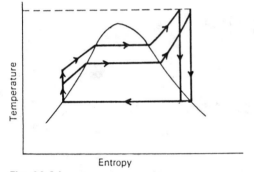

Fig. 32.14

Therefore, reheating is essential for high pressure modern power plants. It also means a longer life for the turbine.

2. Reheating cuts down the mass flowrate of steam input to the turbine by 15–17% and thereby reduces the size of boilers and auxiliaries.

3. Reheating reduces the wetness of exhaust steam to as much as half of the Rankine cycle, thereby saving blade erosion by exhaust steam.

4. Fuel consumption is reduced by as much as 4 to 5% due to adoption of reheat cycle.

5. The turbine condenser size is reduced due to lower quantities of steam and heat load handled. As such cooling water requirement is also reduced.

6. For the required thermal performance, relatively lower steam pressures and temperatures will do. And that reduces the cost of material.

7. Higher thermal efficiency.

8. Lower feed pump power.

Q. *What are the main disadvantages of the reheat cycle in comparison to the Rankine cycle?*

Ans.

1. With the adoption of the reheat cycle, the complexity of operation and control mounts.

2. At low loads, the steam passing over the terminal blades of the turbine just before its entry to the condenser gets severely superheated and that may result in blade overheating.

3. The addition of extra pipe and equipment and control system to the reheat cycle make this cycle more expensive than the simple Rankine cycle.

4. It entails additional design problems as the turbine blades of the intermediate stage immediately downstream of the reheater are appreciably larger than the corresponding stages of the non-reheat system.

5. More floor-space is required.

Problem 32.6 A turbo-generator working on the reheat cycle receives steam at 90 kgf/cm^2 abs. and 480°C. It expands frictionless adiabatically to 15 kgf/cm^2 abs. This exhaust steam is reheated to 15 kgf/cm^2 abs. and 480°C and is then allowed to expand in the turbine to the final exhaust steam pressure 0.05 kgf/cm^2 abs., which is saturated.

Work out
(a) the cycle efficiency
(b) the work output.
Steam feedrate = 12 kg/s.

Solution
Step (I) Sp. Enthalpy vs. Entropy Diagram and Temperature vs. Entropy Diagram

The entire cycyle is represented on a T–φ diagram as shown. (Fig. 32.15A, B)

Step (II) Sp. Enthalpy of Steam

From the **Sp. Enthalpy vs. Entropy chart:**

Sp. enthalpy of steam at 90 kgf/cm^2 abs. and 480°C is $H_C = 798$ kcal/kg

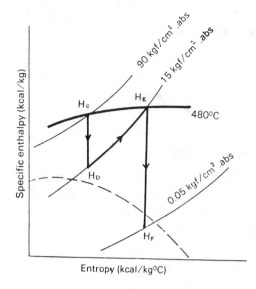

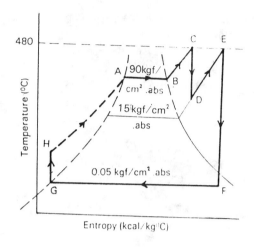

Fig. 32.15 A & B

Sp. enthalpy of steam at 15 kgf/cm² abs. after isentropic expansion is $H_D = 684.8$ kcal/kg

Sp. enthalpy of steam at 15 kgf/cm² abs. and 480°C is $H_E = 820$ kcal/kg

Sp. enthalpy of dry, saturated steam at 0.05 kgf/cm² abs. is $H_F = 546.5$ kcal/kg

Sp. enthalpy at G is $H_G = 32.6$ kcal/kg (from steam tables)

Step (III) Pump Work

Sp. volume of water at 0.05 kgf/cm² abs.

$= 0.001005$ m³/kg

$\therefore$ Pump work $= (90 - 0.05)\,(0.001005) \times 10^4$ kgf m

$= 903.9975$ kgf m

$= 903.9975/427$ kcal $= 2.117$ kcal/kg of steam

Step (IV) Total Work Done

$W_{tot} = (H_C - H_D) + (H_E - H_F) + W_p$

$= (798 - 684.8) + (820 - 546.5) - 2.117$

$= 384.583$ kcal/kg of steam

$\therefore$ Work output per second $= 384.583(12)$

$= 4615$ kcal

Ans.

Step (V) Cycle Efficiency

Net work output $= 384.583$ kcal/kg of steam

Net work input $= (H_C - H_H) + (H_E - H_D)$

$= 798 - (32.6 + 2.117) + (820 - 684.8)$ kcal/kg of steam

$= 898.483$ kcal/kg of steam

Cycle efficiency, $\eta = \dfrac{384.583}{898.483} = 0.4280$

i.e.   42.80%

Ans.

Problem 32.7 Steam at 6000 kN/m² and temperature 450°C is fed to the turbine of a turbo-alternator where it expands isentropically to a pressure 1000kN/m² in the 1st stage. This extraction steam is reheated to 370°C at the same pressure and is again expanded in the 2nd stage of the same turbine where its pressure drops to 200kN/m². This steam is again reheated to a pressure 200 kN/m² and temperature 320°C and is then passed back to the turbine to be expanded to 20kN/m² in the 3rd stage. The finally exhausted steam is then passed to the surface condenser where it is condensed but not undercooled at the same pressure of 20kN/m². The condensate is then recycled to the boiler.

Assuming adiabatic, frictionless expansion in all cases, determine:

(a) the theoretical power available per cycle if the steam flowrate is 10kg/s
(b) the thermal efficiency of the reheat cycle, neglecting feed pump work
(c) the efficiency of the cycle if there were no reheat.

Solution

Step (I) Sp. Enthalpy vs. Entropy Diagram

The entire cycle can be conveniently described in the sp. enthalpy vs. entropy diagram, as shown. (Fig. 32.16)

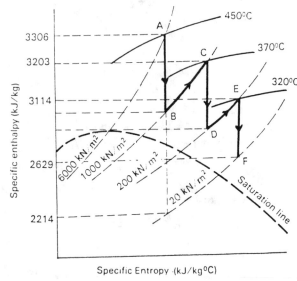

Fig. 32.16

Step (II) Sp. Enthalpies of Steam

From the H–φ diagram on the specific enthalpy vs. entropy chart we find the specific enthalpies of steam at various points.

$H_A = 3306$ kJ/kg; $H_B = 2851$ kJ/kg;

$H_C = 3203$ kJ/kg; $H_D = 2811$ kJ/kg;

$H_E = 3114$ kJ/kg; $H_F = 2629$ kJ/kg;

$H_G = 2214$ kJ/kg

Step (III) Theoretical Power Output

Now net work available per kg of steam expanding through the three stages of the turbine

$$= (H_A - H_B) + (H_C - H_D) + (H_E - H_F)$$

$$= (3306 - 2851) + (3203 - 2811) + (3114 - 2629) \text{ kJ}$$

$$= 1332 \text{ kJ}$$

∴ Theoretical power output per kg of steam passing through the turbine per second

$$= 1332 \text{ kJ/s} = 1332 \text{ kW} = 1.332 \text{ MW}$$

For steam flowrate = 10 kg/s, the theoretical power output per cycle would be = 13.32 MW.

Step (IV) Thermal Efficiency

$$\eta = \frac{\text{Network Output/kg of steam}}{\text{Net Energy supplied/kg of steam}}$$

$$= \frac{(H_A - H_B) + (H_C - H_D) + (H_E - H_F)}{(H_A - h_{w_F}) + (H_C - H_B) + (H_E - H_D)}$$

$$= \frac{1332}{(3306 - 251) + (3203 - 2851) + (3114 - 2811)}$$

$$= 0.3590 \quad \text{i.e.} \quad 36\%$$

Ans.

Step (V) Thermal Efficiency with no Reheat

$$\eta = \frac{H_A - H_G}{H_A - h_{w_G}} = \frac{3306 - 2214}{3306 - 250} = 0.3573$$

i.e. 35.73%

Ans.

Q. *From the above numerical example it is evident that the gain in efficiency is insignificant by the application of reheat. So is it useful to adopt the reheat cycle?*

Ans. Despite the very little gain of thermal efficiency over the non-reheat cycle, the reheat cycle, in the above example, renders the feed steam always superheated minimizing the problem of blade erosion.

Q. *What are the different methods of reheating used in practice?*

Ans.
1. Gas reheating
2. Live steam reheating
3. Combined gas and live steam reheating.

Q. *How is gas reheating carried out?*

Ans. The intermediate extraction from the high pressure steam turbine is reheated in the duct of flue

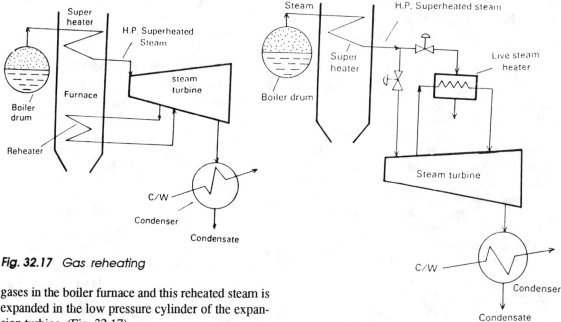

Fig. 32.17 Gas reheating

gases in the boiler furnace and this reheated steam is expanded in the low pressure cylinder of the expansion turbine. (Fig. 32.17)

Usually the steam is reheated to its initial throttle temperature.

Q. *What are the disadvantages of gas reheating?*

Ans.

1. This system requires long and large pipe connections, mounting the cost.
2. With the increase of pipe length, pressure drop of steam increases.
3. It necessitates the provision for expansion and contraction of the steam pipes
4. In case of failure of emergency control, the quantity of steam in the piping and reheater is enough to raise the turbine spinning considerably.

Q. *How is live steam reheating done?*

Ans. This is accomplished by directing a part of the superheated steam from the superheater to reheat the extraction steam from the high-pressure cylinder.

The reheating is carried out in a live steam reheater—a specially designed heat-exchanger. (Fig. 32.18)

Q. *What are the advantages of live steam reheating over gas reheating?*

Fig. 32.18 Live steam reheating

Ans.

1. Large pipe connections are avoided as the reheater can be placed near the turbine
2. Mode of operation is simple
3. Simpler temperature control as the alteration of combustion conditions do not affect the live steam reheater
4. More than one reheating can be employed
5. Wet steam reheating can be carried out.

Q. *What is the chief drawback of live steam reheating?*

Ans. The extraction steam from the high pressure cylinder cannot be reheated to its initial throttle temperature.

Q. *How can this difficulty be avoided?*

Ans. By adopting the combined gas and live steam reheating system.

Q. *How is this combined system carried out?*

Ans. The live steam reheating is hooked up in series with gas reheating. The extraction steam from an H.P.

turbine is first directed to the live steam reheater and it is then further reheated in the gas reheater. Finally, this reheated steam is fed to the L.P. turbine. (Fig. 32.19)

Problem 32.8 A steam turbine working on a reheat cycle receives steam at 100 kgf/cm² abs. and 500°C. The steam expands adiabatically to 10 kgf/cm² abs. with an isentropic efficiency of 82%.

The steam is then reheated to 10 kgf/cm² abs. and 400°C and allowed to expand adiabatically in the lower stage of the turbine upto the condenser pressure 0.1 kgf/cm² abs. with an isentropic efficiency of 85%.

Determine the thermal efficiency of the reheat cycle assuming the pressure loss in the reheater is nil.

If the expansion of steam is carried out in the lower stage of the turine with an isentropic efficiency 80% without reheating, calculate the thermal efficiency of the cycle.

Boiler feedwater pump work may be ignored.

Solution

Step (I) The cycle efficiency can be calculated with the help of specific enthalpy vs. entropy and temperature vs. entropy diagrams (Fig. 32.20B, 20A)

From the **Sp. Enthalpy vs. Entropy chart:**

Sp. enthalpy of steam at 100kgf/cm² abs. and 500°C, $H_A = 808$ kcal/kg

Sp. enthalpy of steam at 10 kgf/cm² abs. after isentropic expansion $H_B' = 666$ kcal/kg

Sp. enthalpy of steam at 10 kgf/cm² abs. and 400°C, $H_B = 780$ kcal/kg

Sp. enthalpy of steam at 10 kgf/cm² abs., after isentropic expansion $H_D' = 565$ kcal/kg

Step (II) Determination of H_B and H_D

$$\frac{H_A - H_B}{H_A - H_B'} = 0.82 \quad \text{or} \quad \frac{808 - H_B}{808 - 666} = 0.82;$$

$$\therefore H_B = 691.56 \text{ kcal/kg}$$

Similarly,

$$\frac{H_C - H_D}{H_C - H_D'} = 0.85 \quad \text{or} \quad \frac{780 - H_D}{780 - 565} = 0.85;$$

$$\therefore H_D = 597.25 \text{ kcal/kg}$$

Step (III) Determination of Cycle Efficiency

$$\eta = \frac{(H_A - H_B) + (H_C - H_D)}{(H_A - h_{w_E}) + (H_C - H_B)}$$

$$= \frac{(808 - 691.56) + (780 - 597.25)}{(808 - 45.4) + (780 - 691.56)}$$

$$= 0.3515 \quad \text{i.e,} \quad 35.15\%$$

Ans.

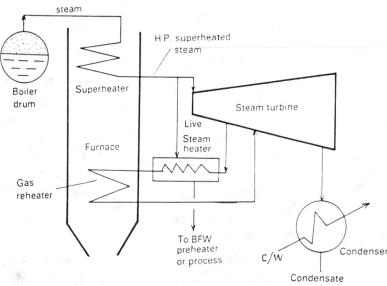

steam

H.P. superheated steam

Boiler drum

Superheater

Furnace

Gas reheater

Live Steam heater

Steam turbine

To BFW preheater or process

C/W

Condenser

Condensate

Fig. 32.19 *In the combined system of steam heating, the live-steam heater is lined up in series with gas reheater*

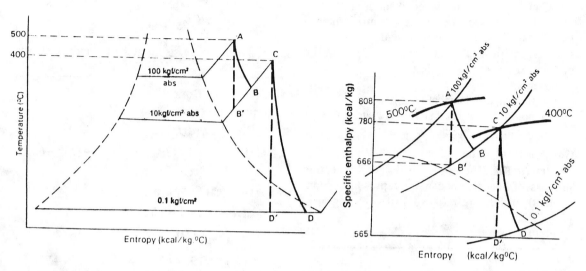

Fig. 32.20 A & B

Step (IV) Cycle Efficiency in Absence of Reheating

If there were no reheating, the isentropic expansion of steam (10 kgf/cm² abs.) in the low pressure cylinder would take place along BF'. (Fig. 32.21)

From the **sp. enthalpy vs. entropy** diagram we find $H_F' = 515$ kcal/kg

Since, **this isentropic expansion $(B) \longrightarrow (F')$** has 80% efficiency, so

$$\frac{H_B - H_F}{H_B - H_F'} = 0.80 \quad \text{or,} \quad \frac{691.56 - H_F}{691.56 - 515} = 0.80;$$

$$\therefore \quad H_F = 550.312 \text{ kcal/kg}$$

Therefore, the thermal efficiency of the cycle without reheating,

$$\eta = \frac{(H_A - H_B) + (H_B - H_F)}{(H_A - h_{w_E})}$$

$$= \frac{H_A - H_F}{H_A - h_{w_E}} = \frac{808 - 550.312}{808 - 45.4}$$

$$= 0.3379 \quad \text{i.e.} \quad 33.79\%$$

Ans.

Problem 32.9 A 3-stage steam turbine working on double reheat cycle receives steam at 200 kgf/cm²

abs. and 550°C. This steam expands in the I stage to 50 kgf/cm² abs. and 350°C. The steam is then reheated to 450°C in the reheater where it suffers a pressure loss of 1 kgf/cm² abs. This reheated steam is again expanded in the II stage of the turbine to 10 kgf/cm²

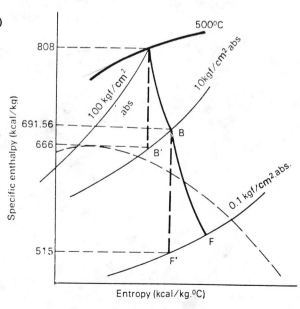

Fig. 32.21

abs. and 280°C. After this the steam is reheated to 350°C in the 2nd reheater where it suffers a pressure loss of 1 kgf/cm² abs. The reheated steam expands in the III stage of the turbine to a condenser pressure 0.5 kgf/cm² abs. The exhaust steam enters the condenser as dry and saturated.

Determine:

(a) thermal efficiency of the cycle
(b) power developed if steam flowrate = 15 kg/s
(c) isentropic efficiency in each stage.

Solution

Step (I) The different expansion and reheating process of the steam can be represented conveniently on **Sp. Enthalpy vs. Entropy diagrams.** (Fig. 32.22)

From the chart:

Point	Sp. Enthalpy
A	$H_A = 814.5 \text{ kcal/kg}$
B	$H_B = 733 \text{ kcal/kg}$
B'	$H_B' = 720 \text{ kcal/kg}$
C	$H_C = 794.5 \text{ kcal/kg}$

D	$H_D = 718.5 \text{ kcal/kg}$
D'	$H_D' = 698 \text{ kcal/kg}$
E	$H_E = 755 \text{ kcal/kg}$
F	$H_F = 632 \text{ kcal/kg}$
F'	$H_F' = 611.5 \text{ kcal/kg}$

Step (II) Work Done

Network done per kg of steam

$$= (H_A - H_B) + (H_C - H_D) + (H_E - H_F)$$
$$= (814.5 - 733) + (794.5 - 718.5) + (755 - 632)$$
$$= 280.5 \text{ kcal/kg}$$

Step (III) Heat Supplied

Input heat energy per kg of steam

$$= (H_A - h_{w_F}) + (H_C - H_B) + (H_E - H_D)$$
$$= (814.5 - 80.9) + (794.5 - 733) + (755 - 718.5)$$
$$= 831.6 \text{ kcal}$$

Step (IV) Thermal Efficiency of the Cycle

$$\eta = \frac{280.5}{831.6} = 0.3373 \quad \text{i.e.} \quad 33.73\% \qquad \text{Ans.}$$

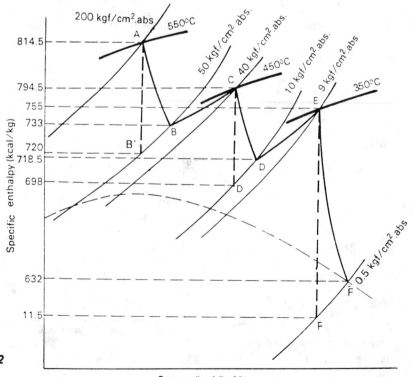

Fig. 32.22

Entropy (kcal/kg°C)

Step (V) Power Developed

Steam flowrate = 15 kg/s
Work output available = 280.5 kcal/kg of steam
Work output per second = 280.5(15) kcal
Power developed = 280.5(15)(5.6918) H.P.
(cf 1 kcal/s = 5.6918 HP)

= 23948 H.P.

Step (VI) Stage Efficiency

Isentropic efficiency of I stage

$$= \frac{H_A - H_B}{H_A - H_B'} = \frac{814.5 - 733}{814.5 - 720}(100) = 86.24\%$$

Ans.

Isentropic efficiency of II stage

$$= \frac{H_C - H_D}{H_C - H_D'} = \frac{794.5 - 718.5}{794.5 - 698}(100) = 78.75\%$$

Ans.

Isentropic efficiency of III stage

$$= \frac{H_E - H_F}{H_E - H_F'} = \frac{755 - 632}{755 - 611.5}(100) = 85.71\%$$

Ans.

Problem 32.10 Steam at 100 kgf/cm² abs. and 500°C expands, in the I stage of a steam turbine to 20 kgf/cm² abs., the isentropic efficiency of expansion being 80% . The steam is then reheated to 450°C in the reheater where it suffers a pressure loss of 1 kgf/cm² abs. The reheated steam then expands to a condenser pressure of 0.05 kgf/cm² abs. The isentropic efficiency of expansion is 85%. The turbine is coupled with a generator generating 45 MW of electricity.

If the transmission efficiency from the turbine to generator is 98% and the generator efficiency is 96%, determine the steam flowrate. Neglect feed pump work.

Solution

Step (I) The expansion and reheating processes are depicted in the **Sp. Enthalpy vs. Entropy chart**. (Fig. 32.23)

From the chart:

H_A = 808 kcal/kg; H_B' = 701 kcal/kg;

H_C = 803 kcal/kg; H_D' = 532.5 kcal/kg

Step (II) Calculation of H_B

Considering the **expansion of steam** $(A) \longrightarrow (B)$

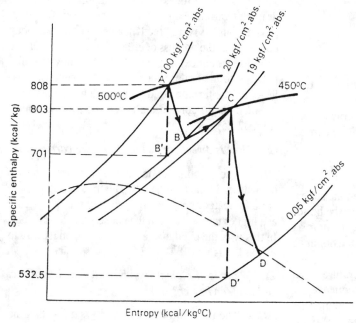

Fig. 32.23

$$\frac{H_A - H_B}{H_A - H_B'} = 0.80 \quad \text{or,} \quad \frac{808 - H_B}{808 - 701} = 0.8;$$

$$\therefore H_B = 722.4 \text{ kcal/kg}$$

Step (III) Calculation of H_D

Considering the **expansion of steam** $(C) \longrightarrow (D)$,

$$\frac{H_C - H_D}{H_C - H_D'} = 0.85 \quad \text{or,} \quad \frac{803 - H_D}{803 - 532.5} = 0.85$$

or $H_D = 573.07$ kcal/kg

Step (IV) Work Done

Work done per cycle $= (H_A - H_B) + (H_C - H_D)$

$= (808 - 722.4) + (803 - 573.07)$

$= 315.53$ kcal/kg of steam

Step (V) Steam Flowrate

Let the mass flowrate of steam be $\dot{M}$ kg/s.

$$\therefore \quad \dot{M}(315.53)(0.98)(0.96) = (45 \times 10^3)(0.2388)$$
$$\text{(cf. 1 kW} = 0.2388 \text{ kcal/s)}$$

$$\therefore \quad \dot{M} = 36.20 \text{ kg/s}$$
Ans.

$= 130.32$ t/h
Ans.

Q. *What is regenerative heating?*

Ans. Also called feed heating, regenerative heating is the process of heating feedwater on its way from the condenser to the boiler by bled steam drawn from various stages of the turbine. The heating is carried out in feed heaters.

Q. *Why is it carried out?*

Ans. There is a large difference between the super-heated steam temperature as supplied from the boiler unit to the turbine and the condensate temperature as it leaves the condenser and is recycled to the boiler as BFW.

Now to increase the condensate temperature on its way back to the boiler and, as a consequence, to increase the thermal efficiency of the plant, the process of regenerative heating (feed heating) is introduced.

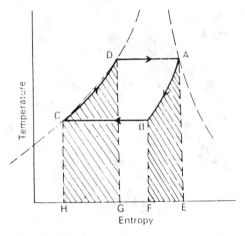

Fig. 32.24

Under ideal conditions, the heat lost by the steam must be equal to the heat gained by the condensate, i.e. (Fig. 32.24)

Area $ABFE$ = Area $DGHC$

This type of heating arrangement results in a cycle efficiency equivalent to Carnot cycle efficiency.

Q. *How is this regenerative heating carried out?*

Ans. In this process, small quantities of steam drawn at various stages of the turbine are allowed to pass through several feed heaters whereupon they condense in the process of either direct heat exchange or indirect heat exchange with the condensate being pumped to the boiler. The bled steam condenses and heats up the turbine condensate which together with the bled-steam-condensate is returned to the boiler. (Fig. 32.25)

Q. *How is direct contact feed heating carried out?*

Ans. The turbine condensate is sprayed onto the bled steam in the feed heater. The mixed condensate, already heated, is again sprayed into the bled steam at a higher temperature in the subsequent feed heater. In this way the condensate temperature, on its way to the boiler, is raised. (Fig. 32.26)

Q. *What is the chief disadvantage of this direct contact feed heater?*

Ans. The main drawback of this system is that the feed pump or recirculation pump has to work with a

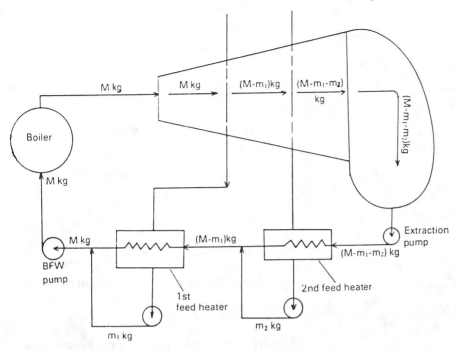

Fig. 32.25 *Regenerative heating of condensate through indirect heat exchange*

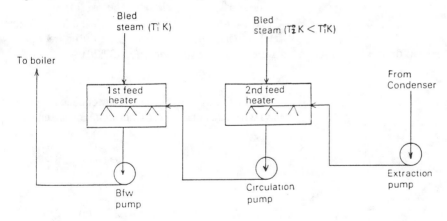

Fig. 32.26 *Direct contact regenerative heating of condensate*

hot feed. That's why this system is rarely used in practice.

Q. *What are the indirect contact methods for feed reheating?*

Ans. There are three methods:
1. Drain Pump method
2. All Drains to Hot Well method

3. Cascade method.

Q. *How does the drain pump method work?*

Ans. In this system the turbine condensate is increasingly heated as it passes through successive feed heaters where bled steam is introduced to heat up the condensate in the process of indirect heat transfer.

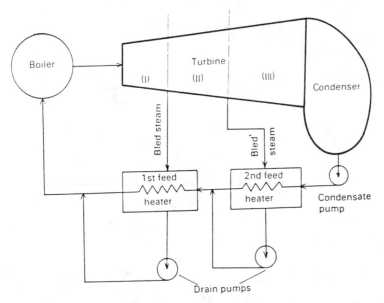

Fig. 32.27 *Regenerative heating of condensate by drain pump method*

The bled steam condensate is added to the feed line by the drain pump. (Fig. 32.27)

Q. *What is an important disadvantage of the drain pump method?*

Ans. The feed pump and the drain pump have to handle a hot feed.

Q. *How does the method of feed heating with all drains to hot well work?*

Ans. In this system the total condensate—sum of turbine condensate and bled steam condensate—is drained to a single well called hot well from which the feed pump delivers the BFW to the boiler through feed heaters where it is heated up by indirect heat exchange with bled steam. (Fig. 32.28)

Q. *How does the cascade system work?*

Ans. In this system, the bled steam condensate from the 1st feed heater is sent to a hot well via the 2nd feed

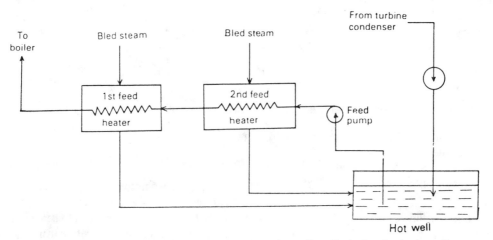

Fig. 32.28 *Regenerative heating of condensate with all drains to hot well*

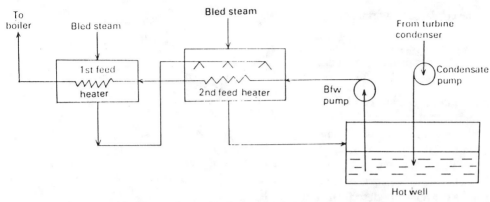

Fig. 32.29 *Regenerative heating of condensate in cascade system*

heater where it cascades down and exchanges heat (indirect heat transfer) with the feed pumped through the tubes from the hot well to the boiler. The bled steam condensate from the 1st feed heater collecting together with the condensate to the bled steam in the 2nd feed heater is discharged to the hot well wherefrom the condensate is pumped back to the boiler through feed heaters. (Fig. 32.29)

Problem 32.11 Superheated steam at a pressure and temperature 7 MPa and 500°C respectively is supplied to a turbine from which steam is bled at pressure 2 MPa and 0.5 MPa for feed heating. In the feed heaters the bled steam is condensed and the liquid enthalpy of feedwater is raised to that of the corresponding bled steam which being condensed is pumped into the feed main.

Determine:
(a) mass of bled steam per unit mass of supply steam
(b) the thermal efficiency of the system as a whole
(c) gain in thermal efficiency over the system without feed heating.

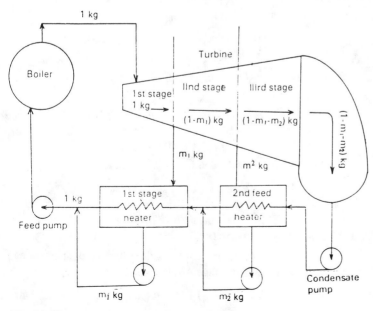

Fig. 32.30

Given: Turbine condenser's working pressure = 0.05 MPa

Steam efficiency of each section of turbine = 80%

No undercooling of bled steam condensate.

Solution

Step (I) The overall arrangement can be diagramatically represented as shown (Fig. 32.30)

Let 1 kg be the supply steam out of which m_1 kg is the bled steam in the 1st feed heater and m_2 kg is the bled steam in the 2nd feed heater.

Step (II) The expansion of steam in successive stages of the turbine can be conveniently described on the Sp. Enthalpy vs. Entropy diagram. (Fig. 32.31)

From the diagram, the specific enthalpies at A, B', C', D' are:

$H_A = 3411$ kJ/kg

$H_{B'} = 3046$ kJ/kg

$H_{C'} = 2791$ kJ/kg

$H_{D'} = 2451$ kJ/kg

Considering expansion AB

$\dfrac{H_A - H_B}{H_A - H_{B'}} = 0.8;$ or $\dfrac{3411 - H_B}{3411 - 3046} = 0.8;$

$\therefore\ H_B = 3119$ kJ/kg

Considering expansion BC

$\dfrac{H_B - H_C}{H_B - H_{C'}} = 0.8;$ or $\dfrac{3119 - H_C}{3119 - 2791} = 0.8;$

$\therefore\ H_C = 2856.6$ kJ/kg

Considering expansion CD

$\dfrac{H_C - H_D}{H_C - H_{D'}} = 0.8;$ or $\dfrac{2856.6 - H_D}{2856.6 - 2451} = 0.8;$

$\therefore\ H_D = 2532.12$ kJ/kg

Step (III) Liquid Enthalpy at B, C, D

From steam tables, specific liquid enthalpies at

2 MPa i.e. $h_{w_B} = 908.5$ kJ/kg

0.5 MPa i.e. $h_{w_C} = 640$ kJ/kg

0.05 MPa i.e. $h_{w_D} = 340.5$ kJ/kg

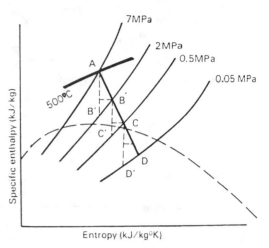

Fig. 32.31

Step (IV) Bled Steam to 1st Feed Heater

Now, the loss of enthalpy by bled steam

= The gain of enthalpy of feedwater

$\therefore\ (m_1)(H_B - h_{w_B}) = (1 - m_1)(h_{w_B} - h_{w_C})$

or,

$m_1 = \dfrac{h_{w_B} - h_{w_C}}{H_B - h_{w_C}} = \dfrac{908.5 - 640}{3119 - 640} = 0.1083$ kg/kg of

supply steam. *Ans.*

Step (V) Bled Steam to 2nd Feed Heater

By energy balance:

$m_2(H_C - h_{w_C}) = (1 - m_1 - m_2)(h_{w_C} - h_{w_D})$

or,

$m_2 = (1 - m_1)(h_{w_C} - h_{w_D})/(H_C - h_{w_D})$

$= \dfrac{(1 - 0.1083)(640 - 340.5)}{2856.6 - 340.5}$

$= 0.1061$ kg/kg of supply steam

Ans.

Step (VI) Total Work Done per kg of Steam Expanding in the Turbine

The theoretical work done by 1 kg of steam due to expansion stage AB, BC and CD is

$= (H_A - H_B) + (1 - m_1)(H_B - H_C) +$
$\qquad (1 - m_1 - m_2)(H_C - H_D)$

$= (3411 - 3119) + (1 - 0.1083)\,(3119 - 2856.6) +$
$(1 - 0.1083 - 0.1061)\,(2856.6 - 2532.12)\ kJ/kg$

$= 780.893\ kJ/kg$

Step (VII) Energy Input

Net energy input

$= H_A - h_{w_B} = 3411 - 908.5 = 2502.5\ kJ/kg$

Step (VIII) Thermal Efficiency

The thermal efficiency of the entire arrangement

$$\eta = \frac{\text{Theoretical Work/kg of steam}}{\text{Energy Input/kg of steam}}$$

$$= \frac{780.893}{2502.5}(100) = 31.20\%$$

Ans.

Step (IX) Without Feed Heating

If there had been no feed heating, the expansion process would be AE as shown. (Fig. 32.32)

$$H_A = 3411\ kJ/kg$$

$$H_E' = 2372\ kJ/kg$$

Considering expansion efficiency 80% as before,

$$\frac{H_A - H_E}{H_A - H_E'} = 0.80$$

or,

$$\frac{3411 - H_E}{3411 - 2372} = 0.8$$

$\therefore \qquad H_E = 2579.8\ kJ/kg$

Thermal efficiency

$$= \frac{H_A - H_E}{H_A - h_{w_E}} = \frac{3411 - 2579.8}{3411 - 340.5} = 0.2707$$

i.e. 27%

Ans.

Hence the gain in thermal efficiency due to feed heating

$$= 31.2 - 27 = 4.2\%$$

Ans.

Q. *What do you mean by topping cycle?*

Ans. It means an additional energy recovery cycle superimposed on a basic power cycle.

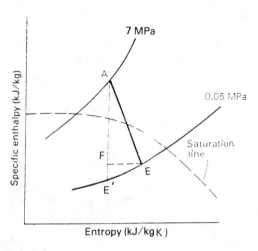

Fig. 32.32

[A cycle is a complete course of operations measured in thermodynamic terms from]
(a) starting point to starting point (CLOSED CYCLE)
(b) starting point to end point (OPEN CYCLE)
   A conventional power plant involves three basic cycles:
   1. Gas (Combustion Process)
   2. Steam
   3. Power]

Q. *What is the utility of a topping cycle?*

Ans. This improves the system heat rate and enhances the utilization of the available heat energy.

Q. *Cite the names of certain topping cycles?*

Ans.
   1. Steam topping cycle
   2. Gas topping cycle
   3. Binary vapour cycle.

Q. *How is a steam topping cycle accomplished?*

Ans. By installing a new boiler plus a turbo-alternator (turbo-generator) that will operate at a higher temperature and pressure than the existing equipment.
   And then feeding the extraction steam of this new turbine into the existing turbo-generator and condenser.

Q. *How is a gas topping cycle established?*

Ans. It is a combination of a gas turbine with a steam cycle.

It can be accomplished in two cycles:

1. SUPERCHARGED BOILER CYCLE—Air is compressed by an air-compressor driven by a gas turbine, combustion taking place at high pressure and the combustion products at elevated pressure and temperature are delivered to the gas turbine to drive the compressor as well as to generate power. (Fig. 32.33)

2. EXHAUST HEAT RECOVERY CYCLE— Gas turbine exhaust gases are used as a source of recoverable heat and of oxygen to supply combustion air to the boiler.

Q. *What is the efficiency gain in a supercharged boiler cycle and exhaust heat recovery cycle?*

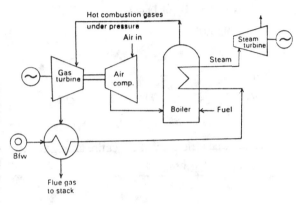

Fig. 32.33 SUPERCHARGED BOILER CYCLE
It combines Gas Turbine and Steam Turbine Cycles

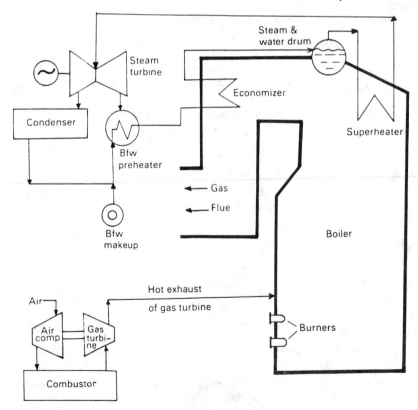

Fig. 32.34 *EXHAUST HEAT RECOVERY CYCLE*
Gas turbine exhaust supplies preheated oxygen to the boiler furnace where steam is generated and superheated and then directed to steam turbine

Ans. Efficiency gain is 4 – 12% in the case of a supercharged boiler cycle and 2–8% in the case of an exhaust heat recovery cycle.

Q. *What are the advantages of a gas topping cycle?*

Ans.

1. Draft fans are not required and that gives a boost to boiler efficiency
2. Neither steam nor electrical energy is expended to deliver combustion air to the burners
3. Air preheater is eliminated
4. Heat is recovered in the form of power
5. Water requirement is reduced
6. Condenser size becomes small
7. Auxiliary equipment sizes also get reduced
8. Improvement of plant efficiency is registered from half-load to full-load.

Q. *What are the advantages of a supercharged boiler?*

Ans.

1. Small furnace volume. Only 25–33% of the heat transfer surface of a conventional boiler is required.
2. Higher heat release rate ($53 \times 10^5 - 62 \times 10^5$ kcal/m^3)
3. Design is symmetrical and compact and so minimum quantity of refractory is required. And that gives an edge in size and weight factor.
4. Centralized, remote control operation is possible
5. Close grouping and compact layout
6. Smaller floor space, smaller building, reduced steel supports are required
7. Rapid startup possible. It requires just 25 minutes to bring a 136 t/h from cold to full steam pressure
8. Lends itself successfully to cogeneration whereupon exhaust gas enthalpy can be extracted in a turbo-generator
9. Removal/replacement of parts for repair is relatively easy
10. Mechanical design flexibility.

Q. *What is the exhaust temperature range of a gas turbine?*

Ans. It ranges from 425–485°C

Q. *What is the oxygen content of a gas turbine exhaust?*

Ans. It contains 15–17% oxygen.
(cf. air 21% oxygen by volume)

Q. *Why is this excess air supplied to the turbine inlet?*

Ans. The purpose is:
(a) to complete the combustion process so that no carbon particles or soot can deposit on turbine bladings
(b) to limit the turbine inlet temperature
(c) to ensure the necessary mass flow of gases to power the turbine.

Note: A gas turbine handles five times the stoichiometric air required to complete the fuel combustion.

Q. *What are the scopes of utilizing the gas turbine exhaust?*

Ans. The heat load and excess oxygen content of the gas turbine exhaust can be used:
(a) to preheat the entire or part of the total combustion air required by the boiler
(b) to supply the primary air to the pulverizer
(c) to preheat boiler feedwater.

Note: The exhaust of a 5MW gas turbine carries enough heat to generate 16–18 ton steam per hour in a pressurized furnace type of waste heat boiler.

Q. *What do you mean by a binary vapour cycle?*

Ans. It means two or more basic single-stage cycles compounded together, such that the heat rejection from the first stage becomes the source of heat for the second stage (or vice versa) while complete separation of heat transfer media is attained.

Q. *What is the most important binary vapour cycle?*

Ans. It is the mercury-steam cycle.

Q. *Briefly describe the mercury-steam binary cycle.*

Ans. It is composed of two cycles—mercury cycle and steam cycle.

MERCURY CYCLE: Liquid mercury is circulated to the evaporator tubes of the boiler converting

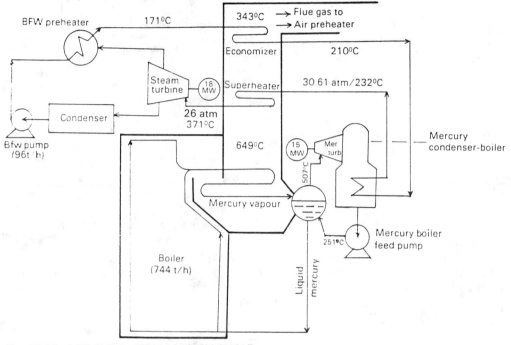

Fig. 32.35 MERCURY—STEAM BINARY CYCLE
Mercury cycle is superimposed on steam cycle whereupon steam is generated in Mercury condenser-boiler where heat is rejected by mercury vapour after being expanded in mercury turbine

the liquid into vapour which is then allowed to expand in the mercury turbine to generate electric power. The exhaust mercury vapour from the turbine is discharged to mercury condenser-boiler where the mercury vapour condenses into liquid releasing its heat of vapourization and pumped back to the boiler. (Fig. 32.35)

STEAM CYCLE: The heat rejected by vapour mercury in the mercury-condenser-boiler is absorbed by the BFW to generate steam at a desired pressure. This steam is superheated and directed to drive the steam turbine to produce an additional power output.

Fuel is burned in the mercury boiler furnace. The heat released by the fuel is utilized to vapourize the mercury, superheat the steam and preheat the combustion air and boiler feedwater.

Q. *Why is the mercury cycle superimposed upon the steam cycle and not the other way around?*

Ans. Mercury boils at a higher temperature than water at any fixed pressure.

Pressure	B.P. of Hg	B.P. of H_2O
1 atm.	629K	373K
12 atm.	811K	462K

Thus mercury has a much higher vapour temperature for any saturated vapour condition than that of water. And therefore, mercury is capable of generating steam at considerable pressure at any condensation point.

That's why the mercury cycle tops the steam cycle.

Note: Mercury at 10 atm. boils at 800K. The mercury vapour at this temp. and press may be allowed to expand in the turbine to produce power and then it may be used to generate steam at 40.8 atm./525K (saturated).

Mercury vapour at 9.5 atm./797K (turbine throttle) is equivalent to steam at 156 atm./838K.

Q. *What are the advantages of the mercury cycle?*
1. High overall plant efficiency
2. High availability (as much as 85%)

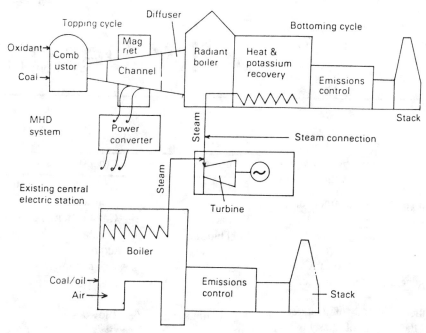

Fig. 32.36 *MHD System superimposed on steam cycle ensures 50% more efficiency than conventional pulverized-coal-fired plants fitted with emission control gadgets*

3. Operation simplicity
4. Acceptable plant maintenance
5. Reasonable plant cost
6. No operational trouble, though mercury is toxic.

Q. *How does a MHD topping cycle work?*

Ans. MHD (magnetohydrodynamics) topping cycle is superimposed on the steam-turbine bottoming cycle.

The topping cycle is a direct-conversion process in which the fuel is burned at high pressure and at a temperature close to 2750°C. Then potassium compounds are added to the combustion gases to produce a seeding effect to enhance the electrical conductivity. The hot gases are allowed to expand, at sonic velocity, through a channel surrounded by a magnetic field. As the ionized gas travels through the magnetic field, it produces direct current which is converted to ac power.

Following this direct conversion process, the combustion gases are passed through a conventional boiler where they transfer their heat to evaporator, superheaters and reheaters to generate high-pressure superheated steam that drives a turbo-alternator. (Fig. 32.36)

Q. *Why are potassium compounds added to the combustion products?*

Ans. The function of potassium is to remove SO_2.

Q. *Why is the potassium recovery unit installed?*

Ans. Potassium seed material is expensive. Economics dictates that it must be regenerated and returned to the combustor.

33

Boiler Design

Q. *What is the basic problem faced by a boiler designer in designing a steam generating unit?*

Ans. In a steam generating unit two distinct fundamental processes take place:
1. Conversion of the potential energy of the fuel into thermal energy
2. Transfer of this liberated thermal energy to the working fluid to generate steam for useful purpose(s).

This being the case, the basic task of a boiler designer is to maximize the output of these two processes simultaneously. And for that purpose he must design the layout of the entire heat-absorbing surface in such a manner that it will receive maximum available heat in the process of fuel combustion.

However, boiler design is an extremely specialized subject. In practical terms, the designers face mounting difficulties as the process of heat transfer becomes complicated due to constant variations of operating parameters, irregularity in the configuration of heating surfaces and a lot of other factors that enter into the equation simultaneously. Finally, there is another paramount problem: How to obtain maximum efficiency at lowest cost.

Q. *What is basically needed for maximum overall economy, i.e., the boiler running at maximum efficiency but at lowest cost?*

Ans. It needs a balanced design. For this, all component parts must be properly designed and accurately proportioned and linked to the other elements with each process being correctly assessed and related to the other processes involved in the unit as a whole.

Q. *What are these components and processes?*

Components	Processes
1. Boiler	1. Air and water preheating
2. Furnace	2. Combustion air supply
3. Fuel burning equipment	3. Combustion products removal
4. Feedwater	4. Ash handling
5. Refractories and baffles	5. Steam separation
6. Foundations and supports	6. Blowdown

Q. *What factors must be taken into account for design consideration of a boiler?*

Ans.
1. Service requirements
2. Load characteristics
3. Fuel characteristics
4. Mode of fuel burning
5. Hydrodynamics of gas flow
6. Feedwater quality
7. Furnace size, shape and material of construction
8. Type of furnace bottom
9. Boiler proper
10. Boiler operation
11. Capital investment.

Q. *What do you mean by service requirements of a steam generating unit?*

Ans. It implies the nature of service it will render, i.e., whether the unit will be for a public utility central station for production of electricity or for an industrial plant for the generation of steam and or electricity for process use.

Q. *What is the difference between an industrial plant boiler and a central station steam generator in the light of service demanded from them?*

Ans. The primary production equipment of a utility plant is the monobloc unit, i.e., boiler and turboalternator taken as a whole. Whereas in the case of an industrial plant the chief equipment are those which are directly involved in the process for the manufacture of end products.

The basic service required from a utility plant is the efficient manufacture of kilowatts whereas in the case of industrial plant boiler, the sole aim is the production of end product.

The industrial plant works under more unfavourable conditions of load, water and fuel than the central power station steam generator. The steam load in an industrial plant may be highly unpredictable, hunting continuously between high and low extremes.

Q. *What should be the prime considerations in designing an industrial boiler?*

Ans.
1. Dependability and operational reliability
2. Capital investment
3. Availability
4. Working efficiency
5. Cost of operation
6. Serviceability.

Q. *What is the economic justification for a large amount of money being invested to instal a public utility central station steam generator?*

Ans. These set-ups are made purely on the basis of the total cost of electricity production. Though the initial capital investment is very high, the cost per unit output of electricity is very low and that justifies the expenditure of large sums when the plant runs at maximum efficiency with assured lowest unit cost.

Q. *What load characteristics are necessary in the design of a steam generator?*

Ans. The boiler must be so designed as to run at high efficiency under normal load and at the same time should be capable of meeting maximum demand as well as fluctuating load characteristics. Hence the boiler designer should essentially consider the following load characteristics:
1. Maximum load, normal load and minimum load
2. Load factor
3. Nature of load—constant or fluctuating
4. Duration time of each load rate.

Q. *Why do fuel characteristics play an important role in boiler design?*

Ans. From these very characteristics, a boiler designer gets the knowledge of the heat value available from the fuel as well as its such specific properties such as:
(a) ash content and the percent of volatile matter
(b) nature of ash and its fusion point
(c) unburnt fuel losses as carbon is lost to ash or escapes through flue gases
(d) the presence of such corrosive agents as sulphur and vanadium that will dictate the flue gas exit temperature as well as the material of construction of the heating surfaces of the boiler to avoid the problem of corrosion and slagging.

It is these characteristics that control the fuel burning mechanism in a boiler, which in turn influences the boiler design.

Q. *How does the mode of fuel burning exert an influence upon boiler design?*

Ans. It is the capacity of the fuel burning device that controls the rate of fuel input which in turn determines the furnace volume and its design specifications.

The design of pulverized coal fired boilers is certainly not the same as stoker or grate fired boilers.

Be it oil fired burners—vapourizing, rotary, gun or steam (or air) atomizing—or gas fired burners, each type dictates its own design peculiarities on the design of the boiler furnace. As the furnace design undergoes a change, so does the layout of the heat absorbing surface of the boiler.

However, it can be safely concluded that the type of fuel burning equipment and the method of firing exercise much greater influence on furnace design than on boiler design.

Q. *How do the gas flow characteristics influence boiler design?*

Ans. The gas flow through the boiler is effected by the differential pressure between the combustion products in the furnace core and the flue gases at the boiler exit. This pressure difference, called draught (draft) may be effected by natural means or by mechanical means to supply the necessary primary and secondary air to sustain and control fuel combustion.

Depending upon whether this draught is produced naturally (by chimney effect) or by mechanical means (by installing induced draught fans, forced draught fans or both), the boiler design is altered accordingly. The quantity of excess air supplied in the form of secondary air influences the boiler capacity as well as furnace temperature.

If the firing rates are high, the draught requirements will be considerably very large and accordingly, the size of the ID and FD fans becomes proportionately much bigger and that increases the size of the boiler unit.

Again, higher boiler efficiency needs combustion air to be preheated and therefore, an air preheater is to be installed almost invariably in the convective shaft of the boiler furnace. And that means a further draft loss which must be taken into account in the overall design of the steam generator.

Q. *In what way does feedwater quality affect boiler design?*

Ans. The presence of dissolved solids and gases, suspended matter and organic contaminants in feedwater cause corrosion, scaling, priming and foaming that effectively impair the performance of a boiler. These undesirable ingredients must be eliminated as much as possible before the feedwater is charged to the boiler. Quality requirements become more stringent with the increase of pressure and temperature of generated steam.

Feedwater quality, together with other factors, influences the design of drum internals, steam separator and steam washer, etc.

Q. *Why is the furnace one of the elements of consideration in boiler design?*

Ans. The furnace volume must be sufficient to maintain the necessary heat release rate and furnace temperature while the combustion space should be sufficient to contain the flame so that it does not directly hit the waterwalls.

The rate of steam generation and the temperature and pressure of output steam dictate the furnace temperature and the heat release rate which in turn govern the size and shape of the furnace, nature and materials of construction of furnace wall and disposition of heat-absorbing surface in the radiant and convective shafts of the furnace.

Q. *Why is furnace bottom design an important part of boiler design?*

Ans. It becomes an essential consideration during designing coal fired boilers. They may be pulverized coal fired or stoker or grate fired. In each case, the removal of ash and slag poses a problem. For pulverized coal firing, the furnace bottom must be cone shaped to drop out all molten slag to be carried off mechanically, pneumatically or by water. In the case of stoker fired boilers, high pressure water jets are directed upon the ash and molten slag as they spill over the chain at the furnace bottom. The lack of proper furnace design for adequate ash removal may result in excessive slagging of waterwalls impairing the heat transfer characteristics and performance of the boiler.

Q. *What factors control the design of the boiler proper?*

Ans.

1. **The operating pressure and temperature**
2. **The Quality of Steam**—Whether the steam required should be wet, dry or superheated. If wet steam is required, the designer may do away with the separator and superheater. If 99.5% dry steam is needed, the designer must opt for suitable steam separators. The incorporation of a superheater or reheater becomes obvious if superheated steam or steam reheating is required downstream, e.g., in the turboalternator.
3. **Layout of Heating Surface**—The prime aim of the boiler designer is to obtain the best disposal of heat-absorbing surface within the limitations of space as dictated by the furnace and other components.

4. **Heating Surface Requirements**—These depend upon the duty of the element heat exchangers such as primary evaporators, secondary evaporators, superheaters—radiant and convective, reheater, economizer and air preheater.

5. **Circulation of steam and water**—natural or forced.

6. **Provision for continuous blowdown**

7. **The capacity of boiler drum**

8. **Materials and methods of construction.**

Q. *Why are boiler operation and maintenance considered essential elements in boiler design?*

Ans. The life of a boiler is extended with good, trouble-free operation and adequate maintenance and cleaning.

The designer must make allowance for efficient removal of:

(a) soot, slag and ash deposited on the fireside of heating surfaces

(b) scale, sludge and silt settled on the waterside of heating surfaces.

Accessibility for operation, maintenance and repairing must be easy and quick to ensure higher operating efficiency and offset the long outage time.

Adequate provision must be made for

(a) soot blowing

(b) tube cleaning—chemically and or mechanically

(c) washing economizer and air preheater surfaces.

Automation should be injected wherever it leads to higher reliability and greater ease in boiler operation.

Q. *What factors are involved in determining the overall capital investment in designing a boiler?*

Ans.

1. Cost of equipment

2. Cost of fuel

3. Cost of labour and materials for operation, maintenance and repairing

4. Cost of the auxiliaries, e.g., cost of running pumps, fans, ash disposal systems, etc.

5. Expected life of the equipment

Q. *What is the most important limitation on the steam generation capacity of a boiler?*

Ans. It is the rate of fuel combustion in the furnace.

Q. *Why?*

Ans. The rate of fuel burning is limited by the firing equipment and the furnace volume. As soon as the highest efficient combustion rate for any particular fuel is established, the only method of increasing the steaming capacity of a boiler is to burn more fuel or to change the fuel burning mechanism for more efficient use of the fuel burnt.

Q. *What other factors impose limitations on the steaming capacity of a boiler?*

Ans. Limitations are imposed by impaired heat transfer characteristics due to

(a) deposition of scale, precipitates and sludge on the waterside of the heat-absorbing surfaces

(b) corrosion, erosion, caustic embrittlement, slagging, etc., on the fireside of the heat-absorbing surfaces.

Also, the boiler design itself, the design pressure and temperature, materials of construction and construction methods impose limitations on the steaming capacity of a boiler.

Q. *What factors determine the efficiency of a steam generating unit?*

Ans.

1. Boiler design characteristics

2. The type of fuel and fuel burning device

3. Operating parameters.

Q. *Can a steam generating unit run with maximum efficiency at a load below the rated capacity?*

Ans. Yes.

Q. *What factors affect the maximum efficiency of a steam generator?*

Ans.

1. Derated capacity utilization

2. Loss of combustibles to flue gases, ash and slag

3. Higher excess air supplied

4. High flue gas exit temperature

5. Lack of proper maintenance

Q. *While investigating fuel characteristics, the type of fuel burning equipment and the method of firing, during a boiler design, what is the other factor pertinent to fuel combustion that a boiler designer should basically consider?*

Ans. It is the flame characteristics of the fuel to be burned in the boiler furnace.

Q. *What is that?*

Ans. A flame appears only when fuel molecules in contact with oxidizer molecules (O_2) are raised to the activation energy so that the chemical reactions between the fuel and the oxidizer take place.

Combustion forms a flame in which the reactions of combustion of fuel constituents take place and evolution of heat occurs. As in modern steam generation unit, the combustion of solid, liquid or gaseous fuels is done by the flame-method, i.e., the fuel and air are supplied into the furnace space in the form of turbulent jets which gradually intermix and the flame assumes a definite length and shape which can be conveniently controlled.

The light and heat given off by the flame as well as its length and shape constitute the flame characteristics.

Q. *How do the flame characteristics influence a boiler design?*

Ans. The luminosity of a flame appears due to the presence of solid particles (coke or foreign matter) heated to incandescense by the burning gas. Non-luminous flames are much hotter than luminous flames.

Anthracite coal—pulverized or lump—burns with a transparent or barely visible flame. A boiler designer must limit the heat-absorbing surface to which heat may radiate in the furnace proper so that the furnace temperature does not drop below the ignition point of anthracite.

There is a marked difference between the flames produced by various fuels. Semianthracite burns with a short, clear flame; bituminous with a long, luminous flame and sub-bituminous and lignite burn with a long, yellow flame.

Luminosity of oil flames being high, combustion of oil contribute to more radiant heat transfer than convection. As such a boiler designer should lay more emphasis on radiant heat transfer and the design of radiant heat absorbing waterwall tubes in the oil-fired furnace core.

As the gaseous fuels are burnt, the major portion of the heat transfer takes place by way of convection after the combustion products leave the furnace proper. Since the radiation heat transfer is at a minimum, greater emphasis is laid on designing and on the best possible disposal of convective heat transfer surfaces (superheaters, economizers and air heaters) than on water-cooled furnace walls.

Q. *Draw the contour lines of temperature at various depths of the furnace for a down fired radiant boiler.*

Ans. Fig. 33.1

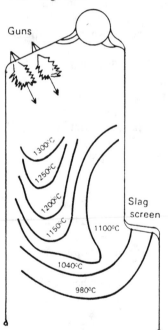

Fig. 33.1 *Isothermal contour lines at various depths of the furnace*

Q. *How do the heat transfer characteristics affect boiler design?*

Ans. All the three modes of heat transfer—radiation, convection and conduction—enter into the equation, either alone or in combination during theoretical calculations of waterwalls, superheaters, economizers and air heaters.

Radiation heat transfer is prevalent in the core (hottest part) of the furnace and the transfer of radiant energy to the boiler tubes is dependent on the luminosity of the flame and the amount of the heat absorbing surface exposed to the flame.

The rate of radiant heat absorbed by the waterwall is

$$q_{rad} = \varepsilon \, AF \, \sigma \, [(T_1)^4 - (T_2)^4]$$

$$= 5.67(\varepsilon) \, AF \left[\left(\frac{T_1}{100} \right)^4 - \left(\frac{T_2}{100} \right)^4 \right], \text{W}$$

where ε = emissivity of the flame

A = area of cross-section of radiant heat absorbing surface, m^2

F = view-factor

σ = Stefan-Boltzmann constant = 5.67×10^{-8} $W/m^2 \, K^4$

T_1 = absolute temperature of flame, °K

T_2 = absolute temperature of radiant heat-absorbing surface, °K

The coefficient of radiative heat transfer is given by

$$h_{rad} = \frac{q_{rad}}{A\Delta\theta} = \frac{5.67 \, \varepsilon \cdot F \left[\left(\frac{T_1}{100} \right)^4 - \left(\frac{T_2}{100} \right)^4 \right]}{\Delta\theta}$$

Radiative Resistance

$$R_{rad} = \frac{1}{h_{rad} \, A} = \frac{\Delta\theta}{5.67 \, \varepsilon \cdot F \cdot A \left[\left(\frac{T_1}{100} \right)^4 - \left(\frac{T_2}{100} \right)^4 \right]}, \text{°K/W}$$

Heat transfer in the second vertical shaft and the horizontal duct of the furnace takes place entirely by convection. The rate of heat transfer from hot flue gases to the heat-absorbing surfaces laid out in this zone is given by

$$q_{conv} = h_{conv} \, A\Delta\theta, \text{W}$$

where, h_{conv} = coefficient of convective heat transfer, W/m^2 °K

A = area of heat-absorbing surface, m^2

$\Delta\theta$ = temperature difference between hot gaseous products of combustion and tube walls

Convective resistance

$$R_{conv} = \frac{1}{h_{conv} \, A}, \text{°K/W}$$

Either natural convection (caused mainly by density differences due to temperature differentials) or forced convection (caused by mechanical force applied to induce motion to the flue gases) is involved and the overall heat transfer coefficient and design of convective heat absorbing elements in the furnace undergo changes accordingly.

Heat transfer by the mode of conduction takes place through the wall thickness of the tubes as well as across the scale or depositions on both the inside and outside of the tube surface. The rate of conductive heat transfer through a wall is given by

$$q_{cond} = k \, A \frac{\Delta\theta}{\Delta x}$$

For a composite wall,

$$q_{cond} = k_1 \, A \left[\frac{\Delta\theta}{\Delta x} \right]_1 + k_2 \, A \left[\frac{\Delta\theta}{\Delta x} \right]_2 + k_3 \, A \left[\frac{\Delta\theta}{\Delta x} \right]_3$$

and conductive resistance

$$R_{cond} = \frac{1}{A} \Sigma \frac{\Delta x}{k}$$

where, k_1 = thermal conductivity of the scale, W/m °K

k_2 = thermal conductivity of the tube wall, W/m °K

k_3 = thermal conductivity of slag deposition, W/m °K

$\left[\dfrac{\Delta\theta}{\Delta x} \right]_1$ = temp. gradient across the scale, °K/m

$\left[\dfrac{\Delta\theta}{\Delta x} \right]_2$ = temp. gradient across the tube wall, K/m

$\left[\dfrac{\Delta\theta}{\Delta x} \right]_3$ = temp. gradient across the slag, K/m

If all the three processes of heat transfer take place simultaneously then overall resistance to the heat flow

$$R = R_{rad} + R_{conv} + R_{cond}$$

Therefore, heat flowrate, $q = \dfrac{\Delta\theta}{\Sigma R} = UA \, \Delta\theta$

where, U is called overall heat transfer coefficient

A boiler designer must consider the internal and external fouling factors while designing the superheaters, waterwalls, economizers, etc.

Therefore, the overall heat transfer coefficient becomes

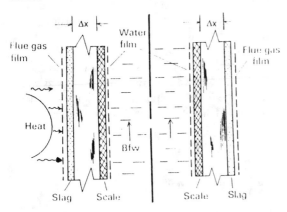

Fig. 33.2

$$U = \cfrac{1}{\cfrac{1}{h_o} + \left[\cfrac{\Delta x}{k} + \cfrac{1}{h_{s_i}} + \cfrac{1}{h_{s_o}}\right] + \cfrac{1}{h_i}} \; \text{W/m}^2 \, {}^{\circ}\text{K}$$

where h_o = coefficient of heat transfer across the flue gas film on the tube surface, W/m^2 °K

h_i = coefficient of heat transfer across the water or air or steam film on the inner surface of the tubes, W/m^2 K

k = thermal conductivity of tube wall, W/m °K

Δx = thickness of tube wall, m

h_{s_i} = coefficient of heat transfer of the scale inside the tube, W/m^2 °K

h_{s_o} = coefficient of heat transfer of the slag outside the tube, W/m^2 °K

Now, if the mode of heat transfer across the hot flue gas film on the outer surface of the tube is both radiative and convective type then the rate of heat transfer. (Fig. 33.3)

$$q = q_{rad} + q_{conv} = (K_r + K_{cv}) \, \Delta\theta$$

where K_r = radiative conductance of hot gas film

$$= \sigma A \, \varepsilon \, F \, (T_1^4 - T_2^4)/\Delta\theta$$

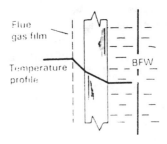

Fig. 33.3

K_{cv} = convective conductance of hot gas film

$$= (h_{cv})_o \, A$$

$$\therefore \; h_o = \sigma \, \varepsilon \, F \, [(T_1^4 - T_2^4)/\Delta\theta] + (h_{cv})_o$$

Determination of View Factor

Problem 33.1 Two rows of waterwell tubes cover the walls of a boiler furnace. The tubes have outer dia = 100 mm and these are arranged in in-line arrangement with two rows on equal pitch (in the plane parallel to the wall).

Pitch = 500 mm

Mean view factor between the surface of the furnace wall and the waterwall tubes is to be determined.

Solution

The required view factor can be determined with the help of the formula:

$$F = 1 - \sqrt{1 - (d/p)^2} + (d/p) \arctan \sqrt{(p/d)^2 - 1}$$

where d = tube dia (OD)

p = tube pitch

$$F = 1 - \sqrt{1 - (100/500)^2} + (100/500) \arctan$$

$$\sqrt{(500/100)^2 - 1} = 0.29$$

Ans.

Determination of Coefficient of Heat Transfer by Radiation

Problem 33.2 It is desired to calculate the coefficient of heat transfer by radiation in a superheater placed in the path of a flue gas so that the flue gas temperature at the inlet to and exit from the superheater are 1470°K and 1080°K respectively.

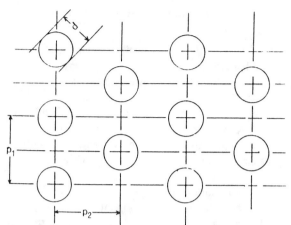

Fig. 33.4

The superheater tubes are staggered in layout as shown with tube pitches:

$p_1 = p_2 = 2d = 2 \times 40$ mm   (Fig. 33.4)

Average temperature of the heat transfer surface may be assumed to be constant and equal to 870°K.

Flue gas analysis : CO_2–12%; H_2O–4.5%
Flue gas pressure = 112.815 kPa (112.815 kN/m²)
Emissivity of tube surface = 0.8

Solution
Step (I) Average Temp. of Flue Gas

$$T = \frac{1}{2}(1470 + 1080) = 1275 \ °K$$

Step (II) Emissivities of CO_2 and H_2O of the Flue Gas

The emissivities of CO_2 and H_2O of flue gas at this average temperature are determined with the help of emissivity curves of CO_2 and H_2O[**L.P$_{CO_2}$ vs. temp.** and **L.P$_{H_2O}$ vs. temp.**]

where P = partial pressure of the corresponding gas
L = mean path of a ray in the intertubular space

$= 1.08d[(p_1 p_2/d) - 0.785]$

$= 1.08(0.04)[2 \times 2 - 0.785] = 0.1388$ m

$LP_{CO_2} = 0.12(112.815 \times 10^3)(0.1388)$ m.Pa

$$= \frac{0.12(112.815 \times 10^3)(0.1388)}{0.981 \times 10^5}$$ m atm.

(cf. 1atm = 0.981×10^5Pa)

= 0.0191 m atm. = 1.91 cm atm.

$LP_{H_2O} = 0.045(112.815 \times 10^3)(0.1388)$ m Pa

$$= 0.045(112.815 \times 10^3) \frac{(0.1388)}{0.981 \times 10^5}$$

= 0.00718 m atm.

= 0.718 cm atm.

From Emissivity Curves

For, $LP_{CO_2} = 1.91$ cm atm and at temp. 1 275 K

$\varepsilon_{CO_2} = 0.059$

$LP_{H_2O} = 0.718$ cm atm and at temp. 1 275 K

$\varepsilon_{H_2O} = 0.0107$

Step (III) Average Emissivity of Flue Gas

Therefore, the emissivity of the flue gas at mean temperature:

$\varepsilon = \varepsilon_{CO_2} + \beta \ \varepsilon_{H_2O} = 0.059 + \beta(0.0107)$

where β = correction factor.

From β **vs. P_{H_2O} curves** we determine for

$LP_{H_2O} = 0.718$ cm atm.

$\beta = 1.04$ corresponding to $P_{H_2O} = 0.05175$ atm.

$$P_{H_2O} = 0.045 \ \frac{112.815 \times 10^3}{0.981 \times 10^5} \ \text{atm.}$$

$$= 0.05175 \ \text{atm.}$$

$\therefore \ \varepsilon = 0.059 + 1.04(0.0107) = 0.0701$

Step (IV) Average Absorptivity of the Flue Gas

This can be determined from the equation:

$$\alpha = \varepsilon_{CO_2} \left[\frac{T}{T_W} \right]^{0.65} + \beta \ \varepsilon_{H_2O}$$

where α = mean absorptivity of the flue gas
T = average temperature of the flue gas, °K
T_w = absolute temperature of the furnace wall, °K
At the tube surface temperature:

$$\varepsilon_{CO_2} = 0.057 \quad ; \quad \varepsilon_{H_2O} = 0.025$$

Therefore, the mean absorptivity of flue gas at the tube surface temperature

$$\alpha = 0.057 \left(\frac{1257}{870}\right)^{0.65} + 1.04(0.025) = 0.0990$$

Step (V) Rate of Radiative Heat Transfer

The rate of heat transfer, by radiation, from the hot flue gas to the superheater is given by

$$q_{rad} = \frac{1}{2} [\varepsilon_w + 1](5.67) \left[\varepsilon \left(\frac{T}{100}\right)^4 - \alpha \left(\frac{T_w}{100}\right)^4 \right],$$
$$\text{W/m}^2$$

$$= \frac{1}{2} (0.8 + 1)(5.67) \left[0.0701 \left(\frac{1275}{100}\right)^4 - 0.099 \left(\frac{870}{100}\right)^4 \right],$$
$$\text{W/m}^2$$

$$= 6\,559.057 \text{ W/m}^2$$

Step (VI) Coefficient of Heat Transfer by Radiation

The coefficient of heat transfer by radiation

$$h_{rad} = \frac{q_{rad}}{\Delta\theta} = \frac{6\,559}{1\,275 - 870} = 16.19 \text{ W/m}^2 \text{ }^\circ\text{K}$$

Ans.

Determination of the Coefficient of Heat Transfer by Convection

Problem 33.3 A waterwall, constructed as a staggered bank of tubes, is exposed to the cross-flow of flue gases of the following composition:

N_2—76%; CO_2—13%; H_2O—11% (by volume)

The gas temperature before the bank of tubes is 1 470 °K and after the bank of tubes is 1 100 °K.

The tube bank is built-up of 4-rows of identical tubes in the direction of the gas flow. (Fig. 33.5)

The transverse pitch, $p_1 = 2.5\ d$
The longitudinal pitch, $p_2 = 2\ d$
The tube diameter, $d = 100$ mm

The average velocity of flue gas in the narrowest cross-section of the bank of tubes $v = 12.5$ m/s.

Ignore the fouling factor.

The mean coefficient of heat transfer by convection from the flue gas to the bank of staggered boiler tubes is to be determined.

Solution The convective heat transfer coefficient from the flue gas to the boiler tubes (clean) disposed in staggered arrangement can be determined by the formula:

$$Nu_f = 0.41(Re_f)^{0.6}(Pr_f)^{0.33} \; \phi_s \qquad (I)$$

Step (I) The Average Temperature of the Flue Gas

$$T = \frac{1}{2} (1470 + 1100) = 1285 \text{ }^\circ\text{K}$$

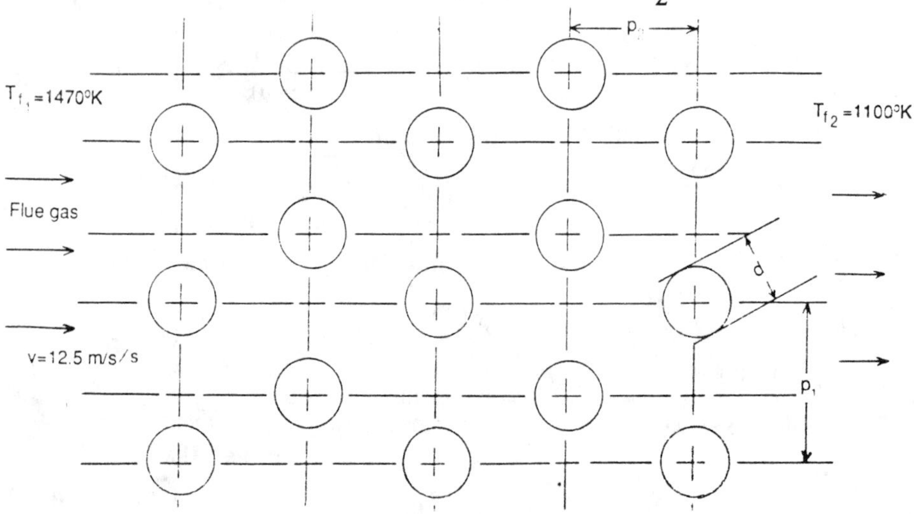

Fig. 33.5

Step (II) Flue Gas Parameters

At this average temperature the flue gas of the said composition has the following physicochemical properties:

$$k = 12.65 \times 10^{-2} \text{ W/m } ^\circ\text{K}$$

$$v = 221 \times 10^{-6} \text{ m}^2/\text{s}$$

$$\text{Pr} = 0.56$$

$$\mu = 53 \times 10^{-6} \text{ Pa.s}$$

$$C_p = 1.341 \text{ kJ/kg } ^\circ\text{K}$$

Step (III) Determination of Reynolds Number

$$\text{Re}_f = d\,u/v_f = (100 \times 10^{-3})\frac{(12.5)}{221 \times 10^{-6}} = 5656$$

Since Re_f lies within $(10^3 - 10^5)$, so we can apply the formula (I) as stated above.

Step (IV) Nusselt No. at Reference Temperature

Re_f	Pr_f	p_1/p_2	$\phi_s = (p_1/p_2)^{1/6}*$	Nu_f
5656	0.56	2.5/2	$(1.25)^{1/6}$	$0.41(5656)^{0.6}$
			$- 1.0378$	$(0.56)^{0.33}$
				(1.0378)
			$*\ \ p_1/p_2 < 2$	$- 62.70$

Step (V) Heat Transfer Coefficient

The heat transfer coefficient referred to the 3rd row of tubes

$$h_3 = \text{Nu}_f\, k_f/d = \frac{62.7(12.65 \times 10^{-2})}{100 \times 10^{-3}}$$

$$= 79.31 \text{ W/m}^2 \, ^\circ\text{K}$$

Step (VI) Mean Heat Transfer Coefficient

$$h = \frac{1}{4} \sum_{x=1}^{x=4} h_x = \frac{1}{4}(h_1 + h_2 + h_3 + h_4)$$

$$= \frac{1}{4}(0.6h_3 + 0.7h_3 + 2h_3)$$

where h_1 = heat transfer coefficient for the first row of the tube banks

$$= 0.6h_3$$

h_2 = heat transfer coefficient for the second row of staggered bank of tubes

$$= 0.7h_3$$

$$h_3 = h_4$$

$$h = \frac{1}{4}(3.3h_3) = \frac{1}{4}(3.3)(79.31) = 65.43 \text{ W/m}^2 \, ^\circ\text{K}$$

Ans.

Design of Double-pas Air Preheater

Problem 33.4 A double-pass air heater is to be designed for a steam generator. It must be capable of handling air at the rate of 90 t/h.

Air is to be heated from 30°C to 274°C.

The flue gas and air are in cross-flow—the flue gas moving through tubes at the rate 72 t/h with a mean velocity 17 m/s and air moves in the shell-side of the tubes with a mean velocity 10 m/s in the narrow cross-section of the tube bank.

ID/OD of the tubes = 50/55 mm

Thermal conductivity of the tube material = 47 W/m°C

The tubes are in staggered arrangement with $p_1 = p_2 = 1.25d_2$

Flue gas temperature inlet to air heater = 387°C

The following parameters are to be determined:

1. total heat transfer surface required
2. the height of tubes in one pass
3. the number of tubes to be arranged across the flow of air
4. the number of tubes along the flow of air.

The flue gas composition being: CO_2—13%; H_2O—11%

Solution

Step (I) Average Temperature of Air

The average temperature of air in the air heater

$$\theta_a = \frac{1}{2}(30 + 274) = 152^\circ\text{C}$$

Step (II) Physichochemical Properties of Air

At this average temp. (152°C), air has the following physicochemical properties:

$$k = 3.58 \times 10^{-2} \text{ W/m } ^\circ\text{C}$$

$$v = 29.174 \times 10^{-6} \text{ m}^2/\text{s}$$

$$\text{Pr} = 0.6792$$

$\mu = 24.18 \times 10^{-6}$ Pa.s

$C_p = 1.0154$ kJ/kg C

$\rho = 0.830$ kg/m^3

Step (III) Heat Load

The amount of heat transferred from the hot flue gas to the air stream

$$Q = \dot{M}_a \, C_p \, (\Delta\theta)_a$$

$$= \left(\frac{90 \times 1000}{3600}\right)(1.0154)\,(274 - 30)$$

$$= 6193.94 \text{ kW}$$

where, $\dot{M}_a$ = mass flowrate of air

$[C_p]_a$ = sp. heat of air at mean temp.

$(\Delta\theta)_a$ = temp. diff. between the air at inlet and outlet

Step (IV) Flue Gas Temp at Air Heater Outlet

This is done through a trial and error method.

Let as the case of first approximation, the mean flue gas temperature in the air heater = 317°C

$[C_p]_{fg} = 1.128$ kJ/kg °C at 317°C

Therefore, flue gas temperature at the air heater outlet

$$[\theta_{fg}]_2 = [\theta_{fg}]_1 - \frac{Q}{\dot{M}_{fg}[C_p]_{fg}}$$

$$= 387 - \frac{6193.94}{\left[\dfrac{72 \times 1000}{3600}\right](1.128)}$$

$$= 112.45°C$$

$[\theta_{fg}]_2$ = flue gas temp. at inlet to air heater

$\dot{M}_{fg}$ = mass flowrate of flue gas.

Accordingly, average temperature of flue gas in air heater comes out to be

$$\theta_{fg} = \frac{1}{2}\,(387 + 112.45) = 249.72°C$$

This figure does not tally with our figure (317°C) assumed.

So as the case of 2nd approximation, let the mean flue gas temperature in the air heater = 250°C. At this temperature:

$[C_p]_{fg} = 1.109$ kJ/kg °C

$$\therefore \ [\theta_{fg}]_2 = 387 - \frac{6193.94}{72 \times \dfrac{1000}{3600} \times 1.109} = 108°C$$

and therefore, $\theta_{fg} = \dfrac{1}{2}(387 + 108) = 247.5°C$

This time the calculated value of θ_{fg} is in close agreement with the assumed value 250°C.

So, $\theta_{fg} = 247°C$

Step (V) Reynolds No. For Flue Gas at Mean Temperature

FLUE GAS   247°C

$\rho_{fg} = 0.6864$ kg/m^3

$[C_p]_{fg} = 1.108$ kJ/kg °C

$k_{fg} = 4.40 \times 10^{-2}$ W/m °C

Pr = 0.66

$v_{fg} = 51.92 \times 10^{-6}$ m^2/s

$$[Re]_{fg} = \frac{d_i \, v_{fg}}{v_{fg}} = \frac{(50 \times 10^{-3})(17)}{51.92 \times 10^{-6}} = 16371$$

Step (VI) Reynolds No. for Air at Mean Temperature

$$Re_a = \frac{d_o \, v_a}{v_a} = \frac{(55 \times 10^{-3})(10)}{29.174 \times 10^{-6}} = 18852$$

Step (VII) Nusselt No. for Flue Gas

$$Nu_{fg} = 0.021 \,(Re_{fg})^{0.8} \,(Pr_{fg})^{0.43}$$

$$= 0.021 \,(16371)^{0.8} \,(0.66)^{0.43} = 41.122$$

Step (VIII) Nusselt No. for Air

For air in crossflow over a clean bank of tubes

$$Nu_a = 0.41 (Re_a)^{0.6} (Pr_a)^{0.33} \phi, \text{ for staggered arrangement}$$

Since, $\dfrac{p_1}{p_2} = 1.25 < 2, \ \phi = (1.25)^{1/6} = 1.0378$

$$\text{Nu}_a = 0.41(18852)^{0.6}(0.6792)^{0.33}(1.0378)$$

$$= 137.616$$

Step (IX) Heat Transfer Coefficient from Tube Walls to Air

$$h_{fg} = \text{Nu}_{fg}\left(\frac{k_{fg}}{d_i}\right) = \frac{41.122(4.40 \times 10^{-2})}{50 \times 10^{-3}}$$

$$= 36.187 \text{ W/m}^2 \, ^\circ\text{C}$$

Step (X) Heat Transfer Coefficient from Tube Walls to Air

$$h_a = \text{Nu}_a\left(\frac{k_a}{d_o}\right) = \frac{137.616}{\dfrac{3.58 \times 10^{-2}}{55 \times 10^{-3}}}$$

$$= 89.575 \text{ W/m}^2 \, ^\circ\text{C}$$

Step (XI) Overall Heat Transfer Coefficient

$$U = \cfrac{1}{\dfrac{1}{h_{fg}} + \Delta\dfrac{x}{k_w} + \dfrac{1}{h_a}}$$

$$= \cfrac{1}{\dfrac{1}{36.187} + \dfrac{2.5 \times 10^{-3}}{47} + \dfrac{1}{89.575}}$$

$$= 25.739 \text{ W/m}^2 \, ^\circ\text{C}$$

Step (XII) Log-Mean-Temperature-Difference

$$[\theta_{fg}]_1 \longrightarrow [\theta_{fg}]_2 \qquad \text{(Fig. 33.6)}$$

$$\frac{[\theta_a]_2 \longleftarrow [\theta_a]_1}{[[\theta_{fg}]_1 - [\theta_a]_2) \quad ([\theta_{fg}]_2 - [\theta_a]_1)}$$

$$\therefore \quad \Delta\theta_{lm} = \frac{([\theta_{fg}]_1 - [\theta_a]_2) - ([\theta_{fg}]_2 - [\theta_a]_1)}{\ln\left[\dfrac{[\theta_{fg}]_1 - [\theta_a]_2}{[\theta_{fg}]_2 - [\theta_a]_1}\right]}$$

$$= \frac{(387 - 274) - (108 - 30)}{\ln\left[\dfrac{387 - 274}{108 - 30}\right]}$$

$$= 94.42\,^\circ\text{C}$$

Step (XIII) Correction Factor

The correction factor (ε) for tubular 2-pass crossflow heat exchanger is determined from the chart depending on the values of the dimensionless parameters P and R where

$$P = \frac{[\theta_a]_2 - [\theta_a]_1}{[\theta_{fg}]_1 - [\theta_a]_1} = \frac{274 - 30}{387 - 30} = 0.683$$

$$R = \frac{[\theta_{fg}]_1 - [\theta_{fg}]_2}{[\theta_a]_2 - [\theta_a]_1} = \frac{387 - 108}{274 - 30} = 1.143$$

From the chart, corresponding to the value of $P = 0.683$ and $R = 1.14$ we get,

$$\varepsilon \approx 0.85$$

$$\Delta\theta = \varepsilon\,\Delta\,\theta_{1m} = 0.85(94.42) = 80.257\,^\circ\text{C}$$

Step (XIV) Heating Surface of Air Heater

$$A = \frac{Q}{U\Delta\theta} = \frac{6193.94 \times 10^3}{(25.739)(80.257)}$$

$$= 2998 \text{ m}^2$$

Step (XV) Total Number of Tubes

$$n = \frac{\dot{M}_{fg}}{\dfrac{\pi}{4}d_1^2\, v_{fg}\, \rho_{fg}} = \frac{(72 \times 1000/3600)}{\dfrac{\pi}{4}(50 \times 10^{-3})^2(17)(0.6864)}$$

$$= 872$$

Step (XVI) Height of Tubes in One Pass

$$L_1 = \frac{A}{2\pi\, d_1\, n} = \frac{2998}{2(\pi)\,(50/10^3)(872)} = 10.94\,\text{m}$$

Since tube-height becomes abnormally large, let us halve the tube-height and increase the number of tubes two-fold. Therefore,

total number of tubes $= 2 \times 872 = 1744$

tube-height in one pass $= \dfrac{1}{2}(10.94) = 5.47$ m

Step (XVII) The Clear Passage for Air Flow

$$[A]_{clear} = \frac{\dot{M}_a}{v_a\,\rho_a} = \frac{(90 \times 1000/3600)}{10 \times 0.83}$$

$$= 3.01 \text{ m}^2$$

Step (XVIII) The Number of Tubes Arranged Across the Flow

$$n_1 = \frac{[A]_{clear}}{L_1(p_1 - d_o)} = \frac{3.01}{5.47(1.25 \times 0.055 - 0.055)} = 40$$

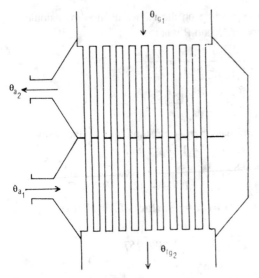

Fig. 33.6 *Air preheater*

Step (XIX) Number of Tubes Exposed to Longitudinal Flow

$$n_2 = \frac{n}{n_1} = \frac{1744}{40} \approx 44$$

Loop Type Economizer Design

Problem 33.5 A loop type economizer is to be designed out of tubes disposed in a staggered arrangement to heat up the BFW from 140°C to 300°C.

Data Given:

1. **Steel Tube Specifications:**

   ID = 45 mm

   OD = 50 mm

   Thermal conductivity = 26 W/m °C

2. **Tube Layout:**

   Transverse pitch : Longitudinal pitch = 1.05

   Longitudinal pitch = Twice the mean dia of tubes

3. **Boiler Feedwater:**

   Flowrate = 200 t/h

   Mean velocity of flow = 0.75 m/s

   Mode of flow : vertical, from bottom to top

4. **Flue Gas:**

   Composition : CO_2—13%; H_2O—11%

Mean velocity of flow in the narrowest cross-section of tube-banks = 15 m/s

Flowrate = 480 t/h

Temp. inlet to economizer = 800°C

Solution

Step (I) Mean Temp of BFW

$$\theta_w = \frac{\theta_{w_1} + \theta_{w_2}}{2} = \frac{140 + 300}{2} = 220°C$$

Step (II) Heat Transfer Rate to BFW

$$\boxed{\text{WATER—220°C}}$$

$\rho = 840 \text{ kg/m}^3$

$C_p = 4.615 \text{ kJ/kg °C}$

$k = 64.5 \times 10^{-2} \text{ W/m °C}$

$\nu = 0.148 \times 10^{-6} \text{ m}^2/\text{s}$

$Pr = 0.89$

The rate of heat transferred to the BFW in the economizer

= Mass flowrate × Sp. heat × Rise in temperature

$$= \left(\frac{200 \times 10^3}{3600}\right)(4.615)(300 - 140) = 41022 \text{ kW}$$

Step (III) Reynolds number for BFW at Mean Temperature

$$Re_1 = d_i \frac{\nu}{\nu} = \frac{(45 \times 10^{-3})(0.75)}{0.148 \times 10^{-6}} = 228040$$

Hence the flow is turbulent. Since the Prandtl number (0.89) is greater than 0.7, we will use the formula

$Nu_1 = 0.021 \, Re_1^{0.8} \, Pr_1^{0.43} \, (Pr_f / Pr_w)^{0.25}$ to determine the Nusselt number. Since the wall temperature is close to the BFW temperature inside the economizer, $(Pr_f/Pr_w)_{outlet}^{0.25} \approx 1$

and the foregoing equation simplifies to

$$Nu_1 = 0.021 \, Re_1^{0.8} \, Pr_1^{0.43}$$

$$= 0.021 \, (228040)^{0.8} \, (0.89)^{0.43} = 386$$

Step (IV) Local Heat Transfer Coefficient for BFW

$$h_1 = \frac{Nu_1 \, k_1}{d_i} = \frac{386(64.5 \times 10^{-2})}{45 \times 10^{-3}}$$

$$= 5532.67 \text{ W/m}^2 \text{ °C}$$

Step (V) Flue Gas Temperature at Economizer Outlet

This is to be determined by trial and error method.

1st Trial Let mean heat capacity of flue gas

$$C_{p_g} = 1.27 \text{ kJ/kg °C}$$

$$Q = \dot{M}_g \, C_{p_g} \, (\theta_{g_1} - \theta_{g_2})$$

or, $41022 = \dfrac{480 \times 10^3}{3600} (1.27) (800 - \theta_{g_2})$;

$$\therefore \; \theta_{g_2} \approx 558°C$$

Therefore, mean temperature of flue gas

$$\theta_g = \frac{1}{2} (\theta_{g_1} + \theta_{g_2}) = \frac{1}{2}(800 + 558) = 679°C$$

At this temperature, the value of heat capacity of the flue gas

$$C_{p_g} = 1.234 \text{ kJ/kg °C}$$

2nd Trial Let $C_{p_g} = 1.234$ kJ/kg °C

$$\therefore \; 41022 = \left(\frac{480 \times 10^3}{3600} \right) (1.234) (800 - \theta_{g_2});$$

$$\therefore \; \theta_{g_2} \approx 551°C$$

Therefore, mean temperature of flue gas

$$\theta_g = \frac{1}{2} (551 + 800) = 675.5°C$$

At this temperature, the value of heat capacity of flue gas
$C_{p_g} = 1.233$ kJ/kg °C which closely tallies with the assumed value.

Hence the mean temperature of flue gas = 675.5°C

Step (VI) Reynolds No. for Flue Gas

FLUE GAS —675.5°C

$\rho = 0.373$ kg/m³

$k = 8.06 \times 10^{-2}$ W/m. °C

$v = 107.57 \times 10^{-6} \text{ m}^2/\text{s}$

Pr = 0.612

$$Re_g = \frac{d_2 \, v_g}{v_g} = \frac{(50 \times 10^{-3}) \, (15)}{107.57 \times 10^{-6}} = 6972$$

Step (VII) Nusselt No. For Flue Gas

Since the economizer tubes are arranged in a staggered arrangement, Nusselt no.

$Nu_g = 0.41 \, Re_g^{0.6} \, Pr_g^{0.33} \, \varepsilon$ (where ε = correction factor = 1.008 for p_1/p_2 = 1.05)

$$= 0.41 \, (6972)^{0.6} \, (0.612)^{0.33} \, (1.008)$$

$$= 71.10$$

Step (VIII) Heat Transfer Coefficient for Flue Gas

$$h_2 = \frac{Nu_g \, k_g}{d_o} = \frac{(71.10) \, (8.06 \times 10^{-2})}{50 \times 10^{-3}}$$

$$= 114.61 \text{ W/m}^2 \text{ °C}$$

In practice, due to fouling, the effective heat transfer coefficient for flue gases is less than its theoretical value. Normally, it is assumed to be 80% of its calculated value.

$$h_2 = 114.61 \left(\frac{80}{100} \right) = 91.688 \text{ W/m}^2 \text{ °C}$$

Step (IX) Coefficient of Heat Transfer by Thermal Radiation

A considerable portion of heat is transferred by radiation from the hot flue gas to the walls of the tubes. The coefficient of this radiant heat transfer is determined as follows:

1. **Mean Beam Length**

$$L = 1.08 \, d_o \, (p_1 p_2 d_o^2 - 0.785)$$

$$= 1.08(50 \times 10^{-3})[(1.05 \times 2 \, d_o)(2 \, d_o)/d_o^2 - 0.785]$$

$$= 0.1844 \text{ m}$$

2. **Product of Mean Beam Length and Partial Pressure**

$$p_{CO_2} L = 0.13 \, (0.1844) = 0.0239 \text{ m.atm.}$$

$$p_{H_2O} L = 0.11 \, (0.1844) = 0.02028 \text{ m. atm.}$$

3. **Average Emissivity of Flue Gas at 675.5°C**

$\varepsilon_{CO_2} = 0.071;$ $\varepsilon_{H_2O} = 0.047;$ $\beta = 1.07$

$\varepsilon_g = \varepsilon_{CO_2} + \beta\varepsilon_{H_2O} = 0.071 + 1.07\,(0.047) = 0.1212$

4. **Average Absorptivity of Flue Gas**

The calculation shows $h_1 \gg h_2$

Therefore, outer wall temperature of economizer tubes can be assumed to be Average Temp. of BFW $+ 30°C \approx 220 + 30 = 250°C$

Absorptivity of flue gas at the surface temperature of the tubes

$$\alpha_g = \varepsilon_{CO_2}\left[\frac{T_g}{T_{wall}}\right]^{0.65} + \beta\cdot\varepsilon_{H_2O}$$

$$= 0.063\left[\frac{675.5 + 273}{250 + 273}\right]^{0.65} + 1.07\,(0.075)$$

$$= 0.1730$$

5. **Effective Emissivity of the Enclosure**

$$\varepsilon = \frac{1}{2}\,[\text{Emissivity of the tubewall} + 1]$$

$$= \frac{1}{2}\,(0.8 + 1) = 0.9$$

6. **Rate of Radiation Heat Transfer**

$$q_{rad} = (5.67)\left[\varepsilon_g\left(\frac{T_g}{100}\right)^4 - \alpha_g\left(\frac{T_{wall}}{100}\right)^4\right]$$

$$= 0.9(5.67)\left[0.1212\left(\frac{675.5 + 273}{100}\right)^4\right.$$

$$\left. - 0.173\left(\frac{250 + 273}{100}\right)^4\right],\ W/m^2$$

$$= 4345.34\ W/m^2$$

7. **The coefficient of Heat Transfer by radiation**

$$h_{rad} = \frac{q_{rad}}{\theta_g - \theta_{wall}} = \frac{4345.34}{675.5 - 250}$$

$$= 10.21\ W/m^2\ °C$$

Step (X) Total Coefficient of Heat Transfer from Flue Gas

The total coefficient of heat transfer from flue gas to the tube walls

$$h_{tot} = h_2 + h_{rad} = 114.61 + 10.21$$

$$= 124.82\ W/m^2\ °C$$

Step (XI) Overall Heat Transfer Coefficient

$$U = \frac{1}{\dfrac{1}{h_1} + \left[\dfrac{\Delta x}{k}\right]_{wall} + \dfrac{1}{h_{tot}}}$$

$$= \frac{1}{\dfrac{1}{5532.67} + \dfrac{2.5 \times 10^{-3}}{26} + \dfrac{1}{123.92}}$$

Step (XII) Log-Mean-Temperature-Difference

$$800°C \xrightarrow{\ \ fluc\ \ \ \ \ \ gas\ \ } 551°C$$

$$300°C \xleftarrow{\ \ \ \ BFW\ \ \ \ } 140°C$$

$$\underline{\text{Diff. } 500°C \qquad\qquad\qquad 411°C}$$

Assuming the flue gas in counterflow with water,

$$\therefore\ \Delta\theta_{lm} = \frac{500 - 411}{\ln(500/411)} = 454°C$$

Step (XIII) Heating Surface of Economizer

$$A = \frac{Q}{U\Delta\theta_{lm}} = \frac{41022 \times 10^3}{(120)\,(454)}$$

$$= 752.973\ m^2 \approx 753\ m^2$$

Step (XIV) Number of Loops in Parallel

Let the number of parallel loops be n.

$$\left(\frac{\pi}{4}\,d_i^2\,v_{water}\right)\rho_{H_2O}\,(n) = \left(\frac{200 \times 10^3}{3600}\right);$$

or $\left(\dfrac{\pi}{4}\right)(45 \times 10^{-3})^2(0.75)(840)(n) = \dfrac{1}{18} \times 10^3$

$\therefore\ n = 55.44$ i.e.   56 number of loops

Step (XV) Length of Individual Loop

$$(\pi\,d_o)n\,L = A$$

or $\pi(50 \times 10^{-3})\,(56)L = 753;\ \therefore\ L = 85.60\ m$

Design of Loop Type Superheater

Problem 33.6 A loop type of superheater of following duty:

Steam to be superheated = 250 t/h
Superheated steam pressure = 10 MN/m²
Superheated steam temperature = 525°C
is to be designed so that flue gas is in crossflow with steam.

The superheater coils are to be designed from steel tubes of

OD/ID = 35/30 mm

Thermal conductivity of the material of the tubes = 25 W/m °C

Average velocity of steam = 18 m/s

Velocity of flue gas in the narrow section of the bank of tubes = 15 m/s

The tubes are to be disposed in the mode of in-line arrangement with longitudinal and transverse pitch ratio 1.2 and transverse pitch = 2.5 × OD of the tubes.

Flue gas composition: CO_2—13%; H_2O—11%
Flue gas flowrate: 600 t/h
Flue gas inlet temp.: 1200°C

Solution

Step (I) Heat Load

	Saturation Temp. (θ_s)	Enthalpy at θ_s	Enthalpy at 525°C
SUPERHEATED STEAM, 10 MN/m²	311°C	2725 kJ/kg	3437 kJ/kg

←——— From steam tables ———→

The rate of superheated steam generation = 250 t/h

Heat load of the superheater

$$= \frac{250 \times 10^3}{3600} (3437 - 2725)\, kW$$

$$= 49444.44\, kW$$

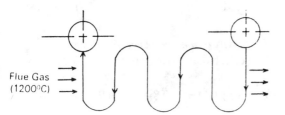

Fig. 33.7 Loop type superheater

Step II Steam Parameters

Average temperature of superheated steam

$$= \frac{1}{2}(311 + 525) = 418°C$$

SUPERHEATED STEAM 418°C, 10 MN/m²	$\rho = 37\ kg/m^3$ $k = 7.1 \times 10^{-2}\ W/m°C$ $v = 0.703 \times 10^{-6}\ m^2/s$ $Pr = 1.1$

Step (III) Coefficient of Heat Transfer from Wall to Steam

1. *Reynolds Number for Steam*

$$Re = d_i \frac{v}{v} = \frac{(30 \times 10^{-3})(18)}{0.703 \times 10^{-6}} = 768136$$

2. *Nusselt Number for Steam*

$$Nu = 0.021\, Re^{0.8}\, Pr^{0.43}$$

$$= 0.021\,(768136)^{0.8}\,(1.1)^{0.43} = 1117.81$$

3. *Coefficient of Heat Transfer from Tubewall to Steam*

$$h_i = Nu\, k/d_i = (1117.81)\frac{(7.1 \times 10^{-2})}{30 \times 10^{-3}}$$

$$= 2645\ W/m^2\, C$$

Step (IV) Flue Gas Temperature at Outlet

In order to determine the flue gas exit temperature, we must go through successive approximations.

1st Approximation: Let $C_{p_g} = 1.31$ kJ/kg°C for flue gas.

Exit temp. of flue gas, $[\theta_2]_g = [\theta_1]_g - \dfrac{Q}{M_g \times C_{p_g}}$

$$[\theta_g]_2 = 1200 - \frac{49444.44}{\frac{600 \times 10^3}{3600} \times 1.31} = 974°C$$

Therefore, average temperature of flue gas

$$= (1/2)(1200 + 974) = 1087°C$$

At this average temperature, the specific heat of flue gas,

$$C_{p_g} = 1.303\ kJ/kg.\ °C$$

which is very close to our assumed value. So we need no further approximation and hence the flue gas temperature at superheater outlet = 974°C

Step (IV) Flue Gas Parameters at Average Temperature

$$\boxed{\text{FLUE GAS } 1087°C}$$

$$\rho = 0.259 \text{ kg/m}^3$$
$$k = 11.64 \times 10^{-2} \text{ W/m}°C$$
$$v = 194 \times 10^{-6} \text{ m}^2/\text{s}$$
$$Pr = 0.571$$

Step (V) Heat Transfer Coefficient from Flue Gas to Superheater Tubewalls

1. *Reynolds No. for Flue Gas*

$$Re = d_o\left(\frac{u}{v}\right)_g = (35 \times 10^{-3})\left(\frac{15}{194 \times 10^{-6}}\right) = 2706$$

2. *Nusselt No. for Flue Gas*

$$Nu = 0.26 \, Re^{0.65} \, Pr^{0.33} \, \varepsilon$$

Where $\varepsilon = (d_o/p_2)^{0.15}$

p_1 = transverse pitch

p_2 = longitudinal pitch

Given $p_1 = 2.5(35 \times 10^{-3})$ m; $\dfrac{p_2}{p_1} = 1.2$

$$\therefore \ p_2 = 1.2(2.5)(35 \times 10^{-3}) \text{ m} = 0.105 \text{ m}$$

$$\therefore \varepsilon = \left(\frac{35 \times 10^{-3}}{0.105}\right)^{0.15} = 0.848$$

$$\therefore Nu_g = 0.26 \, (2706)^{0.65} \, (0.571)^{0.33} \, (0.848) = 31.19$$

3. *Heat Transfer Coefficient*

$$h_g = \frac{(Nu \ k)_g}{d_o} = \frac{(31.19)(0.1164)}{35 \times 10^{-3}}$$

$$= 103.729 \approx 104 \text{ W/m}^2°C$$

This is for clean tubes laid in an in-line arrangement.

Fouling of heat transfer surface of tubes will reduce its value to some extent. So let's assume, the effective heat transfer coefficient is 80% of the heat transfer coefficient calculated for clean tubes.

$$h'_g = 104 \, (0.8) = 83.2 \text{ W/m}^2 \ °C$$

Step (VI) Heat Transfer Coefficient by Radiation

Mean Beam Length

$$L = 1.08 d_o\left[\left(\frac{p_1 p_2}{d_o^2}\right) - 0.785\right]$$

$$= 1.08 \, (0.035)\left[\frac{(2.5 d_o)(1.2 p_1)}{d_o^2} - 0.785\right]$$

$$= 0.0378\left[\frac{(2.5 d_o)(1.2 \times 2.5 d_o)}{d_o^2} - 0.785\right]$$

$$= 0.2538 \text{ m}$$

Products of Mean Beam Length & Partial Pressure

$$L \, p_{CO_2} = 0.2538 \, (0.13) = 0.0329 \text{ m atm.}$$

$$L \, p_{H_2O} = 0.2538 \, (0.11) = 0.0279 \text{ m atm.}$$

Emissivity of the Flue Gas at 1087°C

$$\varepsilon_g = \varepsilon_{CO_2} + + \beta \, \varepsilon_{H_2O} = 0.065 + 1.08 \, (0.038)$$

$$= 0.106$$

Absorptivity of Flue Gas

Let us assume that tubewall temperature is greater than the arithmetic mean temperature of steam by 40 to 42°C.

$$\therefore \ \Theta_w \approx 418 + 42 = 460°C$$

At this temperature, $\varepsilon_{CO_2} = 0.075$; $\beta = 1.06$; $\varepsilon_{H_2O} = 0.07$

$$\alpha = \varepsilon_{CO_2}\left[\frac{T_g}{T_w}\right]^{0.65} + \beta \varepsilon_{H_2O}$$

$$= 0.075\left[\frac{1087 + 273}{460 + 273}\right]^{0.65} + 1.06 \, (0.07) = 0.186$$

Emissivity of the Enclosure

The effective emissivity of the enclosure

$$\varepsilon = \frac{1}{2}(\varepsilon_w + 1) = \frac{1}{2} \, (0.8 + 1) = 0.9$$

Heat Flowrate due to Thermal Radiation

$$q_{rad} = \varepsilon(5.7)\left[\varepsilon_g\left(\frac{T_g}{100}\right)^4 - \alpha_g\left(\frac{T_w}{100}\right)^4\right]$$

$$0.9(5.7)\left[0.106\left(\frac{1087+273}{100}\right)^4 - 0.186\left(\frac{460+273}{100}\right)^4\right]$$

$$= 15848 \text{ W/m}^2$$

The Coefficient of Heat Transfer by Radiation

$$h_{rad} = \frac{q_{rad}}{\Theta_g - \Theta_w} = \frac{15848}{1087-460} = 25.27 \text{ W/m}^2 \text{ }^\circ C$$

The Coefficient of Heat Transfer from Flue Gas to Tubewalls

$$h_o = h'_g + h_{rad} = 83.2 + 25.27 = 108.47 \text{ W/m}^2 \text{ }^\circ C$$

Step (VII) Overall Heat Transfer Coefficient

$$U = \frac{1}{\dfrac{1}{h_i} + \dfrac{\Delta x}{k} + \dfrac{1}{h_o}} = \frac{1}{\dfrac{1}{2645} + \dfrac{2.5 \times 10^{-3}}{25} + \dfrac{1}{108.47}}$$

$$= 102.91 \approx 103 \text{ W/m}^2 \text{ C}$$

Step (VIII) Log-mean - Temp.-Difference

Assuming the flue gas and superheated steam to be in counterflow

FLUE GAS	1200°C ⟶ 974°C	$\therefore \Delta\theta_{lm} = \dfrac{675 - 663}{\ln(675/663)}$
	525°C ⟵ 311°C STEAM	$= 668.98°C \approx 669°C$
Diff.	675°C 663°C	

Step (IX) Check
Outside Surface Temp. of Tubes

$$\Theta_w = \Theta_g - (U\Delta\Theta/h_o) = 1087 - \left(\frac{103 \times 669}{108.47}\right)$$

$$= 452°C$$

During calculation of radiation heat transfer we have assumed $\theta_w = 460°C$ (cf. Step VI.4). Thus we see that there is a close coincidence between the assumed and calculated temperatures of the tube wall (outer). Therefore, we need no further approximation for the tube wall temperature.

Step (X) Heating Surface of Superheater

$$Q = U A \Delta\Theta_{lm}$$

$$\therefore A = \frac{Q}{U \Delta\Theta_{lm}}$$

$$A = \frac{49444.44 \times 10^3}{103 \times 669} = 717.55 \approx 718 \text{ m}^2$$

Ans.

Step (XI) Number of Loops

Let N be the number of loops

$$G_s = \left(\frac{\pi}{4} d_i^2 N\right)(v_s)(\rho_s)$$

or $\dfrac{250 \times 10^3}{3600} = \left(\dfrac{\pi}{4}\right)(30 \times 10^{-3})^2 (N) (18) (37);$

$$\therefore N \approx 148$$

Ans.

Step (XII) Length of Each LOOP

If L be the length of each loop,

$$A = (\pi d_o L)N$$

or $718 = \pi(35 \times 10^{-3})L(148)$

$$\therefore L = 44.12 \text{ m}$$

Ans.

34

Nuclear Steam Generators

Q. *What is nuclear fission?*

Ans. The nucleus of heavy elements like U-233, U-235 and Pu-239 has the ability to capture and absorb neutrons, whereupon it gets converted to a compound nucleus. Sometimes this compound nucleus is highly unstable undergoing spontaneous fragmentation into several approximately equal lighter nuclei plus two or three neutrons. This process is called nuclear fission and is accompanied by enormous amount of heat liberation. (Fig. 34.1)

Q. *How much fission energy is available from 1kg of U-235?*

Ans. It is roughly equivalent to the energy liberated by the combustion of 3086 ton of coal or 1362 m^3 of fuel oil.

Q. *What do you mean by a nuclear chain reaction?*

Ans. It is a self-sustained nuclear fission process which is attained when the production of neutrons equals or exceeds the quantity lost through absorption of non-fissioning materials or leakage.

Q. *What is the difference between fissionable and fertile materials?*

Ans. Fissionable materials are those which are capable of sustaining a fission chain reaction.

U-235 is the only fissionable isotope found in nature.

Fertile materials are non-fissionable materials that can be converted into fissionable materials. U-238 and Th-232 can be converted to fissionable material.

Q. *What do you mean by neutron capture?*

Ans. It means absorption of neutrons.

Neutrons, depending on their energy, may be absorbed or scattered by the target nuclei or reactor materials. Usually a slow neutron or thermal neutron is captured while fast nuetrons are bounced off. (Fig. 34.2)

Q. *What is neutron balance?*

Ans. In any power reactor where a self-perpetuating fission process is going on, the production rate of neutrons must balance the rate of neutron capture [by target nuclei, reactor materials, fission products (poisons, e.g. $_{54}^{135}$Xe) and control rods] plus the rate of neutron loss through leakage.

Q. *What is critical mass?*

Ans. It is the minimum mass of fissionable material that will sustain a fission-chain reaction.

Q. *What is the magnitude of critical mass for naturally occurring fissionable materials?*

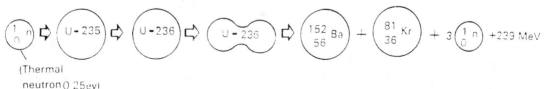

(Thermal
neutron 0 25ev)

Fig. 34.1 *Nuclear fission mechanism*

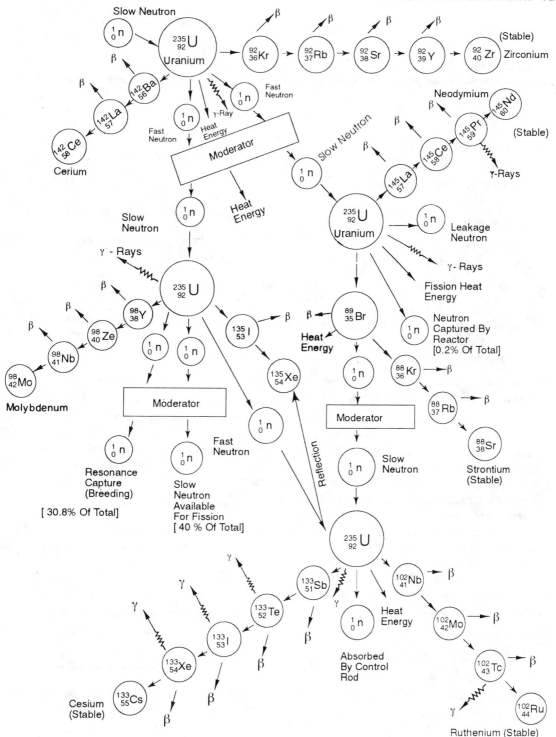

Fig. 34.2 *Typical nuclear fission reactions with U-235 isotope*

Ans. It varies from just 200 gm to as much as half a ton.

Q. *Should the actual fuel loading of a reactor exceed, be equal to or be less than the critical mass?*

Ans. It must exceed the critical mass.

Q. *Why?*

Ans. This is to make-up for fuel burnup and compensate other effects encountered during reactor operation.

Q. *What is fuel burnup?*

Ans. It refers to depletion of fuel in the course of sustained fission reaction in a reactor.

Q. *What do you mean by multiplication factor k for a reactor?*

Ans. It is defined as the ratio of the number of neutrons produced in one generation to the number of neutrons produced in the preceding generation.

Q. *How can a reactor be classified as supercritical, critical or subcritical on the basis of this multiplication factor?*

Ans. For a reactor with k-value greater than one, the neutron population is increasing with time. Since reactor power is directly proportional to the neutron level in the reactor, for $k > 1$, the reactor power is increasing with time. Such a reactor is said to be a supercritical reactor.

If k is exactly equal to unity, a neutron balance is just maintained in the reactor. And such a reactor is said to be a critical reactor.

For k less than unity, the neutron population and the reactor power are decaying with time and the reactor is said to be subcritical.

Q: *What is the reactivity of the reactor?*

Ans. It is defined by the ratio

$$\frac{k-1}{k} = \rho$$

For supercritical reactor, ρ is positive
For critical reactor, ρ is zero
For subcritical reactor, ρ is negative.

Q. *What do you mean by neutron flux?*

Ans. It is the number of neutrons passing through a unit area in unit time

Reactor	Neutron Flux
Low-Powered reactor	$\phi = 10^{15}$ neutrons/m^2 s
High-Powered reactor	$\phi = 10^{20}$ neutrons/m^2 s

Q. *How are reactor neutrons classified?*

Ans. On the basis of their kinetic energy, reactor neutrons can be classified into three basic types:
1. SLOW NEUTRONS having K.E. less than 0.1 eV
2. INTERMEDIATE NEUTRONS having K.E. between 0.1 eV and 0.1 MeV
3. FAST NEUTRONS having K.E. exceeding 0.1 MeV

Slow neutrons are also called thermal neutrons. They move with a velocity of about 2200 m/s at 293K.

Fast neutrons having kinetic energy of about 2 MeV move with a velocity 2×10^7 m/s.

Q. *What do you mean by neutron reaction rate?*

Ans. It is the measure of the rate of target nuclei reacting with the neutrons to produce a given reaction. It is the product of average neutron flux $\bar{\phi}$, the total number of nuclei N reacting with the neutrons and the microscopic neutron cross-section σ for the given reaction, i.e.

Neutron reaction rate $= N \sigma \bar{\phi}$

The fission rate in the reactor core $= N_{fuel} \, \sigma_f \, \bar{\phi}$

σ_f = microscopic neutron cross-section for fission

Q. *What is the microscopic cross-section σ?*

Ans. It corresponds to the area of each target nucleus for producing a given reaction as a result of reacting with projectiles (e.g. neutrons, protons, deutrons, etc.)

Q. *What is the macroscopic cross-section Σ?*

Ans. It is the product of microscopic cross-section (σ) and the isotopic density $(N/V,$ the number of atoms per unit volume)

$$\Sigma = \sigma \frac{N}{V}$$

Note: Neutron reaction rate $= N\sigma\bar{\phi} = \Sigma \cdot V \cdot \bar{\phi}$

Q. *What is the fission rate in a given reactor operating at a power level of P MW_{th} (i.e. megawatt thermal)?*

Ans. Roughly 3.1×10^{16} fissions should occur to generate energy equal to 1 MW · s.

So for a reactor operating at a power level P MW_{th}

Fission rate $= 3 \times 10^{16} \times P$ fissions/s

$$= N_{fuel}\, \sigma_f \bar{\phi} = \Sigma_f V_c \bar{\phi}$$

where N_{fuel} = the number of fuel atoms in the reactor core volume V_c

$\bar{\sigma}_f$ = average microscopic fission cross-section for neutrons

$\bar{\Sigma}_f$ = average macroscopic fission cross-section for neutrons.

Problem 34.1 A 1 000MW nuclear power station operates at a thermal efficiency 33%. The reactor fuel is U-235 and the average thermal neutron flux in the reactor core is 5×10^{17} neutrons/m² s.

If the average fission cross-section is 420 barns, determine the reactor fuel loading in terms of U-235.

If the fuel used is in the form of uranium dioxide containing 2.5% enriched UO_2, determine the fuel loading of UO_2 in this system.

Solution The reactor fuel loading is to be calculated on the basis of the fission rate.

Step (I) Reactor Thermal Power

Reactor thermal efficiency

$$= \frac{\text{Reactor electrical power}}{\text{Reactor thermal power}}$$

∴ Reactor thermal power $= \dfrac{1\,000}{0.33}$ MW_{th}

Step (II) Fission Rate

Fission rate

$$= 3.1 \times 10^{16} \left(\frac{\text{fissions}}{MW_{th}.s}\right) \times \frac{1\,000}{0.33}\ (MW_{th})$$

$$= 9.3939 \times 10^{19}\ \text{fissions/s}$$

Step (III) Number of Fuel Atoms

The number of fuel atoms (N_{fuel}) undergoing this rate of fission in the reactor core is given by

$$\bar{\sigma}_f N_{fuel}\,\bar{\phi} = 9.3939 \times 10^{19}$$

or $420 \times$

$$\left(\frac{\text{fission barns}}{\text{fuel atom neutrons}}\right) \times 10^{-28} \left(\frac{m^2}{\text{barn}}\right) \times N_{fuel} \times$$

$$5 \times 10^{17} \left(\frac{\text{neutron}}{m^2\ s}\right) = 9.3939 \times 10^{19}\ \text{fissions/s}$$

$$\therefore N_{fuel} = \frac{9.3939 \times 10^{19}}{(420)\,(10^{-28})\,(5 \times 10^{17})}$$

$$= 4.473 \times 10^{27}\ \text{atoms of U-235}$$

Mass of U-235

$$= \frac{4.473 \times 10^{27}(\text{atoms}) \times 235(\text{kg/kmol})}{6.023 \times 10^{26}(\text{atoms/kmol})}$$

$$= 1\,745.23\ \text{kg} = 1.745\ \text{tons of U-235}$$

Step (IV) Loading of Enriched UO_2

When the reactor is fueled with 2.5% enriched UO_2, it contains mostly $^{238}UO_2$.

∴ Molecular weight of fuel UO_2

$$= 238 + 2 \times 16 = 270\ \text{kg/kmol}$$

Total mass of uranium

$$= \frac{1.745\ \text{tons of U–235}}{0.025\text{tons of U–235/ton of fuel}}$$

$$= 69.809\ \text{t of uranium}$$

Total mass of UO_2

$$= \frac{69.809\ \text{tons of uranium}}{238\ \text{tons uranium/270 tons of }UO_2}$$

$$= 79.195\ \text{tons of }UO_2$$

Q. *What is a nuclear reactor?*

Ans. It is an equipment specially designed to initiate, maintain and control a nuclear fission chain reaction so that the accompanying heat release can be used fruitfully.

In other words, a nuclear reactor can be portrayed as a furnace in which heat is produced because of a fission chain reaction. This generated heat must be removed and in a power reactor, converted to work.

Q. *What are the essential components of all power reactors?*

Ans.

1. Reactor core containing the fissionable material (fuel) and probably a moderator (fuel dilutent)
2. A system of control rods/plates
3. A coolant for heat removal
4. A neutron source to trigger the fission
5. A neutron reflector to bring escaping neutrons back to the core
6. A shield for operational safety
7. Structural elements
8. Fuel handling equipment.

Q. *With what fuels is the reactor core loaded?*

Ans. The reactor fuel may be of:
   (a) Uranium-235
   (b) Uranium-Thorium mixture
   (c) Uranium-Plutonium mixture
   (d) Thorium ($^{232}_{90}$Th)
   (e) Plutonium ($^{239}_{94}$Pu)

Q. *In which form is the fuel used?*

Ans. The fuel may be of
   (a) Metallic form (molten or solid as pellets)
   (b) Oxide form (e.g. uranium dioxide and plutonium dioxide). Nearly all of the present power reactors employ oxides of fuel isotopes.
   (c) Carbides (uranium carbide and plutonium carbide)
   (d) Salt solution (uranyl nitrate, uranyl sulfate, uranyl phosphate, uranium hexafluoride)
   (e) Molten salts (uranyl fluoride)

Q. *How is fuel distributed in the reactor core?*

Ans. Fuel may be distributed in the reactor in a
   (a) homogeneous manner
   (b) heterogeneous manner.

Q. *In which form is the fuel disbursed in a heterogeneous reactor?*

Ans. In heterogeneous reactors fuel may come in the form of thin rods, hollow tubes, pellets, thin plates sandwiched together or lumps.

The fuel is usually alloyed with (embeded into) a suitable metal (aluminium, zirconium, niobium or molybdenum).

The fuel element is cladded (jacketed) with suitable protective material (aluminum, zirconium or stainless steel) to protect it from corrosive attack.

Q. *Why has only limited success been achieved in the use of metallic fuel alloys?*

Ans. Metallic fissionable elements have proved to be unstable during long periods of irradiation and thermal cycling.

They develop a tendency to warp, grow, elongate or distort under intense radiation.

Q. *How is the fuel element distributed in the heterogeneous reactor core?*

Ans. The fuel element is evenly distributed in the reactor core.

The distribution order follows a definite geometrical pattern (lattices) in the matrix of the moderator.

Q. *In which form is the fuel distributed in the homogeneous reactor?*

Ans. The fuel is alloyed with or homogeneously mixed with moderator, diluent and coolant and the mixture is evenly dispersed in the reactor core in the form of solution, slurry or melt. Aqueous slurries (D_2O or H_2O) and liquid metal (molten bismuth) are used. The container material is stainless steel.

Q. *What are the potential disadvantages of heterogeneous reactors?*

Ans.

1. Instability of metallic fuel elements over long exposure of irradiation and thermal cycling
2. Expensive fuel preparation
3. Removal of cladding for reprocessing the fuel is very complicated
4. Fuel reprocessing is doubly expensive.

Q. *What are the potential advantages of homogeneous reactors?*

Ans.

1. Very good heat transfer characteristics
2. Very good nuclear control characteristics
3. Continuous chemical processing of the fuel solution is possible without the need of a dissolving solid fuel element
4. Expensive fuel fabrication is eliminated
5. Expensive refabrication of radioactive fuel material is eliminated.

Q. *What is a moderator?*

Ans. It is a substance incorporated in the reactor core to slow down the fission neutrons (fast neutrons), i.e., a moderator converts a fast neutron into a thermal neutron by absorbing its kinetic energy and thereby increasing the probability of neutron capture by the fissionable material.

Q. *What substances are used as moderators?*

Ans. Only three elements are actually used to moderate neutrons:

1. Ordinary hydrogen ($_1^1$H) or heavy hydrogen ($_1^2$H).

    As a moderator, ordinary hydrogen is used in the form of
    (a) water
    (b) solid and liquid organic compounds (diphenyl and terphenyls)
    (c) intermetallic compounds viz, zirconium hydride.
    Heavy hydrogen is used as the moderating material in the form of heavy water, D_2O.
2. Beryllium
    It is used as pure metal or as beryllium oxide (BeO).
3. Carbon (high purity graphite).

Q. *What should be the essential qualities of a good moderator?*

Ans. It must have:
(a) High macroscopic neutron-scattering cross-section
(b) Low macroscopic neutron-absorption cross-section
(c) Low atomic mass to get the maximum K.E. transfer from the fast neutrons to the moderator atoms
(d) High resistance to the high radiation level in the reactor core
(e) High resistance to corrosion under high pressure and temperature
(f) Good machinability if it is solid
(g) High melting point if it is solid
(h) High chemical stability
(i) High thermal conductivity for better heat transfer in the reactor core.

Q. *What is a diluent?*

Ans. It is a substance that adds mass to the fuel element. It is required particularly in fast reactors for heat removal and temperature reduction purposes.

Q. *What is a reflector?*

Ans. It is a material used in the form of a layer around the reactor core to prevent the escape of neutrons by bouncing them back into the core.

Not only does the reflector return the leakage neutrons to the core, it also slows down fast and intermediate neutrons with the effect that more neutron flux is available for inducing fission and fission energy.

Q. *What is the effect of adding a reflector to a bare thermal reactor?*

Ans. It effectively raises the neutron flux and hence boosts up the power density at the edges of the core. (Fig. 34.3)

That is, the addition of a reflector reduces the critical mass of a reactor by decreasing the rate of neutron leakage and creates a sharp improvement in the ratio of maximum-to-average power in the reactor core by increasing the neutron flux and hence results in greater power generation at the core boundary.

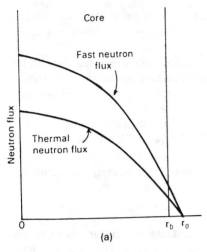

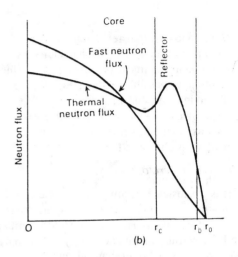

Fig. 34.3 *The thermal neutron flux distribution (a) Bare Reactor Core; (b) Reflector Reactor Core* r_o = *extrapolated core radius* = r_b + *extrapolation distance* r_b = *reactor core radius.*

Q. *What is the value of the maximum-to-average power in a reflected assembly?*

Ans. It varies, but one such typical value is

$$P_{max}/P_{av} = 2.4$$

Q. *What materials are used as reflectors?*

Ans.

1. Heavy water (D_2O)
2. Metallic beryllium or beryllium oxide (BeO)
3. Pure graphite
4. Uranium-238 (in production reactors).

Q. *What are the essential qualities of a good reflector?*

Ans.

1. Low macroscopic neutron-absorption cross-section
2. High macroscopic neutron-scattering cross-section
3. High resistance to oxidation
4. High resistance to irradiation
5. High radiation stability.

Q. *What is a coolant?*

Ans. A coolant is a thermal fluid (gaseous or liquid) used to transport thermal energy from the core.

It is passed through the reactor core to remove the heat that is generated as a result of nuclear fission.

Q. *By how many methods can the coolant remove the heat from the reactor core?*

Ans. A coolant can transport thermal energy from the core by any one of the three methods:

1. **Internal Cooling:** circulating the coolant right through the reactor core. Circulation may be forced or natural
2. **External Cooling:** circulating the liquid fuel through a tube- and-shell type heat exchanger outside the reactor while the coolant abstracts heat from the fuel in the heat exchanger.
3. **Boiling Liquid Cooling:** circulating and allowing the coolant to vapourize within the reactor and to take the reactor heat out.

Q. *Has the coolant any other utility apart from transporting heat from the reactor core?*

As. Yes; it can serve as a moderator as well as a fuel carrier.

Q. *What substances are used as coolants?*

Ans. Coolants may be of two types: Liquid or gas.

Liquid Coolants:
1. Ordinary and Heavy Water
2. Lithium
3. Sodium
4. Sodium-potassium alloy (22 : 78)
5. Lead-bismuth alloy (44.5 : 55.5)
6. Lead-magnesium alloy (97.5 : 2.5)
7. Diphenyls and terphenyls
8. Petroleum oil
9. Sodium hydroxide

Gaseous 1. Helium
Coolants: 2. Carbon dioxide
3. Air
4. Steam

Q. *What should be the essential qualities of a good reactor coolant?*

Ans. It must have
(a) High thermal conductivity
(b) High boiling point, should it be a liquid
(c) High specific heat
(d) High thermal stability
(e) High density
(f) Low viscosity
(g) Low vapour pressure, if it is a liquid
(h) Low melting point
Over and above, it should
(a) Chemically inert to in-core materials
(b) Have low macroscopic neutron-absorption cross-section
(c) Not become highly radioactive
(d) Non-oxidizing and non-toxic.

Q. *What are the disadvantages of using ordinary water in the primary circuit, i.e., as a reactor coolant?*

Ans. Water has a low boiling point and so it cannot be heated to a substantially high temperature. If it is used as reactor coolant, it must be circulated under high pressure and that makes the whole path too complicated and expensive as well.

Q. *Why are reactor shields used?*

Ans. They are used as envelopes (liquid or solid) to absorb all forms of gamma radiation. They also slow down and capture neutrons.

There are two reactor shields:
1. **Inner (Thermal) Shield** comprises 10% of the total reactor shield thickness yet they absorb 90% of the nuclear radiation that tends to escape.
2. **Outer (Biological) Shield** is several feet thick to protect the operating personnel and the environment from irradiation by γ-rays or neutrons.

Q. *Why are thermal shields used?*

Ans. These are installed to protect the vessel from excessive heating due to gamma-ray absorption.

The absorption of high energy radiations by the steel vessel or concrete shields leads to high temperatures and temperature gradients with the effect that considerable thermal stresses are induced in the structures. Moreover, prolonged exposure to the radiations emanating from the reactor core brings about changes in the properties of the materials: Steel becomes brittle and the concrete goes to form aggregates which swell non-uniformly and that tell upon the strength of the concrete and may cause failure.

To avoid this disaster, reactor vessel and the biological shield are protected by thermal shields made of high heat-resistant and radiation-stable materials (iron or steel) absorbing the excessive radiation energy.

Q. *What should be the qualities of a good shield?*

Ans. The shielding material.
(a) Must be a heavy element like lead, lead-cadmium alloy, tantalum, iron, iron oxide or barium sulphate to absorb the harmful gamma radiations.
(b) Should preferably have a low mass number element, e.g., hydrogen in water, concrete, wood for maximum kinetic energy transfer from the fast neutrons.
(c) Should be a good neutron absorber (e.g., boron used as borax, borates or boric acid)

Q. *Each reactor is provided with a reactivity control system. What is the purpose of this system?*

Ans. It is provided to every reactor to serve the following purposes:
1. To startup the reactor
2. To bring it up to the design load and capacity
3. To increase and sustain the specified power output
4. To shutdown as well as cool down the reactor.
5. To render the reactor subcritical under any abnormal conditions.

Q. *What is the difference between reactor cooldown and reactor shutdown?*

Ans. Cooldown is a process which is the aftermath of shutdown. If a reactor operating at full or part load is put out of operation either for routine maintenance and overhauling or to avoid accidents, the phenomenon is called reactor shutdown.

Whereas cooldown refers to the process of reducing the temperature and pressure of the coolants to their normal values following the reactor shutdown.

Q. *What do you mean by the term "scram capability"?*

Ans. It is a special capability endowed to the reactor by its reactivity control system as a result of which the reactor goes subcritical under any abnormal conditions.

Q. *What do you mean by shim rods?*

Ans. These are coarse control rods.

The purpose of their use is to compensate for variations in the reactivity when the reactor changes from the cold to the hot state, i.e., they are designed to compensate for variations in the temperature coefficient of reactivity. They are also used to compensate for fission-product processing and fuel burnup.

Q. *What do you mean by automatic or manual control rods?*

Ans. These are fine control rods, also called regulator rods. They are designed to maintain the reactor capacity and the basic parameters of the coolant at a specified level by slight variations in the reactivity.

Q. *What are scramming rods?*

Ans. These are safety rods to ensure quick shutdown of the reactor, if there arise some abnormalities.

Q. *In which configurational patterns do the control rods appear?*

Ans. They may be
   (a) cylindrical
   (b) prismatic
   (c) cruciform
   (d) rodded cruciform
   (e) spherical.

Q. *What is a chemical shim system?*

Ans. This refers to dosing a neutron-absorbing chemical to the coolant to effect a control of the nuclear fission process in the reactor.

These must have high thermal neutron-capture cross-section. For example, boric acid solution in-

jected into the coolant serves the purpose of reactivity control as boron has high thermal neutron-capture capability.

Q. *How is reactor control achieved?*

Ans.
   1. Introducing or removing the strong neutron absorber called control rods (rods, strips or plates) from the core. This is called absorption control
   2. Fuel control, i.e., by varying the amount of fuel within the reactor core. If the fuel quantity is reduced below the critical value fission chain reaction will cease
   3. Coarse control, i.e., varying the moderator (D_2O) level in the reactor
   4. Repositioning the reflector to vary neutron leakage.

Q. *What are the essential characteristics of a good control rod?*

Ans. The control rod
   (a) should absorb neutrons in the same way as does blotting paper the extra ink, i.e., it should have very high neutron capture cross-section
   (b) should have chemical and radiation stability
   (c) should have high melting point
   (d) should not react with the coolant or moderator in the reactor core.

Q. *Why is every nuclear reactor core fitted with a neutron source?*

Ans. To provide a source of neutrons to bring start-up neutron level in the core.

Q. *Why is the neutron level during start-up period monitored?*

Ans. To prevent any inadvertent power excursions during this interval.

Q. *What is the source of neutrons in the reactor?*

Ans. There are two types of neutron sources
   1. (α, n) source
   2. (γ, n) source
both using beryllium as the target metal.

$$^{9}_{4}Be + {}^{4}_{2}He \longrightarrow {}^{13}_{6}C + {}^{1}_{0}n + \gamma$$

$$\ce{^9_4Be} + \gamma \longrightarrow \ce{^9_4Be} \longrightarrow \ce{^8_4Be} + \ce{^1_0n} + \gamma$$

$$\downarrow$$

$$2\ce{^4_2He} + \ce{^1_0n} + \gamma$$

The source of α–particles (projectiles) is plutonium-239 which is mixed with beryllium and the mixture is used as neutron source.

The source of γ–rays (projectile) is radioactive antimony-124 (half-life 60 days).

Q. *What should be the threshold energy of this radiation to eject a neutron from beryllium after impingement?*

Ans. 1.6 MeV.

Q. *How are reactors classified?*

Ans. Reactors are classified on the basis of
(a) their use
(b) fuel conversion
(c) neutron energy level.

Q. *How are reactors classified according to their use?*

Ans. Reactors can be designed to generate power either as a prime function or as a byproduct:
1. **Power Reactor:** designed to produce power. It may be stationary as in a central station utility plant.
   It may be mobile as encountered in a nuclear powered submarines, oceanliners.
   It may be package type transportable by dismantling.
2. **Production Reactor:** designed to convert fertile materials into fissionable materials to be used elsewhere.
3. **Research Reactor:** primarily meant for scientific research, study and investigation in nuclear engineering and basic physics. Here power is produced as a byproduct.
4. **Material Testing Reactor:** specially designed for case study of
   (a) radiation damage
   (b) tolerance level to irradiation
   (c) beneficial effects of radiation.

Q. *How can reactors be classified according to fuel conversion?*

Ans. Reactors may be
(a) regenerative
(b) non-regenerative type.
A regenerative type reactor converts a fertile material like U-238 into fissionable material like Pu-238 by using a part of the neutrons available in the reactor. This generated fuel (always less than what is consumed) augment the life of the core.

A breeder is another form of regenerative reactor that not only converts fertile materials into fissionable materials but also produces more fissionable material than is consumed.

A non-regenerative reactor depletes the reactor fuel to the point where sustained nuclear fission is no longer possible. Such reactors are activated by loading a fresh charge of fuel.

Q. *How can reactors be classified on the basis of neutron energy level?*

Ans. Depending on the energy level of the neutrons used, the reactors may be classified as:
1. **Thermal Reactor** that uses slow neutrons having kinetic energy less than 0.1 eV.
2. **Intermediate Reactor** a reactor in which a portion of the fast neutrons (fission yields) is moderated down to thermal neutrons.
3. **Fast Reactor** a reactor where fast neutrons are not slowed down.

Q. *How many types of thermal reactors are there?*

Ans.
1. Pressurized Water Reactor using ordinary water under high pressure
2. Heavy Water Reactor (pressurized or unpressurized)
3. Graphite Water Reactor
4. Boiling Water Reactor
5. Organic Cooled and Moderated Reactor
6. Sodium Graphite Reactor
7. Aqueous Homogeneous Reactor
8. Liquid Metal Fuel Reactor
9. Fused Salt Fuel Reactor
10. Dust-fueled Reactor.

Q. *Apart from the above three basis of reactor classification, is there any other basis in vogue?*

Ans. Reactors can also be classified on the basis of

1. Fuel used
   (a) Natural fuel
   (b) Enriched fuel
2. Moderator used
   (a) Ordinary water moderated
   (b) Heavy water (D_2O) moderated
   (c) Graphite moderated
   (d) Beryllium moderated
3. Coolant used
   (a) Ordinary water cooled reactor
   (b) Heavy water cooled reactor
   (c) Gas cooled reactor
   (d) Liquid metal cooled reactor
   (e) Organic liquid cooled reactor.

Q. *Briefly describe a pressurized water reactor (PWR) and its working principle.*

Ans. A pressurized water reactor is a thermal reactor where slightly or highly enriched, heterogeneous, clad uranium is used as fuel and ordianry water under high pressure is used as a moderator, coolant reflector.

The fuel used is usually uranium oxide, UO_2, because it is highly resistant to irradiation damage and well adopted to high burnups. Also it has another big plus point: It is highly resistant to corrosion inflicted by water at high temperature and pressure in case of accidental cladding failure.

Coolant-cum-moderator-cum-reflector is ordinary water which is circulated in the primary circuit by the

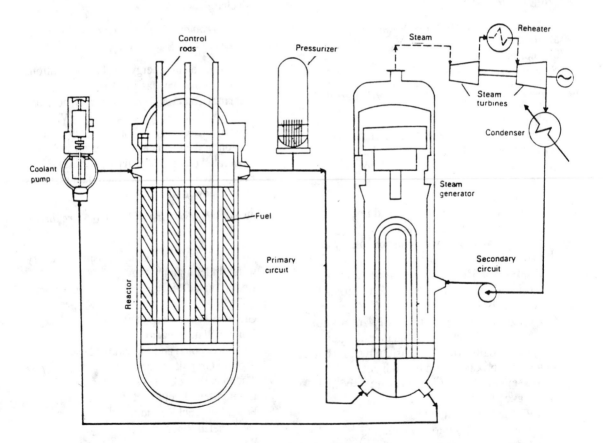

Fig. 34.4 *Pressurized Water Reactor. Nuclear Fission heat is transferred to the reactor coolant while steam is generated in the secondary circuit by indirect heat transfer to the working fluid in the steam generator*

circulating pump pressurizing the liquid to 100–130 atm. Heat carried by the coolant from the reactor transferred to the DM water of the secondary circuit to generate steam in the heat exchanger. The primary circuit including the heat excanger is well shielded as the coolant in the primary circuit becomes radioactive. (Fig. 34.4)

The generated steam is directed to the steam turbine to generate electricity.

Q. *What are the advantages of a pressurized water reactor?*

Ans.

1. The principal advantage of a PWR is that it uses ordinary water as a coolant, moderator and reflector. Since water is available abundantly and cheaply, there is great saving in cost.
2. The reactor is very compact and has high power density, 65 KW/litre. It can be made more compact if enriched fuel is used.
3. Small number of control rods are required. Barely 60 control rods are required in a 1000 MW nuclear power plant operating PWR.
4. Greater energy is extracted per unit weight of fuel loaded. Therefore, PWR is ideally fitted for fuel designed for higher burnups.
5. Safe and stable operation as well as self-regulation are possible because of high negative temperature coefficient of PWR.
6. The power demand coefficient of PWR is positive, i.e., when more power is demanded the reactor spontaneously responds to supply the same. The negative temperature coefficient has made this possible.
7. Optimization of the turbine cycle for the purpose of low heat rate supply is possible because of separation of the secondary circuit from the primary circuit.
8. As steam is not contaminated by radiation, inspection and maintenance of turbine, condenser and feed pump are possible.

Q. *What are the disadvantages of a pressurized water reactor?*

Ans.

1. High capital cost and heavy and strong pressure vessel is required in the primary circuit.

2. Difficulty in obtaining high steam pressure at the turbine.
   Because of low pressure (60–70 kgf/cm^2) in the secondary circuit, the thermodynamic efficiency of this plant is as low as 20%.
3. Corrosion is a severe problem induced by rapidly circulating water under high temperature and pressure. This dictates the use of stainless steel vessel and cladding, that further hikes up the cost.
4. It requires slightly enriched uranium fuel.
5. Fuel reloading takes a couple of months and that means a considerable downtime loss.
6. Deadly and penetrating γ–radiations cause uneven heating of the pressure vessel and thereby inflict thermal stresses in the vessel wall which is already under pressure stresses. This fact further complicates the design.

Q. *What is a fused-salt fuel reactor?*

Ans. It is a homogeneous thermal reactor that uses fused salt (uranium-235 tetrafluoride, UF$_4$) as the fuel while sodium fluoride and beryllium oxide is used as a carrier.

Q. *What is the purpose of using beryllium oxide?*

Ans. It serves as a moderator as well as reflector.

Q. *Can this reactor be used as a breeder reactor?*

Ans. Yes.
In this case lithium flouride or beryllium fluoride is used as a carrier while thorium tetraflouride may be used as reflector.

Q. *What are its advantages over an aqueous homogeneous reactor?*

Ans.

1. High temperature operation (825°C) makes possible greater amount of heat transfer from the reactor core and hence higher efficiency
2. Operating pressure is low. So there is a great saving in the first cost as high pressure vessel is not required
3. Extremely stable
4. Completely responsive to power-demand factor.

Q. *What is a dust-fueled reactor?*

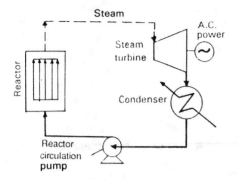

Fig. 34.5 *Single-Circuit Nuclear Power Station*

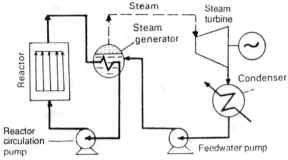

Fig. 34.6 *Two-Circuit Nuclear Power Station*

Ans. It is a thermal reactor wherein fissionable dust is carried in an inert (helium) gas.

It can also act as a breeder reactor if thorium-uranium fuel is used.

Q. *What moderator is used in a dust-fueled reactor?*

Fuel	Moderator
Uranium-carbide (enriched)	Graphite moderator lined with silicon carbide
Uranium-dioxide (enriched)	Beryllium oxide moderator lined with alumina

Q. *How is steam generated in nuclear power stations?*

Ans. The tremendous thermal energy liberated due to controlled nuclear fission of heavy nucleids (uranium-233, uranium-235 or plutonium-239) by slow or thermal neutrons within the reactor core is transferred directly or indirectly to the working fluid, i.e., water to generate steam. That is, a nuclear reactor may be thought of as a boiler furnace—a heat source—to produce steam and superheat it.

Steam is generated directly in a single-circuit nuclear power station operating a boiling type nuclear reactor (Fig. 34.5). Also it may be generated indirectly, as a result of heat transfer from the reactor coolant to the working fluid, in a two-circuit nuclear power station (Fig. 34.6) and in a three-circuit nuclear power station (Fig.34.7).

Q. *Why do nuclear power stations with aqueous coolant usually operate on saturated steam of pressure 5–7 MPa (about 50–70 atm.)?*

Ans. Water is a low-temperature coolant. And it is the pressurized water reactor where aqueous coolant can be subjected to the highest pressure and hence maximum temperature.

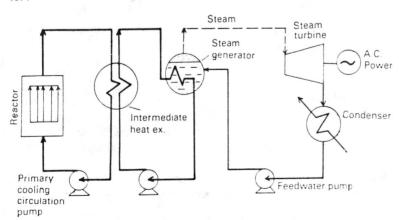

Fig. 34.7 *Three-Circuit Nuclear Power Plant*

Now if it is required to generate steam at a pressure as high as possible, e.g. P = 63 atm. and the temperature of the aqueous coolant at the reactor inlet be θ'_1 = 593 K, the degree of superheating will be only 20K (see Fig. 34.8).

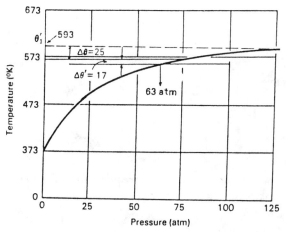

Fig. 34.8 *Selection of Working Fluid Parameters for Generation of Steam in Water Cooled Water Moderated Reactors*

To ensure active heat transfer we can assume that the temperature gradient between the coolant (i.e., pressurized water) and the working fluid to be 25K. nevertheless, the degree of superheating that can be obtained at 63 atm. will be only 17K = $\Delta\theta'$. This explains why nuclear power stations operating on aqueous coolant usually operate on saturated steam at a pressure of 50–70 atm.

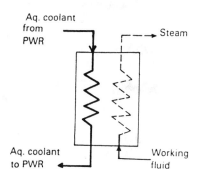

Fig. 34.9

Q. *Why is it essential to superheat the steam, usually by 20K?*

Ans. As the steam expands in the turbine, it increasingly becomes more wet. Wet steam can cause erosion of turbine blades and steam distribution devices and impair the turbine's reliability. Hence to prevent these effects, it is essential to superheat the steam slightly, usually by 20K, so that moisture formation can be avoided.

Steam must be at least 99.8% dry.

Q. *But as the steam expands through the successive stages of the turbine, it will become wet at some stages of the turbine. Therefore, the turbine will operate in the region of wet steam which will reduce the efficiency of the power plant as well as inflict blade erosion on the turbine. What can be done to get rid of this problem?*

Ans. Before the steam becomes wet to the degree at which it becomes inadmissible (generally to the low-pressure cylinder of the turbine) due to the risk of moisture formation and consequent blade erosion, the wet steam (i.e., extraction steam) from the high-pressure cylinder is introduced tangentially to the apparatus, called separating steam superheater, at the top. As it passes through the separator, its moisture is removed and it then sweeps the superheating surfaces heated up by the second-stage heating steam. The steam becomes superheated to 514K and exits from the top. It is then directed to the low-pressure cylinder of the turbine.

Q. *Why can steam of high or supercritical parameters only be produced by using a gas or liquid metal and not water as coolant?*

Ans. Water is a low-temperature coolant while liquid metals or gases are high temperature coolants. In fact, liquid metals and gases have no limitations regarding their temperature at the reactor outlet, while water must be kept under very high pressure to maintain its liquid state to ensure a better heat transfer rate.

Though the gases have poor thermophysical properties, gases with high mass flowrate, i.e., gases under pressure can be used to intensify the heat trans-

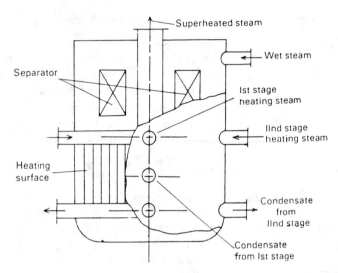

Fig. 34.10 *Separating Superheater*

fer. Liquid metals have very favourable thermophysical properties that make them perfectly suitable for high temperature coolants without taking any special measures to intensify the heat transfer.(Fig. 34.11)

Therefore, by using these high temperature coolants only, high or supercritical pressure steam can be produced.

Q. *How much coolant (gas/liquid metal) tempera-*

ture at the reactor outlet is required to ensure steam superheating to standard parameters e.g., 125–240 atm. and $\theta_{ss} = 818\,K$?

Ans. It has been found in practice that if the temperature of the gas or liquid metal coolant at the reactor outlet is in the range 873–923 K, then it can ensure steam superheating to the parameters 125–240 atm./$\theta_{ss} = 818$ K.

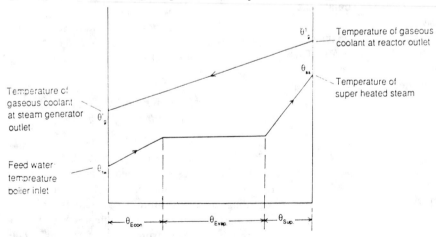

Fig. 34.11 *Temperature v. Heat Transfer diagram in the steam generator of a gas cooled reactor*
  $Q_{ECON} = $ *Quantity of heat transferred in the economizer*
  $Q_{EVAP} = $ *Quantity of heat transferred in the evaporator*
  $Q_{SUP} = $ *Quantity of heat transferred in the superheater*

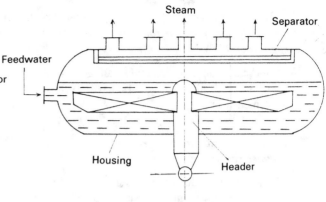

Fig. 34.13 *Horizontal steam generator*

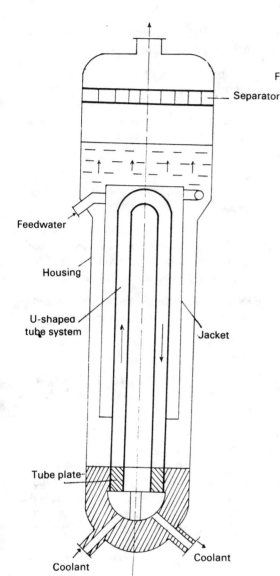

Fig. 34.12 *Vertical steam generator*

Q. *What is the general configuration pattern of steam generators of nuclear power plants?*

Ans.
1. Vertical (Fig. 34.12)
2. Horizontal (Fig. 34.13)

Q. *What type of configuration pattern of steam generators is preferred in the cases of gaseous or liquid metal cooled reactors?*

Ans. Usually vertical.

Q. *What type of configuration pattern of steam generators is selected in the case of water-cooled-water-moderated reactors?*

Ans. Either vertical or horizontal.

Q. *What are the chief advantages of a horizontal steam generator of a nuclear power plant?*

Ans.
1. Design simplicity
2. Ease of manufacture
3. Reliable operation.

Q. *What are the chief advantages of a vertical steam generator of a nuclear power plant?*

Ans.
1. Very compact
2. Better radiation safety
3. Greater unit capacity for a given volume.

Q. *What are the comparable features between horizontal and vertical steam generators?*

Ans.
1. Both are more or less equally costly with respect to total cost of the apparatus
2. Both have roughly equivalent performance characteristics and reliability
3. Both can be assembled and tested at the manufacturing site and transported by railroad.

Q. *Why is it not usually favoured to manufacture a steam generator of the horizontal type for a capacity exceeding 300 MW?*

Ans. The size of the horizontal steam generator of capacity beyond 300 MW will be excessively and inconveniently large.

Q. *How many types of steam generators exist for nuclear power stations where water-cooled-water-moderated reactors are used?*

Ans. As regards the processes occurring in the secondly circuit, the nuclear steam generators of water-cooled-water-moderated reactors can be divided into two main types:
1. Natural circulation steam generators
2. Once-through steam generators.

Q. *How can these two be distinguished on the basis of the mode of steam formation?*

Ans. In the natural circulation steam generators, heat is transferred from a submerged heating surface to the working fluid (water) which boils in the bulk on the surface of the heating elements.

In once-through systems, water flows under pressure through a system of tubes or in the intertubular space of the dense tube bundle and gets evaporated to steam as it abstracts heat from tubular heating elements on its way up from the bottom of the steam generator (vertical).

Q. *What are the heating elements of the steam generators?*

Ans. They are a system of tubes. They come in the form of straight tubes, U-shaped tubes, or helical. They may also be assembled into plain curtains.

Q. *What is the diameter and thickness of the tubes forming the heating elements in both horizontal and vertical steam generators?*

Ans. OD : 12–22 mm
Thickness : 1.2–1.5 mm

Q. *How many such tubes are used in high-capacity steam generators?*

Ans. As many as some ten thousand tubes may form the heating surface of high-capacity steam generators.

Q. *Why are all the elements of the primary circuit of a nuclear steam generator made of stainless steel?*

Ans. All the elements of the primary circuit are made of stainless steel to eliminate corrosion to avoid leakage of reactor coolant from the primary circuit to the secondary circuit. Contamination of the aqueous coolant by corrosion products is inadmissible.

Q. *The ends of the tubes used as heating elements in steam generators are expanded in tube plates and the tightness of the joints should be very high. Why is such high joint-tightness necessary?*

Ans. To avoid the slightest chance of leakage of reactor coolant in the primary circuit, as this might lead to the contamination of working fluid with radioactive material in the secondary circuit.

Q. *Why is the heat transfer agent in steam generators with aqueous coolant circulated in tubes?*

Ans. To minimize the use of metal in the generator housing.

Q. *Why is the intensity of steam generation different in various portions of the drum of a horizontal steam generator?*

Ans. As the coolant emerging from the reactor and loaded with heat passes through the steam generating coils of the horizontal steam generator, it develops a variable temperature along the length of the coils: It is highest at the coolant's inlet to the steam generator and lowest at the coolant's outlet from the generator. This explains why the steam generation intensity is different in various portions of the drum.

Q. *Is the intensity of evaporation from the disengagement surface also different?*

Ans. Yes.

Q. *How can this be equalized?*

Ans. This can be effected by placing a submerged perforated plate or differentiated supply of boiler feedwater to the coils.

Q. *Between the vertical steam generator and horizontal steam generator which one is affected by higher moisture carryover?*

Ans. Vertical steam generator.

Q. *Why ?*

Ans. The intensity of evaporation, in the case of vertical steam generators, from the disengagement surface is much higher than in the horizontal type. As a result, vertical steam generators suffer from higher carryover of moisture by steam.

Q. *How can this moisture carryover be diminished?*

Ans. The steam space height of the drum is substantially increased to fit within it two or three steam separating stages. They will produce steam of satisfactory quality (dryness fraction 99.8%).

Q. *What are the advantages of design of vertical nuclear steam generators?*

Ans.
1. Higher compactness that bestows radiational safety
2. Substantial increase of unit capacity
3. Thermal expansion of U-shaped tube bundles where such tubes are used in vertical steam generators, is self-compensatory. This is a big advantage.

Q. *Is there any typical fabrication problem encountered in vertical steam generators with U-shaped tubes?*

Ans. The ends of the U-ends are expanded in the

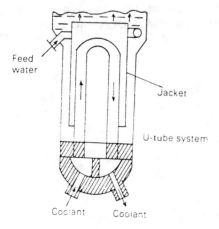

Feed water

Jacket

U-tube system

Coolant   Coolant

Fig. 34.14 *Vertical Steam Generator with U-shaped Tube System. It needs precision fabrication to fit large number of U-tubes*

tube plates. For high-capacity steam generators, the thickness of such tube plates may range from 600 to 700 mm. They must be accurately drilled to produce a large number of deep holes to fit a large number of U-tubes for expanding of tube ends. The manufacture and assembly of such plates is expensive as well as complicated. (Fig. 34.14)

Q. *What is the major drawback in vertical, U-tube type steam generators with a bottom tube plate?*

Ans. Here comes the problem of corrosion. The water in a steam generator may contain sludge (par-

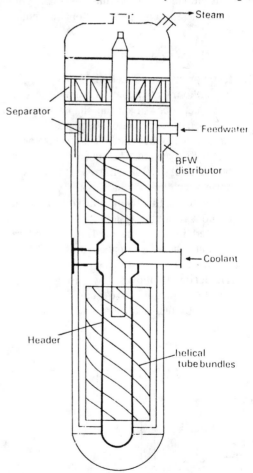

Steam

Separator

Feedwater

BFW distributor

Coolant

Header

helical tube bundles

Fig. 34.15 *Vertical steam generator without bottom plate*

ticularly iron oxide) which tends to accumulate (despite blowdown) in the bottom portion of the tube bundle just near the tube plate. Gradually, the sludge layer is evaporated, increasing the concentration of all water impurities, viz., chloride ions and alkalis within the pores of the sludge layer. Under high temperature and higher concentration, these impurities become potentially corroding and can destroy even stainless steel tubes.

Q. *How can this problem be coped with?*

Ans. This problem can be eliminated by designing vertical-type steam generators without the bottom tube plate.

Q. *How does such a system work?*

Ans. It comprises of two sections:
1. The bottom section that accommodates stainless steel tubes given the shapes of helical coils and expanded into tube plates. The helically coiled tubes are self-compensatory in thermal expansion. The aqueous coolant at a pressure 158–160 atm. moves in the tubes that serve as heat transfer surfaces.

    The working fluid is introduced from the top of the helical tube system which is kept submerged in the working fluid to generate saturated steam at a pressure of 60–63 atm. (Fig 34.15)

    The feedwater moves down, by gravity circulation, through a descending channel formed by the jacket surrounding the helical

tube system. The working fluid moves ascendingly inside the jacket.
2. The upper section forms the steam space that houses a cyclone type separator (for coarse steam separation, 1st stage) and a vertical annular louver-type separator (2nd stage) to produce dry steam.

Q. *What is the basic advantage of a vertical steam generator with helically coiled tubes during temperature variation?*

Ans. The thermal expansion and contraction of helically coiled tubes due to temperature variation is self-compensatory.

Q. *What is the principal drawback of such a system?*

Ans. It involves difficulties in the manufacture of helical coiling since the angle of coiling differs from one tube row to other.

Q. *What is the operating principle of a three-circuit nuclear power station?*

Ans. Steam is generated as a result of indirect heat transfer in this system that employs liquid sodium as the reactor coolant and primary heat transfer agent.

In the first circuit (I), molten sodium is pumped through the reactor to transfer heat from the reactor core to the intermediate heat-transfer agent which circulates in the second-circuit (II). (Fig. 34.16)

The heat-transfer agent of the 2nd-circuit (II) may be sodium or an alloy of sodium-potassium. The flow

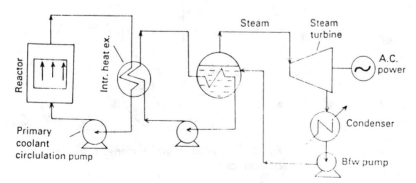

Fig. 34.16 *A Three-Circuit Nuclear Power Station. It involves one intermediate heat exchanger and steam is generated in the second circuit*

of the heat transfer agent in the 2nd-circuit is made at a pressure higher than liquid sodium flowing in the 1st-circuit.

In the third circuit heat is transferred to the working fluid (water) from the heat-transfer agent of the 2nd-circuit. The steam generated is fed to the steam turbine to drive the turboalternator to generate electricity.

Q. *Why is the heat-transfer agent in the 2nd circuit circulated at higher pressure than liquid sodium in the primary circuit?*

Ans. This is a preventive measure to avoid radioactive contamination of the heat-transfer agent of the 2nd-circuit by that of the first circuit, should there arise any leakage in the intermediate heat exchanger.

Molten sodium of the 1st circuit becomes highly radioactive due to its direct contact with nuclear fission and fission products. It emits high energy gamma-radiation. If the shielding in the 2nd circuit becomes loose, radioactive sodium will pass into the 2nd circuit and will render Na-K alloy or sodium of this circuit radioactive. Ultimately, radioactivity will pass into the working fluid and endanger the safety of the working personnel.

To overcome this problem, the heat-transfer agent of the 2nd-circuit is circulated at a higher pressure than liquid Na of the 1st-circuit.

Q. *Why does a nuclear steam generator involving liquid sodium cooled reactor consist of three circuits instead of one?*

Ans. Liquid sodium while passing through the reactor becomes highly radioactive, which renders one heat exchanger insufficient and the whole heat-exchange system becomes complicated. In order to make the steam generator operation safe, the nuclear power station incorporates three circuits and two successive heat exchangers.

Q. *Briefly describe a steam generator with sodium coolant?*

Ans. It is a shell-type steam generator whose shell diameter (OD) is about 3m or less. The heating surfaces are made of U-tubes arranged on the shell periphery. The working fluid is circulated through tubes and the molten sodium circulates in the intertubular space. (Fig. 34.17)

A cylindrical jacket is fitted on the shell to force a countercurrent current flow of the coolant and the working fluid. The shell is prevented from overheating by a thermal shield.

Q. *Why is the shell diameter (OD) of a steam generator with sodium coolant 3m or less?*

Ans. For manufacturing ease and transportation facility, the shell-diameter of a steam generator rarely exceeds 3m.

Q. *In the sodium-coolant steam generator, molten sodium (heat-transfer agent) is circulated in the intertubular space and working fluid through the tubes (U-shaped). Why not the reverse?*

Ans. Since the working fluid is circulated at a much higher pressure than the heat-transfer agent (liquid sodium), the former is taken within the tubes and the latter (Na-coolant) in the intertubular space. (Fig. 34.17).

Q. *Can you mention certain basic parameters of a typical sodium-coolant steam generator?*

Ans. One such typical generator with 1320MW output has the following working parameters:

Temperature of coolant at steam generator inlet

= 833K

Temperature of coolant at steam generator outlet

= 653K

Superheated steam pressure = 16.3/3.4 MN/m^2
Superheated steam temperature = 813K/813K.

Q. *What are the advantages of sodium used as a coolant?*

Ans.

1. It has low neutron absorption cross-section (0.5–2barns). So greater number of slow neutrons will be available for fission process
2. It has low melting point (371K) and high boiling point (1156K). Between these two extreme temperatures, sodium will remain liquid and can be easily pumped and circulated through the reactor in the primary circuit
3. It has high specific heat 0.3 kcal/kg/K

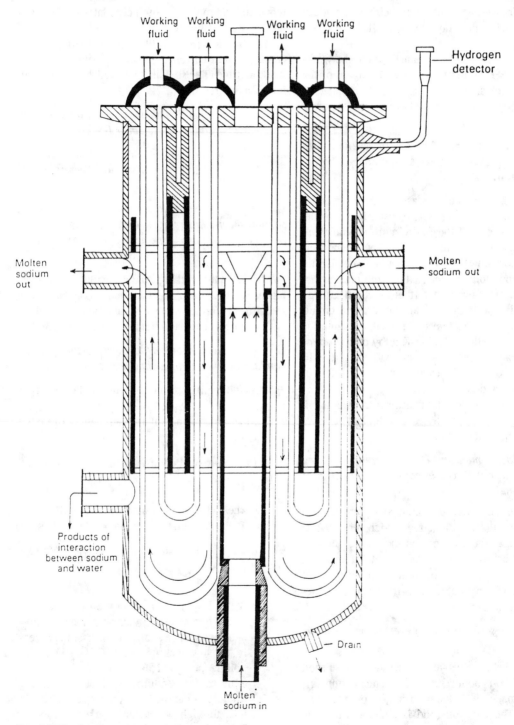

Fig. 34.17 Sodium-coolant steam generator

4. Its thermal conductivity is 120 times the thermal conductivity of water. Hence it is a much more efficient heat transfer agent than aqueous coolant.
5. It is economically available.

Q. *What are the advantages of liquid metal cooled reactors?*

Ans.

1. High thermal efficiency at low cost
2. Low cost of output power
3. High thermal conductivity and high specific heat of liquid metals render a higher heat removal rate at much lower flowrate in comparison to gas cooled or water cooled reactors. Hence liquid metal need not be circulated at high pressure
4. High reactor stability. In other reactors, the reactivity of the reactor increases with the loss of coolant because of the removal of neutron absorber. Hence to avoid overheating, these reactors must be shutdown by inserting control rods. But in the case of liquid metal cooled reactors, the reactivity decreases with the rise of temperature. This is a great advantage from the point of view of safety.
5. The reactor is compact and hence its size is comparatively smaller than other reactors of same power output
6. If the liquid metal used as a coolant be sodium, then it will be best suited for thermal reactors with slightly enriched fuel. This is due to the low neutron absorption cross-section of sodium.

Q. *What are the disadvantages of liquid metal cooled reactors?*

Ans.

1. With the increase of reactor core temperature, the neutron economy falls off, as the hot moderator subjects the high-energy fission-yield-neutrons to resonance and thereby increases the probability of the non-fissionable capture of neutrons
2. Each block of graphite, used as a moderator in sodium cooled reactor must be well cladded to prevent the diffusion of liquid sodium into porous graphite. This sodium-penetration in

graphite layers causes mechanical failure. Hence graphite is to be protected from sodium-penetration by cladding, which further adds to the construction cost
3. The liquid metal, particularly sodium, becomes highly reactive as it extracts heat directly from the core. So to avoid contamination by irradiation steam is generated in the secondary circuit and both primary and secondary circuits are well shielded by thick slabs of concrete
4. Compared to other coolants, the leakage of sodium (primary circuit) is very dangerous as sodium becomes highly radioactive due to irradiation
5. All care must be taken to prevent the leakage of sodium in the water-circuit as it will react exothermally with water, producing enough heat as well as radioactive caustic soda, which increases the risk of further contamination.

Q. *What are the advantages of organic cooled and moderated reactors?*

Ans.

1. Organic liquids viz., diphenyls and triphenyls act both as coolants and very efficient moderators. Hence they're called super-moderators
2. Compact core design is possible due to dual characteristics (coolant and moderator) of liquid hydrocarbons
3. Non-corrosive. And therefore, low-cost mild steel piping can be used in the system. This saves initial capital investment
4. Selection can be made from a wide variety of fuels viz., uranium, uranium oxide, uranium carbide or uranium alloy because of excellent moderating properties of the organic coolants
5. Low operating core pressure
6. Steam of higher parameters of pressure and temperature can be produced.

Q. *What are the basic disadvantages with organic fluids used as moderator as well as coolant?*

Ans.

1. Poor heat transfer characteristics, even less than those of water
2. They undergo thermochemical changes due to radiation and heat. As a result they deposit on

fuel-can surfaces thereby impairing the heat transfer further

3. Organic fluids being inflammable, adequate protection must be taken to guard against accident

Q. *Can you cite an example of a nuclear power plant that uses organic fluid as moderator and coolant?*

Ans. The Piqua Nuclear Power Plant (11.4MW), Piqua city, Ohio, USA is an example.

Q. *What organic fluid does it use as a moderator and coolant?*

Ans. It uses a mixture of orthometa terphenyl and para-terphenyl as a coolant and moderator.

Q. *What fuel and control rods are used there?*

Ans. It uses enriched uranium as fuel.
Boron is used as control rods.

Q. *What basic condition should a nuclear reactor fulfill in order to act as a breeder reactor?*

Ans. A breeder reactor produces a set of new fissionable material in the course of the burn-up of primary fuel. That is, in a breeder reactor, the burnup of primary fuel results in the production of the secondary fuel.

In order to breed, the average number of neutrons produced per neutron absorbed by the fuel ($\eta = v\, \sigma_f / \sigma_a$) must exceed 2— one neutron to continue the fission chain reaction and the other neutron to be absorbed by a fertile nucleus to yield a new fissile (i.e., fissionable) nucleus.

For example, when one U-235 nucleus undergoes fission, by a thermal neutron, it ejects 2.5 fast neutrons. One out of these 2.5 neutrons goes to sustain the nuclear chain reaction of U-235 fission, 0.6 neutron becomes absorbed by moderator, coolant and structural elements and the rest 0.9 neutron is captured by U-238 nucleus to produce a new fissile material Pu-239.

$$^{238}_{92}U + ^{1}_{0}n \longrightarrow ^{239}_{92}Pu$$

This plutonium can undergo fission process and is called secondary fuel.

Q. *So in a breeder reactor, the burnup of primary fuel is compensated to some extent by the production of secondary fuel. How can this compensation be measured?*

Ans. This can be measured in terms of the conversion factor, which is the ratio of the number of secondary fuel nuclei generated to the number of primary fuel nuclei depleted.

Q. *What should be the conversion factor in order to "breed" in a reactor?*

Ans. The conversion factor should be unity or higher.

Q. *Why cannot U-235 or Pu-239 be used as a fuel in a thermal breeder reactor despite the fact that the value of η (the average number of neutrons produced per neutron absorbed) exceeds 2 for all fuels and neutrons of all energy?*

Ans. This is because of the absorption of fission neutrons by non-fuel material and some leakage from the reactor core.

When U-235 is used as fuel, the average number of thermal neutrons emitted is 2.12 of which one goes to maintain the fission chain reaction and the remaining 1.12 are left for breeding. Since the loss of neutron due to leakage and absorption by non-fuel material well exceeds 0.12, so less than one neutron is available for breeding. This exaplains why the breeding mechanism ceases shortly.

With plutonium-239 as the primary fuel, the situation is worse, as the fission neutrons (thermal) available as a result of the fission of one Pu-239 nucleus is only 1.94.

Q. *Why are all the breeder reactors necessarily fast reactors?*

Ans. Fast reactors employ fast neutrons as projectiles. The value of η for fast neutrons is much greater than 2 for U-233, U-235 and Pu-239.

Target	η	Fission Cross Section	Absorption Cross Section
$\left(^{1}_{0}n\right) \Rightarrow$ (> 1.5 MeV) U-233	2.6	1.85 barns	1.884 barns
U-235	2.48	1.28 barns	1.358 barns
Pu-239	3.04	1.95 barns	1.990 barns

With respect to fission cross-section they have a relatively high absorption cross-section for fast neutrons—a fact that increases the breeding possibility considerably. With fast neutrons, the chances of fission are less while the chances of neutron absorption by U-233, U-235, Pu-239 and Th-232 are considerably greater. Therefore, when fissions are induced by high-energy neutrons, high breeding ratios can be achieved with any fuel isotope.

Q. *What is the principal advantage of thermal reactors over fast reactors?*

Ans. The low fuel requirement.

If a fast reactor and a thermal reactor are allowed to operate at the same power with the same average fluxes causing fission, the fission rate

$$= \overline{\sigma}_f \cdot N_{fuel} \cdot \overline{\phi}_f \qquad \text{for the fast reactors}$$

$$= \overline{\sigma}_f \cdot N_{fuel} \cdot \overline{\phi}_{th} \qquad \text{for thermal reactors}$$

whereupon $\overline{\phi}_f$ and $\overline{\phi}_{th}$ become same and equal to $\overline{\phi}$.

Hence the number of fuel atoms, i.e., the total fuel loadings become directly proportional to the microscopic fission cross-sections ($\overline{\sigma}_f$).

The fuel isotope being U-235, the fission cross-section for thermal (0.0253 eV) neutrons is $\overline{\sigma}_f = 582$ while for fast neutrons (> 1.5MeV) the fission cross-section is $\overline{\sigma}_f = 1.28$.

So even if the average flux of fast neutrons exceeds the thermal neutron flux by 10 times, the U-235 fueled fast reactor will have a fuel loading equal to 582/(1.28 × 10) = 45.4 times that of a thermal reactor operating at the same power.

Q. *Why is Pu-239 a more preferred fuel than U-233 in a breeder reactor?*

Ans. The average number of neutrons ejected per fast neutron absorbed (η) is 3 when plutonium-239 is used as fuel and it is 2.5–2.6 when uranium-233 is used as fuel.

That's why Pu-239 is more preferrable in a breeder reactor.

Q. *Why is no moderator used in a breeder reactor?*

Ans. A breeder reactor is essentially a fast reactor. If a moderator is used to slow down the fission neutrons, the value of η (the average number of neutrons produced per neutron absorbed by the fuel nucleus) will be roughly equal to two whereupon enough neutrons will not be available for breeding.

Q. *What is the major advantage of a fast breeder reactor?*

Ans. A fast breeder reactor employs high-energy neutrons as projectiles. And upon fission, the target nuclei eject fast neutrons also. Since there is no moderator, these neutrons are not slowed down. Since the structural materials do not absorb fast neutrons, a wide range of choices of constructional material is possible. This is a major advantage.

Q. *What is the major technical drawback of a breeder reactor?*

Ans. A breeder reactor has a very high power density— as high as 430 MW/m^3 of core volume. This is 13 times greater than a BWR, 40 times that of a CANDU reactor and 200 times that of a gas cooled reactor.

Therefore, it is necessary to remove large quantities of heat from the core and for that special coolants and special arrangements are required.

Q. *What is a boiling water reactor (BWR)?*

Ans. In this reactor, saturated steam is generated in the reactor core instead of a separate steam generator, i.e., in this system, fission heat is directly transferred to the water to produce steam which is dried and sent straightaway to the steam turbine.

Q. *In the early models, BWR was tagged into dual-cycle flow systems. Why was such a dual-cycle system in vogue?*

Ans. The dual-cycle flow system BWR consisted of two circuits—primary circuit that generated most of the turbine steam directly in the reactor core and the secondary circuit that contained a steam generator to produce turbine control steam. (Fig.34.18).

In any BWR, the reactor power is controlled by the amount of steam bubbles in the reactor core. In a BWR, water is used as a coolant as well as the

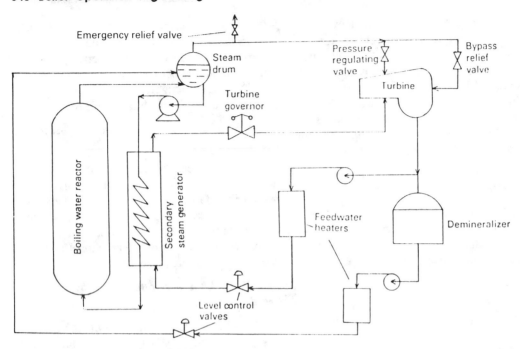

Fig. 34.18 *Schematic Diagram of a Dual-Cycle BWR System. Turbine control steam is produced in the secondary steam generator*

working fluid and moderator. Therefore, the displacement of water from the core by steam following higher steam demand by the turbine is accompanied by a fall of reactor power unless the control rods are withdrawn. If the reactor steam to the turbine is fed through a throttle valve, and the valve is opened to increase the turbine power, the reactor steam pressure will drop immediately. As the steam pressure in the reactor decreases, the steam volume in the core increases and that will bring about a lower reactor power.

This problem is solved by introduction a secondary circuit incorporating a small secondary steam generator and placing the turbine throttle valve on the secondary steam supply line.

Now with the increase of steam demand by the turbine, the throttle valve is opened, as a result of which the steam parameters—both temperature and pressure—will decrease. This will result in an increase of heat transfer from the primary water in the secondary steam generator and as a consequence the reactor inlet temperature of the feedwater will decrease.

As the cold water enters the reactor, it condenses some of the in-core steam increasing thereby the reactor power. This is how the dual-circuit arrangement essentially endows the reactor with an inherent load following capability.

Q. *In the latest BWR steam generator systems, a single-cycle BWR circuit is used. And there is no secondary steam generator. Then how is the reactor power controlled?*

Ans. The reactor power is controlled by altering the recirculation rate through the reactor core with a series of jet pumps. (Fig. 34.19).

With the increase of power demand, the water (working fluid) recirculation rate through the core is increased and that sweeps away the steam bubbles from the core increasing the reactor power level.

Following the fall in reactor power demand, the water recirculation rate through the core with jet pumps is reduced increasing the incore steam volume that downs the reactor power level.

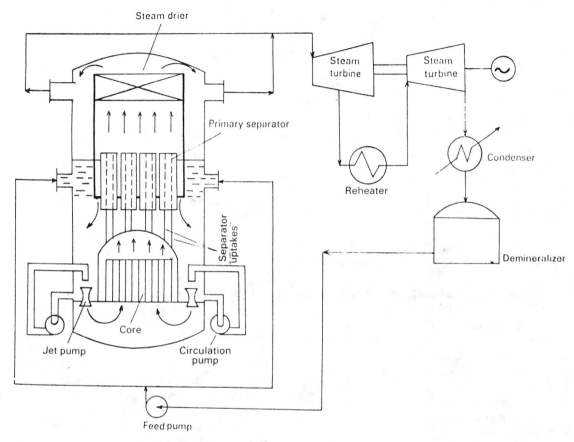

Fig. 34.19 *Boiling Water Reactor. Steam is generated in the reactor core because of direct heat transfer to the working fluid used as coolant also*

Q. *In the BWR system, the control rods enter from the bottom of the BWR vessel. Why not from the top?*

Ans. The top of the BWR is filled with steam, steam separator and steam drier assembly. Hence the control rods are inserted through the bottom of the BWR.

Q. *Is there any difference in the basic operating principle of control rods between PWR and BWR systems?*

Ans. The control rods in a PWR enter from the top of the reactor and they are actuated by an electromagnetic system. During emergency, the reactor is scrammed by switching off power to the electromagnets letting the rods drop into the core under the influence of gravity.

However, BWR systems do not rely on gravity for emergency insertion. Here a hydraulic accumulator system is deployed to store enough energy to drive the scram rods into the core in the event of an emergency shutdown.

Q. *While designing as well as operating a boiling-water reactor care must be taken that the maximum heat flux is below the DNB (departure from nucleate boiling) value.*

What is DNB? And why is it imperative that the maximum heat flux in the reactor core does not exceed the value of DNB?

Ans. Departure from nucleate boiling—a typical heat transfer phenomenon—can be demonstrated with the help of a boiling heat transfer curve.(Fig. 34.20)

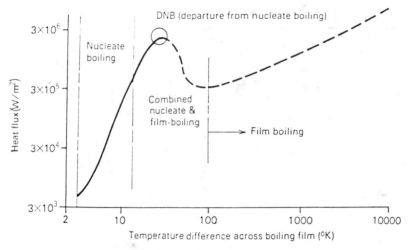

Fig. 34.20 *Typical boiling heat-transfer relationships*

Nucleate boiling is a boiling heat transfer process that takes place at relatively low heat fluxes. In this region vapour bubbles are formed at the surface of the fuel element and break away. Because of high convective heat transfer coefficients between the vapour bubble and the fuel element, a very high heat-transfer rate is established resulting in very marginal temperature difference between the saturation temperature and the fuel-element surface.

If the heat flux is raised further, the rate of steam bubble formation will increase and the smaller bubbles will coalesce to form big bubbles, while simultaneously increasing the heat-transfer rate. Ultimately the big bubbles will merge to form long, continuous vapour bubbles almost completely isolating the liquid film from the surface of the fuel element.

The vapour acts as an insulating layer and when the surface of the fuel element is completely covered with vapour, the surface temperature shoots up dramatically. This mode of heat transfer is called film boiling. The transition from nucleate to film boiling occurs at the departure from nucleate boiling.

Departure from nucleate boiling (DNB) takes place at high heat flux. Until or unless the heat flux is reduced, the fuel element suface will literally melt. And that's why during design and operation of a BWR (or any liquid cooled reactor) care must be taken to ensure that the maximum heat flux is always below the DNB value.

Q. *So it is extremely essential to operate a BWR at a maximum heat flux which is well below the DNB value. How is this accomplished?*

Ans. This is achieved by differential disposition of fuel rods about the control rod. The fuel rods having lower enrichment of fuel are disposed in the regions of high neutron flux to reduce the power generation of these fuel tubes. (Fig. 34.21)

The less enriched fuel tubes are placed along the outer edge next to the water gap where the control rod is inserted. The control rod shown here is in cruciform configuration. As the cruciform control rods are withdrawn, the thermal neutron-flux increases sharply in the water gap produced by the control rod withdrawal. However, the thermal neutrons encounter less-enriched fuel elements, the nuclear fission rate becomes moderate and therefore, the heat flux is kept at bay.

Q. *What is HTGR?*

Ans. It stands for High Temperature Gas-cooled Reactor.

Q. *What gas is used for cooling such reactors?*

Ans.
1. Helium
2. Carbon dioxide.

Q. *Briefly describe one HTGR and its operation.*

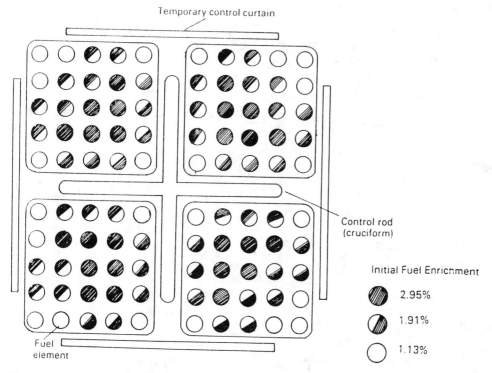

Temporary control curtain

Control rod
(cruciform)

Initial Fuel Enrichment

2.95%

1.91%

1.13%

Fuel
element

Fig. 34.21 *A typical in-core section of a BWR fuel module. Four fuel elements are symmetrically disposed around a cruciform control rod*

Ans. Two gas-cooled commercial reactors are in operation in the USA. One is the 40MW reactor (Peach Bottom-I) of Philadelphia Elec. Co. and the other is the 300MW power reactor of Ft. St. Vrain.

The reactor is enclosed in a prestressed-concrete pressure vessel (PCPV) along with the helium coolant and steam-generating and steam-superheating equipment. (Fig. 34.22)

These systems are cooled by helium and moderated with graphite. One proposed reactor of 3 800 MW will produce steam of high temperature (515°C) and high pressure (16.55 MN/m^2). The design efficiency of the HTGR is about 38% which is comparable to the efficiency of best fossil-fuel fired units. Its operational parameters are:

Helium (coolant) Inlet Temp. 318°C
Helium (coolant) Outlet Temp. 741°C
Helium Pressure 5 MN/m^2
Throttle Steam Temperature 515°C
Throttle Steam Pressure 16.55 MN/m^2

Reheat Steam Temperature 540°C
Reheat Steam Pressure 4 MN/m^2
Core Power 3 800MW
Net Electrical output 1 450MW
Net Thermal Efficiency 38.2%

Another HTGR operating at Sommerset, England has an electrical power output of 248MW with 26% efficiency. The reactor is a spherical vessel made of low carbon steel. Graphite is used as a moderator as well as reflector. The core is a 24-sided prism through which are inserted fuel rods. The whole reactor is encased in a 3m thick concrete shield that also acts as a biological shield. Carbon dioxide is used as coolant and natural uranium is used as fuel.

Fission heat is transferred to the coolant gas, helium or carbon dioxide, from which heat is extracted by the working fluid in the steam generator to produce dry, superheated steam. The steam so obtained is expanded in the turbo-alternator to generate electricity.

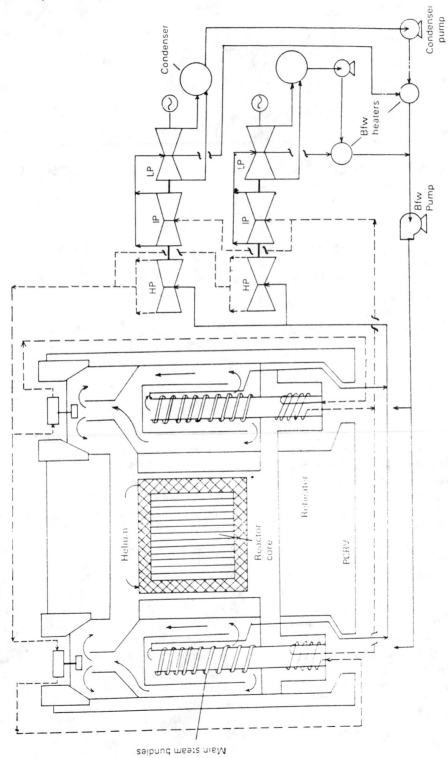

Fig. 34.22 Schematic diagram of a nuclear power system incorporating a helium-cooled and graphite moderated HTGR

Q. *What is the advantage of using graphite to build the core of HTGR?*

Ans. Graphite is a good moderator as well as excellent refractory material. It can tolerate very high temperatures. Besides, graphite has a high heat capacity. Therefore, the graphite-built core of HTGR will absorb large amount of fission heat without the risk of core meltdown.

Q. *What are the advantages of the gas-cooled reactor?*

Ans.
1. Simplicity in fuel processing
   Cladding is not required. Uranium carbide(fuel) and graphite (moderator) are simply ground together and reprocessed.
2. No corrosion problem.
3. Higher stability of graphite at higher temperature and irradiation makes it an excellent high-temperature moderator. Therefore, high reactor core temperature can be obtained without the risk of core meltdown.
4. Both uranium carbide and graphite can withstand high temperatures. Therefore, limiting the fuel element temperature is not that serious a problem as in other types of nuclear reactors.
5. Better neutron economy due to low parasitic absorption
6. CO_2 being a non-corrosive, non-toxic and non-flammable gas, completely eliminates the possibility of explosion in the reactor (cf. this threat is prevalent in water-cooled reactors)
7. Higher efficiency than Light-Water Reactors (LWR) viz., pressurized-water reactor (PWR) and boiling-water reactor (BWR). Thermal efficiency of LWR ranges from 32 to 33% .

Q. *What are the disadvantages of an HTGR?*
Ans.
1. Very low power density
   Gases have very low heat transfer coefficient.
2. Large vessel is required.
   Because of low heat transfer coefficient of coolant gas, a large amount of gas is to be circulated.

3. Fuel loading is more elaborate a process. Besides, it is costly.
4. If helium is used instead of CO_2, leakage of radioactive helium will cause a problem.
5. Circulation of coolant should be higher and that chews up 10 – 20% of generated power.
6. Higher fuel loading is needed, as the critical mass is high.
7. Complicated control mechanism due to low negative coefficient as helium does not absorb neutrons.

Q. *Gas-cooled steam generators should preferably incorporate an economizer as well as a superheater. If no economizer is installed, what will be the limitation?*

Ans. This can be demonstrated with the help of the accompanying figure (Fig. 34.23) where $[\theta_{wf}]_{ex}$ is the exit temperature of the working fluid producing saturated steam and θ_{ss} is the temperature of the superheated steam at the same temperature of the gaseous coolant.

The temperature gradient $\Delta\theta$ on the colder side should not go below 10°C in order that intensive heat transfer from the coolant to the working fluid takes place.

Now if no economizer is used, saturated steam will be generated at the constant temperature of the work-

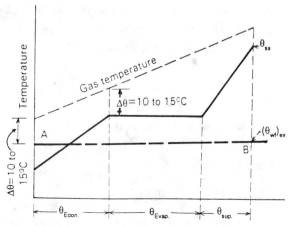

Fig. 34.23 θ vs Q curve in a gas-cooled steam generator fitted with an economizer and a superheater

ing fluid in the steam generator (line *AB*) with the effect that the temperature of the saturated steam will be limited to $[\theta_{wf}]_{ex}$ corresponding to the given value of $\Delta\theta$.

However, if an economizer is incorporated within the gascooled steam generator, it will increase the temperature gradient at the inlet and as a consequence the exit temperature θ_{ss} of steam will rise.

Q. *How is the performance of a nuclear power plant described?*

Ans. It is described in terms of electrical and thermal capacity.

Q. *How can this be expressed?*

Ans. This can be expressed in terms of the utilization factor, K_{ut}, which is the ratio of the energy actually generated (Q_{ac}) during the time in question (it may be a month, quarter of a year or a year) to Q_{av}, the available energy which the plant can generate operating at a 100% installed capacity during the whole period in question

$$K_{ut} = \left[\frac{Q_{ac}}{Q_{av}}\right](100)\%$$

Q. *On which factor or factors, does the electrical capacity of a nuclear power plant depend?*

Ans. Capacity of the steam turbines (that ultimately convert nuclear heat energy into electrical energy).

Q. *On which factor or factors does the thermal capacity of nuclear power plants depend?*

Ans.
1. Fuel loading
2. Design of the reactor.

Q. *What is the thermal capacity of a reactor operating on U-235 as fuel?*

Ans. It is given by the expression

$$N_r = 3 \times 10^{-11} \; \phi \, \rho \, \sigma_f V$$

where 3×10^{-11} J or 190 MeV thermal energy is released upon fission of 1 nucleus of U-235.

ϕ = average neutron flux density

ρ = number of fissile fuel nuclei per unit core volume

V = volume of the reactor core

σ_f = microscopic fission cross-section of U-235.

Q. *What is the difference between pressure-tube reactors and pressure vessel reactors.*

Ans. In pressure-tube reactors, the pressure is taken up by the tubes. The coolant is pumped under pressure through hollow tubes.

In pressure-vessel reactors, the pressure is taken up by the vessel itself where the nuclear fission process is accomplished.

Q. *Why must the water used as a nuclear coolant be thoroughly purified?*

Ans. Water to be used as a nuclear reactor coolant must be thoroughly purified and free from traces of impurities, otherwise these undesired materials will initiate corrosion and erosion of the materials used in the reactor core as well as the entire circuit.

Q. *Ordinary water is an extremely stable, non-toxic, non-explosive heat transfer material. Hence it finds its application as a reactor coolant in nuclear power plants. Is there any chance of explosion when ordinary water is used as coolant?*

Ans. Yes; despite its thermal stability, ordinary water undergoes radiolysis, i.e., radiation decomposing water into hydrogen and oxygen as well as hydrogen peroxide

$$H_2O + h\nu \longrightarrow H + \overset{\bullet}{O}H$$

$$\overset{\bullet}{O}H + \overset{\bullet}{O}H \longrightarrow H_2O_2$$

$$H_2O_2 \longrightarrow H_2O + O_2$$

The mixture of such gases is explosive.

Q. *What factors should be considered in selecting the site for a nuclear power plant?*
1. The land allocating for constructing a nuclear power plant should be of little or no agricultural value.
2. The site should be chosen in the vicinity of a natural body of water, lying on unflooded land so that cooling water required by the

powerplant needs to be lifted for a minimum height.

3. The soil at the erection site must have sufficient bearing capacity to minimize the expensive grouting process.

4. The ground water level should be preferably below the bottom of the projected building basements as well as buried services so as to avoid additional expenses required for ground water lowering during construction.

5. The site should preferably have a flat surface bordered with slopes for the construction of power plant buildings and to ensure sufficient surface drainage.

Besides, nuclear power plants should not be constructed in

(a) localities of active karst
(b) localities with a high risk of massive landslides and mudflow
(c) localities with high risk of avalanches
(d) localities prone to tsunamis or other natural disasters
(e) localities likely to be flooded following a failure of a dam upstream
(f) zones left void after mining
(g) swamped or overdamped areas under constant inflow of pressure ground water
(h) localities near and around health resorts, spa and water supply sources.

35

Energy from Waste

Q. *What is the latest trend in solid waste management?*

Ans. It aims to attain four primary objectives:
1. source reduction
2. recycling
3. waste-to-energy conversion
4. landfilling.

Q. *What is source reduction?*

Ans. It means the reduction of the quantum of waste requiring disposal at the source. This needs a radical change in public attitude and the industrial and commercial processes to minimize the solid waste.

Q. *What do you mean by recycle of solid waste?*

Ans. It means sorting out and return to the point of manufacture of those materials from the solid waste for which it is economically feasible to do so, e.g., waste paper can be sorted out and recycled to the paper industry. Aluminium cans of soft drinks and other foodstuffs can be conveniently recycled. So also use-and-throw polythene cups, dishes, trays and cutlery.

Q. *What degree of recycling is economically feasible?*

Ans. Industry experts opine that recycling is economically sound to the extent of 35%. Beyond this it is unlikely.

Q. *Why?*

Ans.
1. Public attitude has not been fully geared to accepting anything and everything produced from recycled materials

2. Source diversity is another problem that renders sorting and recycling of the specific materials to the point of manufacture difficult
3. Waste itself poses the very characteristics that deny 100% total recycle
4. Method of production of waste and the collection responsibility also stand in the way of total recycle.

Q. *What do you mean be waste-to-energy conversion?*

Ans. Burning of waste to produce a valuable commodity—steam and/or electricity.

It also offers a viable means of source reduction, i.e., reducing the volume of waste.

As much as 47 000 tons of solid wastes were gulped down per day by the 160 existing municipal waste-to-energy facilities in the USA in 1989.

Q. *How does "landfilling" play its role in the integrated approach to solid-waste management?*

Ans. Filling the voids on earth with the wastes/residues leftover materials and energy recovery as well as maintaining enviornmental health complete the landfilling strategy.

Q. *What is the traditional method of burning solid waste to produce thermal energy?*

Ans. Mass-burning of solid waste.

Q. *How is mass burning of solid waste carried out?*

Ans. Solid refuse is trucked in and unloaded into a storage pit capable of holding a 2 to 4-day supply of refuse.

Large, non-combustible objects are sorted out and rejected.

The solid waste is mixed to a more homogeneous mass, picked up from the garbage pit with the help of mobile crane and grapple bucket and unloaded into separate feed chutes that supply each boiler.

As the refuse drops down the chute onto a feed table, a series of hydraulically driven charging rams continually pushes the mass into the furnace grate at a rate dictated by fuel or steam demand. The speed of the rams is controlled by motorized flow-control valves. (Fig. 35.1)

The grate consists of three inclined stepped sections where drying, combustion and burnout take place. Each section is composed of alternating arrays of stationary and moving grate bars. The latter agitate the refuse and force it down the grate by a push-pull mechanism.

Primary combustion air is fed under the grate while the secondary air, also called overfire air, is injected over the top of the fuel bed to ensure optimum combustion and emission control.

Bottom ash representing 80–90% of the total ash is discharged into the ash chute by the final member of pusher-type conveyors arranged in a cascade along the length of the grate. Also a portion of bottom ash falls through the grate-air-slots.

The system normally runs automatically though manual override from the local panel is possible.

Q. *Why is refuse in the tipping house mixed before feeding it to boiler furnace?*

Ans. Mixing ensures homogenization of fuel to a certain extent. This fuel homogenization improves the operation of the furnace and minimizes load swings due to variations in refuse heating value.

Q. *Why are oversize and otherwise undesirable materials removed from the refuse prior to its feeding by crane grapple down to the feed chutes?*

Ans. They may create problems in the feed chute or ash removal system. The feed chute may get clogged or ash discharging conveyor may get jammed or grate-air-slots may get blocked.

Q. *Why are the feed chutes kept full with refuse?*

Ans. To maintain an air-seal.

Q. *How is this safeguard ensured?*

Ans. An alarm is fitted to ensure a steady rate of feed (fuel) supply. If the level in the feed chute drops below a prescribed limit, an alarm sounds.

Q. *How is the charging-throat area protected against burnback and radiant heat?*

Ans. The whole charging-throat area at the furnace interface is lined with refractory brick or cooled with water.

Q. *Why are the furnace grates upon which refuse burns stepped?*

Ans. To promote burnout.

The steppings expose the unburned fuel for combustion better.

Q. *How the specific refuse buring conditions are maintained?*

Ans. Each inclined stepped section of the grate is provided with at least one independent air plenum to supply undergrate primary air. Therefore, the proportion of undergrate combustion air can be varied to fullfill the specific refuse-burning conditions.

Q. *How are grate functions controlled?*

Ans. Signals generated by a discrete stoker controller control the grate functions. The speed of the charging ram as well as the reciprocation rate of the grate and siftings system are determined by the microprocessor-based controller. This also feeds information on the feed chute and shut-off gate and sounds alarms when the grate and siftings sections have stalled.

Q. *How are the refuse-fired boilers designed to combat the corrosive and abrasive nature of the refuse and flue gas generated?*

Ans. The lower furnace (extending up to 10.5m above the grate) is lined with a 25mm thick silicon carbide refractory anchored to the walls with high density pin studs.

Protection is also ensured along the grate line by high-alloy cast blocks. These are replaceable liners. They are bolted to the waterwalls and extend up to 1.2–1.25m above the grate in the charging section and up to 1m above the grate in the discharging section.

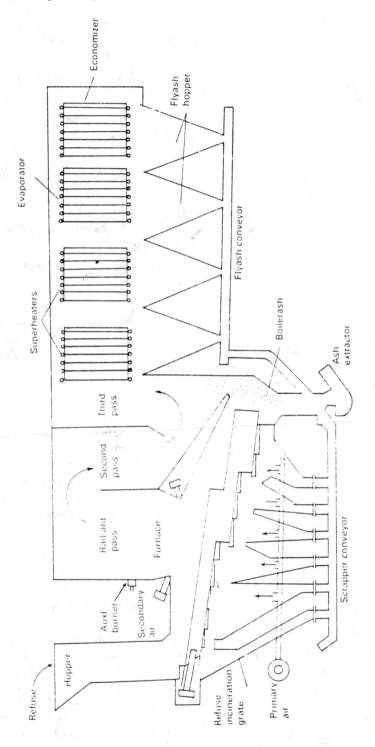

Fig. 35.1 Multipass mass burning system of solid waste municipal solid waste is incinerated on in-clined stepped grate sections to provide high combustion efficiency and low emission

Q. *Why is secondary air introduced?*

Ans. It promotes high turbulence and thereby ensures good mixing of volatile gases. As a result, complete combustion of unburnt particles and partial oxidation products of the hydrocarbon fraction of refuse is achieved.

Q. *Why is it imperative to operate the furnace at the correct level of oxygen?*

Ans. Correct level of excess oxygen introduced in the furnace achieves good combustion. Oxygen deficiency may lead to incomplete destruction of such toxic hydrocarbons as furnas, dioxins, etc. Appropriate amount of excess oxygen reduces the potential for corrosion.

Q. *How much excess air is used?*

Ans. 80% percent excess air is used.

Q. *Why 80%?*

Ans. This has been derived from industry experience which suggests that 80% of excess air is just enough to optimize the combustion of solid refuse in the mass-burning system.

Q. *Frequently it is found that mass-burning furnaces are equipped with one or two auxiliary burners. Why?*

Ans. These burners are installed for startup/shutdown or backup when refuse cannot be fired. Also in response to higher steam demand these auxiliary burners are switched on. Besides, some installations fire these auxiliary burners to ensure the complete destruction of organic compounds formed during combustion.

Q. *How are these auxiliary burners tuned to the complete destruction of organic compounds generated due to the mass-burning of refuse?*

Ans. Mass-burning installations are provided with temperature sensors or some other mechanisms that trigger the firing of auxiliary burners when operating parameters fall below the prescribed limit.

Q. *How are the superheater tubes arranged?*

Ans. Superheaters are of convective type. In a current Foster Wheeler design, parallel-flow superheater tubes are laid out in the convective shaft of the horizontal heat recovery pass.

Q. *Why are these superheater tubes protected from excessive temperature build-up?*

Ans. To avoid the risk of high-temperature corrosion.

Q. *How does the probability of high-temperature corrosion arise?*

Ans. Flyash arrested on superheater tubewalls may gradually build up, impairing the heat transfer from hot flue gas to the steam passing through superheater tubes. As a result, the tube metal temperature will increase and that will fuse the corrosive ash on the external surfaces of the tubes. Under the onslaught of aggressive ash and high temperature, the superheater tubes will fail.

Q. *How can this risk be avoided?*

Ans. The superheater tubes are spaced widely apart to avoid the risk of plugging. And to eliminate ash-buildup they are cleaned by mechanical rappers.

Q. *How does this rapping system work?*

Ans. It employs a series of rapping hammers to impart shock forces to the superheater tubes. (Rapping means to induce vibration by means of mechanical shock impulses.)

Each tube is independently rapped from one or both sides. Hammer blows induce high shock forces to the tubes continually with the effect that the bond between the slag and tubewall is ultimately broken and the tubes get freed from ash buildup.

However a thin layer of ash always remains adhered to the tubewall to impart some degree of protection to the superheater tubes from corrosion.

Q. *Refuse is a heterogeneous mixture of varied kinds of materials having different heating value. Does this variation in refuse heating value, affect the boiler output in any way?*

Ans. Yes. Variations in refuse heating value cause swings in the boiler output. And this is inevitable.

Q. *How to accommodate these swings in boiler output?*

Ans. To counter the swings in boiler output due to variation in the heating value of the refuse, a turbine-following-boiler control scheme is used. This means a constant boiler header pressure is always maintained by turbine controls while the electrical load is allowed to vary in proportion to the actual output.

Besides, there is a combustion control system that works to minimize load swings by responding to changes in the refuse heating value. This control system is pivoted on two loops—oxygen flow-control loop and furnace exit-gas temperature-control loop. These two loops give a feedback on changes in furnace condition due to variations in the heating value of the refuse which changes the specified steam flowrate. The combustion control system then adjusts the flow of air and fuel to maintain the specified steam flowrate and therby minimizes load swings.

Q. *Upon which factors does the efficient operation of a mass-burning system depend?*

Ans.
1. Optimum regulation of fuel bed depth
2. Combustion air distribution.

Q. *How is the fuel bed-depth controlled*
This is effected by regulating
(a) refuse charging rate
(b) grate speed.
However, both these factors are functionally dependent on the nature of the refuse being fired as well as the location of the complete burnout on the grate.

Q. *How is combustion air introduced to the mass-burning combustor?*

Ans. A part of the combustion air is fed undergrate. This is also called primary air. The remaining portion is fed above the bed. It is called overfire air or secondary air.

Depending on the burning pattern and emission characteristics, the primary or undergrate air is distributed to the various air zones where dampers are provided to effect flow control. Usually this air is preheated, but if the moisture in the refuse mass is low air preheating is bypassed.

Overfire ambient air is introduced into the furnace at different levels above the fuel bed.

Q. *How much of the total combustion air is the undergrate air?*

Ans. About 60%.

Q. *Where are the secondary air nozzles located?*

Ans. These high-pressure air nozzles are located at different levels on the front and rear walls.

Q. *What is the basis of the set up of overfire air nozzles at different depths?*

Ans. The following factors provide the basis of the location of overfire air nozzles:
1. NO_x level in the flue gas
2. CO level in the flue gas
3. Observed flame shape
4. Observed flame height
With the change of refuse heating values in different seasons, it requires adjustment of the overfire air system.

Q. *Fireside corrosion of the tube metal of the evaporator is a nagging problem in a mass-burning system. What recent improvement has been done to protect the tubes from the highly corrosive environment?*

Ans. Refractory lining has been found to be effective. However, it demands constant maintenance and measurement of the thickness of the refractory wall. Besides, refractory is prone to slagging, which impairs heat transfer.

In a recent development, a weld-deposited alloy of Inconel-625 is applied to the furnace walls. In the zones of higher gas-temperature, Incoloy-825 is used.

Q. *Why are these alloys used?*

Ans. They have high heat and corrosion resistance.

Q. *Is there any other method to slow down this corrosion process in the boiler furnace?*

Ans. The maximum steam temperature is kept in the range 400–450°C in order to keep the tube metal temperature sufficiently low so that carbon steels and other low alloy steels can be used to reduce the corrosion rate. For the same reason, the temperature of flue gases entering the superheaters is kept at 700°C or less.

Q. *What is the present trend of solid waste burning?*

Ans. Instead of burning untreated garbage in the mass-burn facility, the present trend is mass-burn facility with upstream preparation.

Q. *What is upstream preparation of solid waste subject to mass burning?*

Ans. It includes:
1. Primary shredding
2. Separation of light fraction and heavy fraction by density separation
3. Recovery of such materials as ferrous, aluminium, glass, copper, brass from heavy fraction
4. Secondary shredding of light fraction.
    (a) Burning the classified and shredded refuse directly in a boiler
    (b) Anaerobic digestion of shredded product to produce fuel gas to be burnt in the boiler or gas turbine or internal combustion engine
    (c) Transformation of secondary shreddings into pellets or some other convenient form and then burning them in the boiler
    (d) Pyrolysis of the product of secondary shreddings at elevated temperature and in absence of O_2 whereupon the carbonaceous material are thermally decomposed to liquid or gaseous fuel which is burned in the boiler

Q. *What is the utility of upstream preparation?*

Ans. It offers better combustion control which is particularly necessary to limit the products of incomplete combustion that escape in the flue gas. For example, upstream preparation drastically cuts down the quantities of heavy metals entering the combustor and thereby reduces the emission of metals in the ash or flyash.

Upstream preparation produces refuse-derived fuel (RDF) that can be cofired in the utility boilers with other premium grade fuels. This offers an excellent scope for disposing of large amounts of prepared wastes and at the same time recovering their inherent energy value.

Q. *How many different types of combustor technology convert waste to energy today?*

Ans.
1. Mass-burn units
2. Starved-air modular units
3. Excess-air modular units
4. Fluidized bed combustion
5. Rotary combustor technology
6. Cofiring units that burn wood or coal along with the waste.

Q. *How many different types of emissions occur from municipal waste combustors?*

Ans. Four types:
1. Particulates
2. Acid gases, particularly HCl
3. Metals
4. Toxic organics.

Q. *How does HCl occur in the emissions from municipal waste combustors?*

Ans. It comes from the combustion of plastic materials (cf. divinyl chloride plastic) and paper in the waste mass.

Q. *What metals are emitted?*

Ans. Cadmium, chromium, arsenic and mercury

Q. *What toxic organics occur in the emissions from municipal waste combustors?*

Ans. Mainly dioxins and furans.

Q. *How are these emissions kept at bay?*

Ans. Emissions are minimized by adopting two basic technologies:
1. combustion control
2. downstream add-on controls.

Q. *What do you mean by combustion control?*

Ans. This involves:
    (a) design of combustors
    (b) operation of combustors
    (c) control of combustor performance
    (d) enhancing the destruction of organic stuffs
    (e) reduction of the quantity of uncontrolled particulate matter leaving the furnace
    (f) maintaining minimum furnace temperature to reduce NO_x emission.

Q. *How can a reduction of the quantity of uncontrolled particulate matter leaving the furnace be achieved?*

Ans. This is achieved through proper mixing of combustion air with the wastes in the furnace.

Q. *What do you mean by add-on control?*

Ans. It comes in the form of
(a) ESP (Electro Static Precipitator) introduced downstream of the economizer.
(b) Spray dryers followed by an ESP
(c) Spray dryers followed by a fabric filter
All are added downstream of the economizer and in the flue gas shaft.

Q. *Is there any particular advantage of a spray dryer-fabric filter combination?*

Ans. This combination eliminates multiple pollutants, viz., dioxins, furans, acid gases, particulate matter and trace heavy metals.

Q. *What kinds of particulate emissions take place due to municipal waste combustion?*

Ans. Two types:
1. solid particulates
2. condensible particulates.

Q. *How do solid particulates originate?*

Ans. There are non-combustible materials always present in almost all solid wastes. Under normal combustion conditions in the furnace, these produce solid particulates as flyash in the flue gas.

Q. *How do the condensible particulates originate?*

Ans. These are produced when a certain portion of refuse mass entering the combustion zone gets vapourized instead of burning out into CO_2 and water.

These vapourized constituents eventually get cooled downstream and condense on their way from the boiler to the stack.

Q. *In what form are the trace heavy-metals emitted?*

Ans. They are emitted as vapour mainly in the elemental form. Also they are released as solid particulate matter as metal oxides, chlorides and sulphates.

Q. *What are the plus points and minus points of wet scrubbers?*

Ans. High efficiency and low reagents requirement are its two plus points.
Its disadvantages are:
1. water consumption
2. need to treat scrubber bleed stream
3. it is incompatible with fabric filters.

Q. *Besides the above mentioned pollutants is there any other pollutant emitted by the municipal waste upon which special attention to be focussed?*

Ans. Yes; it is NO_x.

Q. *How is this generated?*

Ans. It results from the oxidation of nitrogen contained in the waste itself. Also it is generated due to high temperature conversion of atmospheric nitrogen with the available oxygen. The latter is called thermal NO_x.

Q. *How can the NO_x emissions from solid-waste incinerators be reduced?*

Ans. This can be implemented through several control measures.
Precombustion Control includes
(a) lowering the nitrogen content of solid-waste introduced into the furnace
(b) minimizing the amount of oxygen available for reaction with the available nitrogen at the point of combustion.
Combustion Control includes
(a) operating the furnace at a lower temperature
(b) selective non-catalytic reduction (SNR) of NO_x by injecting ammonia through large jets fitted in the furnace walls at such locations where temperatures are in the range 870–980°C.
Postcombustion Control includes
(a) selective catalytic reduction of NO_x in the flue gas by ammonia over a catalyst bed downstream
(b) wet scrubbing of flue gases.

Q. *What are the toxic organics that may exit the stack during municipal solid waste (MSW) burning?*

Ans. These are chiefly dioxins and furans.

Q. *In which form do they mainly exit?*

Ans. They may emit into three forms:
1. vapour form
2. condensed form
3. absorbed onto fine particulate form.

Q. *Why do they form?*

Ans. Incomplete combustion of carbon compounds is the genesis of toxic organics.

Q. *What steps should be taken for the efficient destruction of these harmful, incomplete products of combustion?*

Ans. This requires:
(a) adequate control of temperature
(b) supplying appropriate amount of combustion air through appropriate locations of the combustion system.

Q. *What steps should be taken to ensure good furnace operation in a large mass-burning incinarator?*

Ans.
1. Combustible materials are to be separated and non-combustibles to be removed from the refuse feed.
2. Large blocks of combustible materials should be broken into more or less uniform size.
3. Elimination of moisture content of the refuse combustibles should be carried out by air drying to uniformly low levels prior to combustion.
4. Furnace temperature should be controlled to ignite the refuse but it should not be so high as to increase the formation of NO_x or vapourization of metals. Besides, furnace temperature should be below the melting point of flyash.
5. Correct amount of air in the correct location should be injected into the furnace to supply O_2 and ensure uniform turbulence throughout the furnace to permit complete reaction of unburnt matter with oxygen and at the same time minimize the entrainment of bottom ash into the flue gas.
6. Residence time of the combustion products in the combustor should be sufficient to complete combustion.

Q. *Apart from mass burning, what other technologies are available for deriving energy from waste?*

Ans. There are six other technologies available:
1. Fluidized-bed combustion
2. Pyrolysis
3. Supercritical water oxidation
4. Infrared combustion
5. Wet-air oxidation
6. Deep-well wet oxidation.

Q. *What are the prospects of fluidized bed combustion?*

Ans. This offers a clean and efficient method of incinerating a variety of solid waste fuels.

Combustion efficiency is high because of intense mixing of combustion air and combustibles in the turbulent bed.

Boiler efficiencies are comparable to conventional units.

Bed temperature (810–870°C) is substantially low and more uniformly distributed with the effect that NO_x formation is greatly minimized while the destruction efficiency of hazardous organics is as high as 99.99%.

Q. *What types of fluidized-bed combustion are used?*

Ans. Commercially available technologies are:
1. Bubbling-bed units
2. Circulating fluidized-bed (CFB) units.

Both work on atmospheric fluidized-bed combustion.

Q. *Name certain solid wastes that find their way to fluidized bed combustion.*

Ans.
1. Coal mining wastes (e.g., anthracite culm)
2. Petroleum coke
3. Bagasse
4. Bark
5. General wood wastes
6. Sawdust and shavings
7. Nut hulls
8. Rice husk
9. Corn cobs.

Q. *What is culm?*

Ans. Anthracite mining waste-dirt is known as culm. These are waste coals that pile up near anthracite coal fields.

Over the past one and a half century, about 1 billion tons of anthracite culm has been generated in the anthracite mining areas of Pennsylvania, USA. This is a fuel whose ash content is high but heating value is low.

This plus its high moisture content has so far dumped it as waste. However with the advent of the trend of waste-to-energy facilities, 20 projects are operating in Pennsylvania burning culm in fluidized-bed boilers generating 50 MW of power each.

Q. *What is the strongest point in favour of that vote circulating fluidized-bed boilers (CFB) for burning anthracite culm?*

Ans. They can burn practically anything with heat content—with minimum fuel preparation. They can turn waste into energy. For instances, several culm-fired CFB units in Pennsylvania's anthracite coal fields are converting high-grade dirt into electric power. Ash contents of these fuels are as high as 70%.

Municipal solid waste boilers at Alexandria, Virginia gobble up as much as 975 tons of waste per day. In the year 1989, 22 plants in USA alone burned municipal solid wastes at the rate of 32 000 tons/day.

Q. *What are the disadvantages associated with CFBs?*

Ans.

1. Lengthy startup/shakedown period
2. Constant wear and tear of critical auxiliary equipment because of high non-combustible content
3. Very large and complex ash handling system. Power output of the CFB facilities belies the size of the major auxiliary equipments
4. Fuel feed system suffers from improper performances if fuel moisture content, particle size deviate much from design
5. CFBs firing on fuels with high non-combustibles as high as 70% accomplish little regarding solid-waste reduction, for the obvious reason of higher dosage of lime to absorb SO_2

6. Refractory failure is a nagging problem. Refractory failure has been reported in the combustor and cyclone (The Gilberton Power Co., San Diego, California)
7. In many cases fuel preparation is required using heavy-media floatation and centrifuging to separate non-combustibles
8. The tube erosion is a major problem. This has been reported to occur in the economizer and ash cooler (The Westwood Energy Properties, Pennsylvania)
9. Sintering and/or agglomeration are common problems. Agglomeration in the ash cooler has been reported (The Westwood Energy Properties and Frankville Power Corpn.). Many problems stem from moving voluminous quantities of ash.

Q. *How can the refractory failure problem be avoided?*

Ans.

1. Changeover to refractory brick from castable refractory.
2. Increasing the height of the refractory in the combustor zone.

Q. *How do the agglomeration and sintering problems arise in culm-fed CFBs?*

Ans. They arise from two main problems:
1. discharging ash from the combustor and/or fabric filter
2. maintaining the ash flowing properly throughout the entire system.

Q. *How to cope with this problem?*

Ans.

1. Switchover from Vacuum Ash Handling System to Pressurized or Partially-pressurized Ash Handling System. (The Gilberton power Co., California, USA). Using air lance to maintain flow
2. Incorporating Water Cooled Screw Conveyor for continuous removal of oversize ash particles which contribute to the sintering problem
3. Installing control valves on the Fabric-Filter Hoppers and proper adjustment of speeds of

the discharge conveyor to avoid leakage of very fine ash.

Q. *What problems are associated with Fuel Feed Systems of culm fired CFBs?*

Ans.

1. Where fuel processing is required prior to its supply to the CFB, high moisture content and improper fuel particle size contribute to the jamming of the fuel feed conveyors (The Gilberton power Co., California, USA).
2. If the dryer fails to function properly, the above problem aggravates.
3. Where fuel benefication is not employed (The Westwood Energy Properties, Pennsylvania, USA), proper screening of the fuel is the critical criteria for good combustor performance. High moisture levels make this difficult to achieve.
4. Rotary valves are to be avoided in the fuel feed system.

Q. *Is cofiring compatible with fluidized-bed combustion of solid waste?*

Ans. Yes the fluidized-bed combustion technique has enough flexibility to permit simultaneous combustion of solids, sludges, slurries, liquids and gases in the same unit.

Q. *Can you name such a unit operating?*

Ans. A unit in La Crosse, Wisconsin (USA) burns RDF (Refuse Derived Fuel) and wood in equal amounts at a capacity of 400 tons/day.

Q. *What are the disadvantages of fluidized-bed combustion of solid waste?*

1. Feed solid waste stream must be closely monitored and sized properly and more or less uniformly
2. Inorganic salts present in the feed waste must be absent as these salts may melt at the bed temperature or form eutectics causing bed seizure
3. At low loads, combustion efficiency is generally poor
4. Inbed tube erosion is a nagging problem.

Q. *What is pyrolysis?*

Ans. "Pyr" means fire while "lysis" means disintegration into simpler products. Thus pyrolysis means chemical decomposition of a material at high temperature in a closed system in the absence of oxygen. It also means destructive distillation.

Q. *What is the fundamental difference between pyrolysis and incineration?*

Ans. Pyrolysis is an endothermic process that takes place in absence of O_2 whereas incineration is an exothermic process that occurs in the presence of oxygen.

Q. *In what type of equipment, is pyrolysis carried out?*

Ans. This can be carried out in any of the following units:

1. Gasifiers
2. Vertical-shaft reactors
3. Rotary kilns lined with refractory inside
4. Regenerative towers.

Q. *What are the products of pyrolysis?*

Ans.

1. GASES—H_2, CH_4, CO, CO_2
2. TARRY LIQUIDS—water plus organic compounds, viz., methanol, acetic acid.
3. SOLID — carbonaceous char.

The derived fuels—gas is partly burnt to supply heat to the reactor while the surplus gas and liquids are available as fuel by-products. These may be burnt in a boiler unit in a gas turbine to generate electricity.

Q. *What are the advantages of pyrolysis?*

Ans. It is a closed system operation and hence contributes very little to environmental pollution.

Q. *What is supercritical water oxidation?*

Ans. In this process aqueous organic as well as inorganic wastes are mixed with air or oxygen in a reactor at temperatures exceeding 374°C and pressure 225 kgf/cm^2.

At these conditions water remains liquid and lends itself as an excellent solvent for organic substances which whereupon reform to gases like CO, CO_2, H_2 and CH_4 and volatile organic liquids, viz., alcohols,

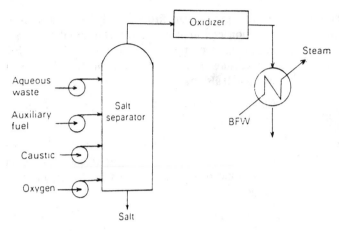

Fig. 35.2 *Supercritical water oxidation*

aldehydes and furans. No char is produced. These are the primary products which ultimately get fully oxidized to CO_2 and H_2O liberating sufficient quantity of heat which is carried out by the exit liquid stream and can be utilized to produce steam in a heat exchange process. (Fig. 35.2)

Q. *Is there any limitation of the supercritical oxidation process?*

Ans. Yes, its application is limited to liquid feed streams. It is also not applicable to wastes which contain more than 15 to 20% salts or liquid wastes whose primary contaminants are heavy metals.

Q. *What is infrared incineration?*

Ans. This is one of the latest combustion techniques. It employs infrared heating elements in the combustion chambers to incinerate solid waste on a moving chain. (Fig. 35.3)

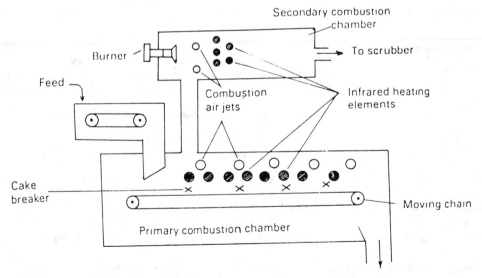

Fig. 35.3 *INFRARED INCINERATOR*
It involves two combustion chambers. Combustion is completed in the secondary combustion chamber by simultaneous firing of gas fired burners & infrared heating elements

There are two combustion chambers to which combustion air is supplied by air jets. While most of the solid waste is incinerated in the primary combustion chamber, gaseous products of partial oxidation of waste are completely burned in the secondary combustion chamber.

Both combustion chambers are ceramic fibre insulated.

Residence time of solid waste in the primary combustion ranges from 5 –180 minutes at an operating temperature of 260°C –1040°C.

Gas residence time in the secondary combustion chamber is about 2 seconds. Here upto 100% excess air is introduced and temperature attained is as much as 1260°C by the simultaneous firing of gas fired burners as well as infrared heating elements.

Q. *What are the advantages of infrared incineration?*

Ans.

1. Lower capital cost
2. Greater availability
3. Total mobility for on-site applications
4. Incineration can be carefully controlled and regulated
5. 99.9999% destruction efficiency for toxic hydrocarbons like dioxins and furans.

Appendix I

MEAN SPECIFIC HEATS OF SELECTED GASES [kcal/m^3°C]

TEMP (°C)	O$_2$	N$_2$	H$_2$	CO	CO$_2$	AIR
300	0.3126	0.3023	0.3060	0.3048	0.4358	0.3049
400	0.3155	0.3048	0.3092	0.3060	0.4475	0.3075
500	0.3169	0.3073	0.3107	0.3085	0.4593	0.3088
600	0.3198	0.3085	0.3128	0.3098	0.4691	0.3113
700	0.3112	0.3110	0.3148	0.3123	0.4789	0.3126
800	0.3226	0.3123	0.3168	0.3148	0.4888	0.3152
900	0.3255	0.3148	0.3189	0.3160	0.4966	0.3165
1 000	0.3269	0.3173	0.3209	0.3185	0.5045	0.3191
1 100	0.3298	0.3185	0.3229	0.3198	0.5123	0.3203
1 200	0.3312	0.3210	0.3249	0.3223	0.5182	0.3229
1 300	0.3326	0.3223	0.3269	0.3248	0.5241	0.3242
1 400	0.3355	0.3248	0.3290	0.3260	0.5300	0.3268
1 500	0.3369	0.3273	0.3310	0.3285	0.5339	0.3294
1 600	0.3397	0.3285	0.3330	0.3310	0.5378	0.3307
1 700	0.3412	0.3310	0.3350	0.3323	0.5418	0.3332
1 800	0.3426	0.3322	0.3371	0.3348	0.5457	0.3345
1 900	0.3469	0.3348	0.3390	0.3360	0.5496	0.3371
2 000	0.3483	0.3372	0.3411·	0.3385	0.5516	0.3384

HEAT CONTENTS OF DIFFERENT GASES AT VARIOUS TEMPERATURES (kcal/kg)

TEMP (°C)	H$_2$O	CO$_2$	SO$_2$	CO	N$_2$	O$_2$	H$_2$
800	380.111	199.068	136.859	201.286	201.286	181.125	2828
900	432.277	227.886	156.672	227.893	227.893	205.031	3190.5
1 000	486.111	257.204	176.828	254.821	254.821	229.219	3567.5
1 100	541.833	286.931	197.265	282.071	282.071	253.687	3949
1 200	599.722	316.931	217.890	309.643	309.643	278.437	4335
1 300	660.055	347.159	238.672	337.536	337.536	303.469	4725.5
1 400	723.277	377.431	259.484	365.750	365.750	328.781	5120.5
1 500	789.555	407.681	280.281	394.286	394.286	354.375	5520

1 600	859.444	437.931	301.078	423.143	423.143	380.250	5924
1 700	933.388	469.090	322.500	452.321	452.321	406.406	6332.5
1 800	1011.611	500.318	343.968	481.821	481.821	432.844	6745.5
1 900	1094.833	531.613	365.484	51·1.643	511.643	459.562	7163
2 000	1183.333	562.954	387.031	541.786	541.786	486.562	7585
2 100	1277.722	594.250	408.547	572.250	572.250	513.844	8011.5
2 200	1378.666	625.590	430.094	603.036	603.036	541.406	8442.5
2 300	1486.444	656.863	451.593	634.143	634.143	569.250	8878
2 400	1602.000	688.091	473.062	665.571	665.571	597.375	9318
2 500	1725.666	719.318	494.531	697.321	697.321	625.781	9762

CONVERSION FACTOR

kcal/kg $\longrightarrow$ kJ/kg Multiply by 4.187

Appendix II

INTERNAL TREATMENT OF BOILER FEEDWATER BY CHEMICALS

Corrective Treatment Required	Chemicals to be Dosed
Adjustment of feedwater pH and maintaining boiler water alkalinity to inhibit scaling and corrosion	Soda ash/ Caustic soda/ Sulphuric acid
Internal softening of BFW to prevent scaling	Soda ash/ Phosphates/ Sodium aluminate/ Sodium silicate
Boiler sludge conditioning	Tannins/ Lignin derivatives/ Starch/ Glucose derivatives
Protecting evaporators, stage heaters and pipelines from scaling from hot water	Polyphosphates/ Glucose derivatives/ Lignin derivatives/ Tannins
Chemical deaeration of BFW to inhibit O_2-corrosion	Hydrazine/ Sulphites/ Ferrous hydroxide/ Tannins/ Glucose derivatives
Inhibition of caustic embrittlement	Phosphates/ Nitrates/ Sodium sulphate/ Tannins
Prevention of foaming of BFW	Polyamides/ Polyalkylene glycols
Protective film formation to prevent corrosion	Tannins/ Lignins/ Glucose derivatives
Prevention of corrosion by condensate	Ammonia/ EDTA

BFW SPECIFICATIONS FOR NATURAL CIRCULATION BOILERS

Water Quality Characteristics	Working Pressure (MN/m^2)		
	up to 4	up to 10	above 10
pH at 298K	9–9.2	9–9.2	9–9.2
Total Hardness (μg-equiv/kg)	10 (max)	3–5	1
Silica (ppm)	—	0.08	0.04
Dissolved Oxygen (ppm)	0.02–0.03	0.02–0.03	0.01–0.02
Dissolved CO_2 (ppm)	Nil	Nil	Nil
Hydrazine (ppm)	0.02–0.06	0.02–0.06	0.02–0.06
Dissolved Ammonia (ppm)	1	1	1
Dissolved NO_2 (ppm)	—	0.02 (max)	0.02 (max)
$NO_2^- + NO_3^-$ (ppm)	—	0.02 (max)	0.02 (max)
Fe (ppm)	0.1–0.2	0.05–0.1	0.02–0.03
Cu (ppm)	0.01–0.02	0.01–0.02	0.005 (max)
Oil (ppm)	1	0.03	0.03

BFW SPECIFICATIONS FOR ONCE-THROUGH BOILER UNITS

Water Quality Characteristics	Amount
pH at 298K	9–9.2
Total Hardness	0.0002
Silica (ppm)	0.015
Dissolved oxygen (ppm)	0.01
Iron (ppm)	0.01
Copper (ppm)	0.005
Hydrazine (excess) (ppm)	0.02–0.06
Oil (ppm)	0.1
Na^+ ions (ppm)	0.005

CHARACTERISTICS OF NATURAL WATER

pH

The pH value of water is defined as the negative logarithm of hydrogen ion concentration to the base 10, the H^+ ion concentration being measured in kg-equivalent per m^3 or gm- equivalent per litre.

$$pH = -\log_{10}[H^+]$$

At 25°C, pure water has a concentration of H^+ ions of 10^{-7} kmol/m^3 and a concentration of OH^- ions of 10^{-7} kmol/m^3. And therefore, the product of $[H^+]$ and $[OH^-]$ ions is 10^{-14} and that means, for pure water

$$pH = 7 \quad \text{and} \quad pOH = 7$$

and

$$pH + pOH = 14$$

In other words, the pH value is a number between 0 and 14 which indicates the degree of acidity or alkalinity of water. A pH value equal to 7 means the water is neutral, since it contains the same concentration of H^+ ions as OH^- ions. When pH is less than 7, the water is distinctly acidic as it contains more H^+ ions than OH^- ions. For example, if pH = 5, then pOH value of the solution will be pOH = 14 − 5 = 9 and therefore $[OH^-] = 10^{-9}$ kmol/m^3, which is less than the OH^- ion concentration when the water is neutral. Similarly if pH is greater than 7, the water is alkaline.

Alkalinity

It is the total load of that much part of total dissolved solids (TDS) in water which contribute to the increase of the hydroxyl ion concentration in water as a result of hydrolysis or dissociation.

The alkalinity of natural waters is usually due to the presence of HCO_3^{2-}, SiO_3^{2-}, $HSiO_3^-$ and sometimes CO_3^{2-} as well as due to the presence of PO_4^{3-}, HPO_4^{2-} and of OH^- ions. These anions of weak acids undergo hydrolysis to add OH^- ions to water

$$SiO_3^{2-} + H_2O \rightleftharpoons HSiO_3^- + OH^-$$

$$PO_4^{3-} + H_2O \rightleftharpoons HPO_4^{2-} + OH^-$$

Depending upon the anion that is present in water, the alkalinity is classified. For example:

bicarbonate alkalinity is due to the presence of HCO_3^- ions

carbonate alkalinity is due to the presence of CO_3^{2-} ions

hydrate alkalinity is due to the presence of OH^- ions.

The alkalinity of water is usually determined by using methyl orange or phenolpthalein as an indicator. Accordingly, M-alkalinity and P-alkalinity have been distinguished. M-alkalinity refers to total alkalinity.

These two alkalinities are interrelated differently depending upon the nature of alkalinity. For example:

1. for a solution of NaOH
   M-value = P-value
2. for a solution of Na_2CO_3
   M-value = 2P-value
3. for a solution of only bicarbonates
   P-value = nil

In some literature, the methyl orange alkalinity, phenolpthalein alkalinity, carbonate alkalinity, bicarbonate alkalinity, etc., are conveniently represented by A_{mo}, A_{ph}, A_c, A_b respectively and accordingly $A_{mo} = A_h + A_c + A_b$

$$A_{ph} = A_h + \frac{1}{2}A_c$$

where

A_h = hydrate alkalinity.

Hardness

The total hardness is defined as the sum of the concentration of calcium (Ca^{2+}) and magnesium (Mg^{2+}) ions present in the water. It is expressed in milligram-equivalents per kilogram (mg-equiv./kg) or microgram-equivalents per kilogram (μg-equiv/kg)

$$H_t = H_{Ca} + H_{Mg}$$

Hardness is due to the presence of calcium and magnesium salts—bicarbonates, carbonates, sulphates, chlorides and nitrates. And sometimes the presence of aluminium, iron and manganese salts causes water to be hard, but normally these substances are not found in appreciable quantities in natural water.

Therefore, it has become customary to divide hardness into two main types = carbonate and non-carbonate hardness.

Carbonate hardness is due to the presence of calcium, magnesium carbonates and bicarbonates.

Non-carbonate hardness is due to the presence of calcium, magnesium sulphates and chlorides

$$H_t = H_c + H_{nc} = H_{Ca} + H_{Mg}$$

where H_c = carbonate hardness

H_{nc} = non-carbonate hardness

Soft water features only a total or residual hardness H_t.

The presence of sodium compounds in water does not contribute to its hardness.

Total Solids

It is the net load of non-volatile substances present in water in a colloidal as well as a molecular dispersed state. It is expressed in mg/kg of water.

Total solids is determined by evaporating an unfiltered sample of water to dryness, heating the leftover residue at 380–390K to constant weight. The amount of these substances related to 1 kg of water is the measure of total solids.

Total Dissolved Solids

It refers to the content of dissolved non-volatile substances in water. It is expressed in mg/kg.

It is determined by evaporating a filtered sample free from any suspended matter and drying the residue to constant weight at 380–390K. The amount of these substances related to 1 kg of water gives the total dissolved solids (TDS).

Total dissolved solids can also be determined by measuring the specific conductance or conductivity, of the water.

Total Suspended Solids

It is the amount of non-volatile substances distributed in water in a colloidal state. It is expressed in mg/kg.

It is determined from the relation

$$TSS = TS - TDS$$

Oxidability

This parameter indicates, to some extent, the degree of contamination of water by organic substances. It is the amount of oxygen required to oxidize, under given conditions, the organic substances contained in a 1 kg water sample and is expressed in milligrams of oxygen/kg of water (mg O_2/kg). Since potassium permanganate is used as the oxidizing agent, oxidability can also be expressed by the quantity of $KMnO_4$, in mg/kg of water, expended to oxidize the organic substances in the water.

It is important to note that oxidability is not identical to the content of organic substances in water, since no complete oxidization of all the organic substances present in water takes place under the conditions in which oxidability is determined.

Transparency

It characterizes the concentration of total suspended solids in raw water, but it gives only a rough estimate of the TDS, which must be determined by the weight method for accuracy.

It is determined in terms of the height of a water column (cm) held in a glass tube through which it is still possible to read printing or discern two crossed black lines 1 mm thick on a white paper placed underneath the glass tube.

Silica Content

It is the concentration of silicic acid H_2SiO_3 (expressed as SiO_2) in raw water. This silicic acid is produced due to hydrolysis of silicates—salts of a weak acid.

$$SiO_3^{2-} + 2H_2O \rightleftharpoons H_2SiO_3 + OH^-$$

and the process is accelerated at low pH. This silicic acid exists predominantly in a colloidal state which cannot be removed by the processes of settling and filtration. It is only partially removed by the precipitation of hydroxides (iron, aluminium, etc.). Because of its slippage, silica is carried over to boiler feedwater and due to its solubility in steam at higher pressures

(4 MN/m^2 and above) and temperature, it is carried over to steam turbines and causes nozzle and blade deposition therein. That's why modern steam generation plants are equipped with demineralization units where silica content in water is reduced to as low as 0.01–0.02 ppm. However, since silicic acid lends itself to removal by the ion exchange process only in the ionic state (HSiO$_3^-$), the presence of colloidal silica complicates the overall desilication process.

ANALYTICAL PROCEDURES

Determination of P-Alkalinity

If the raw water sample contains alkaline hydroxides, carbonates, phosphates and silicates, it will turn pink when a few drops of phenolpthalein indicator are added.

To a measured 100 ml of filtered sample of raw water taken in a flask placed over a white sheet of paper, four drops of neutral phenolpthalein indicator are added. If a pink colour is developed the sample is titrated with 0.02N sulphuric acid solution from a burette. The acid should be added slowly dropwise, the solution well-stirred and the titration carried out till the endpoint is reached, i.e. until the pink colour just disappears.

P-alkalinity (as CaCO$_3$) in ppm

$$= \frac{\text{ml. of Acid}}{\text{ml. of Sample}} \times 100$$

Example: A 100-ml sample of water requires 8 ml of 0.02N sulphuric acid for complete neutralization. Determine it P-alkalinity.

Solution P-value $= \dfrac{8}{100} \times 1000 = 80$ ppm.

Determination of M-Alkalinity If the raw water sample does not develop pink color when phenolpthalein indicator is added but turns yellow when methyl orange indicator is added, then the sample contains no alkaline hydroxides or carbonates but bicarbonates, phosphates, etc. M-alkalinity determines total alkalinity.

To a measured 100 ml of filtered sample taken in a flask placed over a white sheet of paper, 2 to 4 drops of methyl orange indicator are added to get a distinct straw yellow color. The sample is titrated with 0.02N sulphuric acid from the burette. The acid is added dropwise and slowly and the solution is stirred well. The endpoint is marked by the appearance of a faint pink colour.

M-alkalinity (as CaCO$_3$) in ppm

$$= \frac{\text{ml of Acid}}{\text{ml of Sample}} \times 1000$$

Example: A 100-ml sample of water takes 15 ml of 0.02N H$_2$SO$_4$ solution for complete neutralization. Determine its M-alkalinity.

Solution M-alkalinity $= \dfrac{15}{100} \times 1000 = 150$ ppm.

Determination of Hydroxide Alkalinity. If the raw water is practically free of silicates, aluminates and organic matter or present in a small percentage in comparison to hydroxide, the following volumetric method may be used.

To a 100 ml filtered sample taken in a flask placed atop a white sheet of paper, 5 drops of barium chloride solution, [100 gm of BaCl$_2$.2H$_2$O in 1000 ml of water] are added to precipitate out the carbonate and sulphate ions.

Two to three drops of phenolpthalein indicator are added and the sample is titrated with 0.02N HCl solution from a burette until the pink colour just disappears.

OH - alkalinity in ppm. $= \dfrac{\text{ml of Acid}}{\text{ml of Sample}} \times 340$

Example: A 100 ml sample of raw water requires 6 ml of 0.02N HCl for neutralization. Determine its hydroxide alkalinity.

Solution OH-alkalinity $= \dfrac{6}{100} \times 340 = 20.4$ ppm OH

Determination of Total Hardness

To a 50 ml sample of clean raw water neutralized by adding drops of 0.1N NaOH or 0.1N H$_2$SO$_4$ solution as required, standard soap solution is added from a burette in small increments (0.2 ml at a time) and the bottle is shaken vigorously after each addition. When a permanent foam begins to form, soap solution is added in 0.1 ml increments and shaken as before until with further shaking a foam persists over the surface of water for a period of 5 minutes. Soap solution consumed is recorded.

After this point, a test for a false endpoint is done by adding 0.2 ml more of the solution and then shaking. If the foam disappears, the titration is continued to a final endpoint.

Total Hardness (as $CaCO_3$)
= (ml Soap Solution – ml of Foam Factor) × (Hardness Factor)

Standard Soap Solution

100 gms of powered castile soap is dissolved in 1000 ml of enthyl alcohol (80%), the solution is allowed to stand overnight and decanted. A portion of the stock solution is diluted with ethyl alcohol (80%) until, when titrated against $CaCl_2$ solution, 1 ml of resulting dilution is equivalent to 1 ml of the standard $CaCl_2$ solution. Accordingly, the rest of the soap solution is diluted.

1 ml of Soap Solution—Foam Factor = 1 mg $CaCO_3$

Foam Factor of the soap solution is determined by adding small quantities of soap solution to 50 ml of freshly boiled and cooled distilled water until a stable foam is formed after shaking. Milliliters of soap solution used is the foam factor.

Hardness Factor is determined by adding 5 ml of standard hard water to 45 ml of freshly boiled and cooled distilled water. The mixture is adjusted to phenolpthalein endpoint, and then titrated with small increments of soap solution and shaken vigorously after each addition of soap solution. The endpoint is marked when a foam stable for 5 minutes is formed.

Determination of Chlorides is done by titrating a fixed volume of water sample with standard silver nitrate solution using potassium chromate as indicator. A red colour is developed when all the chloride has been precipitated by the silver nitrate:

$$Cl^- + AgNO_3 \longrightarrow AgCl + NO_3^-$$

To a 50 ml filtered sample of water, already neutralized by adding drops of 0.1N NaOH or 0.1N

H_2SO_4 solution as required, 1 ml of potassium chromate indicator [50 gm of K_2CrO_4 dissolved in 100 ml of distilled water] is added. The sample is titrated slowly with stirring by adding a standard solution of silver nitrate from the burette until a stable slight reddish-orange or pink persists after stirring. It marks the endpoint.

The indicator blank is determined by titrating 50 ml of distilled water (free from chloride) with the same silver nitrate standard solution and by using 1 ml of K_2CrO_4 indicator solution.

Cl^- (as NaCl) in ppm

$$= \frac{\text{(ml of } AgNO_3 \text{ solution—Blank)}}{\text{ml of Sample Solution}} \times 1000$$

Determination of Sulphates

The sulphate of water sample is precipitated by benzidine hydrochloride as benzidine sulphate which is dissolved in distilled water and titrated with a standard solution of alkali.

To a 50 ml filtered water sample taken in a 250 ml round bottom flask 10 ml of benzidine solution [11.2 gm of benzidine hydrochloride dissolved in 400 ml of distilled water + 100 ml of 5% HCl solution] is added and thoroughly agitated and allowed to settle for 10 minutes.

The benzidine sulphate is filtered off and to the filtrate a few drops of benzidine hydrochloride is added; if the solution becomes cloudy 2 ml of benzidine solution more is added and shaken and filtered off after settling.

The precipitate is washed with 25 ml of distilled water saturated with benzidine sulphate, carefully transferred to the original flask and dissolved in 25 ml of distilled water. The solution is titrated with 0.1N NaOH from the burette using phenolpthalein as the indicator. The amount of standard NaOH solution consumed is recorded.

Blank is determined by titrating a solution of 10 ml of benzidine hydrochloride solution + 50 ml of distilled water with 0.1N NaOH using phenolpthalein as the indicator. The volume of NaOH consumed is recorded as sulphate blank.

SO_4^{2-} (as Na_2SO_4) = (ml of NaOH for the sample — ml of NaOH for the Blank) × 142, ppm

Determination of Phosphate

Determination of Phosphate as PO_4^{3-} in boiler or evaporator waters which are particularly not too coloured is carried out by the colorimetric method. In case considerable colour is present, the sample is either decolourized or the colour effect is balanced when the colour comparison is made.

This method is based on the formation of phosphomolybdate when molybdic acid is added to the sample water and the production of a blue colour when the phosphomolybdate is reduced by stannous chloride solution. The intensity of blue colour produced is proportional to the quantity of phosphate in the sample.

To a 5 ml of clear, filtered sample taken in a clean test tube 10ml of molybdic acid solution [10 gms of 85% molybdic acid dissolved in 500 ml of distilled water] is added and mixed. Into this solution added is 2.5 ml of dilute stannous chloride solution [12 gms of $SnCl_2 \cdot 2H_2O$ dissolved in 100 ml of conc. HCl and diluted in 1:50 ratio with distilled water] and mixed. The mixture is allowed to stand for 2 minutes to develop the blue colour.

Blank is prepared by adding to a 2nd test sample of 5 ml, 10 ml of molybdic acid + 2.5 ml of distilled water and mixed. It is then transferred to the colour comparison tube of the photometer, the colour filter is placed, the scale reading is set at zero and the galvanometer is balanced to the zero mark by turning the light-adjusting knob.

The blank is replaced by the blue coloured sample in the colour comparison tube and the galvanometer is now zeroed by turning the knob on the colour density scale. At this point the scale reading is recorded.

The PO_4^{3-} concentration (ppm) in the sample is read out from the Standard Calibration Curve in which the scale readings are plotted against ppm of phosphate in standard phosphate solutions using the same light filter.

Determination of pH Value

This is done by the electromagnetic method using glass and calomel electrodes immersed in the sample and a potentiometer (also called pH meter) which is calibrated to read in terms of pH.

Determination of Dissolved Oxygen

This is based on the oxidation of manganous ion into manganic ion by dissolved oxygen. The manganic ion is allowed to react with potassium iodide to liberate iodine. The quantity of iodine liberated is equivalent to the dissolved oxygen. The endpoint is determined by using a starch indicator.

Determination of Total Solids

This is carried out by the gravimetric method. A measured volume of the sample, 100 ml in the case of BFW, 500 ml in the case of raw water is evaporated in an accurately weighed beaker to dryness over a water or steam bath or on a sand bath. Then it is dried in an oven at 105°C for 1h, cooled in a desicator and weighed.

ppm of Total Solid

$$= \frac{\text{Wt of the beaker (in mg)}}{\text{Volume of Sample taken (in litre)}}$$

Determination of Total Dissolved Solids

This is done by measuring the conductivity of a clear, filtered sample and multiplying the result with a conversion factor.

Determination of Total Suspended Solids

It is obtained by subtracting the total dissolved solids from the total solids of the sample.

Determination of Soluble Silica

This can be conveniently done by the colorimetric method. The sample is treated with ammonium molybdate and then acidified whereupon the soluble silica forms silicomolybdate (yellow colour). It is then reduced by sodium sulphite whereupon a blue colour is developed. The intensity of the blue colour is proportional to the concentration of silica. From the standard calibration curve, the silica content in ppm is read out.

To a 10 ml of clear, filtered sample taken in a pyrex glass flask are added 5 ml of HCl [10 ml of conc. HCl diluted to 500 ml with distilled water] and 5 ml of ammonium molybdate [50 gm of ammonium molybdate dissolved in 500 ml of distilled water] solution. The solution is mixed and allowed to stand for 2 min and 10 ml of sodium sulphite solution [140 gms of $Na_2SO_3 \cdot 7H_2O$ dissolved in 500 ml of distilled water] and allowed to stand for 2 min for the full development of the blue colour.

A blank is prepared, the photometer is set at null-point, the comparison tube containing the coloured sample is placed in the photometer, the galvanometer is balanced and the colour intensity is recorded on the photometer.

The silica content (ppm) in the sample is obtained from the Standard Calibration Curve in which the photometer scale readings have been plotted as a function of the ppm of silica.

Appendix III

MATERIALS OF CONSTRUCTION FOR BOILERS & AUXILIARIES

Both carbon steel and alloy steels find their way to the manufacture of boilers.

Basic open-hearth steel is also used for making boiler components.

Though Bessemer steel is not fit for high-pressure boiler components yet "killed" Bessemer (i.e., deoxidized steel) is used for fabricating seamless steel tubes.

High-alloy steels made in electric furnaces are ususaly resistant to corrosion and capable of withstanding high temperatures over a prolonged period. These steels are used for constructing water-wall and superheater tubes, baffles, supports and hanger fittings, soot blower clamps, etc.

The steels that are employed, in general, for the construction of boilers belong to:

(a) Plain carbon steels (rimmed, semikilled and killed)
(b) Low alloy steels (pearlitic class)
(c) High-alloyed austenitic steels
(d) High-alloyed martensitic and martensitic-ferritic class.

Plain Carbon Steels

These are used for boiler construction with an average carbon content of 0.15 to 0.30%.

Carbon steel parts subject to pressure and exposed to fire should contain a maximum of 0.25% carbon for plate thickness 18–20 mm; for higher plate thickness a maximum carbon content of 0.30% can be recommended. Phosphorus should not exceed 0.04% for acid steel and 0.035% for basic steel. Maximum limit of sulphur content: 0.04%.

Carbon steel boiler components subject to pressure but not exposed to fire are made from flange steels which contain the maximum amount of the following elements:

Mn—0.8%; P—(0.04–0.05)%; S—0.05%.

Reasonably resistant to scaling in steam and corrosion by combustion products upto 540°C, carbon steels are sensitive to corrosive oxidizing media at somewhat lower temperatures if these are under high stress. Only killed steels can be recommended for use in the temperature range above 480°C. The fundamental problem inherent in carbon steel is that its carbide phase is precipitated along the grain boundaries upon prolonged exposure to temperatures above 425°C. This leads to intercrystalline cracks and tube failure. Though some manufacturers recommend the use of carbon steel superheaters to 500°C, the use of carbon steel is generally restricted to temperatures below 425°C.

Carbon steels find their application in the manufacture of boiler drums and vessels for pressure 6–12 MN/m^2 and temperatures upto 450°C. These are also used in making tubes of heating surfaces for temperatures upto 480°–500°C.

Low-Alloy Steels

These are superior to carbon steels in that they:

(a) retain greater strength and impact resistance at higher temperature
(b) are more resistant to corrosion
(c) possess high-temperature oxidation resistance and stability.

The low-alloy steels belong to the pearlitic class; they contain a maximum of 4 –5% of all alloying elements.

Low-alloy steels are resistant to creep at temperatures upto 580°C. Hence these are used in making superheater tubes and headers and steam pipelines— the elements which operate under creep conditions. They are also used for making boiler drums for working pressures upto 18 MN/m^2.

Not only are these creep resistant but also resistant to scale formation at temperatures upto 580°C. They are called refractory steels if they can retain the creep resistance and high temperature scale resistance beyond 580°C.

The chief alloying elements being: Chromium, molybdenum, silicon and aluminium.

Molybdenum gets dissolved in ferrite and increases the long-term strength and creep resistance. However, prolonged exposure to temperatures exceeding 470°C causes the transformation of carbide phase into graphite and that restricts the use of plain carbon-molybdenum steel to a maximum temperature of 425°–470°C.

Elements like chromium, silicon and aluminium impart scale resistance to the steel as they form dense refractory oxides of Cr_2O_3, SiO_2 and Al_2O_3 on the surface of the components. Their coefficients of expansion are very close to that of steel and they protect the base metal from further oxidation. In the making of boiler components, low-alloyed Cr-Mo class steels are frequently used because of their better creep resistance at elevated temperatures, less chances for graphitization and good weldability.

Cr	Mo	Applications
0.4–0.6	0.4–0.6	Steam headers and pipelines
0.8–1.1	0.4–0.6	Superheater tubes for temp. 475–550°C

Vanadium has a strong carbide forming tendency and hence the addition of vanadium (0.15–0.35%) to the low-alloyed pearlitic class improves the creep resistance of steel which therefore can be utilized for making boiler elements subjected to higher temperature and pressure.

Cr	Mo	V	Nb	Applications
0.9–1.25	0.25–0.35	0.15–0.35	—	Boiler tubes and steam pipelines operating temperature up to 565°–580°C
1.2–1.5	1–1.25	0.25–0.35	—	perheaters and steam pipelines subject to operating temperature up to 565°–580°C
2–2.5	0.5–0.75	0.25–0.35	0.5–0.75	Superheaters operating at temperatures up to 580°C

Chromium improves the scale resistance of steel. This property can be further boosted by addition of a small quantity of silicon (0.4–0.7%) rendering the pearlitic steel very highly scale resistant. If a very small quantity of boron (0.003–0.005%) is added to it, the steel attains a high degree of refractoriness. However, these steels are very hard and need special care during the heat treatment operation.

Cr	Mo	V	Si	B	Applications
1.5–2	0.5–0.75	0.25–0.35	0.4–0.7	0.005	Superheaters (operating temperatures up to 580°C

Manganese improves strength and hardness of steel. To it, if molybdenum is added, the steel gains higher heat resistance (creep resistance). Nickel also improves the strength and hardness as well as the corrosion resistance of steel. Hence this type of steel is used in making boiler drums and feedwater pipelines for supercritical pressure for continuous operation without the risk of creep.

Mn	Ni	Mo	Applications
0.8–1.2	1–1.35	0.45–0.55	Boiler drums (Working Pressure: 14–18.5 MN/m^2)
0.9–1.4	0.75–1.2	–	Feedwater pipelines (Pressure: up to 38 MN/m^2; Temp. : up to 450°C)

High-Alloyed Austenitic Class Steels

They belong to either high-alloyed chromium-nickel austenite or chromium-nickel-manganese austenite class whose chief alloying elements are chromium and nickel. Together these two elements may exceed 30% of steel composition. Both nickel and chromium substantially increase the austenite stability and improve the strength and hardness without any appreciable loss in ductility and toughness.

Cr	Ni	W	Mo	Ti	Nb	B	Applications
13–15	13–15	2.5–2.75	0.5–0.6	–	–	–	Superheaters and steampipes at temp. up to 650°C
17–19	12–15	–	–	0.65	–	–	-do-
15–17	8–10	–	1.5–2	–	–	–	Steam pipelines at steam temp. up to 650°C
13–15	18–20	2–2.5	–	–	1–1.25	–	Superheater coils and steam pipelines at temp. up to 700°C.
15–18	15–18	2–3	0.5–1	–	0.5–1	0.005	- do-

These have high heat and scale resistance at elevated temperatures and hence find applications in the manufacture of steam pipelines subject to temperature of 650°–700°C.

The austenitic class of steels owe their high scale resistance to the presence of a high concentration of chromium which forms a protective, impervious, hard chromium oxide film over the surface to prevent the base metal from further oxidation. However, these high chromium steels must be stabilized by adding stabilizing elements, viz. titanium (0.5–1%) and niobium (0.6–1%) to prevent intercrystalline corrosion that occurs due to chromium carbide precipitation at the grain boundaries under prolonged exposure to high temperature. These elements (Ti, Nb) being strong carbide formers prevent the formation of chromium carbides.

Austenitic steels have high heat resistance due to chromium which increases the resistance to softening at elevated temperatures. This property can be further enhanced by adding molybdenum and tungsten which form high-disperse strong compounds Fe_2Mo and Fe_2W in the steel matrix to improve the strength and creep resistance of the structure at elevated temperature.

High-alloyed Martensitic and Martensitic-Ferritic Class

These are no-nickel steels having chromium (11–15%) as the chief alloying element together with the additions of tungsten, vanadium and molybdenum for higher heat resistance. These combinations of alloying elements produce a low-carbon martensite or a martensitic-ferrite structure in steel which lacks the following drawbacks of austenitic steel:

1. The austenitic steel is very expensive because of high nickel content.
2. Austenitic steels are liable to stress-corrosion cracking.
3. Austenitic steels are prone to weld-decay—formation of annular cracks closed to welds during heating due to sharp drop of ductility.

Appendix IV

FORMATION OF COAL

Coal originated from prehistoric plants. Buried deep beneath the heaps of earth, the plant bodies underwent morphological changes and gradually converted to coal in the course of millions of years. The conversion was anaerobic and took place under heat and pressure.

The gradual conversion of plant bodies to coal occurred probably via the following steps:

1. Peat Formation

The plant constituents—lignocellulose, vegetable proteins, fats and oils, resins and waxes—underwent fermentation process in peat bog with the evolution of CH_4, CO_2 and H_2O as well as heat.

This was followed by maceration and segregation in the presence of water whereupon the fermented mass converted into colloidal gel and insoluble matter. The product was peat.

2. Conversion to Lignites

Under pressure of the earth's crust, peat underwent further consolidation and lost much of its water.

The consolidated mass gradually took the shape of laminates under high pressure, due to the movement of the earth. The product was lignite.

3. Formation of Bituminous Coal

The lignite or brown coal gradually transformed to bituminous coal over thousands of years under the influence of increasing pressure and rising temperatures (100 –
200°C) through the following steps:
   (a) Progressive dewatering.
   (b) Gradual elimination of oxygen content from the coal mass in the form of CO_2 and H_2O. The coal became lighter.
   (c) Amalgamation and complexation of acidic and alkaline humic substances of decayed vegetable matter.
   (d) Conversion of phenolic substances into coking constituents.

4. Conversion to Anthracite

Bituminous coal underwent anthracitization where conditions were favourable i.e., pressure reached 1500 atm. and temperatures shot up to a great extent. This was the last stage of coal formation. More oxygen content of bituminous was expelled as CO_2 and H_2O and coal became enriched in carbon.

COAL CLEANING

Cleaning of coal means physically separating the ash, and other non-combustibles as well as pyritic sulphur from coal.

Coal cleaning is carried out to ensure better emission control. And almost all coal cleaning is carried out at the mines by the coal companies/authorities. This adds value to their product.

But not all the coal bound for powerplants is clean coal. About 33% of the coal burnt in powerplants undergoes some sort of cleaning.

Coal cleaning involves the following steps:
   1. The raw coal is screened and sized. Standard hammermills, roll crushers and rotary breakers are used to size the raw coal. This is called size reduction which must avoid the creation of a large amount of coal-fines. These are difficult to clean.

Once crushed, the coal is screened into different size ranges.

2. Screened coal is subjected to a separation process to remove impurities from it.

In jig-type units water pulsations are fed to a suspension of raw coal in water to effect a separation of the refuse from the coal by size and density.

In advanced coal-cleaning plants froth floatation is used to recover coal fines. This technique is based on hydrophobic and hydrophilic properties of coal to effect a separation. Sized raw coal is introduced in a vessel containing a dense fluid— usually magnetite suspension. Air is introduced to produce fine bubbles to agitate the slurry

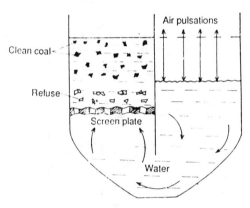

Fig. IV.1 *JIG TYPE UNIT*
Air pulsation is fed to aqueous suspension of raw coal to force a separation from refuse

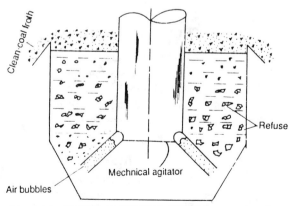

Fig. IV.2 *FROTH-CLEANING SYSTEM*
Raw coal suspension in magnetite slurry is agitated by air bubbles to effect a separation of raw coal from solid refuse

whereupon particles (pyrites and non-combustibles) with specific gravity greater than the medium sink to the bottom and get discharged as refuse while light particles float and separate as clean coal.

REFERENCE FUEL

It is a conditional fuel that liberates 7,000 kcal (29.309 MJ) of heat per kg of fuel burned.

If $\dot{M}$ be the consumption of a particular fuel that upon combustion will release the same amount of heat as the reference fuel will do for its combustion $\dot{M}_{ref}$ then

$$\dot{M} Q = \dot{M}_{ref} Q_{ref}$$

$$\dot{M}_{ref} = \dot{M}\left[\frac{Q}{Q_{ref}}\right]$$

where Q = heating value of fuel under consideration, kcal/kg

Q_{ref} = heating value of reference fuel, kcal/kg

For the sake of simplicity in calculation of the combustion of various fuels in steam boilers, the concept of reference fuel has been introduced.

CONGELATION TEMPERATURE

It is referred to liquid petroleum products and is a characteristic of its fluidity.

It is the temperature at which liquid fuels become so thick that they remain static in a place and do not flow out for a duration of 60s from a test glass held at a 45° inclination.

Fuel oils having high sulphur and paraffin content are characterized by high congelation temperature (300 –315 K).

The usefulness of congelation temperature lies in the selection of the scheme of transportation and storage of fuel oils.

CRITICAL PRESSURE

This is the pressure at which the density of water and steam becomes equal at saturation temperature. This is about 225 kgf/cm² (218 atm.) at temperature 374°C.

Because of zero density difference between water and steam, no natural circulation is possible at this point.

When this condition is reached, the entire absorbed heat goes to convert water to steam of the same volume and no absorption of latent heat occurs.

SUPERCRITICAL PRESSURE BOILERS

These boilers are those which operate at a pressure higher than 225 kgf/cm^2 and temperature exceeding 374°C (saturation temperature at cirtical pressure).

The first supercritical boiler plant that went into operation was PHILO-6 of the Ohio Power Co. It generates steam at 31 MN/m^2 (306 atm.) and temperature 621°C. Its designed steam generation capacity is 306 t/h to drive a 125 MW turbo-generator with a thermal efficiency of 40%. The unit consists of a once-through boiler with three slag-tap bottom cyclone furnaces burning crushed coal.

The second supercritical pressure boiler unit was installed at Philadelphia Electric Co. It is designed to produce steam [34.48 MN/m^2 (340 atm.)/649°C] at the rate of 907 t/h to generate 325 MW of power from turbo-alternators. Fuel burned is bituminous and the boiler is Sulzer type— dry-bottom twin furnaces fired with tangential burners located at the corners.

Another giant is the 1 300 MW monobloc unit of the Gavin Power Station, USA. Its once-through boiler unit— 61m high, 34.4m wide and 15.8m in depth—has the designed capacity of generating steam (269 atm./538–543°C) is 4 438 t/h. The open-type furnace with a dry-bottom hopper is befitted with 112 Nos. of pulverized coal fired burners laid out in four tiers on the front and rear walls of the furnace.

SUPERCHARGED BOILERS

These are those boilers in which combustion takes place under a pressure higher than atmospheric pressure. The term "supercharging" refers to a combustion process which is carried out in the furnace at such a pressure that useful work may be obtained from the hot flue gases after they leave the boiler by virtue of their expansion in the gas turbine.

The furnace is operated under considerable pressure, viz. 3–7 kgf/cm^2.

The supercharged boiler cycle affords the generation of steam economically at any pressure and temperature (even at supercritical pressures).

There are two types of supercharged cycles:
1. self-sustained cycle
2. power cycle.

SELF-SUSTAINED SUPERCHARGED CYCLE

Hot pressurized flue gas drives the gas turbine which is coupled to a air compressor to supply combustion air under pressure. There is no electrical power output. [Fig. IV. 3]

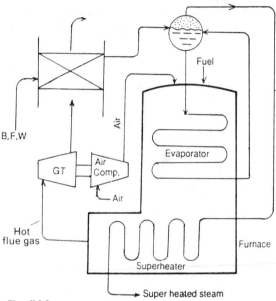

Fig. IV.3

POWER SUPERCHARGED CYCLE

In this cycle the gas turbine also generates electric power apart from driving the air compressor for the supply of combustion air. It has high cycle efficiency. It can be applied to large power generating stations providing a 90% efficiency over a wide load range. [Fig. IV. 4]

THERMAL CHARACTERISTICS OF WATER WALLS

Incident Heat Flux

Heat absorption by waterwalls in a boiler furnace is mainly by the radiant heat transfer from the high-

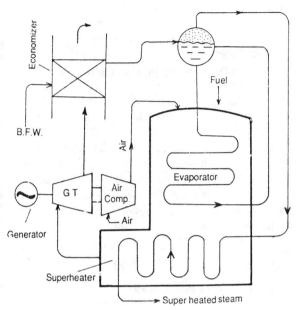

Fig. IV.4

θ_{fl} = flame temperature

The incident heat flux from the flame core in a boiler furnace is usually in the range 400–700 kW/m².

Effective Heat Flux

It is the sum of intrinsic heat flux and reflective heat flux

$$q_{eff} = q_{intr} + q_{refl}$$

The intrinsic heat flux is due to radiation from deposits on the tubes surface and from the refractive lining

$$q_{intr} = \varepsilon_{dep} (56.7 \times 10^{-9}) \left[\frac{\theta_{dep} + 273}{100} \right]^4$$
$$\times 10^{-3} [(1 - \varepsilon_{dep}) q_{inc}] \, x \, kW/m^2$$
$$= \varepsilon_{dep} (1 - \varepsilon_{dep}) \, x \, q_{inc} (\theta_{dep} + 273)^4$$
$$(56.7 \times 10^{-20}) \, kW/m^2$$

And the reflected heat flux is

$$q_{refl} = \varepsilon_{refl} (56.7 \times 10^{-9}) \left[\frac{\theta_{refl} + 273}{100} \right]^4$$
$$\times 10^{-3} (1 - \varepsilon_{refl}) q_{inc} (1 - x) \, kW/m^2$$
$$= \varepsilon_{refl} (1 - \varepsilon_{refl}) (1 - x) q_{inc} (\theta_{refl} + 273)^4 \times$$
$$(56.7 \times 10^{-20}) \, kW/m^2$$

x is a factor that determines the fraction of total radiation which falls on the waterwall surfaces. It is essentially the ratio of the surface of the waterwall absorbing radiant heat to the total surface area of the waterwalls inside the furnace.

temperature gaseous products of combustion to the waterwall tubes. The intensity of heat exchange by radiation is determined by the incident heat flux from the flame core onto the waterwalls in the boiler furnace.

According to Stefan-Boltzmann's law, the incident heat flux from the flame core can be expressed in the following form:

$$q_{inc} = \varepsilon_{fl} (56.7 \times 10^{-9}) \left[\frac{\theta_{fl} + 273}{100} \right]^4 \times (10^{-3}) \, kW/m^2$$

where ε_{fl} = flame emissivity

emissivity of black body = 56.7×10^{-9} W/m². K⁴

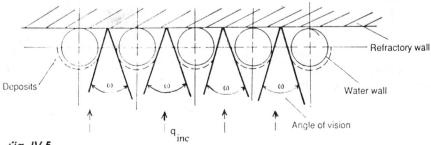

Fig. IV.5

The effective heat flux in boiler furnaces may run as high as 50–60% of the incident heat flux.

Radiant Heat Flux

It is the numerical difference between the incident heat flux and effective heat flux

$$q_{rad} = q_{inc} - q_{eff}$$

It is the radiant heat flux which is absorbed by the waterwalls and transferred to the working fluid from thereof.

Coefficient of Thermal Efficiency of Waterwalls

It is the ratio of the radiant heat flux to the incident heat flux on the waterwall surfaces

$$\phi_{ef} = q_{rad}/q_{inc}$$

This parameter characterizes the fraction of total heat absorbed by the heating surfaces on the furnace walls. It varies with the type of fuel:

Fuel	ϕ_{ef}
Solid	0.4–0.45
Liquid	0.5–0.55
Natural gas	0.65

Higher the value of ϕ_{ef}, higher the operating efficiency of waterwalls, i.e., they will absorb a higher fraction of the total heat.

The efficiency of waterwalls is not constant throughout the boiler furnace: It is higher in the zone of flame core and diminishes with the height of the furnace towards the exit from the furnace.

Angular Coefficient of a Waterwall

It is the fraction of total radiation heat falling on the surface of the waterwalls.

This angular coefficient of waterwalls (*x*) is the ratio of the radiant heat absorbing surface of the waterwalls to the total surface of the waterwalls.

However, calculation of *x* based on the geometry of the waterwall alone will lead to erroneous result, because a part of the incident heat flux absorbed by the refractory wall is re-radiated onto the rear surface of the waterwall tubes and thus transferred to the working fluid. Only a small fraction of the heat flux from the refractory lining within the angle of vision of the flame ($<\omega$) escapes to the furnace volume. (Fig.IV.5)

Hence, *x* is dependent on ω which is a function of tube spacing determined by the tube pitch $p = s/d$. Smaller the tube pitch, i.e., denser the tube spacing, narrower will be the angle of vision, with the effect that a higher fraction of total heat flux falls upon the waterwall surface.

For,

$$p = 1.05 - 1.11, x = 0.97 - 0.99$$

$$p = 1 \qquad x = 1 \text{ (theoretically)}$$

Coefficient of Fouling of Waterwall

It is the fraction of the incident radiant heat absorbed by the waterwall, i.e., it is the ratio of the absorbed radiant heat to the total radiant heat falling on the waterwall surface.

Now, the amount of absorbed radiant heat flux is proportional to the factor $(\theta_{fl} + 273)^4 - (\theta_{dep} + 273)^4$, therefore, the coefficient of fouling of waterwall

$$\varepsilon = [(\theta_{fl} + 273)^4 - (\theta_{dep} + 273)^4]/(\theta_{fl} + 273)^4$$

where

θ_{fl} = flame temperature

θ_{dep} = temperature of deposits on waterwall

ε is always less than unity and is a measure of the relative cleanliness of waterwall tubes. Greater the deposit thickness, larger will be the value of absolute temperature of deposit surface and consequently the coefficient of fouling will be less.

Appendix V

DETERMINATION OF GROSS CALORIFIC VALUE OF COAL

Theory

A weighed quantity of coal is completely burned in an enclosure and the released heat is allowed to increase the temperature of water in the calorimeter. Net heat absorbed by water divided by the weight of the sample gives the Gross Calorific Value of coal.

APPARATUS REQUIRED
1. Mahler Bomb Calorimeter
2. Pelletizer
3. Fine iron wire-piece (10 cm long)
4. Thermometer $20 - 40 \pm 0.001\,^\circ\text{C}$ scale

Method

About 1 gm finely powdered coal sample is pelletized on the loop of the previously weighed iron wire in the pelletizer.

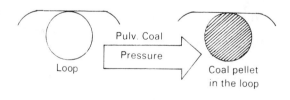

Loop   Pulv. Coal   Pressure   Coal pellet in the loop

It is then carefully removed from the machine, weighed and placed in the crucible of the bomb calorimeter.

The ends of the loop-wire are attached to the terminal rods to ensure electrical contact. The head of the bomb calorimeter is screwed, tightened and sealed.

The outlet valve of the bomb is closed and oxygen gas from the oxygen cylinder is let in the bomb through the inlet valve till the pressure reaches 2–5 MPa. The inlet valve of the bomb and the outlet valve of the O_2 - cylinder are closed.

The bomb is then placed in the calorimeter into which is taken 2000 ml of water. The electrically driven stirrer as well as the thermometer are inserted in the calorimeter.

The stirrer is started and temperature is recorded every 1 minute for 5 minutes. At this point, the ignition switch is pressed and the coal is fired. Temperature readings are taken every 1 minute for another 8 minutes.

Results

    Weight of iron wire = 0.0076 g
    Weight of iron-wire + Coal sample = 0.7523 g
    Water equivalent of the calorimeter = 400 g
    Water used in the calorimeter = 2000 ml

Temperature Recordings

Initial Period	Firing Period	Final Period
0 min. 23.015°C	5 min. 22.980°C	8 min. 24.996°C
1 min. 23.010°C	6 min. 24.940°C	9 min. 24.992°C
2 min. 22.999°C	7 min. 24.975°C	10 min. 24.980°C
3 min. 22.999°C	8 min. 24.996°C	11 min. 24.965°C
4 min. 22.989°C		12 min. 24.955°C
5 min. 22.980°C		13 min. 24.940°C

Calculations

Weight of sample on the loop

$= 0.7523 - 0.0076$

$= 0.7447$ g

Total water equivalent of calorimeter

$= 2000 + 400$

$= 2400$ g

Cooling Rate	Initial Period	Final Period
$\Delta\theta$	$23.015 - 22.980$	$24.996 - 24.940$
t	5	5
	$- 0.007$ °C/min.	$- 0.0112$ °C/min.

∴ Cooling correction = Length of Firing Period × Average of Initial and Final Cooling Rates

$= 3 \times \frac{1}{2} (0.007 + 0.0112) = 0.0273$°C

Therefore, corrected temperature rise

$= (24.996 - 22.980) + 0.0273 = 2.0433$°C

Therefore, Gross Calorific Value of coal

$$= \left[\frac{\begin{array}{c}(\text{Total Water Equiv.} \times \text{Sp. Heat} \times \text{Corrected} \\ \text{Temp.rise}) - (\text{Wt. of Iron Loop Wire} \\ \times \text{Cal. Value of Iron})\end{array}}{\text{Weight of the Coal Sample}} \right]$$

$= (2400 \times 4.187 \times 2.0433 - 0.0076 \times 1600)/$
$\qquad 0.7447$ kJ/kg

$= 27\ 555$ kJ/kg $= 6\ 581$ kcal/kg

Note: The same test can be carried out for the determination of the gross calorific value of liquid fuel which is to be held in the bomb calorimeter in a plastic of known calorific value.

PROXIMATE ANALYSIS OF COAL

The proximate analysis is carried out to determine the moisture content, volatile matter, fixed carbon, coke, ash content and calorific value of a coal sample.

Theory

The moisture content is determined by drying the preweighed coal sample to a constant weight at a temperature slightly above the boiling point of water (100°C) under normal conditions.

In another run, the coke content (equal to fixed carbon + ash content) is determined by driving off the total volatile matter as well as the moisture content from a known amount of coal sample. The volatile-matter-free, dry residue is coke which is calculated out by weight difference.

Finally, the coke residue in the crucible is burnt off till all the fixed carbon burns out and only the ash remains. By difference of weights, the ash content is measured.

From these data, the other parameters are computed as follows:

% Volatile Matter = 100 – [Coke % + Moisture %]

% Fixed Carbon = Coke % – Ash %

Determination of Moisture Content

About 1g of coal sample taken in a previously weighed, dry and clean crucible is weighed.

The crucible is heated at 375 K for about an hour in an oven. At this temperature, the moisture content of coal evaporates. The crucible is allowed to cool in a dessicator and reweighed.

This process is repeated till the final constant.

Result

Weight of crucible = 10.7505 g
Weight of crucible + coal = 11.7601 g
Weight of crucible + dry coal = 11.7411 g

Calculation

Therefore, weight of moisture

$= 11.7601 - 11.7411$ g

$= 0.0190$ g

Weight of coal $= 11.7601 - 10.7505 = 1.0096$ g
Therefore,
moisture content $= 0.019/1.0096$ kg/kg of coal

$= 0.0188$

% moisture content $= 0.0188 \times 100 = 1.88$ %

Determination of Coke Content

About 1g of coal sample is taken in a previously

weighed, dry and clean, crucible. The crucible together with the coal sample is reweighed.

A lid is placed over the crucible which is heated over a bunsen flame for about 7 minutes.

The crucible and contents are reweighed after cooling in a dessicator. The residue is coke.

Result

Weight of crucible = 10.7712 g
Weight of crucible + coal = 11.8351 g
Weight of crucible + coke = 11.4781 g

Calculation

Weight of coke residue = 11.4781 − 10.7712 g

$$= 0.7069 \text{ g}$$

Weight of coal sample = 11.8351 − 10.7712 g

$$= 1.0639 \text{ g}$$

Therefore, coke % $= \dfrac{0.7069}{1.0639} \times 100 = 66.444 \%.$

Determination of Ash Content

The coke residue in the crucible in the above experiment is now heated over a strong bunsen flame with the lid open until all the carbon is burnt off and only the ash remains.

After about 1 h of heating, the crucible is cooled in the dessicator and weighed.

Result

Weight of crucible = 10.7712 g
Weight of crucible + ash = 10.8649 g

Calculation

Weight of ash = 10.8649 − 10.7712 = 0.0937 g
Weight of coal = 11.8351 − 10.7712 = 1.0639 g
Ash % = [0.0937/1.0639] × 100 = 8.80 %

Estimation of Volatile Matter

Volatile Matter % = 100 − (Coke % + Moisture %)

$$= 100 - (66.444 + 1.88) = 31.676\%$$

Estimation of Fixed Carbon

Fixed Carbon % = Coke % − Ash %

$$= 66.444 - 8.80 = 57.644 \%$$

Determination of Calorific Value of Coal

Determination of volatile matter % on a dry, ash-free basis:

$$[VM]_{dry} = \frac{VM \%}{VM \% + FC \%} \times 100$$

$$= \frac{31.676}{31.676 + 57.644} \times 100 = 35.46 \%$$

From the $[VM]_{dry}$ vs. α graph (Fig. V.1) we find

$$\alpha = 90 \text{ for } [VM]_{dry} = 35.46$$

$$GCV = 4.187 \, [(82)(57.644) + (90)(24.756)] \text{ kJ/kg}$$

$$= 31\,727 \text{ kJ/kg}$$

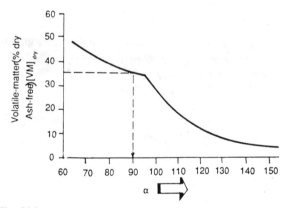

Fig. V.1

$$= 7\,577 \text{ kcal/kg} \qquad \boxed{\text{cf. 1 kcal} = 4.187 \text{ kJ}}$$

Goutal's formula:

$$GCV = 4.187 \, [82 \, (FC) + \alpha \, (VM)_{dry}], \text{ kJ/kg}$$

DETERMINATION OF CALORIFIC VALUE OF A GASEOUS FUEL

Theory

A metered quantity of gaseous fuel is burnt under preset conditions and the heat evolved is allowed to raise the temperature of water flowing at a known and constant rate.

The total heat received by water as determined

from its mass flowrate, specific heat and rise of temperature, divided by the volume of gas (corrected to standard state) burned gives the higher calorific value of the gaseous fuel.

The amount of heat of evaporation of the moisture formed due to combustion of hydrogen in the fuel is subtracted from the higher calorific value to determine the lower calorific value of the gaseous fuel

Apparatus Required

1. Junker's Gas Calorimeter
2. Constant water head tank
3. Thermometers (2 Nos.)
4. Gas meter
5. Gas pressure regulator

Procedure

Thermometers are placed— one at the water inlet to calorimeter and the other at water outlet from the calorimeter. (Fig. V.2)

With the aid of the constant head device, the water flowrate surrounding the annular space of the calorimeter is adjusted to about 2 litres/min.

The gas flowrate is set in the range 0.007–0.0085 m^3/min with the help of the gas pressure regulator.

After these adjustments, the equipment is left for 30– 45 minutes to attain steady state.

The gas is ignited and readings of temperatures and the volume of water passed are noted after every 1 minute for about 12 minutes.

Results

Volume of gas burned = $0.1012 \ m^2$
Pressure of gas supply = $101.25 \ kN/m^2$
Room temperature = 293 K
Effluent gas temperature = 295 K
Temperature of water at inlet = 291 K
Temperature of water at outlet = 306.70 K
Mass of water circulated = 25 kg
Steam condensed = 0.068 kg

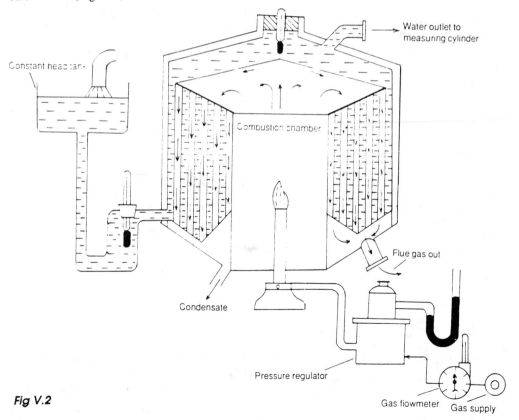

Fig V.2

Calculations

1. *Gas Volume Correction:* As the gas was metered at conditions other than standard conditions, its volume must be corrected to 101.33 kN/m² and 288 K.

Corrected gas volume

$$= 0.1012 \left[\frac{101.25}{101.33}\right]\left[\frac{288}{293}\right] = 0.09939 \text{ m}^3$$

2. *Effluent Gas Correction:* Since the effluent gas usually leaves the calorimeter at a temperature other than room temperature, a correction in the gross calorific value of fuel is necessary by adding or subtracting the figure 8.27 × (Effluent Gas Temp. − Room Temp.) from the calorific value obtained.

In this case, the correction = 8.27 (295 − 293) = (+) 16.57 kJ/m³

3. *Gross Calorific Value of Gas*

GCV =

$$\frac{\text{Mass of Water} \times \text{Specific Heat} \times \text{Rise in Temp.}}{\text{Volume of Gas burned}}$$

+ Correction

$$= \frac{25(4.187)(306.70 - 291)}{0.09939} + 16.57$$

$$= 16\,551 \text{ kJ/m}^3$$

4. *Lower Calorific Value of Gas:* Since the lower calorific value differs from the higher calorific value by the heat of evaporation of the water content of fuel and that produced on combustion of hydrogen in the fuel,

LCV = HCV — (Steam Condensate)(Latent Heat of Steam at 288K)

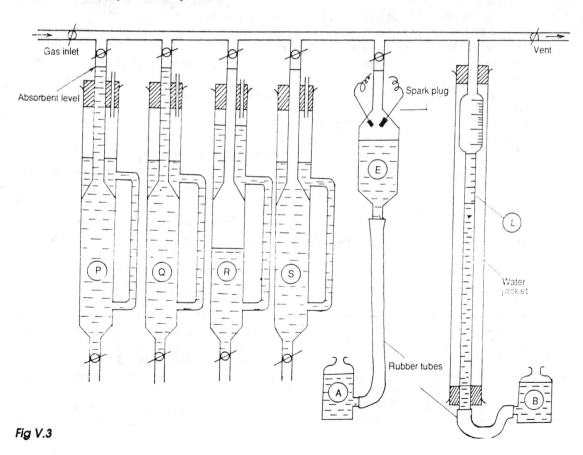

Fig V.3

$$= 16\ 551 - \left[\frac{0.068}{0.09939}\right] (2\ 454) = 14\ 872 \text{ kJ/m}^3$$

> cf. Latent Heat of Steam at 288 K = 2 454 kJ/kg

DETERMINATION OF PERCENTAGE COMPOSITION OF A FUEL GAS & ITS CALORIFIC VALUE

Volumetric analysis of a fuel gas as well as its gross and net calorific values can be determined with the help of an Orsat apparatus.

The gas under consideration contains CO, CO_2, H_2, CH_4, C_nH_{2n} (Unsaturated hydrocarbons), plus some oxygen and nitrogen.

Apparatus

The entire experiment requires the following apparatus set-up [Fig. V.3]

1. Two aspirators A and B for intake of gas in the pipettes and eudiometer
2. Eudiometer tube L for metering intake gas volume and the volume of gas absorbed
3. Pipettes 5 Nos.—P, Q, R, S and E, the last one being provided with a spark plug to ignite the gas-air mixture.

The function of the rest four pipettes being:

Pipette	Gas Absorbed	Reagent	Concentration
P	C_nH_{2n}	Br_2 - water	Saturated Br_2 -water + 2 vols. of water
Q	CO	Cu_2Cl_2	16% Cu_2Cl_2 solution in conc. HCl
R	O_2	Alkaline pyrogallol	Pyrogallic acid 1 part + 3 part water + 24 parts NaOH + 16 parts water
S	CO_2	KOH solution	1 part KOH + 2 parts water by mass.

All these pipettes are filled with some glass rods to increase the area of gas-liquid contact.

Theory

The gas composition is determined volumetrically. The fractions of CO, CO_2, O_2 and unsaturated hydrocarbons in a metered volume of the fuel gas are determined by absorption in liquid reagents (vide above). The hydrogen and methane contents are estimated by exploding a given volume of the gas-air mixture in the explosion pipette. Their concentrations are calculated from the reduction in volume (due to condensation of water vapour formed) and CO_2 left-over.

When the composition of gas is known, its GCV and LCV are computed out from the known calorific values of its constituents.

Procedure

The levels of the reagents in P, Q, R and S and the water level in E are raised to the marks indicated near the taps by lowering B containing water acidified with a few drops of sulphuric acid and reddened by methyl orange indicator.

Keeping L full of liquid, the apparatus is purged off air by the fuel gas.

50 ml of the gas is drawn into L and the gas tap is closed.

The gas is now transferred to pipette S whereupon its CO_2 content is allowed to get absorbed in the KOH solution.

The gas is then transferred back to L by lowering B, the tap on pipette S is closed and the volume of the gas is noted. Every time the reading is taken, the liquids in B and L are brought to the same level so that gas volumes are measured at atmospheric pressure and at a uniform temperature as provided by the jacket water. This operation is repeated to get constant volume readings (at least three readings).

The same procedure is carried out for pipettes R, Q and P.

Now, to the pipette R all but 10 ml gas is transferred and the vent tap is opened and 40 ml of air is sucked into L. And this mixture of 10 ml. gas and 40 ml. of air is delivered to E by lowering A and then transferred back to L. This is repeated several times to ensure thorough mixing of the gas with air.

Finally, the mixture is driven to E and ignited by means of spark plugs. The volume of the leftover gas is measured in L.

Afterwards, the gas is transferred to the pipette S to estimate the volume of CO_2 produced due to combustion and then it is let out to the atmosphere.

This combustion process is carried out three times—each time taking 10 ml of gas from the pipette R.

Results

Intake sample volume = 50 ml.

		Gas Volume after absorption in pipette	
S	R	Q	P
48.6 ml	48.3 ml.	41.5 ml.	41.3 ml.

10 ml. of the left over gas is combustioned with 40 ml. air.

	1st *run*	2nd *run*	Average
Volume after combustion	35.7 ml	35.3 ml	35.5 ml
Volume after absorption in S	32.5 ml	32.7 ml	32.6 ml

Calculations

$\%\ CO_2 = (50 - 48.6)\ 100/50 = 2.8\%$
$\%\ O_2 = (48.6 - 48.3)\ 100/50 = 0.6\%$
$\%\ CO = (48.3 - 41.5)\ 100/50 = 13.6\%$
$\%\ C_nH_{2n} = (41.5 - 41.3)\ 100/50 = 0.4\%$

Combustion Reactions	*Volume Reduction After Condensation*
(i) $CH_4 + 2O_2 \longrightarrow CO_2 + 2H_2O$	2x
(x) (2x) (x) (2x)	
(ii) $H_2 + 0.5\ O_2 \longrightarrow H_2O$	0.5y
(x) (0.5y) (y)	

The CO_2 formed on combustion = 35.5 – 32.6

= 2.9 ml.

And this CO_2 entirely comes from combustion of methane in the 10 ml sample.

$\therefore\quad x = 2.9$ ml

Volume reduction due to combustion = 50 – 35.5

= 14.5 ml

This is equal to the volume of oxygen consumption and water vapour produced condensing out.

$\therefore\ \ 2x + 0.5y + y = 14.5$

or $2\ (2.9) + 1.5y = 14.5$; $\therefore\ y = 5.8$ ml

Gas Sample Volume	CH_4 Content	H_2 Content
10 ml	2.9 ml	5.8 ml
41.3 ml	2.9 (41.3/10)	5.8 (41.3/10)
	= 11.977 ml	= 23.954 ml

Therefore, $CH_4\% = 11.977\ (100/50) = 23.954\%$

$H_2\% = 23.954\ (100/50) = 47.908\%$

Therefore, $N_2\ \% + \text{Error}\ \%$

$= 100 - (2.8 + 0.6 + 13.6 + 0.4 + 23.954 + 47.90)$

$= 10.738\%$

CALCULATION OF CALORIFIC VALUES

1. The *Gross Calorific Value* of the fuel gas sample is calculated from the calorific values of its constituents.

$GCV = [0.47908 \times 12\ 780 + 0.23954 \times 39\ 755 +$

$0.136 \times 12\ 706 + 0.004 \times 62412]\ kJ/m^3$

$= 17\ 623\ kJ/m^3\ (4\ 209\ kcal/m^3)$

Gas	Calorific Value
H_2	12 780 kJ/m^3
CH_4	39 755 kJ/m^3
CO	12 706 kJ/m^3
C_2H_4	62 412 kJ/m^3

2. The *Net Calorific Value* is obtained by deducting the latent heat of evaporation of water (formed during combustion) from gross calorific value.

Assuming the unsaturated hydrocarbon present almost as C_2H_4, the volume of water vapour produced per 100 m^3 of the sample on combustion can be calculated as:

Constituent	Combustion Reaction	Water Produced
H_2 (47.908 ml)	$H_2 + 1/2\ O_2 \longrightarrow H_2O$	47.908 ml
CH_4 (23.954 ml)	$CH_4 + 2O_2 \longrightarrow CO_2 + 2H_2O$	2×23.954 = 47.908 ml
C_2H_4 (0.4 ml)	$C_2H_4 + 4O_2 \longrightarrow 2CO_2 + 2H_2O$	2×0.4 = 0.8 ml
		96.616 ml
		$= \dfrac{96.616}{22.4}$ (18) kg
		= 77.634 kg.

Therefore, the net calorific value of fuel gas

$$= 17\ 623 - \frac{77.634}{100}\ (2\ 455)\ \text{kJ/m}^3$$

(cf. Latent Heat of Steam = 2 455 kJ/kg)

$$= 15\ 717\ \text{kJ/m}^3\ (3\ 754\ \text{kcal/m}^3)$$

SUMMARY

Fuel Gas Composition	% (Volume)
H_2	47.908
CH_4	23.954
CO	13.6
O_2	2.8
CO_2	0.6
C_nH_{2n}	0.4
N_2 + Error	10.738
	100.000

Gross Calorific Value

$$= 17\ 623\ \text{kJ/m}^3\ (4\ 209\ \text{kcal/m}^3)$$

Net Calorific Value

$$= 15\ 711\ \text{kJ/m}^3\ (3\ 754\ \text{kcal/m}^3)$$

Appendix VI

Environmental Pollution and its Control in Boiler Plants

Large boiler units of thermal power plants around the world gobble up as much as 40% of the total global output of fuel, let alone those chemical plants that burn fossil fuels for the generation of steam for process use. And therefore, these fossil - fuel - burning systems pose a substantial threat to the environment by ejecting heat and certain harmful pollutants into the atmosphere and water basins. Since environmental pollution—particularly air and water pollution—knows no political boundaries, it has become a global concern to protect our precious environment from the harmful effluents of industrial plants for the sustained survival of life on earth.

AIR POLLUTION DUE TO FOSSIL-FUEL BURNING

The boiler plants pollute the atmosphere by injecting into it:

1. oxides of nitrogen ($NO_x = NO + NO_2$)
2. oxides of sulphur ($SO_x = SO_2 + SO_3$)
3. solid ash particles as suspended matters or particulates
4. vanadium pentoxide V_2O_5.

However the types of gaseous pollutants depend on the nature of the fuel burned. When coal is burned, the major pollutants are fly ash, soot, sulphur dioxide and NO_x. Whereas in the case of fuel oil burning, SO_2 and NO_x constitute the major gaseous effluents. However most Indian coals are low in sulphur content, except the Assam coal having sulphur content from 2% to as high as 8%.

Over and above the foregoing air pollutants, flue gases may contain carbon-monoxide (CO) and hydrocarbons like CH_4, C_2H_4 and $C_{20}H_{12}$ (benzpyrene) and soot.

Of course, certain components of gaseous effluents from fossil- fuel burning systems are harmless to man. These include water vapour, carbon dioxide, oxygen and nitrogen. Other ingredients as cited above are harmful to the biosphere in varying degrees.

Vanadium pentoxide and benzpyrene are extremely dangerous substances. Vanadium appears in minor quantities in certain fuel oils and when these oils are burned V_2O_5 is formed.

Benzpyrene is formed during any type of fossil-fuel burning under condition of oxygen deficiency and is also formed on the slow decomposition of soot. It is a carcinogen and causes lung cancer.

Carbon monoxide is potentially quite dangerous. It is odourless, colourless and tasteless. The human body can tolerate CO in concentrations upto 10 ppm without a noticeable effect.

Nitrogen dioxide and sulphur trioxide are highly dangerous air pollutants. $NO + NO_2$ are produced in the high-temperature zones of the flame—the yield of NO_2 being 10% (usually) of the total NO_x formation. NO_x may undergo a photochemical reaction with the hydrocarbons in the atmosphere in the presence of sunlight to release some toxic substances in the air.

Sulphur in the fuel burns off to liberate SO_2 which forms SO_3 in the final stage of flame burning when there is an exigency of atomic oxygen. SO_3 is also produced from SO_2 on the surface of superheater deposits that act as a catalyst at elevated temperatures. Sulphur trioxide reacts with atmospheric moisture to form an aerosol of sulphuric acid which rains down as acid rain.

Sulphur dioxide is a moderately harmful substance. But because of its acidic nature, it attacks like SO_3 and NO_2, the delicate membrane lining the eyes, nose and respiratory tract. These air pollutants are known to cause four major types of respiratory damage: Bronchitis, bronchial asthma, emphysema and lung cancer. They affect plants as well. They burn the leaf's chlorophyl causing red or brown blotches on leaves between the veins.

Suspended particulates cause smog and obscure visibility.

TOXIC RATE

It is convenient to compute the risk of exposure of a living being to a harmful ingredient in air as the ratio of the actual concentration of the substance C (expressed in $\mu g/m^3$) to the highest permissible concentration HPC (expressed in $\mu g/m^3$) of that substance in the air at the breathing level:

$$\phi = [C/HPC]_i$$

This represents the toxic rate of the substance and should always be less than unity.

Since te toxicity level in air increases with the number of harmful substances in it, the presence of multiple toxic components renders their concentrations even at the HPC level inadmissible. That is why in such cases, toxic hazard is estimated by the cumulative toxic rates of these substances:

$$[C/HPC]_{SO_2} + [C/HPC]_{SO_3} + [C/HPC]_{NO_2} +$$

$$[C/HPC]_{NO} + [C/HPC]_{CO} < 1$$

The highest permissible concentrations of certain toxic substances in air are as follows:

Pollutant	In Inhabited Zones	
	Highest Single Concentration $(\mu g/m^3)$	Average Daily $(\mu g/m^3)$
SO_2	500	50
SO_3	300	100
NO_2	85	85
NO	250	250

CO	3000	1000
$C_{20}H_{12}$ (benzpyrene)	—	0.001
V_2O_5	—	2
Soot	150	50
Ash particles	500	150

AMBIENT AIR QUALITY STANDARD

The Central Board of Water Pollution (INDIA) has prescribed the following air quality standards (adopted in November 1982).

Area	Category	Suspended Solids $(\mu g/m^3)$	SO_2 $(\mu g/m^3)$	NO_x $(\mu g/m^3)$	CO $(\mu g/m^3)$
A	Industrial and Mixed Use	500	120	120	5000
B	Residential and Rural	200	80	80	2000
C	Sensitive Area	100	30	30	1000

The Environmental Protection Agency (EPA), USA has enforced the following standards for the ambient air quality.

Pollutant	Concentration $(\mu g/m^3)$		Remarks
	Primary	Secondary	
SO_2	80	60	Annual average
	365	260	Max. 24-h concentration not to be exceeded more than once in a year
Suspended Particulates	75	60	Annual geometric mean
	260	150	Max. 24-h concentration
CO	10^4	10^4	Max. 8-h concentration
NO_x	100	100	Annual average

The particulate emission standards for coal fired boilers in India are as cited below:

Boiler Size		Particulate Emission (mg/Nm³)	
	Protected Area	Other Areas	
		Old (before 1979)	New (after 1979)
< 200 MW	150	600	350
≥ 200 MW	150	—	150

AIR POLLUTION MONITORING EQUIPMENT

Equipment	Pollutants	Operating Range	Remarks
1. Air Monitor CEA - 555	SO_2 NO_x	0–0.5 ppm 0–5 ppm	Made in U.S.A.
2. SO_2 Monitor Model SO 261	SO_2	0–200 ppm	Industrial Scientific Corpn., U.S.A.
3. Brink Mist Analyzer	Particulate Concentration		Made in U.S.A.
4. IPAC-2000	CO	0–1000 ppm	Industrial Pump and Compressor, Denver, Colorado, U.S.A.
	H_2S	0–500 ppm	

DETERMINATION OF PARTICULATE MATTER IN FLUE GAS

Particulate matter discharged by thermal power plants and coal fired boiler units is potentially harmful to human health. Hence determination of particulate concentration is necessary to regulate the emission of this pollutant.

It consists of:

(a) isokinetic sampling of the flue gas to extract an uniform sample of gas and particulates
(b) separation of particulate solids from the sample
(c) determination of particulate concentration.

For isokinetic sampling, a stainless steel nozzle (ID = 7mm min) with sharp tapered leading edge is inserted in the gas duct at the transverse points*. The nozzle dia varies with the velocity of flue gas and the amount of sample required.

The other end of the nozzle is inserted into a thimble (made of paper or alundum) set inside a thimble holder. Thimble is a filter media to collect the particulates. Paper thimbles can withstand temperature upto 150°C while alundum can withstand 550°C as well as more corrosive environment. Glass cloth filters made of finely woven glass fibre can also be successfully used on coarse dust and where sampling is to be taken under high dust load.

The thimble holder is connected in series with the condenser (to collect water vapour condensate), a dry gas meter attached with a temperature and vacuum gauge (for metering the conditions of sampled gas) and a vacuum pump or blower or ejector to induce gas flow through the sampling train. (Fig. VI.1)

The sampling is to be continued for at least one hour so that at least 1.6 m³ of dry gas has been withdrawn and the collected mass of particulates amounts to at least 20% of the mass of the filtering medium.

During each run, the gas volume, pressure and temperature as well as condenser temperature are recorded at 5 minute intervals.

Sampling is done for at least 5 minutes at each transverse point, and when it is finished at one point, the sampler is quickly shifted to another point.

When the sampling is over, the sampler is cooled, particulates collected inside the nozzle are carefully brushed into the thimble and weighed. The mass of dust collected in the thimble is found by difference, i.e., by weighing the thimble before and after the run.

CALCULATIONS

1. *GAS FLOWRATE* The sampling rate at the meter

$$V_r = u\, a \left(\frac{T}{T_s} \right) \left[\frac{P_s}{P_a - p - p_{H_2O}} \right] \left(1 - \frac{m_w}{100} \right)$$

[For circular stacks, the area of cross-section is divided into a number of equal annular areas the midpoints of which are the location of transverse points. In the case of rectangular stacks, the cross-section is divided into 12 to 15 equal areas and the centre of each area is a location for a transverse point.]

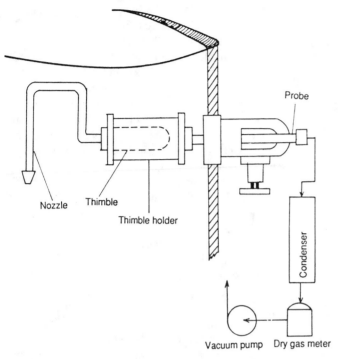

Nozzle

Thimble

Thimble holder

Probe

Condenser

Vacuum pump   Dry gas meter

Fig. VI.1

where V_r = volume flowrate (m³/s) as recorded by gas meter

u = flue gas velocity at sampling point, m/s

a = nozzle area, m²

T = gas meter temperature, K

T_s = temperature of stack gas, K

P_s = stack gas pressure, kPa (i.e. kN/m²)

P_a = atmospheric pressure, kPa

p = suction pressure at gas meter, kPa

P_{H_2O} = vapour pressure of water at condenser exit temperature, kPa

m_w = moisture content of the gas (% by mass)

2. *VOLUME OF GAS SAMPLED* Therefore, the volume of gas sampled

$$= V_r(t), m^3$$

where, t = sampling period, s.

Therefore, the volume of gas sampled (dry basis/298°K/101 kPa)

$$V_d = (V_r t)\frac{298}{T} \cdot \frac{P_a}{101}, m^3$$

And, the volume of gas sampled (wet basis/298°K/101 kPa)

$$V_w = V_d\left[1 + \frac{m_w}{100 - m_w}\right]$$

3. *DUST CONCENTRATION* Dust concentration (dry basis/298 K/101 kPa)

$$[Dust]_{dry} = M/V_d, g/m^3$$

where M = mass of dust collected in the thimble, g.

$$[Dust]_{wet} = M/V_w, g/m^3$$

4. *EMISSION RATE* The dust emission rate

$$\dot{M} = [Dust]_{wet} \times V_{fg}, g/h$$

where V_{fg} = flue gas flowrate, m³/h

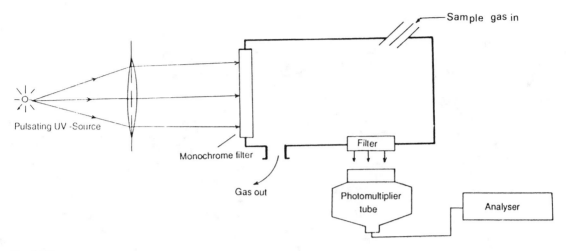

Sample gas in

Pulsating UV -Source

Monochrome filter

Gas out

Filter

Photomultiplier tube

Analyser

Fig. VI.2

MEASUREMENT OF SO₂ CONCENTRATION IN FLUE GAS

This is carried out with the help of a package analyzer which consists of

(a) probe assembly to extract the sample from the mainstream of flue gas (Fig. VI.2)
(b) sample tubing through which the collected sample goes to the instrumentation set-up. The sample tubing is heated electrically to prevent condensation in the lines
(c) gas conditioning unit
(d) analyzer
(e) calibration gas cylinder.

The analyzer which operates on the pulsed fluorescence principle is calibrated to zero calibration, first, by using clean and dry compressed air. Then a series of SO₂/Air mixtures of known concentrations are provided for accomplishing span calibration.

Now, a sample gas is subjected to a source of pulsed UV-light through a monochromatic filter. The high intensity pulsed UV-light energizes SO₂ molecules which emit specific illumination when they return to the ground state.

The emitted light, passing through a filter, impinges on a photomultiplier tube. The intensity of emitted radiation is linearly proportional to the concentration of SO₂ molecules in the sample.

The range of this analyzer is 0–10 000 ppm.

MEASUREMENT OF NOₓ CONCENTRATION

It is done with the help of a package analyzer that comprises of

(a) probe assemblies
(b) sample tubing
(c) gas conditioner
(d) analyzer.

The working technique is based on the chemiluminiscence principle.

When in a mixture of NO and NO₂, adequate quantity of ozone (O₃) is introduced, the latter will react with nitric oxide and convert all NO molecules to NO₂ in the excited state:

$$NO + O_3 \rightarrow NO_2^* + O_2$$

When these excited NO_2^* molecules return to the ground state they emanate photons ($h\nu$) $NO_2^* \longrightarrow NO_2 + h_\nu$ which are measured in the analyzer, thus affording a viable as well as sensitive measurement of NO in the flue gas.

Both NO and NO₂ are measured by this technique.

In one mode, the sample gas is directed to the reaction chamber where only NO content of NOₓ reacts with ozone to produce excited molecules of NO_2^*. The chemiluminiscence of these excited NO_2^* as detected and recorded by the analyzer gives the value

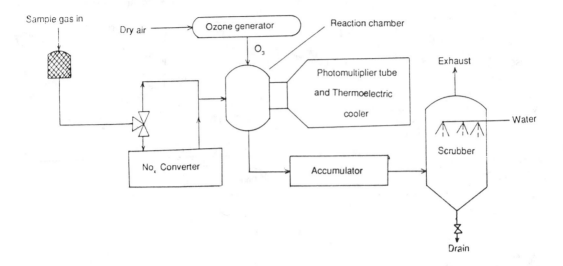

Fig. VI.3

of NO concentration in the sample. (Fig. VI.3)

In the other mode, the entire sample gas is directed to NO_x converter where all NO_2 content of NO_x mixture is converted to NO which is then fed to the reaction chamber and the concentration of excited molecules of NO_2^* recorded by the analyzer gives the total concentration of $NO + NO_2$ in the flue gas.

The difference between these two results is the concentration of NO_2 in the flue gas sampled.

The analyzer records the output on strip-chart and has the range of $0 - 10\,000$ ppm with minimum detectable limit 1 ppm.

AIR POLLUTION CONTROL IN FOSSIL FUEL BURNING SYSTEMS

Control of SO_2 Emission

As much as 75% of total sulfur dioxide ejected into the atmosphere come from thermal power stations. With the increased industrial utilization of high-sulfur coal and fuel oils, it has become a growing concern to reduce the SO_x emission to atmosphere. A number of options are being available in this regard:

(a) Burning low sulfur coals

(b) Desulphurization of coal prior to its combustion
(c) Desulphurization of petroleum
(d) Subjecting fuel oil and high sulfur coal to deep thermal processing to produce gaseous fuel
(e) Post combustion SO_2 removal from waste gases.
(f) Fluidized bed combustion.

Switching over to low sulphur fuels is, however, a short term measure. Whereas desulphurization of coal and liquid fuels, removal of SO_2 from stack gas and adopting advanced technologies of coal combustion are long term measures.

Uptil now, the coal desulphurization is in the pilot plant stage. However, coal gasification of high-sulphur feedstock befitted with sulphur recovery system, according to some experts, may be economically more lucrative. Desulphurization of fuel oils, in the refineries, to produce low-sulphur fuel oil is expensive, yet with steady technological advancement it is increasingly becoming economical.

However, at present the most widely practised method is desulphurization of flue gases. Since the volume concentration of SO_x in the flue gases is rather low (about 0.2%) and the volume of flue gases to be handled is enormous, stripping flue gases off sulphur oxides presents enormous technical difficulties.

There are several methods available for cleaning up flue gases from SO_x:

(a) Dry scrubbing
(b) Wet scrubbing
(c) Absorption of SO_x on coke or activated charcoal bed
(d) Catalytic oxidation of SO_2 to SO_3 and subsequent conversion, as well as removal, of the latter as H_2SO_4.

Dry scrubbing involves absorption and chemical reaction between SO_x and lime/MgO/$CaCO_3$ bed.

Wet scrubbing involves spraying lime slurries or NaOH or caustic soda solution into the flue gas stream.

Scrubbing has added advantages: It removes a substantial quantity of particulate solids.

Wet lime/limestone scrubbing uses lime or limestone slurry as the reactive absorbent. But these systems are beset with the problems of

(a) scaling
(b) downtime loss
(c) corrosion of air preheater tubings
(d) plugging of demister
(e) solid waste disposal.

The solid waste ($CaSO_3$) must be stabilized to destroy its quick-sand effect. SO_2 removal efficiency 70–85%.

Also sulphur dioxide removal efficiency as high as 90% has been achieved by the non-regenerable alkali process using sodium hydroxide as absorbent.

SO_2 removal efficiency of 85–90% is obtained by adopting the double alkali process in which the flue gas is scrubbed with NaOH solution and the scrubber effluent is treated with lime or limestone to regenerate NaOH or Na_2SO_3 and precipitate out calcium sulphite and the sulphates are removed by vacuum filtration.

The MAGOX scrubbing system for flue gas desulphurization uses a fine suspension of magnesium oxide (90% passing through 325 US Standard Mesh) in water as the scrubbing liquid injected in the venturi-throat or just upstream of it. The acceleration of flue gas through venturi atomizes the injected MgO-suspension-water into fine droplets ($10 \mu m$ in size) which form droplet-clouds of 170-300 μm in dia. Because of tremendous turbulence, the SO_2 in the flue gas reacts quickly with MgO - suspension and gets

removed as magnesium sulphite. Stack gas reheating is not always necessary. Efficiency of SO_2 removal is above 90%. Magnesium oxide loss (10–12%), dust emission from dryer erosion and chocking of scrubber by fine crystals of $MgSO_3$ are the chief problems.

The CATOX (catalytic oxidation) approach to FGD involves the oxidation of SO_2 to SO_3 by air over a vanadium pentoxide catalyst. The sulphur trioxide produced is absorbed in dilute sulphuric acid. SO_2 removal efficiency is 85%. However this system requires a high efficiency mist eliminator and stack gas reheating after absorber.

Control of NO_x Emission

As much as 40% of the total quantity of NO_x ejected into the atmosphere comes from thermal power plants. However the removal of nitrogen oxides from flue gas presents greater engineering difficulties than that of SO_2. That is due to

(a) lower concentration of NO_x in flue gases. It amounts to 20 – 50% of SO_2 concentration.
(b) chemical inactivity of NO which is usually present in FG in much larger concentration than NO_2.

Unlike SO_2 whose presence in flue gas is unavoidable until and unless the sulphur content of fuel is eliminated beforehand, the concentration of NO_x in combustion gases largely depends on the combustion technique in the furnace. Nitrogen oxides are produced in two ways during fuel combustion:

1. chemical reaction between aerial nitrogen and oxygen in the high-temperature combustion zone (above 1600°C)
2. oxidation of nitrogen content in fuel. This can take place below 1600°C.

Since the bulk of NO_x produced in the combustion process comes from reaction (1) the chief means of limiting NO_x generation is to lower the temperature in the combustion zone, i.e., to carry out a low-temperature combustion technique as well as reduce the quantity of excess air. This can be achieved by

(a) adopting fluidized bed combustion technique. As this process contains the combustion temperature in the range 800–900°C, the chances of NO_x formation are minimum.
(b) carrying out combustion with minimum ex-

cess air
(c) corner firing
(d) flue gas recirculation
(e) double-stage combustion, particularly with gas fired large boilers.

NO_x emission from steam generation plants can be successfully diminished by
—Selective Non-catalytic Reduction (SNR) of NO_x
—Selective Catalytic Reduction (SCR) of NO_x

SNR system reduces NO_x by injecting ammonia into the furnace gas. Through large jets, ammonia is introduced to the furnace at those points where temperature is in the vicinity of $1140°—1250°K$. NO_x reacts with ammonia to form water vapor and elemental nitrogen

$$NO + NO_2 + 2NH_3 \longrightarrow 2N_2 + 3H_2O$$

Successful operation of SNR system demands stringent combustion control and minimum fluctuations of temperature and oxygen content.

While the same chemical principle is involved in SCR process, it differs from SNR in the aspects that ammonia is injected at a different location and by different means. Through specific ports, ammonia is charged to the flue gas and then the reduction is carried out over a catalyst bed (V_2O_5 based on TiO_2) downstream.

SCR unit is installed downstream of a particulate collecting device (bag filter or ESP) fitted beyond the economizer. This, sometimes necessitates flue gas reheating upstream of catalyst bed.

SO_x and NO_x REMOVAL FROM FLUE GASES
Japanese scientists of Okayama University have been successful in developing a batch of coprecipitated copper-oxide/titanium-oxide catalysts for the simultaneous removal of SO_x and NO_x from coal - and fuel-oil- combustion flue gases.

The catalysts are composed of:
$45 – 47.5$ mol % CuO
$2.5 – 5$ mol % Cr_2O_3
$47.5 – 52.5$ mol % TiO_2

That ensure optimal removal of SO_x and NO_x. These catalysts retain their stability after regeneration (up to 20 times) and are capable of effecting 90%

desulphurization and 100% denitrification from flue gases containing 500 ppm each of SO_2 and NO, 5% O_2, 10% H_2O and N_2 at 350°C.

U.K.'s Humphreys & Glasgow Ltd., and Sweden's Nykomb AB (Stockholm) have developed a quenching technique to burn all types of coal but producing cleaner flue gases than conventional or fluidized combustion processes. It limits SO_2 emissions to less than 5 mg of sulphur, i.e., removing as much as 99% of the sulphur in the fuel per MJ and less than 50 mg of NO_2 per MJ. The figures look certainly promising in comparison to 50 mg S and 100 mg NO_2 per MJ for fluidized bed combustion (or conventional coal firing with after-combustion flue gas clean-up).

The method is based on any high pressure, slagging gasifier in the quench mode. The key is to start with a quench to cool the gas and knock out the pollutants.

Control of Particulate Emission

Several devices are available for the removal of particulate solids from the stack gas of thermal power plants and steam generation plants. These are:
(a) Cyclone separator
(b) Fabric filters
(c) Dust scrubbers
(d) Electrostatic precipitators.

Cyclones afford the cheapest investment while ESPs are the costliest amongst all. However, the efficiency of cyclones goes down with particulate size lower than 10 μ.

ESPs have high collection efficiency, as high as 99.9%. Besides, these are capable of eliminating dust particles as small as 0.01 μ. Draft loss is also low: 0.25–1.25 cm of water. And they can operate at gas pressures upto 40 atm.

Fabric filters are liable to chemical attack; moreover they are susceptible to thermal and mechanical shock. Hence their application to control particulate emission is not favoured in thermal power plants.

WATER POLLUTION AND SOLID WASTE DISPOSAL PROBLEM IN BOILER PLANTS

Large boilers of thermal power stations and of process

industries contribute to water pollution by way of discharging into the water basin the:

(a) boiler blowdown

(b) SO_2 - scrubber waste

(c) cooling waters that mainly cause thermal pollution

(d) waste waters from water treatment plants and demineralizing units

(e) waste waters contaminated with petroleum products

(f) waste waters from hydraulic ash-disposal system.

These discharged waste waters carry a rich load of harmful impurities, viz., heavier metal cations, organic substances and coarse-dispersed solids besides dissolved salts. The toxic substances added to the water basin from boiler plants may adversely affect the hydrobionts—all living organisms inhabiting the water basin. At higher concentrations they will simply perish while at lower concentrations they may suffer from reduced metabolism and growth rate, abnormal change in mutagenesis and reproductive capacity. The maximum permissible concentrations of selected substances in water basins, according to ISI standard are tabulated below.

Effluent Characteristics	General Limits
Temperature, (max)	40°C
pH	5.5–9.0
BOD, mg/1 ; max	30
COD, mg/1 ; max	250
Oils and Grease, mg/1 ; max	10
Ammoniacal Nitrogen (as N) mg/1	50
Phosphates (as P), mg/1 ; max	15
Sulphides (as S), mg/1 ; max	2
Fluorides (as F), mg/1 ; max	2
Cyanides (as CN), mg/1 ; max	0.2
Lead (as Pb), mg/1 ; max	0.1
Zinc (as Zn), mg/1 ; max	5
Copper (as Cu), mg/1 ; max	3
Nickel (as Ni), mg/1 ; max	3
Cadmium (as Cd), mg/1 ; max	2
Chromium (as Cr^{6+}), mg/1 ; max	0.1
Mercury (as Hg), mg/1 ; max	0.01
Arsenic (as As), mg/1 ; max	0.2
Phenolic compounds (as C_6H_5 OH), mg/1 ; max	1.0

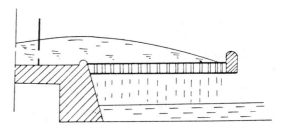

Fig. VI.4 *Open Spillways can substantially cool down the hot discharge water from turbine condenser*

Now the load of impurities discharged into water basins can be decreased by two ways:

1. by purifying the waste waters

2. by reducing the quantitative discharge of impurities from particular technological processes.

Of course, the latter should be given more importance.

THERMAL POLLUTION

The heat released by cooling water (open-circuit) from the turbine condensers of super thermal power stations is enormous. As much as one million kilocalories of heat is dumped per hour in the water basin by cooling water flowing at a rate of 100 000 – 130 000 kg/MWh from a 1000 MW conventional thermal power plant. This creates a huge zone of elevated temperature in the water basin, reduces its dissolved oxygen and thereby impairs the growth and development of aquatic life.

The zone of elevated temperature in a water basin can be substantially reduced by allowing the inflow of hot discharge water into the basin through:

(a) open spillways with

    (i) transverse and side weir bulkheads

    (ii) distributing grill

(b) submerged jet-type spillways

Open-type spillways (Fig. VI. 4) allow the discharge water to get cooled as well as saturated with atmospheric oxygen. Jet-type spillways, disperse hot discharge water over a wide zone of water basin

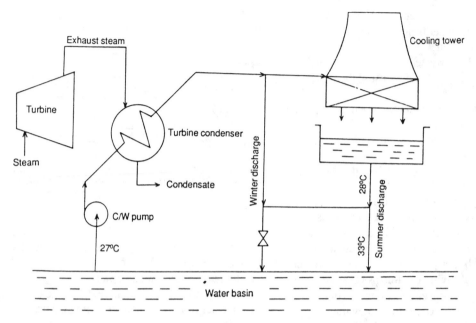

Exhaust steam

Cooling tower

Turbine

Turbine condenser

Steam

Condensate

C/W pump

27°C

Winter discharge

28°C

33°C

Summer discharge

Water basin

Fig. VI.5 *Cooling tower incorporated into the condensate discharge circuit substantially downs the thermal load of the hot discharge water*

through submerged jets intensifying the intermixing of hot water with cold water of the basin.

The thermal load of hot discharge water can also be substantially decreased by incorporating a cooling tower or spray-pond downstream of the surface condenser of the turbine to cool the coolant water before its discharge to the water basin. (Fig. VI. 5)

WASTE WATERS FROM WATER TREATMENT PLANTS AND DM UNITS

All the boilers generally need water-treatment plant and demineralizing unit for the production of make-up cooling water and BFW from raw water. The reagents used in the water-treatment plant and DM-unit and the salts resulting from their chemical action must be discharged off as wastes. As much as 1.5–2 ton of various salts are discharged with waste water every hour from a WT-plant of nominal capacity (2 000 M^3/h).

Waste water of water treatment plants (WTP) contains slime, coarse-dispersed solids, organic substances, magnesium hydroxide, calcium carbonate and salts of iron and aluminium. The composition and concentration of various impurities in waste water depends on the quality of raw water and the methods adopted for water treatment. Where lime treatment is adopted, the clarifier blowdown contains sludge of CaCO$_3$ (mainly) and it has elevated pH value (10–10.2).

In the DM-unit, regeneration of H-cation exchanger and OH-anion exchanger is done by using H$_2$SO$_4$ and NaOH solutions and as a consequence, the disposed waste water becomes respectively acidic and alkaline in nature.

So the principal effects of discharging waste waters of WTP and DM-units in the water basin are to:

(a) increase the salt content of the basin
(b) raise the concentration of suspended solids
(c) alter the pH of the basin
(d) raise the BOD level of the basin.

That is why the direct disposal of waste waters from the water treatment plant and DM-unit to the water basin is not admissible.

It has been estimated that as much as 700 kg of

salts are discharged per hour from a water treatment plant of a boiler producing enough steam to generate 1000 MWh of electrical energy.

1. *Purification:* The waste water of water treatment plant (WTP) and discharge from DM-unit may be disposed as follows:

(a) transferring the waste water into the hydraulic ash handling system of coal fired boiler units

(b) neutralizing the waste water (pH > 9) of WTP with acid-wastes of the DM-unit

(c) subjecting the waste water to slime separation, i.e., slime dewatering in drum-type vacuum filters and recycling the clarified water for washing of mechanical filters.

Waste water of WTP can be discharged into the hydraulic ash-disposal system of coal fired boiler units provided it does not cause to the formation of deposits in the ash-disposal system.

If the purge water of WTP has pH value about 10 (as is the case when lime treatment is employed), it is better to neutralize a part or whole of this alkaline waste water by acid wastes of the DM-unit.

If the pH of waste water is below 6.5 or above 8.5, the water must be neutralized before discharge.

Waste waters from ion-exchanges of DM-units upon mixing together with regenerative waters from cation and anion exchanges have their pH index, in most cases, less than 6. Hence, these waters are neutralized by alkalis, lime, marble and dolomite.

Neutralization by lime is convenient as it does not, unlike other reagents, lead to a sharp increase in the salt concentration in water.

Neutralizing by ammonia is much more convenient. The quantity of ammonia required is not high and the salt content of neutralized water does not exceed that resulting from lime-neutralization process.

Neutralization is carried out in neutralizing tanks. A tank is filled with acidic and alkaline waste waters, a metered quantity of neutralizing solution is charged into it and the liquid is agitated by compressed air at a rate of 0.4–0.8 M^3 per minute/M^2 of the cross-sectional area of the neutralizer.

If the water upon mixing of regenerative wastes is alkaline, then it is better to neutralize it by bubbling flue gases into it. The flue gases contain acidic oxides like CO_2, SO_2, NO_2 which upon dissolution in water will produce acids that neutralize the alkalinity of waste water.

2. *Reducing Waste Water Discharge of WTP and DM-Units:* The amount of impurities discharged into the water basin by waste waters from the WTP and DM-unit can be diminished by adopting techniques that will minimize the use of reagents and water for water treatment and regeneration purposes.

The quantity of water used for regeneration of mechanical filters can be drastically cut down by increasing the filtering capacity of mechanical filters. It has been found that that certain volcanic slag have absorbing capacity 2–2.5 times that of quartz sand. And therefore, by using expanded clay instead of quartz sand in mechanical filters it is possible to use less water for regeneration by a factor 2.5–3.5.

The flowrate of wastes from the DM-unit can be effectively reduced by adopting the process of:

(a) continuous ion-exchange

(b) stepwise counter-current ion-exchange

(c) thermal regeneration (instead of chemical regeneration) of ion-exchanges.

The continuous ion-exchange has certain distinct advantages:

(a) smaller size of apparatus

(b) reduction of the quantity of regenerants down to almost theoretical (stoichiometric) quantity

(c) diminishing the flowrates of waste waters down to half to one-third.

The method of stepwise counter-current ion-exchange cuts down the NaOH consumption for regeneration of OH-anion exchanger by 30–40% and brings the H_2SO_4 consumption for regeneration of H-cation exchanger down to near stoichiometric quantity.

Though thermal regeneration—effected by hot water (about 80°C)—has not been found applicable for all kinds of ion-exchange materias, it is possible to thermally regenerate 42% of borate-form anion exchangers.

The colloidal impurities of waste water can be precipitated down by electrocoagulator using either Fe- or Al-anodes instead of using chemical coagulants $FeSO_4$, $Al_2(SO_4)_3$.

Electrocoagulators are of various types—the most

popular being those with plate type electrodes or with filled-in electrodes.

A plate-type electrocoagulator is essentially a reservoir into which are vertically suspended a series of parallel electrode plates 10–20 mm apart. Waste water is usually charged into the vessel from the bottom and purified water exits from the top. As a direct current is passed through the electrodes the following processes occur:

(a) electrolytic dissolution of Fe or Al f the anodes

$$Fe - 2e \rightarrow Fe^{2+}$$

(b) reduction of dissolved oxygen at cathodes

$$2H_2O + O_2 + 4e \rightarrow 4OH^-.$$

(c) formation of sparingly soluble $Fe(OH)_2$ or $Al(OH)_3$

$$Al^{3+} + 3OH^- \longrightarrow Al(OH)_3$$

(d) oxidation of chloride ions at the anode

$$Cl^- - e \longrightarrow Cl \xrightarrow{Cl} Cl_2$$

(e) oxidation of organic impurities by free chlorine

(f) removal of CO_2 to a great extent

(g) diminishing of HCO_3^- ion concentration

(h) substantial reduction of silicon, iron and oxygen concentration in water.

It has been possible to decrease the organic compounds in water by 70%, silicic acid by 50%, chlorides by 30% by electrocoagulation by using iron anodes. It also reduces the hardness and alkalinity of water.

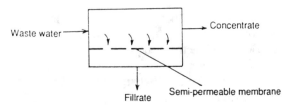

Fig. VI.6 Reverse Osmosis

Stripping waste water off its dissolved salts can be effectively accomplished by a physical process known as reverse osmosis. In this process, the waste water with dissolved salts is forced through a semi-permeable membrane (Fig. VI.6) at a pressure, usually 5–8 MN/m^2, exceeding the osmotic pressure of the solution. The membrane allows only water and a small fraction of salts as ions to pass with the effect that the filtrate contains an appreciably smaller quantum of dissolved impurities.

Cellulose acetate membrane is the most popular. It can operate at a

(a) wide range of pressure 0.6 to 8MN/m^2

(b) wide pH range 3–8

(c) temperature range 0–30°C

Cellulose acetate membranes in the form of hollow fibres have

(a) selectivity upto 90% of NaCl dissolved in water

(b) permeability of water 1.25×10^{-5} $m^3/m^2 \cdot s$ at 1.5–2 MN/m^2

The technique of reverse osmosis, it has been observed, has drastically cut down the consumption of reagents used for water treatment with the effect that the concentration of impurities in discharged waste waters sharply declines. (Fig. VI. 7)

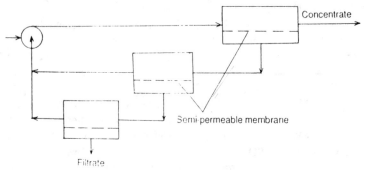

Fig. VI.7 Reverse Osmosis connected in series

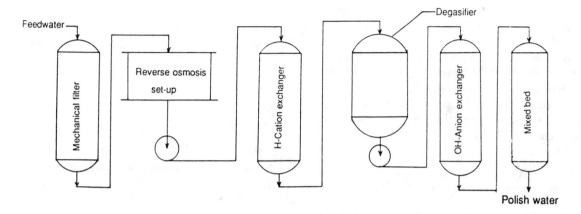

Fig. VI.8 *A typical line-up of reverse osmosis plant with a DM-unit*

The line up of reverse osmosis plant upstream of DM-units (Fig. VI. 8) can slash down the discharge of salt solution by 50% and the cost of desalted water by 25%.

Desalination of waste waters can be successfully carried out by another technique known as electrodialysis. The apparatus consists of an assembly of parallel cation and anion exchange membranes (which are selectively permeable to cations and anions respectively) flanked by a stainless steel cathode and platinum coated steel anode.

Water is delivered to the desalting (even) cells. As the electric field is applied to the system, cations migrate towards the cathode by penetrating the cation-exchange membrane and the anions move, likewise, to the anode through the anion-exchange membrane. As a result of this ionic migration, the water in the even cells gets desalted while the concentration of odd cells (brine) increases. Desalinated water and brine concentrate are tapped out separately.

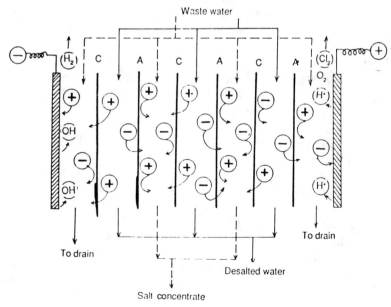

Fig. VI.9 *Schematic representation of electrodialysis*

Electrodialysis substantially reduces the use of reagents for the waste water treatment and consequently the amount of salt discharged to waste waters is also diminished. By installing an electrodialysis plant upstream of the DM-unit, the load on the ion-exchangers is radically dropped and so the quantity of regenerants expended per m^3 of DW water yield and the amount of salt discharged with waste water of regeneration are also substantially decreased.

WASTE WATERS CONTAMINATED WITH PETROLEUM PRODUCTS

The petroleum products like lube oils, fuel oils, kerosene, etc., find their way to the water basin in an emulsified, colloidal or dissolved state. They are particularly dangerous for water basins. The maximum allowable concentration of petroleum products in the water basin is 0.5 mg/kg of water.

They form films on the water surface, inhibit the natural aeration of water and thereby inflict serious harm to aquatic life.

The main sources of oil in water are:
(a) pump cooling water
(b) condensate discharged from surface condensers
(c) ground water contamination with fuel oil due to spillage.

Waste waters may get contaminated with oil (from 2 to 10 mg/kg) from the oil-systems of turbines and bearings of the rotating mechanisms.

In some cases the same cooling water circuit is involved in turbine oil coolers, the bearings of axiliary machines and turbine condensers. Oil ingression to cooling water system due to leakage in oil coolers may occur and rise as high as 25-30 mg/kg in practical cases. Add to this, a considerable quantity of oily water from turbine tanks ad oil regeneration plants.

The flowrate of petroleum-contaminated waste water from a modern thermal plant usually does not exceed 100 m^3/h and oil concentration $\leq$ 50 mg/kg.

1. *Purification:* The oil contaminated waste water is charged to oil traps that separate out efficiently the coarse oil particles of size 80–100 μm or more.

The clarified water is then fed into the flotator where finer oil particles are separated from water at a high rate under pressure flotation.

The purified water is then filtered through a mechanical filter. The former consists of double-layer packing of quartz sand (grain dia 2 mm, bed thickness 700 mm) and anthracite (grain dia 3–15 mm, bed thickness 600 mm). The carbon filter consists of a bed (2500 mm thick) of activated carbon to adsorp oily suspensions.

The final effluent water of this plant is 95% free from oil.

2. *Decreasing the Amount of oil Contaminated Waste Water:* The problem of decreasing the disposal of petroleum contaminated waste-waters can be solved to a great extent if adequate and careful measures are taken:

(a) Special care should be taken to the handling, storage and transportation of fuel oil.

(b) Synthetic detergents should be used, while cleaning emptied oil tank wagons, instead of hot water or steam. This reduces the volume of oily waste water by a factor of more than 10 because of multiple use of washing solution.

(c) For storing fuel oil, steel reservoirs are preferable to reinforced concrete reservoirs. Experience shows concrete reservoirs suffer from substantial oil leakages while steel reservoirs do not.

(d) It is advisable to use tube-in-tube type heat exchanger, while heating fuel oil, to avoid fuel oil leakages.

(e) The problem of turbine oil penetrating into the cooling waters of turbine oil coolers can be prevented by hermetically sealing (with epoxy resins) the joints between the tubes and tube-plates of the oil coolers.

(f) Corrosion of oil cooler tubes often lead to oil contamination of cooling water. Brass tubes, in particular, are susceptible to salt cracking induced by dissolved salts in the cooling water. Hence selection of materials for oil cooler tubes should be made from stainless steel, German silver and titanium to avoid tube leakage due to corrosion.

(g) A good solution to the problem of minimizing oil leakage is to switch-over from shell-and-tube type oil coolers to plate type coolers for more efficient operation, better access to mechanical cleaning, visual inspection and leak detection.

(h) Another viable technique is to set up a separate cooling water circuit to supply cooling water to the turbine condenser, oil coolers, auxiliaries. Oil contamination, it has ben found, of up to 30 mg/kg in cooling water normally does not present any problem to the process of cooling.

(i) The leakage of oil from oil lubricated bearings is also a source of oil contamination of waste water. This oil loss from bearing housing can be minimized or prevented by using thick lubricants and special bearing seals.

WASTE WATER OF HYDRAULIC ASH-DISPOSAL SYSTEM

Hydraulic ash-disposal of solid-fuel fired boilers is an efficient ash removal system in which water is allowed to act upon the ash, a part of which dissolves and the rest forms a pulp (suspension) which is pumped to ash-settling ponds where the coarse impurities settle down and the clarified water can either be discharged directly into a water basin (straight-flow hydraulic system) or recycled for reuse.

The composition and concentration of impurities transferred from ash to water depends largely upon the chemical composition of ash which may vary appreciably:

SiO_2 : 10 – 70% ; Al_2O_3 : 10–40% ;

Fe_2O_3 : 2 – 30%; CaO : 2–70%; MgO : 0 – 10%

$K_2O + Na_2O$: 0 – 10% plus it may contain traces amount of compounds of vanadium, germanium, arsenic, mercury, beryllium and fluorine.

The ash-disposal water may be categorized into four groups:

(a) lime saturated water
(b) calcium sulphate saturated water
(c) lime as well as calcium sulphate saturated water
(d) water with relatively low mineral content.

In straight-flow hydraulic ash-disposal system, as much as 10000–15000 metric tons of salts are discharged to the water basin annually from a coal fired boiler producing enough steam to generate 1.25 MW

power. Obviously such high discharges of salts to water is not permissible. Therefore, a closed-hydraulic system (in which clarified water is recycled) is adopted in modern boiler plants. However certain salts of calcium, e.g., $CaSO_4$, $CaCO_3$ and $Ca(OH)_2$ form supersaturated solutions with water and then precipitate as scale in fluid transport systems (pumps, pipelines) and therefore present a difficulty to the operation of closed hydraulic system.

Since the ash-disposal waters contain both scale formers and salt contaminants, the essential characteristics of clarified water from an ash-disposal system are:

1. Alkalinity ⎤ determine
2. Sulphate ⎥ scale formation
   concentrations ⎦

3. Total dissolved ⎤
   salts ⎥ determine
4. Concentration of ⎥ contamination
   toxic impurities ⎦

The pH of clarified water of a hydraulic ash-disposal system may vary from strongly acidic to strongly alkaline. Therefore, the clarified water before its final discharge to water basin must be treated so that

(a) pH value of water basin does not alter beyond the specified limit 6.5–8.5
(b) salt content within admissible value
(c) concentration of toxic substances is less than their highest permissible concentration.

Decontamination of Ash-Disposal Water

The high flowrate and high concentrations of impurities in the waste water of a hydraulic ash-disposal system prevent purification of the entire bulk of the ash-disposal system. What can be achieved is decontamination of toxic impurities to a safe level and the principal processes involved are as follows:

1. Precipitation as well as co-precipitation of impurities
2. Sorption of impurities
3. Oxidation reduction followed by precipitation.

Removal of impurities by forming sparingly soluble precipitates or by adsorption on the surface of the solid-phase separated in the water mass during the course of the chemical treatment of ash-disposal water is the most popular one. Lime is the most commonly used precipitant, sometimes used with other reagents to enhance decontamination. For example, arsenic is precipitated out from the ash-disposal water by lime as basic arsenate $3Ca(AsO_4)_2 \cdot Ca(OH)_2$. Though it is less soluble among complexes of lime, still it has a solubility 4 mg//kg—a figure which is 18 times the safe level in the water basin.

But if $FeSO_4$ is added together with lime (pH = 9–10), a more insoluble compound, $FeAsO_4$ is precipitated along with some flakes of ferrous hydroxide onto which adsorbed are AsO_4^{2-} ions, thereby reducing the arsenic concentration in the waste water of the hydraulic ash-disposal system down to less than 0.05 mg/kg. Simultaneously, the concentration of harmful chromium is also reduced.

Removal of chromium ions is best achieved by adding Na_2SO_3 together with $FeSO_4$ at pH > 9. The former reduces Cr^{6+} (sexivalent chromium) to Cr^{3+} which separates out as insoluble precipitate $Cr(OH)_3$ and the concentration of chromium ions in clarified water drops to 0.014–0.024 mg/kg. Simultaneously, the concentration of fluoride ions drops by more than 90%.

Fluoride ions (F^-) are precipitated readily by adding $MgCl_2$ to waste water. However optimal removal of fluoride ions is achieved on adding $MgCl_2$ in amounts of 50 mg/kg of fluorine at pH 10.2–10.4. Fluorides precipitate (CaF_2, MgF_2) out together with flakes of $Mg(OH)_2$

A better result is obtained in the coagulation process induced by adding $Al_2(SO_4)_3$ at pH < 7. When aluminium sulphate in an amount 18–24 mg/kg

of fluorine is added to waste water at pH : 4.5–5.5, the concentration of fluoride ions in decontaminated water reduces practically to zero.

The waste water of hydraulic ash-disposal system can be partially purified by sorption of its toxic impurities on the sorbents. Ash is a good sorbent. It contains aluminosilicates—inorganic ion-exchange materials capable of extracting various metallic ions. Unburnt coal particles in ash also serve as good sorbents of organic substances. It has been found that the greatest reduction of concentration of harmful substances in waste water is attained at an ash content 3–5 g/kg of waste water. Under these conditions, concentration of Cu^{2+} ions is reduced to 0.003 mg/kg, Zn^{2+} ions to 0.02 mg/kg, As^{5+} to 0.44 mg/kg, oil products 5.5–6 mg/kg.

SOLID WASTE DISPOSAL

Ash and slag can be pelletized and then stored on abandoned quarries and bogs, etc., or used as building materials.

They can be used as fillers and binders in the
(a) production of concrete and concrete elements
(b) road construction
(c) manufacture of heat insulating materials
(d) manufacture of sound-proofing materials.

They can also be used as the starting material for the manufacture of aluminium as well as sorbents for purification of different types of waste waters.

If the calcium content of ash is high, it'll prove beneficial to the calcination of acid soils.

Appendix VII

Quantity	SI Unit	Conversion Factor
LENGTH	m	$1 \mu m = 10^{-6} m = 10^{-4} cm = 10^{-3} mm$
		$1 A = 10^{-6} m = 10^{-8} cm$
		$1 ft = 0.3048 m$
		$1 inch = 0.0254 m = 2.54 cm$
AREA	m^2	$1 ft^2 = 0.0929 m^2$
		$1 inch^2 = 6.4516 cm^2$
		$= 6.451 \times 10^{-4} m^2$
VOLUME	m^3	$1 lt = 10^{-3} m^3$
		$1 ft^3 = 0.0283 m^3 = 28\ 316 cm^3$
		$1 inch^3 = 16.387 cm^3$
		$= 16.39 \times 10^{-6} m^3$
		$1 gallon = 4.546 lt = 0.1605 ft^3$
		$= 0.00454 m^3$
MASS	kg	$1 MT = 1000 kg$
		$1 lb = 0.45359 kg = 453.59 g$
		$1 slug = 32.1739 lb = 14.593 kg$
DENSITY	kg/m^3	$1 ton/m^3 = 1 kg/dm^3$
		$= 1 g/cm^3 = 1000 kg/m^3$
		$1 kgf.s^2/m^4 = 9.81 kg/m^3$
		$1 lb/ft^3 = 16.02 kg/m^3$
		$= 0.016 kg/cm^3$
		$1 lb/inch^3 = 27.68 \times 10^3 kg/m^3$

Quantity	SI Unit	Conversion Factor
TEMPERA-TURE	K	$\Theta\,^{\circ}C = (\Theta + 273.15)K$
		$\Theta\,^{\circ} F = \left[\frac{5}{9}(t - 32) + 273.15\right] K$
FORCE (WEIGHT)	N	$1 kgf = 9.81 N$
		$1 dyn = 10^{-5} N$
		$1 lbf = 4.481 N$
		$1 kg.m/s^2 = 1 N$
		$1 lb.m/s^2 = 0.13826 N$
		$1 kgf = 2.1887 lbf$
PRESSURE	N/m^2 (Pa)	$1 bar = 10^5 Pa$
		$1 dyn/cm^2 = 0.1 Pa$
		$1 kgf/cm^2 = 9.81 \times 10^4 Pa$
		$1 kgf/m^2 = 9.81 Pa$
		$1 atm = 760 mm\ Hg = 101\ 325 Pa$
		$1 atm = 14.696 lbf/inch^2$
		$1 mm\ H_2O = 9.81 Pa$
		$1 mm\ Hg = 133.33 Pa$
		$1 lbf/inch^2 (psi) = 6\ 894.76 Pa$
		$1 lbf/ft^2 = 47.88 Pa$
DYNAMIC VISCOSITY	Pa s	$1P = 1 dyn.s/cm^2 = 0.1 Pa\ s$
	(poise)	
		$1cP = \frac{1}{98\ 180} kgf\ s/m^2 = 10^{-3} Pa\ s$
		$= 10^{-3} kg/m\ s = 10^{-7} g/cm\ s$
		$1 lb/ft\ s = 148.82 cP = 0.1488 Pa\ s$

Quantity	SI Unit	Conversion Factor
		1 kg/m s – 1 Pa.s
KINEMATIC VISCOSITY m^2/s		1 St – 1 cm^2/s – 10^{-4} m^2/s (Stokes)
		1 ft^2/s – 0.0929 m^2/s – 929 St
		1 ft^2/h = 25.807×10^{-6} m^2/s
		1 m^2/h = 10.7639 ft^2/h
THERMAL CONDUC-TIVITY	W/m K	1 kcal/m h °K – 1.163 W/m K
		1 BTU/ft h °F – 1.73 W/m K
		1 cal/s cm K – 418.4 W/m K
		1 BTU/ft h °F – 1.488 kcal/m h °C
		1 W/m °C – 0.5778 BTU/(h ft^2 °F/ft)
		– 0.2388 cal/ (s m °C)
HEAT TRANSFER COEFFICIENT	W/m^2 K	1 kcal/m^2 h °K – 1.163 W/m^2 K
		1 BTU/ft^2 h °F – 5.678 W/m^2 K
		1 cal/cm^2 s °K – 4.184 W/cm^2 K
		= 7.3686×10^3 BTU/ft^2 h °F
		1 BTU/ft^2 h °F– 4.88 kcal/m^2 h°C
WORK, or, ENERGY	J	1 N m – 1 J
		1 kgf m – 9.81 J
		1 lbf ft – 1.3558 J – 0.324 cal
		1 erg – 10^{-7} J
		1 kWh – 3.6×10^6 J
		1 kcal – 4.1868 kJ – 4.1868×10^3 J
		1 BTU – 251.95 cal – 778.16 lbf ft – 1 055 J
POWER	W	1 J/s – 1 W
		1 kgf.m/s – 9.81 W
		1 erg/s – 10^{-7} W

Quantity	SI Unit	Conversion Factor
		1 kcal/h – 1.163 W
		1 lbf. ft/s – 1.3558 W
		1 HP (FPS) – 550 lbf ft/s – 745.69 W
		– 596.8 kcal/h
		1 HP (MKS) – 75 kgf m/s – 735.75 W
		– 0.1756 kcal/s
ENERGY FLOWRATE	W	1 kcal/h – 1.163 W
		1 cal/s – 4.1868 W
		1 BTU/h – 0.252 kcal/h = 0.293 W
		1 ton of refrigeration – 50 kcal/min
		– 198.45 BTU/min
		– 3.489 kW
ENERGY FLUX	W/m^2	1 kg/s^3 – 1 W/m^2 – 0.317 BTU/h.ft^2
		1 BTU/h ft^2 – 3.1545 W/m^2
		1 kcal/h m^2 – 1.163 W/m^2
		1 cal/s cm^2 – 36 000 kcal/h m^2
SPECIFIC ENTHALPY	J/kg	1 kcal/kg = 4 190 J/kg
		1 BTU/lb – 2 326 J/kg – 0.5551 kcal/kg
SPECIFIC ENTROPY	J/kg K	1 kcal/kg. °K – 4 190 J/kg K
		1 BTU/lb.°F – 4 190 J/kg K
SPECIFIC HEAT CAPACITY	J/kg K	1 kcal/kg °K – 4 190 J/kg K
		1 erg/g °K – 10^{-4} J/kg K
		1 BTU/lb °F – 4 190 J/kg K
SPECIFIC HEAT	J/kg	1 kcal/kg – 1 cal/g – 4 190 J/kg
		1 BTU/1b – 2 326 J/kg

Further Reading References

A. Literature

1. H.M. Spring—Boiler Operators Guide (McGraw-Hill, New York, 1940)
2. O.de Lorenzi—Combustion Engineering (Combustion Engineering Inc., 1947)
3. Babcock & Wilcox Co.—Steam, Its Generation and Use
4. E.B. Woodruff and H.B. Lammers—Steam Plant Operation (McGraw-Hill, 1950)
5. C.D. Shields—Boilers (McGraw-Hill, New York)
6. R.F.Latham—Naval Boilers (U.S. Naval Institute)
7. R.C. King—Practical Marine Engineering (Prentice-Hall, Englewood Cliffs, New Jersey)
8. M.I. Reznikov & Yu M. Lipov—Steam Boilers of Thermal Power Stations (Mir, Moscow)
9. Babcock & Wilcox, USA—Instruction Manual for Watertube Boilers
10. Babcock & Wilcox—Water Treatment
11. A.W. Culp—Principles of Energy Conversion (McGraw-Hill, Kogakusa)
12. S.M. Elonka & A.L. Kohan—Standard Boiler Operators Questions & Answers (McGraw-Hill)
13. Worthington Corporation, USA—Instruction Manual for Turbines
14. F.T. Morse—Power Plant Engineering
15. Strehlow—Combustion Fundamentals, 5e (McGraw-Hill, N.Y.)
16. Ministry of Power—The Efficient Use of Fuels (HMSO, London)
17. O. Lyle—The Efficient Use of Steam (HMSO, London)
18. J.S.S. Brame & J.G. King—Fuels—Solid, Liquid and Gaseous (Edward Arnold, London)
19. R.T. Haslam & R.P. Russel—Fuels and Their Combustion (McGraw-Hill, New York)
20. W. Francis—Fuels and Fuel Technology (Oxford, London)
21. D.A. Williams & G. Jones—Liquid Fuels (Pergamon Press, London)
22. B. Mullins—Spontaneous Ignition of Liquid Fuels (Butterworth, London)
23. G.W. Himus—The Elements of Fuel Technology (Leonard Hill, London)
24. J. Griswold—Fuels, Combustion and Furnaces (McGraw-Hill, New York)
25. H.H. Lowry—Chemistry of Coal Utilization (John Wiley, New York)
26. G.J. Minkoff & C.F.H. Tipper—Chemistry of Combustion Reactions (Butterworth, London)
27. M.W. Thrings—The Science of Flames and Furnaces (Chapman & Hall, London)
28. G. Tine—Gas Sampling and Chemical Analysis in Combustion Processes (Pergamon Press, London)
29. William A. Bone—Coal and its Scientific Uses (Longman, London)
30. K. Sterner—Oil Burners (McGraw-Hill, New York)
31. K. Sterner—Fuel and Fuel Burners (McGraw-Hill, London)
32. J.H. and D.A. Allen—Fuel Science (Oliver & Boyd, London)
33. F.I. Belan—Water Treatment (Mir, Moscow)
34. L.I. Pincus—Practical Water Treatment
35. S. Domkundwar—Power Plant Engineering (Dhanpat Rai & Sons)
36. R. Joel—Basic Engineering Thermodynamics (ELBS, London)
37. A.S. Sarao—Thermal Engineering (Satya Prakashan, New Delhi)
38. K.P. Roy & N. Roy—Problems in Heat Engines (Oxford & IBH Publ. Co.)
39. C. Davies—Calculations in Furnace Technology (Pergamon Press, London)

40. W.J. Peck & A.J. Richmond—Applied Thermodynamic Problems for Engineers (Edward Arnold, London)
41. J.R. Campbell—Methods and Analysis of Fuels and Oils (Constable, London)
42. R.M. Bethea—Air Pollution Control Technology (Van Nostrand Reinhold, New York)
43. R.D. Ross (ed.)—Air Pollution and Industry (Van Nostrand Reinhold, New York)
44. B. Rudomino and Yu. Remzhin—Steam Power Plant Piping Design (Mir, Moscow)
45. H. Spring and A. Kohan—Boiler Operator's Guide (McGraw-Hill, New York)
46. Nalco Chemical Co.—The Nalco Guide to Boiler Failure (McGraw-Hill, N.Y.)
47. Tyler G. Hicks—Power Generation Calculations Reference Guide (McGraw-Hill, N.Y.)
48. David Lindsley—Boiler Control Systems
49. K.G. Oliver—Industrial Boiler Management—An Operator's Guide
50. E.B. Woodruff—Steam Plant Operation (5th Edition)
51. C.D. Schroeder—Solutions to Boiler & Cooling Water Problems (The Fairmont Press)

B. Periodicals

1. Steam Generation (Power, Dec. 1946)
2. Boiler Factors Affecting the Part Load Efficiency (Energy World, 1977)
3. Boilers (Energy World, Aug. & Sept., 1975)
4. Boilers (Energy World, Feb. 1976)
5. Furnace Design of Large Steam Generators (Power, July & Aug., 1951)
6. Chemical Engineering, Vol. 93. 16
7. Chemical Engineering, Vol. 93, No. 21
8. Chemical Engineering, Vol. 94, No. 8
9. Chemical Engineering, Vol. 94, No. 17
10. Pulverized Coal Firing (Energy World, March, 1975)
11. Modern Ash Handling System (Power Engineering, Jan. 21 & Feb. 18, 1971)
12. The Basis of Dust Collection (Plant Engineering, Jan. 21 & Feb. 18, 1971)
13. Selecting a Boiler Water Treatment System (Plant Engineering, March 9, 1971)
14. Stack Gas Scrubbing (Combustion, Oct. 1975)
15. Blue Gas Desulfurization (Combustion, Oct. 1975)
16. Sulphur Oxide Removal by Wet Scrubbing Application to Unit Boilers (Combustion, Oct. 1978)
17. Central Industrial Boiler Emission (Power, April, 1976)
18. Air Pollution Control of Coal Fired Boilers (Power, April, 1976)
19. Economizer Preheat Cycle Raises Boiler Plant Efficiency (Power, Feb. 1966)
20. Waste Heat Recovery (Power Engineering, June, 1967)
21. Power, Feb. 1982
22. Power, Oct. 1984
23. Power, March, 1988
24. Power, June, 1988
25. Power, Feb. 1989
26. Power, March, 1989
27. Active Conservation Technique, Nov., 1988
28. Active Conservation Technique, June, 1989
29. Chemical Industry Digest (4th Quarter issue), 1988
30. Mechanical Engineering, Dec., 1988.

Index